Introduction to Finite and Infinite Dimensional Lie (Super)algebras

Introduction to Finite and Infinite Dimensional Lie (Super)algebras

N. Sthanumoorthy

Ramanujan Institute for Advanced Study in Mathematics, University of Madras
Chennai, India

ELSEVIER

AMSTERDAM • BOSTON • HEIDELBERG • LONDON
NEW YORK • OXFORD • PARIS • SAN DIEGO
SAN FRANCISCO • SINGAPORE • SYDNEY • TOKYO
Academic Press is an imprint of Elsevier

Academic Press is an imprint of Elsevier
125 London Wall, London, EC2Y 5AS, UK
525 B Street, Suite 1800, San Diego, CA 92101-4495, USA
50 Hampshire Street, 5th Floor, Cambridge, MA 02139, USA
The Boulevard, Langford Lane, Kidlington, Oxford OX5 1GB, UK

Library of Congress Cataloging-in-Publication Data
A catalog record for this book is available from the Library of Congress

British Library Cataloguing in Publication Data
A catalogue record for this book is available from the British Library

ISBN: 978-0-12-804675-3

For information on all Academic Press publications
visit our website at http://www.elsevier.com

Publisher: Nikki Levy
Acquisition Editor: Graham Nisbet
Editorial Project Manager: Susan Ikeda
Production Project Manager: Poulouse Joseph
Designer: Matthew Limbert

In memory of
my beloved parents
R. Neelakanda Pillai and L. Anantham Neelakanda Pillai.

Contents

About the author **xi**
Preface **xv**
Acknowledgments **xxi**

1 Finite-dimensional Lie algebras **1**
1.1 Basic definition of Lie algebras with examples and structure constants **2**
1.2 Subalgebras of Lie algebras and different classes of subalgebras of $gl(n, \mathbb{C})$ **8**
1.3 Ideals, quotient Lie algebras, derived sub Lie algebras, and direct sum **10**
1.4 Simple Lie algebras, semisimple Lie algebras, solvable and nilpotent Lie algebras **12**
1.5 Isomorphism theorems, Killing form, and some basic theorems **14**
1.6 Derivation of Lie algebras **19**
1.7 Representations of Lie algebras and representations of $sl(2, \mathbb{C})$ **21**
1.8 Rootspace decomposition of semisimple Lie algebras **28**
1.9 Root system in Euclidean spaces and root diagrams **30**
1.10 Coxeter graphs and Dynkin diagrams **34**
1.11 Cartan matrices, ranks, and dimensions of simple Lie algebras **36**
1.12 Weyl groups and structure of Weyl groups of simple Lie algebras **39**
1.13 Root systems of classical simple Lie algebras and highest long and short roots **40**
1.14 Universal enveloping algebras of Lie algebras **44**
1.15 Representation theory of semisimple Lie algebras **49**
1.16 Construction of semisimple Lie algebras by generators and relations **50**
1.17 Cartan-Weyl basis **51**
1.18 Character of a finite-dimensional representation and Weyl dimension formula **52**
1.19 Lie algebras of vector fields **53**

2 Kac-Moody algebras **59**
2.1 Basic concepts in Kac-Moody algebras **60**
2.2 Classification of finite, affine, hyperbolic, and extended-hyperbolic Kac-Moody algebras and their Dynkin diagrams **66**
2.3 Invariant bilinear forms **75**
2.4 Coxeter groups and Weyl groups **78**
2.5 Real and imaginary roots of Kac-Moody algebras **81**
2.6 Weyl groups of affine Lie algebras **85**
2.7 Realization of affine Lie algebras **86**

2.8 Different classes of imaginary roots (special imaginary roots, strictly imaginary roots, purely imaginary roots) in Kac-Moody algebras **94**
2.9 Representations of Kac-Moody algebras, integrable highest weight modules, Verma modules, and character formulas **100**
2.10 Graded Lie algebras and root multiplicities **105**

3 Generalized Kac-Moody algebras **135**
3.1 Borcherds-Cartan matrix (BCM), Generalized Generalized Cartan matrix (GGCM), Borcherds Kac-Moody (BKM) algebras, and Generalized Kac-Moody (GKM) algebras **136**
3.2 Dynkin diagrams of GKM algebras **139**
3.3 Root systems and Weyl groups of GKM algebras **144**
3.4 Special imaginary roots in GKM algebras and their complete classifications **150**
3.5 Strictly imaginary roots in GKM algebras and their complete classifications **164**
3.6 Purely imaginary roots in GKM algebras and their complete classifications **169**
3.7 Representations of GKM algebras **180**
3.8 Homology modules and root multiplicities in GKM algebras **187**

4 Lie superalgebras **203**
4.1 Basic concepts in Lie superalgebras with examples **204**
4.2 Coloring matrices, θ-colored Lie superalgebras, and examples **212**
4.3 Subsuperalgebras, ideals of Lie superalgebras, abelian Lie superalgebras, solvable, and nilpotent Lie superalgebras **215**
4.4 General linear Lie superalgebras **216**
4.5 Simple and semisimple Lie superalgebras and bilinear forms **216**
4.6 Representations of Lie superalgebras **218**
4.7 Different classes of classical Lie superalgebras **220**
4.8 Universal enveloping algebras of Lie superalgebras and θ-colored Lie superalgebras **224**
4.9 Cartan subalgebras and root systems of Lie superalgebras **230**
4.10 Killing forms on Lie superalgebras **237**
4.11 Dynkin diagrams of Lie superalgebras **239**
4.12 Lie superalgebras over an algebraically closed field of characteristic zero **241**
4.13 Classification of non-classical Lie superalgebras **245**
4.14 Lie superalgebras of vector fields **248**

5 Borcherds Kac-Moody Lie superalgebras **257**
5.1 BKM supermatrices and BKM Lie superalgebras **258**
5.2 Dynkin diagrams of BKM Lie superalgebras and in particular Kac Moody Lie superalgebras and some examples **264**

5.3 Domestic type and Alien type imaginary roots in BKM Lie superalgebras 272
5.4 Special imaginary roots in BKM Lie superalgebras and their complete classifications 281
5.5 Strictly and Purely imaginary roots in BKM Lie superalgebras and complete classification of BKM Lie superalgebras possessing purely imaginary roots 298
5.6 BKM Lie superalgebras possessing purely imaginary property but not strictly imaginary property 308
5.7 Complete classification of BKM Lie superalgebras possessing strictly imaginary property (SIM property) 315
5.8 Borcherds superalgebras and root supermultiplicities 325
5.9 Root supermultiplicities of Borcherds superalgebras which are extensions of Kac-Moody algebras and some combinatorial identities 351
5.10 Description of finite and infinite dimensional Lie algebras and Lie superalgebras and their different classes of root systems 368

6 Lie algebras of Lie groups, Kac-Moody groups, supergroups, and some specialized topics in finite- and infinite-dimensional Lie algebras 375
6.1 Lie groups and Lie algebras of Lie groups 376
6.2 Kac-Moody groups, supergroups, and some applications 386
6.3 Homogeneous spaces, corresponding Lie algebras, and spectra of some differential operators on homogeneous spaces 390
6.4 Spectral invariants of zeta function of the Laplace-Beltrami operator 394
6.5 Generalization of Macdonald's identities for some Kac-Moody algebras 402
6.6 Some special infinite dimensional Lie algebras 406
6.7 Hirota bilinear differential operators and soliton solutions for KdV equation 411
6.8 Principal vertex operator construction of basic representation and homogeneous vertex operator construction of the basic representation 415
6.9 Principal vertex operators and the corresponding Hirota bilinear equations for $A_2^{(1)}, A_2^{(2)}, C_2^{(1)}$, and $B_3^{(1)}$ 426
6.10 Fermionic Fock space, Clifford algebra, Bosonic Fock space, and Boson-Fermion correspondence 459
6.11 Remark on Quantum groups, String theory and Mathematical Physics with some references 460

Appendix 463
Bibliography 471
Index 483

About the author

N. Sthanumoorthy has more than 45 years of teaching and research experience. He obtained Ph.D. degree in Mathematics, with a highly commended thesis, from the University of Madras. He is presently working as the Principal Investigator of a Book-Writing Project funded by "Science and Engineering Research Board (SERB)—Department of Science and Technology (DST), Government of India" and was formerly working as Professor and Professor Emeritus (granted by University Grants Commission (UGC), Government of India) in the Ramanujan Institute for Advanced Study in Mathematics (RIASM), University of Madras, Chennai, India. UGC awarded Visiting Associate-ship to him for two years during 1994–1996. He also served as a "Visiting Professor" offered by NBHM during 2004 and a "Visiting Fellow" in the Tata Institute of Fundamental Research, Mumbai during 1986. Among his 36 research publications on topics closely related to the title of the present book, two articles were published in Science Encyclopedia and four articles were published in Source Books brought out in "Algebra," "Applied Algebra," and "Real Analysis" (Part 1 and Part 2). He was the convener and coordinator of "Ramanujan International Symposium" on "Kac-Moody Lie Algebras and Applications" held in RIASM, during 2002 and was an editor of "Contemporary Mathematics (AMS), Vol. 343 – Kac-Moody Algebras and Related Topics" which was the proceedings of this symposium. As the coordinator of a refresher course on Differential Geometry, organized by UGC-Academic Staff College, University of Madras, he brought out a reading material on "Differential Geometry" in 2005. He supervised and guided six research fellows for the Ph.D. degree and 35 research fellows for the M.Phil. degree. He successfully completed four Major Research Projects, two of them funded by DST, Government of India, rest funded by UGC, Government of India and one minor research project, also funded by the UGC. He delivered lectures in many international institutions located in countries like United States, Germany, Italy, China, and India. Awards and Honors conferred on the author: (i) he was a recipient of Performance Linked Cash Incentive award with highest credit in his department in the University of Madras for 2 years (2002–2004); (ii) he was a subject of biographical record for outstanding achievement published by "Marquis 'Who's Who' in the World" during 2009; (iii) he was a recipient of "Vijay Shree Award" with "Certificate of Excellence" from the India International Friendship Society during 2006 for outstanding services, series of achievements and contributions and also selected for "Rajiv Gandhi Excellence Award"; and (iv) he received "Best Citizen of India Award" from "International Publishing House," during 2005. He refereed papers for many journals, evaluated several Ph.D. theses, reviewed research papers for Mathematical Reviews like AMS (USA) and Zentralblatt fur Mathematik (Germany), evaluated many research projects and served as an expert committee member for UGC. He was a member of many Professional societies. Now he is an Associate Editor of "International Journal of Theoretical and Computational Mathematics (ISSN: 2395–6607)."

Catalysed and Supported by the
Science & Engineering Research Board,
Department of Science & Technology, Government of India
Under its Utilisation of Scientific Expertise of Retired Scientists Scheme

Preface

The aim of writing this book is to bring the essence in finite and infinite-dimensional Kac-Moody Lie algebras, Generalized Kac-Moody (GKM) algebras, Lie superalgebras, and Borcherds Kac-Moody superalgebras and some applications of Lie groups and Kac-Moody algebras to "Number theory, Differential geometry, and Differential equations" in one book. The present book may not contain the proofs of some important theorems proved in other standard books, instead will give many examples to illustrate the standard results. This makes the point that we can cover so much material by not giving all the detailed proofs of general theorems (which can already be found in other textbooks). The idea is to present enough examples so that the readers can really understand the main concepts behind the theory. Expected readers may be graduate students, research scholars, and teachers. It is hoped that the graduate students can learn the fundamentals from this book, whereas the researchers will have many options to choose and will get much working knowledge in the concerned topics after reading the book and teachers may find it interesting to teach the topics.

First I briefly explain the historical development of the topics mentioned above.

The creators of the Lie theory considered Lie group as a group of symmetries of an algebraic or geometric object and Lie algebra as the set of infinitesimal transformations. Lie [5–7] considered the problem of classification of GL_n and infinite-dimensional group of transformations. Starting from the works of Sophus Lie, Wilhelm Killing, and Elie Cartan, the theory of finite-dimensional Lie groups and Lie algebras developed in depth and scope. As stated in Kac [8], the following are four classes of infinite-dimensional Lie groups and Lie algebras that underwent intensive study: (i) Lie algebras of vector fields and corresponding groups of diffeomorphisms on manifolds, (ii) Lie groups and Lie algebras of smooth mappings on a given manifold into finite-dimensional Lie groups or Lie algebras, respectively, (iii) classical Lie groups and algebras of operators in a Hilbert's space or Banach space, and (iv) infinite-dimensional Lie algebras, namely, Kac-Moody algebras. Using the researches of Sophus Lie and Wilhelm Killing, Cartan [9] in his 1894 thesis, completed the classification of finite-dimensional simple Lie algebras over C. The nine types of this classification (consisting of the four classes of classical simple Lie algebras and five exceptional simple Lie algebras) correspond to the nine types of finite Cartan matrices and to the nine types of Dynkin Diagrams [10, 11]. Wilhelm Killing and Elie Cartan had developed a process from finite-dimensional simple Lie algebras to finite Cartan matrices, whereas, Chevalley [12] and Harish-Chandra [13] constructed a scheme that began with a finite Cartan matrix and produced finite-dimensional simple Lie algebra. During 1976, Serre [14–16] proved the defining relations on the generators and Cartan integers (elements of the Cartan matrix) of the finite-dimensional complex semi-simple Lie algebras. On the other hand, Moody [17, 18] in Canada and Kac [19–21] in Russia worked simultaneously and independently to extend the construction that Jacobson [22] had presented in Chapter 7 of his book to infinite-dimensional setting. Both of them gave the classification of the generalized Cartan matrices of affine type and gave standard realizations of them. Kac-Moody Lie algebras, in general, appear as infinite-dimensional generalizations of semi-simple Lie algebras over generalized Cartan matrices. For a finite-dimensional complex semisimple Lie algebra, all the corresponding principal minors of the Cartan matrix are positive whereas Kac and Moody suggested to drop this condition and enlarge the corresponding abelian Lie algebra (Cartan subalgebra), so that the linear functions on the linear span of the abelian subalgebra will be linearly independent. Hence for a Kac-Moody algebra, all the principal minors of the corresponding generalized Cartan matrix need not be positive but may be also singular and the corresponding Kac-Moody algebra is in general infinite dimensional. It was in their paper on "Lie algebra Homology and Macdonald-Kac Formulas" by Garland and Lepowsky [23], the term "Kac-Moody Lie algebras" was first used in the literature. Within a decade, this subfield of mathematics had found many surprising physical applications. In the theory of the finite-dimensional complex semi-simple Lie algebras, there are four main tools: the root string (or more generally, the weight string), the Weyl group, the Killing form, and the Casimir operator. All of them were generalized to Kac-Moody algebras by Kac and Moody, namely, the root string and the Weyl group were generalized to arbitrary Kac-Moody algebras whereas the Killing form and the Casimir operator were generalized to symmetrizable Kac-Moody Lie algebras. The algebraic work of Moody and Kac led to deep results in Physics and won for them, the prestigious Wigner Medal in 1994. To classify the Kac-Moody algebras, it is sufficient to classify the generalized Cartan matrices. The generalized Cartan matrices of finite type are simply the Cartan matrices, and their classification is well-known. The classification of the generalized Cartan matrices of affine type were achieved by Kac and Moody independently. In the class of indefinite type, there are hyperbolic types [8], extended hyperbolic types [24], and other indefinite types. For details regarding the history of the early development of Kac-Moody algebras, one can refer to Berman and Parshall [25] and Macdonald [128]. For details regarding Flag varieties and conjugacy theorems, one can refer Kac and Peterson [26].

GKM algebras, also known as Borcherds algebras, are a natural generalization of Kac-Moody algebras allowing imaginary simple roots. They were constructed by Borcherds [27] in his study of vertex algebras and Conway and Norton's moonshine conjectures [28] for the Monster sporadic simple group. GKM algebras contain as a subclass, the symmetrizable Kac-Moody algebras, and in particular, the affine and finite-dimensional simple Lie algebras. Additionally here, the generalized Cartan matrix may be infinite. The structure and the representation theory of GKM algebras are very similar to those of Kac-Moody algebras, and many basic facts about Kac-Moody algebras can be extended to GKM algebras. But there are some differences, too. For example, GKM algebras may have imaginary simple roots with norms ≤ 0 and multiplicities can be >1. GKM algebras can also be obtained as quotients of vertex algebras [29–33]. GKM algebras were originally described in terms of generators and relations which are certain generalizations of the defining conditions for Kac-Moody algebras.

Lie superalgebras appeared as Lie algebras of certain generalized groups, nowadays called Lie supergroups, whose function algebras are algebras with commuting and anticommuting variables. Recently, a satisfying theory, similar to Lie's theory, has been developed on the connection between Lie supergroups and Lie superalgebras. If G is a finite-dimensional Lie superalgebra, then G contains a unique maximal solvable ideal R (the solvable radical). The Lie superalgebra G/R is semisimple (that is, it has no solvable ideals). Therefore, the theory of finite-dimensional Lie superalgebras is reduced, in a certain sense, to the theories of semisimple and solvable Lie superalgebras. The main fact in the theory of solvable Lie algebras is Lie's theorem, which asserts that every finite-dimensional irreducible representation of a solvable Lie algebra over C is one-dimensional. For Lie superalgebras this is not true, in general. Next, it is well known that a semisimple Lie algebra is a direct sum of simple ones. This is by no means true for Lie superalgebras. However, there is a construction that allows us to describe finite-dimensional semisimple Lie superalgebras in terms of simple ones. So we come to the fundamental problem of classifying the finite-dimensional simple Lie superalgebras. The principal difficulty lies in the fact that the Killing form may be degenerate, which cannot happen in the case of simple Lie algebras. Therefore, the classical technique of Killing-Cartan is not applicable here. A detailed exposition of the theory of Lie superalgebras can also be seen in Kac [34] and Scheunert [35]. Kac and Wakimoto [36] further developed the theory of Lie superalgebras and presented some interesting applications of affine Kac-Moody superalgebras to number theory.

The theories of Lie superalgebras, Kac-Moody superalgebras, and GKM algebras were combined to give rise to the theory of Borcherds Kac-Moody Lie superalgebras. Wakimoto [3, 4] introduced the definitions of Borcherds Kac-Moody supermatrix (BKM supermatrix) and Borcherds Kac-Moody Lie superalgebra (BKM Lie superalgebra) by imposing an additional structure in the index set in the definition of Borcherds Cartan matrix. Kang and Kim [37] introduced the definition of Borcherds superalgebras by introducing a coloring matrix to the Borcherds Cartan matrix. Using this coloring matrix, one can define a bimultiplicative form θ [38] which in turn will give a θ-colored Lie superalgebra to make it finally as the Borcherds superalgebra. On the other hand, the additional structure due to the introduction of a subset (of the index set in the definition of Borcherds Cartan matrix) corresponding to odd roots of the BKM Lie superalgebra coincides with the introduction of the coloring matrix in the definition of Borcherds superalgebra.

During the last more than four decades, the theory of finite and infinite-dimensional Lie algebras and Lie superalgebras attracted researchers from different areas of Mathematics and Physics because of its close connections with combinatorics, differential equations, group theory, modular forms, singularities, knot theory, statistical mechanics, quantum field theory, and string theory to name a few.

Below I give briefly the chapter-wise description of this book. To understand the fundamentals about all types of finite- and infinite-dimensional Lie algebras and Lie superalgebras, it is better that one knows the fundamentals of finite-dimensional Lie algebras. So in this volume, I start with the finite-dimensional Lie algebras.

Chapter 1 is about finite-dimensional Lie algebras: Starting from the basic concepts in Lie algebras, I go up to representations of Lie algebras, in particular, that of semisimple Lie algebras and that of $sl(2, \mathbb{C})$. I also explain the structure of Weyl groups of simple Lie algebras, root systems of classical Lie algebras, Cartan-Weyl basis, universal enveloping algebras, and character of finite-dimensional representations. Moreover, Lie algebras of vector fields are also explained in this chapter.

Chapter 2 deals with the infinite-dimensional Kac-Moody Lie algebras: Starting from the basic concepts, various classes of Kac-Moody algebras and different classes of imaginary roots are explained with examples. Weyl groups of affine Lie algebras, realization of affine Lie algebras, representation theory of Kac-Moody algebras, character formulae, graded algebras, and root multiplicities are given in detail with examples. In particular, recently obtained results on different classes of root systems and root multiplicities are also described in this chapter.

In Chapter 3, I discuss about GKM algebras: Starting from the definition of Borcherds Cartan Matrices and Dynkin Diagrams, different classes of root systems of GKM algebras are given with examples. Homology modules, root multiplicities, and representations of GKM algebras are discussed in this chapter. Newly obtained results on complete classification theorems on special, strictly and purely imaginary roots, and dimension formulae are also given in this chapter. Monstrous Lie algebras are also explained here.

Chapter 4 describes about Lie superalgebras: Lie superalgebras are defined here basically in two different ways with examples. Then the basic concepts in Lie superalgebras, coloring matrices, θ-colored Lie superalgebras, the representations of Lie superalgebras, universal enveloping algebras of Lie superalgebras, and θ-colored Lie superalgebras are explained. Moreover, in this chapter, the Killing forms, Dynkin diagrams, classifications of Lie superalgebras and nonclassical Lie superalgebras, and Lie superalgebras of vector fields are also given.

Chapter 5 is about Borcherds Kac-Moody Lie superalgebras: All basic concepts in BKM Lie superalgebras, their Dynkin diagrams along with all types of imaginary roots, namely, domestic, alien, purely, strictly, and special imaginary roots and their complete classifications are explained. Root supermultiplicities of Borcherds superalgebras along with recently found out superdimension-formulae and corresponding combinatorial identities are also given. Diagrammatic description of finite and infinite dimensional Lie algebras, Lie superalgebras and their different classes of root systems is also given.

Chapter 6 is about Lie groups and Lie algebras, Kac-Moody groups, supergroups, some specialized topics, and some applications of finite- and infinite-dimensional Lie algebras to Differential Geometry, Number theory, and Differential Equations. First I discuss about Lie algebras of Lie groups, Kac-Moody groups, supergroups, and some applications. Then spectra of some differential operators on some homogeneous spaces and the spectral invariants of the zeta function of the Laplacian on 1-forms on the above spaces and 2-forms on $(4r-1)$-dimensional sphere and eta function and spectral asymmetry of the operator $B = \pm({}^*d - d^*)$ acting on 2-forms on $(4r-1)$-dimensional sphere along with some generalizations of McDonald's identities for some Kac-Moody algebras are given. Some special infinite-dimensional Lie algebras are also discussed. Also, in this chapter, Hirota bilinear differential operators and soliton solutions for KdV and KP equations and many results on principal vertex operators and Hirota bilinear equations for $A_2^{(1)}$ and $A_2^{(2)}$, Hirota bilinear equations and super-bilinear equations for $B_3^{(1)}$, and super-bilinear equations for $C_2^{(1)}$ are also explained. Moreover, homogeneous vertex operators, in particular for $A_1^{(1)}$, and corresponding nonlinear differential equations are discussed in this chapter. Some preliminaries on Fermionic Fock space, Clifford algebra, Bosonic Fock space, and Boson-Fermion correspondence and a remark on Quantum groups, String theory and Mathematical Physics with some references are also given.

In the Appendix, basic definitions of differentiable manifold along with some examples, C^∞ functions on differentiable manifold, product manifold, Lie group $SO(3)$ and Lie algebra $so(3)$, pseudo Riemannian manifold, Riemannian manifold, Riemannian symmetric space with examples, Riemann surface, symmetries of heat equation, Lie algebra realized as operators, Maya diagram, Young tableau, presheaf and sheaf of sets, symmetric group with an example are given. Finally definitions and Dynkin diagrams of quasi finite, quasi affine and quasi hyperbolic Kac-Moody algebras are also given here.

N. Sthanumoorthy
Ramanujan Institute for Advanced Study in Mathematics
September 18, 2015

Acknowledgments

I am grateful to my research guide and supervisor T.S. Bhanu Murthy for his research guidance during 1978–1982 to me in Ramanujan Institute for Advanced Study in Mathematics, University of Madras for my research leading to Ph.D. degree. I am thankful to the authorities who were responsible for my academic accomplishments at the International level: (i) visits to Abdus Salam International Center for Theoretical Physics, Trieste, Italy to attend international conferences during April 17–28, 1989; March 15 to April 2, 1993; August 3 to September 1, 1995; and September 4–15, 1995; (ii) invited talk delivered in the International Congress of Mathematicians (ICM 2002) held in Beijing, China, during August 20–28, 2002; (iii) lectures given in the Max Plank Institute Fur Gravitational Physics, Albert Einstein Institute, Germany during July 4 to August 2, 2004; (iv) visit to School of Mathematics, University of Minnesota, Minneapolis, USA on September 29, 2006; (v) lectures given in the Department of Mathematics, University of Texas at Arlington, Texas, USA on December 28, 2006; (vi) lectures given in the Department of Mathematics, University of Nebraska, Lincoln, USA on May 6, 2010; and (vii) participation and lectures given in many conferences in India. My visits to Tata Institute of Fundamental Research (TIFR), Mumbai, India, several times, in particular, as a Visiting Fellow (offered by TIFR) and Visiting Professor offered by National Board for Higher Mathematics (NBHM) were also useful to me. I also acknowledge American Mathematical Society (AMS) for my work as an editor of *Contemporary Mathematics* (Vol. 343), published by AMS.

I thank the authorities of the University of Madras for their approval to do this book project, sanctioned and financially supported by SERB, Department of Science and Technology (DST), Government of India and also thank the Director and Head, entire teaching and non-teaching staff of Ramanujan Institute for Advanced Study in Mathematics for the necessary facilities offered to me to complete this book project.

I am thankful to Alex Feingold, State University of New York, Binghamton, New York, for his encouragement and best wishes for the successful completion of this book.

I thank the authorities of Elsevier Academic Press, in particular, Cathleen Sether (Publishing Director), Nikki Levy (Publisher), Graham Nisbet (Senior Acquisitions Editor), Steven Mathews (Editorial Project Manager), Susan Ikeda (Editorial Project Manager), Poulouse Joseph (Senior Project Manager), and Matthew Limbert (Designer) for their kind cooperation in bringing out this volume on time.

I thank A. Uma Maheswari for her valuable assistance in the proofreading and also thank K. Priyadharsini for her skillful technical assistance and sincere role in typesetting the manuscript of the book.

I am thankful to my wife, Anantha Radha Sthanumoorthy, for her support and encouragement throughout my career in teaching and research. I thank my sons, Neelakantan Sthanumoorthy and Raja Sthanumoorthy, for their constant encouragement during the preparation of this book. I also thank all other members of my family for their best wishes. Last but not least, I would like to mention that the warmth and affection of my grand children, Riya, Shrihan and Ishan were great source of inspiration to me.

This book is dedicated to the memory of my beloved parents, R. Neelakanda Pillai (father) and L. Anantham Neelakanda Pillai (mother) and to the memory of my sister, A. Visalakshi. My late father-in-law, M. Subramania Pillai, would be immensely happy to know the publication of this book.

Above all, I thank the Almighty, for his grace and kindness to complete this work.

N. Sthanumoorthy
September 18, 2015

Finite-dimensional Lie algebras

1

Chapter Outline

1.1 Basic definition of Lie algebras with examples and structure constants 2
1.2 Subalgebras of Lie algebras and different classes of subalgebras of $gl(n, \mathbb{C})$ 8
1.3 Ideals, quotient Lie algebras, derived sub Lie algebras, and direct sum 10
1.4 Simple Lie algebras, semisimple Lie algebras, solvable and nilpotent Lie algebras 12
1.5 Isomorphism theorems, Killing form, and some basic theorems 14
1.6 Derivation of Lie algebras 19
1.7 Representations of Lie algebras and representations of $sl(2, \mathbb{C})$ 21
1.8 Rootspace decomposition of semisimple Lie algebras 28
1.9 Root system in Euclidean spaces and root diagrams 30
1.10 Coxeter graphs and Dynkin diagrams 34
1.11 Cartan matrices, ranks, and dimensions of simple Lie algebras 36
1.12 Weyl groups and structure of Weyl groups of simple Lie algebras 39
1.13 Root systems of classical simple Lie algebras and highest long and short roots 40
1.14 Universal enveloping algebras of Lie algebras 44
1.15 Representation theory of semisimple Lie algebras 49
1.16 Construction of semisimple Lie algebras by generators and relations 50
1.17 Cartan-Weyl basis 51
1.18 Character of a finite-dimensional representation and Weyl dimension formula 52
1.19 Lie algebras of vector fields 53

Abstract

History of the development of finite-dimensional Lie algebras is described in the preface itself. Lie theory has its name from the work of Sophus Lie [6], who studied certain transformation groups, that is, the groups of symmetries of algebraic or geometric objects that are now called Lie groups. Using the researches of Sophus Lie and Wilhelm Killing, Cartan [9] in his 1894 thesis, completed the classification of finite-dimensional simple Lie algebras over C. The nine types of this classification (consisting of the four classes of classical simple Lie algebras and five exceptional simple Lie algebras) correspond to the

Introduction to Finite and Infinite Dimensional Lie (Super)algebras. http://dx.doi.org/10.1016/B978-0-12-804675-3.00001-7

nine types of finite Cartan matrices and to the nine types of Dynkin Diagrams [10, 11]. Chevalley [12] and Harish-Chandra [13] constructed a scheme that began with a finite Cartan matrix and produced finite-dimensional simple Lie algebra. During 1976, Serre [14–16] proved the defining relations on the generators and Cartan integers (elements of the Cartan matrix) of the finite-dimensional complex semi-simple Lie algebras.

In this chapter, we give all preliminaries in finite-dimensional Lie algebras which are necessary to develop all other chapters in this book. We start from the definition of Lie algebras, structure constants, subalgebras, ideals, quotient Lie algebras, simple Lie algebras, and semisimple Lie algebras with examples. Homomorphism and isomorphism theorems, Killing form, derivation, representations of semisimple Lie algebras and also that of $sl(2,\mathbb{C})$ are also explained. Moreover, the rootspace decomposition of semisimple Lie algebras, Dynkin diagrams, Cartan matrices, rank and dimensions of simple Lie algebras, Weyl groups, universal enveloping algebras, construction of semisimple Lie algebras by generators and relations, Cartan-Weyl basis, and character of finite-dimensional representations are also explained. Cartan matrices of all classical simple algebras, structures of Weyl groups of simple algebras, Weyl groups, root systems highest short and long roots of all classical simple algebras are given. Basic properties of Lie algebras of vector fields are also given.

Throughout this chapter, all Lie algebras are finite dimensional unless otherwise stated.

1.1 Basic definition of Lie algebras with examples and structure constants

Definition 1. Let $\mathcal{G}$ be a vector space over a field $\mathbb{F}$. We define an operation, $\mathcal{G}\times\mathcal{G}\to\mathcal{G}$, denoted by $(x,y)\to[x,y]$, called the Lie bracket of x and y. Then $\mathcal{G}$ is called a *Lie algebra* over $\mathbb{F}$, if the following axioms are satisfied:

(1) The bracket operation [,] is bilinear.
(2) $[x,x]=0$ for all $x\in\mathcal{G}$ (skew symmetry).
(3) $[x,[y,z]]+[y,[z,x]]+[z,[x,y]]=0$, for all $x,y,z\in\mathcal{G}$ (Jacobi identity).

Remark 1. Note that (1) and (2) applied to $[x+y,x+y]$ implies (2′) $[x,y]=-[y,x]$, which is anticommutativity.

Conversely, if the characteristic of $\mathbb{F}$, char $\mathbb{F}\neq 2$, then apply $y=x$ in (2′) $\Rightarrow$ $[x,x]=-[x,x]\Rightarrow 2[x,x]=0\Rightarrow[x,x]=0\Rightarrow$ (2).

A Lie algebra can also be defined starting from the definition of an algebra

Definition 2. An algebra $\mathcal{G}$ is a vector space over a field $\mathbb{F}$, endowed with a multiplication ab, which is bilinear, that is,

$$a(\lambda b+\mu c)=\lambda ab+\mu ac$$
$$(\lambda b+\mu c)a=\lambda ba+\mu ca,$$

for all $a,b,c\in\mathcal{G}$ and $\lambda,\mu\in\mathbb{F}$.

Definition 3. A Lie algebra is an algebra $\mathcal{G}$ with product $[a,b]$, called the bracket of a and b, subject to the following two axioms:

- anticommutativity: $[a,a]=0$ or skew symmetry
- Jacobi identity: $[a,[b,c]]+[b,[c,a]]+[c,[a,b]]=0$.

Definition 4. A Lie algebra $\mathcal{G}$ is *abelian* if $[a,b]=0$ for all $a,b\in\mathcal{G}$.

Definition 5. An algebra $\mathcal{G}$ is associative if $(ab)c = a(bc)$ for all $a,b,c\in\mathcal{G}$. In particular, in *an associative algebra* $\mathcal{G}$ with bilinear multiplication ab, define the Lie bracket $[,]$ to be the commutator $[a,b] = ab - ba$. With this $[,]$, $\mathcal{G}$ becomes a Lie algebra. If the associated algebra is also commutative, then $ab = ba \Rightarrow [a,b] = 0$. So a commutative associative algebra is an *abelian Lie algebra*.

Definition 6. A *homomorphism*, say φ, from a Lie algebra $\mathcal{G}$ to another Lie algebra $\mathcal{G}_1$ is a *linear map* $\varphi\colon \mathcal{G}\to\mathcal{G}_1$ that preserves bracket: $\varphi([X,Y]) = [\varphi(X),\varphi(Y)]$. (If $\mathcal{G}_1=\mathcal{G}$, we speak of an *endomorphism*.)

The image of such a homomorphism is $\Im\,\phi\equiv\{\phi(a)|a\in\mathcal{G}_1\}$, which is a subset of $\mathcal{G}_1$. The *Kernel of a homomorphism* $\varphi\colon\mathcal{G}\to\mathcal{G}_1$ is the set $\{X\in\mathcal{G}|\varphi(X)=0\}$. We prove in the Proposition 6 that the Kernel is an ideal in $\mathcal{G}$ and we write $\ker\varphi$ for it. A homomorphism ϕ is called an *isomorphism* if it is bijective, that is if $\ker\phi=0$ and $\Im\,\phi=\mathcal{G}_1$.

We give below the formal definition of isomorphism.

Definition 7. We say that two Lie algebras $\mathcal{G}',\mathcal{G}$ over $\mathbb{F}$ are *isomorphic* if there exists a vector space isomorphism $\phi\colon\mathcal{G}\to\mathcal{G}'$ satisfying $\phi([x,y])=[\phi(x),\phi(y)]$ for x,y in $\mathcal{G}$.

An isomorphism of a Lie algebra with itself is an *automorphism*.

Example 1 (General linear Lie algebras). Let V be a finite-dimensional vector space over a field $\mathbb{F}$ with $\dim V=n$. Lie algebras arise in nature as *vector spaces of linear transformations on* V endowed with a new operation [,], called the commutator $[f,g]=f\circ g-g\circ f$ (which is neither commutative nor associative) in terms of the associative composition operation $\circ$ in $End(V)=\{f\colon V\to V|f \text{ is linear}\}$, for $f,g\in End(V)$. Actually this $End(V)$ is an associative algebra. As a vector space over $\mathbb{F}$, $End(V)$ has dimension n^2. It can be directly verified that this bracket $[f,g]$ for all $f,g\in End(V)$ satisfies the axioms required for a Lie algebra. With this operation, $End(V)$ becomes a Lie algebra over $\mathbb{F}$.

In order to distinguish this new algebra structure from the old associative one, we write $gl(V)$ for $End(V)$ viewed as a Lie algebra and call it, the *general linear algebra*, $gl(V)$.

Any subalgebra of $gl(V)$ is called *a linear Lie algebra*.

There is an isomorphism of $End(V)$ or $gl(V)$ with the Lie algebra $gl(n,\mathbb{F})$ of $n\times n$ matrices over $\mathbb{F}$ after a choice of basis is made for V. The standard basis of $gl(V)$ consists of the matrices (e_{ij}) (having 1 in the (i,j)-th position and 0 elsewhere). Since $e_{ij}e_{kl}=\delta_{jk}e_{il}$, we have $\left[e_{ij},e_{kl}\right]=\delta_{jk}e_{il}-\delta_{li}e_{kj}$.

If $\mathbb{F}=\mathbb{C}$ and $\dim V=n$, then $gl(V)$ is denoted as $gl(n,\mathbb{C})$.

Remark 2. In particular, if V is a Lie algebra $\mathcal{G}$, then we can define $gl(\mathcal{G})$ and we get the following homomorphism "*ad*" as follows. For a Lie algebra $\mathcal{G}$, we define

$ad\colon \mathcal{G} \to gl(\mathcal{G})$ by $(ad\ x)(y) := [x, y]$ for $x, y \in \mathcal{G}$. Using Jacobi identity, we can easily verify that

$$ad([x, y]) = ad\ x \circ ad\ y - ad\ y \circ ad\ x \quad \text{for all } x, y \in \mathcal{G}.$$

Theorem 1. *Prove that the three-dimensional vector space with cross product is a Lie algebra which is isomorphic to the algebra of the set of all* 3×3 *skew-symmetric matrices over* $\mathbb{C}$.

Proof. Let V_3 be a three-dimensional vector space over $\mathbb{C}$ and $\{e_1, e_2, e_3\}$ be a basis for V_3. For any two elements $x = x_1e_1 + x_2e_2 + x_3e_3$ and $y = y_1e_1 + y_2e_2 + y_3e_3$ in V_3, define

$$[x, y] = (x_2y_3 - x_3y_2)e_1 + (x_3y_1 - x_1y_3)e_2 + (x_1y_2 - x_2y_1)e_3,$$

for $x_1, x_2, x_3, y_1, y_2, y_3 \in \mathbb{C}$. Then V_3 becomes a Lie algebra with this bracket. The bracket here is nothing but the cross product of vectors x and y in V_3. The required axioms for Lie algebras can be directly verified.

This algebra is isomorphic to the Lie algebra $o(3) = \{A \in gl(3, \mathbb{C}) \mid A + A^t = 0\}$, *the set of all* 3×3 *skew-symmetric matrices over* $\mathbb{C}$.

Any element of $o(3)$ can be taken as

$$\begin{pmatrix} 0 & -d & e \\ d & 0 & -f \\ -e & f & 0 \end{pmatrix}$$

with a basis $\{R_x, R_y, R_z\}$, where

$$R_x = \begin{pmatrix} 0 & 0 & 0 \\ 0 & 0 & -1 \\ 0 & 1 & 0 \end{pmatrix}, \ R_y = \begin{pmatrix} 0 & 0 & 1 \\ 0 & 0 & 0 \\ -1 & 0 & 0 \end{pmatrix} \text{ and } R_z = \begin{pmatrix} 0 & -1 & 0 \\ 1 & 0 & 0 \\ 0 & 0 & 0 \end{pmatrix}.$$

with $[R_x, R_y] = R_z$, $[R_y, R_z] = R_x$, and $[R_z, R_x] = R_y$. Actually, R_x, R_y, and R_z will form the infinitesimal rotations around x-, y-, and z-axis, respectively.

On the other hand, if we take

$$X = x_1R_x + x_2R_y + x_3R_z \text{ and } Y = y_1R_x + y_2R_y + y_3R_z,$$
$$\text{then } [X, Y] = XY - YX = (x_2y_3 - x_3y_2)R_x + (x_3y_1 - y_3x_1)R_y + (x_1y_2 - x_2y_1)R_z.$$

Now one can establish an isomorphism between V_3 and $o(3)$ through the correspondence $e_1 \to R_x$, $e_2 \to R_y$, $e_3 \to R_z$. □

Remark 3.

(1) One can define a Lie algebra structure on a three-dimensional vector space with a basis $\{x, y, z\}$ and Lie brackets satisfying the following relations, $[x, y] = z, [y, z] = x, [x, z] = y$. All the three axioms for Lie algebra can be easily verified. This algebra is isomorphic to the algebras V_3 and $o(3)$. Moreover, the Lie algebra satisfying the above relationships can also be considered as a *Real Lie algebra with three elements in the basis*. This is also *angular momentum algebra in three dimensions*, equivalent to $so(3)$.

Below we give some more examples to Lie algebras.

(2) Every finite-dimensional Lie algebra is isomorphic to a linear Lie algebra.

(i) If $\mathcal{G}$ is an arbitrary finite-dimensional vector space over $\mathbb{F}$, we make $\mathcal{G}$ as a Lie algebra by setting $[x, y] = 0$ for all $x, y \in \mathcal{G}$. Such an algebra having trivial Lie multiplication is *Abelian*.

(ii) *Lorentz algebra*: The Lie algebra satisfying

$$[X_1, X_2] = X_3, \quad [X_2, X_3] = -X_1, \quad [X_3, X_1] = X_2$$

is a real Lie algebra with three elements in the basis, called the *Lorentz algebra* in 2+1 dimension, denoted as $so(2, 1)$.

(iii) Let $\mathcal{G}_1$ and $\mathcal{G}_2$ be two Lie algebras over the same field K. The vector space $\mathcal{G}_1 \times \mathcal{G}_2$ becomes a Lie algebra over K if we define the bracket as

$$[(X, Y), (X', Y')] = ([X, X'], [Y, Y']).$$

This Lie algebra is called the product of Lie algebras $\mathcal{G}_1$ and $\mathcal{G}_2$. The sets $\{(X, 0): X \in \mathcal{G}_1\}, \{(0, Y): Y \in \mathcal{G}_2\}$ are ideals in $\mathcal{G}_1 \times \mathcal{G}_2$ and $\mathcal{G}_1 \times \mathcal{G}_2$ is the direct sum of these ideals.

(iv) In general, if $M \in gl(n, \mathbb{C})$, the set of all complex $n \times n$ matrices X satisfying $XM + MX^t = 0$ is linear Lie algebra. This set is denoted as $gl(n, M, \mathbb{C})$. It is known that $gl(n, M_1, \mathbb{C})$ and $gl(n, M_2, \mathbb{C})$ are isomorphic whenever M_1 and M_2 are congruent. In particular, if M is nonsingular and symmetric, then M is congruent to the identity matrix. So in this case,

$$\begin{aligned} gl(n, M, \mathbb{C}) &\cong gl(n, I, \mathbb{C}) \\ &= \{X \in gl(n, \mathbb{C})/X + X^t = 0\} \\ &= so(n, \mathbb{C}). \end{aligned}$$

Proposition 1. *Let $\mathcal{G}$ be any Lie algebra with a basis $\{x_1, \ldots, x_n\}$, such that $[x_i, x_j] = \sum_{k=1}^{n} a_{ij}^k x_k$. Then above structure constants a_{ij}^k satisfy the following relations:*

$$a_{ij}^k = -a_{ji}^k \tag{1.1}$$

(or equivalently $a_{ii}^k = 0$)

$$\text{and} \quad \sum_{k,m} (a_{ij}^k a_{kl}^m + a_{jl}^k a_{ki}^m + a_{li}^k a_{kj}^m) = 0. \tag{1.2}$$

Conversely, if $a_{ij}^k (i, j, k = 1, \ldots, n)$ are n^3 constants satisfying Equations (1.1) and (1.2), and if one can choose a basis $\{X_1, \ldots, X_n\}$ for a vector space $\mathcal{G}$ such that $[X_i, X_j] = \sum_{k=1}^{n} a_{ij}^k X_k$, then $\mathcal{G}$ becomes a Lie algebra.

Proof. We have

$$
\begin{aligned}
&[x_i + x_j, x_i + x_j] = 0 \\
\Rightarrow &[x_i, x_j] + [x_j, x_i] = 0 \\
\Rightarrow &\sum_k (a_{ij}^k + a_{ji}^k) x_k = 0 \\
\Rightarrow &a_{ij}^k = -a_{ji}^k \text{ or equivalently } a_{ii}^k = 0 \text{ as } x_k \text{ are linearly independent.}
\end{aligned}
$$

So (1) is proved.

Moreover,

$$
\begin{aligned}
&[[x_j, x_l], x_i] + [[x_l, x_i], x_j] + [[x_i, x_j], x_l] = 0 \\
\Rightarrow &\sum_m \left(\left(\sum_k a_{jl}^k a_{ki}^m \right) + \left(\sum_k a_{li}^k a_{kj}^m \right) + \left(\sum_k a_{ij}^k a_{kl}^m \right) \right) x_m = 0 \\
\Rightarrow &\sum_m \left(\sum_k a_{jl}^k a_{ki}^m + a_{li}^k a_{kj}^m + a_{ij}^k a_{kl}^m \right) x_m = 0,
\end{aligned}
$$

which proves the result (Equation 1.2) as the set $\{x_m\}_{m=1}^n$ is linearly independent.

Conversely if $a_{ij}^k (i, j, k = 1, \ldots, n)$ are n^3 constants satisfying Equations (1.1) and (1.2), we can choose a basis $\{X_1, \ldots, X_n\}$ for a vector space $\mathcal{G}$ such that $[X_i, X_j] = \sum_{k=1}^n a_{ij}^k X_k$. Hence $\mathcal{G}$ becomes a Lie algebra. Moreover the entire multiplication table of $\mathcal{G}$ can be recovered from the structure constants a_{ij}^k, which occur in the expressions. □

Remark 4. If two Lie algebras are isomorphic with respect to some suitable bases, then they have same structure constants and conversely the Lie algebras with same set of structure constants are isomorphic.

Proposition 2. *If $a_{ij}^k (i, j, k = 1, \ldots, n)$ and $b_{ij}^k (i, j, k = 1, \ldots, n)$ are two sets of structure constants occurring in the equations $[X_i, X_j] = \sum_{k=1}^n a_{ij}^k X_k$ and $[Y_i, Y_j] = \sum_{k=1}^n b_{ij}^k Y_k, 1 \le i, j \le n$, then linear combination of one set of structure constants can be written as linear combination of another set of structure constants. Here $\{X_1, X_2, \ldots, X_n\}$ and $\{Y_1, Y_2, \ldots, Y_n\}$ are two different sets of basis.*

Proof. If $\{Y_1, \ldots, Y_n\}$ is another basis such that

$$
[Y_i, Y_j] = \sum_{k=1}^n b_{ij}^k Y_k \quad 1 \le i, j \le n
$$

with

$$
Y_i = \sum_{j=1}^n c_i^j X_j \text{ and } Y_j = \sum_{k=1}^n c_j^k X_k \quad \text{where } c_i^j \neq 0 \text{ and } c_k^j \neq 0.
$$

Then

$$[Y_i, Y_j] = \left[\sum_l c_i^l X_l, \sum_m c_j^m X_m\right] = \sum_{l,m} c_i^l c_j^m [X_l, X_m]$$
$$= \sum_{l,m} c_i^l c_j^m \sum_s a_{lm}^s X_s = \sum_{l,m,s} c_i^l c_j^m a_{lm}^s X_s. \tag{1.3}$$

Hence

$$[Y_i, Y_j] = \sum_{k=1}^{n} b_{ij}^k \, Y_k \quad (1 \le i, j \le n) = \sum_k b_{ij}^k \sum_s c_s^k X_s = \sum_{k,s} c_s^k b_{ij}^k X_s. \tag{1.4}$$

Equating the coefficients of X_s on both sides, we have

$$\sum_{k,s=1}^{n} b_{ij}^k c_s^k = \sum_{l,m,s=1}^{n} c_i^l c_j^m a_{lm}^s \quad \text{for } 1 \le i, j, l \le n.$$

□

Remark 5. So two Lie algebras are isomorphic if their structure constants a_{ij}^k and b_{ij}^k satisfy the above equation, where the matrix $(c_i^j)_{i,j=1}^n$ is nonsingular.

Proposition 3. *In the three-dimensional vector space* V_3 *with cross product with a basis* $\{e_1, e_2, e_3\}$ *with* $e_1 = (1,0,0), e_2 = (0,1,0)$, *and* $e_3 = (0,0,1)$, *the structure constants are* $0, 1, -1$.

Proof. We have the following structure table:

$e_i \times e_j$	e_1	e_2	e_3
e_1	0	e_3	$-e_2$
e_2	$-e_3$	0	e_1
e_3	e_2	$-e_1$	0

Here $e_i \times e_j = [e_i, e_j]$. We have $[e_i, e_j] = \sum_{k=1}^{3} a_{ij}^k e_k$.

We have all the structure constants are zero except $a_{12}^3 = a_{23}^1 = a_{31}^2 = 1$ and $a_{13}^2 = a_{21}^3 = a_{32}^1 = -1$. □

Lie algebras of one, two, and three dimensions and their structure constants

Proposition 4. *We find all structure constants of Lie algebras upto dimension three.*

Proof.

(i) One-dimensional Lie algebra: If there is only one element X in the basis of a Lie algebra then it should satisfy the relation $[X, X] = 0$. It is an abelian Lie algebra. Here all structure constants are zero.

(ii) Two-dimensional Lie algebras: Let X_1, X_2 be generators of a two dimensional Lie algebra. Then the following are the two possibilities:

(1) $[X_1, X_2] = 0$ (Here all structure constants are zero.)
(2) $[X_1, X_2] = X_1$ (Here $a^1_{12} = 1, a^1_{21} = -1$, and all other structure constants are zero.)
(or)
$[X_2, X_1] = X_2$ (Here $a^2_{21} = 1, a^2_{12} = -1$, and all other structure constants are zero.)

In the case (1), the algebra is abelian. In the case (2), $\{X_1\}$ (or $\{X_2\}$) is an ideal of the Lie algebra.

(iii) **Three-dimensional Lie algebras:** Let X_1, X_2, and X_3 be generators of a three dimensional Lie algebra, then following are the four possibilities:

(1) $[X_1, X_2] = [X_2, X_3] = [X_3, X_1] = 0$. Here all structure constants are zero and this Lie algebra is abelian.

(2) $[X_1, X_2] = X_3; [X_1, X_3] = [X_2, X_3] = 0$. Here $a^3_{12} = 1, a^3_{21} = -1$, and all other structure constants are zero.
(or)
$[X_1, X_3] = X_2; [X_1, X_2] = [X_2, X_3] = 0$. Here $a^2_{13} = 1, a^2_{31} = -1$, and all other structure constants are zero.

(3) $[X_1, X_2] = 0; [X_3, X_1] = \alpha X_1 + \beta X_2; [X_3, X_2] = \gamma X_1 + \delta X_2$ where the matrix $\begin{pmatrix} \alpha & \beta \\ \gamma & \delta \end{pmatrix}$ is nonsingular. Here $a^1_{31} = \alpha, a^2_{31} = \beta$, $a^1_{13} = -\alpha, a^2_{13} = -\beta$, $a^1_{32} = \gamma, a^2_{32} = \delta, a^1_{23} = -\gamma, a^2_{23} = -\delta$, and all other structure constants are zero.

(4) **(i)** $[X_1, X_2] = X_3; [X_2, X_3] = X_1; [X_3, X_1] = X_2$,
(ii) $[X_1, X_2] = X_3; [X_2, X_3] = -X_1; [X_3, X_1] = X_2$. In both the above cases the structure constants can be directly written.

□

1.2 Subalgebras of Lie algebras and different classes of subalgebras of $gl(n, \mathbb{C})$

Definition 8. Let $\mathcal{G}$ be a Lie algebra. A *sub Lie algebra* or a *sub algebra of a Lie algebra* $\mathcal{G}$ is a subspace, say Q, (the subspace of the underlying vector space) of $\mathcal{G}$ that is closed under the bracket operations (That is, $[Q, Q] \subset Q$). So Q becomes a Lie algebra with the bracket operations inherited from that of $\mathcal{G}$.

Remark 6. The intersection of two sub Lie algebras is again a sub Lie algebra. Moreover if A is a sub Lie algebra and B is an ideal of $\mathcal{G}$, then $A \cap B$ is an ideal of A if $A \cap B$ is nonempty.

1.2.1 Different subalgebras of $gl(n, \mathbb{C})$

Four families of classical Lie algebras, namely, A_n, B_n, C_n, and D_n and their bases

(i) $A_n = sl((n+1), \mathbb{C}) = \{A \in gl(n+1, \mathbb{C})$ such that trace $A = 0\}$. The dimension of $A_n = n + (n+1)^2 - (n+1) = n^2 + 2n$.

A basis of A_n: The set $\{e_{ij} (i \neq j)$ and $h_i = e_{ii} - e_{i+1,i+1}$ for $1 \leq i \leq n,\ 1 \leq i,j \leq n\}$. Here one can note that all trace zero diagonal matrices form an n-dimensional abelian subalgebra of A_n. The set $\{e_{ii} - e_{i+1,i+1}$ for $1 \leq i \leq n\}$ forms a basis of all trace zero diagonal matrices. In fact, for $1 \leq i,j \leq n, i \neq j$

$$[e_{ii} - e_{i+1,i+1}, e_{jj} - e_{j+1,j+1}] = -\delta_{i,j+1} e_{i,j+1} + \delta_{j+1,i} e_{j+1,i} - \delta_{i+1,j} e_{i+1,j} + \delta_{j,1+i} e_{j,i+1}. \tag{1.5}$$

If $i = j + 1$ and $i + 1 \neq j$, Equation (1.5) becomes

$$-e_{j+1,j+1} + e_{j+1,j+1} = 0.$$

If $i + 1 = j$ and $i \neq j + 1$, Equation (1.5) becomes

$$-e_{jj} + e_{jj} = 0. \tag{1.6}$$

As Equations (1.5) and (1.6) are true, the set $\{e_{ii} - e_{i+1,i+1}$ for $1 \leq i \leq n\}$ forms a basis for A_n with dimension n.

(ii) $B_n = o(2n + 1, \mathbb{C})$, the complex skew-symmetric matrices of order $2n + 1$. That is, $B_n = \{X \in gl(2n + 1, \mathbb{C}) \mid X + X^t = 0\}$.

In fact, if we take

$$S = \begin{pmatrix} 1 & 0 & 0 \\ 0 & 0 & I_n \\ 0 & I_n & 0 \end{pmatrix} \text{ and } X = \begin{pmatrix} a & b & c \\ d & u & v \\ e & p & q \end{pmatrix},$$

where I_n is the identity matrix and all u, v, p, q are matrices of the type $n \times n$ and a, b, c, d, e are corresponding row and column vectors, then B_n will consist of matrices of the form $X \in B_n$ such that $SX + X^t S = 0$. Using the condition $SX + X^t S = 0$, we will get

$$X = \begin{pmatrix} a & b & c \\ -c^t & u & v \\ -b^t & p & -u^t \end{pmatrix} \quad \text{with } a = 0,\ v^t = -v,\ p^t = -p.$$

A basis for B_n: Diagonal elements: $\{e_{ii} - e_{n+i,n+i}, 2 \leq i \leq n+1\}$ will give n elements.
Elements b and c: $\{e_{1,n+i+1} - e_{i+1,1}; e_{1,i+1} - e_{n+i+1,1}, 1 \leq i \leq n\}$ will give $2n$ elements.
Element u: $\{e_{i+1,j+1} - e_{n+j+1,n+i+1}, 1 \leq i \neq j \leq n\}$ will give $n^2 - n$ elements.
Element v: $\{e_{i+1,n+j+1} - e_{j+1,n+i+1}, 1 \leq i < j \leq n\}$ will give $\frac{1}{2}n(n-1)$ elements.
Element p: $\{e_{i+n+1,j+1} - e_{j+n+1,i+1}, 1 \leq j < i \leq n\}$ will give $\frac{1}{2}n(n-1)$ elements.
The union of the above sets will give a basis for B_n with dimension $2n^2 + n$.

(iii) $C_n = sp(n, \mathbb{C})$, the set of all symplectic matrices in $gl(2n, \mathbb{C})$.

The condition for symplectic matrices X will be $SX + X^t S = 0$ for $X = \begin{pmatrix} Z_1 & Z_2 \\ Z_3 & Z_4 \end{pmatrix}$ and $S = \begin{pmatrix} 0 & I_l \\ -I_l & 0 \end{pmatrix}$. So $sp(n, \mathbb{C}) = \left\{ \begin{pmatrix} Z_1 & Z_2 \\ Z_3 & -Z_1^t \end{pmatrix} \mid Z_i (i = 1, 2, 3) \text{ are complex } n \times n \text{ matrices with } Z_2 \text{ and } Z_3 \text{ being symmetric} \right\}$.

A basis for C_n: For Z_1,

$$\{\{e_{ii} - e_{n+i,n+i}, 1 \leq i \leq n\}, \quad \{e_{ij} - e_{n+j,n+i} \ (1 \leq i \neq j \leq n)\}\}$$

as basis elements will give totally, $n + n^2 - n = n^2$ elements.

For $Z_2 = Z_2^t$:

$$\{\{e_{i,n+i}(1 \le i \le n)\}, \{e_{i,n+j} + e_{j,n+i}(1 \le i < j \le l)\}\}$$

as basis elements will give totally, $n + \frac{1}{2}n(n-1)$ elements.

For $Z_3 = Z_3^t$:

$$\{\{e_{i,n+i}(1 \le i \le n)\}, \{e_{n+i,j} + e_{n+i,j}(1 \le i < j \le l)\}\}$$

as basis elements will give totally, $n + \frac{1}{2}n(n-1)$ elements.

The union of the above sets will give a basis for C_n. $\dim C_n = 2n^2 + n$.

(iv) $D_n = o(2n, \mathbb{C})$, the complex skew-symmetric matrices of order $2n$.

Let $S = \begin{pmatrix} 0 & I_n \\ I_n & 0 \end{pmatrix}$. Then D_n will consist of $2n \times 2n$ matrices $X \in gl(2n, \mathbb{C})$ satisfying $XS + SX^t = 0$. Hence

$$\begin{aligned} \dim D_n &= \dim o(2n, \mathbb{C}) \\ &= 2n^2 - n \end{aligned}$$

Equivalently, D_n consists of the matrices of the form $\begin{pmatrix} A & B \\ C & D \end{pmatrix}$, where A, B, C, and D are $n \times n$ matrices such that $B^t = -B$, $C = -C^t$, and $D = -A^t$.

A basis for D_n:

$A = -D^t$: $\{\{e_{ii} - e_{n+i,n+i}, 1 \le i \le n\} \cup \{e_{ij} - e_{n+j,n+i}(1 \le i \ne j \le n)\}\}$ as basis elements will give totally $n + (n^2 - n) = n^2$ elements.

$B = -B^t$: $\{-e_{i,n+j} + e_{j,n+i}\}(1 \le i < j \le n)$ as basis elements will give $\frac{1}{2}n(n-1)$ elements.

$C = -C^t$: $\{e_{n+i,j} - e_{j+n,i}\}(1 \le i < j \le n)$ as basis elements will give $\frac{1}{2}n(n-1)$ elements.

The union of the above sets will give a basis for D_n with dimension $2n^2 - n$. It can be directly verified that A_n, B_n, C_n, D_n are subalgebras of $gl(n, \mathbb{C})$.

Remark 7. All diagonal matrices in $gl(n, \mathbb{C})$ form an n-dimensional abelian subalgebra of $gl(n, \mathbb{C})$ as $[e_{ii}, e_{jj}] = 0$, for all $1 \le i, j \le n$.

Definition 9.

(i) Let $\mathcal{G}$ be a Lie algebra. The *centralizer* $\mathcal{G}_S$ of a subset S ($\subset \mathcal{G}$) is $\{X \in \mathcal{G} |$ for all $Y \in S$, $[X, Y] = 0\}$, which is also a Lie subalgebra. If $S = \mathcal{G}$ then $\mathcal{G}_S$ of $\mathcal{G}$ is the center of $\mathcal{G}$.

(ii) Let A be subalgebra of a Lie algebra $\mathcal{G}$. The *normalizer* of the sub Lie algebra $A = \{X \in \mathcal{G}$ with $[X, A] \subset A\}$, is a sub Lie algebra of $\mathcal{G}$, which contains A as an ideal and it is the largest sub Lie algebra of $\mathcal{G}$ with this property.

1.3 Ideals, quotient Lie algebras, derived sub Lie algebras, and direct sum

Definition 10. *A subspace* I of a Lie algebra $\mathcal{G}$ is called an *ideal* of $\mathcal{G}$ if for any $x \in \mathcal{G}$ and $y \in I$, $[x, y] \in I$. Obviously 0 (additive identity of underlying vector space) and $\mathcal{G}$ themselves are ideals of $\mathcal{G}$.

A *sub Lie algebra* $\mathcal{Q}$ *is an ideal* of a Lie algebra $\mathcal{G}$ if $[\mathcal{G},\mathcal{Q}] \subset \mathcal{Q}$ (ie, if $X \in \mathcal{G}$ and $Y \in \mathcal{Q}$, then $[X,Y] \in \mathcal{Q}$). By the skew-symmetry property of the bracket, it is clear that the Lie bracket [,] of ideals are automatically two-sided. That is, $[\mathcal{G},\mathcal{Q}] = [\mathcal{Q},\mathcal{G}]$.

Example 2.

(i) A trivial example: Every subspace of an abelian Lie algebra is an ideal.
(ii) $A_n = sl(n+1,\mathbb{C})$ is an ideal of $gl(n+1,\mathbb{C})$. Clearly for $x \in A_n$ and $y \in gl(n+1,\mathbb{C})$, trace $[x,y] = 0$ and hence $[x,y] \in A_n$. In fact, $gl(n+1,\mathbb{C})$ is the direct sum of A_n and one-dimensional Lie algebra $s(n+1,\mathbb{C})$ formed by scalar matrices in $gl(n+1,\mathbb{C})$. We have $gl(n+1,\mathbb{C}) = sl(n+1,\mathbb{C}) \oplus s(n+1,\mathbb{C})$ with $[s(n+1,\mathbb{C}), gl(n+1,\mathbb{C})] = 0$ and hence the center of $gl(n+1,\mathbb{C})$ is $s(n+1,\mathbb{C})$.

Definition 11. If $\mathcal{Q}$ is an ideal in a Lie algebra $\mathcal{G}$, then the quotient space $\mathcal{G}/\mathcal{Q}$ (whose elements are the linear cosets $X+\mathcal{Q}$) carries an induced [,] operation, defined by $[X+\mathcal{Q}, Y+\mathcal{Q}] = [X,Y]+\mathcal{Q}$. Moreover one can verify that this is well defined; that is, it does not depend on the choice of the representatives X,Y in $\mathcal{Q}$. So with this operation, $\mathcal{G}/\mathcal{Q}$ becomes a Lie algebra, the *quotient Lie algebra* of $\mathcal{G}$ by $\mathcal{Q}$.

The *Center* of a Lie algebra $\mathcal{G}$ is $Z(\mathcal{G}) = \{z \in \mathcal{G} \mid [x,z] = 0 \text{ for all } x \in \mathcal{G}\}$.

The Jacobi identity shows that $Z(\mathcal{G})$ is actually an ideal. It is clear that a Lie algebra $\mathcal{G}$ is *abelian* if and only if $Z(\mathcal{G}) = \mathcal{G}$.

Definition 12. The *derived* sub Lie algebra of the Lie algebra $\mathcal{G}$ is the ideal $\mathcal{G}' = [\mathcal{G},\mathcal{G}]$, spanned by all $[X,Y]$ for all $X,Y \in \mathcal{G}$. It corresponds to the commutator subgroup of a group. The quotient $\mathcal{G}/\mathcal{G}'$ is abelian and $\mathcal{G}'$ is the unique minimal ideal of $\mathcal{G}$.

Definition 13. The (external) *direct sum* of two algebras $\mathcal{G}_1$, $\mathcal{G}_2$, written as $\mathcal{G}_1 \oplus \mathcal{G}_2$, has the obvious definition; it is the vector space direct sum with [,] defined "componentwise," that is, $[(X_1,Y_1),(X_2,Y_2)] = ([X_1,X_2],[Y_1,Y_2])$.

In particular, if $\mathcal{G}_1$ and $\mathcal{G}_2$ are two commuting Lie algebras with $\mathcal{G}_1 \cap \mathcal{G}_2 = \phi$, satisfying the relations

$$[X_\rho, X_\sigma] = \sum_\tau c^\tau_{\rho\sigma} X_\tau$$

$$[Y_\rho, Y_\sigma] = \sum_\tau c^\tau_{\rho\sigma} Y_\tau$$

$$[X_\rho, Y_\sigma] = 0,$$

where $X_\rho, X_\sigma \in \mathcal{G}_1$, $Y_\sigma, Y_\rho \in \mathcal{G}_2$, then the set $\{\mathcal{G}_1 \cup \mathcal{G}_2\}$ of all the elements in $\mathcal{G}_1$ and $\mathcal{G}_2$ forms an algebra $\mathcal{G}$ called the direct sum $\mathcal{G} = \mathcal{G}_1 \oplus \mathcal{G}_2$.

For two subalgebras $\mathcal{G}_1$ and $\mathcal{G}_2$ of $\mathcal{G}$, the two summands $\mathcal{G}_1$ and $\mathcal{G}_2$ (ie, the set of elements of the form $(X,0)$ and $(0,Y)$) are ideals in $\mathcal{G}_1 \oplus \mathcal{G}_2$. The direct sum has intersection 0 and "nullify" each other that is, $[\mathcal{G}_1,\mathcal{G}_2] = 0$.

Conversely, if A and B are two ideals in $\mathcal{G}$ that span $\mathcal{G}$ linearly (ie, $A+B = \mathcal{G}$) and have intersection 0, then the map $(X,Y) \to X+Y$ is an isomorphism of $A \oplus B$ with $\mathcal{G}$ (thus $\mathcal{G}$ is *internal* direct sum of A and B).

This uses the fact that $[A,B]$ is contained in $A \cap B$ which is 0 here. Here A and B are called complementary ideals.

An ideal A is a *direct summand* if there exists a complementary ideal or equivalently, if there exists a "retracting" homomorphism $\rho: \mathcal{G} \to A$ with $\rho \circ i = id_A$ (here $i: A \subset \mathcal{G}$ is the identity on A).

Example 3. It is possible to write the algebra $so(4)$ as direct sum of other algebras as $so(4) \cong so(3) \oplus so(3) \cong su(2) \oplus su(2) \cong sp(2) \oplus sp(2)$.

1.4 Simple Lie algebras, semisimple Lie algebras, solvable and nilpotent Lie algebras

Definition 14. If the Lie algebra $\mathcal{G}$ has no ideals except itself and 0 and if $[\mathcal{G}, \mathcal{G}] \neq 0$, then $\mathcal{G}$ is called *simple*. Here the condition $[\mathcal{G}, \mathcal{G}] \neq 0$ (ie, $\mathcal{G}$ is nonabelian) is imposed in order to avoid undue prominence to one-dimensional algebra. Clearly $\mathcal{G}$ is simple $\Rightarrow Z(\mathcal{G}) = 0$ and $\mathcal{G} = [\mathcal{G}, \mathcal{G}]$. In case, a Lie algebra is not simple (and not one dimensional), it is possible to factor out a nonzero ideal I in $\mathcal{G}$ and thereby we obtain a Lie algebra (of smaller dimension), namely $\mathcal{G}/I$ which is a *quotient Lie algebra*.

Any simple finite-dimensional Lie algebra is either isomorphic to one of the linear classical Lie algebras $sl(n, \mathbb{C}), so(n, \mathbb{C})$ or $sp(2n, \mathbb{C})$ or one of the five exceptional Lie algebras E_6, E_7, E_8, F_4, and G_2. It is natural to study a Lie algebra $\mathcal{G}$ through its ideals.

Basic properties of A_n, B_n, C_n, and D_n were given in Section 1.3. More details about these algebras regarding the root systems, etc., and the details regarding the five exceptional Lie algebras E_6, E_7, E_8, F_4, and G_2 will be given in Section 1.16 of this chapter.

Remark 8. For classification of simple Lie algebras over a field of nonzero characteristic, one can refer Kac [39].

Definition 15. Let $\mathcal{G}$ be a Lie algebra. We define a sequence of ideals of $\mathcal{G}$ *(called the derived series)* by $\mathcal{G}^{(0)} = \mathcal{G}$, $\mathcal{G}^{(1)} = [\mathcal{G}, \mathcal{G}]$, $\mathcal{G}^{(2)} = [\mathcal{G}^{(1)}, \mathcal{G}^{(1)}], \ldots, \mathcal{G}^{(i)} = [\mathcal{G}^{(i-1)}, \mathcal{G}^{(i-1)}]$. We call $\mathcal{G}$ *solvable*, if $\mathcal{G}^{(n)} = 0$ for some n.

If $\mathcal{G}$ is an arbitrary Lie algebra, there exists a unique maximal solvable ideal, called the radical of $\mathcal{G}$, denoted by *Rad* $\mathcal{G}$.

Example 4. Below we give some examples of solvable Lie algebras and nonsolvable Lie algebras:

(1) The Heisenberg algebra (defined in Chapter 6) is solvable.
(2) Any two-dimensional Lie algebra is solvable.
(3) $sl(2, \mathbb{C})$ is not solvable.
(4) Abelian algebras are solvable whereas simple algebras are not solvable.
(5) The algebra $t(n, \mathbb{F})$ of upper triangular matrices is solvable.

In the following proposition, sum will always mean sum as vector spaces:

Proposition 5. *The following results can be directly verified:*

(1) *Let $\mathcal{G}$ be a Lie algebra.*
  (a) *The sum of two ideals of $\mathcal{G}$ is an ideal.*
  (b) *The sum of an ideal and a subalgebra is a subalgebra.*

(2) (a) *If a Lie algebra $\mathcal{G}$ is solvable then every subalgebra and all homeomorphic images of $\mathcal{G}$ are solvable.*
(b) *If a Lie algebra $\mathcal{G}$ has an ideal I such that I and $\mathcal{G}/I$ are solvable then $\mathcal{G}$ is solvable.*
(c) *The sum of finitely many solvable ideals of $\mathcal{G}$ is a solvable ideal. In particular, the sum of a collection of solvable ideals of a finite-dimensional Lie algebra is a solvable ideal.*

Proof.

(1) (a) Let $\mathcal{G}$ be a Lie algebra. Let I and J be two ideals of $\mathcal{G}$. Then $I+J = \{x+y \mid x \in I,\ y \in J\}$. Let $z \in \mathcal{G}$, $x+y \in I+J$. Then

$$[x+y,z] = [x,z]+[y,z] \subseteq I+J.$$

Hence $I+J$ is also an ideal.
(b) Let $\mathcal{G}$ be a Lie algebra. Let I be an ideal and Q be a subalgebra of $\mathcal{G}$. Then $I+Q = \{x+q \mid x \in I,\ q \in Q\}$. Let $y \in \mathcal{G}$, $q_1,q_2 \in I+Q$. Then

$$[x+q_1,y+q_2] = [x,y]+[x,q_2]+[q_1,y]+[q_1,q_2] \subseteq I+Q.$$

Hence $I+Q$ is also a subalgebra.
(2) (a) Let K be a subalgebra of solvable Lie algebra $\mathcal{G}$, then $K^{(i)} \subset \mathcal{G}^{(i)}$ (by definition). So K is solvable.
Let ϕ: $\mathcal{G}_1 \to \mathcal{G}_2$ be an algebra epimorphism, then using induction on i, it is clear that $\phi(\mathcal{G}_1^{(i)}) = \mathcal{G}_2^{(i)}$. Hence $\phi(\mathcal{G}_1)$ is also solvable.
(b) Let I be a solvable ideal of $\mathcal{G}$ satisfying the condition that $(\mathcal{G}/I)$ is solvable.
Then $(\mathcal{G}/I)^{(n)} = 0$ for some particular n. Consider the map ϕ: $\mathcal{G} \to \mathcal{G}/I$, by part (2)(a), we get $\phi(\mathcal{G}^{(n)}) = 0$ or $\mathcal{G}^{(n)} \subset I = \mathit{kernel}\ \phi$.
If I is solvable, $I^{(m)} = 0$, then $(\mathcal{G}^{(n)})^{(m)} \subseteq I^{(m)} = 0$ for some m. As $(\mathcal{G}^{(n)})^{(m)} = \mathcal{G}^{(n+m)}$, $\mathcal{G}$ is solvable.
(c) Let I and J be two solvable ideals of $\mathcal{G}$. By second isomorphism theorem, there is an isomorphism between $(I+J)/J$ and $I/(I \cap J)$.
Since homomorphic image of I is solvable, $(I+J)/J$ is solvable. Then by part (2)(b), $I+J$ is also solvable.

□

Definition 16. If $\mathcal{G} \neq 0$ and $\mathit{Rad}\ \mathcal{G} = 0$, then $\mathcal{G}$ is called *semisimple*. A simple Lie algebra is semisimple.

Example 5.

(1) Note that if $\mathcal{G}$ is not solvable, that is, $\mathcal{G} \neq \mathit{Rad}\ \mathcal{G}$, then $\mathcal{G}/\mathit{Rad}\ \mathcal{G}$ is semisimple.
(2) The Lie algebra $so(3)$ with generators X_1, X_2, and X_3 satisfying the relations:

$$[X_1,X_2] = X_3,\ [X_2,X_3] = X_1, \text{ and } [X_3,X_1] = X_2$$

is semisimple.
(3) The Lie algebra $e(2)$ defined by $[X_1,X_2] = X_3$, $[X_1,X_3] = -X_2$, $[X_2,X_3] = 0$ is not semisimple.

Remark 9. The following results can be easily verified.

(1) A semisimple Lie algebra has a center $\{0\}$.
(2) A semisimple Lie algebra $\mathcal{G}$ is a direct sum $\mathcal{G} = \mathcal{G}_1 + \mathcal{G}_2 + \cdots + \mathcal{G}_r$, where each $\mathcal{G}_i$ is a simple ideal in $\mathcal{G}$. Moreover, any ideal in $\mathcal{G}$ is the direct sum of some $\mathcal{G}_i$'s.

Definition 17. We define another sequence of ideals of $\mathcal{G}$, called the *descending central series or lower central series* by $\mathcal{G}^0 = \mathcal{G}$, $\mathcal{G}^1 = [\mathcal{G}, \mathcal{G}](= \mathcal{G}^{(1)}), \ldots, \mathcal{G}^i = [\mathcal{G}, \mathcal{G}^{i-1}]$. $\mathcal{G}$ is called *nilpotent*, if $\mathcal{G}^n = 0$ for some n.

Clearly, nilpotent algebras are solvable. Thus $\mathcal{G}^r$ is spanned by *iterated* or *long* brackets $[X_1[X_2[\cdots X_{r+1}]\cdots]$ (which we abbreviate as $[X_1 X_2 \cdots X_{r+1}]$). Again $\mathcal{G}^r$ are characteristic ideals and the relation $\mathcal{G}^{r+1} \subset \mathcal{G}^r$ holds. The following properties can be proved:

(a) If a Lie algebra $\mathcal{G}$ is nilpotent then any subalgebra of $\mathcal{G}$ is also nilpotent.
(b) For a Lie algebra $\mathcal{G}$, if $\mathcal{G}/Z(\mathcal{G})$ is nilpotent then $\mathcal{G}$ is nilpotent where $Z(G)$ is center of $\mathcal{G}$.

Example 6.

(i) Any abelian algebra is *nilpotent*.
(ii) The Lie algebra $\eta(n, \mathbb{C})$ of strictly upper triangular matrices over the field $\mathbb{C}$ is nilpotent.

The proof of the following standard theorem can be directly verified.

Theorem 2 (Engel's theorem). *First version: Let V be a vector space; let $\mathcal{G}$ be a sub Lie algebra consisting of entirely nilpotent operators in general linear Lie algebra $gl(V)$. Then $\mathcal{G}$ is a nilpotent Lie algebra.*

Second version: If $\mathcal{G}$ is a Lie algebra such that all operators ad X (with X in $\mathcal{G}$) are nilpotent, then $\mathcal{G}$ is nilpotent.

Remark 10.

(1) If $\mathcal{G}$ is nilpotent, then there exists a basis of V in which every element of $\mathcal{G}$ can be represented by a strictly upper triangular matrix.
(2) Lie algebra $\mathcal{G}$ is nilpotent if and only if for all $x \in \mathcal{G}$, a linear map *ad* x: $\mathcal{G} \to \mathcal{G}$ is nilpotent.
(3) All nilpotent algebras are solvable where as all solvable Lie algebras need not be nilpotent. The best example is set $t(n, \mathbb{C})$ of all upper triangular matrices $n \geq 2$ is solvable but not nilpotent.
(4) A two-dimensional nonabelian Lie algebra is solvable which is not nilpotent.

Moreover, the algebra $e(2)$ defined in Example 5 is not nilpotent.

1.5 Isomorphism theorems, Killing form, and some basic theorems

Proposition 6. *Let ϕ: $\mathcal{G} \to \mathcal{G}_1$ be a homomorphism of a Lie algebra $\mathcal{G}$ into another Lie algebra $\mathcal{G}_1$. Then*

(a) $\ker\phi$ *is an ideal in* $\mathcal{G}$.
(b) $\Im\phi = Im\,\phi$ *is a subalgebra of* $\mathcal{G}$.
(c) *The Lie algebra $Im\,\phi$ is isomorphic to the Lie algebra $\mathcal{G}/\ker\phi$ (first isomorphism theorem).*

Proof.

(a) Given $a \in \ker\phi, b \in \mathcal{G}, \phi([a,b]) = [\phi(a), \phi(b)] = [0, \phi(b)] = 0 \Rightarrow [\ker\phi, \mathcal{G}] \subset \ker\phi \Rightarrow$ $\ker\phi$ is an ideal.
(b) $a, b \in \Im\phi \Rightarrow a = \phi(a'), b = \phi(b') \Rightarrow \phi(\lambda a' + \nu b') = \lambda a + \nu b$ (by the linearity of ϕ) $\Rightarrow \lambda a + \nu b \in \Im\phi \Rightarrow \Im\phi$ is a subspace. We have

$$[a,b] = [\phi(a'),\phi(b')] = \phi([a',b'])$$
$$\Rightarrow [a,b] \in \Im\phi$$
$$\Rightarrow \Im\phi \text{ is a subalgebra of } \mathcal{G}$$

(c) **(i)** Define $\psi: \mathcal{G}/\ker\phi \to \Im\phi$ by $\psi(a+\ker\phi) = \phi(a)$ for $a \in \mathcal{G}$. So $\psi(\mathcal{G}/\ker\phi) = \Im\phi$. Hence ψ is surjective.

(ii) We have $a + \ker\phi = \{a + b | b \in \ker\phi\}$. Let $a+b = c$ and $a + b' = d, b, b' \in \ker\phi$. Choosing two representatives $c, d \in a + \ker\phi$, let us prove that both yield $\phi(a)$ under ψ, that is $\psi(c) = \psi(d) = \phi(a)$. Now $\psi(c) = \psi(d)$ implies $\psi(a+b) = \psi(a+b')$ both are equal to $\phi(a)$. Hence ψ is well defined.

(iii) $\psi(a+\ker\phi) = \psi(b+\ker\phi) \Rightarrow \phi(a) = \phi(b) \Rightarrow a+d = b+c, c, d \in \ker\phi \Rightarrow a + \ker\phi = b + \ker\phi \Rightarrow \psi$ is bijective.

One can easily verify that ψ is a homomorphism. From (i), (ii), and (iii), it follows that ψ is an isomorphism.

□

Example 7.

(1) Consider $su(2)$, the set of all skew-Hermitian matrices of trace 0. The matrices $S_x = 1/2\begin{pmatrix} 0 & i \\ i & 0 \end{pmatrix}$, $S_y = 1/2\begin{pmatrix} 0 & -1 \\ 1 & 0 \end{pmatrix}$, $S_z = 1/2\begin{pmatrix} i & 0 \\ 0 & -i \end{pmatrix}$ will form a basis for $su(2)$. It can be easily proved that $su(2)$ and $o(3)$ are isomorphic, via the map $S_x \to R_x$, etc., where R_x, R_y, and R_z are as in Theorem 1.

(2) The nonabelian two-dimensional Lie algebra $\mathcal{G}_2$ with $[x,y] = x$ has the table

[,]	x	y
x	0	x
y	$-x$	0

On the other hand, the set of all matrices of the form $\left\{\begin{pmatrix} -a & b \\ 0 & a \end{pmatrix} \Big/ a, b \in \mathbb{R}\right\}$ is a nonabelian Lie algebra X_2, which is a sub algebra of $sl(2,\mathbb{R})$, with a basis,

$$\left\{X = \begin{pmatrix} 0 & 1 \\ 0 & 0 \end{pmatrix} \text{ and } h = -\frac{1}{2}\begin{pmatrix} -1 & 0 \\ 0 & 1 \end{pmatrix}\right\}.$$

It is clear that isomorphism between $\mathcal{G}_2$ and X_2 can be established by the correspondence

$$x \mapsto X \text{ and } y \mapsto h.$$

Remark 11.

(1) The proofs of the isomorphism theorems are similar to those for vector spaces. Hence they can be directly verified. One can also refer Humphreys [40] and Jacobson [22].

(2) Another version of the first isomorphism theorem: Let $\mathcal{G}$ and $\mathcal{G}_1$ be Lie algebras and K be the kernel of the homomorphism $\varphi: \mathcal{G} \to \mathcal{G}_1$. Then the induced map $\varphi': \mathcal{G}/K \to \mathcal{G}_1$ sets up an isomorphism of $\mathcal{G}/K$ with the image, Lie algebra $\varphi(\mathcal{G})$.

Example 8. Consider a map *trace* tr: $gl(n,\mathbb{C}) \to \mathbb{C}$ sending a matrix to its trace. This is a Lie algebra homomorphism. We have

$$tr[x,y] = tr(xy - yx) = tr\, xy - tr\, yx = 0.$$

But $[tr\ x, tr\ y] = tr\ x\ tr\ y - tr\ y\ tr\ x = 0$. Therefore $tr[x, y] = [tr\ x, tr\ y] = 0$. This map tr is onto and its kernel is $sl(n, \mathbb{C})$. By first isomorphism theorem, the map $gl(n, \mathbb{C})/sl(n, \mathbb{C}) \to \mathbb{C}$ will be $x + sl(n, \mathbb{C}) \mapsto tr\ (x)$.

We state below the following isomorphism theorems.

The second isomorphism theorem: If A is a subalgebra and B is an ideal in a Lie algebra $\mathcal{G}$ with $A \cap B$ being nonempty, then natural map of A into $A + B$ induces an isomorphism of $A/(A \cap B)$ with $(A + B)/B$.

The third isomorphism theorem: If A and B are ideals of a Lie algebra $\mathcal{G}$ such that $A \subseteq B$ then, B/A is an ideal of $\mathcal{G}/A$ and $(\mathcal{G}/A)/(B/A)$ is isomorphic to $\mathcal{G}/B$.

Definition 18. Let $\mathcal{G}$ be a Lie algebra. For a symmetric bilinear form $\beta(x, y)$, we can define its radical by $S = \{x \in \mathcal{G} \mid \beta(x, y) = 0 \text{ for } y \in \mathcal{G}\}$. It is clear that $\beta(x, y)$ is *nondegenerate* if and only if $S = 0$.

Definition 19. Let $\mathcal{G}$ be a Lie algebra. For $x, y \in \mathcal{G}$, we can define an associative, symmetric bilinear form κ on $\mathcal{G}$ called the *Killing form* as $\kappa(x, y) = trace\ (ad\ x \circ ad\ y)$, where $(ad\ x)y = [x, y]$.

Killing form (named after W. Killing; in the literature), can also be defined as follows:

Definition 20 (Killing form defined through adjoint representation). Let $\mathcal{G}$ be a finite-dimensional Lie algebra and ϕ be a representation of $\mathcal{G}$ in V. We define an invariant bilinear form B_V on $\mathcal{G}$ as $B_V(x, y) := tr_V(\phi(x)\phi(y))$, where ϕ: $\mathcal{G} \to Gl(V)$ and $\phi(x)$: $V \to V$. If V is $\mathcal{G}$ itself then the bilinear form associated to the adjoint representation of $\mathcal{G}$ in $\mathcal{G}$ is the bilinear form

$$B(x, y) := tr_{\mathcal{G}}(ad(x)ad(y)), \quad \text{for all } x, y \in \mathcal{G},$$

called the *Killing form* and denoted as κ.

The *Killing form* κ is associative in the sense, $\kappa([x, y], z) = \kappa(x, [y, z])$. It can be established that $tr([A, B], C) = tr(A, [B, C])$, for endomorphisms A, B, C of a finite-dimensional vector space. $\kappa(X, Y)$ can be thought of as an inner product on $\mathcal{G}$, attached to the Lie algebra structure on $\mathcal{G}$. Similarly any representation φ gives rise to the symmetric bilinear *trace form* t_φ, defined by

$$t_\varphi(X, Y) = tr(\varphi(X) \circ \varphi(Y)).$$

If α is an automorphism on $\mathcal{G}$ then we have

$$\kappa(\alpha(X), \alpha(Y)) = \kappa(X, Y)$$

for all X, Y in $\mathcal{G}$. So the Killing form is *invariant* under all automorphisms of $\mathcal{G}$: If I is an ideal of $\mathcal{G}$ and κ is a Killing form of $\mathcal{G}$ and κ_I is the Killing form of I (viewed as Lie algebra) then $\kappa_I = \kappa|_{I \times I}$.

Definition 21. The Killing form κ is nondegenerate if for some X_0 in $\mathcal{G}$ the value $\kappa(X_0, Y)$ is 0 for all Y in $\mathcal{G}$, then X_0 is 0. For any subspace S of a Lie algebra $\mathcal{G}$ one can define, $S^\perp$ by

$$S^{\perp} = \{x \in \mathcal{G} \mid \kappa(x,s) = 0 \quad \text{for all } s \in S\}.$$

Here if $\mathcal{G}^{\perp} = 0$, we say that Killing form is nondegenerate. If I is an ideal of $\mathcal{G}$, then $I^{\perp}$ with respect to Killing form is also an ideal of $\mathcal{G}$.

Below we give the different Killing forms for classical Lie algebras:

(i) For $sl(n+1,\mathbb{C})$, the Killing form is $\kappa(X,Y) = 2(n+1)tr(XY)$.
(ii) For $so(2n+1,\mathbb{C})$, the Killing form is $\kappa(X,Y) = (2n-1)tr(XY)$.
(iii) For $sp(2n,\mathbb{C})$, the Killing form is $\kappa(X,Y) = 2(n+1)tr(XY)$.
(iv) For $so(2n,\mathbb{C})$, the Killing form is $\kappa(X,Y) = 2(n-1)tr(XY)$.

Here X and Y are elements in the corresponding classical Lie algebras.

Examples for the matrix of the Killing form

(1) One can compute the Killing form of $sl(2,\mathbb{C})$ using the standard basis x, h, and y as defined before. Defining the Killing form κ in the usual sense, the matrix of this bilinear form κ on $sl(2,\mathbb{C})$ can be found out to be,

$$\begin{pmatrix} \kappa(x,x) & \kappa(x,h) & \kappa(x,y) \\ \kappa(h,x) & \kappa(h,h) & \kappa(h,y) \\ \kappa(y,x) & \kappa(y,h) & \kappa(y,y) \end{pmatrix} = \begin{pmatrix} 0 & 0 & 4 \\ 0 & 8 & 2 \\ 4 & 0 & 0 \end{pmatrix}.$$

(2) Let $\mathcal{G}$ be a two-dimensional nonabelian Lie algebra with basis $x,\ y$ defined by $[x,y] = x$ for all $x, y \in \mathcal{G}$. It can be easily verified that matrix κ of the Killing form here is $\begin{pmatrix} 0 & 0 \\ 0 & 1 \end{pmatrix}$.

We prove the following Theorem 3 and Theorem 4 [40, 41, 43].

Theorem 3 (Cartan's first criterion). *A Lie algebra $\mathcal{G}$ is solvable if and only if its Killing form κ vanishes identically on the derived Lie algebra $\mathcal{G}'$. That is, let $\mathcal{G}$ be a Lie algebra over $\mathbb{C}$ then $\mathcal{G}$ is solvable if and only if $\kappa(x,y) = 0$ for all $x \in \mathcal{G}$ and $y \in [\mathcal{G},\mathcal{G}]$.*

Proof. It is sufficient to prove that $\mathcal{G}' = [\mathcal{G},\mathcal{G}]$ is nilpotent. Hence we have to prove all x in $[\mathcal{G},\mathcal{G}]$ are nilpotent endomorphisms. With out loss of generality, we can assume that $\mathcal{G}$ be an subalgebra of $gl(V)$.

Let $\mathcal{G}' = [\mathcal{G},\mathcal{G}]$, $\mathcal{G}_1 = \mathcal{G}$ and $M = \{x \in gl(V) | \ [x,\mathcal{G}] \subset [\mathcal{G},\mathcal{G}]\}$. It is clear that $\mathcal{G} \subset M$. Since $Tr(xy) = 0$, for all $x \in [\mathcal{G},\mathcal{G}]$ and $y \in \mathcal{G}$, to prove that for each $x \in [\mathcal{G},\mathcal{G}]$ is nilpotent, it is enough to prove $Tr(xy) = 0$, for all $x \in [\mathcal{G},\mathcal{G}]$ and $y \in M$. Let $[x,y] \in [\mathcal{G},\mathcal{G}]$ and $z \in M$, then it can be directly verified that $Tr([x,y]z) = Tr(x[y,z]) = Tr([y,z]x)$. Using the definition of M, $[y,z] \in [\mathcal{G},\mathcal{G}]$, $Tr([y,z]x) = 0$. Hence $\mathcal{G}$ is solvable. □

Theorem 4 (Cartan's second criterion). *Let $\mathcal{G}$ be a (nonzero) Lie algebra over $\mathbb{C}$. Then $\mathcal{G}$ is semisimple if and only if its Killing form κ is nondegenerate.*

Proof. Let $\mathcal{G}$ be semisimple. Then $Rad\ \mathcal{G} = 0$. Let R be the radical of Killing form κ. So $Tr(ad\ x \circ ad\ y) = 0$ for all $x \in R, y \in \mathcal{G}$.

Using Cartan's first criterion, $ad_{\mathcal{G}}R$ is solvable, hence R is solvable. But R is an ideal of $\mathcal{G}$, so $R \subseteq Rad\mathcal{G} = 0$. This implies κ is nondegenerate. Conversely, let $R = 0$. Now, we have to prove $\mathcal{G}$ is semisimple. For that it is enough to prove that every abelian ideal I of $\mathcal{G}$ is included in R. Let $x \in I, y \in \mathcal{G}$. Then consider $ad\ x \circ ad\ y$: $\mathcal{G} \to I$ and $(ad\ x \circ ad\ y)^2$: $\mathcal{G} \to [I, I]$. But $[I, I] = 0$. Hence $ad\ x \circ ad\ y$ is nilpotent. Hence $0 = Tr(ad\ x \circ ad\ y) = \kappa(x, y)$. So $I \subset R = 0$. Hence $\mathcal{G}$ is a semisimple Lie algebra. □

Definition 22 (Semisimple Lie algebra using Killing form). A Lie algebra $\mathcal{G}$ over a field of characteristic 0 is called *semisimple* if the Killing form κ of $\mathcal{G}$ is nondegenerate. We shall call a Lie algebra $\mathcal{G} \neq \{0\}$ is simple if it is semisimple and has no ideals except $\{0\}$ and $\mathcal{G}$.

The following result can be easily proved.

Remark 12. Let $\mathcal{G}$ be a semisimple Lie algebra and J be an ideal in $\mathcal{G}$. Let $J^{\perp}$ denote the set of elements $X \in \mathcal{G}$ which are orthogonal to J with respect to Killing form. Then J is semisimple, $J^{\perp}$ is an ideal and $\mathcal{G} = J + J^{\perp}$.

Definition 23 (Abstract Jordan decomposition). Let $\mathcal{G}$ be a complex semisimple Lie algebra. An *abstract Jordan decomposition* of an element $a \in \mathcal{G}$ is a decomposition of the form $a = a_s + a_n$, for a_n, $a_s \in \mathcal{G}$ where $ad(a_s)$ is a semisimple operator and $ad(a_n)$ is a nilpotent operator of $\mathcal{G}$ and $[a_s, a_n] = 0$. If a_n =0 then a is semisimple.

Example 9. If $\mathcal{G} = gl_{\mathbb{N}}(\mathbb{F})$, where $\mathbb{F}$ is algebraically closed, then for $A \in \mathcal{G}$, $A = A_s + A_n$ (the concrete Jordan decomposition) is also an abstract Jordan decomposition, as $ad(A_s)$ is semisimple, $ad(A_n)$ is nilpotent, and $[ad(A_s), ad(A_n)] = ad[A_s, A_n] = 0$.

Remark 13. Note that $A = A_s' + A_n'$ is another abstract Jordan decomposition with $A_s' = A_s + \lambda I, A_n' = A_n - \lambda I$ for any $\lambda \in \mathbb{C}$. Thus we see that the abstract Jordan decomposition is not unique in itself. The uniqueness fails in this case because I is a central element.

For the following Theorem 5 and Corollary 1, one can refer Humphreys [40].

Theorem 5. *Let $\mathcal{G}$ be semisimple. Then there exist ideals $\mathcal{G}_1, \ldots, \mathcal{G}_t$ of $\mathcal{G}$, which are simple (as Lie algebras), such that $\mathcal{G} = \mathcal{G}_1 \oplus \cdots \oplus \mathcal{G}_t$. Every simple ideal of $\mathcal{G}$ coincides with one of the $\mathcal{G}_i$. Moreover, the Killing form of $\mathcal{G}_i$ is the restriction of κ to $\mathcal{G}_i \times \mathcal{G}_i$.*

Proof. We prove this theorem, by induction on $\dim \mathcal{G}$. If $\mathcal{G}$ has no proper nonzero ideal, then $\mathcal{G}$ is itself simple. Hence the proof.

Otherwise, consider a minimal nonzero ideal $\mathcal{G}_1$ of $\mathcal{G}$. Then $\mathcal{G}_1^{\perp} = \{x \in \mathcal{G} \mid \kappa(x, y) = 0$, for all $x \in \mathcal{G}_1\}$. This implies that $\mathcal{G} = \mathcal{G}_1 \oplus \mathcal{G}_1^{\perp}$. Any ideal of $\mathcal{G}_1$ is an ideal of $\mathcal{G}$. By minimality property, $\mathcal{G}_1$ is simple and hence semisimple. So $\mathcal{G}_1^{\perp}$ is also semisimple. Hence it is clear that $\mathcal{G}_1^{\perp}$ can be split into simple ideals.

Now we have to prove that the simple ideals are unique. Let I be a simple ideal of $\mathcal{G}$, then $[I, \mathcal{G}]$ is also an nonzero ideal of I. This implies $[I, \mathcal{G}] = I$. On the other hand, $[I, \mathcal{G}] = [I, \mathcal{G}_1] \oplus \cdots \oplus [I, \mathcal{G}_t]$, so all but one summand must be 0. Let $[I, \mathcal{G}_i] = I$. Then $I \subset \mathcal{G}_i$ and $I = \mathcal{G}_i$. Hence the proof. □

Corollary 1. *If $\mathcal{G}$ is semisimple, then $\mathcal{G} = [\mathcal{G}, \mathcal{G}]$ and all ideals and, homomorphic images of $\mathcal{G}$ are semisimple. Moreover, each ideal of $\mathcal{G}$ is a sum of certain simple ideals of $\mathcal{G}$.*

Remark 14. A semisimple ideal in a Lie algebra is a direct summand. All simple Lie algebras are semisimple, but there exist semisimple Lie algebras which are not simple. It is easy to see both solvability and vanishing of κ on $\mathcal{G}'$ remain unchanged under complexification for a real $\mathcal{G}$; thus we may take $\mathcal{G}$ to be complex.

Example 10. If $\mathcal{G} = sl(n, \mathbb{F})$ and H is the set of all diagonal matrices in $\mathcal{G}$, then $\mathcal{G}$ is the direct sum of the subspaces $\mathcal{G}_\alpha = \{x \in \mathcal{G} | [h, x] = \alpha(h)x$ for all $h \in H\}$, where α ranges over H^*. In this case we have a root space decomposition (or Cartan decomposition) $\mathcal{G} = H \oplus \mathcal{G}_\alpha$, $\alpha \in \Phi$. The restriction of the Killing form κ to H is nondegenerate as $H = C_{\mathcal{G}}(H)$.

Remark 15. Root system decomposition in general will be explained in Section 1.11 of this chapter.

1.6 Derivation of Lie algebras

Definition 24. Let A be an algebra over a field K. A *derivation* of A is a mapping D: $A \to A$ such that

(i) $D(\alpha f + \beta g) = \alpha Df + \beta Dg$ for $\alpha, \beta \in K, f, g \in A$ and
(ii) $D(fg) = f(Dg) + (Df)g$ for $f, g \in A$.

In fact, for any algebra A, an endomorphism D is called a derivation if the Leibnitz rule holds, that is,

$$D(ab) = (Da)b + a(Db) \quad \text{for all } a, b \in A.$$

Definition 25. A *derivation* of a Lie algebra $\mathcal{G}$ is an operator D: $\mathcal{G} \to \mathcal{G}$ that satisfies $D([X, Y]) = [DX, Y] + [X, DY]$ for all X, Y in $\mathcal{G}$.

Given an algebra A, denote by $Der\ A(\subset End(A))$, the vector space of derivations of A. Clearly, $Der\ A$ is a vector subspace of $gl(A)$. In fact, it is a Lie subalgebra of $gl(A)$.

Moreover, the derivation D is the infinitesimal version of an automorphism: If $\alpha(s)$ is a differentiable family of automorphisms with $\alpha(0) = id$, one finds on differentiating (using Leibnitz's rule) the relation $\alpha(s)([X, Y]) = [\alpha(s)(X), \alpha(s)(Y)]$, $\alpha'(0)$ (the derivative at 0), which is a derivation. In other words, the first-order term in the expansion $\alpha(s) = id + sD + \cdots$ is a derivation. Conversely, if D is a derivation, then all e^{sD} are automorphisms, as one can verify again by differentiating.

Moreover, the Killing form is (infinitesimally) invariant under any derivation D of $\mathcal{G}$, that is, we have $\kappa(DX, Y) + \kappa(X, DY) = 0$ for all X, Y in $\mathcal{G}$.

Remark 16. Recall that if $\mathcal{G}$ is any F-algebra of finite dimension, then $Der\ \mathcal{G}$ contains the semisimple and nilpotent parts in $End(\mathcal{G})$ of all its elements. In particular, since $Der\ \mathcal{G}$ coincides with $ad\ \mathcal{G}$, such that $\mathcal{G} \to ad\ \mathcal{G}$ is $1-1$, each $x \in \mathcal{G}$ determines unique elements $s, n \in \mathcal{G}$ such that $ad\ x = ad\ s + ad\ n$ (the usual Jordan decomposition of $ad\ x$ in $End\ \mathcal{G}$). This means that if $x = s + n$, with $[s, n] = 0$, then s is ad-semisimple (ie, $ad\ s$ is semisimple) and n is ad-nilpotent (ie, $ad\ n$ is nilpotent). We write $s = x_s, n = x_n$, and call s and n as the *semisimple* and *nilpotent parts* of x, respectively.

Proposition 7. *Commutator* $[\delta_1, \delta_2]$ *of two derivations* δ_1 *and* δ_2 *of an* F*-algebra is a derivation.*

Proof. Denote by δ, the derivation of an F-algebra $\mathcal{A}$ to $\mathcal{A}$ as δ: $\mathcal{A} \to \mathcal{A}$ such that $\delta\ (ab) = \delta(a)b + a\delta(b)$. Denote by $Der\ \mathcal{A}$ the set of all derivations of $\mathcal{A}$. As $Der\ \mathcal{A}$ is a subspace of $End(\mathcal{A})$, we define the commutator

$$[\delta_1, \delta_2] = \delta_1\delta_2 - \delta_2\delta_1$$

$$\begin{aligned}\text{We have } [\delta_1, \delta_2](ab) &= (\delta_1\delta_2 - \delta_2\delta_1)(ab)\\ &= (\delta_1\delta_2)(ab) - (\delta_2\delta_1)(ab)\\ &= \delta_1(\delta_2(ab)) - \delta_2(\delta_1(ab))\\ &= \delta_1(\delta_2(a)b + a\delta_2(b)) - (\delta_2(\delta_1(a)b + a\delta_1(b)))\\ &= \delta_1(\delta_2(a))b + \delta_2(a)\delta_1(b) + \delta_1(a)\delta_2(b) + a\delta_1(\delta_2(b))\\ &\quad - \delta_2(\delta_1(a))b - \delta_1(a)\delta_2(b) - \delta_2(a)\delta_1(b) - a\delta_2(\delta_1(b))\\ &= (\delta_1\delta_2(a))b - (\delta_2\delta_1(a))b + a(\delta_1\delta_2)b - a(\delta_2\delta_1)(b)\\ &= ((\delta_1\delta_2 - \delta_2\delta_1)a)b + a(\delta_1\delta_2 - \delta_2\delta_1)(b)\\ &= ([\delta_1, \delta_2]a)b + a[\delta_1, \delta_2]b.\end{aligned}$$

Hence $[\delta_1, \delta_2]$ is a derivation. □

Definition 26. Given an element $a \in \mathcal{G}$, define the operator *ad a* (adjoint) on $\mathcal{G}$ by:

$$(ad\ a)b = [a, b].$$

Jacobi's identity implies that *ad a* is a derivation of the Lie algebra $\mathcal{G}$. It is called an *inner derivation.*

The *ad X*'s are the *inner derivations* of $\mathcal{G}$, similar to the inner automorphisms of a group.

Remark 17. It can be verified that, if I is an ideal of a Lie algebra $\mathcal{G}$, κ is a Killing form of $\mathcal{G}$ and κ_I is the Killing form of I, then $\kappa_I = \kappa|_{I\times I}$.

For the proof of the following theorem, one can also refer Humphreys [40].

Theorem 6. *For a semisimple Lie algebra* $\mathcal{G}$, $Der\ \mathcal{G} = ad\ \mathcal{G}$.

Proof. Let $\mathcal{G}$ be a semisimple Lie algebra. Then center of $\mathcal{G}, Z(\mathcal{G}) = 0$. Hence $\mathcal{G} \to ad\ \mathcal{G}$ is an isomorphism of Lie algebras. If $ad\ \mathcal{G}$ is semisimple, then it has a nondegenerate Killing form.

If $ad\ \mathcal{G}$ is an ideal in $Der\ \mathcal{G}$, then $[Der\ \mathcal{G}, ad\ \mathcal{G}] \subset ad\ \mathcal{G}$. By the Remark 17, $\kappa_{ad\ \mathcal{G}}$ is the restriction to $ad\ \mathcal{G} \times ad\ \mathcal{G}$ of the Killing form $\kappa_{Der\ \mathcal{G}}$ of $Der\ \mathcal{G}$. In particular, if $I = ad\ \mathcal{G}^{\perp}$, then $I \cap ad\ \mathcal{G} = 0$ as $\kappa_{ad\ \mathcal{G}}$ is nondegenerate such that $ad\ \mathcal{G} \cup ad\ \mathcal{G}^{\perp} = Der\ \mathcal{G}$. As both I and $ad\ \mathcal{G}$ are ideals of $Der\ \mathcal{G}$, $[I, ad\ \mathcal{G}] = 0$. If $\delta \in I$, then

$$[\delta, ad\ x] = ad(\delta x), \quad x \in \mathcal{G}, \quad \delta \in Der\ \mathcal{G}.$$

If $\delta \in I$, then $ad\ (\delta x) = 0$ for all $x \in \mathcal{G}$. This implies $\delta x = 0$ $(x \in \mathcal{G})$, because *ad* is one-one and $\delta = 0$. So this implies $I = 0$. Hence $Der\ \mathcal{G} = ad\ \mathcal{G}$. □

Proposition 8. *If* $\mathcal{G}$ *is a Lie algebra, the map ad x*: $\mathcal{G} \to \mathcal{G}$ *is a derivation of* $\mathcal{G}$, *then*

$$(ad\ x)[y, z] = [x, [y, z]].$$

Proof. By Jacobi identity, we have

$$[x,[y,z]] = [[x,y],z] + [y,[x,z]] = [(ad\, x)y, z] + [y, (ad\, x)z]$$
$$= ad\, x[y,z] \quad \text{for all } x, y, z \in \mathcal{G}.$$

□

1.7 Representations of Lie algebras and representations of $sl(2,\mathbb{C})$

Definition 27. Let $\mathcal{G}$ be a Lie algebra over complex numbers $\mathbb{C}$. A *representation* of a Lie algebra $\mathcal{G}$ on a vector space V is a homomorphism φ from $\mathcal{G}$ to the general linear algebra $gl(V)$ of V. So φ assigns to each X in $\mathcal{G}$, an operator $\varphi(X)$: $V \to V$ (or, if one wants to use a basis of V, a matrix), depending linearly on X, that is,

(i) $\varphi(aX + bY) = a\varphi(X) + b\varphi(Y)$ and
(ii) $\varphi([X,Y]) = [\varphi(X), \varphi(Y)] = \varphi(X)\varphi(Y) - \varphi(Y)\varphi(X)$ ("preservation of brackets").

If we allow the possibility of $\mathcal{G}$ to be real, one considers $gl(V)$ as a real Lie algebra, by "restriction of scalars."

Example 11. Let $\mathcal{G}$ be a Lie subalgebra of $gl(V)$. Then the inclusion map $\mathcal{G} \hookrightarrow gl(V)$ is Lie algebra homomorphism which is the restriction of the identity map on $gl(V)$ to $\mathcal{G}$. The corresponding representation is a natural representation of $\mathcal{G}$. For example, the set $sl(n,\mathbb{C})$ of trace 0 matrices in $gl(n,\mathbb{C})$ has the natural representation in $gl(n,\mathbb{C})$. Similarly, the set $bl(n,\mathbb{C})$ of upper triangular matrices has a natural representation.

Definition 28. The dimension of the representation φ is the dimension of the vector space V. For $V = \mathbb{F}^n$, the elements $\varphi(X)$ are $n \times n$ matrices and we get the notion of *matrix representation*. A representation φ is *faithful* if the kernel of map φ, namely, $\ker \varphi = 0$, that is, if the only X with $\varphi(X) = 0$ is 0 itself. If φ has kernel K, it induces a faithful representation of $\mathcal{G}/K$ in the standard way. The *trivial representation* is the representation on a one-dimensional space, with all representing operators being 0. As a matrix representation it assigns to each element of $\mathcal{G}$, the matrix [0].

Let $\mathcal{G}$ act on V via a representation φ. An *invariant or stable* subspace is a subspace W of V with $\varphi(X)(W) \subset W$ for all X in $\mathcal{G}$. There is an obvious induced representation of $\mathcal{G}$ in W. Furthermore, there is an induced representation on the quotient space V/W and the canonical quotient map $V \to V/W$ is equivariant.

For proof of the following Lie's theorem, one can refer to Erdman and Wildon [41] and Samelson [43].

Theorem 7 (Lie's theorem). *We give below two equivalent forms of the Lie's theorem:*

Theorem *A***:** *Let $\mathcal{G}$ be a solvable Lie algebra, acting on the vector space V by a representation φ, all over $\mathbb{C}$. Then there exists a "joint eigen-vector"; that is, there is a nonzero vector v_0 in V that satisfies $\varphi(g)v_0 = \lambda(g)v_0$, where $\lambda(g)$ is a complex number (depending on g), for all g in $\mathcal{G}$. $\lambda(g)$ depends of course linearly on g, that is, λ is a linear function on V.*

Theorem A'**:** *Any complex representation of a complex solvable Lie algebra is equivalent to a triangular one, that is, to one with all (upper-)triangular matrices. The above result can be stated as follows.*

Let V be a n-dimensional complex vector space and $\mathcal{G}$ be a solvable subalgebra of $gl(V)$. Then there exists a basis of V by which every element of $\mathcal{G}$ can be represented by an upper-triangular matrix.

Definition 29. Let φ_1, φ_2 be two representations of $\mathcal{G}$ on the vector spaces V_1, V_2, respectively. A linear map T: $V_1 \to V_2$ is *equivariant* with respect to φ_1 and φ_2 (or intertwines φ_1 and φ_2), if it satisfies the relation $T \circ (\varphi_1(X)) = (\varphi_2(X)) \circ T$ for all X in $\mathcal{G}$.

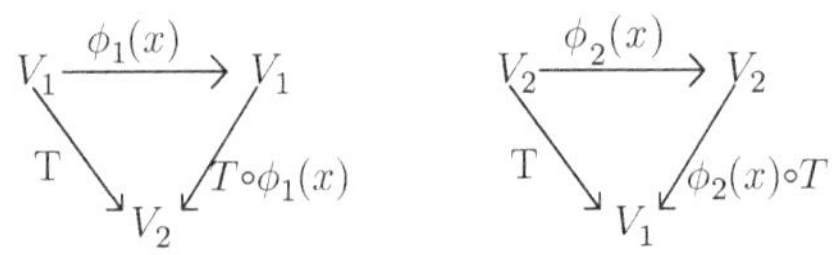

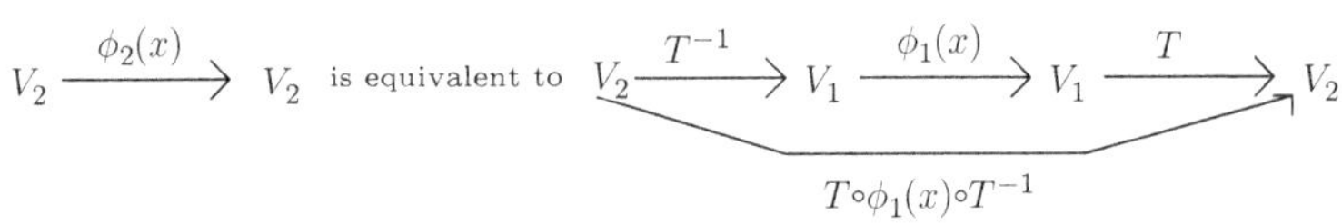

If T is an isomorphism, then φ_1 and φ_2 are equivalent and we have $\varphi_2(X) = T \circ \varphi_1(X) \circ T^{-1}$ for all X in $\mathcal{G}$.

Definition 30. The representation φ and corresponding vector space V are said to be *irreducible* or *simple* if there is no nontrivial (ie, different from 0 and V) invariant subspace of ϕ. Moreover, φ and V are *completely reducible* or *semisimple*, if every invariant subspace of V is a direct sum of irreducible subspaces.

Definition 31. Let φ_1, φ_2 be two representations of $\mathcal{G}$ on V_1 and V_2, respectively. Then the representation $\varphi_1 \oplus \varphi_2$ on $V_1 \oplus V_2$ is defined in the usual sense: $(\varphi_1 \oplus \varphi_2)(X)(v_1, v_2) = (\varphi_1(X)(v_1), \varphi_2(X)(v_2))$.

There is also the *tensor product* $(\varphi_1 \otimes \varphi_2)$ of representations ϕ_1 and ϕ_2 on the tensor product $V_1 \otimes V_2$, defined by $(\varphi_1 \otimes \varphi_2)(X)(v_1 \otimes v_2) = \varphi_1(X)(v_1) \otimes v_2 + v_1 \otimes \varphi_2(X)(v_2)$. This is the infinitesimal version of the tensor product of operators.
Let T_1, T_2 be operators on V_1, V_2, then taking the derivative of $\exp(sT_1) \otimes \exp(sT_2)$ at $s = 0$, one can get $T_1 \otimes id + id \otimes T_2$.

Remark 18. Note that $(\varphi_1 \otimes \varphi_2)(X)$ is not the tensor product of the two operators $\varphi_1(X)$ and $\varphi_2(X)$; it is the *infinitesimal tensor product or tensor sum* and one can use the notation, for example, $\varphi_1 \sharp \varphi_2(X)$.

Definition 32. A very important representation of a Lie algebra $\mathcal{G}$ is the *adjoint representation*, denoted by "*ad*." It is just the (left) regular representation of $\mathcal{G}$. The vector space on which it operates is $\mathcal{G}$ itself. The operator $ad\, X$, assigned to X, is given by $ad\, X(Y) = [X, Y]$ for all Y in $\mathcal{G}$("$ad\, X = [X, -]$"). The representation condition $ad\, [X, Y] = ad\, X \circ ad\, Y - ad\, Y \circ ad\, X$ for any X, Y in $\mathcal{G}$ turns out to be just the

Jacobi condition (plus skew symmetry). The kernel of ad is the center of $\mathcal{G}$ and ideals of $\mathcal{G}$ are the same as ad-invariant subspaces.

Proposition 9. *For the Lie algebra $sl(2,\mathbb{C})$ with an ordered basis*

$$x=\begin{pmatrix}0&1\\0&0\end{pmatrix},\ h=\begin{pmatrix}1&0\\0&-1\end{pmatrix},\ \text{and } y=\begin{pmatrix}0&0\\1&0\end{pmatrix},$$

one can find the matrices of ad x, ad y, and ad h. If

$$((\exp ad\ x)(\exp ad\ (-y))(\exp ad\ x))=\sigma,$$

then order of automorphism of σ is 2.

Proof. Consider $sl(2,\mathbb{C})$ with an ordered basis

$$x=\begin{pmatrix}0&1\\0&0\end{pmatrix},\ h=\begin{pmatrix}1&0\\0&-1\end{pmatrix},\ \text{and } y=\begin{pmatrix}0&0\\1&0\end{pmatrix},$$

respectively. We have $[h,x]=2x,\ [h,y]=-2y,\ [x,y]=h.$

Since,

$$\begin{aligned}(ad\ x)(x) &= [x,x]=0=0\cdot x+0\cdot h+0\cdot y,\\ (ad\ x)(h) &= [x,h]=-2x=-2\cdot x+0\cdot h+0\cdot y, \quad \text{and}\\ (ad\ x)(y) &= [x,y]=h=0\cdot x+1\cdot h+0\cdot y.\end{aligned}$$

$$\text{Hence the matrix of } ad\ x=\begin{pmatrix}0&-2&0\\0&0&1\\0&0&0\end{pmatrix}.$$

$$\text{Similarly, the matrix of } ad\ y=\begin{pmatrix}0&0&0\\-1&0&0\\0&2&0\end{pmatrix} \quad \text{and}$$

$$\text{the matrix of } ad\ h=\begin{pmatrix}2&0&0\\0&0&0\\0&0&-2\end{pmatrix}.$$

Moreover, we can find out an automorphism σ of order 2 on $sl(2,\mathbb{C})$. We have

$$(ad\ x)^2=\begin{pmatrix}0&0&-2\\0&0&0\\0&0&0\end{pmatrix} \text{ and } (ad\ x)^3=\begin{pmatrix}0&0&0\\0&0&0\\0&0&0\end{pmatrix}, \text{etc.}$$

Hence

$$\exp ad\ x=1+ad\ x+\frac{(ad\ x)^2}{2!}+0 \quad \text{and}$$

$$\begin{aligned}(\exp ad\ x)x &= x + (ad\ x)(x) + \frac{(ad\ x)^2}{2!}x \\ &= x.\end{aligned}$$

Apply $\exp ad\ (-y)$, we find

$$\begin{aligned}((\exp ad\ (-y))(\exp ad\ x))(x) &= (\exp ad\ (-y))x \\ &= x + ad\ (-y)x + \frac{(ad\ (-y))^2 x}{2!} + 0 \\ &= x + [-y, x] + \frac{1}{2}[-y, [-y, x]] \\ &= x + h + \frac{1}{2}[-y, h] \\ &= x + h - y.\end{aligned}$$

Now applying $\exp ad\ x$ again, we find

$$((\exp ad\ x)(\exp ad\ (-y))(\exp ad\ x))x = -y.$$

Assuming

$$((\exp ad\ x)(\exp ad\ (-y))(\exp ad\ x)) = \sigma,$$

we find $\sigma(x) = -y$, similarly, $\sigma(y) = -x$ and $\sigma(h) = -h$. $\sigma^2(x) = x$, $\sigma^2 = id$ and hence σ is an automorphism of order 2. □

Definition 33 (Lie algebra modules). Let $\mathcal{G}$ be a Lie algebra over a field $\mathbb{F}$. It is often convenient to use the language of modules along with the (equivalent) language of representations. As in other algebraic theories, there is a natural definition. A vector space V, endowed with an operation $\mathcal{G} \times V \to V$ (denoted by $(x, v) \mapsto x \cdot v$ or just xv) is called a $\mathcal{G}$-module if the following conditions are satisfied:

(1) $(ax + by) \cdot v = a(x \cdot v) + b(y \cdot v)$,
(2) $x \cdot (av + bw) = a(x \cdot v) + b(x \cdot w)$, and
(3) $[x, y] \cdot v = x \cdot y \cdot v - y \cdot x \cdot v$, for all $x, y \in \mathcal{G}; u, w \in V; a, b \in \mathbb{F}$.

For example, if ϕ: $\mathcal{G} \to gl(V)$ is a representation of $\mathcal{G}$, then V may be viewed as an $\mathcal{G}$-module via the action $x \cdot v = \phi(x)(v)$. Conversely, given a $\mathcal{G}$-module V, the equation $\phi(x)v = x \cdot v$ defines a representation ϕ: $\mathcal{G} \to gl(V)$.

Lie modules and representations are two ways to describe the same structure. One can write $X \cdot \upsilon$ or simply $X\upsilon$ instead of $\varphi(X)(\upsilon)$ (the image of the vector υ under the operator $\varphi(X)$); preservation of bracket appears then in the form $[X, Y]\upsilon = XY\upsilon - YX\upsilon$.

Quotient or factor module: If W is a submodule of the $\mathcal{G}$-module V, we can construct the factor module V/W by setting

$$x \cdot (v + W) = (x \cdot v) + W \quad \text{for } x \in \mathcal{G} \text{ and } v \in V.$$

Definition 34 (Complete reducibility of representations). Let V and W be two $\mathcal{G}$-modules. A *homomorphism of $\mathcal{G}$-modules* is a linear map ϕ: $V \to W$ such that $\phi(x \cdot v) = x \cdot \phi(v)$. The kernel of such a homomorphism is then an $\mathcal{G}$-submodule of V. If ϕ is an isomorphism of vector spaces, we call it an *isomorphism* of $\mathcal{G}$-modules. In this case, the two modules will give *equivalent* representations of $\mathcal{G}$. A $\mathcal{G}$-module V is called *irreducible* if it has precisely only two $\mathcal{G}$-submodules, namely, itself and 0. We do not regard a zero-dimensional vector space as an irreducible $\mathcal{G}$-module. We also allow a one-dimensional space on which $\mathcal{G}$ acts (perhaps trivially) to be called irreducible. V is called *completely reducible* if V is a direct sum of irreducible $\mathcal{G}$-submodules, or equivalently if each $\mathcal{G}$-module W of V has a complement W' (an $\mathcal{G}$-submodule such that $V = W \oplus W'$).

For the following Lemma 1, one can refer Humphreys [40] and Karin and Wildon [41].

Lemma 1 (Schur's lemma). *Let ϕ: $\mathcal{G} \to gl(V)$ be irreducible. Then the only endomorphisms of V commuting with all $\phi(x)(x \in \mathcal{G})$ are the scalars.*

Proof. Consider the irreducible representation ϕ: $\mathcal{G} \to gl(V)$. Let M be the corresponding simple $\mathcal{G}$-module and ψ: $M \to M$ be a $\mathcal{G}$-module homomorphism. So ψ is a linear map on the complex vector space M and so it should have an eigen value, say λ. This implies, $\psi - \lambda 1_V$ is also an $\mathcal{G}$-module homomorphism. Hence $\ker(\psi - \lambda 1_V)$ contains λ-eigen vector for ψ and ker $(\psi - \lambda 1_V)$ is a nonzero submodule of M. As M is irreducible, $M =$ ker $(\psi - \lambda 1_V)$. Hence, $\psi = \lambda 1_V$. □

Remark 19. Converse of the above statement in Schur's lemma is also true, if all endomorphisms of V commuting with all $\phi(x)(x \in \mathcal{G})$ are the scalars, that is $f(\phi(x)) = (\phi(x))f$, for $f = \lambda$. Then ϕ: $\mathcal{G} \to gl(V)$ is irreducible.

Remark 20. We briefly explain how to give a module structure to the tensor product of two modules for a Lie algebra $\mathcal{G}$. On the generators $v \otimes w$, require $g \cdot (v \otimes w) = g \cdot v \otimes g \cdot w$. For Lie algebras the correct definition can be got by differentiating: $x \cdot (v \otimes w) = x \cdot v \otimes w + v \otimes w \cdot w$. We have

$$\begin{aligned}([x,y]) \cdot (v \otimes w) &= ([x,y] \cdot v) \otimes w + v \otimes ([x,y] \cdot w),\\ &= (x \cdot y \cdot v - y \cdot x \cdot v) \otimes w + v \otimes (x \cdot y \cdot w - y \cdot x \cdot w)\\ ([x,y]) \cdot (v \otimes w) &= ([x,y]) \cdot (v \otimes w)(x \cdot y \cdot v \otimes w + v \otimes x \cdot y \cdot w)\\ &\quad - (y \cdot x \cdot v \otimes w + v \otimes y \cdot x \cdot w). \end{aligned} \tag{1.7}$$

Similarly,

$$(x \cdot y - y \cdot x) \cdot (v \otimes w) = (x \cdot y \cdot v \otimes w + v \otimes x \cdot y \cdot w) - (y \cdot x \cdot v \otimes w + v \otimes y \cdot x \cdot w). \tag{1.8}$$

By Equations (1.7) and (1.8), we have $([x,y]) \cdot (v \otimes w) = (x \cdot y - y \cdot x) \cdot (v \otimes w)$.

Theorem 8 (Weyl). *[40] Let ϕ: $\mathcal{G} \to gl(V)$ be a (finite-dimensional) representation of a semisimple Lie algebra with $\mathcal{G} \neq 0$. Then ϕ is completely reducible.*

Theorem 9. *[40] Let $\mathcal{G} \subset gl(V)$ be a semisimple linear Lie algebra (V is finite dimensional). Then $\mathcal{G}$ contains the semisimple and nilpotent parts in $gl(V)$ of all its elements. In particular, the abstract and usual Jordan decompositions in $\mathcal{G}$ coincide.*

Definition 35. If $\mathcal{G}$ is a Lie algebra and $\pi\colon \mathcal{G} \to gl(V)$ is its representation, and $\lambda \in \mathcal{G}^*$, then the subspace $V_\lambda = \{v \in V \mid \pi(a)v = \lambda(a)v \text{ for all } a \in \mathcal{G}\}$ is called a weight space of $\mathcal{G}$ attached to the weight λ.

Remark 21. Weights and weight spaces will be explained in Section 1.15 for semisimple Lie algebra.

Representation of $sl(2,\mathbb{C})$ in an $(n+1)$-dimensional vector space

Proposition 10. *The representation of $sl(2,\mathbb{C})$ in $n+1$-dimensional vector space in terms of matrices.*

Proof. Let ρ_n denote a representation of $sl(2,\mathbb{C})$ in V_{n+1}, $n+1$-dimensional vector space over $\mathbb{C}$. In the usual notation, $sl(2,\mathbb{C})$ is spanned by e, f, and h. It is well known that the vector space V_{n+1} is isomorphic to the subspace of $\mathbb{C}[X,Y]$ (all polynomials with complex coefficients in two variables X, Y) consisting of homogeneous polynomials of degree n with basis $X^n, X^{n-1}Y, \dots, XY^{n-1}, Y^n$. Consider the representation ρ_n of $sl(2,\mathbb{C})$ in V_{n+1} defined by

$$\rho_n(e) = X\frac{\partial}{\partial Y};\ \rho_n(f) = Y\frac{\partial}{\partial X} \text{ and } \rho_n(h) = X\frac{\partial}{\partial X} - Y\frac{\partial}{\partial Y}.$$

One can verify the following:

$$\rho_n(h)X^aY^b = (a-b)X^aY^b \quad \text{and}$$
$$[\rho_n(e), \rho_n(f)](X^aY^b) = (a-b)X^aY^b = \rho_n(h)X^aY^b.$$

In particular, $[\rho_n(e), \rho_n(f)](X^n) = nX^n = \rho_n(h)X^n$ and

$[\rho_n(e), \rho_n(f)](Y^n) = -nY^n = \rho_n(h)Y^n$.

Also $[e,f] = h$.

So $\rho_n[e,f] = \rho_n(h)$.

Hence $\rho_n[e,f] = [\rho_n(e), \rho_n(f)] = \rho_n(h)$.

Similarly, $[\rho_n(h), \rho_n(e)] = \rho_n([h,e]) = \rho_n(2e) = 2\rho_n(e)$ and

$[\rho_n(h), \rho_n(f)] = \rho_n([h,f]) = -\rho_n(2f) = -2\rho_n(f)$.

So ρ_n gives a representation of $sl(2,\mathbb{C})$ in V_{n+1}, the $n+1$-dimensional vector space over $\mathbb{C}$. □

Remark 22. In particular, for $n = 1$, the basis elements themselves are the images of these elements under the map $\rho_1\colon sl(2,\mathbb{C}) \to gl(2,\mathbb{C})$ with dimension of V being 2 and $\rho_1(h) = \begin{pmatrix} 1 & 0 \\ 0 & -1 \end{pmatrix}$, $\rho_1(e) = \begin{pmatrix} 0 & 1 \\ 0 & 0 \end{pmatrix}$, and $\rho_1(f) = \begin{pmatrix} 0 & 0 \\ 1 & 0 \end{pmatrix}$. For $n = 2$ and $\rho_2\colon sl(2,\mathbb{C}) \to gl(3,\mathbb{C})$ (with dimension of V being 3), the following can be directly proved:

$$\rho_2(h) := \begin{pmatrix} 2 & 0 & 0 \\ 0 & 0 & 0 \\ 0 & 0 & -2 \end{pmatrix}, \rho_2(e) := \begin{pmatrix} 0 & 2 & 0 \\ 0 & 0 & 1 \\ 0 & 0 & 0 \end{pmatrix}, \rho_2(f) := \begin{pmatrix} 0 & 0 & 0 \\ 1 & 0 & 0 \\ 0 & 2 & 0 \end{pmatrix}.$$

Similarly, for $n = 3$, ρ_3 will become

$$\rho_3(h) := \begin{pmatrix} 3 & 0 & 0 & 0 \\ 0 & 1 & 0 & 0 \\ 0 & 0 & -1 & 0 \\ 0 & 0 & 0 & -3 \end{pmatrix}, \rho_3(e) := \begin{pmatrix} 0 & 3 & 0 & 0 \\ 0 & 0 & 2 & 0 \\ 0 & 0 & 0 & 1 \\ 0 & 0 & 0 & 0 \end{pmatrix},$$

$$\rho_3(f) := \begin{pmatrix} 0 & 0 & 0 & 0 \\ 1 & 0 & 0 & 0 \\ 0 & 2 & 0 & 0 \\ 0 & 0 & 3 & 0 \end{pmatrix}.$$

Finally in general for any n, ρ_n will become

$$\rho_n(h) := \begin{pmatrix} n & 0 & \cdots & \cdots & 0 \\ 0 & n-2 & \cdots & \cdots & 0 \\ \vdots & \vdots & \vdots & \vdots & \vdots \\ 0 & 0 & \cdots & -n+2 & 0 \\ 0 & 0 & \cdots & \cdots & -n \end{pmatrix}, \rho_n(e) := \begin{pmatrix} 0 & n & \cdots & \cdots & 0 \\ 0 & 0 & n-1 & \cdots & 0 \\ \vdots & \vdots & \vdots & \vdots & \vdots \\ 0 & 0 & \cdots & \cdots & 1 \\ 0 & 0 & \cdots & \cdots & 0 \end{pmatrix},$$

$$\rho_n(f) := \begin{pmatrix} 0 & 0 & \cdots & \cdots & 0 \\ 1 & 0 & 0 & \cdots & 0 \\ \vdots & \vdots & \vdots & \vdots & \vdots \\ 0 & 0 & \cdots & n & 0 \end{pmatrix}.$$

General theory of the representation of $sl(2,\mathbb{C})$. Throughout this section $\mathcal{G}$ denotes $sl(2,\mathbb{C})$

Let V be an arbitrary $\mathcal{G}$-module. Now h acts diagonally on V. ($\mathbb{C}$ is algebraically closed ensures that all the required eigen values already lie in $\mathbb{C}$.) This yields a decomposition of V as direct sum of eigen spaces $V_\lambda = \{v \in V \mid h.v = \lambda v, \lambda \in \mathbb{C}\}$. Of course, the subspace V_λ still makes sense (and is 0) when λ is not an eigen value for the endomorphism of V which represents h. Whenever $V_\lambda \neq 0$, we call λ a *weight* of h in V and V_λ, a *weight space*.

We prove the following two preliminary results, given in Propositions 11 and 12 (Humphreys [40]).

Proposition 11. *If $v \in V_\lambda$, then prove that $x \cdot v \in V_{\lambda+2}$ and $y \cdot v \in V_{\lambda-2}$.*

Proof. Let $v \in V_\lambda$ then $h \cdot v = \lambda \cdot v$. We have $h \cdot (x \cdot v) = [h,x] \cdot v + x \cdot h \cdot v = 2x \cdot v + \lambda x \cdot v = (\lambda+2)x \cdot v \Rightarrow x \cdot v \in V_{\lambda+2}$ and similarly $h \cdot (y \cdot v) = [h,y] \cdot v + y \cdot h \cdot v = -2y \cdot v + \lambda y \cdot v = (\lambda-2)y \cdot v \Rightarrow y \cdot v \in V_{\lambda-2}$. □

Remark 23. The above result implies that x and y are represented by nilpotent endomorphisms of V.

Definition 36. Let V be a finite-dimensional $\mathcal{G} = sl(2,\mathbb{C})$-module. Then $V = \coprod_{\lambda \in \mathbb{C}} V_\lambda$ with $V_\lambda \neq 0$ such that $V_{\lambda+2} = 0$. In this case for any nonzero vector $v \in V_\lambda$, we have $x \cdot v = 0$. Here x is called *maximal vector of weight* λ.

Let V be an irreducible $\mathcal{G}$-module with $v_0 \in V_\lambda$, $v_{-1} = 0$, $v_n = (1/n!)y^n \cdot v_0 (n \geq 0)$.

Proposition 12. *In the above notation,*

$$(a)\ h \cdot v_n = (\lambda - 2n)v_n,$$

$$(b)\ y \cdot v_n = (n+1)v_{n+1},$$

$$(c)\ x \cdot v_n = (\lambda - n + 1)v_{n-1} \quad (n \geq 0).$$

For proof, one can refer Humphreys [40].

Remark 24. Let V be an irreducible module such that dim $V < \infty$. As per our earlier result $h \cdot v_n = (\lambda - 2n)v_n$, v_i's are linearly independent. Then there exists a smallest integer m, such that $v_m \neq 0, v_{m+1} = 0$. The earlier result shows that there is a submodule of $\mathcal{G}$ with basis $(v_0, v_1, \ldots, v_m)$ different from 0. This is equal to V itself.

It is clear that for $i = m + 1$, left side of (a) in the above proposition is 0 and right side of (a) in the above proposition is $(\lambda - m)v_m$. This implies $\lambda = m$. In other words, the weight of a maximal vector is a nondegenerate integer, called as highest weight of V. Moreover, each weight μ occurs with multiplicity one.

In particular, as V determines λ uniquely, $\lambda = \dim V - 1$. The following theorem is the direct consequence of above remarks.

The results given below (Humphreys [40]) are basic for an irreducible $sl(2,\mathbb{C})$-module.

Theorem 10. *Let V be an irreducible module for $\mathcal{G} = sl(2,\mathbb{C})$.*

(a) *Relative to h, V is the direct sum of weight spaces V_μ, where $\mu = m, m-2, \ldots, -(m-2), -m$, with $m + 1 = \dim V$ and $\dim V_\mu = 1$ for each μ.*

(b) *V has (up to nonzero scalar multiples) a unique maximal vector, whose weight (called the highest weight of V) is m.*

(c) *The action of $\mathcal{G}$ on V is given explicitly by the above formulas, if the basis is chosen in the prescribed fashion. In particular, there exists at most one irreducible $\mathcal{G}$-module (up to isomorphism) of each possible dimension $m + 1, m \geq 0$.*

(d) *Let V be any (finite-dimensional) $\mathcal{G}$-module with $\mathcal{G} = sl(2, \mathbb{C})$. Then the eigen values of h on V are all integers and each occurs along with its negative (an equal number of times). Moreover, in any decomposition of V into direct sum of irreducible submodules, the number of summands is precisely* $\dim V_0 + \dim V_1$.

1.8 Rootspace decomposition of semisimple Lie algebras

Let $\mathcal{G}$ be a semisimple Lie algebra. If $\mathcal{G}$ is not consisting of entirely nilpotent elements, then it consists of elements of the form x whose semisimple part x_s in the abstract Jordan decomposition is not zero. In this case, $\mathcal{G}$ has subalgebras consisting of semisimple elements and such a subalgebra is called *toral subalgebra.* A toral subalgebra is abelian.

Now fix a *maximal toral subalgebra* H of $\mathcal{G}$ (a toral subalgebra not properly included in any other). Since H is abelian, in the usual notation, $ad_{\mathcal{G}}H$ is a commuting family of semisimple endomorphisms of $\mathcal{G}$. $\mathcal{G}$ is the direct sum of the subspaces $\mathcal{G}_\alpha = \{x \in \mathcal{G} \mid [h,x] = \alpha(h)x \text{ for all } h \in H\}$, where α ranges over H^*. $\mathcal{G}_0 = \{x \in G \mid [h,x] = 0 \text{ (or) } x = 0\}$ is $C_{\mathcal{G}}(H)$, the centralizer of H in $\mathcal{G}$; *it includes* H. The set of all nonzero $\alpha \in H^*$ for which $\mathcal{G}_\alpha \neq 0$ is denoted by Φ; the elements of Φ are called the *roots of* $\mathcal{G}$ *relative to* H (and are finite in number). With this notation, we have a *root space decomposition* (or *Cartan decomposition*) of $\mathcal{G}$ as $\mathcal{G} = C_{\mathcal{G}}(H) + \coprod_{\alpha\in\Phi}\mathcal{G}_\alpha$.

Basic properties of root systems

In the usual notations the following results can be verified [40].

I. Orthogonality properties of root system:
  - **(a)** Φ spans H^*.
  - **(b)** If $\alpha \in \Phi$, then $-\alpha \in \Phi$.
  - **(c)** Let $\alpha \in \Phi, x \in \mathcal{G}_\alpha, y \in \mathcal{G}_{-\alpha}$, then $[x,y] = \kappa(x,y)t_\alpha$.
  - **(d)** If $\alpha \in \Phi$, then $[\mathcal{G}_\alpha, \mathcal{G}_{-\alpha}]$ is one dimensional with basis t_α.
  - **(e)** $\alpha(t_\alpha) = \kappa(t_\alpha, t_\alpha) \neq 0$, for $\alpha \in \Phi$.
  - **(f)** If $\alpha \in \Phi$ and x_α is any nonzero element of $\mathcal{G}_\alpha$, then there exists $y_\alpha \in \mathcal{G}_{-\alpha}$ such that $x_\alpha, y_\alpha, h_\alpha = [x_\alpha, y_\alpha]$ span a three-dimensional simple subalgebra of $\mathcal{G}$ isomorphic to $sl(2,\mathbb{F})$ through $x_\alpha \mapsto \begin{pmatrix} 0 & 1 \\ 0 & 0 \end{pmatrix}, y_\alpha \mapsto \begin{pmatrix} 0 & 0 \\ 1 & 0 \end{pmatrix}, h_\alpha \mapsto \begin{pmatrix} 1 & 0 \\ 0 & -1 \end{pmatrix}$.
  - **(g)** $h_\alpha = \frac{2t_\alpha}{\kappa(t_\alpha,t_\alpha)}$; $h_\alpha = -h_{-\alpha}$.

II. Integrality properties of root system:
  - **(a)** If $\alpha \in \Phi$, then $\dim \mathcal{G}_\alpha = 1$. In particular, for $S_\alpha = \mathcal{G}_\alpha + \mathcal{G}_{-\alpha} + H_\alpha$ ($H_\alpha = [\mathcal{G}_\alpha, \mathcal{G}_{-\alpha}]$), and for given nonzero $x_\alpha \in \mathcal{G}_\alpha$, there exists a unique $y_\alpha \in \mathcal{G}_{-\alpha}$ satisfying $[x_\alpha, y_\alpha] = h_\alpha$.
  - **(b)** If $\alpha \in \Phi$, the only scalar multiples of α which are roots are α and $-\alpha$.
  - **(c)** If α and $\beta \in \Phi$, then $\beta(h_\alpha) \in \mathbb{Z}$ and $\beta - \beta(h_\alpha)\alpha \in \Phi$. (The numbers $\beta(h_\alpha)$ are called *Cartan integers.*)
  - **(d)** If $\alpha, \beta \in \Phi, \alpha + \beta \in \Phi$, then $[\mathcal{G}_\alpha, \mathcal{G}_\beta] = \mathcal{G}_{\alpha+\beta}$.
  - **(e)** Let $\alpha, \beta \in \Phi$ and $\beta \neq \pm\alpha$. Let r, q be (respectively) the largest integers for which $\beta - r\alpha, \beta + q\alpha$ are roots. Then all $\beta + i\alpha \in \Phi (-r \leq i \leq q)$, for which $\beta(h_\alpha) = r - q$.
  - **(f)** $\mathcal{G}$ is generated (as Lie algebra) by the root spaces $\mathcal{G}_\alpha$.

Remark 25. In general, if $\mathcal{G}$ is a semisimple Lie algebra over $\mathbb{C}$ and H is a maximal toral subalgebra with $\Phi \subset H^*$, the set of roots of $\mathcal{G}$, then we have $\mathcal{G} = H \bigoplus_{\alpha\in\Phi}\mathcal{G}_\alpha$ is the root space decomposition.

We can transform the Killing form κ to H^*, by letting

$$(\alpha, \beta) = \kappa(t_\alpha, t_\beta) \quad \forall\, \alpha, \beta \in H^*.$$

Here, for $\alpha \in \Phi, x \in \mathcal{G}_\alpha, y \in \mathcal{G}_{-\alpha}$, we have $[x,y] = \kappa(x,y)t_\alpha$.

Moreover, to each $\alpha \in H^*$, there corresponds a unique $t_\alpha \in H$ satisfying $\alpha(h) = \kappa(t_\alpha, h)$ for all $h \in H$. In particular, Φ corresponds to the subset $\{t_\alpha; \alpha \in \Phi\}$ of H. Φ spans H^*. If $(\alpha_1, \dots, \alpha_l)$ is a basis of H^*, since the form (α_i, α_j) is nondegenerate, the matrix $(\alpha_i, \alpha_j)_{1 \le i,j \le l}$ is nonsingular.

Root space decomposition and properties of Killing form

For proof of the following results ((1) - (5)), one can refer Humphreys [40].

(1) For all $\alpha, \beta \in H^*, [\mathcal{G}_\alpha, \mathcal{G}_\beta] \subset \mathcal{G}_{\alpha+\beta}$. If $x \in \mathcal{G}_\alpha, \alpha \neq 0$, then *ad* x is nilpotent. If $\alpha, \beta \in H^*$ and $\alpha + \beta \neq 0$, then $\mathcal{G}_\alpha$ is orthogonal to $\mathcal{G}_\beta$, relative to the Killing form κ of $\mathcal{G}$.
(2) The restriction of the Killing form to $\mathcal{G}_0 = C_{\mathcal{G}}(H)$ is nondegenerate.
(3) If x, y are commuting endomorphisms of a finite-dimensional vector space, with y nilpotent, then xy is also nilpotent; in particular, $tr(xy) = 0$.
(4) Let H be a maximal toral subalgebra of $\mathcal{G}$. Then $H = C_{\mathcal{G}}(H)$.
(5) The restriction of the Killing form to $H \times H$ is nondegenerate.
Remark:
  (1) For each linear form on H, there exists a $h \in H$ such that $B(h, h_\alpha) = \alpha(h)$ for all $h \in H$.
  (2) If $\alpha \in \Delta$, then $-\alpha \in \Delta$ and $[g_\alpha, g_{-\alpha}] \in C(H_\alpha)$.
  (3) With respect to the Killing form κ on $\mathcal{G}$, $\mathcal{G}_\alpha$, and $\mathcal{G}_\beta$ are said to be orthogonal if $\kappa(\mathcal{G}_\alpha, \mathcal{G}_\beta) = 0$, where $\alpha + \beta \neq 0$.
  (4) $\kappa|_{\mathcal{G}_{-\alpha}+\mathcal{G}_\alpha}$ is a nondegenerate bilinear form. In particular, $\kappa|_{\mathfrak{h}}$ is nondegenerate and κ defines a nondegenerate pairing of $\mathcal{G}_\alpha$ and $\mathcal{G}_{-\alpha}$. $\mathfrak{h}$ is a maximal abelian subalgebra of $\mathcal{G}$. $\mathfrak{h}$ consists of semisimple elements (ie, $ad(h)$ is semisimple for all $h \in \mathfrak{h}$).

1.9 Root system in Euclidean spaces and root diagrams

Let $\mathcal{G}$ be a semisimple Lie algebra (over algebraically closed field of characteristic 0), H be a maximal toral subalgebra, $\Phi = \{\alpha_1, \alpha_2, \dots, \alpha_n\} \subset H^*$ be the set of roots of $\mathcal{G}$ and $\mathcal{G} = H + \coprod_{\alpha \in \Phi} \mathcal{G}_\alpha$ be the root space decomposition. Let Q be the set of all rational numbers, $\mathbb{R}$ be the set of all real numbers, and E_Q be the Q-subspace of H^* spanned by all roots. We have $n = \dim H^*$. If Q is the base field, then we can extend the base field to $\mathbb{R}$ with E being the corresponding real vector space. That is, $E = \mathbb{R} \otimes_Q E_Q$. Hence E is an Euclidean space. Φ contains a basis of E and dimension of E is n.
The following results can be established [40]:

(a) Φ spans E and 0 does not belong to Φ.
(b) If $\alpha \in \Phi$ then $-\alpha \in \Phi$, but no other scalar multiple of α is a root.
(c) If $\alpha, \beta \in \Phi$, then $\beta - \frac{2(\beta,\alpha)}{(\alpha,\alpha)}\alpha \in \Phi$.
(d) If $\alpha, \beta \in \Phi$, then $\frac{2(\beta,\alpha)}{(\alpha,\alpha)} \in \mathbb{Z}$.

Definition 37. A *reflection* in an Euclidean space E is an invertible linear transformation leaving pointwise fixed some hyperplane (subspace of codimension one) and sending any vector orthogonal in that hyperplane into its negative.

A reflection preserves the inner product on E. Any nonzero vector α determines a reflection σ_α with *reflecting hyperplane* $P_\alpha = \{\beta \in E | (\beta, \alpha) = 0\}$.

So we get the following:

$$\sigma_\alpha(\beta) = \beta - \frac{2(\beta,\alpha)}{(\alpha,\alpha)}\alpha.$$

Denote by $\langle \beta, \alpha \rangle = \frac{2(\beta,\alpha)}{(\alpha,\alpha)}$. Then $\sigma_\alpha(\beta) = \beta - \langle \beta, \alpha \rangle \alpha$.
A subset Φ of the Euclidean space E is called a *root system* in E [40] if

(1) Φ is finite, spans E, and does not contain 0.
(2) If $\alpha \in \Phi$, the only multiplies of α in Φ are $\pm\alpha$.
(3) If $\alpha \in \Phi$, the reflection σ_α leaves Φ invariant.
(4) If $\alpha, \beta \in \Phi$, then $\langle \beta, \alpha \rangle = \frac{2(\beta,\alpha)}{(\alpha,\alpha)} \in Z$

Let Φ denote a *root system of rank n in an Euclidean space E*. A subset Δ of Φ is called a *base* if

(1) Δ is a basis of E.
(2) Each root β can be written as $\beta = \sum_\alpha k_\alpha \alpha (\alpha \in \Delta)$ with all integral coefficients k_α being nonnegative or all nonpositive.

The roots in Δ are called *simple*. The height of the above root β is $\sum_{\alpha\in\Delta} k_\alpha$. A root system Φ is called *irreducible* if it cannot be partitioned into the union of two proper subsets such that each root in one set is *orthogonal* to each root in other set.

Remark 26.

(1) Let V be an n-dimensional vector space over a field F. The dual space of V, denoted by V^*, is the set of all linear maps from V to F. If $f, g \in V^*$ then $f + g$ and λf for $\lambda \in F$ are defined by $(f + g)v = f(v) + g(v)$ for $v \in V$ and $(\lambda f)(v) = \lambda f(v)$. Given a basis $\{v_1, v_2, \ldots, v_n\}$ of a vector space V, one can define associated dual basis of V as follows: Let f_i: $V \to F$ be the linear map defined by

$$f_i(v_j) = \begin{cases} 1 & \text{for } i = j \\ 0 & \text{for } i \neq j. \end{cases}$$

One can check that $\{f_1, \ldots, f_n\}$ is a basis of V^* and it is a dual to the basis $\{v_1, v_2, \ldots, v_n\}$ of V.
(2) *Dual root system*: Let E be an inner product space and R be the root system. Then one can verify that $\hat{R} = \left\{\frac{2\alpha}{(\alpha,\alpha)} : \alpha \in R\right\}$ is also a root system in E. Also one can verify that Cartan matrix of $\hat{R}$ is the transpose of Cartan matrix of R. Here $\hat{R}$ is the dual root system to R. One can prove that Weyl group of R and $\hat{R}$ are isomorphic.

Consider the space $\mathbb{R}^{n+1}$ with the Euclidean inner product. Let ϵ_i be the vector in the Euclidean space with ith entry 1 and all other entries are zero. Now define

$$R = \{\pm(\epsilon_i - \epsilon_j) : 1 \leq i < j \leq l + 1\}.$$

One can show that R is an root system in E. Let R be a root system in the real inner product space E. Let $\alpha, \beta \in R$ with $\beta \neq \pm\alpha$. Then it can be proved that

$$\langle \alpha, \beta \rangle \langle \beta, \alpha \rangle \in \{0, 1, 2, 3\}.$$

Now one can show that there are only few possibilities for $\langle\alpha,\beta\rangle$. If we take two roots α,β in R with $\alpha\neq\pm\beta$ and $\langle\beta,\beta\rangle\geq\langle\alpha,\alpha\rangle$, then we have

$$|\langle\beta,\alpha\rangle|=\frac{|2(\beta,\alpha)|}{(\alpha,\alpha)}\geq\frac{|2(\alpha,\beta)|}{(\beta,\beta)}=|\langle\alpha,\beta\rangle|.$$

Let α and β be two roots in E. Cosine of the angle θ between vectors $\alpha,\beta\in E$ is given by the formula, $\|\alpha\|\ \|\beta\|\cos\theta=(\alpha,\beta)$.

So

$$\begin{aligned}\langle\beta,\alpha\rangle&=\frac{2(\beta,\alpha)}{(\alpha,\alpha)}\\&=\frac{2\|\alpha\|\ \|\beta\|\cos\theta}{\|\alpha\|^2}=\frac{2\|\beta\|}{\|\alpha\|}\cos\theta.\end{aligned}$$

Similarly

$$\langle\alpha,\beta\rangle=\frac{2\|\alpha\|}{\|\beta\|}\cos\theta.$$

Hence

$$\langle\alpha,\beta\rangle\langle\beta,\alpha\rangle=4\cos^2\theta,$$

which is a nonnegative integer. As $0\leq\cos^2\theta\leq1$, $\langle\alpha,\beta\rangle$ and $\langle\beta,\alpha\rangle$ have same sign.

So when $\alpha\neq\pm\beta$ and $\|\beta\|\geq\|\alpha\|$, we have Table 1.1.

Properties of Root diagrams:

We have the following properties of the root system Φ.

Let $\alpha,\beta\in\Phi$.

(a) If the angle between α and β is strictly obtuse, then $\alpha+\beta\in\Phi$.
(b) If the angle between α and β is strictly acute and $(\beta,\beta)\geq(\alpha,\alpha)$, then $\alpha-\beta\in\Phi$.

Let $E=R$ with the Euclidean inner product:

Table 1.1 **Angles between root vectors as explained above**

$\langle\alpha,\beta\rangle$	$\langle\beta,\alpha\rangle$	$4\cos^2\theta=<\alpha,\beta><\beta,\alpha>$	θ	$\frac{\langle\beta,\beta\rangle}{\langle\alpha,\alpha\rangle}=\frac{\|\beta\|^2}{\|\alpha\|^2}$
0	0	**0**	$\pi/2$	Undetermined
1	1	**1**	$\pi/3$	1
−1	−1	**1**	$2\pi/3$	1
1	2	**2**	$\pi/4$	2
−1	−2	**2**	$3\pi/4$	2
1	3	**3**	$\pi/6$	3
−1	−3	**3**	$5\pi/6$	3

(1) The rank of the root system l is 1. Now there is only one possibility, namely, (A_1) as shown below.

$$-\alpha \longleftrightarrow \alpha$$

(2) Now we consider rank $l = 2$. In this case, there are four possibilities.

(i) Let $\theta = \pi/2$. Using the properties of root diagrams, we can find Φ, which contains four roots as shown below. In this case, root system is said to be of type $A_1 \times A_1$.

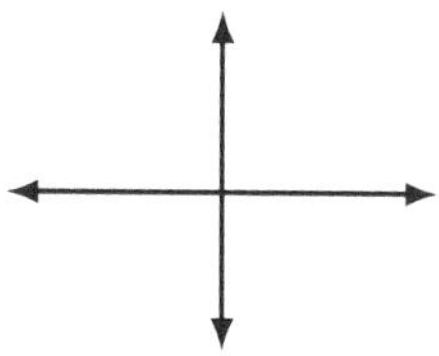

(ii) Let $\theta = 2\pi/3$. Using the properties of root diagrams, we can find Φ, which contains six roots as shown below. In this case, root system is said to be of type A_2.

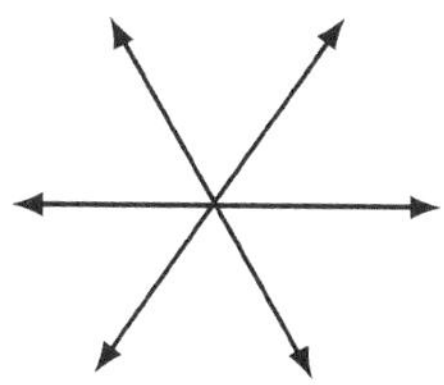

(iii) Let $\theta = 3\pi/4$. Using the properties of root diagrams, $\alpha + \beta$ and $2\alpha + \beta$ are roots. Root diagram is shown below. In this case, root system is of type B_2.

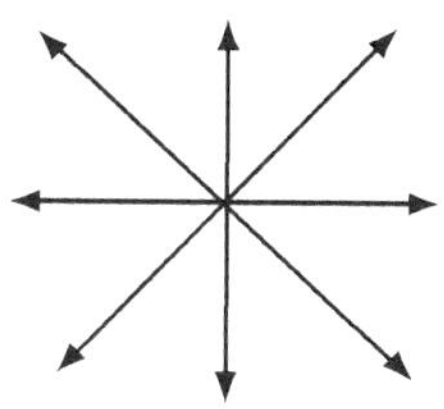

(iv) Let $\theta = 5\pi/6$. The positive root system is $\{\alpha, \beta, \alpha + \beta, \alpha - \beta, 2\alpha + \beta, \alpha + 2\beta\}(\{\alpha, \beta, \alpha + \beta, 2\alpha + \beta, -\alpha + \beta, -(\alpha + 2\beta)\})$. Root diagram is shown below. This root system is of type G_2.

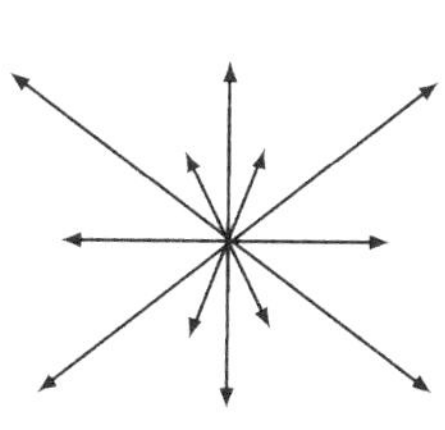

Remark 27. In each case, one can check the axioms directly and determine Weyl group W.

Definition 38 (Reduced root system and Chevalley's normalization). Let V be a vector space and V^* be its dual. Then one can define a symmetry S_α to be an automorphism (i.e., $S_\alpha : V \to V$) of V such that

(i) $S_\alpha(\alpha) = -\alpha$ and
(ii) the set H of elements of V fixed by S_α is a hyperplane of V. That is, $H = \{\alpha \in V | S_\alpha(\alpha) = \alpha\}$ is a hyperplane of V.

A subset R of a vector space V is said to be a root system in V if

(a) R is finite, spans V, and does not contain 0,
(b) The orthogonal transformations $S_\alpha(\beta) = \beta - \frac{2<\beta,\alpha>}{|\alpha|^2}\alpha$, for $\alpha \in R$ transforms R to itself.
(c) $\frac{2(\beta,\alpha)}{|\alpha|^2}$ is an integer if α and β are in R.

This root system R is said to be reduced if for each $\alpha \in R, \alpha$ and $-\alpha$ are only roots proportional to α, (hence $2\alpha \notin R$).

The following properties of the reduced system can be verified: Let R be a reduced root system.

(1) For a semisimple Lie algebra $\mathcal{G}$, the root system is isomorphic to R.
(2) For $\mathcal{G}$ to be a simple Lie algebra, it is necessary and sufficient that R should be irreducible.
(3) *Chevalley's normalization* [15]: For each $\alpha \in R$, choose a nonzero element $x_\alpha \in \mathcal{G}_\alpha$, such that

$$[x_\alpha, x_\beta] = \begin{cases} N_{\alpha,\beta} x_{\alpha+\beta} & \text{if } \alpha+\beta \in R \\ 0 & \text{if } \alpha+\beta \notin R \text{ and } \alpha+\beta \neq 0. \end{cases}$$

Here $N_{\alpha,\beta}$ is a nonzero scalar. The coefficients $N_{\alpha,\beta}$ determine the multiplication table of $\mathcal{G}$.

One can choose the elements x_α so that

$$\begin{array}{lll} [x_\alpha, x_{-\alpha}] = H_\alpha & \text{for all } \alpha \in R & \text{and} \\ N_{\alpha\beta} = -N_{-\alpha,-\beta} & \text{for } \alpha, \beta, \alpha+\beta \in R & \end{array}$$

In this case, for $\alpha, \beta \in R$ with $\alpha + \beta \in R$, let p be the greatest integer such that $\beta - p\alpha \in R$. Then $N_{\alpha,\beta} = \pm(p+1)$.

Remark 28. One can also refer to Cartan-Weyl basis in Section 1.17.

1.10 Coxeter graphs and Dynkin diagrams

Definition 39. If α, β are distinct positive roots, then by Table 1.1, $\langle\alpha, \beta\rangle \langle\beta, \alpha\rangle \in \{0, 1, 2, 3\}$.

We define the *Coxeter graph* of Φ to be a graph having n vertices, the ith vertex joined to jth vertex ($i \neq j$) by $\langle\alpha_i, \alpha_j\rangle \langle\alpha_j, \alpha_i\rangle$ edges.

Example 12.

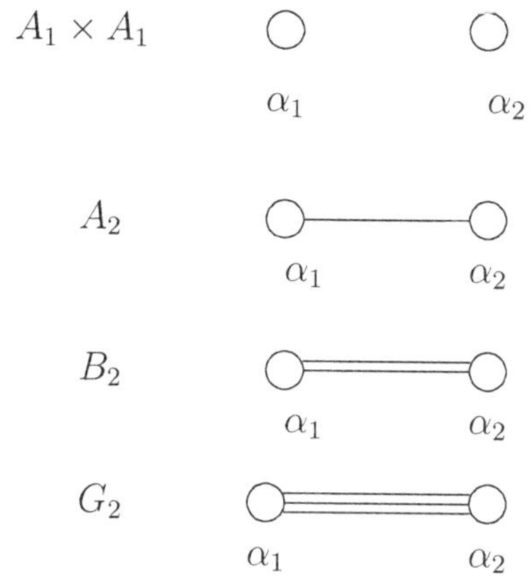

The Coxeter graph determines the numbers $\langle \alpha_i, \alpha_j \rangle$, in case all the roots have equal length, $\langle \alpha_i, \alpha_j \rangle = \langle \alpha_j, \alpha_i \rangle$. In the case of more than one root length (for example, B_2 or G_2) the graph fails to tell us which of a pair of vertices should correspond to a short simple root, which to a long (in case these vertices are joined by two or three edges).

Definition 40. Whenever a double or triple edge occurs in the Coxeter graph Φ, we can add an *arrow* pointing to the shorter of the two roots. The resulting diagram is the *Dynkin diagram* which allows us to determine Cartan integers.

Remark 29. If n is the number of vertices (the rank of the corresponding Lie algebra which is the dimension of the Cartan subalgebra) and Φ is an irreducible root system of rank n, then its Dynkin diagram is one of the following cases:

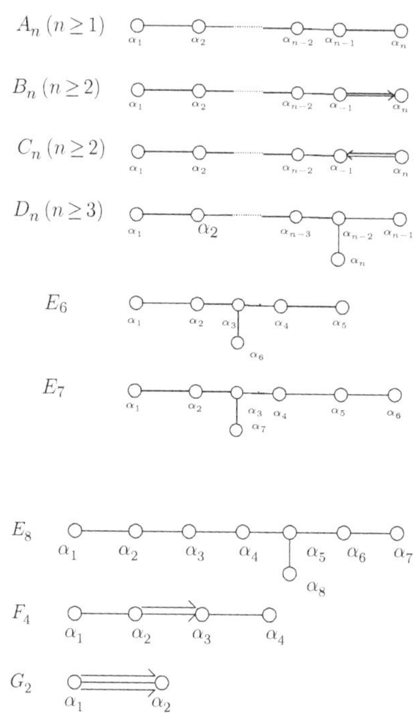

Here,

$$\begin{aligned} A_n &= sl(n+1,\mathbb{C}) \\ B_n &= so(2n+1,\mathbb{C}) \\ C_n &= sp(n,\mathbb{C}) \\ D_n &= so(2n,\mathbb{C}). \end{aligned}$$

The dimension of the Euclidean space E is $\dim E = n$, the rank of the root system Φ. By basic property of root system, if $\alpha \in \Phi$, then the only multiples of α in Φ are $\pm\alpha$. When $n \leq 2$, we can describe Φ by simply drawing pictures as shown below.

(i) For the case $n = 1$, there is only one possibility, which can be labeled as

(A_1): ○

(ii) For the case $n = 2$, we have

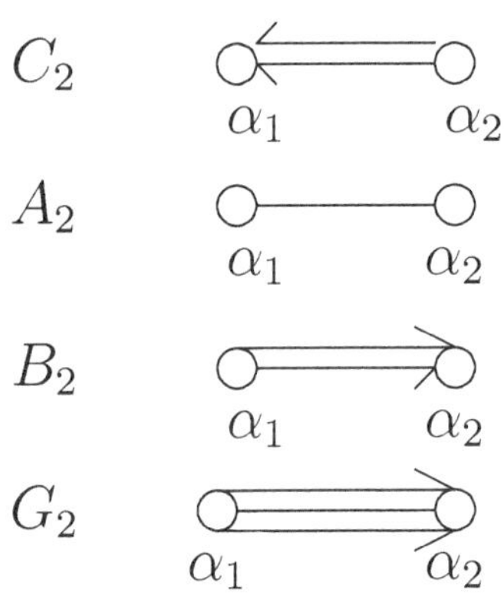

Remark 30. Dynkin diagrams of A_1, B_1, C_1 are identical and hence they are isomorphic. Similarly, the Dynkin diagrams of B_2 and C_2 are identical and hence they are isomorphic. Moreover, Dynkin diagrams of A_3 and D_3 are identical and hence they are isomorphic.

1.11 Cartan matrices, ranks, and dimensions of simple Lie algebras

Definition 41. Let Φ be a root system of rank n and $\triangle$ be a base of Φ. Fix an ordering $(\alpha_1, \ldots, \alpha_n)$ of the *simple roots*. The matrix $(\langle \alpha_i, \alpha_j \rangle)_{i,j=1}^n$ is called the *Cartan matrix* of Φ. Its entries are called *Cartan integers*. Dimension of the Cartan subalgebra is called the *rank* of the corresponding Lie algebra.

For the systems of rank 2, the Cartan matrices for $A_1 \times A_1, A_2, B_2$, and G_2 are

$$\begin{pmatrix} 2 & 0 \\ 0 & 2 \end{pmatrix}, \begin{pmatrix} 2 & -1 \\ -1 & 2 \end{pmatrix}, \begin{pmatrix} 2 & -2 \\ -1 & 2 \end{pmatrix}, \text{ and } \begin{pmatrix} 2 & -1 \\ -3 & 2 \end{pmatrix},$$

respectively.

Cartan matrices of classical simple Lie algebras

$$\text{Cartan matrix of } A_n: \begin{pmatrix} 2 & -1 & 0 & & & \cdots & & 0 \\ -1 & 2 & -1 & 0 & & \cdots & & 0 \\ 0 & -1 & 2 & -1 & 0 & \cdots & & 0 \\ \cdot & \cdot & \cdot & \cdot & \cdot & \cdots & \cdot & \cdot \\ 0 & 0 & 0 & 0 & & \cdots & -1 & 2 \end{pmatrix};$$

$$\text{Cartan matrix of } B_n: \begin{pmatrix} 2 & -1 & 0 & & \cdots & & & 0 \\ -1 & 2 & -1 & 0 & \cdots & & & 0 \\ \cdot & \cdot & \cdot & \cdot & \cdots & \cdot & \cdot & \cdot \\ 0 & 0 & 0 & & \cdots & -1 & 2 & -2 \\ 0 & 0 & 0 & & \cdots & 0 & -1 & 2 \end{pmatrix};$$

$$\text{Cartan matrix of } C_n: \begin{pmatrix} 2 & -1 & 0 & & \cdots & & & 0 \\ -1 & 2 & -1 & & \cdots & & & 0 \\ 0 & -1 & 2 & -1 & \cdots & & & 0 \\ \cdot & \cdot & \cdot & \cdot & \cdots & \cdot & \cdot & \cdot \\ 0 & 0 & 0 & & \cdots & -1 & 2 & -1 \\ 0 & 0 & & & \cdots & 0 & -2 & 2 \end{pmatrix};$$

$$\text{Cartan matrix of } D_n: \begin{pmatrix} 2 & -1 & 0 & \cdots & & & & & 0 \\ -1 & 2 & -1 & \cdots & & & & & 0 \\ \cdot & \cdot & \cdot & \cdots & \cdot & \cdot & \cdot & \cdot & \cdot \\ 0 & 0 & & \cdots & -1 & 2 & -1 & 0 & 0 \\ 0 & 0 & & \cdots & & -1 & 2 & -1 & -1 \\ 0 & 0 & & \cdots & & 0 & -1 & 2 & 0 \\ 0 & 0 & & \cdots & & 0 & -1 & 0 & 2 \end{pmatrix};$$

$$\text{Cartan matrix of } E_6: \begin{pmatrix} 2 & 0 & -1 & 0 & 0 & 0 \\ 0 & 2 & 0 & -1 & 0 & 0 \\ -1 & 0 & 2 & -1 & 0 & 0 \\ 0 & -1 & -1 & 2 & -1 & 0 \\ 0 & 0 & 0 & -1 & 2 & -1 \\ 0 & 0 & 0 & 0 & -1 & 2 \end{pmatrix};$$

$$\text{Cartan matrix of } E_7: \begin{pmatrix} 2 & 0 & -1 & 0 & 0 & 0 & 0 \\ 0 & 2 & 0 & -1 & 0 & 0 & 0 \\ -1 & 0 & 2 & -1 & 0 & 0 & 0 \\ 0 & -1 & -1 & 2 & -1 & 0 & 0 \\ 0 & 0 & 0 & -1 & 2 & -1 & 0 \\ 0 & 0 & 0 & 0 & -1 & 2 & -1 \\ 0 & 0 & 0 & 0 & 0 & -1 & 2 \end{pmatrix};$$

$$\text{Cartan matrix of } E_8: \begin{pmatrix} 2 & 0 & -1 & 0 & 0 & 0 & 0 & 0 \\ 0 & 2 & 0 & -1 & 0 & 0 & 0 & 0 \\ -1 & 0 & 2 & -1 & 0 & 0 & 0 & 0 \\ 0 & -1 & -1 & 2 & -1 & 0 & 0 & 0 \\ 0 & 0 & 0 & -1 & 2 & -1 & 0 & 0 \\ 0 & 0 & 0 & 0 & -1 & 2 & -1 & 0 \\ 0 & 0 & 0 & 0 & 0 & -1 & 2 & -1 \\ 0 & 0 & 0 & 0 & 0 & 0 & -1 & 2 \end{pmatrix};$$

$$\text{Cartan matrix of } F_4: \begin{pmatrix} 2 & -1 & 0 & 0 \\ -1 & 2 & -2 & 0 \\ 0 & -1 & 2 & -1 \\ 0 & 0 & -1 & 2 \end{pmatrix}; \text{ and}$$

$$\text{Cartan matrix of } G_2: \begin{pmatrix} 2 & -1 \\ -3 & 2 \end{pmatrix}.$$

Table 1.2 shows the set of all simple Lie algebras and the dimension of the corresponding Cartan subalgebras.

Table 1.2 Set of all simple Lie algebras and the dimension of the corresponding Cartan subalgebras (Helgason [42])

Name	Description	Rank (dimension of the CSA)	Dimension (dimension of the Lie algebra)
A_n	$sl(n+1,\mathbb{C})(n \geq 1)$	n	$n(n+2)$
B_n	$o(2n+1,\mathbb{C})(n \geq 2)$	n	$n(2n+1)$
C_n	$sp(n,\mathbb{C})(n \geq 3)$	n	$n(2n+1)$
D_n	$o(2n,\mathbb{C})(n \geq 4)$	n	$n(2n-1)$
G_2	–	2	14
F_4	–	4	52
E_6	–	6	78
E_7	–	7	133
E_8	–	8	248

1.12 Weyl groups and structure of Weyl groups of simple Lie algebras

Definition 42. Let $\mathcal{G}$ be a simple Lie algebra and Δ and $\Delta^\vee$ be the root system and coroot system of $\mathcal{G}$, respectively. For any root $\alpha \in \Delta$, we can define a transformation w_α (in the weight space) called *Weyl reflection* for a weight λ as

$$w_\alpha(\lambda) = \lambda - \frac{2(\alpha, \lambda)}{(\alpha, \alpha)}\alpha = \lambda - \langle \alpha^\vee, \lambda \rangle \alpha.$$

The set of all *Weyl reflections* with respect to all the roots of $\mathcal{G}$ forms a finite group W called the *Weyl group* of $\mathcal{G}$. If the Weyl reflection $w_\alpha(\beta) = \beta$, then β is a vector in the hyperplane orthogonal to α. Moreover, $w_\alpha(\alpha) = -\alpha$.

Definition 43. The *Weyl group* W is generated by the set of all Weyl reflections w_{α_i} $(1 \leq i \leq r)$ and the identity element. Here α_i are simple positive roots for $1 \leq i \leq r$. In other words, the group of all invertible linear transformations of Euclidean space, generated by the reflection σ_α for $\alpha \in \Phi$ is called the *Weyl group* of Φ and is denoted by W or $W(\Phi)$.

For any element $w \in W$, length of w, namely, $l(w)$, is the minimum number of reflections w_{α_i} such that $w = \prod_i w_{\alpha_i}$. The *parity of* w, namely, $\epsilon(w)$ is equal to $(-1)^{l(w)}$.

For a finite-dimensional Lie algebra, the following can be proved directly [40].

The following statements are equivalent.

(a) A is a generalized Cartan matrix of finite type.
(b) The restriction of the standard bilinear form $(\, , \,)$ to $\mathfrak{h}_R$ is positive definite.
(c) $|W| < \infty$.
(d) $|\Delta| < \infty$.
(e) $\mathcal{G}(A)$ is a simple finite-dimensional Lie algebra.
(f) There exists an $\alpha \in \Delta$ such that $\alpha + \alpha_i \notin \Delta$ for all $i = 1, \ldots, n$.

Remark 31. W is a normal subgroup of $Aut\ \Phi$. Let $\Gamma = \{\sigma \in Aut\ \Phi \mid \sigma(\Delta) = \Delta\}$, where Δ is a *fixed basis of* Φ. Then Γ is a subgroup of $Aut\ \Phi$. $Aut\ \Phi$ is the semidirect product of Γ and W. Γ may be identified with the group of diagram automorphism (Table 1.3).

Definition 44. Weyl group is a normal subgroup of the group of all automorphisms of Δ, the root system of simple Lie algebra $\mathcal{G}$. If $\{\alpha_1, \alpha_2, \ldots, \alpha_r\}$ is a simple root system of $\mathcal{G}$, then the element $c = w_{\alpha_1} \ldots w_{\alpha_r}$ is called the *Coxeter element* of W.

Definition 45. The vector $\rho = \frac{1}{2} \sum_{\alpha \in \Delta^+} \alpha$ is called the *Weyl vector* of the simple Lie algebra $\mathcal{G}$ and $\rho^\vee = \frac{1}{2} \sum_{\alpha \in \Delta^+} \alpha^\vee$ is called the *dual Weyl vector* of $\mathcal{G}$, where Δ^+ is the set of all positive simple roots.

Table 1.3 Structure of Weyl group for simple Lie algebras (Humphreys [40])

Type	Number of positive roots	Order of W	Structure of W	Γ
A_n	$\binom{n+1}{2}$	$(n+1)!$	S_{n+1}	$Z/2Z(n \geq 2)$
B_n, C_n	n^2	$2^n n!$	$(Z/2Z)^n \rtimes S_n$	1
D_n	$n^2 - n$	$2^{n-1} n!$	$(Z/2Z)^{n-1} \rtimes S_n$	$\begin{cases} S_3(n=4) \\ Z/2Z(n>4) \end{cases}$
E_6	36	$2^7 \cdot 3^4 \cdot 5$		Z/2Z
E_7	63	$2^{10} \cdot 3^4 \cdot 5 \cdot 7$		1
E_8	120	$2^{14} \cdot 3^5 \cdot 5^2 \cdot 7$		1
F_4	24	$2^7 3^2$	$S_3 \rtimes (S_4 \rtimes (\mathbb{Z}/2\mathbb{Z})^2)$	1
G_2	6	$2^2 3$	D_6, Dihedral group	1

1.13 Root systems of classical simple Lie algebras and highest long and short roots

The set of all root vectors, Cartan matrices, and Weyl groups for different classes of finite-dimensional Lie algebras are given below. In particular, the set of all roots [8], positive roots, fundamental root system, fundamental Weyl chamber (refer Chapter 2, definition 88), maximal root, and fundamental coroots for different classes of finite-dimensional Lie algebras are given below ([15], [40], [41], [43]):

Let $\epsilon_1, \ldots, \epsilon_n$ denote the usual unit orthonormal vectors which form a basis in R^n and I denote the $\mathbb{Z}$-span of this basis, which is a lattice.

Class (1): $A_n(n \geq 1)$: Consider the vector $\epsilon_1 + \cdots + \epsilon_{n+1}$ in $\mathbb{R}^{n+1}$. Let E be the n-dimensional subspace of $\mathbb{R}^{n+1}$ orthogonal to the vector $\epsilon_1 + \cdots + \epsilon_{n+1}$. Now consider the set $I' = I \cap E$. Let Φ be the set of all vectors $\alpha \in I'$ such that $(\alpha, \alpha) = 2$. Now it is clear that $\Phi = \{\varepsilon_i - \varepsilon_j, i \neq j\}$. Moreover the vectors $\alpha_i = \varepsilon_i - \varepsilon_{i+1} (1 \leq i \leq n)$ are independent. Now one can write $\varepsilon_i - \varepsilon_j = (\varepsilon_i - \varepsilon_{i+1}) + (\varepsilon_{i+1} - \varepsilon_{i+2}) + \cdots + (\varepsilon_{j-1} - \varepsilon_j)$, for $i < j$. Hence $\{\alpha_i | 1 \leq i \leq n\}$ is a basis for Φ. Now the Cartan matrix of A_n is $(\langle \alpha_i, \alpha_j \rangle)$. The Weyl reflection of β with respect to α is

$$\sigma_\alpha(\beta) = \beta - \frac{2(\beta, \alpha)}{(\alpha, \alpha)} \alpha.$$

Now one can note that the reflection σ_{α_i} with respect to α_i permutes the subscripts $i, i+1$ and leaves all other subscripts fixed. Hence σ_{α_i} corresponds to the transposition $(i, i+1)$ and all these transpositions $(i, i+1)$ generate the symmetric group on $n+1$ symbols, namely, S_{n+1}. Hence Weyl group of A_n is S_{n+1}.

The Cartan subalgebra, $\mathfrak{h}$ is the space of all diagonal matrices $h = diag(a_1, a_2, \ldots, a_{n+1})$ such that $\sum a_i = 0$. Now $[h, e_{ij}] = (a_i - a_j) e_{ij}$.

We have the following description about root systems.

The set of roots is $\{\epsilon_i - \epsilon_j \text{ with } i \neq j\}$.
The fundamental root system is $\{\epsilon_1 - \epsilon_2, \epsilon_2 - \epsilon_3, \epsilon_3 - \epsilon_4, \dots, \epsilon_n - \epsilon_{n+1}\}$.
The set of all positive roots is $\{\epsilon_i - \epsilon_j\}$ with $i < j$.
The fundamental Weyl chamber consists of the elements h, where
$h = diag(a_1, a_2, \dots, a_{n+1})$ with $a_1 > a_2 > \cdots > a_{n+1}$.
The maximal root is $\epsilon_1 - \epsilon_2 + \epsilon_2 - \epsilon_3 + \epsilon_3 - \epsilon_4 + \cdots + \epsilon_n - \epsilon_{n+1} = \epsilon_1 - \epsilon_{n+1}$.
The fundamental coroots are $h_1 = e_{11} - e_{22}, h_2 = e_{22} - e_{33}, \dots, h_n = e_{nn} - e_{n+1,n+1}$.
We have the following root space decomposition of S_{n+1}

$$sl(n+1, \mathbb{C}) = \mathfrak{h} \oplus_{i \neq j} \mathcal{G}_{\alpha_i}.$$

The Cartan matrix is $(\langle \alpha_i, \alpha_j \rangle)$ where

$$\langle \alpha_i, \alpha_j \rangle = \alpha_i(h_{\alpha_j}) = \begin{cases} 2 & i = j \\ -1 & |i - j| = 1 \\ 0 & \text{otherwise.} \end{cases}$$

Class (2): $B_n (n \geq 2)$: Here $E = \mathbb{R}^n$, $\Phi = \{\alpha \in I | (\alpha, \alpha) = 1 \text{ or } 2\}$. Then $\Phi = \{\pm\varepsilon_i$ (of squared length 1), $\pm(\varepsilon_i \pm \varepsilon_j), i \neq j$ (of squared length 2)$\}$. The n vectors $\varepsilon_1 - \varepsilon_2, \varepsilon_2 - \varepsilon_3, \dots, \varepsilon_{n-1} - \varepsilon_n, \varepsilon_n$ are independent. Short root is $\varepsilon_i = (\varepsilon_i - \varepsilon_{i+1}) + (\varepsilon_{i+1} - \varepsilon_{i+2}) + \cdots + (\varepsilon_{n-1} - \varepsilon_n) + \varepsilon_n$. Long root is $\varepsilon_i - \varepsilon_j$ or $\varepsilon_i + \varepsilon_j$. Weyl group W is isomorphic to the semidirect product of $(\mathbb{Z}/2\mathbb{Z})^n$ and S_l.

We have the following description about root systems:

For $o(2n+1, \mathbb{C})$, the Cartan subalgebra consists of the elements of the form
$h = diag(0, a_1, -a_1, a_2, -a_2, \dots, a_n, -a_n)$. The roots and root elements are given below.
For roots ϵ_i, $1 \leq i \leq l$, *root elements are* $\sqrt{2}(e_{2i-1,0} - e_{0,2i})$.
For roots $-\epsilon_i$, $1 \leq i \leq l$, *root elements are* $\sqrt{2}(e_{0,2i-1} - e_{2i,0})$.
For roots $\epsilon_i - \epsilon_j$, $i \neq j$, *root elements are* $e_{2i-1,2j-1} - e_{2j,2i}$.
For roots $\epsilon_i + \epsilon_j$, $i < j$, *root elements are* $e_{2j-1,2i} - e_{2i-1,2j}$.
For roots $-\epsilon_i - \epsilon_j$, $i < j$, *root elements are* $e_{2i,2j-1} - e_{2j,2i-1}$.

The fundamental root system is $\{\epsilon_1 - \epsilon_2, \epsilon_2 - \epsilon_3, \dots, \epsilon_{n-1} - \epsilon_n, \epsilon_n\}$.
The fundamental Weyl chamber is given by $a_1 > a_2 > \cdots > a_l > 0$.
The maximal root is $\epsilon_1 + \epsilon_2 = \epsilon_1 - \epsilon_2 + 2(\epsilon_2 - \epsilon_3 + \cdots + \epsilon_{n-1} - \epsilon_n + \epsilon_n)$.
Fundamental coroots are $h_1 = e_{11} - e_{22}, \dots, h_{n-1} = e_{n-1,\,n-1} - e_{ll}$, $h_n = 2e_{nn}$.
The Cartan matrix differs from that of A_l only by having -2 as $(n-1, n)$-th entry.

Class (3): $C_n (n \geq 3)$: (Here $B_2 = C_2$.) Here $E = R^n$. The root system of C_n is the set $\{\pm 2\varepsilon_i, \pm(\varepsilon_i \pm \varepsilon_n)$, for all i such that $i \neq j\}$ with a basis $\{\varepsilon_1 - \varepsilon_2, \dots, \varepsilon_{n-1} - \varepsilon_j, 2\varepsilon_n\}$. The Weyl group is isomorphic to that of B_n.

For $sp(n, \mathbb{C})$, the Cartan subalgebra $\mathfrak{h}$ is the set of matrices of the form

$$h = diag(a_1, -a_1, a_2, -a_2, \dots, a_l, -a_l).$$

We have $\mathfrak{h}_0 = \mathbb{R}^l$.

The roots and root elements are given below [43].
For roots $\epsilon_i - \epsilon_j$, $i \neq j$, *root elements are* $e_{2i-1,2j-1} - e_{2j,2i}$.
For roots $\epsilon_i + \epsilon_j$, $i < j$ *root elements are* $e_{2i-1,2j} + e_{2j-1,2i}$.
For roots $-\epsilon_i - \epsilon_j$, $i < j$, *root elements are* $e_{2i,2j-1} + e_{2j,2i-1}$.
For roots $2\epsilon_i$, *root elements are* $e_{2i-1,2i}$.
For roots $-2\epsilon_i$, *root elements are* $e_{2i,2i-1}$.

The fundamental root system is $\{\epsilon_1 - \epsilon_2, \epsilon_2 - \epsilon_3, \dots, \epsilon_{n-1} - \epsilon_n, 2\epsilon_n\}$.
The set of all positive roots is $\{\epsilon_i - \epsilon_j, \epsilon_i + \epsilon_j, 2\epsilon_i \text{ with } i < j\}$.
Fundamental Weyl chamber is $a_1 > a_2 > \cdots > a_n$.
Maximal root is $2\epsilon_1 = 2(\epsilon_1 - \epsilon_2 + \epsilon_2 - \epsilon_3 + \cdots + \epsilon_{n-1} - \epsilon_n) + 2\epsilon_n$.
Fundamental coroots are $h_1 = e_{11} - e_{22}, \dots, h_{n-1} = e_{n-1,n-1} - e_{nn}, h_n = e_{nn}$.
The Cartan matrix is the transpose of that for B_n.

Class (4): $D_n (n \geq 4)$: Here $E = R^n$, $\Phi = \{\alpha \in I | (\alpha, \alpha) = 2\} = \{\pm(\epsilon_i \pm \varepsilon_j), i \neq j\}$. One can take $\{\varepsilon_1 - \varepsilon_2, \dots, \varepsilon_{n-1} - \varepsilon_n, \varepsilon_{n-1} + \varepsilon_n\}$ set of all independent vectors as a basis. The Weyl group W is isomorphic to the semidirect product of $(\mathbb{Z}/2\mathbb{Z})^{n-1}$ and S_n.

For $o(2n, \mathbb{C})$, the Cartan subalgebra $\mathfrak{h}$ is the set of all

$$h = diag(a_1, -a_1, a_2, -a_2, \dots, a_n, -a_n).$$

We have the following description about root systems.

The roots and root elements are given below [43].
For roots $\epsilon_i - \epsilon_j$, $i \neq j$, *root elements are* $e_{2i-1,2j-1} - e_{2j,2i}$.
For roots $\epsilon_i + \epsilon_j$, $i < j$, *root elements are* $e_{2i-1,2j} - e_{2j-1,2i}$.
For roots $-\epsilon_i - \epsilon_j$, $i < j$, *root elements are* $e_{2i,2j-1} - e_{2j,2i-1}$.

The set of all positive roots is $\{\epsilon_i - \epsilon_j \text{ and } \epsilon_i + \epsilon_j \text{ with } i < j\}$.
Fundamental root system is $\{\epsilon_1 - \epsilon_2, \epsilon_2 - \epsilon_3, \dots, \epsilon_{n-1} - \epsilon_n, \epsilon_{n-1} + \epsilon_n\}$.
Fundamental Weyl chamber is $a_1 > a_2 > \cdots > a_{l-1} > |a_n|$.
Maximal root is $\epsilon_1 + \epsilon_2 =$
$\epsilon_1 - \epsilon_2 + 2(\epsilon_2 - \epsilon_3 + \epsilon_3 - \epsilon_4 + \cdots + \epsilon_{n-2} - \epsilon_{n-1}) + (\epsilon_{n-1} - \epsilon_n) + (\epsilon_{n-1} + \epsilon_n)$.
Fundamental coroots are
$h_1 = e_{11} - e_{22}, \dots, h_{n-1} = e_{n-1,n-1} - e_{nn}, h_n = e_{n-1,n-1} + e_{nn}$.
The Cartan matrix differs from that of A_n by having $a_{n-1,n} = a_{n,n-1} = 0$ and $a_{n-2,n}$ and $a_{n,n-2}$ equal to -1.

Class (5): E_6, E_7, E_8: It is clear that E_6, E_7 can be identified canonically with subsystems of E_8, hence one can construct E_8 directly. Here $E = R^8$ and $I' = I + Z((\varepsilon_1 + \cdots + \varepsilon_8)/2)$, $I'' =$ subgroup of I' consisting of all elements $\sum c_i \varepsilon_i + \frac{c}{2}(\varepsilon_1 + \cdots + \varepsilon_8)$ for which $\sum c_i$ is an even integer.

Define $\Phi = \{\alpha \in I'' | (\alpha, \alpha) = 2\} = \pm(\varepsilon_i \pm \varepsilon_j)$ for $i \neq j \cup \frac{1}{2}\sum_{i=1}^{8}(-1)^{\kappa(i)}\varepsilon_i$ (where the $k(i) = 0, 1$, add up to an even integer). All inner products here are in Z. A basis of Φ, $\{\frac{1}{2}(\varepsilon_1 + \varepsilon_8 - (\varepsilon_2 + \cdots + \varepsilon_7)), \varepsilon_1 + \varepsilon_2, \varepsilon_2 - \varepsilon_1, \varepsilon_3 - \varepsilon_2, \varepsilon_4 - \varepsilon_3, \varepsilon_5 - \varepsilon_4, \varepsilon_6 - \varepsilon_5, \varepsilon_7 - \varepsilon_6\}$. The order of the Weyl group can be shown to be equal to $2^{14}3^5 5^2 7$.

Case (1): For E_6, $\mathfrak{h}$ is $\mathbb{C}^6$ and $\mathfrak{h}_0$ is $\mathbb{R}^6$. The roots are $\epsilon_i - \epsilon_j$, $\pm(\epsilon_i + \epsilon_j + \epsilon_k)$ with $i < j < k$, and $\pm(\epsilon_1 + \epsilon_2 + \cdots + \epsilon_6)$.

We have the following description about root systems.

The set of all positive roots is $\{\epsilon_i - \epsilon_j$ with $i < j, \epsilon_i + \epsilon_j + \epsilon_k$ with $i < j < k, \epsilon_1 + \cdots + \epsilon_6\}$.
Fundamental root system is $\{\alpha_1 = \epsilon_1 - \epsilon_2, \alpha_2 = \epsilon_2 - \epsilon_3, \ldots, \alpha_5 = \epsilon_5 - \epsilon_6,$
$\alpha_6 = \epsilon_4 + \epsilon_5 + \epsilon_6\}$.
Fundamental Weyl chamber is $a_1 > a_2 > \cdots > a_6, a_4 + a_5 + a_6 > 0$.
Maximal root is $\epsilon_1 + \epsilon_2 + \cdots + \epsilon_6 = \alpha_1 + 2\alpha_2 + 3\alpha_3 + 2\alpha_4 + \alpha_5 + 2\alpha_6$.
Fundamental coroots are $h_1 = e_{11} - e_{22}, \ldots, h_5 = e_{55} - e_{66},$
$h_6 = 1/3(-e_{ll} - e_{22} - e_{33} + 2e_{44} + 2e_{55} + 2e_{66})$.

Case (2): For E_7, $\mathfrak{h}$ is $\mathbb{C}^7$ and $\mathfrak{h}_0$ is $\mathbb{R}^7$. The roots are $\epsilon_i - \epsilon_j$, $\pm(\epsilon_i + \epsilon_j + \epsilon_k)$ with $i < j < k$, and $\pm \sum_{r \neq i} \epsilon_r$.

We have the following description about root systems.

The set of all positive roots is $\epsilon_i - \epsilon_j$ with $i < j, \epsilon_i + \epsilon_j + \epsilon_k$ with $i < j < k, \sum_{r \neq i} \epsilon_r$.
Fundamental root system is $\{\alpha_1 = \epsilon_1 - \epsilon_2, \ldots, \alpha_6 = \epsilon_6 - \epsilon_7, \alpha_7 = \epsilon_5 + \epsilon_6 + \epsilon_7\}$.
Fundamental Weyl chamber is $a_1 > a_2 > \cdots > a_7, a_5 + a_6 + a_7 > 0$.
Maximal root is $\epsilon_1 + \cdots + \epsilon_6 = \alpha_1 + 2\alpha_2 + 3\alpha_3 + 4\alpha_4 + 3\alpha_5 + 2\alpha_6 + 2\alpha_7$.
Fundamental coroots are $h_1 = e_{11} - e_{22}, \ldots, h_6 = e_{66} - e_{77},$
$h_7 = 1/3(-e_{11} - e_{22} - e_{33} + 2e_{55} + 2e_{66} + 2e_{77})$.

Case (3): For E_8. For $\mathfrak{h}$, we take the subspace $\epsilon_1 + \epsilon_2 + \cdots + \epsilon_9 = 0$, of $\mathbb{C}^9$, with $\mathfrak{h}_0 = \mathfrak{h} \cap \mathbb{R}^9$. Roots: $\{\epsilon_i - \epsilon_j$, with $i \neq j, \pm(\epsilon_i + \epsilon_j + \epsilon_k)$ with $1 \leq i < j < k \leq 9\}$.

We have the following description about root systems.

The set of all positive roots is $\{\epsilon_i - \epsilon_j$ with $i < j, \epsilon_i + \epsilon_j + \epsilon_k$ with $i < j < k < 9$, and $-\epsilon_i - \epsilon_j - \epsilon_9$ with $i < j < 9\}$.
Fundamental root system is $\{\alpha_1 = \epsilon_1 - \epsilon_2, \ldots, \alpha_7 = \epsilon_7 - \epsilon_8, \alpha_8 = \epsilon_6 + \epsilon_7 + \epsilon_8\}$.
Maximal root is $\epsilon_1 - \epsilon_2 = 2\alpha_1 + 3\alpha_2 + 4\alpha_3 + 5\alpha_4 + 6\alpha_5 + 4\alpha_6 + 2\alpha_7 + 3\alpha_8$.
Fundamental coroots are
$h_1 = e_{11} - e_{22}, \ldots, h_6 = e_{66} - e_{77}, h_7 = e_{77} - e_{88}, h_8 = e_{66} + e_{77} + e_{88} - \frac{1}{3}e$, with $e = (1, \ldots, 1)$.
There is an alternative description of E_8, with $\mathfrak{h} = \mathbb{C}^8$.
The set of all positive roots is $\{\pm\epsilon_i - \epsilon_j$ with $1 \leq i < j < 8, \pm\epsilon_i + \epsilon_8$ with $1 \leq i < 8, 1/2(\pm e_{11} \pm e_{22} \pm \cdots \pm e_{77} + e_{88})\}$ (even number of negative signs).
Fundamental root system is $\alpha_1 = 1/2 \sum \epsilon_i, \alpha_2 = -\epsilon_1 - \epsilon_2, \alpha_3 = \epsilon_2 - \epsilon_3, \alpha_4 = \epsilon_1 - \epsilon_2, \alpha_5 = \epsilon_3 - \epsilon_4, \alpha_6 = \epsilon_4 - \epsilon_5, \alpha_7 = \epsilon_5 - \epsilon_6, \alpha_8 = \epsilon_6 - \epsilon_7$.

Remark 32. The Cartan matrices for E_7 and E_6 in this system can be obtained from that for E_8 by omitting the last row and column or the last two rows and columns, respectively.

Class (6): For F_4. Here $E = R^4, I' = I + Z((\varepsilon_1 + \varepsilon_2 + \varepsilon_3 + \varepsilon_4)/2)$, $\Phi = \{\alpha \in I' | (\alpha, \alpha) = 1 \text{ or } 2\}$. Explicitly, $\Phi = \{\pm\varepsilon_j, \pm(\varepsilon_i \pm \varepsilon_j), i \neq j, \pm\frac{1}{2}(\varepsilon_1 \pm \varepsilon_2 \pm \varepsilon_3 \pm \varepsilon_4)\}$. A basis of $\Phi = \{\varepsilon_2 - \varepsilon_3, \varepsilon_3 - \varepsilon_4, \varepsilon_4, \frac{1}{2}(\varepsilon_1 - \varepsilon_2 - \varepsilon_3 - \varepsilon_4)\}$. Here W has order 1152.

For F_4, the Cartan subalgebra $\mathfrak{h}$ is C^4 and $\mathfrak{h}_0$ is $\mathbb{R}^4$. The roots are $\pm\epsilon_i$ and $\pm(\epsilon_i \pm \epsilon_j)$ with $i, j = 1, 2, 3, 4$ and $i < j$, and $1/2(\pm\epsilon_1 \pm \epsilon_2 \pm \epsilon_3 \pm \epsilon_4)$.

We have the following description about root systems.

Table 1.4 Highest long roots and short roots of classical Lie algebras (Humphreys [40])

Type	Long	Short
A_n	$\alpha_1 + \alpha_2 + \cdots + \alpha_n$	
B_n	$\alpha_1 + 2\alpha_2 + 2\alpha_3 \cdots + 2\alpha_n$	$\alpha_1 + \alpha_2 + \cdots + \alpha_n$
C_n	$2\alpha_1 + 2\alpha_2 + \cdots + 2\alpha_{n-1} + \alpha_n$	$\alpha_1 + 2\alpha_2 + \cdots + 2\alpha_{n-1} + \alpha_n$
D_n	$\alpha_1 + 2\alpha_2 + \cdots + 2\alpha_{n-2} + \alpha_{n-1} + \alpha_n$	
E_6	$\alpha_1 + 2\alpha_2 + 2\alpha_3 + 3\alpha_4 + 2\alpha_5 + \alpha_6$	
E_7	$2\alpha_1 + 2\alpha_2 + 3\alpha_3 + 4\alpha_4 + 3\alpha_5 + 2\alpha_6 + \alpha_7$	
E_8	$2\alpha_1 + 3\alpha_2 + 4\alpha_3 + 6\alpha_4 + 5\alpha_5 + 4\alpha_6 + 3\alpha_7 + 2\alpha_8$	
F_4	$2\alpha_1 + 3\alpha_2 + 4\alpha_3 + 2\alpha_4$	$\alpha_1 + 2\alpha_2 + 3\alpha_3 + 2\alpha_4$
G_2	$3\alpha_1 + 2\alpha_2$	$2\alpha_1 + \alpha_2$

The set of all positive roots is $\{\epsilon_i, \epsilon_i + \epsilon_j$ and $\epsilon_i - \epsilon_j$ with $i < j, \frac{1}{2}(\epsilon_1 \pm \epsilon_2 \pm \epsilon_3 \pm \epsilon_4)\}$.
Fundamental root system is $\{\alpha_1 = 1/2(\epsilon_1 - \epsilon_2 - \epsilon_3 - \epsilon_4), \alpha_2 = \epsilon_4, \alpha_3 = \epsilon_3 - \epsilon_4,$
$\alpha_4 = \epsilon_2 - \epsilon_3\}$.
Maximal root is $\epsilon_1 + \epsilon_2, = 2\alpha_1 + 4\alpha_2 + 3\alpha_3 + 2\alpha_4$.
Fundamental Weyl chamber is $a_1 > a_2 + a_3 + a_4, a_4 > 0, a_3 > a_4, a_2 > a_3$.

Class (7): For G_2, E is the subspace of R^3 orthogonal to $\varepsilon_1 + \varepsilon_2 + \varepsilon_3$. $I' = I \cap E, \Phi = \{\alpha \in I' | (\alpha, \alpha) = 2 \text{ or } 6\}$. So $\Phi = \pm\{\varepsilon_1 - \varepsilon_2, \varepsilon_2 - \varepsilon_3, \varepsilon_1 - \varepsilon_3, 2\varepsilon_1 - \varepsilon_2 - \varepsilon_3, 2\varepsilon_2 - \varepsilon_1 - \varepsilon_3, 2\varepsilon_3 - \varepsilon_1 - \varepsilon_2\}$. A basis of Φ is the set $\varepsilon_1 - \varepsilon_2, -2\varepsilon_1 + \varepsilon_2 + \varepsilon_3$.

Again for G_2, the Cartan subalgebra $\mathfrak{h}$ is the subspace of $\mathbb{C}^3$ with equation $\epsilon_1 + \epsilon_2 + \epsilon_3 = 0$. $\mathfrak{h}_0$ is the corresponding subspace of $\mathbb{R}^3$. The roots are the restriction to $\mathfrak{h}$ of $\pm\epsilon_i$ and $\pm(\epsilon_i - \epsilon_j)$.

We have the following description about root systems.

The set of all positive roots is $\{\epsilon_1, \epsilon_2 - \epsilon_3, \epsilon_1 - \epsilon_2, \epsilon_2, -\epsilon_3, \epsilon_1 - \epsilon_3\}$.
Fundamental root system is $\{\epsilon_2, \epsilon_1 - \epsilon_2\}$.
Fundamental Weyl chamber is $a_1 > a_2 > 0$.
Maximal root is $\epsilon_1 - \epsilon_3 = 3\epsilon_2 + 2(\epsilon_1 - \epsilon_2)$.
If the set of all root system Φ is irreducible with two distinct root lengths, then we get long and short roots. If all roots are of equal length we call them as long (Table 1.4).

1.14 Universal enveloping algebras of Lie algebras

Definition 46. Let $\mathcal{G}$ be a Lie algebra over $\mathbb{C}$. Let $T(\mathcal{G})$ be a tensor Lie algebra defined by $T(\mathcal{G}) = \bigoplus_{k=0}^{\infty} T^k(\mathcal{G})$. Here $T^0(\mathcal{G}) = \mathbb{C}$ is the corresponding field. Moreover, $T^1(\mathcal{G}) = \mathcal{G}, \ldots, T^k(\mathcal{G}) = \mathcal{G} \otimes \mathcal{G} \otimes \cdots \otimes \mathcal{G}$ (a k-fold tensor product is a universal k-multilinear map). Let J be the two-sided ideal in $T(\mathcal{G})$ generated by all $X \otimes Y - Y \otimes X - [X, Y]$ with $X, Y \in T^1(\mathcal{G})$. Let $U(\mathcal{G}) = T(\mathcal{G})/J$. Then $U(\mathcal{G})$ is an associative algebra with identity, called the *universal enveloping algebra of* $\mathcal{G}$. Denote by i, the canonical map from $\mathcal{G}$ to $U(\mathcal{G})$ defined by the embedding $\mathcal{G}$ to $T^1(\mathcal{G})$. This map i satisfies

$$i[X,Y] = i(X)i(Y) - i(Y)i(X) \quad \text{for } X \text{ and } Y \text{ in } \mathcal{G}.$$

The above definition can also be written as follows

Let $\mathcal{G}$ be a Lie algebra over a field K. The rule of composition $(X,Y) \to [X,Y]$ is rarely associative. We shall now assign to $\mathcal{G}$ an associative algebra with unit element, the universal enveloping algebra $U(\mathcal{G})$ of $\mathcal{G}$. This algebra is defined as the factor algebra $T(\mathcal{G})/J$, where $T(\mathcal{G})$ is the tensor algebra over $\mathcal{G}$ (considered as a vector space) and J is the two-sided ideal in $T(\mathcal{G})$ generated by the set of all elements of the form $X \otimes Y - Y \otimes X - [X,Y]$, where $X, Y \in \mathcal{G}$.

For a finite-dimensional Lie algebra over a field F with a basis $\{X_1, X_2, \ldots, X_n\}$, let the structure constants a_{ij}^k satisfy

$$[X_i, X_j] = \sum_k a_{ij}^k X_k \quad \text{for } 1 \le i,j \le n.$$

Then the universal enveloping algebra $U(\mathcal{G})$ of $\mathcal{G}$ is an associative algebra with unity generated by $\{X_1, X_2, \ldots, X_n\}$ satisfying

$$X_iX_j - X_jX_i = \sum_{k=1}^{n} a_{ij}^k X_k \quad \text{for } 1 \le i,j \le n.$$

The universal enveloping algebra $U(\mathcal{G})$ does not depend on the choice of the basis. That is, for two different bases of $\mathcal{G}$ the corresponding $U(\mathcal{G})$ are isomorphic.

Remark 33. In fact, a universal enveloping algebra of $\mathcal{G}$ is a pair $(U(\mathcal{G}), i)$, where $U(\mathcal{G})$ is an associative algebra with 1 over F and i is a linear mapping $i\colon \mathcal{G} \to U(\mathcal{G})$ satisfying

$$i([x,y]) = i(x)i(y) - i(y)i(x)$$

for $x, y \in \mathcal{G}$. Moreover for any other associative algebra $\mathfrak{U}$ over F with unity and any other linear map $j\colon \mathcal{G} \to \mathfrak{U}$ satisfying

$$j([x,y]) = j(x)j(y) - j(y)j(x),$$

there exists a unique homomorphism $\phi\colon U(\mathcal{G}) \to \mathfrak{U}$ such that $\phi \circ i = j$.

The universal mapping property

Let $U(\mathcal{G})$ be the universal enveloping algebra of a Lie algebra $\mathcal{G}$ over $\mathbb{C}$. Consider the canonical map $i\colon \mathcal{G} \to U(\mathcal{G})$. If A is a complex associative algebra with identity and

$\phi: \mathcal{G} \to A$ is a linear mapping such that $\phi(X)\phi(Y) - \phi(Y)\phi(X) = \phi[X, Y]$ for all X and Y in $\mathcal{G}$, then there exists a unique algebra homomorphism $\tilde{\phi}: U(\mathcal{G}) \to A$ such that $\tilde{\phi}(1) = 1$ and the diagram commutes.

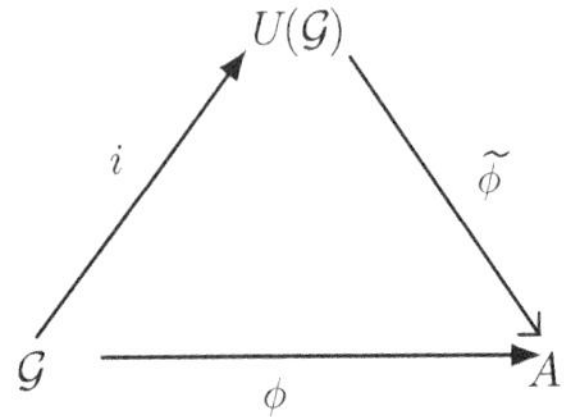

The above property is called the *Universal mapping property.*

For any Lie algebra $\mathcal{G}$ over $\mathbb{F}$ generated by e_i, f_i, and $\mathfrak{h}$, the corresponding universal enveloping algebra $U(\mathcal{G})$ of $\mathcal{G}$ is the associative algebra over $\mathbb{F}$ with unity and generated by $e_i, f_i (i \in I)$, and $\mathfrak{h}$ and satisfying the following conditions:

(1) $hh' = h'h$ for $h, h' \in \mathfrak{h}$;
(2) $e_i f_j - f_j e_i = \delta_{ij} h_i$ for all $i, j \in I$;
(3) $he_i - e_i h = \alpha_i(h) e_i$ for $h \in \mathfrak{h}, i \in I$;
(4) $hf_i - f_i h = -\alpha_i(h) f_i$ for $h \in \mathfrak{h}, i \in I$;
(5) $\sum_{k=0}^{1-a_{ij}} (-1)^k \binom{1-a_{ij}}{k} e_i^{1-a_{ij}-k} e_j e_i^k = 0$ for $i \neq j$; and
(6) $\sum_{k=0}^{1-a_{ij}} (-1)^k \binom{1-a_{ij}}{k} f_i^{1-a_{ij}-k} f_j f_i^k = 0$ for $i \neq j$.

Let U^+ (respectively, U^0 and U^-) be the subalgebra of $U = U(\mathcal{G})$ generated by the elements e_i (respectively, by h and f_i) for $i \in I$. We also define the root spaces to be

$$U_\beta = \{u \in U \mid hu - uh = \beta(h)u, \quad \text{for all } h \in \mathfrak{h}\}, \quad \beta \in Q.$$

Theorem 11 (Poincare-Birkhoff-Witt theorem). *Let $\mathcal{G}$ be a Lie algebra over $\mathbb{C}$ with basis $\{X_j\}_{j \in A}$ (A – a index set with simple ordering). Then the set of all monomials*

$$(iX_{j_1})^{k_1} \ldots (iX_{j_n})^{k_n}, \quad (j_1, \ldots, j_2 \in A)$$

with $j_1 < \cdots < j_n$ and $k_1 < \cdots < k_n$ is a basis of $U(\mathcal{G})$ where $i: \mathcal{G} \to U(\mathcal{G})$ is a universal mapping which is one-to-one. For the proof of the above theorem, one can refer [40, 43, 22, 174].

Remark 34.

(1) In fact for a Lie algebra $\mathcal{G}$ over a basis $\{X_1, X_2, \ldots, X_n\}$, the set $\{X_1^{m_1}, X_2^{m_2}, \ldots : m_i \in \mathbb{Z}_{\geq 0}\}$ (all but a finite number of m_i's are 0) is a basis of $U(\mathcal{G})$.
(2) If $\mathcal{G}_1, \mathcal{G}_2, \ldots, \mathcal{G}_n$ are subalgebras of $\mathcal{G}$ satisfying $\mathcal{G} = \mathcal{G}_1 \oplus \cdots \oplus \mathcal{G}_n$, then the corresponding universal enveloping algebra is

$$U(\mathcal{G}) = U(\mathcal{G}_1) \ldots U(\mathcal{G}_n).$$

Moreover, we will get the following isomorphism of vector spaces:

$$U(\mathcal{G}) \cong U(\mathcal{G}_1) \otimes \cdots \otimes U(\mathcal{G}_n).$$

(3) By Poincare-Birkhoff-Witt (PBW) theorem, the universal enveloping algebra has the triangular decomposition.

$$U(\mathcal{G}) \simeq U^- \otimes U^\circ \otimes U^+.$$

Here

U^+**:** subalgebra of $U(\mathcal{G})$, generated by e.

U^0**:** subalgebra of $U(\mathcal{G})$, generated by h.

U^-**:** subalgebra of $U(\mathcal{G})$, generated by f.

Moreover, we have the following root space decomposition:

$$U(\mathcal{G}) = \oplus_{\beta \in Q} U_\beta$$
$$U(I) = \oplus_{\beta \in Q_I} U^I_\beta.$$

For the triangular decomposition of $\mathcal{G} = \mathcal{G}_+ \oplus \mathfrak{h} \oplus \mathcal{G}_-$, we have

$$U(\mathcal{G}) = U(h) \oplus (\mathcal{G}_+ U(\mathcal{G}) + U(\mathcal{G})\mathcal{G}_-).$$

Example 13.

(1) Let $\mathcal{G} = \{x\}$(span of x) be a one-dimensional abelian Lie algebra over a field F satisfying $[x,x] = 0$. The corresponding universal enveloping algebra $U(\mathcal{G})$ is the associative algebra generated by a single element X satisfying $XX - XX = 0$. So, $U(\mathcal{G})$ here is the polynomial algebra $F[X]$.

(2) Let $\mathcal{G}$ be an n-dimensional abelian Lie algebra with a basis $\{x_1, x_2, \ldots, x_n\}$. $U(\mathcal{G})$ is isomorphic to the polynomial algebra in n variables with all structure constants a^k_{ij} being 0.

(3) Let $\mathcal{G} = sl(2,\mathbb{C})$.
We have

$$e = \begin{pmatrix} 0 & 1 \\ 0 & 0 \end{pmatrix}, f = \begin{pmatrix} 0 & 0 \\ 1 & 0 \end{pmatrix}, \text{ and } h = \begin{pmatrix} 1 & 0 \\ 0 & -1 \end{pmatrix},$$

with

$$[h,e] = 2e,\ [h,f] = -2f, \text{ and } \quad [e,f] = h.$$

Its universal enveloping algebra $U(sl(2,\mathbb{C}))$ is the associative algebra over $\mathbb{C}$ with 1 generated by the elements e, f, and h satisfying $he - eh = 2e$, $hf - fh = -2f$, and $ef - fe = h$. Here we have

$$\mathcal{G}_\alpha = \{x \in \mathcal{G} | [h,x] = \alpha(h)x\} = \mathbb{C}e$$
$$\mathcal{G}_{-\alpha} = \{x \in \mathcal{G} | [h,x] = -\alpha(h)x\} = \mathbb{C}f \quad \text{and}$$
$$\mathcal{G}_0 = \{x \in \mathcal{G} | [h,x] = 0\} = \mathbb{C}h.$$

This will give the roots space decomposition.

The triangular decomposition of $\mathcal{G}$ is equal to triangular decomposition of $sl(2,\mathbb{C}) = \mathbb{C}f \oplus \mathbb{C}h \oplus \mathbb{C}e$.

U^+: subalgebra of $U(sl(2,\mathbb{C}))$ generated by e.

U^0: subalgebra of $U(sl(2,\mathbb{C}))$ generated by h.

U^-: subalgebra of $U(sl(2,\mathbb{C}))$ generated by f.

We have $U(sl(2,\mathbb{C})) \simeq U^- \otimes U^0 \otimes U^+$ is the triangular decomposition of $U(sl(2,\mathbb{C}))$.

Applying PBW theorem, it is clear that $\{e^i f^j h^k\}(i,j,k \geq 0)$ will form a basis of $U(sl(2,\mathbb{C}))$. The following relations can be directly proved for $U(sl(2,\mathbb{C}))$ for any $p \geq 0, q \geq 0$

(i) $e^p h^q = (h-2p)^q e^p$,
(ii) $f^p h^q = (h+2p)^q f^p$,
(iii) $[e, f^p] = pf^{p-1}(h-p+1) = p(h+p-1)f^{p-1}$, and
(iv) $[e^p, f] = pe^{p-1}(h+p-1) = p(h-p+1)e^{p-1}$.

Remark 35. The element $C = XY + YX + H^2/2 \in U(sl(2,\mathbb{C}))$ is called the Casimir element of the corresponding universal enveloping algebra of $sl(2,\mathbb{C})$.

(4) Consider $\mathcal{G} = sl(n,F)$, where $sl(n,F) = \{X \in M_{n\times n}(F) |$ trace $X = 0\}$. The set $\{E_{i,i} - E_{i+1,i+1}\}(i = 1,\ldots,n)$ is a basis of $sl(n,F)$. We have $e_i = E_{i,i+1}, f_i = E_{i+1,i}, h_i = E_{i,i} - E_{i+1,i+1}(1 \leq i \leq n-1)$.

$$[e_i, f_j] = \delta_{ij} h_i,$$

$$[h_i, e_j] = \begin{cases} 2e_j & \text{if} \quad i = j \\ -e_j & \text{if} \quad |i-j| = 1 \\ 0 & \text{if} \quad |i-j| > 1 \end{cases} \quad [h_i, f_j] = \begin{cases} -2f_j & \text{if} \quad i = j \\ f_j & \text{if} \quad |i-j| = 1 \\ 0 & \text{if} \quad |i-j| > 1 \end{cases}$$

Let $A = (a_{ij})_{i,j,=1,\ldots,n-1}$ be a square matrix such that $a_{ii} = 2, a_{ij} = -1$ if $|i-j| = 1$, and $a_{ij} = 0$ otherwise. The generators e_i, f_i, and h_i satisfy the additional relations $(ad\ e_i)^{1-a_{ij}}(e_j) = (ad\ f_i)^{1-a_{ij}} f_j = 0$ for $i \neq j$.

We have $\mathcal{G}_\alpha = FE_{ij}$, and $\mathcal{G}_{-\alpha} = FE_{ij} = \{x \in sl_n(F) | [h,x] = (\epsilon_i - \epsilon_j)(h)(x)$ for all $h \in \mathfrak{h}\}$

$$\mathcal{G} = sl_n(F) = \left(\bigoplus_{\alpha\in\Phi_-} \mathcal{G}_\alpha\right) \oplus \mathfrak{h} \oplus \left(\oplus_{\alpha\in\Phi_+} \mathcal{G}_\alpha\right)$$

$$\mathcal{G}_+ = \oplus_{\alpha\in\Phi_+} \mathcal{G}_\alpha = \oplus_{i<j} FE_{ij}$$

$$\mathcal{G}_- = \oplus_{\alpha\in\Phi_-} \mathcal{G}_\alpha = \oplus_{i>j} FE_{ij}$$

$U^- = U(\mathcal{G}_-), U^0 = U(h), U^+ = U(\mathcal{G}_+)$. U^+ is generated by e_i. U^- is generated by f_i and U^0 is generated by $\mathfrak{h}$.

Remark 36. For bilinear form on universal enveloping algebras of semisimple Lie algebras, one can refer Shapovalov [44].

1.15 Representation theory of semisimple Lie algebras

Let $\mathcal{G}$ be a semisimple Lie algebra over an algebraic closed field of characteristic 0, $\mathfrak{h}$ be a Cartan subalgebra of $\mathcal{G}$, and Π be the root system with $\Delta = \{\alpha_1, \ldots, \alpha_n\}$ a basis of Φ. Let W be the Weyl group of $\mathcal{G}$.

Let V be a finite-dimensional $\mathcal{G}$-module. Then $\mathfrak{h}$ acts diagonally on V such that $V = \bigoplus_{\lambda \in \mathfrak{h}^*} V_\lambda$, where $V_\lambda = \{v \in V | h \cdot v = \lambda(h)v\}$. If $V_\lambda \neq 0$, then V_λ is called a *weight space* and λ is called *weight* of V_λ. (Already explained for any Lie algebras and in particular for $sl(2, \mathbb{C})$ also.) In the usual notations, the following results are direct consequences of the definition.

(a) $\mathcal{G}_\alpha$ maps V_λ into $V_{\lambda+\alpha} (\lambda \in \mathfrak{h}^*, \alpha \in \Phi)$.
(b) $V' = \bigoplus_{\lambda \in \mathfrak{h}^*} V_\lambda$ is direct sum and V' is an $\mathcal{G}$-submodule of V.
(c) If $\dim V < \infty$, then $V = V'$.

A *maximal vector* in an $\mathcal{G}$-module is an nonzero vector $v^+ \in V_\lambda$ such that $\mathcal{G}_\alpha \cdot v^+ = 0, \alpha \in \Delta$.

Definition 47. If $U(\mathcal{G})$ is the universal enveloping algebra of $\mathcal{G}$ and if $V = U(\mathcal{G}) \cdot v_\lambda$ for a maximal vector v_λ, then V is called *standard cyclic* and we call λ the *highest weight of V*.

Let V be a highest weight module with maximal weight λ. Then V has the following properties:

(a) $V = U(\mathcal{G}) \cdot v_\lambda = U(\eta_-) \cdot v_\lambda$ for any weight vector v_λ of weight λ.
(b) $\dim V_\lambda = 1$.
(c) The set of all weights, $P(V) \subset D(\lambda) = \{\lambda - \alpha | \alpha \in Q_+\}$.
(d) $\mathcal{G}_\alpha V_\lambda = 0$ for any $\alpha \in \Delta^+$.
(e) V has a unique *maximal submodule* hence a unique simple quotient module.
(f) If V'' is a nonzero homomorphic image of V, then V'' is again a highest weight module.

For proofs of the following results, one can refer Humphreys [40].

Theorem 12. *Let V be a standard cyclic $\mathcal{G}$-module, with maximal vector $v^+ \in V_\lambda$ and $\Phi^+ = \{\beta_1, \beta_2, \ldots, \beta_m\}$. Then:*

(a) *V is spanned by the vectors $y_{\beta_1}^{i_1} y_{\beta_2}^{i_2} \cdots y_{\beta_m}^{i_m} \cdot v^+ (i_j \in \mathbb{Z}^+)$ and V is the direct sum of its weight spaces.*
(b) *The weights of V are of the form $\mu = \lambda - \sum_{i=1}^{l} k_i \alpha_i (k_i \in \mathbb{Z}^+)$, that is, all weights satisfy $\mu < \lambda$.*
(c) *For each $\mu \in \mathfrak{h}^*$, $\dim V_\mu < \infty$, and $\dim V_\lambda = 1$.*
(d) *Each submodule of V is the direct sum of its weight spaces.*
(e) *V is an indecomposable $\mathcal{G}$-module with a unique maximal submodule and a corresponding unique irreducible quotient.*
(f) *Every nonzero homomorphic image of V is also a standard cyclic module of weight λ.*
(g) *If V is an irreducible $\mathcal{G}$-module, then v^+ is the unique maximal vector in V, up to nonzero scalar multiples.*

(h) *If V, W are two standard cyclic irreducible modules of highest weight λ, then they are isomorphic. For $\lambda \in \mathfrak{h}^*$, there exists an irreducible standard cyclic module $V(\lambda)$ of weight λ.*

Definition 48 (Finite-dimensional modules). *Necessary and sufficient condition for existence of finite-dimensional irreducible module*: If V is a finite-dimensional irreducible $\mathcal{G}$-module of highest weight λ, then $\lambda(h_i)$ is an nonnegative integer for each i. If $\lambda \in \mathfrak{h}^*$ is dominant integral, then the necessary and sufficient condition for existence of the irreducible $\mathcal{G}$-module $V = V(\lambda)$ is that V is of finite dimension and its set of weights $P(V)$ is permuted by W, with $\dim V_\mu = \dim V_{\sigma\mu}$ for $\sigma \in W$. The map $\lambda \mapsto V(\lambda)$ induces a one-to-one correspondence between Λ^+ and the isomorphism classes of finite-dimensional irreducible $\mathcal{G}$-modules. Also, the following identities hold in $U(\mathcal{G})$, for $k \geq 0, 1 \leq i,j \leq l$: Let $\{x_i, y_i\}$ be standard generators of $\mathcal{G}$. Then

(a) $[x_j, y_i^{k+1}] = 0$ when $i \neq j$ and
(b) $[h_j, y_i^{k+1}] = -(k+1)\alpha_i(h_j)y_i^{k+1}$
(c) $[x_i, y_i^{k+1}] = -(k+1)y_i^k(k \cdot 1 - h_i)$.

1.16 Construction of semisimple Lie algebras by generators and relations

Using the definition of root system of Euclidean space (Section 1.12), we construct below semisimple Lie algebras by relations and generators. Let V be a complex vector space and Φ be a root system in a complex vector space V, in the sense of root system in E. Let V^* be the dual of V with $V^* = \mathfrak{h}$. Let $S = \{\alpha_1, \ldots, \alpha_n\}$ be a basis of Φ and let $H_1, \ldots, H_n$ (the inverses of $\alpha_1, \ldots, \alpha_n$, respectively) $\in \mathfrak{h}$.

Let $\mathcal{G}_{\alpha_i}$ be the rootspace generated by α_i, $\mathcal{G}_{-\alpha_i}$ be the rootspace generated by $-\alpha_i$. Choose $X_i \in \mathcal{G}_{\alpha_i}$, $Y_i \in \mathcal{G}_{-\alpha_i}$ such that $[X_i, Y_i] = H_i$. Let $a_{ij} = \alpha_j(H_i)$. In fact, $a_{ij} = \langle \alpha_j, H_i \rangle$. The matrix formed by the numbers a_{ij} is the Cartan matrix of the given system.

Let

η^+ be generated by X_i and
η^- be generated by Y_i.

Let $\mathcal{G}$ be the Lie algebra defined by $3n$ generators X_i, Y_i, H_i satisfying the relations

$$\begin{aligned}
[H_i, H_j] &= 0,\\
[X_i, Y_j] &= H_i, \quad [X_i, Y_j] = 0 \quad \text{if } i \neq j,\\
[H_i, X_j] &= a_{ij}X_j, \quad [H_i, Y_j] = -a_{ij}Y_j,\\
ad(X_i)^{-n_{ij}+1}(X_j) &= 0 \quad \text{if } i \neq j,\\
ad(Y_i)^{-n_{ij}+1}(Y_j) &= 0 \quad \text{if } i \neq j.
\end{aligned}$$

Then $\mathcal{G}$ is semisimple Lie algebra with the Cartan subalgebra $\mathfrak{h}$ generated by the elements H_i. Now Φ is the root system of $\mathcal{G}$.

The following properties can be verified:

1. There is an automorphism σ of $\mathcal{G}$ which is equal to Id on $\mathfrak{h}$, such that $\sigma(X_i) = -Y_i$, $\sigma(Y_i) = -X_i$ for all i and $\sigma^2 = 1$.
2. Two semisimple Lie algebras corresponding to isomorphic root systems are isomorphic: Let $\mathcal{G}$ and $\mathcal{G}'$ be semisimple Lie algebras with Cartan subalgebras $\mathfrak{h}$ and $\mathfrak{h}'$, respectively. Let S and S' be the corresponding bases of the root system with $\gamma\colon S \to S'$ be a bijection sending the Cartan matrix of S to that of S'. For each $i \in S$ and $j \in S'$, let X_i and X'_j be nonzero elements of $\mathcal{G}_i$ and of $\mathcal{G}'_j$, respectively. Then there is a unique isomorphism $f\colon \mathcal{G} \to \mathcal{G}'$ sending H_i to $H'_{\gamma(i)}$, X_i to $X'_{\gamma(i)}$, $\forall\, i \in S$.

1.17 Cartan-Weyl basis

Let $\mathcal{G}$ be a simple complex Lie algebra of dimension n. The Cartan-Weyl basis of $\mathcal{G}$ is generated by h_i $(i = 1, \ldots, r)$ and $n - r$ generators $e_{\pm\alpha}$ satisfying the following conditions

$$[h_i, h_j] = 0; \quad [h_i, e_\alpha] = \alpha_i e_\alpha$$

$$[e_\alpha, e_{-\alpha}] = \sum_{i=1}^{r} \alpha_i h_i$$

and

$$[e_\alpha, e_\beta] = N_{\alpha\beta} e_{\alpha+\beta}, \quad \text{if } \alpha + \beta \text{ is a nonzero root.}$$

The r-dimensional vector $\alpha = (\alpha_1, \ldots, \alpha_r)$ of $\mathbb{R}^r$ is a root associated to the generator e_α. For a pair of roots α and β, the coefficient $N_{\alpha\beta}$ satisfies

$$N_{\alpha\beta}^2 = \frac{1}{2} k(k' + 1)|\alpha|^2.$$

Here k and k' are integers, so that $\alpha + k\beta$ and $\alpha - k'\beta$ are roots. Moreover, the coefficients $N_{\alpha\beta}$ will satisfy

$$N_{\alpha\beta} = -N_{\beta\alpha} = -N_{-\alpha-\beta}.$$

The roots are normalized such that

$$\sum_{\alpha\neq 0} \alpha_i \alpha_j = \delta_{ij} \Rightarrow \sum_{\alpha\neq 0} \alpha^2 = r$$

with respect to the Killing form $\kappa(X, Y)$ on $\mathcal{G}$ (bilinear form associated to the adjoint representation of $\mathcal{G}$) defined by

$$\kappa(X, Y) = tr(ad\, X ad\, Y), \quad \forall\, X, Y \in \mathcal{G}.$$

The matrix of this Killing form [129] is

$$(g_{ij}) = \begin{pmatrix} \delta_{ij} & & & & & & \\ & 0 & 1 & \dots & 0 & & \\ & 1 & 0 & \ddots & & & \\ & & \ddots & \ddots & \ddots & & \\ & 0 & & \ddots & 0 & 1 & \\ & & & & 1 & 0 & \end{pmatrix},$$

where δ_{ij} is a $r \times r$ matrix.

1.18 Character of a finite-dimensional representation and Weyl dimension formula

Let $\mathcal{G}$ be a finite-dimensional Lie algebra and V be a $\mathcal{G}$-module. If V is $\mathfrak{h}$-diagonalizable with Cartan subalgebra $\mathfrak{h}$, then define the character,

$$ch(V) = \sum_{\lambda \in P(V)} (\dim V_\lambda) e^\lambda. \tag{1.9}$$

Here $P(V)$ is the set of all weights of the representation ϕ of $\mathcal{G}$ in V. Actually, the above character is an element of a ring. The value of the character $ch\ V$ at $h \in \mathfrak{h}$ is

$$(ch\ V)(h) := \sum_{\lambda \in P(V)} (\dim V_\lambda) e^{\lambda(h)}.$$

This value may converge for some $h \in \mathfrak{h}$.

The set of all weights of the representation (ϕ, V) is $P(V) \subset \mathfrak{h}^*$. As $V = \oplus_{\lambda \in P(V)} V_\lambda$ is the eigen space decomposition with respect to $\phi(h)$, $\phi(h)$ is the scalar operation and $\phi(h)|_{(V_\lambda)} = \lambda(h) Id_{V_\lambda}$. Taking exponential on both sides, we get

$$e^{\phi(h)}|_{V_\lambda} = e^{\lambda(h)} Id_{V_\lambda}.$$

So, $\sum (\dim V_\lambda) e^\lambda = tr(e^{\phi(h)})$. Hence Equation (1.9) becomes

$$ch(V) = \sum_{\lambda \in P(V)} tr_V(e^{\phi(h)}).$$

Remark 37. More details, regarding Weyl Character formula for Kac-Moody algebra in general, are in Section 2.9 of Chapter 2.

Definition 49 (Weyl Character formula). Let $\mathcal{G}$ be a semisimple or a simple Lie algebra. The character $\chi(\Lambda)$ of the irreducible representation $\pi(\Lambda)$ with highest weight Λ of $\mathcal{G}$ is given by

$$\chi(\Lambda) = ch(V) = \frac{\sum\limits_{w \in W} \epsilon(w) e^{w(\Lambda+\rho)}}{\sum\limits_{w \in W} \epsilon(w) e^{w(\rho)}}.$$

This is called the classical Weyl Character formula. Here W is the Weyl group of $\mathcal{G}$, $e(w)$ is the parity of w, and ρ is the sum of positive roots. Here the parity $e(w) = (-1)^{l(w)}$, where $l(w)$ is the minimum number of reflections w_{α_i} such that $w = \prod_i w_{\alpha_i}$.

1.19 Lie algebras of vector fields

Classification of Lie algebras of vector fields under local diffeomorphism became important because of its applications to the integration of differential equations with infinitesimal symmetries. Lie classified Lie algebras of vector fields in one real variable and one, two variables [5, 45–47], or three complex variables [7]. For basic details, one can also refer Singer and Thorpe [241].

Definition 50. Let M be a differentiable manifold and $C^\infty(M) = C^\infty(M, \mathbb{R})$ be the set of all real-valued C^∞ functions on M. The tangent space at $m \in M$ is $T_m(M) = \{f\colon C^\infty(M) \to R \mid f \text{ is linear and } X(fg)(m) = X(f)g(m)+f(m)X(g) \text{ for } f, g \in C^\infty(M)\}$. Any element of $T_m(M)$ is called a *tangent vector* at $m \in M$.

Let $T_m(M)$ be the tangent space to M at m. Let $T(M) = \cup_{m\in M} T_m(M)$ and $T^*(M) = \cup_{m \in M} T^*_m(M)$. Here $T(M)$ is called the *tangent bundle* and $T^*(M)$ is called the *cotangent bundle*.

One can define a projection map $\pi\colon T(M) \to M$ as follows:

$v \in T(M)$ means that $v \in T_m(M)$ for a unique $m \in M$ such that $\pi(v) = m$.

Similarly one can define a projection map from $T^*(M)$ onto M, denoted by same π.

Definition 51. A *vector field* V is a map from $M \to T_m(M)$ such that for each point $m \in M$, $V|_m = V_m \in T_m(M)$. For a local coordinate system $\{x_1, \ldots, x_n\}$, the vector field V can be expressed as

$$V_m(f) = \sum_{k=1}^{n} a_k(m) \frac{\partial f}{\partial v_k},$$

where a_k's are real-valued functions. A vector field is called smooth if the coefficient functions a_k's are smooth. We have $V(f)|_m = V_m(f)$. Clearly $V(f)$ is a smooth function if V is a smooth vector field. Hence a smooth vector field V is a map from $C^\infty(M) \to C^\infty(M)$ such that

$$V(fg) = fV(g) + V(f)g.$$

This shows that a vector field is a derivation of the algebra of smooth functions.

That is, a vector field V on a C^∞ manifold M is a derivation of the algebra $C^\infty(M)$. Let $D^1(M)$ denote the set of all vector fields on M. If $f \in C^\infty(M)$ and $X, Y \in D^1(M)$, then fX and $X + Y$ denote the vector fields defined by,

$$f\,X: g \to f(Xg), \quad \text{for } g \in C^\infty(M),$$

and

$$X + Y: g \to Xg + Yg, \quad \text{for } g \in C^\infty(M),$$

respectively.

This turns $D^1(M)$ into a module over the ring $\mathcal{F} = C^\infty(M)$.

If V and W are two smooth vector fields on manifold M, then it can be easily proved that the commutator $[V, W]$ is also a vector field.

For $f, g \to C^\infty(M)$,

$$\begin{aligned}
[V, W](fg) &= (V \circ W - W \circ V)(fg) = V(W(fg)) - W(V(fg)) \\
&= V(W(g) + fV(g)) + W(V(f)g + fV(g)) \\
&= V \circ W(f)g + W(f)V(g) + V(f)W(g) + fV \circ W(g) \\
&\quad - W \circ V(f)g - V(f)W(g) - W(f)V(g) - fW \circ V(g) \\
&= [V, W](f)g + f[V, W](g).
\end{aligned}$$

Hence it satisfies Leibnitz rule.

Definition 52. Let M be a differentiable manifold and $D^1(M)$ be the set of all vector fields on M. If $X, Y \in D^1(M)$, then $XY - YX$, denoted as $[X, Y]$, is also a derivation of $C^\infty(M)$. Denote by $\theta(X)Y$, the derivative of $[X, Y]$. Here $\theta(X)Y$ is called the *Lie derivative* of Y with respect to X. It can be easily verified that this bracket satisfies Jacobi identity. That is, $[X, [Y, Z]] + [Y, [Z, X]] + [Z, [X, Y]] = 0$, otherwise $\theta(X)[Y, Z] + \theta(Y)[Z, X] + \theta(Z)[X, Y] = 0$.

Definition 53. We say that two vector fields are commuting if their Lie bracket is the zero vector field. It is clear that a vector field of M is a map $V: M \to T(M)$ such that $\pi \circ V = i_X$. A differential 1-form is a map $\omega: M \to T^*(M)$ such that $\omega(V) \in C^\infty(M, R^1)$ with $(\omega(V))(x) = \omega(x)(V(x))$. We denote the set of all smooth vector fields on M by $C^\infty(M, T(M))$ and set of smooth 1-forms by $C^\infty(M, T^*(M))$. For $f \in C^\infty(M, R^1)$, $df \in C^\infty(M, T^*(M))$, and $V \in C^\infty(M, T(M))$, we will have $df(V) = Vf \in C^\infty(M, R^1)$.

Proposition 13. *Prove that set of all smooth vector fields, namely, $C^\infty(M, T(M))$ is a Lie algebra.*

Proof. Let M be a smooth manifold and V and W be smooth vector fields on M. So $V, W \in C^\infty(M, T(M))$. One can define a bracket $[V, W]$ of V and W by

$$[V, W](f) = V(Wf) - W(Vf) \quad \text{for all } f \in C^\infty(M, R^1).$$

Let ϕ be a local coordinate system with domain U and coordinate functions $(x_1, \dots, x_n)$. Then one can write $V|_U = \sum_{i=1}^n a_i(\frac{\partial}{\partial x_i})$ and $W|_U = \sum_{i=1}^n b_i(\frac{\partial}{\partial x_i})$ for some $a_i, b_i \in C^\infty(U, R)$. It can be proved that $[V, W]$ is a vector field.

Since $[V, W]$ is bilinear, in order to check that $[V, W]$ is a vector field, it is sufficient to assume $V = a(\partial/\partial x_i)$ and $W = b(\partial/\partial x_j)$.

We have

$$[V, W](f) = a\left(\frac{\partial}{\partial x_i}\right)\left(b\left(\frac{\partial}{\partial x_j}\right)(f)\right) - b\left(\frac{\partial}{\partial x_j}\right)\left(a\left(\frac{\partial}{\partial x_i}\right)(f)\right)$$

After expanding and simplifying the RHS we get,

$$[V, W](f) = \left[a\frac{\partial}{\partial x_i}(b)\frac{\partial}{\partial x_j} - b\frac{\partial}{\partial x_j}(a)\frac{\partial}{\partial x_i}\right](f).$$

Since $a\left(\frac{\partial}{\partial x_i}\right)(b)$ and $b\frac{\partial}{\partial x_j}(a) \in C^\infty(M, R^1)$, $[V, W]$ is a smooth vector field.

It can be easily verified that the bracket $[V, W]$ satisfies the following properties:

(1) $[V, W] = -[W, V]$
(2) $[V_1 + V_2, W] = [V_1, W] + [V_2, W]$
(3) $[cV, W] = c[V, W]$ for $c \in R^1$
(4) $[[V, W], Z] + [[W, Z], V] + [[Z, V], W] = 0$

Hence $C^\infty(M, T(M))$ is a Lie algebra under bracket multiplication, called *Lie algebra of vector fields.* □

The following proposition can be directly proved.

Proposition 14. *Let $\{\theta_t \mid t \in \mathbb{R}\}$ be a one-parameter family of diffeomorphisms generated by the vector field V and $\{\phi_t \mid t \in \mathbb{R}\}$ be a one-parameter family of diffeomorphisms generated by the vector field W satisfying $[V, W] = 0$. Then for each point p of the manifold, there exists an ϵ depending on p such that for any two real numbers $s, t \in \mathbb{R}$ with $|s| + |t| \leq \epsilon$, $\theta_s(\phi_t(p))$ and $\phi_t(\theta_s(p))$ are defined and will satisfy $\theta_s(\phi_t(p)) = \phi_t(\theta_s(p))$. That is, the diffeomorphisms θ_s and ϕ_t commute.*

Some basic properties of Lie algebras of vector fields

Let M be a differential manifold and $\partial_1 = \frac{\partial}{\partial x_1}, \partial_2 = \frac{\partial}{\partial x_2}, \dots, \partial_n = \frac{\partial}{\partial x_n}$ be the associated coordinate vector fields. Then it can be directly proved that these vector fields satisfy the following conditions:

(i) $[\partial_i, \partial_j] = 0$.
(ii) $[fV, gW] = fg[V, W] + fV(g)W - gW(f)V$ for each $V, W \in C^\infty(M, T(M))$ and f and $g \in C^\infty(M, R)$.
(iii) If $V = \sum_{i=1}^n f_i\partial_i$, $W = \sum_{i=1}^n g_i\partial_i$ are arbitrary vector fields, then

$$[V, W] = \sum_{i=1}^n (V(g_i) - W(f_i))\partial_i = \sum_{i=1}^n \left(\sum_{j=1}^n f_j\frac{\partial g_i}{\partial x_j} - g_j\frac{\partial f_i}{\partial x_j}\right)\partial_i.$$

Definition 54 (Vector bundle). A vector bundle is a topological construction that makes precise the idea of a family of vector spaces parameterized by another space X, where X is a topological space or a manifold or an algebraic variety. Hence to every point $x \in X$, one can associate a vector space $V(x)$ such that these vector spaces together will form another space as that of X, which is then called a vector bundle over X.

Remark 38.

(1) Finite-dimensional Lie algebras of vector fields in the real plane can be classified depending on whether they are locally primitive algebras or imprimitive algebras. For details, one can refer Gonzalez-Lopez et al. [48].

(2) For classification of Lie algebras of first-order differential operators in two complex variables, based on Lie's classification of Lie algebras of vector fields in two complex variables, one can refer Gonzalez-Lopez et al. [49].

(3) For details regarding Riemann surface, its genus, stable vector bundles, holomorphic vector bundles, irreducible unitary representations, Fuchsian groups, etc., one can refer Narasimhan and Seshadri [50].

(4) For details regarding Lie algebras of differential operators, one can refer Kamran and Olver [51].

(5) For differential manifolds, integration on differential manifolds, etc., one can refer Ramanan [235].

(6) For Drinfeldians and Yangians for Lie algebras $A_{l-1}, l \geq 3$, $B_l, l \geq 3$, $C_l, l \geq 2$ $D_l, l \geq 4$, one can refer Tolstoy [53].

Exercises

1.1 Let $\mathcal{G}$ be a Lie algebra with a basis $\{X_1, X_2, X_3\}$, satisfying the relations $[X_i, X_j] = \sum a_{ij}^k x_k 1 \leq i,j,k \leq 3$. If $[Y_i, Y_j] = \sum_{k=1}^{3} b_{ij}^k\, Y_k\ 1 \leq i,j \leq 3$, where

$$Y_1 = (1,0,0),\ Y_2 = (1,1,0),\ \text{and}\ Y_3 = (1,1,1),$$

then compute b_{ij}^k for $1 \leq i,j \leq 3$ and also find out the relations between b_{ij}^k and a_{ij}^k for $1 \leq i,j \leq 3$.

1.2 Find the structure constants for $sl(2,\mathbb{C})$ with respect to the basis $\left\{ e = \begin{pmatrix} 0 & 1 \\ 0 & 0 \end{pmatrix}, f = \begin{pmatrix} 0 & 0 \\ 1 & 0 \end{pmatrix}, h = \begin{pmatrix} 1 & 0 \\ 0 & -1 \end{pmatrix} \right\}$.

1.3 Compute the determinant of the Killing form κ, relative to the standard basis of $sl(3,\mathbb{C})$.

1.4 For the Lie algebra $sl(2,\mathbb{C})$, compute the basis dual to the standard basis with respect to the Killing form.

1.5 Prove that the Lie algebra $\mathcal{G}$ is solvable if and only if $[\mathcal{G},\mathcal{G}]$ lies in the radical of the Killing form.

1.6 Let x be an element of Lie algebra $\mathcal{G}$. Prove that set of all inner derivations $ad\, x$ is an ideal of $Der\, \mathcal{G}$.

1.7 Prove that the derived algebra of $gl(n,\mathbb{C})$ is $sl(n,\mathbb{C})$.

1.8 For a three-dimensional Lie algebra $\mathcal{G}$ with $\mathcal{G} = [\mathcal{G}, \mathcal{G}]$, prove that $\mathcal{G}$ is simple.

1.9 Prove that the Lie algebra $\mathcal{G}$ is

(i) nilpotent if and only if $ad\ \mathcal{G}$ is nilpotent and

(ii) solvable if and only if $ad\ \mathcal{G}$ is solvable.

1.10 If $\mathcal{G} = sl(n, C)$, prove that $Rad\ \mathcal{G} = Z(\mathcal{G})$, the center of $\mathcal{G}$.

1.11 Prove that the root system of $sl(2, C)$ is same as that of every three-dimensional semisimple Lie algebra.

1.12 Prove that a Lie algebra $\mathcal{G}$ is semisimple if and only if it has no nonzero abelian ideal.

1.13 Let $\mathcal{G} = \mathcal{G}_1 \oplus \mathcal{G}_2$ be the direct sum of two Lie algebras. Prove that $\{(x_1, 0)\colon x_1 \in \mathcal{G}_1\}$ is an ideal of $\mathcal{G}$ isomorphic to $\mathcal{G}_1$ and $\{(0, x_2)\colon x_2 \in \mathcal{G}_2\}$ is an ideal of $\mathcal{G}$ isomorphic to $\mathcal{G}_2$. Prove also that the projections $p_1(x_1, x_2) = x_1$ and $p_2(x_1, x_2) = x_2$ are Lie algebra homomorphisms.

1.14 Find a basis and root system of $su(3)$ and prove that $su(2)$ is a subalgebra of $su(3)$. Find also the Cartan-Weyl basis of $su(2)$.

1.15 Prove that the set of all derivations of a Lie algebra $\mathcal{G}$ has the structure of a Lie algebra.

1.16 Prove that any semisimple Lie algebra $\mathcal{G}$ can be written as a direct sum of two vector spaces, where one vector space is a compact subalgebra C of $\mathcal{G}$ (ie, $[C, C] \subset C$).

1.17 Prove that for the Lie algebra $so(5)$, eight nonzero roots are of the form $\pm\epsilon_i \pm \epsilon_j$ and $\pm\epsilon_i (i, j = 1, 2, i \neq j)$, where (ϵ_1, ϵ_2) forms an orthonormal basis in the vector space $\mathbb{R}^2$. Find also the simple roots, positive roots, roots of three different types, and highest root.
(Ans: Simple roots: $\alpha_1 = \epsilon_1$ and $\alpha_2 = \epsilon_2 - \epsilon_1$.
Positive roots: $\alpha_1, \alpha_2, \alpha_1 + \alpha_2, 2\alpha_1 + \alpha_2$.)

1.18 Find the adjoint representation of $sl(3, \mathbb{C})$.

1.19 Let $\frac{\partial}{\partial x}$ and $\frac{\partial}{\partial y}$ be the two coordinate vector fields on $\mathbb{R}^2$ determined by the identity mapping

$$X_1 = x\left(\frac{\partial}{\partial x}, \frac{\partial}{\partial y}\right) = x\frac{\partial}{\partial x} + y\frac{\partial}{\partial y}.$$

$$X_2 = x\left(\frac{\partial}{\partial x}, \frac{\partial}{\partial y}\right) = y\frac{\partial}{\partial x} - x\frac{\partial}{\partial y}.$$

Compute $[X_1, X_2]$.

Kac-Moody algebras

2

Chapter Outline

2.1 Basic concepts in Kac-Moody algebras 60
2.2 Classification of finite, affine, hyperbolic, and extended-hyperbolic Kac-Moody algebras and their Dynkin diagrams 66
2.3 Invariant bilinear forms 75
2.4 Coxeter groups and Weyl groups 78
2.5 Real and imaginary roots of Kac-Moody algebras 81
2.6 Weyl groups of affine Lie algebras 85
2.7 Realization of affine Lie algebras 86
2.8 Different classes of imaginary roots (special imaginary roots, strictly imaginary roots, purely imaginary roots) in Kac-Moody algebras 94
2.9 Representations of Kac-Moody algebras, integrable highest weight modules, Verma modules, and character formulas 100
2.10 Graded Lie algebras and root multiplicities 105

Abstract

For many articles on Kac-Moody algebras and related topics, one can refer Sthanumoorthy and Misra [240]. The real and imaginary roots [25] for Kac-Moody algebras were introduced by Kac [20]. The notion of imaginary roots has no counter part in the finite-dimensional theory. The imaginary roots for affine Kac-Moody Lie algebras had been completely and explicitly described by Kac [8, 21] and their characterization to a certain extent for symmetrizable hyperbolic indefinite Kac-Moody algebras was done by Moody [55]. Kac [56] also introduced the notion of strictly imaginary roots. Moody [17] introduced real roots, called Weyl roots. Casperson [57] gave a complete characterization of Kac-Moody algebras possessing the strictly imaginary property and Bennett [54] has shown the existence of special imaginary roots.

In this chapter, we start with all basic concepts regarding different classes (finite, affine, and indefinite) of Kac-Moody algebras and their corresponding Dynkin diagrams. Herein, the classification of finite, affine, hyperbolic, and extended-hyperbolic Kac-Moody algebras and other indefinite Kac-Moody algebras are explained with examples. Moreover, Dynkin diagrams, Weyl groups, and realizations of affine Lie algebras [4] are explained with examples. Different classes of imaginary roots, namely, special, strictly,

Introduction to Finite and Infinite Dimensional Lie (Super)algebras. http://dx.doi.org/10.1016/B978-0-12-804675-3.00002-9

and purely imaginary roots and their complete classifications with examples, wherever necessary, are given [54, 57, 79]. Moreover, representations of Kac-Moody algebras [58, 59] along with integrable highest weight modules, Verma modules, and character formulae, general construction of graded Lie algebras and root multiplicities are also given here with examples [60].

The problem of finding root multiplicities for affine Kac-Moody algebras has been completely solved [3, 8]. Berman and Moody [61] derived a closed form root multiplicity formula for all symmetrizable Kac-Moody algebras. Using Euler-Poincare principle and Kostant formula, Kang obtained two root multiplicity formulas for all symmetrizable Kac-Moody algebras [62, 63]. One of them is in recursive form, and the other is in closed form. Feingold and Frenkel [64] computed level 2 root multiplicities for the hyperbolic Kac-Moody algebra $HA_1^{(1)}$. Kang [65–67] obtained root multiplicities for $HA_1^{(1)}$ upto level 5 and for $HA_2^{(2)}$ upto level 3. Root multiplicities of some indefinite type Kac-Moody Lie algebras $HA_n^{(1)}$ are obtained in Kang and Melville [68] and in Hontz and Misra [69] using two different view points. Kac, Moody, and Wakimoto computed level 2 root multiplicities for $HE_8^{(1)}$ [70]. Computation of root multiplicities of many Kac-Moody algebras and generalized Kac-Moody algebras (Chapter 3) can also be seen in Kass et al. [71], Kang [63, 66, 72, 73], Frenkel and Kac [29], Feingold and Frenkel [64], Feingold and Lepowsky [74], Feingold [75], Kac and Wakimoto [36], Hontz and Misra [76], Misra [77, 78], Sthanumoorthy and Uma Maheswari [24], Sthanumoorthy et al. [80, 81], and Sthanumoorthy and Lilly [82]. Root multiplicities, realizations, and homology modules for some extended-hyperbolic Kac-Moody algebras for $EHA_1^{(1)}$ and $EHA_2^{(2)}$ [60] are also discussed in detail in this chapter. For a general study of finite and affine Lie algebras, one can refer Carter [94].

2.1 Basic concepts in Kac-Moody algebras

Definition 55. Let $A = (a_{ij})_{i,j=1}^{n}$ be a complex $n \times n$ matrix of rank l and $\mathfrak{h}$ be a complex vector space of finite dimension. Let $\Pi = \{\alpha_1, \dots, \alpha_n\} \subset \mathfrak{h}^*$ and $\Pi^\vee = \{\alpha_1^\vee, \dots, \alpha_n^\vee\} \subset \mathfrak{h}$ be indexed sets.

A realization of a square matrix A is a triple $(\mathfrak{h}, \Pi, \Pi^\vee)$ such that

(i) Π and $\Pi^\vee$ are linearly independent and
(ii) $\langle \alpha_i^\vee, \alpha_j \rangle = a_{ij}$ for all $i,j = 1, \dots, n$, where $\langle \alpha_i^\vee, \alpha_j \rangle$: $\mathfrak{h} \times \mathfrak{h}^* \to \mathbb{C}$ is the pairing $\langle \alpha_i^\vee, \alpha_j \rangle = \alpha_j(\alpha_i^\vee)$.

Remark 39. In general, if $(\mathfrak{h}, \Pi, \Pi^\vee)$ is a realization of a matrix A, then dimension $\mathfrak{h} \geq 2n - l$.

Definition 56. Two realizations $(\mathfrak{h}, \Pi, \Pi^\vee)$ and $(\mathfrak{h}_1, \Pi_1, \Pi_1^\vee)$ with $\Pi = \{\alpha_1, \alpha_2, \dots, \alpha_n\}$, $\Pi^\vee = \{\alpha_1^\vee, \alpha_2^\vee, \dots, \alpha_n^\vee\}$, $\Pi_1 = \{\beta_1, \beta_2, \dots, \beta_n\}$, and $\Pi_1^\vee = \{\beta_1^\vee, \beta_2^\vee, \dots, \beta_n^\vee\}$ are *isomorphic* if there exists a vector space isomorphism ϕ: $\mathfrak{h} \to \mathfrak{h}_1$ such that $\phi(\alpha_i^\vee) = \beta_i^\vee$ and $\phi^*(\beta_i) = \alpha_i$ $(i = 1, 2, \dots, n)$, where ϕ^*: $\mathfrak{h}_1^* \to \mathfrak{h}^*$ is an isomorphism induced by ϕ with $\langle \phi(h), \beta \rangle =< h, \phi^*(\beta) >$ for $h \in \mathfrak{h}$ and $\beta \in \mathfrak{h}^*$.

Definition 57. There exists a realization $(\mathfrak{h}, \Pi, \Pi^\vee)$ of A with $\dim \mathfrak{h} = 2n - l$, *called minimal realization*, which is unique up to an isomorphism. This isomorphism is unique if and only if $det\, A \neq 0$.

Remark 40.

(1) Note that if $(\mathfrak{h}, \Pi, \Pi^\vee)$ is a minimal realization of A then $(\mathfrak{h}^*, \Pi^\vee, \Pi)$ is a minimal realization of A^t, the transpose of matrix A.

(2) The following elementary results can be proved:

  (i) For two matrices A and B, the corresponding realizations are isomorphic if B can be obtained from A by permutation of the index set.

  (ii) *Minimal realization of a direct sum*: Let A_1 and A_2 be two complex square matrices with minimal realizations $(\mathfrak{h}_1, \Pi_1, \Pi_1^\vee)$ and $(\mathfrak{h}_2, \Pi_2, \Pi_2^\vee)$. Then we have the direct sum $\begin{pmatrix} A_1 & 0 \\ 0 & A_2 \end{pmatrix}$ of these two matrices A_1 and A_2 with a minimal realization:

$$((\mathfrak{h}_1 \oplus \mathfrak{h}_2),\ \Pi_1 \times \{0\} \cup \{0\} \times \Pi_2,\ \Pi_1^\vee \times \{0\} \cup \{0\} \times \Pi_2^\vee).$$

Definition 58 (Construction of a Lie algebra $\mathcal{G}(A)$, given a matrix A). Let $A = (a_{ij})_{i,j=1}^n$ be a complex $n \times n$ matrix of rank l and $(\mathfrak{h}, \Pi, \Pi^\vee)$ be a minimal realization of A, where $\Pi = \{\alpha_1, \alpha_2, \ldots, \alpha_n\} \subset \mathfrak{h}^*$ and $\Pi^\vee = \{\alpha_1^\vee, \alpha_2^\vee, \ldots, \alpha_n^\vee\} \subset \mathfrak{h}$. Hence $\langle \alpha_i^\vee, \alpha_j \rangle = a_{ij}$ for $1 \leq i,j \leq n$. Let $\tilde{\mathcal{G}}(A)$ be *an auxiliary Lie algebra* generated by e_i, f_i $(i = 1, \ldots, n)$ and $\mathfrak{h}$ satisfying: $[h, h'] = 0$, for $h, h' \in \mathfrak{h}$; $[e_i, f_j] = \delta_{ij}\alpha_i^\vee$, for $i, j = 1, \ldots, n$; and $[h, e_i] = < h, \alpha_i > e_i, [h, f_i] = - < h, \alpha_i > f_i$ for $h \in \mathfrak{h}, i = 1, \ldots, n$. It is clear that $\tilde{\mathcal{G}}(A)$ depends (up to isomorphism) only on the matrix A.

Definition 59. Let $\tilde{\eta}_+$ be the subalgebra of $\tilde{\mathcal{G}}(A)$ generated freely by $e_1, \ldots, e_n$ and $\tilde{\eta}_-$ be the subalgebra of $\tilde{\mathcal{G}}(A)$ generated freely by $f_1, \ldots, f_n$. Then $\tilde{\mathcal{G}}(A) = \tilde{\eta}_- \oplus \mathfrak{h} \oplus \tilde{\eta}_+$ (direct sum of vector subspaces). The maps $e_i \to -f_i, f_i \to -e_i$ *and* $h \to -h (h \in \mathfrak{h})$ *can be uniquely extended to an involutive automorphism (of order 2)* $\tilde{\omega}$ *of* $\tilde{\mathcal{G}}(A)$. Here Π is called a *root basis* (elements are *simple roots*) and $\Pi^\vee$ is called the *coroot basis* (elements are simple *coroots*). Set $Q = \sum_{i=1}^n \mathbb{Z}\alpha_i$, called the *root lattice*. Let $Q_+ = \sum_{i=1}^n \mathbb{Z}_+\alpha_i$.

Introduce a partial ordering $\geq$ on $\mathfrak{h}^*$ by $\lambda \geq \mu$ if $\lambda - \mu \in Q_+$.

Definition 60. Let M be an abelian group and $\mathcal{G}$ be a Lie algebra. Then $\mathcal{G} = \bigoplus_{\alpha \in M} \mathcal{G}_\alpha$ (the direct sum of vector subspaces $\mathcal{G}_\alpha$ of $\mathcal{G}$) is called the *M-gradation* if $[\mathcal{G}_\alpha, \mathcal{G}_\beta] \subseteq \mathcal{G}_{\alpha+\beta}$ for $\alpha, \beta \in M$.

Now $\mathcal{G}(A)$ is also a Q-graded Lie algebra with $\mathcal{G}(A) = \oplus_{\alpha \in Q}\mathcal{G}_\alpha$, where $\mathcal{G}_\alpha \simeq \tilde{\mathcal{G}}_\alpha / \tau_\alpha$, $\tau_\alpha = \tau \cap \tilde{\mathcal{G}}_\alpha$ with $[\mathcal{G}_\alpha,\ \mathcal{G}_\beta] \subset \mathcal{G}_{\alpha+\beta}$. Here $\mathcal{G}_\alpha = \{x \in \mathcal{G}(A) | [h, x] = < h, \alpha > x\ \forall\, h \in \mathfrak{h}\}$ is finite dimensional.

Root space decomposition: For a Q-graded Lie algebra $\tilde{\mathcal{G}}(A)$, one has the root space decomposition:

$$\tilde{\mathcal{G}}(A) = \left(\bigoplus_{\substack{\alpha \in Q_+ \\ \alpha \neq 0}} \tilde{\mathcal{G}}(A)_{-\alpha} \right) \oplus \mathfrak{h} \oplus \left(\bigoplus_{\alpha \in Q_+, \alpha \neq 0} \tilde{\mathcal{G}}(A)_\alpha \right),$$

where

$$\tilde{\mathcal{G}}(A)_\alpha = \{x \in \tilde{\mathcal{G}}(A) | [h, x] = \alpha(h)x\}.$$

We have $\dim \tilde{\mathcal{G}}(A)_\alpha < \infty$ and $\tilde{\mathcal{G}}(A)_\alpha \subseteq \tilde{\eta}_+$ for $\alpha \neq 0$ and $[\tilde{\mathcal{G}}_\alpha, \tilde{\mathcal{G}}_\beta] \subseteq \tilde{\mathcal{G}}_{\alpha+\beta}$.

In fact,

$$\tilde{\mathcal{G}}_\alpha = \sum_{\alpha_{i_1}+\cdots+\alpha_{i_r}=\alpha} \mathbb{C}[e_{i_1}, \ldots, e_{i_r}] \text{ and } \tilde{\mathcal{G}}_{-\alpha} = \sum_{\alpha_{i_1}+\cdots+\alpha_{i_r}=\alpha} \mathbb{C}[f_{i_1}, \ldots, f_{i_r}],$$

where $[e_{i_1}, e_{i_2}, \ldots, e_{i_n}] = [[e_{i_1}, e_{i_2}], \ldots, e_{i_n}]$ and $[f_{i_1}, f_{i_2}, \ldots, f_{i_n}] = [[f_{i_1}, f_{i_2}], \ldots, f_{i_n}]$. There exists a unique maximal ideal τ of $\tilde{\mathcal{G}}(A)$ which intersects $\mathfrak{h}$ trivially.

Let

$$\mathcal{G}(A) = \tilde{\mathcal{G}}(A)/\tau.$$

Then $\mathcal{G}(A)$ *is the Lie algebra associated with the complex matrix* A.

The above maximal ideal $\tau \in \tilde{\mathcal{G}}(A)$ satisfies

$$\tau = (\tau \cap \tilde{\eta}_-) \oplus (\tau \cap \tilde{\eta}_+).$$

We denote the elements $e_i, f_i, \mathfrak{h}$ by the same notation in both $\mathcal{G}(A)$ and $\tilde{\mathcal{G}}(A)$.

Definition 61. The subalgebra $\mathfrak{h}$ of $\mathcal{G}(A)$ is called the *Cartan subalgebra* and the elements e_i, f_i are called *Chevalley generators.*

Definition 62 (Positive and negative roots and triangular decomposition). We have

$$\mathcal{G}_0 = \tilde{\mathcal{G}}_0 = \mathfrak{h}, \quad \mathcal{G}_{\alpha_i} = \mathbb{C}e_i, \quad \mathcal{G}_{-\alpha_i} = \mathbb{C}f_i \quad (i = 1, \ldots, n).$$

If $\mathcal{G}_\alpha \neq 0$, then $\alpha \neq 0$, with $\alpha > 0$ or $\alpha < 0$.

If $\alpha > 0$, then $\mathcal{G}_\alpha$ is the linear span of $[e_{i_1}, \ldots, e_{i_r}]$.

If $\alpha < 0$, $\mathcal{G}_\alpha$ is the linear span of $[f_{i_1}, \ldots, f_{i_r}]$ such that

$$\alpha_{i_1} + \cdots + \alpha_{i_r} = \alpha \text{ or } (= -\alpha).$$

If $\mathcal{G}_\alpha \neq 0$ and $\alpha \neq 0$, then α is a root of $\mathcal{G}(A)$, with multiplicity $\operatorname{mult}\alpha := \dim \mathcal{G}_\alpha$

The roots $\alpha > 0$ are called *positive roots* and $\alpha < 0$ are called *negative roots*. We have $\mathcal{G}(A) = \oplus_\alpha \mathcal{G}_\alpha$ is the *root space decomposition.*

Let η_+ be the subalgebra of $\mathcal{G}(A)$ generated by $e_1, \ldots, e_n$ and η_- be the subalgebra of $\mathcal{G}(A)$ generated by $f_1, \ldots, f_n$. Now

$$\eta_+ = \oplus_{\alpha>0}\mathcal{G}_\alpha \simeq \tilde{\eta}_+/(\tau \cap \tilde{\eta}_+) \quad \text{and}$$
$$\eta_- = \oplus_{\alpha<0}\mathcal{G}_\alpha \simeq \tilde{\eta}_-/(\tau \cap \tilde{\eta}_-).$$

We have the *triangular decomposition*:

$$\mathcal{G}(A) = \eta_- \oplus \mathfrak{h} \oplus \eta_+.$$

All ideals of $\mathcal{G}(A)$ are Q-graded and $\mathcal{G}(A)$ contains no nonzero ideal which intersects $\mathfrak{h}$ trivially.

Definition 63. Let $\tilde{\omega}$ be an involutive automorphism of $\tilde{\mathcal{G}}(A)$ as defined earlier. This $\tilde{\omega}$ induces an involutive automorphism ω of $\mathcal{G}(A)$ satisfying $\omega(e_i) = -f_i, \omega(f_i) = -e_i (i = 1, \ldots, n)$ and $\omega(h) = -h (h \in \mathfrak{h})$. Here ω is called the *Chevalley involution* of $\mathcal{G}(A)$. We have $\omega(\mathcal{G}_\alpha) = \mathcal{G}_{-\alpha}$, *mult* α = *mult* $(-\alpha)$ and ω interchanges η_+ and η_-.

Definition 64. Let $A = (a_{ij})_{i,j=1}^n$ be a complex $n \times n$ matrix such that

(i) $a_{ii} = 2$ for all $i, a_{ij} \leq 0$ (for $i \neq j$) are nonpositive integers.
(ii) $a_{ij} = 0 \Rightarrow a_{ji} = 0$. Then A is called a *Generalized Cartan matrix* (GCM).

If A is a GCM, then corresponding Lie algebra $\mathcal{G}(A)$ *(constructed above) is called the Kac-Moody algebra associated with A.*

Definition 65. Two complex $n \times n$ matrices A and B are said to be *equivalent*, if after reordering the indices of A (ie, a permutation of the rows of A and the same permutation of its columns) we obtain B.

Definition 66. Given two $n \times n$ matrices A_1 and A_2, there exists *an isomorphism of the associated quadruples* $(\mathcal{G}_1, A_1, \Pi_1, \Pi_1^\vee)$ and $(\mathcal{G}_2, A_2, \Pi_2, \Pi_2^\vee)$ if and only if A_2 can be obtained from A_1 by reordering the index set.

It can be proved that if two complex $n \times n$ matrices A and B are equivalent, then their corresponding Lie algebras $\mathcal{G}(A)$ and $\mathcal{G}(B)$ are isomorphic.

Definition 67 (Kac-Moody algebra in terms of generators associated with symmetrizable GCM). Let $A = (a_{ij})_{i,j=1}^n$ be a complex $n \times n$ matrix. A is called *symmetrizable* if there exists a nonsingular diagonal matrix $D = diag(\epsilon_1, \ldots, \epsilon_n)$ (ϵ_i are positive rational numbers) and a symmetric matrix $B = (b_{ij})_{i,j=1}^n (b_{ij} = b_{ji}$ for all $i,j)$ such that $A = DB$.
It can be directly verified that A is *symmetrizable* if and only if $a_{ij} = 0 \Rightarrow a_{ji} = 0$ and $a_{i_1 i_2} a_{i_2 i_3} \cdots a_{i_k i_1} = a_{i_2 i_1} a_{i_3 i_2} \cdots a_{i_1 i_k} \ \forall \ i_1, \ldots, i_k$.

Let $\tilde{\mathcal{G}}(A)$ be the auxiliary Lie algebra defined in Definition 58. Here the corresponding Kac-Moody algebra is $\mathcal{G}(A) = \tilde{\mathcal{G}}(A)/\tau$ with $\tau = \tau_+ \oplus \tau_-$, $\tau_\alpha = \tau \cap \tilde{\mathcal{G}}_\alpha$. Here the ideal τ_+ is generated as an ideal in η_+ by those τ_α for which $\alpha \in Q_+ \backslash \Pi$ and $2(\rho, \alpha) = (\alpha, \alpha)$. Similarly, the ideal τ_- is generated as an ideal in η_- by those $\tau_{-\alpha}$ for which $\alpha \in Q_+ \backslash \Pi$ and $2(\rho, \alpha) = (\alpha, \alpha)$. Let A be a symmetrizable Cartan matrix, then one can also prove that for the corresponding Kac-Moody algebra $\mathcal{G}(A) = \tilde{\mathcal{G}}(A)/\tau$, $(ad\ e_i)^{1-a_{ij}} e_j$ and $(ad\ f_i)^{1-a_{ij}} f_j$ $(i,j = 1, 2, \ldots, n; i \neq j)$ generate the ideals τ_+ and τ_-, respectively, with $\tau = \tau_+ \cup \tau_-$.

Hence for a symmetrizable Cartan matrix, one can give the following definition for Kac-Moody algebra

Let $(\mathfrak{h}, \Pi, \Pi^\vee)$ be a realization of a symmetrizable square matrix A. Then the *associated Kac-Moody algebra* $\mathcal{G}(A)$ with generators e_i, f_i $(i = 1, 2, \ldots, n)$, and $\mathfrak{h}$ satisfy the following defining relations:

$$[h, h'] = 0 \text{ for all } h, h' \in \mathfrak{h}, \quad [e_i, f_j] = \delta_{ij}\alpha_i^\vee$$
$$[h, e_i] = \langle \alpha_i, h\rangle e_i, \quad [h, f_i] = -\langle \alpha_i, h\rangle f_i,$$
$$(ad\ e_i)^{1-a_{ij}} e_j = 0, \quad (ad\ f_i)^{1-a_{ij}} f_j = 0 \quad (i, j = 1, 2, \ldots, n;\ i \neq j).$$

Remark 41.

(i) For a Kac-Moody algebra $\mathcal{G}(A)$, $\mathcal{G}'(A) = [\mathcal{G}(A), \mathcal{G}(A)]$ satisfies

$$\mathcal{G}(A) = \mathcal{G}'(A) + \mathfrak{h}.$$

Moreover $\mathcal{G}(A) = \mathcal{G}'(A)$ if and only if $det\ A \neq 0$.

(ii) For a GCM, A and for the associated Kac-Moody algebra $\mathcal{G}(A)$, the following defining relations are called Serre's relations (Serre's theorem):

$$[h, h'] = 0 \text{ for } h, h' \in \mathfrak{h}\text{: } [e_i, f_j] = \delta_{ij}\alpha_i^\vee \text{ for } i, j = 1, 2, \ldots, n;$$
$$[h, e_i] = < h, \alpha_i > e_i; \quad [h, f_i] = - < h, \alpha_i > f_i$$
$$[e_i, f_i] = \alpha_i^\vee, [\alpha_i^\vee; e_i] = 2e_i, [\alpha_i^\vee, f_i] = -2f_i.$$
$$(ad\ e_i)^{1-a_{ij}} e_j = 0; \quad (ad\ f_i)^{1-a_{ij}} f_j = 0, \text{ if } i \neq j$$

for $h \in \mathfrak{h}, i = 1, 2, \ldots, n$.

(iii) For defining relations of certain infinite dimensional Lie algebras, one can refer Gabber and Kac [83].

Definition 68 (Construction of a GCM given a Lie algebra $\mathcal{G}$). For a given Lie algebra $\mathcal{G}$, we can find the matrix A and the corresponding Lie algebra $\mathcal{G}(A)$ such that $\mathcal{G}(A) \cong \mathcal{G}$ as described below. Let $\mathcal{G}$ be a Lie algebra and $\mathfrak{h} \subset \mathcal{G}$, a commutative subalgebra. Let $e_1, \ldots, e_n, f_1, \ldots, f_n \in \mathcal{G}$ and $\Pi^\vee = \{\alpha_1^\vee, \ldots, \alpha_n^\vee\} \subset \mathfrak{h}$ and $\Pi = \{\alpha_1, \ldots, \alpha_n\} \subset \mathfrak{h}^*$ be linearly independent sets such that

$$[e_i, f_j] = \delta_{ij}\alpha_i^\vee,$$
$$[h, e_i] = < h, \alpha_i > e_i, \text{ and}$$
$$[h, f_i] = - < h, \alpha_i > f_i \quad (h \in \mathfrak{h}, i = 1, \ldots, n).$$

Suppose e_i, f_i, and $\mathfrak{h}$ generate $\mathcal{G}$ as a Lie algebra and $\mathcal{G}$ has no nonzero ideals which intersect $\mathfrak{h}$ trivially. Set $A = (\langle \alpha_i^\vee, \alpha_j\rangle)_{i,j=1}^n$. We have $\dim \mathfrak{h} = 2n - rank\ A$. This is the *GCM associated with the generalized Kac-Moody algebra.* Then it can be directly established that, $\mathcal{G}$ is isomorphic to Lie algebra $\mathcal{G}(A)$ associated with A. Thus one can find out the Cartan matrix $A = (a_{ij})_{i,j=1}^n = (\langle \alpha_i, \alpha_j^\vee\rangle)_{i,j=1}^n$ of a given Lie algebra by their generators and their relations.

Example 14. For the Lie algebra generated by $\{e_1, e_2, f_1, f_2, h_1, h_2\}$ with the relations

$$[h_1, e_1] = 2e_1; \quad [h_1, e_2] = -2e_2;$$
$$[h_2, e_1] = -2e_1 \text{ and } [h_2, e_2] = 2e_2,$$

we obtain the Cartan matrix $\begin{pmatrix} 2 & -2 \\ -2 & 2 \end{pmatrix}$.

Remark 42. This is a Lie algebra $A_1^{(1)}$ which will be described in the class of affine Lie algebras. For construction of the affine Lie algebra $A_1^{(1)}$, one can refer Lepowsky and Wilson [84].

Definition 69 (Construction of Lie algebra given a Cartan matrix). Suppose we are given a Cartan matrix $A = (a_{ij})_{i,j=1}^{n}$, then we can immediately write down the Lie algebra generated by $\{e_i, f_i, h_i\}$ such that $\langle \alpha_i^\vee, \alpha_j \rangle = a_{ij}$, $[e_i, f_j] = \delta_{ij}\alpha_i^\vee$, $[h, e_i] = < h, \alpha_i > e_i$, and $[h, f_i] = - < h, \alpha_i > f_i$.

Example 15. Let $A = \begin{pmatrix} 2 & -1 & 0 \\ -1 & 2 & -1 \\ 0 & -3 & 2 \end{pmatrix}$ be a Cartan matrix. Then we can *construct Lie algebra of A* generated by $\{e_1, e_2, e_3, f_1, f_2, f_3, h_1, h_2, h_3\}$ which satisfies the relations:

$$
\begin{aligned}
&[h_i, h_j] = 0 \quad \text{for all } i, j; \\
&[e_1, f_1] = \alpha_1^\vee; \quad [e_2, f_1] = 0; \quad [e_3, f_1] = 0 \\
&[e_1, f_2] = 0; \quad [e_2, f_2] = \alpha_2^\vee; \quad [e_3, f_2] = 0 \\
&[e_1, f_3] = 0; \quad [e_2, f_3] = 0; \quad [e_3, f_3] = \alpha_3^\vee \\
&[h_1, e_1] = 2e_1; \quad [h_2, e_1] = -e_1; \quad [h_3, e_1] = 0 \\
&[h_1, e_2] = -e_2; \quad [h_2, e_2] = 2e_2; \quad [h_3, e_2] = -e_2 \\
&[h_1, e_3] = 0; \quad [h_2, e_3] = -3e_3; \quad [h_3, e_3] = 2e_3 \\
&[h_1, f_1] = -2f_1; \quad [h_1, f_2] = f_2; \quad [h_1, f_3] = 0 \\
&[h_2, f_1] = f_1; \quad [h_2, f_2] = -2f_2; \quad [h_2, f_3] = f_3 \\
&[h_3, f_1] = 0; \quad [h_3, f_2] = 3f_2; \quad [h_3, f_3] = -2f_3.
\end{aligned}
$$

Definition 70. For any root $\alpha = \sum_{i=1}^{n} k_i\alpha_i$, *support* of $\alpha = \{i \in \{1, \ldots, n\} | k_i \neq 0\}$. The *height of* α is *ht* $\alpha = \sum_{i=1}^{n} k_i$.

Example 16. Let $\alpha = 2\alpha_1 + 3\alpha_2 + 5\alpha_3$ be a root of a KM algebra $\mathcal{G}(A)$, then

$$
\begin{aligned}
Supp(\alpha) &= \{1, 2, 3\}, \\
ht(\alpha) &= \sum_{n}^{i=1} k_i = 5 + 2 + 3 = 10.
\end{aligned}
$$

Definition 71. A square matrix A is said to be *decomposable* if, after reordering the indices, A decomposes into a nontrivial direct sum of square matrices. If a matrix is not decomposable, it is called *indecomposable.* If a matrix is decomposable then the corresponding minimal realization is also decomposable. The following can be proved: Let A be a complex $n \times n$ matrix. If after reordering the indices, one can decompose A into a direct sum of indecomposable matrices, then this decomposition is unique in the following sense: After reordering the indices in two different ways, if $A = A_1 + \cdots + A_m = B_1 + \cdots + B_{m'}$, are two decompositions of A into direct sums of indecomposable constituents then $m = m'$. Moreover after reordering of $B_1, \ldots, B_m, A_i$ and B_i are equivalent, for $i = 1, \ldots, m$.

Definition 72. Let $A = (a_{ij})_{i,j=1}^{n}$ be a complex $n \times n$ matrix of rank l and $J \subset \{1, \ldots, n\}$. Then $A_J = (a_{ij})_{i,j \in J}$ is a principal submatrix of A. Its determinant is called a principal minor. The *Lie algebra $\mathcal{G}(A_J)$ associated with the principal submatrix A_J* can be regarded as a subalgebra of $\mathcal{G}(A)$.

Definition 73. For the Lie algebra $\mathcal{G}(A)$, $[\mathcal{G}(A), \mathcal{G}(A)]$ is the *derived algebra* and is denoted as $\mathcal{G}'(A)$. It is clear that $\mathcal{G}'(A)$ is the subalgebra of $\mathcal{G}(A)$ generated by the Chevalley generators e_i's and f_i's. We have

$$\mathcal{G}(A) = \mathcal{G}'(A) \oplus \mathfrak{h} \quad \text{and}$$
$$\mathcal{G}(A) = \mathcal{G}'(A) \quad \text{if and only if } det(A) \neq 0.$$

Let $\mathfrak{h}' = \sum_{i=1}^{n} \mathbb{C}\alpha_i^\vee$. Then we have $\mathcal{G}(A) \cap \mathfrak{h} = \mathfrak{h}'$ and $\mathcal{G}'(A) \cap \mathcal{G}_\alpha = \mathcal{G}_\alpha$ if $\alpha \neq 0$.

Definition 74. As in the case of finite-dimensional Lie algebras, a Kac-Moody algebra $\mathcal{G}$ is simple if it has no ideals except itself and $\{0\}$ and $[\mathcal{G}, \mathcal{G}] \neq 0$.

Properties of simple Lie algebras: The following can be proved.

(i) Let A be an $n \times n$ complex matrix and $\mathcal{G}(A)$ be the Lie algebra associated to it. Suppose for each pair (i, j), there exists indices $i_1, \ldots, i_s$ such that,

$$a_{ii_1} a_{i_1 i_2}, \ldots, a_{i_s j} \neq 0 \in \mathcal{G}(A), \tag{2.1}$$

then *every ideal* of $\mathcal{G}(A)$ either contains $\mathcal{G}'(A)$ or is contained in the center. Here $\mathcal{G}'(A)$ is the derived algebra of $\mathcal{G}(A)$.

(ii) Suppose for the complex $n \times n$ matrix A, for each pair of indices (i, j) there exists indices $i_1, \ldots, i_s$ satisfying the above condition (2.1) and $det(A) \neq 0$ then $\mathcal{G}(A)$ is *simple*. Conversely, $\mathcal{G}(A)$ is *simple* if the above condition (2.1) holds and $det(A) \neq 0$.

2.2 Classification of finite, affine, hyperbolic, and extended-hyperbolic Kac-Moody algebras and their Dynkin diagrams

Theorem 13 (Vinberg's classification theorem). *Let A be an indecomposable real $n \times n$ GCM satisfying*

(1) $a_{ij} \leq 0$ *if* $i \neq j$ *and*
(2) $a_{ij} = 0 \Rightarrow a_{ji} = 0$.

Then one and only one of the following three possibilities holds for A and its transpose A^t.

(Fin) $det\, A \neq 0$, $\exists\, u > 0$ *such that* $Au > 0$; $Au \geq 0 \Rightarrow u > 0$ *or* $u = 0$.
(Aff) $corank\, A = 1$, $\exists\, u > 0$ *such that* $Au = 0$; $Au \geq 0 \Rightarrow Au = 0$.
(Ind) $\exists\, u > 0$ *such that* $Au < 0$; $Au \geq 0$ *and* $u \geq 0 \Rightarrow u = 0$.

Definition 75. Let A be a real indecomposable $n \times n$ matrix satisfying the above conditions (1) and (2). Then

(a) A is of *finite type* if and only if there exists $u > 0$ such that $Au > 0$.
(b) A is of *affine type* if and only if there exists $u > 0$ such that $Au = 0$.
(c) A is of *indefinite type* if and only if there exists $u > 0$ such that $Au < 0$.

Remark 43. If A is an $n \times n$ matrix of affine type, then any proper principal submatrix of A decomposes into direct sum of matrices of finite types.

Definition 76. We introduce below the *Dynkin diagrams* associated to generalized Cartan matrices. We associate to each GCM A, a graph $S(A)$, called the Dynkin diagram of A as follows: The vertex set of $S(A)$ is $\{1, 2, \ldots, n\}$. If $a_{ij}a_{ji} \leq 4$, the vertices i and j are connected by max $(|a_{ij}|, |a_{ji}|)$ edges and these edges are equipped with an arrow pointing towards i if $|a_{ij}| > 1$. If $a_{ij}a_{ji} > 4$, the vertices i and j are connected by a bold lined edge (dark line) equipped with an ordered pair of integers $(|a_{ij}|, |a_{ji}|)$. We say that $S(A)$ *is of finite, affine, or indefinite type if A is of that type.*

Definition 77 (Dynkin diagrams of finite and affine types). The enumeration of the vertices of the Dynkin diagrams of a GCM A can be explained as follows:
In the Figure *Finite*, vertices are enumerated by symbols $\alpha_1, \alpha_2, \ldots, \alpha_l$. Each Dynkin diagram of $X_l^{(1)}$ in Figure *Affine 1* is obtained from the Dynkin diagram of Finite type X_l by adding one vertex enumerated by the symbol α_0. The enumeration of the rest of the vertices in Figure *Affine 1* will be kept as in Figure *Finite.* Similarly the vertices of the Dynkin diagrams of Figure *Affine 2* and *Affine 3* are enumerated by the symbols $\alpha_0, \alpha_1, \ldots, \alpha_l$.

Properties of Dynkin diagrams

In each Dynkin diagram of finite type, we write within parentheses the determinant of corresponding GCM, A (Chapter 1). Here in the Figures *Affine*, the coefficients of a linear dependence between the columns of A are denoted by the numerical labels. We have the following properties regarding Dynkin diagrams:

Let A be a GCM. Then

(a) A is indecomposable if and only if $S(A)$ is a connected graph. Let A be an indecomposable GCM. Then
  - **(i)** If A is of *finite or affine type*, then any proper subdiagram of $S(A)$ is a union of connected Dynkin diagrams of finite types.
  - **(ii)** If A is of *finite type*, then $S(A)$ contains no cycles.
  - **(iii)** If A is of *affine type* and if $S(A)$ contains a cycle, then $S(A)$ is the cycle $A_l^{(1)}$ from Figure *Affine* 1.

(b) A is determined completely by its Dynkin diagram $S(A)$ and an enumeration of its vertices.

(c) The Dynkin diagrams of all generalized Cartan matrices of *finite types* were listed in Figure *Finite* (Chapter 1).

(d) The Dynkin diagrams of all generalized Cartan matrices of *affine types* will be listed in Figures *Affine* 1–3 (all of them have $l + 1$ vertices).

(e) The labels in Figures *Affine* 1–3 are the coordinates of the unique vector $\delta = (a_0, a_1, \ldots, a_l)$ such that $A\delta = 0$ and the a_i's $(i = 1, \ldots, l)$ are relatively prime positive integers.

Definition 78. We shall say that an edge (i,j) is *simply laced* if $A|_{(i,j)} = \begin{pmatrix} 2 & -1 \\ -1 & 2 \end{pmatrix}$ and the corresponding Dynkin diagram is *simply laced* if all its edges are simply laced. Hence for a simply laced Cartan matrix the entries are 2, 0, and -1 only.

Dynkin diagrams for affine types

Affine 1:

$A_1^{(1)}$

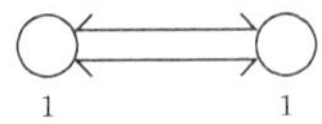

$A^1_{(l)}$

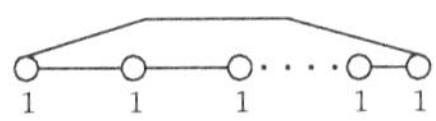

$B_l^{(1)}$ $(l \geq 3)$

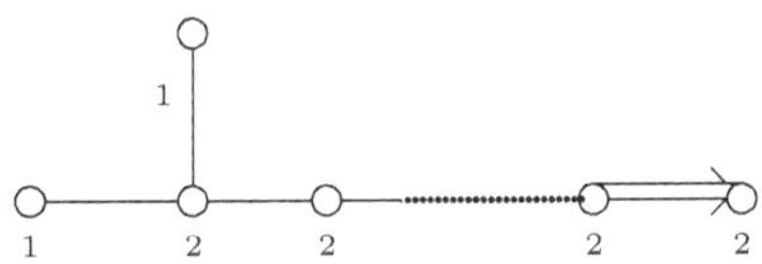

$C_l^{(1)}$ $(l \geq 2)$

$D_l^{(1)}$ $(l \geq 4)$

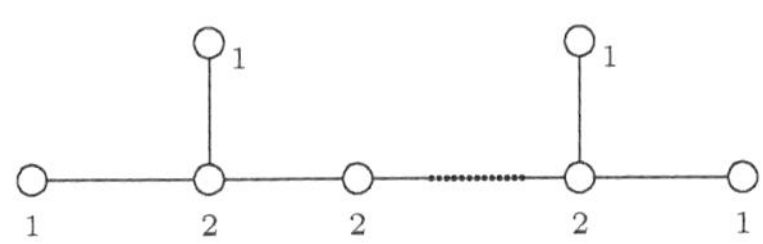

$G_2^{(1)}$

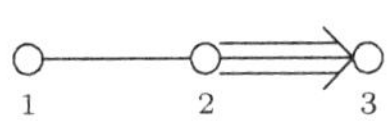

$F_4^{(1)}$

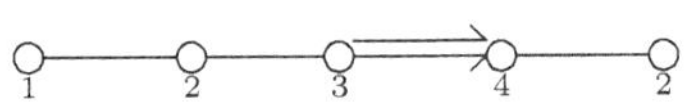

$E_6^{(1)}$

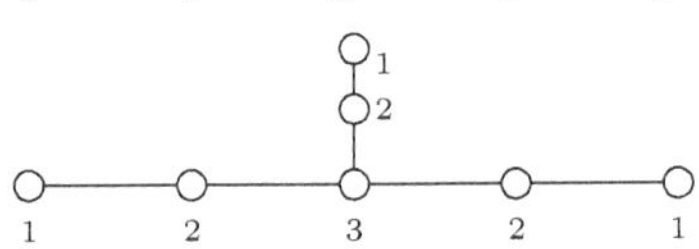

$E_7^{(1)}$

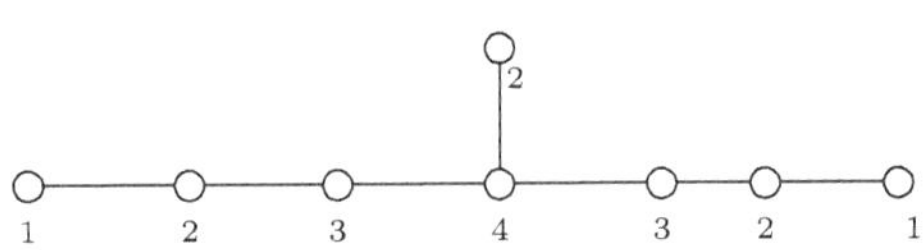

$E_8^{(1)}$

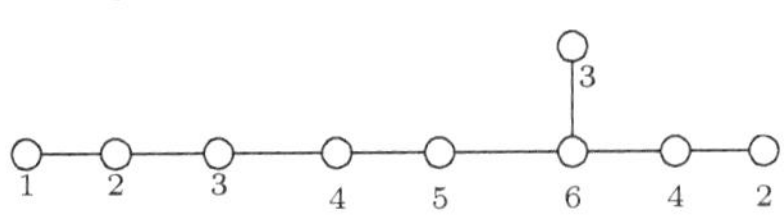

Affine 2:

$A_2^{(2)}$

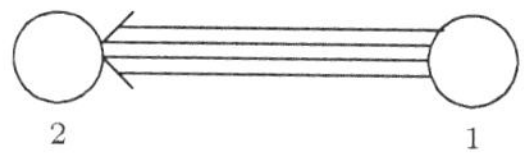

$A_{2l}^{(2)}\ (l \geq 2)$

$A_{2l-1}^{(2)}\ (l \geq 3)$

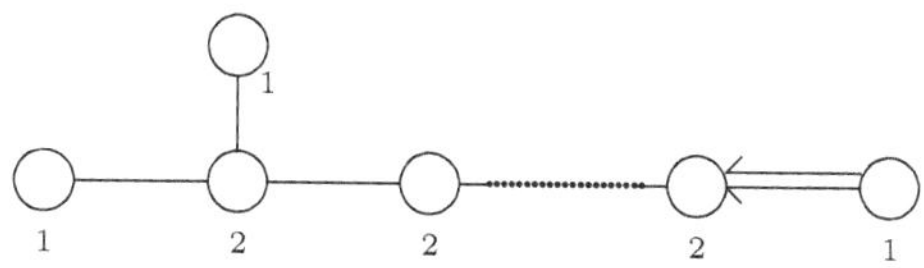

$D_{(l+1)}^{(2)}\ (l \geq 2)$

$E_6^{(2)}$

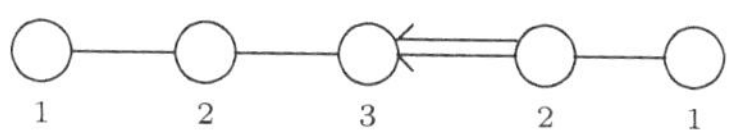

Affine 3:

$D_4^{(3)}$

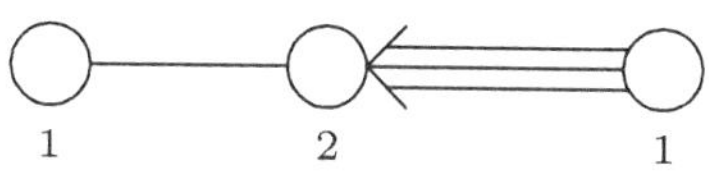

Remark 44.

(1) Let (i,j) be an edge of $S(A)$. We say (i,j) is *an edge of finite type* or a finite edge if $A|_{(i,j)}$ is a *matrix of finite type*. Nonfinite edges are defined similarly. A *nonfinite edge* means an edge of *affine or indefinite type.*

(2) Notion of an indecomposable matrix becomes that of a connected diagram only. If A is a GCM of finite type, than all of its indecomposable principal submatrices are also finite type. All the indecomposable *proper* principal submatrices of affine types are *finite type only.*

Properties of matrices of finite and affine types

In general, we have the following results.

Let A be an indecomposable GCM. Then

(a) A is of *finite type* if and only if all principal minors of A are positive.

(b) A is of *affine type* if and only if all proper principal minors of A are positive and $det\, A = 0$.

(c) If A is of *finite or affine type* then any proper subdiagram of $S(A)$ is a union of connected Dynkin diagrams of finite type.

(d) A is of *affine type* if and only if there exists $\delta > 0$ such that $A\delta = 0$ and such a δ is unique up to a constant factor.

(e) Let $A = (a_{ij})_{i,j=1}^{n}$ be a $n \times n$ matrix of *finite or affine type*. Then any proper principal submatrix of A decomposes into direct sum of matrices of *finite types*.

(f) Let $A = (a_{ij})_{i,j=1}^{n}$ be a *symmetric indecomposable matrix satisfying*

  (i) $a_{ij} \leq 0$ for all $i \neq j$.

  (ii) $a_{ij} = 0$ if and only if $a_{ji} = 0$.

Then A is of finite type if and only if A is positive definite and A is of affine type if and only if A is positive semidefinite and of rank $n - 1$. If $n = 1$, then A is of finite type.

For a GCM of affine type, Dynkin diagrams of A and A^t

We consider the case that A is a GCM of affine type of order $l + 1$ and rank l. The complex Lie algebra $\mathcal{G}(A)$ associated with A is called an affine Lie algebra. Let $a_0, a_1, \ldots, a_l$ be the numerical labels of $S(A)$, in the Dynkin diagram of A in Figure *Affine*. Then $a_0 = 1$ unless A is of type $A_{2l}^{(2)}$. If A is of affine type $A_{2l}^{(2)}$ then $a_0 = 2$. Let $S(A^t)$ be the Dynkin diagram of A^t. Then $S(A^t)$ can be obtained from $S(A)$ by reversing the directions of all arrows and keeping the same enumeration of the vertices. We denote by $a_0^{\vee}, a_1^{\vee}, \ldots, a_l^{\vee}$, the labels of $S(A^t)$. Here $a_0^{\vee} = 1$ for all cases. If A is a GCM of affine type of order $l + 1$ and rank l, then A is symmetrizable and $A = diag(a_0 a_0^{\vee^{-1}}, a_1 a_1^{\vee^{-1}}, \ldots, a_l a_l^{\vee^{-1}})B$, where $B^t = B$.

Definition 79. An indecomposable GCM, A is said to be of *hyperbolic type* if it is of indefinite type and every proper indecomposable principal submatrix (subdiagram of $S(A)$) of A is of *finite* or *affine type*.

An indecomposable GCM, A is said to be *strictly hyperbolic type* if it is of indefinite type and any connected proper subdiagram of $S(A)$ is of finite type.

Remark 45. There exist GCMs of hyperbolic types that are *not symmetrizable*.

In the case of 2×2 *matrices the following can be directly verified.*

Let $A = \begin{bmatrix} 2 & -a_{12} \\ -a_{21} & 2 \end{bmatrix}$ be an indecomposable GCM of order 2, where a_{12} and a_{21} are positive integers. Then

(a) A is of *finite type* if and only if $a_{12}a_{21} \leq 3$;

(b) A is of *affine type* if and only if $a_{12}a_{21} = 4$; and

(c) A is of *strictly hyperbolic type* if and only if $a_{12}a_{21} > 4$.

Some properties of Dynkin diagrams of hyperbolic types

(i) For a hyperbolic matrix A of order n, if a proper connected subdiagram of $S(A)$ is of affine type then it has $(n - 1)$ vertices.

(ii) Let $A = (a_{ij})_{i,j=1}^{n}$ be a hyperbolic matrix. Suppose $a_{ij} \neq 0$ and let S_2 be the subdiagram of $S(A)$ consisting of the vertices i and j.

  (a) If $n = 3$, then S_2 is one of the following diagrams:

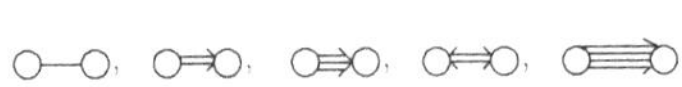

(b) If $n = 4$, then S_2 is one of the following three diagrams:

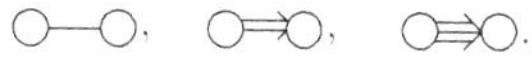

(c) If $n \geq 5$, then S_2 is one of the following two diagrams:

Some examples of Cartan matrices of hyperbolic types and their Dynkin diagrams

Remark 46. For $2 \leq n \leq 10$, the Dynkin diagrams of some of the hyperbolic matrices are listed below. Below we give some examples of hyperbolic type.

Hyp.2

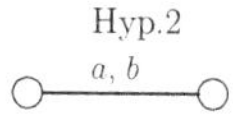

$a, b \in N, ab > 4, a \geq b$

Hyp.3

$$\begin{pmatrix} 2 & -1 & -2 \\ -1 & 2 & -1 \\ -2 & -1 & 2 \end{pmatrix}$$

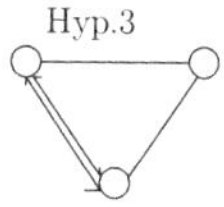

$$\begin{pmatrix} 2 & -1 & 0 \\ -1 & 2 & -4 \\ 0 & -1 & 2 \end{pmatrix}$$

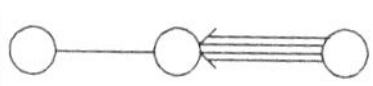

Hyp.4

$$\begin{pmatrix} 2 & -1 & -1 & -1 \\ -1 & 2 & -1 & -1 \\ -1 & -1 & 2 & -1 \\ -1 & -1 & -1 & 2 \end{pmatrix}$$

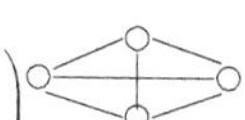

$$\begin{pmatrix} 2 & -1 & 0 & -1 \\ -2 & 2 & -1 & 0 \\ 0 & -2 & 2 & -2 \\ -1 & 0 & -1 & 2 \end{pmatrix}$$

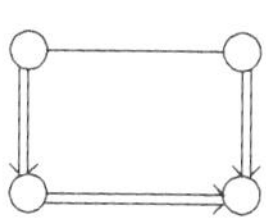

$$\begin{pmatrix} 2 & -2 & 0 & 0 \\ -1 & 2 & -1 & 0 \\ 0 & -2 & 2 & -1 \\ 0 & 0 & -2 & 2 \end{pmatrix}$$

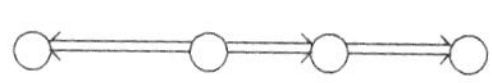

Hyp.5

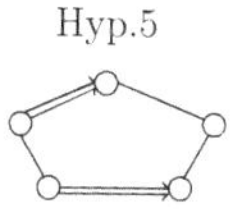

$$\begin{pmatrix} 2 & -2 & 0 & 0 & -1 \\ -1 & 2 & -1 & 0 & 0 \\ 0 & -1 & 2 & -1 & 0 \\ 0 & 0 & -2 & 2 & -1 \\ -1 & 0 & 0 & -1 & 2 \end{pmatrix}$$

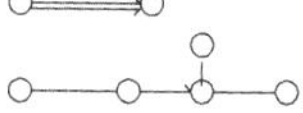

$$\begin{pmatrix} 2 & -1 & 0 & 0 & 0 \\ -1 & 2 & -1 & 0 & 0 \\ 0 & -2 & 2 & -1 & -1 \\ 0 & 0 & -1 & 2 & 0 \\ 0 & 0 & -1 & 0 & 2 \end{pmatrix}$$

For the following cases, we draw only Dynkin diagrams.

Table Hyp.6

$H_3^{(6)}$

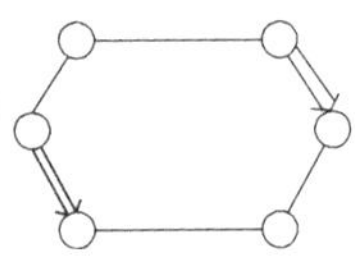

$H_{16}^{(6)}$

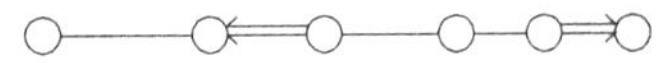

Table Hyp.7

$H_3^{(7)}$

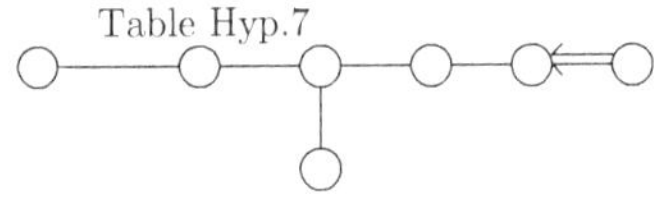

Table Hyp.8

$H_3^{(8)}$

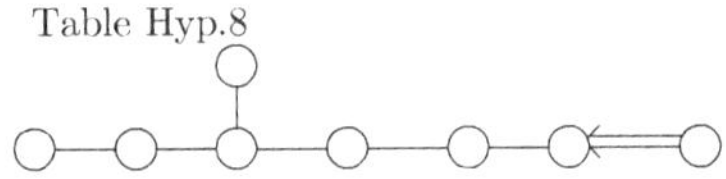

Table Hyp.9

$H_5^{(9)}$

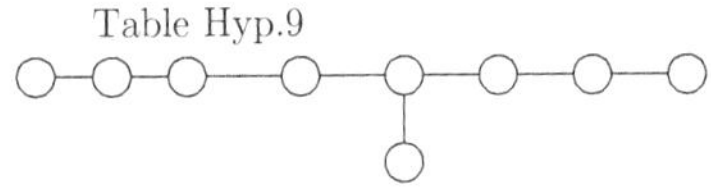

Table Hyp.10

$H_2^{(10)}$

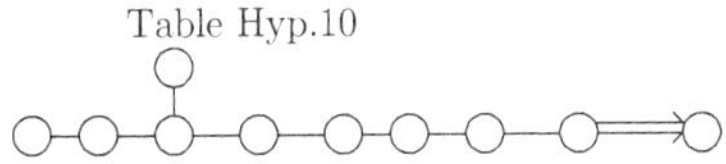

Following can be directly verified:

Remark 47.

(a) The order of a strictly hyperbolic matrix is ≤ 5.
(b) The order of a hyperbolic matrix is ≤ 10.

Definition 80. We define an indefinite nonhyperbolic GCM, $A = (a_{ij})_{i,j=1}^{n}$ to be *extended-hyperbolic type*, if it is an extension of hyperbolic type and every proper connected subdiagram of $S(A)$ is of finite, affine, or hyperbolic type. We denote the Dynkin diagram *associated to an extended-hyperbolic* GCM, $A = (a_{ij})_{i,j=1}^{n}$, by $EHA^{(n)}, n \geq 3$ (for $n \leq 2$, clearly any GCM, $A = (a_{ij})_{i,j=1}^{n}$ is of finite, affine, or hyperbolic type only). Here $HA^{(n)}$ stands for the corresponding Dynkin diagram of the hyperbolic type of order n.

Dynkin diagrams of GCM of extended-hyperbolic types

For $n = 3$, the Dynkin diagram of extended-hyperbolic type contains a bold lined edge with an ordered pair of integers (a, b) with $ab \geq 5$. Diagram can be drawn in each case by definition.

For $n = 4 (n = 5)$, to each Dynkin diagram of hyperbolic types with three vertices (respectively, to each of the 53 Dynkin diagrams of hyperbolic types with four vertices), we will have finitely many associated Dynkin diagrams of extended-hyperbolic type with four (respectively, five) vertices. Dynkin diagrams of extended-hyperbolic type with four vertices (respectively, five vertices) can be drawn in each case following the general principle.

Some examples for the Dynkin diagrams of extended-hyperbolic types

(1) $EHA^{(3)}$:

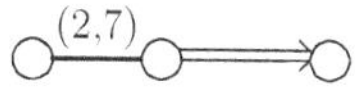

(2) $EHA^{(4)}$:

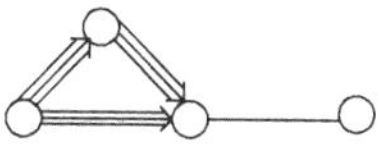

(3) $EHA^{(5)}$:

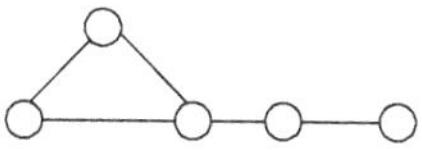

In these extended-hyperbolic diagrams, every subdiagram is of either finite or affine or strictly hyperbolic.

Definition 81. We say that the Kac-Moody algebra associated with the GCM A (respectively, $S(A)$) is of *extended-hyperbolic type,* if A is of extended-hyperbolic type.

Remark 48. Complete set of Kac-Moody algebras $\mathcal{G}(A)$ among the extended-hyperbolic types whose Dynkin diagrams are *completely connected* (ie, $a_{ij} \neq 0$ for all $i, j \in N$) are denoted by $EHA_c^{(n)}$.

Some examples of completely connected extended-hyperbolic type Dynkin diagrams

(i) One example for $EHA_c^{(3)}$: It consists of any indecomposable GCM, $A = (a_{ij})_{i,j=1}^{3}$ such that $\forall i, j \in N, a_{ij} \neq 0$ and $a_{ij}a_{ji} \geq 5$ for atleast one pair of integers (i, j).

$EHA_c^{(3)}$:

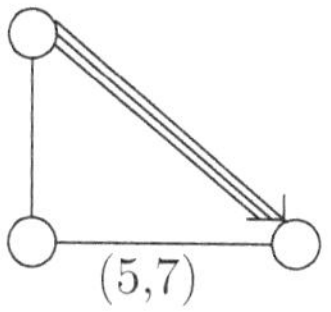

(ii) One example for $EHA_c^{(4)}$: It consists of any indecomposable $GCM, A = (a_{ij})_{i,j=1}^{4}$ of type $EHA_c^{(4)}$ such that $a_{ij} \neq 0$ for all $i, j \in \{1, 2, 3, 4\}$.

$EHA_c^{(4)}$:

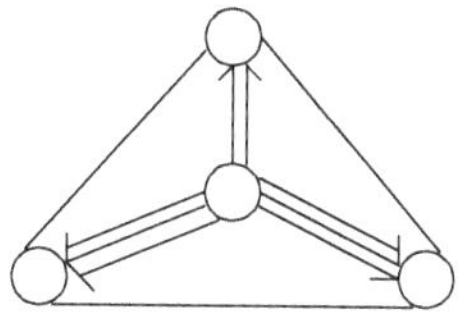

(iii) One example for $EHA_c^{(5)}$: There is exactly one diagram in this class given by

$EHA_c^{(5)}$:

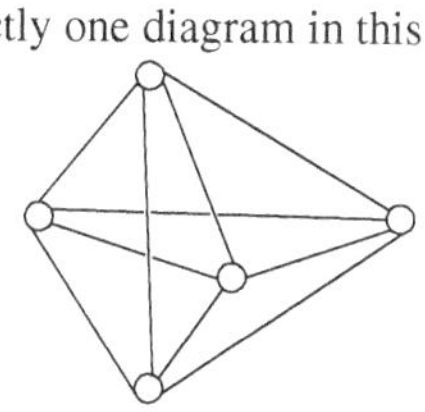

For n ≥ 6, some examples of extended-hyperbolic type Dynkin diagrams

Table ext-hyp 6

EH_1^6

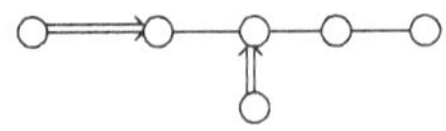

EH_2^6

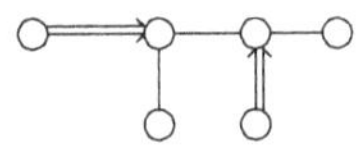

Table ext-hyp 6

EH_3^6

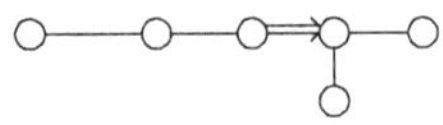

EH_4^6

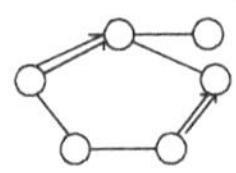

EH_5^6

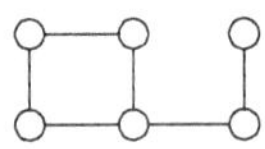

EH_6^6

Table ext-hyp 7

EH_1^7

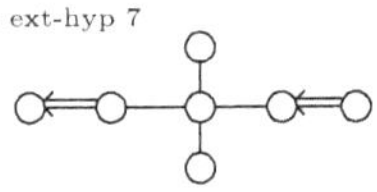

EH_2^7

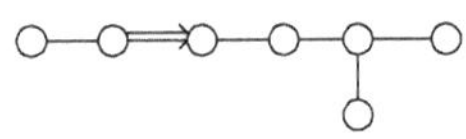

Table ext-hyp 8

EH_1^8

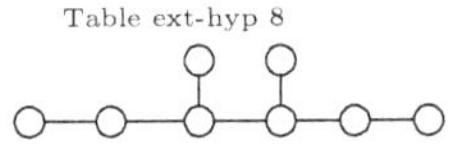

EH_2^8

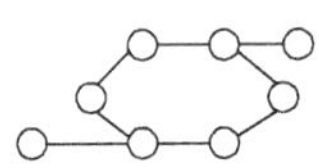

EH_1^8

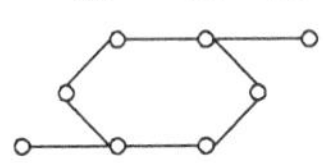

Table ext-hyp 9

EH_1^9

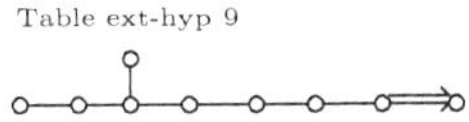

EH_2^9

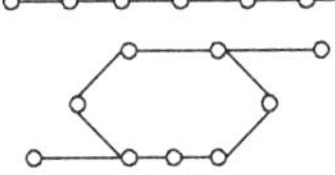

Table ext-hyp 10

EH_1^{10}

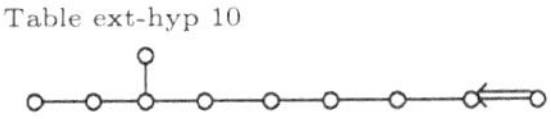

EH_2^{10}

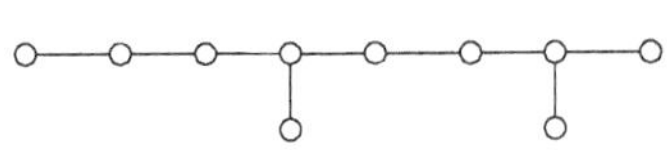

Table ext-hyp 11

EH_1^{11}

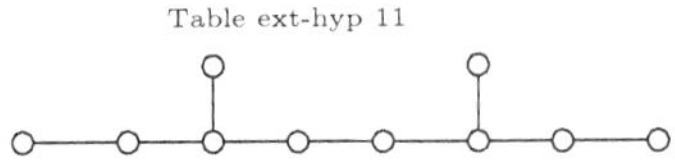

Remark 49. There is no completely connected Dynkin diagram for $EHA^{(n)}, n > 5$.

Remark 50. Any GCM $A = (a_{ij})_{i,j=1}^{n}$ of extended-hyperbolic type has $n \geq 3$.

Proposition 15. *If $A = (a_{ij})_{i,j=1}^{n}$ is an indecomposable GCM of extended-hyperbolic type, then $S(A)$ contains a subdiagram with $n-2$ vertices of affine type and a connected subdiagram with $n-3$ vertices of finite type.*

Proof. Since any hyperbolic diagram of n vertices contains an affine diagram with $n-1$ vertices and any affine diagram with k vertices contains a finite subdiagram with $k-1$ vertices, the proof is immediate. □

Lemma 2. *If $A = (a_{ij})_{i,j=1}^{n}$ is a GCM of extended-hyperbolic, then $n \leq 11$.*

Proof. Since A is of extended-hyperbolic type, there exists a subdiagram of $S(A)$ of hyperbolic type containing $n-1$ vertices. As the maximum number of vertices in a hyperbolic diagram is 10, the maximum number of vertices in a extended-hyperbolic diagram is 11. □

Remark 51. The following properties are clear.

Let $A = (a_{ij})_{i,j=1}^{n}$ be a GCM. Then A is not extended-hyperbolic type if one of the following holds:

(i) There exists a connected subdiagram with $n-3$ vertices of nonfinite type (or)
(ii) There exists a connected subdiagram of nonaffine, nonfinite type with $n-2$ vertices (or)
(iii) There exists a connected subdiagram of nonaffine, nonfinite, nonhyperbolic type with $n-1$ vertices.

From the complete classification of indefinite hyperbolic diagrams, we can easily get a complete classification of indecomposable extended-hyperbolic type Dynkin diagrams.

2.3 Invariant bilinear forms

Properties of invariant nondegenerate symmetric bilinear form

(1) Let A be an $n \times n$ complex matrix. Then A is symmetrizable if and only if there exists a invariant nondegenerate symmetric bilinear C-valued form $(\, , \,)$ on $\mathcal{G}(A)$. Here $(\, , \,)$ is invariant means, $([x,y],z) = (x,[y,z])$ for all $x, y, z \in \mathcal{G}(A)$.

(2) If A is symmetrizable, then $$(\, , \,)|_{\mathcal{G}_\alpha \times \mathcal{G}_\beta} = \begin{cases} 0, & \text{if } \alpha, \beta \in \Delta \cup \{0\} \text{ and } \alpha + \beta \neq 0, \\ \text{nondegenerately paired}, & \text{if } \alpha, \beta \in \Delta \cup \{0\} \text{ and } \alpha + \beta = 0. \end{cases}$$

In particular, the restriction of $(\, , \,)$ to $\mathfrak{h}$ is also nondegenerate.

(3) Let A be a symmetrizable complex $n \times n$ matrix of rank l and $\mathcal{G}(A)$ be the associated complex Lie algebra. Denote by $(\, , \,)$, the invariant nondegenerate symmetric bilinear form on $\mathcal{G}(A)$. As the restriction of the bilinear form to $\mathfrak{h}$ is nondegenerate for each $h \in \mathfrak{h}$, we can define an element $\nu(h)$ of $\mathfrak{h}^*$ such that

$$\langle h_1, \nu(h) \rangle = (h_1, h) \quad \text{for all } h_1 \in \mathfrak{h}.$$

Now we can define a map

$$\nu \colon \mathfrak{h} \to \mathfrak{h}^*$$

such that

$$h \mapsto \nu(h).$$

This is an isomorphism of vector spaces. Again, the nondegenerate symmetric bilinear form (,) on $\mathfrak{h}$ (ie, the restriction of (,) to $\mathfrak{h}$) induces in a natural way a nondegenerate symmetric bilinear form on $\mathfrak{h}^*$ given by

$$(\lambda, \mu) = (\nu^{-1}(\lambda), \nu^{-1}(\mu)) \text{ for } \lambda, \mu \in \mathfrak{h}^*,$$

where ν^{-1} denotes the inverse map of ν. Hence

$$\left.\begin{array}{lll} \nu(\alpha_i^\vee) & = \alpha_i \varepsilon_i, & i = 1, \dots, n \\ (\alpha_i, \alpha_j) & = b_{ij}, & i,j = 1, \dots, n \\ (\alpha_i, \alpha_j) & = a_{ij}\varepsilon_i^{-1}, & i,j = 1, \dots, n \end{array}\right\}. \tag{2.2}$$

We can prove that for $\alpha \in \Delta, x \in \mathcal{G}_\alpha, y \in \mathcal{G}_{-\alpha}$,

$$[x, y] = (x, y)\nu^{-1}(\alpha).$$

Moreover, if $(\, , \,)_1$ and $(\, , \,)_2$ are two invariant nondegenerate symmetric bilinear forms on $\mathcal{G}(A)$ and if the restrictions of $(\, , \,)_1$ and $(\, , \,)_2$ to $\mathfrak{h}$ coincide, then $(\, , \,)_1 = (\, , \,)_2$ on whole of $\mathcal{G}(A)$.

Expression of symmetrizable GCM in terms of invariant bilinear form

Let $A = (a_{ij})_{i,j=1}^n$ be a symmetrizable GCM and a decomposition of A be

$$A = diag(\varepsilon_1, \dots, \varepsilon_n)B, \tag{2.3}$$

where $\varepsilon_1, \dots, \varepsilon_n$ are positive rational numbers and B is a rational symmetric $n \times n$ matrix. If A is indecomposable, then the matrix $diag(\varepsilon_1, \dots, \varepsilon_n)$ is uniquely determined by Equation (2.3) up to a constant factor. Fix such a decomposition of A and a complementary subspace $\mathfrak{h}''$ to $\mathfrak{h}'$ in $\mathfrak{h}$. We define a nondegenerate symmetric bilinear form (,) by

$$\left.\begin{array}{ll} (\alpha_i^\vee, \alpha_j^\vee) & = \varepsilon_i b_{ij} \epsilon_j = a_{ij}\epsilon_j \ (i,j = 1, \dots, n) \\ (h, \alpha_j^\vee) & = (\alpha_j^\vee, h) = \langle h, \alpha_j^\vee \rangle \varepsilon_j \ \ (j = 1, \dots, n) \\ (h', h'') & = 0 \text{ for } h', h'' \in \mathfrak{h} \end{array}\right\}. \tag{2.4}$$

Since $a_{ii} = 2$ and $\varepsilon_i > 0$ for $i = 1, \dots, n$, we deduce the following from Equation (2.2):

$$(\alpha_i, \alpha_i) > 0 \quad \text{for } i = 1, \ldots, n,$$
$$(\alpha_i, \alpha_j) \leq 0 \quad \text{for } i \neq j, \text{ and}$$
$$\alpha_i^\vee = \frac{2}{(\alpha_i, \alpha_j)} \nu^{-1}(\alpha_i).$$

Hence the matrix A is

$$A = \left(\frac{2(\alpha_i, \alpha_j)}{(\alpha_i, \alpha_i)} \right)_{i,j=1}^{n},$$

which is the same as the usual expression of the Cartan matrix.

Then we extend the bilinear form (,) on $\mathfrak{h}$ to an invariant nondegenerate symmetric bilinear form (,) on the entire Kac-Moody algebra $\mathcal{G}(A)$. Such an extension is unique. The extended bilinear form (,) satisfies $(\alpha_i, \alpha_i) > 0$ for $i = 1, \ldots, n$ and is called a *standard invariant form*.

Definition 82 (Canonical central element, Coxeter number, dual coxeter number, and scaling element). In the usual notation, assume that $\mathcal{G}$ is a affine Lie algebra associated with GCM, A of affine type. Let $\mathfrak{h}$ be its Cartan subalgebra, $\Pi = \{\alpha_0, \alpha_1, \ldots, \alpha_l\} \subset \mathfrak{h}^*$ be a system of simple roots, $\Pi^\vee = \{\alpha_0^\vee, \alpha_1^\vee, \ldots, \alpha_l^\vee\} \subset \mathfrak{h}$ be a system of simple coroots with Δ, the total root system. Here $\mathbb{Q}$ and $\mathbb{Q}^\vee$ denote root lattice and coroot lattice, respectively. *Then the center of $\mathcal{G}(A)$ is one dimensional and is spanned by*

$$c = \sum_{i=0}^{l} a_i^\vee \alpha_i^\vee,$$

called the, *canonical central element*. Similarly we define the element,

$$\delta = \sum_{i=0}^{l} a_i \alpha_i \in \mathbb{Q}.$$

Now $h = \sum_{i=0}^{l} a_i$ is called *Coxeter number* and $g = \sum_{i=0}^{l} a_i^\vee$ is called the *dual Coxeter number* for the GCM of affine type.

The following can be proved:

(i) There exists an element $d \in \mathfrak{h}$ which satisfies the conditions

$$\langle d, \alpha_0 \rangle = 1 \text{ and } \langle d, \alpha_i \rangle = 0 \quad \text{for } i = 1, \ldots, l.$$

This element d is uniquely determined up to a summand which is a multiple of c.

(ii) $\{\alpha_0^\vee, \alpha_1^\vee, \ldots, \alpha_l^\vee, d\}$ is a basis of $\mathfrak{h}$. The element $d \in \mathfrak{h}$ is called a *scaling element*, which satisfies the conditions

$$\langle d, \alpha_0 \rangle = 1 \text{ and } \langle d, \alpha_i \rangle = 0 \quad \text{for } i = 1, \ldots, l.$$

Let $\mathfrak{h}'' = \mathbb{C}d$, be the element complementary to $\mathfrak{h}'$ in $\mathfrak{h}$. Now we can define a *nondegenerate symmetric bilinear form* (,) satisfying the following conditions:

$$(\alpha_i^\vee, \alpha_j^\vee) = a_{ij}a_j a_j^{\vee^{-1}}, \quad (i,j = 0, 1, 2, \ldots, l)$$
$$(d, \alpha_i^\vee) = 0, \quad (i = 1, 2, \ldots, l)$$
$$(d, \alpha_0^\vee) = a_0, \text{ and}$$
$$(d, d) = 0.$$

Remark 52. The above-defined bilinear form is positive semi-definite on $\sum_{i=0}^{l} \mathbb{R}\alpha_i^\vee$, which can be uniquely extended to a nondegenerate invariant symmetric bilinear form (,) on $\mathcal{G}(A)$, called *the normalized standard form* on $\mathcal{G}(A)$.

2.4 Coxeter groups and Weyl groups

Definition 83 (Coxeter group). A group with generators $r_1, \ldots, r_n$ with defining relations $r_i^2 = 1 (1 \leq i \leq n)$ and $(r_i r_j)^{m_{ij}} = 1$ $(1 \leq i,j \leq n)$, with m_{ij} being positive integers or ∞ (assume for any $x^\infty = 1$), is called a *coxeter group.*

Definition 84. Let A be a GCM and $\mathcal{G}(A)$ be the corresponding Kac-Moody algebra. In the usual notation, for $1 \leq i \leq n$, define r_i: $\mathfrak{h}^* \to \mathfrak{h}^*$ by $r_i(\lambda) = \lambda - \langle \alpha_i^\vee, \lambda \rangle \alpha_i$ for $\lambda \in \mathfrak{h}^*$. Here r_i is a fundamental reflection r_{α_i}. Now we have

$$\begin{aligned} r_i(\alpha_j) &= \alpha_j - \langle \alpha_i^\vee, \alpha_j \rangle \alpha_i, \\ &= \alpha_j - \alpha_j(\alpha_i^\vee)\alpha_i, \\ &= \alpha_j - a_{ij}\alpha_i \in Q. \\ \text{So} \quad r_i(\alpha_i) &= \alpha_i - a_{ii}\alpha_i = \alpha_i - 2\alpha_i = -\alpha_i. \end{aligned}$$

In particular, $r_i^2 = Id$. Also r_i fixes elementwise the hyperplane $\{\lambda \in \mathfrak{h}^* | \langle \alpha_i^\vee, \lambda \rangle = 0\}$. Here each r_i is called *a fundamental reflection*.

Definition 85. Let W be the group of automorphisms of $\mathfrak{h}^*$ generated by $r_1, \ldots, r_n$. Then $W \subset GL(\mathfrak{h}^*)$ and W *is called the Weyl group of* $\mathcal{G}(A)$. We can write $W = W(A)$.

Here the subgroup of $GL(\mathfrak{h})$ generated by $r_1^*, \ldots, r_n^*$ is the Weyl group $W(A^t)$, called the Weyl group of the Kac-Moody algebra associated to the GCM, A^t. $W(A)$ and $W(A^t)$, Weyl groups of $\mathcal{G}(A)$ and $\mathcal{G}(A^t)$, respectively, are contragradiant linear groups.

Definition 86. Let $\mathcal{G}(A)$ be the Kac-Moody algebra and Δ^{re} be the set of all real roots. For any root $\alpha \in \Delta^{re}$, we define $r_\alpha \in GL(\mathfrak{h}^*)$ by

$$r_\alpha(\lambda) = \lambda - \langle \alpha^\vee, \lambda \rangle \alpha \quad \text{for } \lambda \in \mathfrak{h}^*.$$

As $\langle \alpha^\vee, \alpha \rangle = 2$, $r_\alpha(\alpha) = \alpha - 2\alpha = -\alpha$. Hence r_α is a *reflection with respect to real root* α. If $\alpha = w(\alpha_i)$, then $r_\alpha = w r_i w^{-1} \in W$.

Table 2.1 **Weyl group elements of G_2**

Weyl group element	Image of α_1	Image of α_2
1	α_1	α_2
r_1	$-\alpha_1$	$\alpha_1 + \alpha_2$
r_2	$\alpha_1 + 3\alpha_2$	$-\alpha_2$
$r_1 r_2$	$2\alpha_1 + 3\alpha_2$	$-(\alpha_1 + \alpha_2)$
$r_2 r_1$	$-(\alpha_1 + 3\alpha_2)$	$\alpha_1 + 2\alpha_2$
$r_2 r_1 r_2$	$2\alpha_1 + 3\alpha_2$	$-(\alpha_1 + 2\alpha_2)$
$r_1 r_2 r_1$	$-(2\alpha_1 + 3\alpha_2)$	$\alpha_1 + 2\alpha_2$
$(r_1 r_2)^2$	$\alpha_1 + 3\alpha_2$	$-(\alpha_1 + 2\alpha_2)$
$(r_2 r_1)^2$	$-(2\alpha_1 + 3\alpha_2)$	$\alpha_1 + \alpha_2$
$(r_2 r_1)^2 r_2$	α_1	$-(\alpha_1 + \alpha_2)$
$(r_1 r_2)^2 r_1$	$-(2\alpha_1 + 3\alpha_2)$	α_2
$(r_1 r_2)^3$	$-\alpha_1$	$-\alpha_2$

This fundamental reflection r_i induces an automorphism r_i^* of $\mathfrak{h}$ such that

$$< r_i^*, \lambda > = < h, r_i(\lambda) >, \quad h \in \mathfrak{h}, \ \lambda \in \mathfrak{h}^*.$$

Example 17. One can find the Weyl group of G_2 as follows:
Weyl group of G_2: The GCM of G_2 is $\begin{pmatrix} 2 & -1 \\ -3 & 2 \end{pmatrix}$ and $\Pi = \{\alpha_1, \alpha_2\}$. We have

$$r_1(\alpha_1) = -\alpha_1, \quad r_1(\alpha_2) = \alpha_2 + \alpha_1,$$
$$r_2(\alpha_1) = \alpha_1 + 3\alpha_2, \quad r_2(\alpha_2) = -\alpha_2$$

and we get Table 2.1 on the actions of $r_i, r_i r_j, r_i r_j r_k, i, j, k = 1, 2$ on α_1, α_2.
Hence the Weyl group of G_2 is

$$W = \{1, r_1, r_2, r_1 r_2, r_2 r_1, r_1 r_2 r_1, r_2 r_1 r_2, (r_1 r_2)^2, (r_2 r_1)^2, (r_2 r_1)^2 r_2, (r_1 r_2)^2 r_1, (r_1 r_2)^3\}.$$

Some properties of Weyl groups

The following results can be proved:

(a) **(i)** W permutes the roots, that is, $w\Delta = \Delta$ for every $w \in W$.
(ii) For any $\alpha \in \Delta$, *mult* α = *mult* $w(\alpha)$ for every $w \in W$.
(b) If $r_1, \ldots, r_n$ are fundamental reflections, then

$$r_i^2 = 1 \text{ and } (r_i r_j)^{m_{ij}} = 1 (i \neq j, 1 \leq i \leq n), \tag{2.5}$$

where $m_{ij} \in [2, \infty)$. The values of m_{ij} are given below in terms of the GCM, $(a_{ij})_{i,j=1}^n$.

Moreover,

$a_{ij}a_{ji}$	0	1	2	3	≥ 4
m_{ij}	2	3	4	6	∞

Hence W is a Coxeter group with generators $r_1, \ldots, r_n$ satisfying the above relations (Equation 2.5). Moreover, W acts faithfully on Δ.

(c) $ad\, e_i$ and $ad\, f_i$ $(1 \leq i \leq n)$ are locally nilpotent derivations of $\mathcal{G}(A)$.

(d) **(1)** If $\alpha \in \Delta_+, \alpha \neq \alpha_i$, then $r_i(\alpha) \in \Delta_+$. Hence r_i permutes $\Delta_+ \backslash \{\alpha_i\}$.

(2) Let $w \in W$ be such that $w(\alpha_i) = \alpha_j$. Then

(i) $w(\alpha_i^\vee) = \alpha_j^\vee$ and $wr_i = r_j w$.

(ii) If α_i is a simple root with $(r_{i_1} \ldots r_{i_n})(\alpha_i) < 0$, then there exists a $s(1 \leq s \leq r)$ such that

$$r_{i_s} r_{i_{s+1}} \ldots r_{i_r} = r_{i_{s+1}} \ldots r_{i_r} r_{i_s}.$$

Example for properties of Weyl group

(a) W permutes the roots:

Let $\Delta = \{\alpha_1, \alpha_2\}$ for B_2 with GCM $= \begin{pmatrix} 2 & -1 \\ -2 & 2 \end{pmatrix}$. Here $W = \{1, r_1, r_2, r_1r_2, r_2r_1, r_1r_2r_1, r_2r_1r_2, (r_1r_2)^2\}$.

$$1(\alpha_1) = \alpha_1; \;\; 1(\alpha_2) = \alpha_2;$$
$$r_1(\alpha_1) = \alpha_1 - 2\alpha_1 = -\alpha_1; \;\; r_2(\alpha_1) = \alpha_2 + \alpha_1; \;\; r_1(\alpha_2) = \alpha_2 + 2\alpha_1;$$
$$r_2(\alpha_2) = -\alpha_2; \;\; r_1r_2(\alpha_1) = \alpha_1 + \alpha_2; \;\; r_2r_1(\alpha_1) = -(\alpha_1 + \alpha_2).$$

Thus $w(\alpha) \in \Delta$, hence $w\Delta \subseteq \Delta$.

(b) $r_i^2 = 1$: We have

$$r_1^2(\alpha_1) = r_1(-\alpha_1) = \alpha_1$$
$$r_2^2(\alpha_1) = r_2(\alpha_1 + \alpha_2) = \alpha_1$$
$$r_1^2(\alpha_2) = r_1(\alpha_2 + 2\alpha_1) = \alpha_2.$$

Definition 87. The expression $w = r_{i_1}, \ldots, r_{i_s}$ is called *reduced*, if s is minimal among all representations of $w \in W$ as a product of the r_i. Then s is called the *length of* w denoted by $l(w)$. As $det\, r_i = -1, det\, w = (-1)^{l(w)}$ for $w \in W$.

Definition 88 (Fundamental Weyl chamber and the Tits cone). For a real $n \times n$ matrix of rank ℓ, $A = (a_{ij})_{i,j=1}^n$. Let $(\mathfrak{h}_\mathbb{R}, \Pi, \Pi^\vee)$ be the minimum realization. Here $\mathfrak{h}_\mathbb{R}$ is a real vectorspace of dimension $2n - \ell$ and $\Pi = \{\alpha_1, \alpha_2, \ldots, \alpha_n\} \subset \mathfrak{h}_\mathbb{R}^*$, $\Pi^\vee = \{\alpha_1^\vee, \alpha_2^\vee, \ldots, \alpha_n^\vee\} \subset \mathfrak{h}_\mathbb{R}$.

Let

$$C = \{h \in \mathfrak{h}_R \mid \langle \alpha_i, h \rangle \geq 0 \quad \text{for } i = 1, \ldots, n\}.$$

Here C is called the *fundamental Weyl chamber.* The sets $w(C), w \in W$, are called chambers, and their union

$$\chi = \bigcup_{w \in W} w(\mathbb{C})$$

is called the *Tits cone.* One can have the corresponding dual notions, $C^\vee, \chi^\vee$ in $\mathfrak{h}_R^*$. Here

$$C^\vee = \{\lambda \in \mathfrak{h}_R^* \mid \langle \alpha_i^\vee, \lambda \rangle \geq 0 \quad \text{for } i = 1, \ldots, n\} \text{ and}$$
$$\chi^\vee = \bigcup_{w \in W} w(C^\vee).$$

It can be proved that $|W| < \infty$; $\chi = \mathfrak{h}_R$; $|\Delta| < \infty$; and $|\Delta^\vee| < \infty$ are equivalent conditions.

We state the following theorem, for proof one can refer Wan [3].

Some properties of fundamental Weyl chamber and Tits cone

Theorem.

(i) *The fundamental chamber is* $C = \{h \in \mathfrak{h}_R \mid \textit{for every } w \in W, h - w(h) = \sum_i c_i \alpha_i^\vee, \textit{ where } c_i \geq 0\}$.

(ii) *The Tits cone is* $\chi = \{h \in \mathfrak{h}_R \mid \langle h, \alpha \rangle < 0 \textit{ for only a finite number of } \alpha \in \Delta_+\}$.

Remark. We denote by $\overset{\circ}{\chi} = \{h \in \mathfrak{h}_R \mid \langle h, \alpha \rangle \leq 0$ only for a finite number of $\alpha \in \Delta_+\}$.

2.5 Real and imaginary roots of Kac-Moody algebras

Basic properties of real and imaginary roots

Let A be the GCM and $\mathcal{G}(A)$ be the corresponding Kac-Moody algebra. In the usual notation, let Δ be the root system and W be the Weyl group. A root $\alpha \in \Delta$ is called *real*, if there exists $w \in W$ such that $w(\alpha)$ is a simple root. A root which is not real is called an *imaginary root.* We denote by Δ_+, Δ_-, Δ^{re}, Δ^{im}, Δ_+^{re}, and Δ_+^{im}, the set of all positive roots, negative roots, real roots, imaginary roots, positive real and positive imaginary roots, respectively. We have the following disjoint unions.

$$\Delta = \Delta_+ \cup \Delta_-, \quad \text{where } \Delta_- = -\Delta_+.$$
$$\Delta = \Delta^{re} \cup \Delta^{im}, \quad \text{where } \Delta^{re} = \Delta_+^{re} \cup (-\Delta_+^{re}) \text{ and } \Delta^{im} = \Delta_+^{im} \cup (-\Delta_+^{im}).$$

Moreover, if $\alpha \in \Delta^{re}$, then $\alpha = w(\alpha_i)$ for some $\alpha_i \in \Pi, w \in W$. For a simple root α_i, if $\alpha = w(\alpha_i)$, then $w(\alpha) = w^2(\alpha_i) = \alpha_i$ as $w^2 = 1$.

For the dual real root $\alpha^\vee \in \Delta^\vee$, we define $\alpha^\vee = w(\alpha_i^\vee)$ which is an element of dual of Δ^{re}. So $\alpha^\vee = w(\alpha_i^\vee) \in \Delta^{\vee^{re}}$.

W-equivariant bijection between real roots and their duals

There is a canonical W-equivariant bijection $\Delta^{re} \to \Delta^{\vee^{re}}$, by $\alpha \mapsto \alpha^\vee$, because $w(\alpha_i^\vee) = w'(\alpha_j^\vee)$. This realization $\alpha \mapsto \alpha^\vee$ gives a *W-equivariant bijection.*

Elementary properties of real roots of Kac-Moody algebras

Let α be a real root of a Kac-Moody algebra of $\mathcal{G}(A)$. Then the following properties are true.

(i) Multiplicity of $\alpha = 1$. $k\alpha$ is a root if and only if $k = \pm 1$ and $\langle \alpha^\vee, \alpha \rangle = 2$.
(ii) If $\beta \in \Delta$, there exist nonnegative integers p and q such that $\beta + k\alpha \in \Delta \cup \{0\}$ for $k \in \mathbb{Z}$ if and only if $-p \leq k \leq q$. Here p and q are related by $p - q = \langle \alpha^\vee, \beta \rangle$.
(iii) If $\pm\alpha \notin \Pi$, then there exists an i such that $|ht\ r_i(\alpha)| < |ht\ \alpha|$.
(iv) If A is a *symmetric* indecomposable GCM, then all real roots have the same length.
(v) If A is a symmetrizable GCM and (,) is a standard invariant nondegenerate symmetric bilinear form on $\mathcal{G}(A)$, then for $\alpha \in \Delta^{re}$, we have $(\alpha, \alpha) > 0$ and $\alpha^\vee = 2v^{-1}(\alpha)/(\alpha, \alpha)$.

Definition 89 (Short and long roots). Let A be a symmetrizable indecomposable GCM and (,) be a standard invariant bilinear form.

Let α be a real root, with $\alpha = w(\alpha_i)$ for a root $\alpha_i \in \Pi$ and $w \in W$.

We have $|\alpha|^2 = (\alpha, \alpha) = (\alpha_i, \alpha_i) = |\alpha_i|^2$ for some simple root α_i.

We call a root α *short* if $(\alpha, \alpha) = \min_i(\alpha_i, \alpha_i)$ and *long* if $(\alpha, \alpha) = \max_i(\alpha_i, \alpha_i)$.

The following properties can be proved.

(i) Let $a = \min |\alpha^2|, \alpha \in Q, |\alpha|^2 > 0$. Then a exists and is a positive integer.
(ii) Let A be a symmetizable GCM of finite, affine, or hyperbolic type. If $\alpha \in Q$ and $|\alpha|^2 \leq a$ then either α or $-\alpha \in Q_+$.
(iii) For a symmetrizable GCM A of finite, affine, and hyperbolic type,
  (a) the set of all *short real* roots is

$$\{\alpha \in Q \mid |\alpha|^2 = a = \min_i |\alpha_i|^2\}.$$

  (b) the set of all *long real roots* is $\{\alpha = \sum_j k_j \alpha_j \in Q \mid |\alpha|^2 = a = \max_i |\alpha_i|^2$ and $k_j |\alpha_j|^2 / |\alpha|^2 \in \mathbb{Z}$ for all $j\}$.

If A is a symmetrizable indecomposable GCM but not symmetric and its Dynkin diagram $S(A)$ is equipped with m arrows pointing in the same direction, then there are real roots of exactly $m + 1$ different square lengths. If A is in Figure *Finite* Chapter 1, every real root is either *short* or *long*. If A is in Figure *Affine* and $A \neq A_{2l}^{(2)}$ with $l > 1$, every real root is either *short* or *long*. If $A = A_{2l}^{(2)}$ with $l > 1$, there are real roots of *three different square lengths.*

Definition 90 (Roots of maximal height). Let $\theta = \delta - a_0\alpha_0$. We have $|\theta|^2 = 2a_0$, $\theta = a_0\nu(\theta^\vee)$, $|\theta^\vee|^2 = 2a_0^{-1}$.
Let $(\mathring{\Delta}_+)_l$, $(\mathring{\Delta}_+)_s$ be the sets of all positive long roots and short roots, respectively, of the corresponding finite-dimensional Lie algebra and h be the coxeter number. Then one can establish the following properties:

(i) Let A be of Affine 1 type or Affine $A_{2l}^{(2)}$ type. Then $\theta \in (\mathring{\Delta}_+)_l$ and θ is the unique root in $\mathring{\Delta}$ of *maximal height* $(h-1)$ or $(h-2)$, respectively.
(ii) Let A be of Affine 2 or 3 type and not of type $A_{2l}^{(2)}$. Then $\theta \in (\mathring{\Delta}_+)_s$ and is the unique root in $\mathring{\Delta}_s$ of *maximal height* $h-1$.
(iii) We have the following descriptions of Π and $\Pi^\vee$. $\Pi = \{\alpha_0 = a_0^{-1}(\delta - \theta), \alpha_1, \dots, \alpha_l\}$, $\Pi^\vee = \{\alpha_0^\vee = c - a_0\theta^\vee, \alpha_1^\vee, \dots, \alpha_l^\vee\}$.

Some properties of root systems of Kac-Moody algebras in general

Let A be the GCM and $\mathcal{G}(A)$ be the Kac-Moody algebra associated with it. If $\mathcal{G}_\alpha(A) = \{v \in \mathcal{G}(A) \mid [h, v] = \langle\alpha, h\rangle v \text{ for } h \in \mathfrak{h}\}$, then $\mathcal{G}(A) = \bigoplus_{\alpha\in Q} \mathcal{G}_\alpha(A)$ is called the root space decomposition of $\mathcal{G}$, where Q is the root lattice.

We have $\Delta = \{\alpha \in Q\backslash\{0\} \mid \dim \mathcal{G}_\alpha \neq 0\}$ is the set of all roots. We have

(1) $\Pi \subset \Delta_+ \subset Q_+$ and $2\alpha_i \notin \Delta_+$ for $\alpha_i \in \Pi$, where $\Delta_+ = \Delta \cap Q_+$.
(2) $\alpha \in \Delta_+\backslash\Pi$, then $\alpha - \alpha_i$ is in Δ_+ for some $\alpha_i \in \Pi$.
(3) A root $\alpha \in \Delta$ is a real root and write $\alpha \in \Delta^{re}$ if there exists $\alpha \in W$, such that $W \cdot \alpha \in \Pi$.
(4) For $\alpha \in \Delta_+^{im}$, there is exactly one element in the set $(W \cdot \alpha) \cap K$, where $K = \{\alpha \in Q_+\backslash\{0\} \mid \langle\alpha, \alpha_i^\vee\rangle \leq 0 \text{ for all } i\}$.
(5) For GCM of finite type, $\Delta = \Delta^{re}$. For all other (affine, indefinite) indecomposable matrices both Δ^{re} and Δ^{im} are *infinite sets*.

Description of root systems of affine Lie algebras

Let A be a GCM of order $l+1$. Let $\mathcal{G}(A)$ be the affine Lie algebra of type k, $\mathfrak{h}$ be its Cartan subalgebra, Π be the set of all simple roots $\{\alpha_0, \alpha_1, \dots, \alpha_l\} \subset \mathfrak{h}^*$, and $\Pi^\vee$ be the set of all simple coroots $\{\alpha_0^\vee, \alpha_1^\vee, \dots, \alpha_l^\vee\}$. We have $c = \sum_{i=0}^{l} a_i^\vee \alpha_i^\vee$, the canonical center element. Dually $\delta = \sum_{i=0}^{l} a_i\alpha_i$. Then there exists a unique $\Lambda_0 \in \mathfrak{h}^*$ such that $\langle\alpha_i^\vee, \Lambda_0\rangle = \delta_{0i}$ for $0 \leq i \leq l$ and $\langle d, \Lambda_0\rangle = 0$. The set $\{\alpha_0^\vee, \alpha_1^\vee, \dots, \alpha_l^\vee, d\}$ is a basis of $\mathfrak{h}$ and $\{\alpha_0, \alpha_1, \dots, \alpha_l, \Lambda_0\}$ is a basis of $\mathfrak{h}^*$.

We have a normalized invariant nondegenerate symmetric bilinear form (,) on $\mathcal{G}$. Let $\mathring{\mathcal{G}}$ be the Kac-Moody algebra associated with matrix $\mathring{A}$ obtained from A by deleting 0th row and 0th column. Then $\mathring{\mathcal{G}}$ is a finite-dimensional Lie algebra generated by Chevalley generators $e_i, f_i (1 \leq i \leq l)$ and $\mathring{\mathfrak{h}} = \mathring{\mathcal{G}} \cap \mathfrak{h}$. Here $\mathring{\mathfrak{h}}$ is its Cartan subalgebra.

$\overset{\circ}{\Pi} = \{\alpha_1, \ldots, \alpha_l\}$ is the root basis of $\overset{\circ}{\mathcal{G}}$. $\overset{\circ}{\Pi}{}^{\vee} = \{\alpha_1^{\vee}, \ldots, \alpha_l^{\vee}\}$ is the coroot basis of $\overset{\circ}{\mathcal{G}}$. The set $\overset{\circ}{\Delta} = \Delta \cap \overset{\circ}{\mathfrak{h}}{}^{*}$ is the root system of $\overset{\circ}{\mathcal{G}}$. $\overset{\circ}{\Delta}$ is *finite and consists of real* roots.

Denote by $\overset{\circ}{\Delta}_{+} = \Delta_{+} \cap \overset{\circ}{\Delta}$, the set of all positive roots of $\overset{\circ}{\mathcal{G}}$, $\overset{\circ}{\Delta}_s$, the set of all short roots, and $\overset{\circ}{\Delta}_l$, the set of all long roots. The set of all imaginary roots of $\mathcal{G}$ is $\Delta^{im} = \{n\delta | n \in \mathbb{Z},\ n \neq 0\}$ and the set of all positive imaginary roots is $\Delta_{+}^{im} = \{n\delta | n = 1, 2, \ldots\}$.

The real roots of the affine Lie algebra $\mathcal{G} = \mathcal{G}(A)$ ($X_l^{(k)}$ for $X_l = \overset{\circ}{\mathcal{G}}$) are given below.

(i) $\Delta^{re} = \{\alpha + n\delta | \alpha \in \overset{\circ}{\Delta}, n \in \mathbb{Z}\}$ if $k = 1$.

(ii) $\Delta^{re} = \{\alpha + n\delta | \alpha \in \overset{\circ}{\Delta}_s, n \in \mathbb{Z}\} \cup \{\alpha + nk\delta | \alpha \in \overset{\circ}{\Delta}_l, n \in \mathbb{Z}\}$, if $k = 2$ or 3 but A is not of type $A_{2l}^{(2)}$.

(iii) $\Delta^{re} = \{\frac{1}{2}(\alpha + (2n-1)\delta) | \alpha \in \overset{\circ}{\Delta}_l, n \in \mathbb{Z}\} \cup \{\alpha + n\delta | \alpha \in \overset{\circ}{\Delta}_s, n \in \mathbb{Z}\} \cup \{\alpha + 2n\delta | \alpha \in \overset{\circ}{\Delta}_l, n \in \mathbb{Z}\}$ if A is of type $A_{2l}^{(2)}$.

(iv) $\Delta_{+}^{re} = \{\alpha \in \Delta^{re}, n > 0\} \cup \overset{\circ}{\Delta}_{+}$.

Elementary properties of imaginary roots of Kac-Moody algebras

(1) The set Δ_{+}^{im} is W-invariant. Moreover, for $\alpha \in \Delta_{+}^{im}$, there exists a unique root $\beta \in W \cdot \alpha$ such that $\beta \in -C^{\vee}$ (ie, $\langle \beta, \alpha_i^{\vee} \rangle \leq 0$). Here β is the element with minimum height in $W \cdot \alpha$.

(2) If A is a symmetrizable GCM and (,) is a standard invariant nondegenerate symmetric bilinear form on $\mathcal{G}(A)$, then a root α is imaginary if and only if $(\alpha, \alpha) \leq 0$. For an indecomposable GCM A of order n, the set of all *imaginary roots* is $\{\alpha \in Q \backslash \{0\}$, such that $|\alpha|^2 \leq 0\}$.

(3) Let $K = \{\alpha \in Q_{+} \backslash \{0\} | \ \langle \alpha, \alpha_i^{\vee} \rangle \leq 0$ for all i and *supp* α is connected$\}$. Then $K \subset \Delta_{+}^{im}$. It can be proved that

$$\Delta_{+}^{im} = \bigcup_{w \in W} w(K).$$

(4) If $\alpha \in \Delta_{+}^{im}$ and $r \neq 0$ is a rational number such that $r\alpha \in Q$, then $r\alpha \in \Delta^{im}$. In particular, $n\alpha$ is also an imaginary root for $n \in \mathbb{Z} \backslash \{0\}$.

(5) For an indecomposable GCM, A of order n,

(i) If A is of finite type, then Δ^{im} is empty.

(ii) If A is of affine type, then

$$\Delta_{+}^{im} = \{n\delta | n = 1, 2, \ldots\}.$$

Here $\delta = \sum_{i=0}^{l} a_i \alpha_i$ where a_i are the labels of the vertices of $S(A)$ in Figure *Affine*.

(iii) If A is of indefinite type, there exists a positive imaginary root $\alpha = \sum_i k_i \alpha_i$ such that $k_i > 0$ and $\langle \alpha_i^{\vee}, \alpha \rangle < 0$, for $i = 1, \ldots, n$.

2.6 Weyl groups of affine Lie algebras

Relation between Weyl groups of affine Lie algebras and Weyl groups of corresponding finite-dimensional Lie algebras

The Weyl group W of the affine Lie algebra $\mathcal{G}$ is generated by fundamental reflections $w_0, w_1, \ldots, w_l$, where w_i acts on $\mathfrak{h}^*$ by

$$w_i(\lambda) = \lambda - \langle \alpha_i^\vee, \lambda \rangle \alpha_i \quad (\lambda \in \mathfrak{h}^*).$$

Denote by $\mathring{W}$, the subgroup of W generated by $w_1, \ldots, w_l$. We can identify $\mathring{W}$ with the Weyl group of the finite-dimensional simple Lie algebra $\mathring{\mathcal{G}}$, operating on $\mathring{\mathfrak{h}}^*$. Hence the group $\mathring{W}$ is finite. The map ν: $\mathfrak{h} \to \mathfrak{h}^*$ is defined by $(\cdot|\cdot)$. Moreover, we have

$$a_i^\vee \nu(\alpha_i^\vee) = a_i \alpha_i; \quad \nu(c) = \delta; \quad \nu(d) = a_0 \Lambda_0.$$

Consider a lattice $M \subset \mathring{\mathfrak{h}}^*_{\mathbb{R}}$.
Let $\mathbb{Z}(\mathring{W} \cdot \theta^\vee)$ be the lattice in $\mathring{\mathfrak{h}}_{\mathbb{R}}$ spanned over $\mathbb{Z}$ by the finite set $\mathring{W} \cdot \theta^\vee$ and

$$M = \nu(Z(\mathring{W} \cdot \theta^\vee)).$$

Then

$$M = \begin{cases} \overline{Q} = \mathring{Q} & \text{if } A \text{ is symmetric or } r > a_0 \\ \nu(\bar{Q}^\vee) = \nu(\mathring{Q}^\vee) & \text{otherwise.} \end{cases}$$

Moreover, in the usual notation the lattice M considered as an abelian group, operates faithfully on $\mathfrak{h}^*$ by

$$t_\alpha(\lambda) = \lambda + \langle c, \lambda \rangle \alpha - \left((\alpha, \lambda) + \frac{1}{2} |\alpha|^2 \langle c, \lambda \rangle \right) \delta \quad \text{for } \lambda \in \mathfrak{h}^*.$$

Denote the corresponding subgroup of $GL(\mathfrak{h}^*)$ by T, the group of translations. It can be proved that T acts faithfully on $\mathfrak{h}^*$ with $M \simeq T$ and $W = \mathring{W} \ltimes T$.

Moreover, one can prove that W is isomorphic to $W_{(af)}$, the group of affine transformations of $\mathring{\mathfrak{h}}^*_{\mathbb{R}}$. Here (af) denotes the following isomorphism. (af) is a map (af): $W \to W_{(af)}$ defined by

$$(af)(w)(\bar{\lambda}) = \overline{w(\lambda)} \quad \text{for } \lambda \in \mathfrak{h}^*.$$

Again this map (af) satisfies the following properties:

(i) $(af)(w) = w$ for $w \in \overset{\circ}{W}$.

(ii) $(af)(r_{\alpha_0})(\lambda) = r_\theta(\lambda) + \nu(\theta^\vee)$ for $\lambda \in \overset{\circ}{\mathfrak{h}}{}^*_{\mathbb{R}}$.

(iii) $(af)(t_\alpha)(\lambda) = \lambda + \alpha$ for $\lambda \in \overset{\circ}{\mathfrak{h}}{}^*_{\mathbb{R}}$ and $\alpha \in M$.

Here $W_{(af)}$ is the affine Weyl group of finite-dimensional complex Lie algebra $\overset{\circ}{\mathcal{G}}$.

Moreover, one can prove that

$$W_{(af)} = \overset{\circ}{W} \ltimes (af)(T).$$

For proof, one can refer Kac [8] and Wan [3].

Remark 53. Any element $w \in W$ can be uniquely written as $w = \overset{\circ}{w} t_\alpha$, where $\overset{\circ}{w} \in \overset{\circ}{W}$ and $t_\alpha \in af(T)$.

2.7 Realization of affine Lie algebras

Construction of the GCMs of untwisted affine Lie algebras of type $X_l^{(1)}$ from that of finite type X_l

Let $A = (a_{ij})_{i,j=0}^{l}$ be the GCM of type $X_l^{(1)}$. That is, A is the extended Cartan matrix of the simple finite-dimensional complex Lie algebra $\mathcal{G}(\overset{\circ}{A})$ whose Cartan matrix $\overset{\circ}{A} = (a_{ij})_{i,j=1}^{l}$ is a matrix of finite type X_l, obtained from A by removing the 0th row and 0th column. Let $a_0 = 1, a_1, \ldots, a_l$ be positive integers attached with vertices $0, 1, \ldots, l$ of the Dynkin diagram $S(A)$ of A, respectively. Then $\theta = \sum_{i=1}^{l} a_i\alpha_i$ is the highest root of the simple finite-dimensional complex Lie algebra $\mathcal{G}(\overset{\circ}{A})$.

Let $\overset{\circ}{A} = (a_{ij})_{i,j=1}^{l}$ be a Cartan matrix of type X_l (where $X = A, B, \ldots,$ or G), θ be the highest root of the simple finite-dimensional complex Lie algebra $\mathcal{G}(\overset{\circ}{A})$ and $\theta^\vee = a_1^\vee\alpha_1^\vee + \cdots + a_l^\vee\alpha_l^\vee$ be the highest short root of the Lie algebra $\mathcal{G}(\overset{\circ}{A}{}^t)$. Put $a_{00} = 2, a_{i0} = -(a_1a_{i1} + a_2a_{i2} + \cdots + a_la_{il}), a_{0j} = -(a_1^\vee a_{1j} + a_2^\vee a_{2j} + \cdots + a_l^\vee a_{lj})$. Then $A = (a_{ij})_{i,j=0}^{l}$ is a GCM of type $X_l^{(1)}$.

Example 18. Below we give construction of Cartan matrix of $C_3^{(1)}$ from that of C_3.

We know that the Cartan matrix of C_3 is $\overset{\circ}{A} = \begin{pmatrix} 2 & -1 & 0 \\ -1 & 2 & -2 \\ 0 & -1 & 2 \end{pmatrix} = (a_{ij})_{i,j=1}^{3}$ and its highest root is $2\alpha_1 + 2\alpha_2 + \alpha_3$.

Now $A^t = \begin{pmatrix} 2 & -1 & 0 \\ -1 & 2 & -1 \\ 0 & -2 & 2 \end{pmatrix}$, which is same as the Cartan matrix of B_3 and the highest short root of B_3 is $\alpha_1 + \alpha_2 + \alpha_3$.

Let the corresponding Cartan matrix of $C_3^{(1)}$ be $\begin{pmatrix} a_{00} & a_{10} & a_{20} & a_{30} \\ a_{01} & & & \\ a_{02} & A & & \\ a_{03} & & & \end{pmatrix}$. Here $a_0 = 1, a_1 = 2, a_2 = 2, a_3 = 1, a_1^{\vee} = 1, a_2^{\vee} = 1, a_3^{\vee} = 1$.

We already know that $a_{i0} = -(a_1 a_{i1} + a_2 a_{i2} + \cdots + a_l a_{il})$ and $a_{0j} = -(a_1^{\vee} a_{1j} + a_2^{\vee} a_{2j} + \cdots + a_l^{\vee} a_{lj})$.

Applying the values of a_{ij}'s, a_i's, $a_i^{\vee}$'s in the above equations for a_{i0} and a_{0j}, we have

$$\begin{aligned}
a_{10} &= -(a_1 a_{11} + a_2 a_{12} + a_3 a_{13}) = -(2(2) + 2(-1) + 1(0)) = -2, \\
a_{20} &= -(a_1 a_{21} + a_2 a_{22} + a_3 a_{23}) = -(2(-1) + 2(2) + 1(-2)) = 0, \\
a_{30} &= -(a_1 a_{31} + a_2 a_{32} + a_3 a_{33}) = -(2(0) + 2(-1) + 1(2)) = 0, \\
a_{01} &= -(a_1^{\vee} a_{11} + a_2^{\vee} a_{21} + a_3^{\vee} a_{31}) = -(1(2) + 1(-1) + 1(0)) = -1, \\
a_{02} &= -(a_1^{\vee} a_{12} + a_2^{\vee} a_{22} + a_3^{\vee} a_{32}) = -(1(-1) + 1(2) + 1(-1)) = 0, \\
a_{03} &= -(a_1^{\vee} a_{13} + a_2^{\vee} a_{23} + a_3^{\vee} a_{33}) = -(1(0) + 1(-2) + 1(2)) = 0.
\end{aligned}$$

Hence Cartan matrix of $C_3^{(1)}$ is $\begin{pmatrix} 2 & -1 & 0 & 0 \\ -2 & 2 & -1 & 0 \\ 0 & -1 & 2 & -2 \\ 0 & 0 & -1 & 2 \end{pmatrix}$.

Construction of the nontwisted affine Lie algebras from finite-dimensional Lie algebras

Below we give an explicit construction of the nontwisted affine Lie algebra $\mathcal{G}(A)$, where A is a GCM of $X_l^{(1)}$, where $X = A$ or B or ... or G.

(i) Construction of the Loop algebra: Let $L = \mathbb{C}[t, t^{-1}]$ be the algebra of Laurent polynomials in t and $\mathring{\mathcal{G}} = \mathcal{G}(\mathring{A})$ be the simple finite-dimensional complex Lie algebra with Cartan matrix $\mathring{A}$ of finite type. Now let $L(\mathring{\mathcal{G}}) = L \otimes_c \mathring{\mathcal{G}}$. Then $L(\mathring{\mathcal{G}})$ is an infinite-dimensional complex Lie algebra with bracket $[\ ,\ ]_0$ defined by

$$[t^m \otimes x, t^n \otimes y]_0 = t^{m+n} \otimes [x, y], \qquad \text{for } m, n \in \mathbb{Z};\ x, y \in \mathring{\mathcal{G}}$$

with linearity. If we write $t^n x$ for $t^n \otimes x$, then the above bracket $[\ ,\ ]_0$ becomes $[t^m x, t^n y]_0 = t^{m+n}[x, y]$. Here $[x, y]$ is the Lie algebra bracket in $\mathring{\mathcal{G}}$. Moreover, for a finite-dimensional semisimple complex Lie algebra, there exists an invariant nondegenerate symmetric bilinear $\mathbb{C}$-valued form $(\ ,\ )$ on $\mathring{\mathcal{G}}$ and is unique upto constant multiple. We extend this form to $L(\mathring{\mathcal{G}})$ by $(t^m x, t^n y)_0 = \delta_{m+n,0}(x, y)$ and by linearity. We have

$$([t^m x, t^n y]_0, t^l z)_0 = (t^{m+n}[x,y], t^l z)_0 = \delta_{m+n+l,0}([x,y],z),$$
$$(t^m x, [t^n y, t^l z])_0 = (t^m x, t^{n+l}[y,z])_0 = \delta_{m+n+l,0}(x,[y,z]).$$

Hence $(\, , \,)_0$ is invariant nondegenerate symmetric bilinear form on $L(\overset{\circ}{\mathcal{G}})$. Define a derivation $d = t\frac{d}{dt}$ on $L(\overset{\circ}{\mathcal{G}})$ by $d(t^m x) = mt^m x (m \in \mathbb{Z}, x \in \overset{\circ}{\mathcal{G}})$.

(ii) Construction of an one-dimensional central extension $\tilde{L}(\overset{\circ}{\mathcal{G}})$ of $L(\overset{\circ}{\mathcal{G}})$.

In general, if $\mathcal{G}$ is a Lie algebra, then one can introduce ψ, a 2-cocycle on $\mathcal{G}$, by which, we mean a bilinear $\mathbb{C}$ valued function ψ, ψ: $\mathcal{G} \times \mathcal{G} \to \mathbb{C}$ satisfying the following two conditions:

$$\psi(a,b) = -\psi(b,a), \; a,b \in \mathcal{G} \text{ and}$$
$$\psi([a,b],c) + \psi([b,c],a) + \psi([c,a],b) = 0 \text{ for } a,b,c \in \mathcal{G}.$$

Similarly, for $L(\overset{\circ}{\mathcal{G}})$, one can define ψ: $L(\overset{\circ}{\mathcal{G}}) \times L(\overset{\circ}{\mathcal{G}}) \to \mathbb{C}$ by $\psi(a,b) = (da,b)_0$ $(a,b \in L(\overset{\circ}{\mathcal{G}}))$, where d is defined as in the above construction of Loop algebra. ψ is a 2-cocycle on $L(\overset{\circ}{\mathcal{G}})$ because ψ satisfies the above two conditions.

Now we define an one-dimensional central extension $\tilde{L}(\overset{\circ}{\mathcal{G}})$ of $L(\overset{\circ}{\mathcal{G}})$ by $\tilde{L}(\overset{\circ}{\mathcal{G}}) = L(\overset{\circ}{\mathcal{G}}) \oplus \mathbb{C}c$, where c is the central element.

Using the cocycle ψ defined on $L(\overset{\circ}{\mathcal{G}})$, one can define the Lie bracket on $\tilde{L}(\overset{\circ}{\mathcal{G}})$ as $[a + \lambda c, b + \mu c] = [a,b]_0 + \psi(a,b)c$, for $a, b \in L(\overset{\circ}{\mathcal{G}})$ and $\lambda, \mu \in \mathbb{C}$, where $[\, , \,]_0$ is in $L(\overset{\circ}{\mathcal{G}})$.

It can be proved that $\tilde{L}(\overset{\circ}{\mathcal{G}})$ becomes a Lie algebra with respect to the above-defined bracket.

(iii) Adjoining the derivation d to $\tilde{L}(\overset{\circ}{\mathcal{G}})$: First we construct the semi-direct product by

$$\hat{L}(\overset{\circ}{\mathcal{G}}) = \tilde{L}(\overset{\circ}{\mathcal{G}}) \oplus \mathbb{C}d = L(\overset{\circ}{\mathcal{G}}) \oplus \mathbb{C}c \oplus \mathbb{C}d,$$

where d acts on $L(\overset{\circ}{\mathcal{G}})$ as the derivation defined as above and kills c. The bracket in $\hat{L}(\overset{\circ}{\mathcal{G}})$ becomes $[a + \lambda c + \lambda_1 d, b + \mu c + \mu_1 d] = [a,b]_0 + \psi(a,b)c + \lambda_1 db - \mu_1 da$, where $a, b \in L(\overset{\circ}{\mathcal{G}})$; $\lambda, \lambda_1, \mu, \mu_1 \in \mathbb{C}$.

In particular,

$$[t^m x + \lambda c + \lambda_1 d, t^n y + \mu c + \mu_1 d] = t^{m+n}[x,y] + m\delta_{m+n,0}(x,y)c + \lambda_1 nt^n y - \mu_1 mt^m x$$

for $x, y \in \overset{\circ}{\mathcal{G}}, m, n \in \mathbb{Z}, \lambda, \lambda_1, \mu, \mu_1 \in \mathbb{C}$.

One can directly verify all the required axioms for a Lie algebra $\hat{L}(\overset{\circ}{\mathcal{G}})$. Hence $\hat{L}(\overset{\circ}{\mathcal{G}}) = L(\overset{\circ}{\mathcal{G}}) + \mathbb{C}c + \mathbb{C}d$ is a Lie algebra.

Let $\hat{L}(\overset{\circ}{\mathcal{G}})' = \tilde{L}(\overset{\circ}{\mathcal{G}}) = L(\overset{\circ}{\mathcal{G}}) \oplus \mathbb{C}c$. This implies $\hat{L}(\overset{\circ}{\mathcal{G}})'/\mathbb{C}c = L(\overset{\circ}{\mathcal{G}})$.

Now the following theorem can be directly established.

Theorem 14. *Let $\overset{\circ}{\mathcal{G}}$ be a simple finite-dimensional complex Lie algebra and $A = (a_{ij})_{i,j=0}^n$ be its* extended Cartan matrix *by adding first row and first column in the*

Cartan matrix of $\overset{\circ}{\mathcal{G}}$. Then $\hat{L}(\overset{\circ}{\mathcal{G}})$ is isomorphic to the affine Lie algebra $\mathcal{G}(A)$ associated with the affine matrix A.

Proof. Below we give only an outline of the proof of the above theorem.

(i) Let $\overset{\circ}{\mathfrak{h}}$ be a CSA of $\overset{\circ}{\mathcal{G}}$ and $\overset{\circ}{\Delta} \subset \overset{\circ}{\mathfrak{h}}^*$ be the root system of $\overset{\circ}{\mathcal{G}}$. We have $\overset{\circ}{\mathcal{G}} = \bigoplus_{\alpha \in \overset{\circ}{\Delta} \cup \{0\}} \overset{\circ}{\mathcal{G}}_\alpha$, is the root space decomposition of $\overset{\circ}{\mathcal{G}}$.

We can get $\overset{\circ}{\mathcal{G}} = \eta_-^\circ \oplus \overset{\circ}{\mathfrak{h}} \oplus \eta_+^\circ$, the triangular decomposition of $\overset{\circ}{\mathcal{G}}$, where $\overset{\circ}{\eta}_+ = \bigoplus_{\alpha \in \overset{\circ}{\Delta}_+} \overset{\circ}{\mathcal{G}}_\alpha$, $\eta_-^\circ = \bigoplus_{\alpha \in \overset{\circ}{\Delta}_-} \mathcal{G}_\alpha$.

Let $\{\alpha_1, \alpha_2, \ldots, \alpha_l\}$ be a root basis of $\overset{\circ}{\mathcal{G}}$ and $\{\alpha_1^\vee, \alpha_2^\vee, \ldots, \alpha_l^\vee\}$ be a coroot basis satisfying $\alpha_i^\vee = \frac{2\nu^{-1}(\alpha_i)}{(\alpha_i, \alpha_j)}$. Let θ denote the highest root of $\overset{\circ}{\mathcal{G}}$. Then one can choose $e_\theta \in \overset{\circ}{\mathcal{G}}_\theta$ and $f_\theta \in \overset{\circ}{\mathcal{G}}_{-\theta}$ such that $[e_\theta, f_\theta] = \theta^\vee$, where $\theta^\vee = \frac{2\nu^{-1}(\theta)}{(\theta, \theta)}$ and $2\nu^{-1}(\theta) \in \overset{\circ}{\mathfrak{h}}$ is uniquely determined by the condition $(h, \nu^{-1}(\theta)) = < h, \theta >$ for all $h \in \overset{\circ}{\mathfrak{h}}$. It can be proved that the elements $e_1, \ldots, e_l$ and f_θ generate $\overset{\circ}{\mathcal{G}}$.

(ii) We have $\mathfrak{h} = \overset{\circ}{\mathfrak{h}} + \mathbb{C}c + \mathbb{C}d$ is a $(l+2)$-dimensional commutative subalgebra of $\hat{L}(\overset{\circ}{\mathcal{G}})$. We extend $\lambda \in \overset{\circ}{\mathfrak{h}}^*$ to a linear function on $\mathfrak{h}$ by setting $\langle c, \lambda \rangle = \langle d, \lambda \rangle = 0$. Hence $\overset{\circ}{\mathfrak{h}}^*$ is identified with a subspace in $\mathfrak{h}^*$.

Let δ be a linear function on $\mathfrak{h}$ such that $\delta|_{\overset{\circ}{\mathfrak{h}} + \mathbb{C}c} = 0$, and $\langle d, \delta \rangle = 1$.

Let $e_0 = t \otimes f_\theta = tf_\theta$ and $f_0 = t^{-1} \otimes e_\theta = t^{-1} e_\theta$.

Then it can be proved that

$$[e_0, f_0] = \frac{2}{(\theta,\theta)} c - \theta^\vee \quad \text{and}$$
$$[e_0, f_i] = [e_i, f_0] = 0 \quad \text{for all } i > 0.$$

(iii) Let $\alpha_0 = \delta - \theta$ and $\alpha_0^\vee = \frac{2}{(\theta,\theta)} c - \theta^\vee$.

One can prove that

$$[e_i, f_j] = \delta_{ij} \alpha_i^\vee \qquad (0 \le i, j \le l),$$
$$[h, e_i] = < h, \alpha_i > e_i \qquad (h \in \mathfrak{h} \text{ and } 0 \le i \le l), \text{ and}$$
$$[h, f_i] = - < h, \alpha_i > f_i \qquad (h \in \mathfrak{h} \text{ and } 0 \le i \le l).$$

(iv) Let $\Delta = \Delta^{re} \cup \Delta^{im}$ where $\Delta = \{\alpha + n\delta | \alpha \in \overset{\circ}{\Delta}, n \in \mathbb{Z}\} \cup \{n\delta | n \in \mathbb{Z} \setminus \{0\}\}$ be the root system of $\hat{L}(\overset{\circ}{\mathcal{G}})$. Then it can be proved that $\hat{L}(\overset{\circ}{\mathcal{G}})$ has the following root space decomposition with respect to $\mathfrak{h}$:

$$\hat{L}(\overset{\circ}{\mathcal{G}}) = \mathfrak{h} \oplus \bigoplus_{\alpha \in \Delta} L(\overset{\circ}{\mathcal{G}})_\alpha.$$

Also $L(\overset{\circ}{\mathcal{G}})_{\alpha + n\delta} = t^n \otimes \overset{\circ}{\mathcal{G}}_\alpha$, and $L(\overset{\circ}{\mathcal{G}})_{n\delta} = t^n \otimes \overset{\circ}{\mathfrak{h}}$.

(v) Let $\Pi = \{\alpha_0 = \delta - \theta, \alpha_1, \ldots, \alpha_l\}$ and $\Pi^\vee = \{\alpha_0^\vee = \frac{2}{(\theta,\theta)} c - \theta^\vee, \alpha_1^\vee, \ldots, \alpha_l^\vee\}$. Then $(\mathfrak{h}, \Pi, \Pi^\vee)$ is a minimal realization of A and A is of order $l+1$ and rank l.

We have $2n - l = 2(l+1) - l = l + 2 = \dim \mathfrak{h}$. After computing

$$\langle \alpha_0^\vee, \alpha_0 \rangle = 2,$$
$$\langle \alpha_i^\vee, \alpha_0 \rangle = a_{i0} \quad \text{for } i > 0,$$
$$\langle \alpha_0^\vee, \alpha_j \rangle = a_{0j} \quad \text{for } j > 0,$$

one can get

$$(\langle \alpha_i^\vee, \alpha_j \rangle)_{i,j=0}^{l} = A.$$

So $(\mathfrak{h}, \Pi, \Pi^\vee)$ is a minimal realization of A.

It can be further established that $\hat{L}(\overset{\circ}{\mathcal{G}})$ has no ideals intersecting $\mathfrak{h}$ trivially and $e_i, f_i (0 \leq i \leq l)$ and $\mathfrak{h}$ generate $\hat{L}(\overset{\circ}{\mathcal{G}})$. Hence, it is clear that $\hat{L}(\overset{\circ}{\mathcal{G}})$ is isomorphic to Kac-Moody affine Lie algebra $\mathcal{G}(A)$ for the affine matrix A.

□

Extension of bilinear form from $\overset{\circ}{\mathcal{G}}$ to $\hat{L}(\overset{\circ}{\mathcal{G}})$

Let $\overset{\circ}{\mathcal{G}}$ be a simple finite-dimensional complex Lie algebra. Normalize the Killing form $(\, , \,)$ of $\overset{\circ}{\mathcal{G}}$ to an invariant nondegenerate symmetric bilinear from $(\, , \,)$ by the condition $(\theta, \theta) = 2$. Extend this bilinear form $(\, , \,)$ to a symmetric bilinear form on $\hat{L}(\overset{\circ}{\mathcal{G}})$ by

$$(t^m x, t^n y) = \delta_{m+n,0}(x, y) \quad \text{for } m, n \in \mathbb{Z}, \; x, y \in \overset{\circ}{\mathcal{G}}.$$

Moreover $(\mathbb{C}c + \mathbb{C}d, L(\overset{\circ}{\mathcal{G}})) = 0, (c, c) = (d, d) = 0, (c, d) = 1$. Then $(\, , \,)$ is also invariant and nondegenerate and coincides with the standard invariant nondegenerate symmetric bilinear form on $L(\overset{\circ}{\mathcal{G}})$ defined as follows:

$$(\alpha_i^\vee, \alpha_j^\vee) = a_{ij} a_j \alpha_j^{\vee^{-1}} \quad (i, j = 0, 1, \ldots, l) \text{ and}$$
$$(d, \alpha_i^\vee) = 0 \quad (i = 1, 2, \ldots, l)$$
$$(d, \alpha_0^\vee) = a_0$$
$$(d, d) = 0 \quad \text{with } A = (a_{ij})_{i,j=1}^{l}.$$

Here c is the canonical central element and d is a scaling element.

Triangular decomposition of $\hat{L}(\overset{\circ}{\mathcal{G}})$

$$\hat{L}(\overset{\circ}{\mathcal{G}}) = \eta_- \oplus \mathfrak{h} \oplus \eta_+, \text{ where}$$
$$\eta_- = (t^{-1}\mathbb{C}[t^{-1}] \otimes (\overset{\circ}{\eta}_+ + \overset{\circ}{\mathfrak{h}})) + \mathbb{C}[t^{-1}] \otimes \overset{\circ}{\eta}_-,$$
$$\eta_+ = (t\mathbb{C}[t] \otimes (\overset{\circ}{\eta}_- + \mathfrak{h}^\circ)) + \mathbb{C}[t] \otimes \overset{\circ}{\eta}_+.$$

Chevalley involution of $\hat{L}(\overset{\circ}{\mathcal{G}})$

Let $\overset{\circ}{\omega}$ be the Chevalley involution of $\mathcal{G}^\circ$. Hence $\overset{\circ}{\omega}$ is the involutive automorphism of $\overset{\circ}{\mathcal{G}}$ with $\overset{\circ}{\omega}(e_i) = -f_i, \overset{\circ}{\omega}(f_i) = -e_i (1 \leq i \leq l)$ and $\overset{\circ}{\omega}(h) = -h$ $(h \in \mathfrak{h})$. Then the Chevalley involution ω of $\mathcal{G}(A)$ can be expressed as follows:

$$\omega(p(t) \otimes x + \lambda c + \mu d) = p(t^{-1}) \otimes \overset{\circ}{\omega}(x) - \lambda c - \mu d,$$

where $p(t) \in \mathbb{C}[t, t^{-1}], x \in \overset{\circ}{\mathcal{G}}, \lambda, \mu \in \mathbb{C}$.

Construction of twisted affine Lie algebras from that of finite-dimensional Lie algebras

Let $\mathcal{G}$ be a finite-dimensional complex Lie algebra and σ be the finite-order automorphism of $\mathcal{G}$ satisfying $\sigma^k = 1$ for a positive integer k. Set $\epsilon = \exp(\frac{2\pi i}{k})$, the primitive kth root of unity. Then each eigen value of σ is of the form ϵ^j, for $j \in \mathbb{Z}/k\mathbb{Z}$ and σ is diagonalizable. We have $\mathcal{G} = \bigoplus_{j\in\mathbb{Z}/k\mathbb{Z}} \mathcal{G}_j$ where $\mathcal{G}_j = \{x \in \mathcal{G} \mid \sigma(x) = \epsilon^j x\}$.

Since $[\mathcal{G}_j, \mathcal{G}_i] \subset \mathcal{G}_{j+i(\text{mod } k)}$, we get the $\mathbb{Z}/k\mathbb{Z}$ gradation of $\mathcal{G}$ as given above.

Conversely, if a $\mathbb{Z}/k\mathbb{Z}$ gradation for the Lie algebra $\mathcal{G}$ is given, then the linear transformation of $\mathcal{G}$ defined by multiplying the elements of $\mathcal{G}_j$ by ϵ^j is an automorphism σ of $\mathcal{G}$ of order k.

The following results can be directly proved:

(1) Let $\mathcal{G}$ be a simple finite-dimensional complex Lie algebra with an invariant nondegenerate symmetric bilinear form (,) on $\mathcal{G}$. This bilinear form is also invariant by every automorphism of $\mathcal{G}$, that is $(\sigma(x), \sigma(y)) = (x, y)$.

(2) If σ is the automorphism of finite-order k, then $(\mathcal{G}_i, \mathcal{G}_j) = 0$ if $i + j \not\equiv 0$ (mod$k$) and $\mathcal{G}_i$ and $\mathcal{G}_j$ are nondegenerately paired if $i + j = 0$ (mod$k$). Thus $\dim \mathcal{G}_i = \dim \mathcal{G}_{k-i}$.

Below we explain the diagram automorphism and properties of twisted affine Lie algebra.

(i) Diagram automorphism:

Let $\mathcal{G}$ be a simple finite-dimensional complex Lie algebra of type X_N (where $X = A, B, \ldots$ or G) and $\mathfrak{h}'$ be a Cartan subalgebra of $\mathcal{G}$. Let $\Pi = \{\alpha'_1, \ldots, \alpha'_N\} \subset \mathfrak{h}'^*$ be a root base and $\Pi^\vee = \{\alpha_1^{\vee'}, \ldots, \alpha_N^{\vee'}\} \subset \mathfrak{h}'$ be the corresponding coroot base of $\mathcal{G}$, respectively. We have $\alpha_i^{\vee'} = 2\nu^{-1}(\alpha'_i)/(\alpha'_i, \alpha'_i)$. One can choose Chevalley generators $e'_i, f'_i (i = 1, \ldots, N)$ such that $[e'_i, f'_i] = \alpha_i^{\vee'}$ $(i = 1, \ldots N)$. Let $\bar{\mu}$ be an automorphism of the Dynkin diagram of $\mathcal{G}$, of order $k = 1, 2, 3$. Let μ be an automorphism of $\mathcal{G}$ of order k for $k = 1, 2, 3$ defined by

$$\mu(e'_i) = e'_{\bar{\mu}(i)}, \quad \mu(f'_i) = f'_{\bar{\mu}(i)}, \quad \mu(\alpha_i^{\vee'}) = \alpha_{\bar{\mu}(i)}^{\vee'} \quad (i = 1, \ldots, N).$$

The existence and uniqueness of μ follows from the theory of finite-dimensional semisimple complex Lie algebras. This μ is the *diagram automorphism of* $\mathcal{G}$.

For $X_N = A_l, B_l, \ldots$ or G_2 and $k = 1$, let $E_i = e'_i, F_i = f'_i, H_i = \alpha_i^{\vee'}$.

For $X_N = A_{2l}, A_{2l-1}, D_{l+1}, E_6$ or D_4 and $k = 2, 2, 2, 2$ or 3, respectively, we have the corresponding $\mathbb{Z}/k\mathbb{Z}$-gradation namely,

$$\mathcal{G} = \mathcal{G}_{\bar{0}} \oplus \mathcal{G}_{\bar{1}} \quad \text{if } k = 2,$$
$$\mathcal{G} = \mathcal{G}_{\bar{0}} \oplus \mathcal{G}_{\bar{1}} \oplus \mathcal{G}_{\bar{2}} \quad \text{if } k = 3.$$

Let $\bar{s}$ denote the residue class of $s \in \mathbb{Z}$ mod k. Let $\mathfrak{h}_0$ be the fixed point set of $\mathcal{G}$ in $\mathfrak{h}' = \sum_{i=1}^{n} \mathbb{C}\alpha_i^{\vee'}$. We describe below the representation of $\mathcal{G}_{\bar{0}}$ on $\mathcal{G}_{\bar{1}}$ and $\mathcal{G}_{\bar{2}}$. For proof of the following proposition, one can refer Wan [3].

Proposition 16.

(a) *The Lie algebra $\mathcal{G}_{\bar{0}}$ is simple, with Chevalley generators $E_i, F_i (i \in I)$ and Cartan subalgebra $\mathfrak{h}_0$.*

(b) *The representation of $\mathcal{G}_{\bar{0}}$ on $\mathcal{G}_{\bar{1}}$ is irreducible and is equivalent to the representation on $\mathcal{G}_{\bar{2}}$ if $k = 3$.*

(c) *There exists a unique $\theta_\circ \in \mathfrak{h}_{\bar{0}}^*$ which is a weight of the $\mathcal{G}_0$-module $\mathcal{G}_{\bar{1}}$ such that $\theta_0 + \alpha_i$ is not a weight for all $i \in I$ (θ_0 is called the highest weight of the $\mathcal{G}_{\bar{0}}$-module $\mathcal{G}_{\bar{1}}$).*

(d) *The types of $\mathcal{G}_{\bar{0}}$ and the decomposition $\theta_0 = \sum_{i \in I} a_i \alpha_i$ are listed in the following table:*

k	$\mathcal{G}$	$\mathcal{G}_{\bar{0}}$	**Coefficients a_i's of α_i in θ_0 =(coefficients in the Dynkin diagram of $\mathcal{G}_{\bar{0}}$)**
2	$A_{2l}, l \geq 2$	B_l	$(2, 2, \ldots, 2, 2)$
2	A_2	A_1	(2)
2	$D_{l+1}, l \geq 2$	B_l	$(1, 1, \ldots, 1)$
2	E_6	F_4	$(1, 2, \ldots, 3, 2)$
2	$A_{2l-1} (l \geq 2)$	C_l	$(1, 2, \ldots, 2, 1)$
3	D_4	G_2	$(1, 2)$

Remark 54. In the following $\overset{\circ}{\mathcal{G}}, c, d$ are changed to $\mathcal{G}, c_0, d_0$.

(ii) Extension of diagram automorphism.

Let $\mathcal{G}$ be the finite-dimensional simple complex Lie algebra. Let σ be an automorphism of $\mathcal{G}$ of order k. Extend σ to an automorphism of $L(\mathcal{G})$, denoted by σ again, by setting

$$\sigma(t^j \otimes x) = \epsilon^{-j} t^j \otimes \sigma(x) \quad (j \in \mathbb{Z}, x \in \mathcal{G}),$$

where $t = e^{2\pi i/k}$. Again extend σ to an automorphism $\hat{\sigma}$ of $\hat{L}(\mathcal{G})$ by setting,

$$\hat{\sigma}|_{L(\mathcal{G})} = \sigma, \quad \hat{\sigma}(c_0) = c_0, \quad \hat{\sigma}(d_0) = d_0.$$

Denote the fixed point sets of σ and $\hat{\sigma}$ of $L(\mathcal{G})$ and $\hat{L}(\mathcal{G})$ by $L(\mathcal{G}, \sigma)$ and $\hat{L}(\mathcal{G}, \sigma)$, respectively. Then

$$L(\mathcal{G}, \sigma) = \bigoplus_{j \in \mathbb{Z}} t^j \otimes \mathcal{G}_{\bar{j}}$$

$$\hat{L}(\mathcal{G}, \sigma) = L(\mathcal{G}, \sigma) \oplus \mathbb{C}c_0 \oplus \mathbb{C}d_0.$$

The derived algebra of $\hat{L}(\mathcal{G},\sigma)$ is

$$\tilde{L}(\mathcal{G},\sigma) = L(\mathcal{G},\sigma) \oplus \mathbb{C}c_0.$$

It is clear that $L(\mathcal{G},1) = L(\mathcal{G}), \tilde{L}(\mathcal{G},1) = \tilde{L}(\mathcal{G})$, and $\hat{L}(\mathcal{G},1) = \hat{L}(\mathcal{G})$. It is clear that the decomposition above is a $\mathbb{Z}$-gradation of $L(\mathcal{G},\sigma)$. If we set $\deg c_0 = \deg d_0 = 0$, then we get a $\mathbb{Z}$-gradation of $\hat{L}(\mathcal{G},\sigma)$ as follows:

$$\hat{L}(\mathcal{G},\sigma) = \bigoplus_{j\in\mathbb{Z}} \hat{L}(\mathcal{G},\sigma)_j.$$

Here

$$\hat{L}(\mathcal{G},\sigma)_0 = 1 \otimes \mathcal{G}_{\bar{0}} \oplus \mathbb{C}c_0 \oplus \mathbb{C}d_0 \quad \text{and}$$
$$\hat{L}(\mathcal{G},\sigma)_j = t^j \otimes \mathcal{G}_{\bar{j}} \quad (j \neq 0).$$

The following theorem was proved in Wan [3].

Theorem 15. *Let $\mathcal{G}$ be a finite-dimensional simple complex Lie algebra of type $A_{2l}, A_{2l-1} (l \geq 2), D_{l+1} (l \geq 2), E_6$ or D_4 and σ be a diagram automorphism μ of $\mathcal{G}$ of order 2, 2, 2, 2 or 3, respectively. Then $\hat{L}(\mathcal{G},\mu)$ is isomorphic to the twisted affine Lie algebra $\mathcal{G}(A)$ associated with the affine matrix A of type $A_{2l}^{(2)}, A_{2l-1}^{(2)}, D_{l+1}^{(2)}, E_6^{(2)}$ or $D_4^{(3)}$, respectively.*

We summarize the general construction of the realizations of affine Lie algebras of both nontwisted and twisted cases as follows:

General construction of the affine Lie algebras from that of finite-dimensional Lie algebras (twisted and nontwisted cases)

Let $\mathcal{G}$ be a finite-dimensional, complex, simple Lie algebra with a symmetric, invariant nondegenerate bilinear form $(\cdot,\cdot)$, $\mathfrak{h}$ be a Cartan subalgebra of $\mathcal{G}$ and Δ be the set of roots of $\mathcal{G}$ with respect to $\mathfrak{h}$. Let $\{\alpha_1,\ldots,\alpha_n\}$ be the set of simple roots and $x_\alpha (\alpha \in \Delta)$ be the root vector corresponding to the root α. Let σ be a Dynkin diagram automorphism of $\mathcal{G}$ of order $k (k = 1, 2$ or $3)$. We can extend σ to $\mathfrak{h}$ by linearity and to an automorphism of $\mathcal{G}$ by defining $\sigma(x_\alpha) = x_{\sigma(\alpha)}, \alpha \in \Delta$.

Let $\mathbb{C}[t,t^{-1}]$ be the algebra of Laurent polynomials in the indeterminate t.

Let

$$L(\mathcal{G},\sigma) = \begin{cases} \mathbb{C}[t,t^{-1}] \otimes \mathcal{G} & \text{if order of } \sigma \text{ is } 1; \\ \oplus_{i\in\mathbb{Z}} (t^i \otimes \mathcal{G}_{[i] \bmod k}) & \text{if order of } \sigma \text{ is 2 or 3,} \end{cases}$$

where $\mathcal{G}_{[i]} = \{x \in \mathcal{G}, \sigma(x) = \epsilon^i x\}$, ϵ being the primitive kth root of unity.

Let $\tilde{L}(\mathcal{G},\sigma) = L(\mathcal{G},\sigma) \oplus \mathbb{C}c$ be the one-dimensional central extension of $L(\mathcal{G},\sigma)$, where c is a nonzero central element of $\tilde{L}(\mathcal{G},\sigma) \cdot \tilde{L}(\mathcal{G},\sigma)$ is a Lie algebra with bracket

$$[t^i \otimes x + \lambda c, t^j \otimes y + \mu c] = t^{i+j} \otimes [x,y] + i\delta_{i+j,0}(x,y)c,$$

where $x, y \in \mathcal{G}$ and $\lambda, \mu \in \mathbb{C}$.

Let $\hat{L}(\mathcal{G},\sigma)$ be the Lie algebra obtained by adjoining to $\tilde{L}(\mathcal{G},\sigma)$ a derivation d, which acts on $\hat{L}(\mathcal{G},\sigma)$ as $t\frac{d}{dt}$ and kills c. Therefore, $\hat{L}(\mathcal{G},\sigma) = L(\mathcal{G},\sigma) \oplus \mathbb{C}c \oplus \mathbb{C}d$ with the bracket defined as follows:

$$[t^k \otimes x + \lambda c + \mu d, t^{k_1} \otimes y + \lambda_1 c + \mu_1 d] = t^{k+k_1} \otimes [x,y] + (\mu k_1 y \otimes t^{k_1} - \mu_1 k x \otimes t^k) + k\delta_{k,-k_1}(x,y)c,$$

where $x, y \in \mathcal{G}$, with $\lambda, \mu, \lambda_1, \mu_1 \in \mathbb{C}$. The Lie algebra $\hat{L}(\mathcal{G},\sigma)$ is the affine Kac-Moody Lie algebra associated to the affine GCM, A of type $X_N^{(k)}$ from table $k(k = 1, 2$ or $3)$. Hence we can state the following results. For proof, one can refer Kac [8]:

Theorem 16.

(1) *Let $\mathcal{G}$ be a finite-dimensional complex simple Lie algebra of type $X_N(X = A, B, \ldots, G)$. Let σ be the Dynkin diagram automorphism of $\mathcal{G}$ of order $k(k = 1, 2$ or $3)$. Then the Lie algebra $\hat{L}(\mathcal{G},\sigma)$ is the affine Kac-Moody Lie algebra associated to the affine GCM A of type $X_N^{(k)} = \mathcal{G}^{(k)}$ of order $k(k = 1, 2$ or $3)$ and $\tilde{L}(\mathcal{G},\sigma)$ is isomorphic to $\mathcal{G}'(A)$ and $L(\mathcal{G},\sigma)$ to $\mathcal{G}'(A)/\mathbb{C}c$.*

(2) *Let $\mathcal{G}(A)$ be an affine Lie algebra of rank $l+1$ and let A be of type $X_N^{(k)}$. Then the multiplicity of the imaginary root $nk\delta$ is equal to l and the multiplicity of the imaginary root $s\delta$ for $s \not\equiv 0(\text{mod} k)$ is equal to $(N-l)/(k-1)$.*

(3) *Normalize the Killing form $(\ ,\ )$ of $\mathcal{G}$ to an invariant nondegenerate symmetric bilinear form $(\ ,\ )$ by the condition that the square length of a long root is $2k$. Then*

$$(\theta_0, \theta_0)_0 = 2a_0.$$

Extend $(\ ,\ )_0$ to a symmetric bilinear form on $\hat{L}(\mathcal{G},\mu)$ by

$$(t^m x, t^n y)_0 = \delta_{m+n,0}(x,y), \quad (m, n \in \mathbb{Z}, x \in \mathcal{G}_m, y \in \mathcal{G}_n)$$

$$(\mathbb{C}c_0 + \mathbb{C}d_0, L(\mathcal{G},\mu))_0 = 0; \quad (c_0, c_0)_0 = (d_0, d_0)_0 = 0; \quad (c_0, d_0)_0 = a_0.$$

Moreover, c_0 is the canonical central element and d_0 is a scaling element. Then $(\ ,\ )_0$ is also invariant and nondegenerate and coincides with the standard invariant nondegenerate symmetric bilinear form on $\mathcal{G}(A)$ introduced earlier and defined as in nontwisted case.

Remark 55. Similar to the construction of realization of affine Lie algebras, one can construct toroidal Lie algebras. For details regarding the toroidal Lie algebras and their representations, one can refer Eswara Rao [85–87].

2.8 Different classes of imaginary roots (special imaginary roots, strictly imaginary roots, purely imaginary roots) in Kac-Moody algebras

2.8.1 Special imaginary roots in Kac-Moody algebras [54]

Definition 91. In a Kac-Moody algebra, an imaginary root α will be called *special*, if $(\alpha, \alpha) \neq 0$ and the following conditions are satisfied

(i) The reflection r_α of the root lattice fixes the lattice Δ, Δ^{re}, and Δ^{im} and
(ii) r_α preserves root multiplicities.

Let W be the Coxeter group associated to A, where A is the GCM of indefinite type. In order to define r_α, it is necessary to assume that A is symmetrizable. The reflection r_α must be an element of $-W$ or an element of $-W$ multiplied by a diagram automorphism. Moreover, if $r_\alpha \in -W$ and α is an imaginary root, then α is special.

Theorem 17. *On special imaginary roots, Bennett [54] found two large classes of Kac-Moody algebras having special imaginary roots.*

(1) *Given an $n \times n$ GCM A of indefinite type with associated Kac-Moody Lie algebra $\mathcal{G}(A)$, if σ is a permutation of $\{1, \ldots, n\}$ fixing n and the matrix A (where σ acts on A by permuting rows and columns), and if in addition A_n (the submatrix of A obtained by removing the nth row and nth column) is a Cartan matrix of finite type with $\sigma|_{\{1,\ldots,n-1\}}$ equal to the permutation of $\{1, \ldots, n-1\}$ induced by the longest word ω_0 of the Coxeter group associated to A_n, then there exists a special imaginary root α of $\mathcal{G}(A)$. Moreover, α can be chosen so that $r_\alpha = -\sigma\omega_0$, where ω_0 is the element of maximal length in the Coxeter group associated to A_n.*
(2) *Let $\mathcal{D}$ be a Dynkin diagram on n vertices of indefinite type such that the subdiagram $\mathcal{D}'$ of $\mathcal{D}$ generated by the elements of $\{1, \ldots, n-2\}$ is spherical (finite). Suppose, further, that ω_0 is the longest element of the Coxeter group associated to $\mathcal{D}'$, then if the diagram automorphism $\sigma = -\omega_0$ of $\mathcal{D}'$ extends to an automorphism $\tilde{\sigma}$ of $\mathcal{D}$ interchanging $n-1$ and n, there exists a special imaginary root α of $\mathcal{G}(\mathcal{D})$ such that the reflection over α is equal to $-\tilde{\sigma}\omega_0$.*
(3) *If A is a GCM of weak hyperbolic type with associated Dynkin diagram not equal to*  *and having no infinity subdiagrams, then $\mathcal{G}(A)$ has special imaginary roots.*
(4) *Complete classification of special imaginary roots: Of course, by the definition, if α is special, so is $-\alpha$ (in fact, any root of the form $\kappa\alpha$), hence the following result gives a complete classification.*

If α is a positive special imaginary root of the Kac-Moody Lie algebra $\mathcal{G}(A)$ (with A indecomposable) then, α is W-equivalent to a special imaginary root that arises either as in Result 1 or as in Result 2 of Theorem 17.

Example 19.

(1) $\alpha = a\alpha_1 + 2\alpha_2$ is a special imaginary root for $\mathcal{G}(A)$, where $A = \begin{pmatrix} 2 & -a \\ -b & 2 \end{pmatrix}$.
(2) $\alpha = 2\alpha_1 + 2\alpha_2 + \alpha_3 + \alpha_4$ is a special imaginary root for $\mathcal{G}(A)$, where $A = \begin{pmatrix} 2 & -1 & -1 & -1 \\ -1 & 2 & -1 & -1 \\ -1 & -1 & 2 & -1 \\ -1 & -1 & -1 & 2 \end{pmatrix}$.

2.8.2 Strictly imaginary roots in Kac-Moody algebras

The definition of strictly imaginary roots was introduced by Kac [8] for which, a complete characterization of all Kac-Moody algebras possessing strictly imaginary property was given by Casperson [57].

Definition 92. A root $\gamma \in \Delta^{im}$ is said to be strictly imaginary if for every $\alpha \in \Delta^{re}$, either $\gamma+\alpha$ or $\gamma-\alpha$ is a root. The set of all strictly imaginary roots is denoted by Δ^{sim}.

One can directly prove the following properties regarding Δ^{sim}:

(a) If $\alpha \in \Delta_+$ and $\langle \alpha_i^\vee, \alpha \rangle < 0$ for all i, then $\alpha \in \Delta_+^{sim}$.
(b) If $\alpha \in \Delta_+^{im}$, $r_\gamma(\alpha) \neq \alpha$ for all $\gamma \in \Delta^{re}$, then $\alpha \in \Delta_+^{sim}$.
(c) If $\alpha \in \Delta_+^{sim}$ and $\langle \alpha_i^\vee, \alpha \rangle \leq 0$ for all i, then $\alpha + \beta \in \Delta_+$ for all $\beta \in \Delta_+$.
(d) If $\alpha \in \Delta_+^{sim}, \beta \in \Delta_+^{im}$, then $\alpha + \beta \in \Delta_+^{im}$.
(e) Δ_+^{sim} is a semigroup.

Remark 56. We state below some known results, namely, Theorems 18, 19, and 20, Corollary 2, and Proposition 17, proved in Casperson [57].

Theorem 18. *If $\gamma \in K \setminus \Delta_+^{sim}$, then for some simple root $\alpha_i \in \Pi$, neither $\gamma + \alpha_i$ nor $\gamma - \alpha_i$ is a root.*

Definition 93. A GCM, A is said to have SIM property, if

$$\Delta_+^{sim}(A) = \Delta_+^{im}(A).$$

A is said to satisfy $NC1$, if there exists no subsets $S, T \subset \{1, \ldots, n\}$ such that $A|_S$ is affine or indefinite type and $A|_{S \cup T}$ is decomposable.

Theorem 19. *A GCM lies in SIM if and only if it satisfies the condition NC1 and has no principal submatrix contained in the following list:*

(1) *The 2×2 matrices of the form $\begin{pmatrix} 2 & -a \\ -b & 2 \end{pmatrix}$ with $ab \geq 4$ and $a = 1$ or $b = 1$.*
(2) *The matrices of the following twisted Kac-Moody algebras of affine type: $A_2^{(2)}, A_{2n}^{(2)}, A_{2n+1}^{(2)}, D_n^{(2)}, E_n^{(2)}$, and $D_4^{(3)}$.*
(3) *The strictly hyperbolic 3×3 matrices associated with the Dynkin diagrams of the form:*

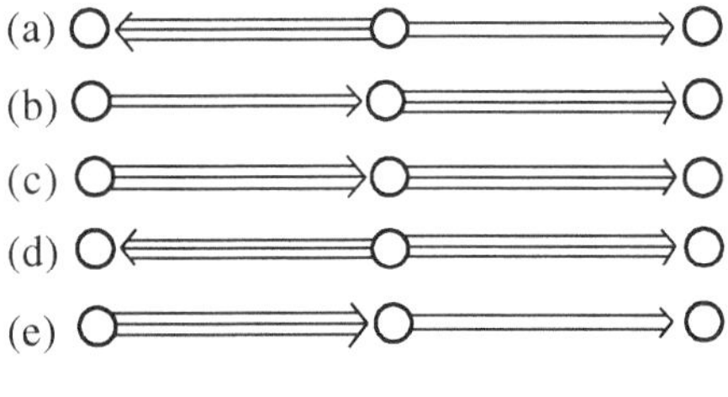

(4) *The hyperbolic 4×4 matrices associated with the Dynkin diagrams of the form:*

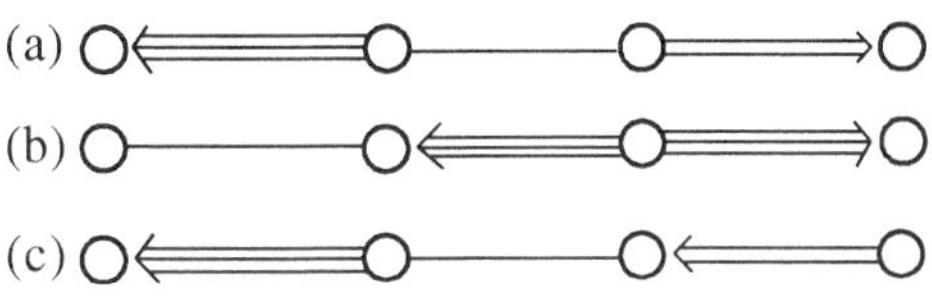

If an algebra is affine, we have that $\Delta^{sim} = \{nk\delta | n \in \mathbb{Z} \setminus \{0\}\}$ and hence $\Delta^{im} \setminus \Delta^{sim} = \{n\delta |\ n \in \mathbb{Z} \setminus k\mathbb{Z}\}$, where δ is the unique minimal positive imaginary root and k is the order of the diagram automorphism used to construct the algebra.

The following theorem gives a complete description of the nonstrictly imaginary roots of 2×2 matrices.

Theorem 20. *Suppose, for the GCM,* $A = \begin{pmatrix} 2 & -a \\ -b & 2 \end{pmatrix}$, $(a, b \in \mathbb{Z}_+)$, *that* $\Gamma_1(A) = \{\gamma \in K_A\colon \gamma \pm \alpha_1 \notin \Delta(A)\} \neq \emptyset$.
Then, either

(1) $A = \begin{pmatrix} 2 & -1 \\ -4 & 2 \end{pmatrix}$, *and* $\Gamma_1(A) = \{(2n+1)(\alpha_1 + 2\alpha_2) | \, n \geq 0\}$ *(or)*

(2) $A = \begin{pmatrix} 2 & -1 \\ -b & 2 \end{pmatrix}$, *where* $b > 4$ *and* $\Gamma_1(A) = \{\alpha_1 + 2\alpha_2\}$.

Corollary 2. *The GCM* $\begin{pmatrix} 2 & -a \\ -b & 2 \end{pmatrix}$ *is not in SIM if and only if* $ab \geq 4$ *and either* $a = 1$ *or* $b = 1$.

Proposition 17. *A GCM of affine type is a member of SIM if and only if it is of nontwisted affine type.*

2.8.3 Purely imaginary roots in Kac-Moody algebras

The definition of purely imaginary roots in Kac Moody algebras was introduced in Sthanumoorthy and Uma Maheswari [79] and therein complete classification of Kac-Moody algebras possessing purely imaginary property was also given.

From the property (*d*) of strictly imaginary roots, we have $\Delta_+^{sim} + \Delta_+^{im} \subset \Delta_+^{im}$. But it can be easily established that $\Delta_+^{im} + \Delta_+^{im} \not\subset \Delta_+^{im}$. Here "+" between the sets denotes the usual addition of roots from the sets.

Example 20. In the following diagram

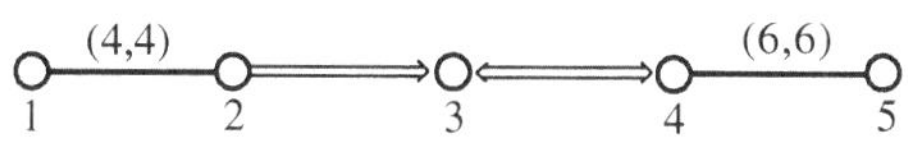

we can easily check that $\alpha_1 + \alpha_2$ and $\alpha_4 + \alpha_5$ are imaginary roots whereas $\alpha_1 + \alpha_2 + \alpha_4 + \alpha_5$ is not a root, because *supp* $(\alpha_1 + \alpha_2 + \alpha_4 + \alpha_5)$ is not connected.

Remark 57. Note that if α and β are imaginary roots and if $\alpha + \beta$ is a root then $\alpha + \beta$ must be imaginary. Hence, one can try to collect all those imaginary roots for a given Kac-Moody algebra having the property that when added with another imaginary root gives an imaginary root. All such imaginary roots are called *purely imaginary roots.*

Definition 94. Let $\alpha \in \Delta_+^{im}$. Then α is called a *purely imaginary*, if for any $\beta \in \Delta_+^{im}$, $\alpha + \beta \in \Delta_+^{im}$. Similarly, a negative root $\gamma \in \Delta_-^{im}$ is purely imaginary if $-\gamma$ is a purely imaginary root. Denote by $\Delta_+^{pim}(A) = \Delta_+^{pim} = \{\alpha \in \Delta_+^{im} | \, \alpha$ is purely imaginary$\}$ and $\Delta_-^{pim}(A) = \Delta_-^{pim} = \{\alpha \in \Delta_-^{im} | \, \alpha$ is purely imaginary$\}$. The set of all purely imaginary roots is

$$\Delta^{pim} = \Delta_+^{pim} \cup \Delta_-^{pim}.$$

Remark 58.

(1) The symmetry of the root system means that we need to prove any results regarding purely imaginary roots only for positive imaginary roots.
(2) From the property (*d*) of the strictly imaginary roots, it is clear that every positive strictly imaginary root is a purely imaginary root. But the converse is not true.

Example 21. Consider the hyperbolic Kac-Moody algebra associated with the GCM,

$$\begin{pmatrix} 2 & -1 \\ -5 & 2 \end{pmatrix}.$$

Here, $\alpha = \alpha_1 + 2\alpha_2$ is a purely imaginary root, whereas α is not a strictly imaginary root because neither $\alpha + \alpha_1$ nor $\alpha - \alpha_1$ is a root.

Definition 95. We say that a GCM A satisfies the *purely imaginary property*, if $\Delta_+^{pim}(A) = \Delta_+^{im}(A)$. We say that the Kac-Moody algebra $\mathcal{G}(A)$ has the *purely imaginary property*, if A satisfies the purely imaginary property.

Properties of Kac-Moody algebras possessing purely imaginary roots

For the proof of the following results and Theorems 21, 22, and 23, one can refer Sthanumoorthy and Uma Maheswari [79].

Results:

(1) Δ_+^{pim} is a semigroup.
(2) $W \cdot \Delta_+^{pim} = \Delta_+^{pim}$.
(3) Let $A = (a_{ij})_{i,j=1}^n$ be an indecomposable GCM. If $\mathcal{G}(A)$ is an affine or a symmetrizable hyperbolic Kac-Moody algebra, then $\Delta_+^{im}(A) = \Delta_+^{pim}(A)$.
(4) Let A be an indecomposable, symmetrizable GCM. Let $\alpha,\ \beta \in \Delta_+^{im}$ be such that *supp* $(\alpha + \beta)$ is connected. Then, $\alpha + \beta \in \Delta_+^{im}$.
(5) If $\alpha \in \Delta_+^{im}, \alpha = \sum_{i=1}^n k_i\alpha_i$, then atleast two of the k_i's are > 0.
(6) Let $A = (a_{ij})_{i,j=1}^n$ be an indecomposable symmetrizable GCM. Suppose $\alpha \in \Delta_+^{im}$ is such that *supp* α contains atleast $n-2$ vertices, then $\alpha \in \Delta_+^{pim}$.
(7) Let $A = (a_{ij})_{i,j=1}^n$ be an indecomposable symmetrizable GCM with $n \leq 4$. Then, $\Delta_+^{pim}(A) = \Delta_+^{im}(A)$.
(8) Let $A = (a_{ij})_{i,j=1}^n$ be an indecomposable symmetrizable GCM such that $a_{ij} \neq 0$ for all $i,j \in \mathbb{N}$. Then, $\Delta_+^{pim} = \Delta_+^{im}$.

In the following theorem, we give a necessary and sufficient condition for any $\alpha \in \Delta_+^{im}$, to be purely imaginary.

Theorem 21. *Let A be an indecomposable symmetrizable GCM and let $\alpha \in \Delta_+^{im}(A)$. Then, $\alpha \in \Delta_+^{pim}$ if and only if every connected subdiagram of $S(A)$, which is of affine or indefinite type, is connected to atleast one of the vertices of supp α.*

The following theorem gives a complete classification of all those Kac-Moody algebras possessing purely imaginary property.

Theorem 22. *An indecomposable symmetrizable GCM, $A = (a_{ij})_{i,j=1}^n$ satisfies the purely imaginary property if and only if for any two subdiagrams $S_1(A)$ and $S_2(A)$ of $S(A)$, each one being a connected subdiagram of nonfinite type, there vertices $i \in S_1(A)$ and $j \in S_2(A)$ such that i and j are connected.*

Remark 59. For all indecomposable, symmetrizable GCM, $A = (a_{ij})_{i,j=1}^n$, we have the following properties:

(i) When $n \leq 4$, $\Delta_+^{im}(A) = \Delta_+^{pim}(A)$.
(ii) When $n \geq 5$, there exist algebras which satisfy the purely imaginary property and there also exist algebras which do not satisfy the purely imaginary property.

For $n = 5$, Theorems 21 and 22 give algebras possessing the purely imaginary property and the following diagram gives an algebra not possessing purely imaginary property.

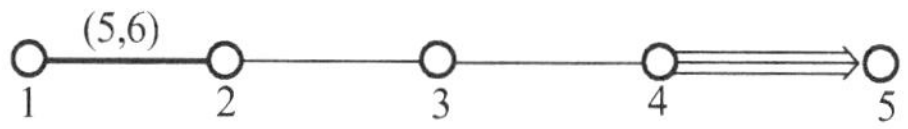

Here, $\alpha_1 + \alpha_2$ and $\alpha_4 + \alpha_5$ are imaginary roots whereas $\alpha_1 + \alpha_2 + \alpha_4 + \alpha_5$ is not a root.

Theorem 23. *Every indecomposable, symmetrizable extended-hyperbolic Kac-Moody algebra $\mathcal{G}(A)$ satisfies the purely imaginary property.*

Complete classification of Kac-Moody algebras possessing purely imaginary property

In Theorems 21 and 22 we have given a necessary and sufficient condition for an imaginary root to be purely imaginary and thereby giving a complete classification of those Kac-Moody algebras with the purely imaginary property, which includes, in particular, the following:

(1) Kac-Moody algebras $\mathcal{G}(A)$ of finite type ($\Delta_+^{im}(A) = \Delta_+^{pim}(A) = \emptyset$).
(2) Affine Kac-Moody algebras.
(3) Symmetrizable hyperbolic Kac-Moody algebras.
(4) Symmetrizable GCM, $A = (a_{ij})_{i,j=1}^n$ for $n \leq 4$.
(5) Symmetrizable extended-hyperbolic, indefinite type Kac-Moody algebras.
(6) Symmetrizable Kac-Moody algebras whose Dynkin diagrams are completely connected.

Remark 60. From the complete classification, one can understand that the Kac-Moody algebras which do not possess the purely imaginary property are in the class of indefinite Kac-Moody algebras other than the hyperbolic and extended-hyperbolic algebras and can also identify the algebras possessing and not possessing this property.

2.9 Representations of Kac-Moody algebras, integrable highest weight modules, Verma modules, and character formulas

Representation of a Kac-Moody algebra $\mathcal{G}(A)$ and a $\mathcal{G}(A)$-module V

Let A be a complex $n\times n$ matrix and $\mathcal{G}(A)$ be the complex Lie algebra over $\mathbb{C}$ associated with A. Let V be a vector space over $\mathbb{C}$ of finite or infinite dimension and $gl(V)$ be the set of all linear transformations on V. The Lie algebra homomorphism $\pi\colon \mathcal{G}(A) \to gl(V)$ is called a representation of $\mathcal{G}$ in V or we say that (V, π) is a representation of $\mathcal{G}(A)$.

If we define, $x\cdot v = \pi(x)v$, for $x \in \mathcal{G}, v \in V$, then V is a $\mathcal{G}(A)$-module, that is, V is a vector space over $\mathbb{C}$ with a module operation $\mathcal{G}(A) \times V \to V$: $(x, v) \to x\cdot v$, defined as

$$(ax+by)\cdot v = a(x\cdot v) + b(y\cdot v) \quad \text{for } x, y \in \mathcal{G}(A),\ a, b \in \mathbb{C},\ v \in V \text{ and}$$

$$[x,y]\cdot v = x\cdot(y\cdot v) - y\cdot(x\cdot v) \quad \text{for } x, y \in \mathcal{G}(A),\ v \in V.$$

Conversely, for a $\mathcal{G}(A)$-module V, one can obtain a representation $\pi\colon \mathcal{G}(A) \to gl(V)$, if we define $\pi(x)v = x\cdot v (x \in \mathcal{G}(A), v \in V)$.

Definition 96. A $\mathcal{G}(A)$-module V is called *h diagonalizable* if it admits a weightspace decomposition $V = \bigoplus_{\lambda\in\mathfrak{h}^*} V_\lambda$ by weight spaces V_λ. Here $V_\lambda = \{v \in V | h\cdot v =< h, \lambda > v = \lambda(h)v, \forall h \in \mathfrak{h}\}$. Clearly, V_λ is a subspace of V. A $\mathfrak{h}$-diagonalizable $\mathcal{G}(A)$-module is called *integrable*, if all e_i and $f_i (i = 1, \ldots, n)$ are locally nilpotent on V. One says that an element $x \in \mathcal{G}$ is locally nilpotent on V if for any $v \in V$, there exists an integer N such that $(x)^N(v) = 0$.

Definition 97. If $V_\lambda \neq 0, \lambda$ is called a *weight* of the $\mathcal{G}(A)$-module V. V_λ is the *weight space* corresponding to the weight λ and $\dim V_\lambda$ is the *multiplicity* of the weight λ, denoted by $mult_V\lambda$. Denote by $P(V) = \{\lambda \in \mathfrak{h}^* | V_\lambda \neq 0\}$, the set of weights of the $\mathcal{G}(A)$-module V.

For any $\lambda \in \mathfrak{h}^*$, let $D(\lambda) = \{\lambda - \alpha | \alpha \in Q_+\}$. For any subset $F \subset \mathfrak{h}^*$, let $D(F) = \cup_{\lambda\in F} D(\lambda)$. Here F is always finite. Now define a partial ordering $\geq$ on $\mathfrak{h}^*$ by $\lambda \geq \mu \Leftrightarrow \lambda - \mu \in Q_+ \Leftrightarrow \mu \in D(\lambda)$.

Definition 98 (Category $\mathcal{O}$-modules). Category $\mathcal{O}$ is the category whose objects are $\mathcal{G}(A)$ modules V satisfying the following conditions

(1) V is $\mathfrak{h}$-diagonalizable, that is $V = \oplus_{\lambda\in\mathfrak{h}^*} V_\lambda$,
(2) $\dim V_\lambda < \infty$ for all $\lambda \in \mathfrak{h}^*$, and
(3) there exists a finite set $F \subset \mathfrak{h}^*$ such that $P(V) \subset D(F)$ and whose morphisms are $\mathcal{G}(A)$-module homomorphisms.

It can be established that every nonzero $\mathcal{G}(A)$-module in $\mathcal{O}$ has at least one maximal weight.

Let A be a complex $n \times n$ matrix and $\mathcal{G}(A)$ be the associated Lie algebra with triangular decomposition $\mathcal{G}(A) = \eta_- \oplus \mathfrak{h} \oplus \eta_+$, where $\eta_-, \mathfrak{h}, \eta_+$ are generated by e_i, h_i, and f_i, respectively. Then the corresponding triangular decomposition of universal enveloping Lie algebra is $U(\mathcal{G}(A)) = U(\eta_-) \oplus U(\mathfrak{h}) \oplus U(\eta_+)$.

Definition 99 (Highest weight modules). A $\mathcal{G}(A)$-module V is called *a highest weight module*, if

(i) V has a unique maximal weight λ and
(ii) V as a $\mathcal{G}(A)$-module, is generated by some $v_\lambda \in V_\Lambda$.

Hence a *weight module* V is a *highest weight module* of highest weight $\lambda \in \mathfrak{h}^*$, if there exists a nonzero vector $v_\lambda \in V$, called *highest weight vector* such that $e_i v_\lambda = 0$ for all i, $hv_\lambda = \lambda(h)v_\lambda$ for $h \in \mathfrak{h}$ and $V = U(\mathcal{G})v_\lambda$, where $U(\mathcal{G})$ is the universal enveloping algebra of $\mathcal{G}$. A vector $v \in V_\lambda$ is called *primitive* if there exists a submodule U in V such that $v \notin U$, $\eta_+(v) \subset U$. Here λ is called a primitive weight. For the proof of the following results, one can refer Kac [8].

Proposition 18. *Let V be a nonzero module from the category $\mathcal{O}$. Then*

(a) *V contains a nonzero weight vector v such that* $\eta_+(v) = 0$.
(b) *The following are equivalent:*
  (i) *V is irreducible.*
  (ii) *V is a highest weight module and any primitive vector of V is highest weight vector.*
(c) *V is generated by its primitive vectors as* $\mathcal{G}(A)$*-modules.*

Properties of highest weight module with maximal weight

A highest weight module V with maximal weight $\wedge$ has the following properties (for proof, one can refer Wan [3]):

(i) $V = U(\mathcal{G}(A)) \cdot v_\wedge = U(\eta_-) \cdot v_\wedge$ for any weight vector $v_\wedge$ of weight $\wedge$.
(ii) $V \in \mathcal{O}$.
(iii) $\dim V_\wedge = 1$.
(iv) $P(V) \subset D(\wedge)$; that is $P(V) \subset \{\wedge - \sum_{i=0}^{l} m_i\alpha_i | m_i \in \mathbb{Z}_{\geq 0}\}$.
(v) $\mathcal{G}_\alpha \cdot V_\wedge = (0)$ for any $\alpha \in \Delta_+$.
(vi) V has a unique maximal submodule, hence a unique simple quotient module.
(vii) If V'' is a nonzero homomorphic image of V, then V'' is again a highest weight module.

Definition 100 (Verma module). A $\mathcal{G}(A)$-module $M(\wedge)$ with highest weight $\wedge$ is called a *Verma module* if every $\mathcal{G}(A)$ module with highest weight $\wedge$ is a quotient of $M(\wedge)$. Fix $\wedge \in \mathfrak{h}^*$. Let $J(\wedge)$ be the left ideal of $U(\mathcal{G})$ generated by all e_i, $(i \in I)$ and $h - \wedge(h)$ $(h \in \mathfrak{h})$. Let $M(\wedge) = U(\mathcal{G})/J(\wedge)$.

Alternatively, one can define Verma module as follows [3]: Let $\wedge \in \mathfrak{h}^*$ and $V_\wedge = \mathbb{C}u_\wedge$ be the one-dimensional $\mathfrak{h}$-module corresponding to $\wedge$. So it is a one-dimensional complex vector space with the action of $\mathfrak{h}$ on $V_\wedge$, defined by $h \cdot u_\wedge = < h, \wedge > u_\wedge$ for all $h \in \mathfrak{h}$. If $b = \mathfrak{h} \oplus \eta_+$, then η_+ is an ideal of b and the sequence

$$0 \to \eta_+ \to b \to \mathfrak{h} \to 0$$

is exact. So $V_\wedge$ is a $\mathfrak{h}$-module with η_+ acting on it trivially, that is $x \cdot v_\wedge = 0$ for all $x \in \eta_+$. Now define

$$M(\wedge) = Ind_{\mathfrak{h}}^{\mathcal{G}(A)} V_{\wedge} = U(\mathcal{G}(A)) \otimes_{U(h)} V_{\wedge}.$$

So $M(\wedge)$ is a $\mathcal{G}(A)$-module, called *the Verma module* corresponding to the weight $\wedge$.

Verma module has the following properties

(1) $M(\Lambda)$ is a highest weight $\mathcal{G}$-module with highest weight Λ and highest weight vector $v_{\wedge} = 1 + \xi(\wedge)$. Every highest weight $\mathcal{G}$(A)-module with highest weight Λ is a homomorphic image of $M(\Lambda)$.
(2) $M(\wedge)$ has a unique maximal submodule with simple quotient $L(\wedge)$. If V is an irreducible highest weight module with highest weight $\wedge$, then $V \simeq L(\wedge)$. $L(\wedge) \simeq L(\wedge')$ if and only if $\wedge = \wedge'$.
(3) As a $U(\eta_-)$-module, $M(\wedge)$ is free of rank 1, generated by the highest weight vector.
(4) $M(\wedge)$ has a unique proper maximal submodule.
(5) If V is a nonzero module from the category $\mathcal{O}$, then $V \simeq L(\wedge)$ for some $\wedge \in \mathfrak{h}^*$.

For the proof of the above properties, one can refer to Wan [3] and Kac [8].

Remark 61. Among the modules with highest weight $\wedge$, there is a unique irreducible one, namely, the module $L(\wedge) = M(\wedge)/M'(\wedge)$. Clearly $L(\wedge)$ is a quotient of any module with highest weight $\wedge$.

Definition 101 (Filtration). Let $V \in \mathcal{O}$ and $\lambda \in \mathfrak{h}^*$. Then there exists a filtration by a sequence of submodules $V = V_t \supset V_{t-1} \supset \cdots \supset V_1 \supset V_0 = 0$ and a subset $J \subset \{1, \ldots, t\}$ such that

(a) if $j \in J$, then $V_j/V_{j-1} \simeq L(\lambda_j)$ for some $\lambda_j \geq \lambda$ and
(b) if $j \notin J$, then $(V_j/V_{j-1})_\mu = 0$ for every $\mu \geq \lambda$.

Every irreducible module in the category $\mathcal{O}$ is isomorphic to $V(\lambda)$ for some $\lambda \in \mathfrak{h}^*$.

Let $V \in \mathcal{O}$ and $\lambda \in \mathfrak{h}^*$ (a weight of V). Construct the filtration of V given by $V = V_t \supset V_{t-1} \supset \cdots \supset V_1 \supset V_0 = 0$. Denote by $[V\colon L(\lambda)]$, the number of times λ appears among $\{\lambda_j | j \in J\}$]. Here $[V\colon L(\lambda)]$ is called *the multiplicity* of $L(\lambda) \in V$.

Then the following results can be established [3].

(a) The number $[V\colon L(\lambda)]$, defined above, is independent of the filtration of V: $V = V_t \supset V_{t-1} \supset \cdots \supset V_0 = 0$ with the property of filtration as defined above.
(b) $L(\lambda)$ has a nonzero multiplicity in V if and only if λ is a primitive weight.

Definition 102. Let V be a $\mathcal{G}$(A)-module belonging to the category $\mathcal{O}$. *The formal character* of V is defined by

$$ch\ V = \sum_{\lambda \in P(V)} (\dim V_\lambda) e^\lambda,$$

where e^λ are formal exponentials with $e^\lambda \circ e^\mu = e^{\lambda+\mu} (\lambda, \mu \in \mathfrak{h}^*)$ and $e^0 = 1$.

Some properties of formal characters

(1) If $\mathcal{O} \to V' \to V \to V''$ is an exact sequence in $\mathcal{O}$, then $chV = chV' + chV''$.
(2) Character of Verma module $M(\Lambda)$: Character of $M(\wedge)$ for $\wedge \in \mathfrak{h}^*$, is $ch\, M(\wedge) = e^{\wedge}/\Pi$, where $\Pi = \prod_{\alpha \in \Delta_+} (1 - e^{\alpha})^{m_\alpha}$, with $m_\alpha = \dim \mathcal{G}_\alpha$.

For the proof of the following properties, one can refer Kac [8].

(1) Let V be a $\mathcal{G}(A)$ module from the category $\mathcal{O}$. Then $ch\ V = \sum_{\lambda \in \mathfrak{h}^*} [V{:}\ L(\lambda)] ch\ L(\lambda)$.
(2) Let V be a $\mathcal{G}(A)$ module with highest weight $\wedge$. Then

$$ch\ V = \sum_{\substack{\lambda \leq \wedge \\ |\lambda+\rho|^2 = |\wedge+\rho|^2}} c_\lambda\ ch\ M(\lambda),$$

where $c_\lambda \in \mathbb{Z}$ and $c_\Lambda = 1$.

Properties of integrable highest weight modules

Let A be a GCM and $\mathcal{G}(A)$ be the corresponding KM algebra with Weyl group W. If V is an integrable $\mathcal{G}(A)$-module then $\dim V_\lambda = \dim V_{w(\lambda)}$ for all $\lambda \in \mathfrak{h}^*$ and $w \in W$. Moreover, set of weights of V is also W-invariant. Let $V(\lambda)$ be a highest weight $\mathcal{G}(A)$-module with v_λ a highest weight vector. Then $V(\lambda)$ is integrable if and only if $f_i^{N_i} \cdot v_\lambda = 0$ for some $N_i > 0$ $(i = 1, 2, \ldots, n)$.

Definition 103. For a Kac-Moody algebra $\mathcal{G}(A)$ of rank n with Cartan subalgebra $\mathfrak{h}$, one can define the following sets:

$$\begin{aligned} P &= \{\lambda \in \mathfrak{h}^* | \langle \alpha_i^\vee, \lambda \rangle \in \mathbb{Z} \quad (i = 1, 2, \ldots, n)\}, \\ P_+ &= \{\lambda \in P | \langle \alpha_i^\vee, \lambda \rangle \geq 0 \quad (i = 1, 2, \ldots, n)\}, \quad \text{and} \\ P_{++} &= \{\lambda \in P | \langle \alpha_i^\vee, \lambda \rangle > 0 \quad (i = 1, 2, \ldots, n)\}. \end{aligned}$$

Elements in P are called *integrals or integral weights*, elements in P_+ are called *dominant integrals or dominant integral weights* and elements in P_{++} are called *regular dominant integrals or regular dominant weights*.

Remark 62. It can be proved that the irreducible highest weight $\mathcal{G}(A)$-module $L(\Lambda)$ is integrable if and only if $\lambda \in P_+$.

Definition 104 (Weyl character formula). Let $\mathcal{G}(A)$ be the Kac-Moody algebra associated with a symmetrizable GCM A and $L(\Lambda)$ be the irreducible highest weight $\mathcal{G}(A)$-module with highest weight Λ. Then in the usual notation one can prove that

$$ch\ L(\Lambda) = \frac{\sum_{w \in W} (det\ w) e^{w(\Lambda+\rho)-\rho}}{\prod_{\alpha \in \Delta_+} (1 - e^{-\alpha})^{mult(\alpha)}},$$

called the *Weyl-Kac character formula*. Here $\rho \in \mathfrak{h}^*$ is defined by $\langle \rho, \alpha_i^\vee \rangle = \frac{1}{2} a_{ii}$ for all $i = 1, 2, \ldots, n$. Taking

$$R := \prod_{\alpha\in\Delta_+} (1 - e^{-\alpha})^{mult(\alpha)},$$

we get

$$ch\, L(\Lambda) = \frac{\sum\limits_{w\in W} (det\ w) e^{w(\Lambda+\rho)-\rho}}{R}.$$

Applying this for the trivial representation (for $\wedge = 0$, $ch\, L(0) = e^0 = 1$), we get

$$R = \sum_{w\in W} (det\ w) e^{w(\Lambda+\rho)-\rho}$$

$$\prod_{\alpha\in\Delta_+} (1 - e^{-\alpha})^{mult(\alpha)} = \sum_{w\in W} (det\ w) e^{w(\wedge+\rho)-\rho}.$$

Hence

$$e^\rho \prod_{\alpha\in\Delta_+} (1 - e^{-\alpha})^{mult(\alpha)} = \sum_{w\in W} (det\ w) e^{w(\rho)} \quad (\text{for } \wedge = 0).$$

This is the *Weyl-Kac denominator identity.* Again for any symmetrizable GCM (from the above equation), we get

$$\prod_{\alpha\in\Delta_+} (1 - e^{-\alpha})^{mult(\alpha)} = \sum_{w\in W} (det\ w) e^{w(\rho)} e^{-\rho}.$$

Hence

$$\sum_{w\in W} (det\ w) e^{w(\rho)-\rho} = \prod_{\alpha\in\Lambda_+} (1 - e^{(-\alpha)})^{mult(\alpha)}.$$

Substituting the right-hand side of the above formula in the Weyl-Kac character formula, we get another form of character formula, namely,

$$ch\, L(\Lambda) = \frac{\sum\limits_{w\in W} (det\ w) e^{w(\Lambda+\rho)}}{\sum\limits_{w\in W} (det\ w) e^{w(\rho)}}.$$

This is called *Denominator formula.*

Definition 105. Applying the above Weyl-Kac denominator identity to the Kac Moody algebra $A_1^{(1)}$, one can get

$$\prod_{n\in Z_{>0}} (1 - u^n v^n)(1 - u^{n-1} v^n)(1 - u^{n+1} v^n) = \sum_{n\in Z} (-1)^n u^{n(n+1)/2} v^{n(n-1)/2},$$

called the *Jacobi's triple product identity.* For proof, one can refer Kac [3].

Remark 63. Most important highest weight modules are $L(\wedge)$ with $\wedge \in P_+$. For these modules, it is known that the actions of all e_i's and f_i's are locally finite and so $\exp(e_i)$ and $\exp(f_i)$ are defined on $L(\wedge)$. For this reason, an irreducible highest weight module with highest weight $\wedge \in P_+$ is called integrable. In this case, $\gamma_i := \exp(f_i)\exp(e_i)\exp(f_i)$ for $0 \leq i \leq l$ will give the action of Weyl group on $L(\wedge)$. So its character is W-invariant. It can be verified that $ch\ M(\wedge) = e(\wedge)\Pi_{\alpha\in\Delta_+}(1 - e(-\alpha))^{-mult(\alpha)}$.

2.10 Graded Lie algebras and root multiplicities

General construction of graded Lie algebras of Kac-Moody type

We first recall some results on the general construction of graded Lie algebras of Kac-Moody type [88–90]. Let G be a Lie algebra over a field of characteristic zero. Let V, V' be two G-modules. Let ψ: $V \otimes V' \to G$ be a G-module homomorphism. Let $G_0 = G, G_{-1} = V, G_1 = V'$. Let $G_+ = \Sigma_{n\geq 1}G_n$ (respectively, $G_- = \Sigma_{n\geq 1}G_{-n}$) denote the free Lie algebra generated by V' (respectively, V). Then G_n for $n > 1$ is the space of all products of n vectors from V' (respectively, V). $G = \Sigma_{n=-\infty}^{\infty}G_n$ can be given a Lie algebra structure by defining the Lie bracket [,] as follows:

For $a, b \in G, v \in V, w \in V'$, define $[a, v] = a \cdot v = -[v, a]$ and $[a, w] = a \cdot w = -[w, a]$. For $a, b \in G$, let $[a, b]$ denote the bracket operation in G. For $w \in V', v \in V, [w, v] = \psi(w \otimes v) = -[v, w]$. Extending this bracket operation, $G = \Sigma_{n\in\mathbb{Z}}G_n$ becomes a graded Lie algebra which is generated by its local part $G_{-1} + G_0 + G_1$. For $n \geq 1$, define the subspaces $I_{\pm n} = \{x \in G_{\pm n} \mid [y_1, [\ldots [y_{n-1}, x]] \ldots] = 0$ for all $y_1, \ldots, y_{n-1} \in G_1\}$. Set $I_+ = \Sigma_{n>1}I_n$, $I_- = \Sigma_{n>1}I_{-n}$. Then I_+ and I_- are ideals of G and the ideal $I = I_- + I_+$ is the largest graded ideal of G trivially intersecting $G_{-1} + G_0 + G_1$. For $n > 1$, let $L_{\pm n} = G_{\pm n}/I_{\pm n}$. Let

$$\begin{aligned} L = L(G, V, V', \psi) &= G_-/I_- \oplus G_0 \oplus G_+/I_+ \\ &= \cdots \oplus L_{-2} \oplus L_{-1} \oplus L_0 \oplus L_1 \oplus L_2 \oplus \cdots, \end{aligned}$$

where $L_0 = G_0, L_1 = G_1, L_{-1} = G_{-1}$. Then $L = \oplus_{n\in\mathbb{Z}}L_n$ is a graded Lie algebra generated by its local part $V \oplus G \oplus V'$ with $L = G/I$. Also $L = \bigoplus_{n\geq 1} L_n$ is a graded Lie algebra generated by the subspace $L_1 = G_1$. Let $J = I/[I, I]$. J is an L-module via adjoint action generated by the subspace J_m. As vector spaces, $J_n \cong I_n$ for $m \leq n < 2m$. Suppose that I_m and G_1 are modules over a Kac-Moody algebra for $\mathcal{G}(A)$. Then G_n has a $\mathcal{G}(A)$-module structure such that $x \cdot [v, w] = [x \cdot v, w] + [v, x \cdot w]$ for $x \in \mathcal{G}(A), v \in G, w \in G_{-1}$. I_n also has a similar module structure.

Now we give below a construction of a graded Lie algebra of Kac-Moody type associated with a Cartan matrix

Let K be an algebraically closed field of characteristic zero. Let $\mathcal{G} = \mathcal{G}(A)$ be a symmetrizable Kac-Moody algebra over K with $A = (a_{ij})_{i,j=1}^{n}$ ($a_{ij} \in \mathbb{Z}$) having rank l.

Let $A = DB$ (where $D = diag(\epsilon_1, \epsilon_2, \ldots, \epsilon_n)$, a nonsingular matrix and $B = (b_{ij})_{i,j=1}^n$, a symmetric matrix) be the decomposition of A, with the minimal realization $(\mathfrak{h}, \Pi^\vee, \Pi)$. Let Z be the center of $\mathcal{G}(A)$.

Let $H = \{h_1, h_2, \ldots, h_l\}$ be a maximal subset of $\Pi^\vee$ independent of Z in $\mathfrak{h}$. Let $\{z_1, z_2, \ldots, z_{n-l}\}$ denote a basis for Z and $\{d_1, \ldots, d_{n-l}\}$ be linearly independent of Z in $H^\perp$ so that $(d_i, z_j) = \delta_{ij}$ for $i, j = 1, 2, \ldots, n-l$. Let $V = V_1 \oplus V_2 \oplus V_3 \oplus \cdots \oplus V_m$, for some integer $m \geq n-l$ each V_i being a faithful, be the irreducible highest weight module of $\mathcal{G}$ of highest weight λ_i. Here λ_i's are chosen such that $(\lambda_i(z_j))_{i,j=1}^{n-l}$ is nonsingular. Let $V^* = V_1^* \oplus V_2^* \oplus V_3^* \oplus \cdots \oplus V_m^*$ be the finite dual of V. Let $\mathcal{G}^e = \mathcal{G} \oplus Kc_{n-l+1} \oplus \cdots \oplus Kc_m$, where the elements $c_{n-l+1}, \ldots, c_m$ act centrally in $\mathcal{G}^e$. Extend the above action of $\mathcal{G}$ on V to $\mathcal{G}^e$ by letting each c_i act trivially. Now using the basis elements z_i and d_i, we can build an orthonormal basis $y_1, \ldots, y_{2n-l}$ for $\mathfrak{h}$. We can extend this basis to a basis $\{x_j\}_{j\in J}$ of $\mathcal{G}$ consisting of $y_1, \ldots, y_{2n-l}$ and root vectors. Using this basis of $\mathcal{G}$, we can also form the dual basis $\{x_j^*\}_{j\in J}$ of $\mathcal{G}$. For each $\mu \in \mathfrak{h}^*$, there exists a unique element $t_\mu = \Sigma_{j=1}^{2n-l} \mu(y_j) y_j$, such that $\mu(h) = (h, t_\mu)$ for all $h \in \mathfrak{h}$. The form $(.\,,.)$ induces a nondegenerate bilinear form, also denoted as $(.\,,.)$ on the dual space $\mathfrak{h}^*$, where $(\mu, \nu) = (t_\mu, t_\nu)$. Let us assume $(\lambda_i, \lambda_i) \neq 0$ for all i. For $w^* = w_1^* + \cdots + w_m^* \in V^*$ and $v = v_1 + v_2 + \cdots + v_m \in V$ where $w_i^* \in V_i^*, v_i \in V_i$, define

$$\Psi(w^* \otimes v) = \Sigma_{i=1}^m \frac{-2}{(\lambda_i, \lambda_i)} \Sigma_{j\in J} < w_i^* | x_j v_i > x_j^* + \Sigma_{i=n-l+1}^m < w^* | v_i > c_i,$$

where $\langle . | . \rangle$ denotes the usual pairing $\langle w^* | v_i \rangle = w_i^*(v_i)$ between V_i^* and V_i. Form the graded Lie algebra $L = L(\mathcal{G}^e, V, V^*, \Psi)$ as defined earlier. Then we have the following theorem proved in Benkart et al. [88]:

Theorem 24. $L = L(\mathcal{G}^e, V, V^*, \Psi)$ *is a* $\mathbb{Z}^{n+m}$ *-graded algebra. Setting for* $i = 1, \ldots, m$, *form the matrix* $C = (\langle \alpha_i, \alpha_j \rangle)_{i,j=1}^{n+m}$, *where* $\langle \alpha_i, \alpha_j \rangle = \frac{2(\alpha_i, \alpha_j)}{(\alpha_j, \alpha_j)}$. *The matrix* C *has rank* $2n - l$ *and* C *is symmetrizable.*

Example 22. Consider the affine Kac-Moody algebra $A_1^{(1)}$ associated with the Cartan matrix $A = (a_{ij})_{i,j=1}^2 = \begin{pmatrix} 2 & -2 \\ -2 & 2 \end{pmatrix}$. Now we can consider the hyperbolic Kac-Moody algebra $HA_1^{(1)}$, associated with the GCM

$$C = (a_{ij})_{i,j=1}^3 = \begin{pmatrix} 2 & -2 & 0 \\ -2 & 2 & -1 \\ 0 & -1 & 2 \end{pmatrix}$$

as the extension of $A_1^{(1)}$. Let V be a representation of G with highest weight λ. Let V^* be the contragradiant of V and ψ be the map defined as above. Now we can form the graded Lie algebra $L(G, V, V^*, \psi)$. Then $L \cong G(C)$ with $C = \begin{pmatrix} 2 & -2 & 0 \\ -2 & 2 & -1 \\ 0 & -1 & 2 \end{pmatrix}$.

Let $\tilde{A}(C)$ be the free Lie algebra on generators $E_i, F_i, H_i, i = 1, \ldots, n+m$ and $\tilde{I}(C)$ be the ideal generated by the homogeneous elements

$$[H_i, H_j],\ [H_i, E_j] - \langle \alpha_j, \alpha_i \rangle E_j,\ [H_i, F_j] + \langle \alpha_j, \alpha_i \rangle F_j \text{ and } [E_i, F_j] - \delta_{ij} H_i.$$

Let $A(C) = \tilde{A}(C)/\tilde{I}(C)$. Let $I(C)$ be the largest graded ideal of $A(C)$ trivially intersecting the span of $H_1, \ldots, H_{n+m}$.

We state the following theorem, the proof of which is in Benkart et al. [88].

Theorem 25. *Let ϕ: $A(C) \longrightarrow L$ be the Lie algebra homomorphism sending $E_i \longrightarrow e_i, F_i \longrightarrow f_i, H_i \longrightarrow h_i$. Then ϕ has kernel $I(C)$ and ϕ: $A(C)/I(C) \longrightarrow L$ is an isomorphism.*

Homology of Lie algebra

We now recall the definition of homology of a Lie algebra [23] and Hochschild-Serre spectral sequence [65]. Let G be a Lie-algebra and V be any module over G. Define the space $C_q(G, V)$ for $q > 0$ of Q-dimensional chains of the Lie algebra G with coefficients in V to be $\Lambda^q(G) \otimes V$. The differential d_q: $C_q(G, V) \longrightarrow C_{q-1}(G, V)$ is defined by

$$\begin{aligned} d_q(g_1 \wedge \cdots \wedge g_q \otimes v) = {} & \Sigma_{1 \le s < t \le q}(-1)^{s+t-1}(([g_s, g_t]) \wedge g_1 \wedge \cdots \wedge \hat{g}_s \wedge \cdots \wedge \\ & \hat{g}_t, \ldots, \wedge g_q) \otimes v \\ & + \Sigma_{1 \le s \le q}(-1)^s (g_1 \wedge \cdots \wedge \hat{g}_s \wedge \cdots \wedge g_q) \otimes g_s \cdot v, \\ & \text{for } v \in V, g_1, \ldots, g_q \in G. \end{aligned}$$

For $q < 0$, define $C_q(G, V) = 0$ and $d_q = 0$. Then $d_q \circ d_{q+1} = 0$. The homology of the complex $(C, d) = \{C_q(G, V), d_q\}$ is called the homology of the Lie algebra G with coefficients in V and is denoted by $H_q(G, V)$. If $V = \mathbb{C}$, we simply write $H_q(G)$ for $H_q(G, \mathbb{C})$.

(i) Horchschild-Serre five-term exact sequence [65]

Assume now that $G, V, C_q(G, V)$ are completely reducible modules in the category $\mathcal{O}$ over a Kac-Moody algebra $\mathcal{G}(A)$ (associated with the GCM A) with d_q being $\mathcal{G}(A)$-module homomorphisms. Let I be an ideal of G and $L = G/I$. Define a filtration $\{K_p = K_p C\}$ of the complex $\{C, d\}$ by

$$K_p C_{p+q} = \{g_1 \wedge g_2 \wedge \cdots \wedge g_{p+q} \otimes v | g_i \in I \quad \text{for } p+1 \le i \le p+q\}.$$

This gives rise to a spectral sequence $\{E^r_{p,q}, d_r\colon E^r_{p,q} \longrightarrow E^r_{p-1,q+r-1}\}$ such that $E^2_{p,q} \cong H_p(L, H_q(I, V))$, where $E^r_{p,q}$'s are determined by

$$E^{r+1}_{p,q} = \ker(d_r\colon E^r_{p,q} \longrightarrow E^r_{p-r,q+r-1}) / \Im(d_r\colon E^r_{p+r,q-r+1} \longrightarrow E^r_{p,q})$$

with boundary homomorphisms $d_{r+1}\colon E^r_{p,q} \longrightarrow E^r_{p-r-1,q+r}$. The modules $E^r_{p,q}$ become stable for $r > \max(p, q+1)$ for each (p, q) and the stable module is denoted by $E^\infty_{p,q}$. The spectral sequence $\{E^r_{p,q}, d_r\}$ converges to $H_n(G, V)$ in the following sense:

$$H_n(G,V) = \bigoplus_{p+q=n} E^{\infty}_{p,q}.$$

Then we have the following Horchschild-Serre five-term exact sequence [65]:

$$\begin{aligned} H_2(G,V) &\longrightarrow H_2(L,H_0(I,V)) \longrightarrow H_0(L,H_1(I,V)) \longrightarrow H_1(G,V) \\ &\longrightarrow H_1(L,H_0(I,V)) \longrightarrow 0. \end{aligned}$$

Now taking $G = \bigoplus_{n\geq 1} G_n$ be the free Lie algebra generated by the subspace G_1 and $I = \bigoplus_{n\geq m} I_n$ be the graded ideal of G generated by the subspace I_m for $m \geq 2$. Consider the quotient algebra $L = G/I$. Then $L = \bigoplus_{n\geq 1} L_n$ is also a graded Lie algebra generated by the subspace $L_1 = G_1$. Let $J = I/[I,I]$. J is an L-module via adjoint action generated by the subspace J_m. As vector spaces, $J_n \cong I_n$ for $m \geq n < 2m$. Suppose that I_m and G_1 are modules over a Kac-Moody algebra $\mathcal{G}(A)$. Then G_n has a $\mathcal{G}(A)$-module structure such that $x\cdot[v,w] = [x\cdot v,w]+[v,x\cdot w]$ for $x \in \mathcal{G}(A), v \in G, w \in G_{n-1}$. I_n also has a similar module structure. We also have the induced module structure of the homogeneous subspaces L_n and J_n. Then we have the following theorem proved in Kang [65].

Theorem 26. *There is an isomorphism of $\mathcal{G}(A)$-modules $H_j(L,J) \cong H_{j+2}(L)$, for $j \geq 1$. In particular $I_{m+1} \cong (G_1 \otimes I_m)/H_3(L)_{m+1}$.*

(ii) Spectral sequence converging to $H^*(L^{(j)})$:
Now for arbitrary $j \geq m$, set $I^{(j)} = \Sigma_{n\geq j} I_n$. Then $I^{(j)}$ is an ideal of G generated by the subspace I_j. We consider the quotient algebra $L^{(j)} = G/I^{(j)}$. Let $N^{(j)} = I^{(j)}/I^{(j-1)}$. In this notation $L = L^{(m)}$. Then we have an important relation: $I_{j+1} \cong (G_1 \otimes I_j)/H_3(L^{(j)})_{j+1}$. There exists a spectral sequence $\{E^r_{p,q}, d_r\colon E^r_{p,q} \longrightarrow E^r_{p-1,q+r-1}\}$ converging to $H^*(L^{(j)})$ [65] such that $E^2_{p,q} \cong H_p(L^{(j-1)}) \otimes \Lambda^q(I_{j-1})$ and

$$H_3(L^{(j)}) \cong E^{\infty}_{3,0} \oplus E^{\infty}_{2,1} \oplus E^{\infty}_{1,2} \oplus E^{\infty}_{0,3}.$$

The proof of the following lemma is in Kang [65].

Theorem 27. *In the above notation, $H_2(L) \cong I_m$.*

Kostant's formula for symmetrizable Kac-Moody algebras

Let $A = (a_{ij})^n_{i,j=1}$ be a symmetrizable GCM. Let $\Delta \subset \mathfrak{h}^*$ be root system of $\mathcal{G}(A)$ and Δ^+, Δ^- denote the systems of positive and negative roots, respectively, of $\mathcal{G}(A)$. We have the triangular decomposition:

$$\mathcal{G}(A) = \eta^- \oplus \mathfrak{h} \oplus \eta^+, \quad \text{where } \eta^{\pm} = \bigoplus_{\alpha\in\Delta^{\pm}} \mathcal{G}_{\alpha}.$$

Let $S = \{1,\ldots,s\}$ be a subset of $N = \{1,\ldots,n\}$ and $\mathcal{G}_S$ be the subalgebra of $\mathcal{G}(A)$ generated by the elements $e_i, f_i, i = 1,\ldots,s$ and $\mathfrak{h}$. Let Δ^+_S denote the set of positive roots generated by $\alpha_1,\ldots,\alpha_S$ and $\Delta^-_S = -\Delta^+_S$. Then $\mathcal{G}_S$ has the corresponding triangular decomposition:

$$\mathcal{G}_S = \eta_S^- \oplus \mathfrak{h} \oplus \eta_S^+,$$

where $\eta_S^\pm = \oplus_{\alpha\in\Delta_S^\pm}\mathcal{G}_\alpha$ and $\Delta_S = \Delta_S^+ \cup \Delta_S^-$ is the root system of $\mathcal{G}_S$. Let $\Delta^\pm(S) = \Delta^\pm \backslash \Delta_S^\pm$ and $\eta^\pm(S) = \bigoplus_{\alpha\in\Delta^\pm(S)} \mathcal{G}_\alpha$. Then $\mathcal{G}(A) = \eta^-(S) \oplus \mathcal{G}_S \oplus \eta^+(S)$. Let $W(S) = \{w \in W | w\Delta^- \cap \Delta^+ \subset \Delta^+(S)\}$, where W is the Weyl group of $\mathcal{G}(A)$. In the following lemma, w is an element of the Weyl group and r_j is a fundamental reflection and $l(w)$ is length of w, one can see Kang [65] for the proof of the following lemma. The following lemma is useful in the actual computation of the homology modules.

Lemma 3. *Suppose $w = w'r_j$ and $l(w) = l(w') + 1$. Then $w \in W(S)$ if and only if $w' \in W(S)$ and $w'(\alpha_j) \in \Delta^+(S)$.*

For $\lambda \in \mathfrak{h}^*$, denote by $\tilde{V}(\lambda)$, the irreducible highest weight module over $\mathcal{G}(A)$ and $V(\lambda)$, the irreducible highest weight module over $\mathcal{G}_S$.

Then we have the following *Kostant's formula,* proved in Garland and Lepowsky [23] and Liu [91].

Lemma 4. $H_j(\eta^-(S), \tilde{V}(\lambda)) \cong \bigoplus_{\substack{w\in W(S)\\ l(w)=j}} V(w(\lambda+\rho)-\rho).$

2.10.1 The combinatorial part of the crystal base theory of quantum affine Lie algebras

For a fixed positive integer l, define

$$B = \{(b_0, b_1) | b_0, b_1 \in \mathbb{Z}_+,\ b_0 + b_1 = l\} \subset \mathbb{Z} \times \mathbb{Z}.$$

A path $p = (p_k)_{k\geq 1}$ is an infinite sequence of elements of B. For $b = (b_0, b_1) \in B$ define $\bar{wt}(b) = (b_1 - b_0)\lambda_0 + (b_0 - b_1)\lambda_1$, where λ_0 and λ_1 are the fundamental weights for the corresponding affine Kac-Moody algebra ($A_1^{(1)}$ or $A_2^{(2)}$). We define the energy function H: $B \times B \longrightarrow \mathbb{Z}$ by

$$H(b \otimes b') = \max\{b_0', b_0' + b_1' - b_1\},$$

where $b = (b_0, b_1)$ and $b' = (b_0', b_1')$ are elements of B. Let $V(\lambda)$ denote the integrable, highest weight module over the affine Lie algebra ($A_1^{(1)}$ or $A_2^{(2)}$) with highest weight λ, where λ is a dominant integral of level l. Define the ground-state path $\eta = (\eta_k)_{k\geq 1}$ of weight λ, a periodic sequence given by $\eta = ((k_0, k_1), (k_1, k_0), (k_0, k_1), \ldots)$. A λ-path $p = (p_k)_{k\geq 1}$ is a path such that $p_k = \eta_k$ for $k \gg 0$. Let $\mathcal{P}(\lambda)$ denote the set of all λ-paths. $\mathcal{P}(\lambda)$ is given a crystal graph structure and $V(\lambda)$ has a basis which is parameterized by λ-paths. For a λ-path $p = (p_k)_{k\geq 1}$ in $\mathcal{P}(\lambda)$, we define the weight function as

$$wt(p) = \lambda + \Sigma_{k=1}^{\infty}(\bar{wt}(p_k) - \bar{wt}(n_k)) - \Sigma_{k=1}^{\infty} k(H(p_{k+1} \otimes p_k) - H(n_{k+1} \otimes n_k))\delta,$$

where $\delta = \alpha_0 + 2\alpha_1$. Let $\mathcal{P}_{\lambda,\mu} = \sharp\{p \in \mathcal{P}(\lambda) | wt(p) = \mu\}$. Then we have following theorem proved in Kang et al. [92, 93].

Theorem 28. *dim* $V(\lambda)_\mu = \mathcal{P}_{\lambda,\mu}$.

2.10.2 Structure and root multiplicities for an extended-hyperbolic KM algebra $EHA_1^{(1)}$

(i) Realization and homology modules for $EHA_1^{(1)}$: Let us denote by $EHA_1^{(1)}$, the class of extended-hyperbolic Kac-Moody algebras whose associated GCM is of the form $\begin{pmatrix} 2 & -2 & -c \\ -2 & 2 & -a \\ -d & -b & 2 \end{pmatrix}$, that is the class of all 3×3 GCM of extended-hyperbolic type obtained from the affine algebra $A_1^{(1)}$ associated with the GCM $\begin{pmatrix} 2 & -2 \\ -2 & 2 \end{pmatrix}$. Here $a, b, c, d \in \mathbb{Z}^+$ and $ab > 4$ or $cd > 4$. In Sthanumoorthy and Uma Maheswari [24], the structure and the root multiplicities were computed for $EHA_1^{(1)}$ for a particular case and the same was considered for $EHA_1^{(1)}$ for more cases in Sthanumoorthy et al. [80]. Here we give the structure and compute the root multiplicities for $EHA_1^{(1)}$ for all cases in general [60]. Consider the affine Kac-Moody algebra associated with the GCM, $A = \begin{pmatrix} 2 & -2 \\ -2 & 2 \end{pmatrix}$. Let $(\mathfrak{h}, \Pi, \Pi^\vee)$ be the realization of A with simple roots $\Pi = \{\alpha_1, \alpha_2\}$ and simple coroots $\Pi^\vee = \{\alpha_1^\vee, \alpha_2^\vee\}$. We have the following bilinear relations:

$$(\alpha_1, \alpha_1) = 2, \quad (\alpha_1, \alpha_2) = -2, \quad (\alpha_2, \alpha_2) = 2;$$

Let α_3' be the element in $\mathfrak{h}^*$ such that $\alpha_3'(\alpha_1^\vee) = 0, \alpha_3'(\alpha_2^\vee) = 1$ and hence

$$(\alpha_3', \alpha_1) = 0, \quad (\alpha_3', \alpha_2) = 1.$$

Let a, b, p be three nonnegative integers such that $ab > 4$ or $(bp^2/4a) > 4$ and $(bp/4a)$ is an integer with $a \neq 0$. Let us define

$$\lambda = \frac{bp}{2a} + \left(\frac{(4a - bp^2)}{(4a(a+p))}\right)\alpha_1 + \left(\frac{(4a - bp^2)}{4a(a+p)}\right)\alpha_2 + \left(b + \frac{bp}{a}\right)\alpha_3'.$$

Set $\alpha_3 = -\lambda$. Then

$$\langle \alpha_3, \alpha_1 \rangle = -bp/a, \quad \langle \alpha_1, \alpha_3 \rangle = -p;$$
$$\langle \alpha_3, \alpha_2 \rangle = -b, \quad \langle \alpha_2, \alpha_3 \rangle = -a, \quad \langle \alpha_3, \alpha_3 \rangle = 2;$$
$$\langle \alpha_2, \alpha_1 \rangle = -2 \text{ and } \langle \alpha_1, \alpha_2 \rangle = -2.$$

Form the matrix $C = (\langle \alpha_i, \alpha_j \rangle)_{i,j=1}^3$. Then $C = \begin{pmatrix} 2 & -2 & -p \\ -2 & 2 & -a \\ -bp/a & -b & 2 \end{pmatrix}$ is a symmetrizable GCM of extended-hyperbolic type. Let V be the integrable highest weight irreducible module over G with the highest weight λ as defined above. Let V^* be the contragradient of V and ψ be the mapping as defined earlier. Let G be the Kac-Moody algebra associated with the GCM $\begin{pmatrix} 2 & -2 \\ -2 & 2 \end{pmatrix}$. Form the graded Lie algebra $L(G, V, V^*, \psi)$. Then $L \cong G(C)$ with $C = \begin{pmatrix} 2 & -2 & -p \\ -2 & 2 & -a \\ -bp/a & -b & 2 \end{pmatrix}$. Thus L is a symmetrizable Kac-Moody algebra of extended-hyperbolic type associated with Cartan matrix which is an extension of the Cartan-matrix $A_1^{(1)}$. In general, we denote the class of extended-hyperbolic Kac-Moody algebra associated with the above GCM by $EHA_1^{(1)}$. Let us find out the homology modules. Let $S = \{1, 2\} \subset N = \{1, 2, 3\}$. The $\mathcal{G}_s$ is the affine Lie algebra $A_1^{(1)}$. Here $\Delta^+(S) = \{k_1\alpha_1 + k_2\alpha_2 + k_3\alpha_3 \in \Delta^+ | k_3 \neq 0\}$. Let Δ_s be the root system of $\mathcal{G}_s$. We consider general case for $p \geq 0$. The only element length 1 in $W(S)$ is r_3. Hence by the Kostant's formula $H_1(L_-) \cong V(-\alpha_3)$, as $r_3\rho - \rho = -\alpha_3$. The only elements of length 2 in $W(S)$ are r_3r_1 and r_3r_2. We have

$$r_3r_2\rho - \rho = -(b+1)\alpha_3 - \alpha_2$$
$$r_3r_1\rho - \rho = -(bp/a+1)\alpha_3 - \alpha_1.$$

Hence $H_2(L_-) \cong V(-(bp/a+1)\alpha_3 - \alpha_1) \oplus V(-(b+1)\alpha_3 - \alpha_2)$. The elements of length 3 in $W(S)$ are $r_3r_2r_3, r_3r_2r_1, r_3r_1r_2, r_3r_1r_3$. We have

$$r_3r_2r_3\rho - \rho = -(ab+b)\alpha_3 - (a+1)\alpha_2 \text{ and}$$
$$r_3r_2r_1\rho - \rho = -\alpha_1 - 3\alpha_2 - (bp/a+3b+1)\alpha_3$$
$$r_3r_1r_2\rho - \rho = -(1+b+3bp/a)\alpha_3 - \alpha_2 - 3\alpha_1$$
$$r_3r_1r_3\rho - \rho = -(p+1)\alpha_1 - (p+1)bp/a\alpha_3.$$

Hence

$$\begin{aligned} H_3(L_-) \cong\ & V(-(ab+a)\alpha_3 - (a+1)\alpha_2) \oplus V(-(1+3b+bp/a)\alpha_3 - 3\alpha_2 - \alpha_1) \\ & \oplus V(-(1+b+3bp/a)\alpha_3 - \alpha_2 - 3\alpha_1) \\ & \oplus V(-bp(p+1)\alpha_3/a - (p+1)\alpha_1). \end{aligned}$$

Similarly we can compute the other homology modules $H_4(L_-), H_5(L_-), H_6(L_-), \ldots$. Using the homological approach together with the representation theory of affine Kac-Moody algebra $A_1^{(1)}$, below we explain some of the boundary homomorphisms and deduce some new structural information on maximal ideals in $EHA_1^{(1)}$.

(ii). Structure of the maximal ideal in $EHA_1^{(1)}$: In this section, we study the structure of the maximal graded ideal I. Here we consider the general case for $p \geq 0$. We know that the ideal I_- of G_- is generated by the homological subspace I_{-2} and hence we may write $I_- = I_-^{(2)}$. Similarly, for $j \geq 2$, we write $I_-^{(j)} = \Sigma_{n \geq j} I_{-n}, L_-^{(j)} = G/I_-^{(j)}$ and $N_-^{(j)} = I_-^{(j)}/I_-^{(j+1)}$. By the homological theory described above, we have in general that $I_{-(j+1)} \cong (V \otimes I_{-j})/H_3(L_-^{(j)})_{-(j+1)}$ for $j \geq 2$. Since G_- is free and I_- is

generated by the subspace I_{-2} form the Hochschild-Serre five-term exact sequence: $I_{-2} \cong H_2(L_-)$. But $H_2(L_-) \cong V(-(bp/a+1)\alpha_3 - \alpha_1) \oplus V(-(b+1)\alpha_3 - \alpha_2)$. Hence $I_{-2} \cong V(-(bp/a+1)\alpha_3 - \alpha_1) \oplus V(-(b+1)\alpha_3 - \alpha_2)$. If $j = 2$, $L_-^{(2)}$ coincides with the subspace $\eta^-(S)$ for $S = \{1,2\}$ and therefore we can compute $H_3(L_-^{(2)})$ using the Kostant's formula. We have

$$\begin{aligned} H_3(L_-^{(2)}) \cong\; & V(-(3b+bp/a+1)\alpha_3 - 3\alpha_2 - \alpha_1) \oplus V(-(ab+b)\alpha_3 - (a+1)\alpha_2) \\ & \oplus V(-(1+b+3bp/a)\alpha_3 - \alpha_2 - 3\alpha_1) \\ & \oplus V(-(p+1)bp/a\alpha_3 - (p+1)\alpha_1). \end{aligned}$$

Thus

$$H_3(L_-^{(2)})_{-3} \cong \begin{cases} V(-3\alpha_3 - 3\alpha_2) & \text{if } b = 1, a = 2 \\ V(-(p+1)bp/a\alpha_3 - (p+1)\alpha_1) & \text{if } bp/a = 1, p = 2, a > 3 \\ = 0, & \text{otherwise} \end{cases}$$

and hence we obtain

$$I_{-3} \cong (V \otimes I_{-2})/H_3(L_-^{(2)})_{-3} \cong \begin{cases} (V \otimes I_{-2})/V(-3\alpha_3 - 3\alpha_2) & \text{if } b = 1, a = 2 \\ (V \otimes I_{-2})/V(-3\alpha_3 - 3\alpha_1) & \text{if } bp/a = 1, \\ & \quad p = 2, a = 3 \\ (V \otimes I_{-2}) & \text{otherwise.} \end{cases}$$

To determine the higher components, we combine the homological techniques and spectral sequences theory. We have $I_{-4} \cong (V \otimes I_{-3})/H_3(L_-^{(3)})_{-4}$. First let us compute $H_3(L_-^{(3)})_{-4}$. Consider the short exact sequence $0 \longrightarrow N_-^{(2)} \longrightarrow L_-^{(3)} \longrightarrow L_-^{(2)} \longrightarrow 0$ and the corresponding spectral sequence $\{E_{p,q}^r\}$ converging to $H*(L_-^{(3)})$ such that $E_{p,q}^r \cong H_p(L_-^{(2)}) \otimes \Lambda^q(I_{-2})$. Consider $0 \longrightarrow E_{2,0}^2 \longrightarrow E_{0,1}^2 \longrightarrow 0$. Note that

$$H_1(L_-^{(3)}) \cong \frac{L_-^{(3)}}{[L_-^{(3)}, L_-^{(3)}]} \cong L_{-1} = V.$$

Since the spectral sequence converges to $H*(L_-^{(3)})$, we have

$$H_1(L_-^{(3)}) \cong E_{1,0}^\infty \oplus E_{0,1}^\infty.$$

But $E_{1,0}^\infty = E_{0,1}^\infty \cong H_1(L_-^{(2)}) \cong L_-^{(2)}/[L_-^{(2)}, L_-^{(2)}] \cong L_{-1} = V$ which implies $E_{0,1}^\infty = E_{0,1}^3 = 0$. Hence the homomorphism d_2 is surjective. Since $E_{2,0}^2 \cong I_{-2}$ and $E_{2,0}^2 \cong I_{-2}$, d_2 must be an isomorphism. Thus $E_{2,0}^3 = 0$ and hence $E_{2,0}^\infty = 0$. Now consider the following sequence

$$0 \longrightarrow E^2_{3,0} \longrightarrow E^2_{1,1} \longrightarrow 0.$$

From the previous equations we get

$$\begin{aligned} E^2_{3,0} &\cong H_3(L^{(2)}_-) \cong V(-(3b+bp/a+1)\alpha_3 - \alpha_1 - 3\alpha_2) \\ &\quad \oplus V(-((3b+bp)/a+1)\alpha_3 - \alpha_2 - 3\alpha_1) \\ &\quad \oplus V(-(b+ba)\alpha_3 - (a+1)\alpha_2) \oplus V(-((p+1)bp/a)\alpha_3 - (p+1)\alpha_1) \\ E^2_{1,1} &\cong H_1(L^{(2)}_-) \otimes I_{-2} \cong V \otimes I_{-2}. \end{aligned}$$

Then $V \otimes I_{-2}$ is a direct sum of irreducible modules over $A^{(2)}_2$ of level $b+2$ and $1+(1+bp/a)$.

Let us assume that $b > 1$. Then by comparing the levels of both the terms $E^2_{3,0}$ and $E^2_{1,1}$ in d_2: $E^2_{3,0} \longrightarrow E^2_{1,1}$, we see that d_2 is trivial. So $E^3_{3,0} = E^2_{3,0}$, and $E^\infty_{1,1} = E^3_{1,1} = E^2_{1,1} \cong V \otimes I_{-2}$. Since $I^{(3)}_-$ is generated by I_{-3}, we get $H_2(L^{(3)}_-) \cong I_{-3}$. But we have $H_2(L^{(3)}_-) \cong E^\infty_{2,0} \oplus E^\infty_{1,1} \oplus E^\infty_{0,2}$. Hence it follows that $E^\infty_{0,2} = E^4_{0,2} = 0$. Therefore the homomorphism d_3: $E^3_{3,0} \longrightarrow E^3_{0,2}$ is surjective.

As $E^\infty_{0,2}$ is a submodule of $E^3_{0,2} \cong \Lambda^2(I_{-2})$, we see that d_3 is trivial. This implies $E^\infty_{3,0} = E^4_{3,0} = E^3_{3,0} = E^2_{3,0} \cong H_3(L^{(2)}_-)$ and $E^3_{0,2} = 0$. Thus the homomorphism d_2: $E^2_{2,1} \longrightarrow E^2_{0,2}$ is surjective in the following sequence $0 \longrightarrow E^2_{4,0} \longrightarrow E^2_{2,1} \longrightarrow E^2_{0,2} \longrightarrow 0$.

Again by comparing levels, we deduce that the homomorphism d_2: $E^2_{4,0} \longrightarrow E^2_{2,1}$ must be trivial. Therefore $E^3_{4,0} = E^2_{4,0}$ and

$$\begin{aligned} E^\infty_{2,1} &= E^3_{2,1} = \ker(d_2\colon E^2_{2,1} \longrightarrow E^2_{0,2})/\Im(d_2\colon E^2_{4,0} \longrightarrow E^2_{2,1}) \\ &= \ker(d_2\colon E^2_{2,1} \longrightarrow E^2_{0,2}). \end{aligned}$$

Since d_2: $E^2_{2,1} \longrightarrow E^2_{0,2}$ is surjective, we have $\Lambda^2(I_{-2}) \cong E^2_{0,2} \cong E^2_{2,1}/\ker d_2 \cong I_{-2} \otimes I_{-2}/\ker d_2$, where $\ker d_2 \cong S^2(I_{-2})$. Therefore $E^\infty_{2,1} \cong S^2(I_{-2})$. If $E^3_{0,2}$ is nonzero and d_3: $E^3_{3,0} \longrightarrow E^3_{0,2}$ is surjective, then since $E^3_{3,0} = E^2_{3,0}$ is irreducible, d_3: $E^3_{3,0} \longrightarrow E^3_{0,2}$ is an isomorphism.

Thus, $E^\infty_{3,0} = E^4_{3,0} = 0$ and $H_3(L^{(2)}_-) \cong E^3_{3,0} \cong E^3_{0,2} \cong E^2_{0,2}/\Im(d_2\colon E^2_{2,1} \longrightarrow E^2_{0,2}) \cong \Lambda^2(I_{-2})/\Im(d_2\colon E^2_{2,1} \longrightarrow E^2_{0,2})$. Since all the modules involved here are completely reducible over $A^{(1)}_1$, we have $\Im(d_2\colon E^2_{2,1} \longrightarrow E^2_{0,2}) \cong \Lambda^2(I_{-2})/H_3(L^{(2)}_-)$. We have seen that the homomorphism d_2: $E^2_{4,0} \longrightarrow E^2_{2,1}$ is trivial.

Thus

$$\begin{aligned} E^\infty_{2,1} &\cong E^3_{2,1} \cong \ker(d_2\colon E^2_{2,1} \longrightarrow E^2_{0,2})/\Im(d_2\colon E^2_{4,0} \longrightarrow E^2_{2,1}) \\ &= \ker\,(d_2\colon E^2_{2,1} \longrightarrow E^2_{0,2}). \end{aligned}$$

Since $\Im\,(d_2) \cong \Lambda^2(I_{-2})/H_3(L_-^{(2)}) \cong E^2_{2,1}/\ker\, d_2 \cong (I_{-2} \otimes I_{-2})/\ker\, d_2$, we have $\ker\, d_2 \cong S^2(I_{-2}) \oplus H_3(L_-^{(2)})$. Now consider the sequence $0 \longrightarrow E^2_{5,0} \longrightarrow E^2_{3,1} \longrightarrow E^2_{1,2} \longrightarrow 0$. By comparing levels, we see that the homomorphism $d_2\colon E^2_{3,1} \longrightarrow E^2_{1,2}$ is trivial. Thus $E^3_{1,2} = E^2_{1,2} \cong V \oplus \Lambda^2(I_{-2})$. Again by comparing the levels of the terms in the sequence $0 \longrightarrow E^3_{4,0} \longrightarrow E^3_{1,2} \longrightarrow 0$, we see that d_3 is trivial. Therefore $E^\infty_{1,2} = E^4_{1,2} = E^3_{1,2} = E^2_{1,2} \cong V \otimes \Lambda^2(I_{-2})$. Finally, since $E^\infty_{0,3}$ is a submodule of $E_{0,3} \cong \Lambda^3(I_{-2})$, we get

$$H_3(L_-^{(3)}) \cong H_3(L_-^{(2)}) \oplus S^2(I_{-2}) \oplus V \otimes \Lambda^2(I_{-2}) \oplus M,$$

where M is a direct sum of irreducible highest weight representation of $A_1^{(1)}$ of level >4. Therefore we obtain

$$H_3(L_-^3)_{-4} = \begin{cases} S^2 \oplus V(-4\alpha_3 - 3\alpha_2) & \text{if } a = 2, b = 1, p = 3 \\ V(-4\alpha_3 - 2\alpha_2) & \text{if } a = 1, b = 2 \\ V(-4\alpha_3 - 3\alpha_2) & \text{if } a = 2, b = 1 \\ 0 & \text{otherwise.} \end{cases}$$

Hence, we have $I_-^{(4)} \cong (V \otimes I_{-3})/H_3(L_-^{(3)})_{-4}$, where $H_3(L_-^{(3)})_{-4}$ is as given above. Now we shall determine I_{-5}. We start with the sequence

$$0 \longrightarrow N_-^{(3)} \longrightarrow L_-^{(4)} \longrightarrow L_-^{(3)} \longrightarrow 0$$

and the corresponding spectral sequence $\{E^r_{p,q}\}$ converging to $H^*(L_-^{(4)})$ such that $E^2_{p,q} \cong H_p(L_-^{(3)}) \otimes \Lambda^q(I_{-3})$. We will compute $H_3(L_-^{(4)})_{-5}$ from this spectral sequence. Now $d_2\colon E^2_{2,0} \longrightarrow E^2_{0,1}$ is an isomorphism and $E^\infty_{2,0} = 0$. Consider the sequence $0 \longrightarrow E^2_{3,0} \longrightarrow E^2_{1,1} \longrightarrow 0$. $E^2_{3,0} \cong H_3(L_-(3))$ and $E^2_{1,1} \cong H_1(L_-^{(3)}) \otimes I_{-3} \cong V \otimes I_{-3}$. By elementary linear algebra, we have $E^2_{3,0} \cong \ker\, d_2 \oplus \Im\, d_2$. Consider $d_2\colon E^2_{3,0} \longrightarrow E^2_{1,1}$. Comparing the levels of the above terms, we get the following cases:

(i) Let $b = 1, a = 2, p = 3$. Then d_2 maps $V(-(ab+b)\alpha_3-(a+1)\alpha_2)\oplus(V\otimes\Lambda^2(I_-^{(2)}))\oplus M$ *to* 0. This implies that $\Im\, d_2$ is isomorphic to a submodule of $S^2(I_-^{(2)}) \oplus V(-4\alpha_3 - 3\alpha_2)$.

(ii) Let $b > 1$. Then d_2 maps $E^2_{3,0}$ to 0. This implies that $\Im\, d_2 = 0$.

Thus

$$H_2(L_-^{(4)}) \cong I_{-4} \cong \begin{cases} (V \otimes I_{-3})/(S^2(I_{-2}) \oplus V(-4\alpha_3 - 3\alpha_2)) & \text{if } a = 2, b = 1, p = 3 \\ (V \otimes I_{-3})/V(-4\alpha_3 - 2\alpha_1) & \text{if } a = 1, b = 2, p^2 > 2 \\ (V \otimes I_{-3})/V(-4\alpha_3 - 3\alpha_1) & \text{if } a = 2, b = 1, p^2 > 2. \end{cases}$$

We have $E^{\infty}_{1,1} = E^3_{1,1} = E^2_{1,1}/\Im\, d_2 \cong V \otimes I_{-3}/\Im\, d_2$. If $b = 1$, $(V \otimes I_{-3})/\Im\, d_2$ is a direct sum of $H_2(L^{(4)}_{-})$. Therefore we conclude that

$$\Im\, d_2 \cong \begin{cases} S^2(I_{-2}) \oplus V(-4\alpha_3 - 3\alpha_2) & \text{if } a = 2, b = 1, p = 3 \\ 0 & \text{if } b > 1 \end{cases}$$

$$E^3_{3,0} \cong \begin{cases} V \otimes \Lambda^2(I_{-2}) & \text{if } a = 2, b = 1, p^2 > 3 \\ E^2_{3,0} & \text{if } b > 1 \end{cases}$$

$$E^{\infty}_{1,1} = E^3_{1,1} \cong I_{-4},$$

which implies that $E^{\infty}_{0,2} = E^4_{0,2} = 0$. Thus the homomorphism d_3 is surjective in the sequence $0 \longrightarrow E^3_{3,0} \longrightarrow E^3_{0,2} \longrightarrow 0$. Since $E^3_{0,2}$ is a submodule of $E^3_{0,2} \cong E^2_{0,2}$ by comparing levels of $E^3_{3,0}$ and $E^3_{0,2}$, we see that ker d_3 must contain $V \otimes \Lambda^2(I_{-2})$. Hence it follows that $E^{\infty}_{3,0} = E^4_{3,0} \cong (V \otimes \Lambda^2(I_{-2})) \oplus M'$, where M' is a direct sum of level ≥ 6 irreducible highest weight representations of $A^{(1)}_1$. So we have

$$(E^{\infty}_{3,0})_{-5} = \begin{cases} V \otimes \Lambda^2(I_{-2}) & \text{if } b = 1 \\ 0 & \text{if } b > 1. \end{cases}$$

Similarly we have $(E^{\infty}_{3,0})_{-5} = (E^{\infty}_{1,2})_{-5} = (E^{\infty}_{0,3})_{-5} = 0$. Therefore we get

$$H_3(L^{(4)}_{-})_{-5} = \begin{cases} V \otimes \Lambda^2(I_{-2}) & \text{if } b = 1 \\ 0 & \text{if } b > 1. \end{cases}$$

Thus we get

$$I_{-5} \cong (V \otimes I_{-4})/H_3(L^{(4)}_{-})_{-5} = \begin{cases} (V \otimes I_{-4})/(V \otimes \Lambda^2(I_{-2})) & \text{if } b = 1, a = 2 \\ (V \otimes I_{-4}) & \text{if } b > 1. \end{cases}$$

Thus we have proved the following theorem which gives the structure of the components of the maximal ideal I (upto level 5) in the extended-hyperbolic Kac-Moody algebra $EHA^{(1)}_1$.

Theorem 29. *With the usual notations, let* $L = \oplus_{n \in \mathbb{Z}} L_n$ *be the realization of* $EHA^{(1)}_1$ *associated with the GCM* $\begin{pmatrix} 2 & -2 & -p \\ -2 & 2 & -a \\ -pb/a & -b & 2 \end{pmatrix}$. *Then, we have the following:*

(i) $I_{-2} \cong V((-pb/a + 1)\alpha_3 - \alpha_1) \oplus V(-(b+1)\alpha_3 - \alpha_2)$

(ii) $I_{-3} \cong (V \otimes I_{-2})/H_3(L^{(2)}_{-})_{-3}$ *where*

$$H_3(L_-^{(2)})_{-3} \cong \begin{cases} V(-3\alpha_3 - 3\alpha_1) & \textit{if } bp/a = 1, p = 2 \\ V(-3\alpha_3 - 2\alpha_2) & \textit{if } b = 1, a = 2 \\ 0, & \textit{otherwise.} \end{cases}$$

(iii) $I_{-4} \cong (V \otimes I_{-3})/H_3(L_-^{(3)})_{-4}$ *where*

$$H_3(L_-^{(3)})_{-4} \cong \begin{cases} S^2(I_{-2} \oplus V(-4\alpha_3 - 3\alpha_2)) & \textit{if } a = 2, b = 1, p = 3 \\ V(-4\alpha_3 - 2\alpha_2) & \textit{if } a = 1, b = 2 \\ V(-4\alpha_3 - 3\alpha_2) & \textit{if } a = 2, b = 1 \\ 0, & \textit{otherwise.} \end{cases}$$

(iv) $I_{-5} \cong (V \otimes I_{-4})/H_3(L_-^{(4)})_{-5}$ *where*

$$H_3(L_-^{(4)})_{-5} \cong \begin{cases} V \otimes \Lambda^2(I_{-2}) & \textit{if } b = 1 \\ 0, & \textit{otherwise.} \end{cases}$$

(iii) Root multiplicities in EH $A_1^{(1)}$: We now find multiplicity of roots upto level 3 for $EHA_1^{(1)}$ associated with the GCM, $\begin{pmatrix} 2 & -2 & -p \\ -2 & 2 & -a \\ \frac{-pb}{a} & -b & 2 \end{pmatrix}$ with $ab > 4$ or $\frac{pb^2}{4a} > 4$.

Let $V = V(\lambda)$ be the integrable, highest weight module over $A_1^{(1)}$ with the highest weight

$$\lambda = \left[\frac{bp}{2a} + \frac{(4a - bp^2)}{(4a(a+p))}\right]\alpha_1 + \frac{(4a - bp^2)}{(4a(a+p))}\alpha_2 + \left(b + \frac{bp}{a}\right)\alpha_3'.$$

Let $\{\tau_i | i = 1, 2, 3, \ldots\}$ be an enumeration of all weights of V. For $\tau \in \mathfrak{H}$, set

$$T(\tau) = \{(n) = (n_1, n_2, \ldots) | n_i \in \mathbb{Z}_{\geq 0}, \Sigma n_i \tau_i = \tau\}.$$

Consider the Birkoff-Witt function

$$B(\tau) = \Sigma_{(n) \in T(\tau)} \frac{(\Sigma n_i - 1)!}{\prod n_i!} \prod (\dim V_{\tau_i})^{n_i}.$$

Any dominant weight μ of $V(\lambda)$ μ will be of the form $\mu = \lambda - n\delta$. Also $\dim V_\mu = P(n)$, where P is the classical partition function. Every column of weights in $V(\lambda)$ is W-conjugate to the dominant column and the multiplicities in any column are the values of the partition function P. We now compute the form (μ, μ):

$$(\mu, \mu) = \frac{2b}{a} - 2n\left(\frac{bp}{a} + b\right)$$

Therefore $n = \frac{1}{a+p} - \frac{a(\mu,\mu)}{2b(a+p)}$, which implies

$$\dim V_{\tau_i} = P\left(\frac{1}{a+p} - \frac{(\tau_i,\tau_i)a}{(p+a)2b}\right).$$

So $B(\tau)$ becomes

$$B(\tau) = \Sigma_{(n)\in T(\tau)} \frac{(\Sigma n_i - 1)!}{\prod n_i!} \prod \left(P\left(\frac{1}{a+p} - \frac{(\tau_i,\tau_i)a}{(a+p)2b}\right)\right)^{n_i}.$$

By the generalization of Witt formula, we get $\dim(G_-)_\alpha = \Sigma_{\tau|\alpha} \mu'\left(\frac{\alpha}{\tau}\right)\frac{\tau}{\alpha}B(\tau)$, where μ' is the classical Mobius function. We use the notation $\frac{\alpha}{k}|\alpha$ for the root $\frac{\alpha}{k}$ if it divides α and k is any positive integer.

$$\dim(L_{-1})_\alpha = P\left(\frac{1}{a+p} - \frac{(\alpha,\alpha)a}{2(a+p)b}\right).$$

Next we shall determine the multiplicity of a root α of level 2. The possible roots which can divide α are α and $\alpha/2$. Hence by the Witt formula,

$$\dim(G_{-2})_\alpha = \begin{cases} B(\alpha) & \text{if } (\alpha/2) \nmid \alpha \\ B(\alpha) - \frac{1}{2}B(\alpha/2) & \text{if } (\alpha/2)|\alpha. \end{cases}$$

In terms of the partition function,

$$\dim(G_{-2})_\alpha = \begin{cases} B(\alpha) & \text{if } (\alpha/2) \nmid \alpha \\ B(\alpha) - \frac{1}{2}P\left(\frac{1}{a+p} - \frac{(\alpha,\alpha)a}{8b(a+p)}\right) & \text{if } (\alpha/2)|\alpha \end{cases} \tag{2.6}$$

Similarly, since $\mu'(3) = -1$ and the possible roots which can divide a root α of level 3 are α and $\alpha/3$, by the Witt formula,

$$\dim(G_{-3})_\alpha = \begin{cases} B(\alpha) & \text{if } (\alpha/3) \nmid \alpha \\ B(\alpha) - \frac{1}{2}P\left(\frac{1}{a+p} - \frac{(\alpha,\alpha)a}{18b(a+p)}\right) & \text{if } (\alpha/3)|\alpha. \end{cases} \tag{2.7}$$

Next we shall determine the multiplicity of a root in I_{-2} and I_{-3}.

$$I_{-2} \cong V(-(b+1)\alpha_3 - \alpha_2) \oplus V\left(-\left(1+\frac{bp}{a}\right)\alpha_3 - \alpha_1\right).$$

Consider any root β of the form $\beta = -(b+1)\alpha_3 - \alpha_2 - m\delta$. So

$$(\beta,\beta) = -\,2mb(b+1)\left(1+\frac{p}{a}\right) - 2b(b+1)\left(1 - \frac{b+1}{a}\right) + 2$$

$$2m = \begin{cases} \dfrac{1}{2\left(1+\dfrac{p}{a}\right)}\left(-(\beta,\beta)-4\left(1-\dfrac{2}{a}\right)+2\right) & \text{if } b=1 \\ \dfrac{1}{b(b+1)\left(1+\dfrac{p}{a}\right)}\left(-(\beta,\beta)-2b(b+1)\left(1-\dfrac{b+1}{a}\right)+2\right) & \text{if } b>1. \end{cases}$$

Therefore,

$$\dim(I_{-2})_\alpha = \begin{cases} E(2m) & \text{if } b=1 \\ \mathcal{P}_{(b+1)\lambda-\alpha_2,\alpha} & \text{if } b>1, \end{cases} \tag{2.8}$$

where the function E is recursively defined as follows:
For $m>0, \Sigma_{i\in\mathbb{Z}}(-1)^i E(m-i(2i+1)) = 0$ with $E(0)=1, E(-m)=0$ and $\Sigma_{i\in\mathbb{Z}_+} E(i)q^i = \prod_{j\geq 1}(1-q^{4j})^{-1}(1-q^{4j-1})^{-1}(1-q^{4j-3})^{-1}$, where $E(i)$ is the number of ways of partitioning i into parts congruent to 2 mod 4.

$$I_{-3} \cong \begin{cases} (V\otimes I_{-2})/V(-3\alpha_3-(a+1)\alpha_2) & \text{if } a=2, b=1 \\ (V\otimes I_{-2})/V(-3\alpha_3-3\alpha_1) & \text{if } \dfrac{bp}{a}=1, p=2 \\ (V\otimes I_{-2}) & \text{otherwise.} \end{cases}$$

We now decompose the tensor product $(V\otimes I_{-3})$ into a direct sum of irreducible modules over $A_1^{(1)}$ of level 3. Now consider $\beta = -\left(1+\frac{bp}{a}\right)\alpha_3-\alpha_1-m\delta$. So

$$\begin{aligned} (\beta,\beta) &= 2-2m\left(b+\frac{bp}{a}\right)\left(1+\frac{bp}{a}\right)-\frac{bp}{a}\left(1+\frac{bp}{a}\right) \\ &\quad -\left(1+\frac{bp}{a}\right)\left(\frac{-2b}{a}\left(1+\frac{bp}{a}\right)+\frac{bp}{a}\right) \\ &\quad \text{if } \frac{bp}{a}=1, \\ 2m &= \frac{1}{2(1+b)}\left(-(\beta,\beta)-2\left(1-\frac{4b}{a}\right)\right). \end{aligned}$$

$$\dim(I_{-2})_\alpha = \begin{cases} E(2m) & \text{if } \frac{bp}{a}=1, \text{ where } 2m \text{ is as given above,} \\ \mathcal{P}_{(b+1)\lambda-\alpha_2,\alpha} & \text{if } \frac{bp}{a}>1. \end{cases} \tag{2.9}$$

Next we shall find the multiplicity of a level 3 root in I_3. We have

$$I_{-3} \cong \begin{cases} (V\otimes I_{-2})/V(-3\alpha_3-(a+1)\alpha_2) & \text{if } a=2, b=1 \\ (V\otimes I_{-2})/V(-3\alpha_3-3\alpha_1) & \text{if } \dfrac{bp}{a}=1, p=2 \\ (V\otimes I_{-2}) & \text{otherwise.} \end{cases}$$

Next we decompose the tensor product $V \otimes I_{-2}$ into a direct sum of irreducible modules over $A_1^{(1)}$ of level 3. In terms of the fundamental weights $\lambda = \frac{bp}{a}w_1 + bw_2$.

Case 1:

$$\tau = \left(\frac{bp}{a}(b+1)+2\right)w_1 + (b(b+1)-2)w_2.$$

or $w = (r_1r_2)^j (j \in \mathbb{Z})$, we have

$$\tau + \rho - w(\tau+\rho) = j(j\left(\frac{bp}{a}(b+1)+(b+1)b+2\right) + (b(b+1)-1))\alpha_1$$
$$+ j(j\left(\frac{bp}{a}(b+1)+b(b+1)+2\right) - \left(\frac{bp}{a}(b+1)+3\right))\alpha_2$$

For $w = r_2(r_1r_2)^j$, we have

$$\tau + \rho - w(\tau+\rho) = j\left(j\left(\frac{bp}{a}(b+1)+(b+1)b+2\right)+b(b+1)-1\right)\alpha_1$$
$$+ (j+1)\left(j\left(\frac{bp}{a}(b+1)+b(b+1)+2\right)+b(b+1)-1\right)\alpha_2.$$

We know that for a linear combination $a\alpha_1 + b\alpha_2$ in $A_1^{(1)}$, the abscissa is $a-b$, ordinate is $a+b$ [74]. Distinct w-conjugates of λ are $w(\lambda)$.

$$\lambda - w(\lambda) = j\left(j\left(\frac{bp}{a}+b\right)\alpha_1\right) + j\left(j\left(\frac{bp}{a}+b\right)-\frac{bp}{a}\right)\alpha_2$$
$$w(\lambda) = \lambda - j\left(j\left(\frac{bp}{a}+b\right)\alpha_1\right) + j\left(j\left(\frac{bp}{a}+b\right)-\frac{bp}{a}\right)\alpha_2$$
$$\text{Abscissa of } w(\lambda) = j\left(bj+\frac{bp}{a}j\right) = bj^2\left(1+\frac{p}{a}\right)$$
$$\text{Ordinate of } w(\lambda) = 2j^2\left(\frac{bp}{a}+b\right) + bj^2 - \frac{bp}{a}j^2.$$

If μ is a weight of $V(\lambda)$ such that $\mu + \tau$ is dominant, then $(\mu+\tau)\alpha_i^\vee \geq 0$, for $i = 1, 2$ and μ is of the form $\mu = \lambda - m\delta$ or $\lambda - \alpha_1 - m\delta$. For a general expression $\mu = \lambda - k_1\alpha_1 - m\delta$, let us find the multiplicity of μ, using the results from Feingold and Lepowsky [74]. Now μ belongs to a unique column of weights which has a top-most weight μ_T which is conjugate to λ. The nonnegative integer m' such that $\mu = \mu_T - m'\delta$, where $mult\ \mu = P(m')$ is to be determined. μ has abscissa k_1 and ordinate $2m + k_1$. μ_T has abscissa $bj^2(1+\frac{bp}{a}) = k_1$. Ordinate of μ_T is

$$2j^2\left(\frac{bp}{a}+b\right) + bj^2 - \frac{bp}{a}j^2 = 2k_1 + \frac{k_1}{(1+\frac{p}{a})} - \frac{p}{a}\frac{k_1}{(1+\frac{p}{a})}.$$

Difference between the ordinates of μ and μ_T is

$$2m + k_1 - \left(2k_1 + \frac{k_1}{(1+\frac{p}{a})} - \frac{p}{a}\frac{k_1}{(1+\frac{p}{a})}\right).$$

Therefore $\mu - \mu_T$ has ordinate $2m'$, where $m' = m - k_1\left(1 + \frac{a}{a+p} - \frac{ap}{a+p}\right)$. Hence

$$mult_\lambda \mu = P(m') = P\left(m - \frac{k_1}{a+p}(2a+p-ap)\right).$$

Let $w = (r_1 r_2)^j$. Then

$$\begin{aligned}\mu + \tau + \rho - w(\tau+\rho) &= \mu + j\left(j\frac{bp}{a}(b+1) + b(b+1) + 2\right) + (b(b+1)-1)\alpha_1\\ &\quad + j\left(j\left(\frac{bp}{a}(b+1) + b(b+1) + 2\right) - \left(\frac{bp}{a}(b+1)+3\right)\right)\alpha_2\\ Let\ A &= j\left(j\frac{bp}{a}(b+1) + b(b+1) + 2\right) + (b(b+1)-1)\\ B &= j\left(j\left(\frac{bp}{a}(b+1) + b(b+1) + 2\right) - \left(\frac{bp}{a}(b+1)+3\right)\right).\end{aligned}$$

Therefore,

$$\mu + \tau + \rho - w(\tau+\rho) = \lambda - \delta(m - A - B) - \alpha_1(k_1 + B - A).$$

Hence

$$mult\ \mu + \tau + \rho - w(\tau+\rho) = P(m - k_1 - 2(B+A)).$$

Let $w = r_2(r_1 r_2)^j$ then

$$\begin{aligned}\mu + \tau + \rho - w(\tau+\rho) &= \mu + j\left(j\frac{bp}{a}(b+1) + b(b+1) + 2\right) + (b(b+1)-1)\alpha_1\\ &\quad + (j+1)\left(j\left(\frac{bp}{a}(b+1) + b(b+1) + 2\right)\right.\\ &\quad \left. + b(b+1) - 1\right)\alpha_2.\end{aligned}$$

Let

$$C = j\left(j\frac{bp}{a}(b+1) + b(b+1) + 2\right) + (b(b+1)-1) \quad \text{and}$$

$$D = (j+1)\left(j\left(\frac{bp}{a}(b+1) + b(b+1) + 2\right) + b(b+1) - 1\right).$$

Then

$$\begin{aligned}\mu + \tau + \rho - w(\tau + \rho) &= \mu + C\alpha_1 + D\alpha_2 \\ &= \lambda - k_1\alpha_1 - m\delta + C\alpha_1 + D\alpha_2 - D\alpha_1 + D\alpha_1 \\ &= \lambda - \delta(m - D) - \alpha_1(k_1 - C + D).\end{aligned}$$

Hence

$$mult\ \mu + \tau + \rho - w(\tau + \rho) = P\left(m - D - \frac{(k_1 - C + D)^2}{b}\right).$$

Case 2: Let $\tau = \left(\frac{bp}{a} + 2\right)\lambda - \alpha_1$. In terms of fundamental weights

$$\tau = \left[\left(\frac{bp}{a} + 1\right)\frac{bp}{a} - 2\right]w_1 + \left\{2 + \left(\frac{bp}{a} + 1\right)\right\}w_2$$

Now to find $(\tau + \rho) - w(\tau + \rho)$:

$$\text{we have } \tau + \rho = \left(\left(\frac{bp}{a} + 1\right)\frac{bp}{a} - 1\right)w_1 + \left\{3 + b\left(\frac{bp}{a} + 1\right)\right\}w_2$$

Let $w = (r_1r_2)^j$. Then

$$\begin{aligned}(\tau + \rho) - w(\tau + \rho) = j&\left[j\left(\frac{bp}{a}\left(\frac{bp}{a} + 1\right) + 2 + b\left(\frac{bp}{a} + 1\right)\right) + 3 + \left(\frac{bp}{a} + 1\right)b\right]\alpha_1 \\ &+ j\left[j\left(\frac{bp}{a}\left(\frac{bp}{a} + 1\right) + 2 + b\left(\frac{bp}{a} + 1\right)\right)\right. \\ &\left.- \left(\frac{bp}{a} + 1\right)\frac{bp}{a} + 1\right]\alpha_2 \\ =\ j&\left[j\left(\left(b + \frac{bp}{a}\right)\left(\frac{bp}{a} + 1\right) + 2\right) + 3 + \left(\frac{bp}{a} + 1\right)b\right]\alpha_1 \\ &+ j\left[j\left(\left(b + \frac{bp}{a}\right)\left(\frac{bp}{a} + 1\right) + 2\right) - \left(\frac{bp}{a} + 1\right)\frac{bp}{a} + 1\right]\alpha_2.\end{aligned}$$

Let $w' = r_2(r_1r_2)^j$. Then

$$\begin{aligned}(\tau + \rho) - w'(\tau + \rho) = (j + 1)&\left[j\left(\left(b + \frac{bp}{a}\right)\left(\frac{bp}{a} + 1\right) + 2\right) + 3 + \left(\frac{bp}{a} + 1\right)b\right]\alpha_1 \\ &+ (j + 1)\left[j\left\{\left(\frac{bp}{a} + 1\right)\frac{bp}{a} + 2 + b\left(\frac{bp}{a} + 1\right)\right\}\right. \\ &\left.+3 + b\left(\frac{bp}{a} + 1\right)\right]\alpha_2.\end{aligned}$$

$$\text{Let } G = j\left[j\left(\left(b+\frac{bp}{a}\right)\left(\frac{bp}{a}+1\right)+2\right)+3+\left(\frac{bp}{a}+1\right)b\right],$$
$$H = j\left[j\left(\left(b+\frac{bp}{a}\right)\left(\frac{bp}{a}+1\right)+2\right)-\left(\frac{bp}{a}+1\right)\frac{bp}{a}+1\right],$$
$$I = (j+1)\left[j\left(\left(b+\frac{bp}{a}\right)\left(\frac{bp}{a}+1\right)+2\right)+3+\left(\frac{bp}{a}+1\right)b\right],$$
$$J = (j+1)\left[j\left\{\left(\frac{bp}{a}+1\right)\frac{bp}{a}+2+b\left(\frac{bp}{a}+1\right)\right\}+3+b\left(\frac{bp}{a}+1\right)\right].$$

Let $\mu = \lambda - k_1\alpha_1 - m\delta$; For $w = (r_1r_2)^j$,

$$\begin{aligned}\mu + (\tau+\rho) - w(\tau+\rho) = \mu &+ j\left[j\left(\left(b+\frac{bp}{a}\right)\left(\frac{bp}{a}+1\right)+2\right)\right.\\ &\left.+3+\left(\frac{bp}{a}+1\right)b\right]\alpha_1\\ &+j\left[j\left(\left(b+\frac{bp}{a}\right)\left(\frac{bp}{a}+1\right)+2\right)\right.\\ &\left.-\left(\frac{bp}{a}+1\right)\frac{bp}{a}+1\right]\alpha_2.\end{aligned}$$

For $w' = r_2(r_1r_2)^j$. Then

$$\begin{aligned}\mu + (\tau+\rho) - w'(\tau+\rho) = \mu &+ (j+1)\left[j\left(\left(b+\frac{bp}{a}\right)\left(\frac{bp}{a}+1\right)+2\right)\right.\\ &\left.+3+\left(\frac{bp}{a}+1\right)b\right]\alpha_1\\ &+(j+1)\left[j\left\{\left(\frac{bp}{a}+1\right)\frac{bp}{a}+2+b\left(\frac{bp}{a}+1\right)\right\}\right.\\ &\left.+3+b\left(\frac{bp}{a}+1\right)\right]\alpha_2\end{aligned}$$

$$\begin{aligned}\mu &+ j\left[j\left(\left(b+\frac{bp}{a}\right)\left(\frac{bp}{a}+1\right)+2\right)+3+\left(\frac{bp}{a}+1\right)b\right]\alpha_1\\ &+j\left[j\left(\left(b+\frac{bp}{a}\right)\left(\frac{bp}{a}+1\right)+2\right)-\left(\frac{bp}{a}+1\right)\frac{bp}{a}+1\right]\alpha_2\\ &= \lambda - k_1\alpha_1 - m\delta + G\alpha_1 + H\alpha_2\\ &= \lambda - \alpha_1(k_1 - G + m) + \alpha_2(H-m) + \alpha_1(H-m) - \alpha_1(H-m)\\ &= \lambda - \alpha_1(k_1 - G + m + H - m) + (H-m)\delta\\ &= \lambda - \alpha_1(k_1 - G + H) + (H-m)\delta.\end{aligned}$$

Therefore,

$$mult\ \mu + \tau + \rho - w(\tau + \rho) = P((m - H) - (k_1 - G + H)).$$

For $w' = r_2(r_1 r_2)^j$, we have

$$\begin{aligned}
\mu + (\tau + \rho) - w'(\tau + \rho) &= \mu + (j+1)\left[j\left(\left(b + \frac{bp}{a}\right)\left(\frac{bp}{a} + 1\right) + 2\right)\right.\\
&\quad \left. +3 + \left(\frac{bp}{a} + 1\right) b\right]\alpha_1\\
&\quad + (j+1)\left[j\left\{\left(\frac{bp}{a} + 1\right)\frac{bp}{a} + 2 + b\left(\frac{bp}{a} + 1\right)\right\}\right.\\
&\quad \left. +3 + b\left(\frac{bp}{a} + 1\right)\right]\alpha_2\\
&= \lambda - k_1\alpha_1 - m\delta + I\alpha_1 + J\alpha_2\\
&= \lambda - (k_1 - I + m)\alpha_1 + (J - m)\alpha_2\\
&\quad + (J - m)\alpha_1 - (J - m)\alpha_1\\
&= \lambda - (k_1 - I + J)\alpha_1 + (J - m)\delta.
\end{aligned}$$

Therefore

$$mult\ \mu + (\tau + \rho) - w'(\tau + \rho) = P\left((m - J) - \frac{1}{b}(k_1 - I + J)^2\right).$$

We have the tensor product decomposition of $V \otimes I_{-2}$ as

$$\begin{aligned}
V \otimes I_{-2} &= \Sigma_{m \geq 0} c_m V((b+2)\lambda - (m+1)\delta) + d_m V((b+2)\lambda - \alpha_2 - m\delta)\\
&\quad \oplus c'_m V\left(\left(\frac{bp}{a} + 2\right)\lambda - 2\alpha_1 - m\delta\right) + d'_m V\left(\left(\frac{bp}{a} + 2\right)\lambda - \alpha_1 - m\delta\right),
\end{aligned}$$

where

$$\begin{aligned}
c_m &= p(m - 2(B + A)) - P\left(m - D - \frac{(-C + D)^2}{b}\right)\\
d_m &= p(m - 1 - 2(B + A)) - P\left(m - D - \frac{(1 - C + D)^2}{b}\right)\\
c'_m &= p(m - H - (H - G)) - P\left(m - J - \frac{(J - I)^2}{b}\right)\\
d'_m &= p(m - H - (1 + H - G)) - P\left(m - J - \frac{(1 - I + J)^2}{b}\right).
\end{aligned} \tag{2.10}$$

Let us find the multiplicity of root in each of the components. Let

$$\beta = (b+2)\lambda - (m+m'+1)\delta - \alpha_2;$$
$$\begin{aligned}(\beta,\beta) &= (b+2)\left\{\frac{2b}{a}(b+2) - (m+m'+1)\left(\frac{bp}{a}+b\right) - b\right\}\\ &\quad - (m+m'+1)\left\{(b+2)\left(\frac{bp}{a}+b\right)\right\} - b(b+2)+2\\ &= \left[(b+2)\left(\frac{2b}{a}(b+2) - (m'+1)\left(\frac{bp}{a}+b\right) - b\right)\right.\\ &\quad \left. -(m'+1)\left((b+2)\left(\frac{pb}{a}+b\right)\right) - b(b+2)+2\right]\\ &\quad - m\left\{(b+2)\left\{\frac{2b}{a}(b+2) - \left(\frac{2b}{a}+b\right) - b - \left(\frac{bp}{a}+b\right)(b+2)\right\}\right\}.\end{aligned}$$

We have

$$\begin{aligned}3m &= \left\{\frac{a}{12}\left\{3\left(\frac{6}{a} - (m'+1)\left(1+\frac{p}{a}\right) - 1\right)\right\}\right.\\ &\quad - (m'+1)\left\{3\left(1+\frac{p}{a}\right) - 1\right\} - (\beta,\beta) \text{ if } b=1\\ 3m+2 &= \left\{\frac{a}{12}\left\{3\left(\frac{6}{a} - (m'+1)\left(1+\frac{p}{a}\right) - 1\right)\right\}\right.\\ &\quad - (m'+1)\left\{3\left(1+\frac{p}{a}\right) - 1\right\} - (\beta,\beta) + 2b = 1.\end{aligned}$$

$$\dim V((b+2)\lambda - (m+1)\delta - \alpha_2)_\alpha = \begin{cases} F(3m+2) & \text{if } b=1 \text{ where } 3m+2 \\ & \text{is given as above} \\ = \mathcal{P}_{(b+2)\lambda-(m+1)\delta,\alpha} & \text{if } b \geq 1 \end{cases}$$

$$\dim V((b+2)\lambda - (m+1)\delta - \alpha_2)_\alpha = \begin{cases} H(3m) & \text{where } 3m \text{ is given} \\ & \text{as above} \\ = \mathcal{P}_{(b+2)\lambda-(m+1)\delta,\alpha} & \text{if } b > 1, \end{cases}$$

where the functions F and H are defined recursively as follows: For $m > 0$,

$$0 = \Sigma_{i\in\mathbb{Z}}(-1)H\left(3m - \frac{1}{2}i(5i+3)\right)$$
$$0 = \Sigma_{i\in\mathbb{Z}}(-1)H\left(3m - 1 - \frac{1}{2}i(5i+1)\right)$$

with the initial conditions $H(0) = 1, H(-3m) = H(-3m+2) = 0$ and $F(m) = H(3m+2)$.

Let

$$\beta = \left(\frac{bp}{a} + 2\right)\lambda - k\alpha_1 - (m + m' + 1)\delta$$

$$\begin{aligned}
\text{Then } (\beta, \beta) &= \left(\frac{pb}{a} + 2\right)\left[\frac{bp}{a}\left(\frac{bp}{a} - k - m - m' - 1\right) + \frac{4b}{a} - b(m + m' + 1)\right] \\
&\quad - k\frac{bp}{a}\left(\frac{pb}{a} + 2\right) + 2k^2 - (m + m' + 1)\left(\left(\frac{pb}{a} + 2\right)\left(\frac{pb}{a} + b\right)\right) \\
&= m\left\{-\left(\frac{pb}{a} + 2\right)\frac{bp}{a} - b - \left(\frac{pb}{a} + 2\right)\left(\frac{pb}{a} + b\right)\right\} \\
&\quad + \left(\frac{pb}{a} + 2\right)\left[\frac{bp}{a}\left(\frac{2b}{a} - k - m' - 1\right) + \frac{4b}{a} - b(m' + 1)\right] \\
&\quad - k\frac{bp}{a}\left(\frac{pb}{a} + 2\right) + 2k^2 - (m' + 1)\left(\frac{pb}{a} + 2\right)\left(\frac{pb}{a} + b\right).
\end{aligned}$$

$$\begin{aligned}
(\beta, \beta) &= \left(\frac{pb}{a} + 2\right)\left[\frac{bp}{a}\left(\frac{2b}{a} - k - m' - 1\right) + \frac{4b}{a} - b(m' + 1)\right] \\
&\quad - k\frac{bp}{a}\left(\frac{pb}{a} + 2\right) + 2k^2 - (m' + 1)\left(\frac{pb}{a} + 2\right)\left(\frac{pb}{a} + b\right) \\
&= m\left\{-\left(\frac{pb}{a} + 2\right)\frac{bp}{a} - b - \left(\frac{pb}{a} + 2\right)\left(\frac{pb}{a} + b\right)\right\}
\end{aligned}$$

$$\begin{aligned}
3m &= 3\frac{1}{\frac{bp}{a}\left\{2 + \frac{bp}{a}\right\} + b\left(3 + \frac{bp}{a}\right)}\left\{\frac{bp}{a} + 2\left[\frac{bp}{a}\left(\frac{2b}{a} - k - m' - 1\right)\right.\right. \\
&\quad \left. + \frac{4b}{a} - b(m' + 1)\right] - (\beta, \beta) \\
&\quad \left. - k\frac{bp}{a}\left(\frac{bp}{a} + 2\right) + 2k^2 - (m' + 1)\left(\frac{bp}{a} + 2\right)\left(\frac{bp}{a} + b\right)\right\} \\
&= 3\frac{1}{\frac{p}{a}\{2 + \frac{p}{a}\} + (3 + \frac{p}{a})}\left\{\frac{p}{a} + 2\left[\frac{p}{a}\left(\frac{2}{a} - k - m' - 1\right)\right.\right. \\
&\quad \left. + \frac{4}{a} - (m' + 1)\right] - (\beta, \beta) \\
&\quad \left. - k\frac{p}{a}\left(\frac{p}{a} + 2\right) + 2k^2 - (m' + 1)\left(\frac{p}{a} + 2\right)\left(\frac{p}{a} + 1\right)\right\} \text{ if } b = 1
\end{aligned}$$

$$\dim V\left(\left(\frac{bp}{a} + 2\right)\lambda - k\alpha_1 - (m + 1)\delta\right)_\alpha = \begin{cases} H(3m) & \text{if } b = 1 \\ \mathcal{P}_{\left(\frac{bp}{a} + 2\right)\lambda - k\alpha_1 - (m+1)\delta, \alpha} & \text{if } b > 1, \end{cases}$$

where $3m$ is given in the above equation. Let

$$\beta = (b+2)\lambda - k\alpha_2 - (m+m')\delta$$

$$(\beta,\beta) = (b+2)\left[\frac{2b}{a}(b+2) - b - (m+m')\left(\frac{bp}{a}+b\right)\right]$$
$$- k((b+2)b - 2k) - (m+m')(b+2)\left(\frac{bp}{a}+b\right)$$

$$(\beta,\beta) - (b+2)\left[\frac{2b}{a}(b+2) - b - m'\left(\frac{bp}{a}+b\right)\right] + k((b+2)b-2k) + m'(b+2)$$
$$\left(\frac{bp}{a}+b\right) = -m\left\{2\left(b+\frac{bp}{a}\right)\right\}$$

$$3m = \frac{3}{2\left(b+\frac{bp}{a}\right)}\left\{-(\beta,\beta) + (b+2)\left[\frac{2b}{a}(b+2) - b - m'\left(\frac{bp}{a}+b\right)\right]\right.$$
$$\left. - k((b+2)b-2k) - m'(b+2)\left(\frac{bp}{a}+b\right)\right\}$$

$$3m+2 = \frac{3}{2\left(b+\frac{bp}{a}\right)}\left\{-(\beta,\beta) + (b+2)\left[\frac{2b}{a}(b+2) - b - m'\left(\frac{bp}{a}+b\right)\right]\right.$$
$$\left. - k((b+2)b-2k) - m'(b+2)\left(\frac{bp}{a}+b\right)\right\} + 2$$

$$\dim V((b+2)\lambda - m\delta)_\alpha = \begin{cases} F\left(\dfrac{3}{2\left(1+\frac{p}{a}\right)}\right)\left\{-(\beta,\beta) + 3\left(\frac{6}{a} - 1 - m'\left(1+\frac{p}{a}\right)\right)\right. \\ \qquad \left. -k(3-2k) - m'3\left(1+\frac{p}{a}\right)\right\} \\ \qquad \text{if } b=1 \\ \mathcal{P}_{(b+2)\lambda - m\delta,\alpha} \quad \text{if } b>1. \end{cases}$$

$$\begin{aligned} \dim(I_{-3})_\alpha = {} & \Sigma_{m\geq 0} c_m H(3m) + d_m F(3m+2) + c'_m H(3m) + d'_m F(3m+2) \\ & + \Sigma c_m \mathcal{P}_{(b+2)\lambda-\alpha_2-m\delta,\alpha} + \Sigma d_m \mathcal{P}_{(b+2)\lambda-m\delta,\alpha} \\ & + c'_m \mathcal{P}_{\left(\left(\frac{bp}{a}+2\right)\lambda - 2\alpha_1 - m\delta_1\alpha\right)} + d'_m \mathcal{P}_{\left(\left(\frac{bp}{a}+2\right)\lambda - \alpha_1 - m\delta\right)} \end{aligned} \tag{2.11}$$

here c_m, d_m are given by the Equation (2.10). Hence we have proved the following theorem.

Theorem 30. *With the usual notations, let* $L = \bigoplus_{n\in\mathbb{Z}} L_n$ *be the realization of the extended-hyperbolic Kac-Moody algebra* $EHA_1^{(1)}$ *associated with the GCM* $\begin{pmatrix} 2 & -2 & -p \\ -2 & 2 & -a \\ \frac{-bp}{a} & -b & 2 \end{pmatrix}$ *with* $ab > 4$ *or* $\frac{pb^2}{4a} > 4$. *Then the multiplicities of roots upto level 3 are given as follows:*

(i) $\dim(L_{-1})_\alpha = P\left(\frac{1}{a+p} - \frac{(\alpha,\alpha)a}{2b(a+p)}\right)$
(ii) $\dim(L_{-2})_\alpha = \dim(G_{-2})_\alpha - \dim(I_{-2})_\alpha$
(iii) $\dim(L_{-3})_\alpha = \dim(G_{-3})_\alpha - \dim(I_{-3})_\alpha$,

where $\dim(G_{-2})_\alpha$ *and* $\dim(G_{-3})_\alpha$ *are given by Equations (2.6) and (2.7) and* $\dim(I_{-2})_\alpha$ *and* $\dim(I_{-3})_\alpha$ *are given by Equations (2.8), (2.9), and (2.11).*

2.10.3 Structure and root multiplicities for an extended Kac-Moody algebra $EHA_2^{(2)}$

(i) Realization and homology modules for $EHA_2^{(2)}$: Let us denote by $EHA_2^{(2)}$, the class of extended-hyperbolic Kac-Moody algebras whose associated GCM is of the form $\begin{pmatrix} 2 & -4 & -c \\ -1 & 2 & -a \\ -d & -b & 2 \end{pmatrix}$, that is, the class of all 3×3 GCM of extended-hyperbolic type obtained from the affine algebra $A_2^{(2)}$ associated with the GCM $\begin{pmatrix} 2 & -4 \\ -1 & 2 \end{pmatrix}$. Here $a, b, c, d \in \mathbb{Z}^+$ and $ab > 4$ or $cd > 4$.

First we give a realization for $EHA_2^{(2)}$ which was done earlier in Sthanumoorthy et al. [81], as a graded Lie algebra of Kac-Moody type.

Consider the affine Kac-Moody algebra associated with the GCM $A = \begin{pmatrix} 2 & -4 \\ -1 & 2 \end{pmatrix}$. Let $(h, \Pi, \Pi^\vee)$ be the realization of A with $\Pi = \{\alpha_1, \alpha_2\}$ and $\Pi^\vee = \{\alpha_1^\vee, \alpha_2^\vee\}$. Then we have the following bilinear relations:

$$(\alpha_1, \alpha_1) = 4, \quad (\alpha_1, \alpha_2) = -2, \quad (\alpha_2, \alpha_2) = 1.$$

Let α_3' be the element in $\mathfrak{h}^*$ such that $\alpha_3'(\alpha_1^\vee) = 0, \alpha_3'(\alpha_2^\vee) = 1, \alpha_3'(\alpha_3'^\vee) = 0$. Then $(\alpha_3', \alpha_1) = 0, (\alpha_3', \alpha_2) = \frac{1}{2}$. Let a, b, p be three nonnegative integers such that $ab > 4$ or $(bp^2/4a) > 4$ and $(bp/4a)$ is an integer with $a \neq 0$. Let us define

$$\lambda = \frac{16a + 4abp + bp^2}{16a(2a+p)}\alpha_1 + \frac{16a - bp^2}{8a(2a+p)}\alpha_2 + \left(b + \frac{bp}{2a}\right)\alpha_3'.$$

Set $\alpha_3 = -\lambda$. Form the matrix $C = (\langle\alpha_i, \alpha_j\rangle)_{i,j=1}^3$. Then $C = \begin{pmatrix} 2 & -4 & -p \\ -1 & 2 & -a \\ -bp/4a & -b & 2 \end{pmatrix}$ is a symmetrizable GCM of extended-hyperbolic type. Let V be the integrable highest weight irreducible module over G with the highest weight λ.

Let $V*$ be the contragradient of V and ψ be the mapping as defined earlier. Let G be the Kac-Moody algebra associated with the GCM $\begin{pmatrix} 2 & -4 \\ -1 & 2 \end{pmatrix}$. Form the graded Lie algebra $L(G, V, V^*, \psi)$. Then $L \cong G(C)$ with $C = \begin{pmatrix} 2 & -4 & -p \\ -1 & 2 & -a \\ -bp/4a & -b & 2 \end{pmatrix}$. Thus L is a symmetrizable Kac-Moody algebra of extended-hyperbolic type associated with Cartan matrix which is an extension of the Cartan-matrix $A_2^{(2)}$. In general, we denote the class of extended-hyperbolic Kac-Moody algebras associated with the above GCM by $EHA_2^{(2)}$. Homology modules upto level 3 for $p \geq 0$ were computed in Sthanumoorthy et al. [81]. From the realization of $L = EHA_2^{(2)}$ as $L = L_- \oplus L_0 \oplus L_+ = G/I$ and using the involutive automorphism, it suffices to study about the negative part $L_- = G_-/I_-$.

Let $S = \{1, 2\} \subset N = \{1, 2, 3\}$. Here $\mathcal{G}_s$ is the affine Lie algebra $A_2^{(2)} \cdot \Delta^+(S) = \{k_1\alpha_1 + k_2\alpha_2 + k_3\alpha_3 \in \Delta^+ | k_3 \neq 0\}$. Let Δ_s be the root system of $\mathcal{G}_s$. The only element of length 1 in $W(S)$ is r_3. Hence $H_1(L-) \cong V(-\alpha_3)$. By the Kostant's formula we see that the elements of length 2 in $W(S)$ are r_3r_1 and r_3r_2. We have $r_3r_1\rho - \rho = -\alpha_1 - (1+bp/4a)\alpha_3$ and $r_3r_2\rho - \rho = -(b+1)\alpha_3 - \alpha_2$. Hence $H_2(L) \cong V(-(b+1)\alpha_3 - \alpha_2) \oplus V(-(1+bp/4a)\alpha_3 - \alpha_1)$. Elements of length 3 in $W(S)$ are $r_3r_2r_3, r_3r_2r_1, r_3r_1r_2$, and $r_3r_1r_3$. We have

$$r_3r_2r_3\rho - \rho = -(a+1)\alpha_2 - (ab+a)\alpha_3,$$
$$r_3r_2r_1\rho - \rho = -\alpha_1 - 2\alpha_2 - \left(2b + 1 + \frac{bp}{4a}\right)\alpha_3,$$
$$r_3r_1r_2\rho - \rho = -5\alpha_1 - \alpha_2 - \left(\frac{4a + 4ab + 5bp}{4a}\right)\alpha_3,$$
$$r_3r_1r_3\rho - \rho = -(1+p)\alpha_1 - \frac{pb}{4a}(p+1)\alpha_3.$$

Hence

$$\begin{aligned} H_3(L_-) \cong\ & V(-(ab+a)\alpha_3 - (a+1)\alpha_2) \oplus V(-\alpha_1 - 2\alpha_2 - (2b + 1 + bp/4a)\alpha_3) \\ & \oplus V(-5\alpha_1 - \alpha_2 - ((4a + 4ab + 5bp)/(4a))\alpha_3) \\ & \oplus V(-(1+p)\alpha_1 - (pb/4a)(p+1)\alpha_3). \end{aligned}$$

Similarly we can compute the other homology modules $H_4(L_-), H_5(L_-), H_6(L_-), \ldots$. Using the homological approach together with the representation theory of affine

Kac-Moody algebra $A_2^{(2)}$ below we give some of the boundary homomorphisms and deduce some new structural information on maximal ideals in $EHA_2^{(2)}$.

(ii). Structure of the maximal ideal in $EHA_2^{(2)}$: In this section, we study for the general case $p \geq 0$ ($p = 0$ was done earlier in Sthanumoorthy et al. [81]). First we study the structure of the maximal graded ideal I. We know that the ideal I_- of G_- is generated by the homological subspace I_{-2} and hence we may write $I_- = I_-^{(2)}$. Similarly for $j \geq 2$, we write $I_-^{(j)} = \Sigma_{n \geq j} I_{-n}, L_-^{(j)} = G/I_-^{(j)}$, and $N_-^{(j)} = I_-^{(j)}/I_-^{(j+1)}$. By the homological theory described earlier, we have in general that $I_{-(j+1)} \cong (V \otimes I_{-j})/H_3(L_-^{(j)})_{-(j+1)}$, for $j \geq 2$. Since G_- is free and I_- is generated by the subspace I_{-2} from the Hochschild-Serre five-term exact sequence we see that $I_{-2} \cong H_2(L_-)$. But we get $H_2(L_-) \cong V(-(b+1)\alpha_3 - \alpha_2) \oplus V(-(1 + bp/4a)\alpha_3 - \alpha_1)$. Hence $I_{-2} \cong V(-(b+1)\alpha_3 - \alpha_2) \oplus V(-(1 + bp/4a)\alpha_3 - \alpha_1)$. When $j = 2$, $L_-^{(2)}$ coincides with the subspace $\eta^-(S)$ for $S = \{1, 2\}$ and therefore we can compute $H_3(L_-^{(2)})$ using the Kostant formula. In this case of $EHA_2^{(2)}$, we have

$$\begin{aligned} H_3(L_-^{(2)}) \cong\ & V(-(ab + a)\alpha_3 - (a + 1)\alpha_2) \oplus V(-\alpha_1 - 2\alpha_2 - (2b + 1 \\ & + bp/4a)\alpha_3) \oplus V(-5\alpha_1 - \alpha_2 - ((4a + 4ab + 5bp)/(4a))\alpha_3) \\ & \oplus V(-(1 + p)\alpha_1 - (pb/4a)\alpha_3). \end{aligned}$$

Thus

$$H_3(L_-^{(2)})_{-3} = \begin{cases} V(-3\alpha_3 - 3\alpha_2) & \text{if } b = 1, a = 2 \\ V(-3\alpha_3 - 3\alpha_1) & \text{if } b = 2a, p = 2 \\ = 0 & \text{otherwise} \end{cases}$$

and hence we obtain $I_{-3} \cong (V \otimes I_{-2})/H_3(L_-^{(2)})_{-3}$ where $H_3(L_-^{(2)})_{-3}$ is given above. To determine the higher components, we combine the theory of homological techniques and spectral sequences. We have $I_{-4} \cong (V \otimes I_{-3})/H_3(L_-^{(3)})_{-4}$. First let us compute $H_3(L_-^{(3)})_{-4}$. Consider the short exact sequence $0 \longrightarrow N_-^{(2)} \longrightarrow L_-^{(3)} \longrightarrow L_-^{(2)} \longrightarrow 0$ and the corresponding spectral sequence $\{E_{p,q}^r\}$ converging to $H^*(L_-^{(3)})$ such that $E_{p,q}^r \cong H_p(L_-^{(2)}) \otimes \Lambda^q(I_{-2})$. Consider

$$0 \longrightarrow E_{2,0}^2 \xrightarrow{d_2} E_{0,1}^2 \longrightarrow 0.$$

Note that $H_1(L_-^{(3)}) \cong L_-^{(3)}/[L_-^{(3)}, L_-^{(3)}] \cong L_{-1} = V$. Since the spectral sequence converges to $H_*(L_-^{(3)})$, we have $H_1(L_-^{(3)}) \cong E_{1,0}^\infty \oplus E_{0,1}^\infty$. But

$$E_{1,0}^\infty = E_{0,1}^\infty \cong H_1(L_-^{(2)}) \cong \frac{L_-^{(2)}}{[L_-^{(2)}, L_-^{(2)}]} \cong L_{-1} = V,$$

which implies $E^{\infty}_{1,0} = E^{3}_{0,1} = 0$. Hence the homomorphism d_2 is surjective. Since $E^{2}_{2,0} \cong I_{-2}$ and $E^{2}_{2,0} \cong I_{-2}$, d_2 must be an isomorphism. Thus $E^{3}_{2,0} = 0$. Hence $E^{\infty}_{2,0} = 0$. Now consider the sequence $0 \longrightarrow E^{2}_{3,0} \xrightarrow{d_2} E^{2}_{1,1} \longrightarrow 0$. We have

$$\begin{aligned} E^{2}_{3,0} &\cong H_3(L^{(2)}_{-}) \cong V(-(2b+1+bp/4a)\alpha_3 - \alpha_1 - 2\alpha_2) \\ &\quad \oplus V(-(ab+b)\alpha_3 - (a+1)\alpha_2) \\ &\quad \oplus V(-(bp/4a(p+1))\alpha_3 - (p+1)\alpha_1) \\ &\quad \oplus V(-(4a+4ab+5bp)/(4a)\alpha_3 - 5\alpha_2 - 5\alpha_1) \text{ and} \\ E^{2}_{1,1} &\cong H_1(L^{(2)}_{-}) \otimes I_{-2} \cong V \otimes I_{-2}. \end{aligned}$$

Then $V \otimes I_{-2}$ is a direct sum of irreducible modules over $A^{(2)}_{2}$ of level $b + 3 + bp/4a$. Then by comparing the levels of both the terms $E^{2}_{3,0}$ and $E^{2}_{1,1}$ in d_2: $E^{2}_{3,0} \longrightarrow E^{2}_{1,1}$ we see that d_2 is trivial. So $E^{3}_{3,0} = E^{2}_{3,0}$ and $E^{\infty}_{1,1} = E^{3}_{1,1} = E^{2}_{1,1} \cong V \otimes I_{-2}$. Since, $I^{(3)}_{-}$ is generated by I_{-3}, we get $H_2(L^{(3)}_{-}) \cong I_{-3}$. But we have $H_2(L^{(3)}_{-}) \cong E^{\infty}_{2,0} \oplus E^{\infty}_{1,1} \oplus E^{\infty}_{0,2}$. Hence it follows that $E^{\infty}_{0,2} = E^{4}_{0,2} = 0$. Hence the homomorphism d_2: $E^{3}_{3,0} \longrightarrow E^{3}_{0,2}$ is surjective. As $E^{\infty}_{0,2}$ is a submodule of $E^{3}_{0,2} \cong \Lambda^2(I_{-2})$, we see that d_3 is trivial. This implies $E^{\infty}_{3,0} = E^{4}_{3,0} = E^{3}_{3,0} = E^{2}_{3,0} \cong H_3(L^{(2)}_{-})$ and $E^{3}_{0,2} = 0$. Thus the homomorphism d_2: $E^{2}_{2,1} \longrightarrow E^{2}_{0,2}$ is surjective in the following sequence:

$$0 \longrightarrow E^{2}_{4,0} \longrightarrow E^{2}_{2,1} \longrightarrow E^{2}_{0,2} \longrightarrow 0.$$

Again by comparing levels, we deduce that the homomorphism d_2: $E^{2}_{4,0} \longrightarrow E^{2}_{2,1}$ must be trivial. Therefore $E^{2}_{4,0} = E^{3}_{4,0}$ and

$$\begin{aligned} E^{\infty}_{2,1} = E^{3}_{2,1} &= \ker\,(d_2\colon E^{2}_{2,1} \longrightarrow E^{2}_{0,2})/\Im(d_2\colon E^{2}_{4,0} \longrightarrow E^{2}_{2,1}) \\ &= \ker\,(d_2\colon E^{2}_{2,1} \longrightarrow E^{2}_{0,2}). \end{aligned}$$

We have $d_2 \cong S^2(I_{-2})$. Therefore, $E^{\infty}_{2,1} \cong S^2(I_{-2})$. Now consider the sequence $0 \longrightarrow E^{2}_{5,0} \longrightarrow E^{2}_{3,1} \longrightarrow E^{2}_{1,2} \longrightarrow 0$. By comparing levels we see that the homomorphism d_2: $E^{2}_{3,1} \longrightarrow E^{2}_{1,2}$ is trivial. Thus, $E^{3}_{1,2} = E^{2}_{1,2} \cong V \otimes \Lambda^2(I_{-2})$. Again by comparing the levels of the terms in the sequence $0 \longrightarrow E^{3}_{4,0} \xrightarrow{d_2} E^{3}_{1,2} \longrightarrow 0$, we see that d_3 is trivial. Therefore $E^{\infty}_{1,2} = E^{4}_{1,2} = E^{3}_{1,2} = E^{2}_{1,2} \cong V \otimes \Lambda^2(I_{-2})$. Finally since $E^{\infty}_{0,3}$ is a submodule of $E^{2}_{0,3} \cong \Lambda^3(I_{-2})$, we get

$$\begin{aligned} H_3(L^{(3)}_{-}) &\cong V(-(2b+1)\alpha_3 - \alpha_1 - 2\alpha_2) \oplus V(-(ab+b)\alpha_3 - (a+1)\alpha_2) \\ &\quad \oplus V(-(4a+4ab+5bp)\alpha_3)\alpha_2 - 5\alpha_1) \\ &\quad \oplus V(-(bp/4a(p+1))\alpha_3 - (p+1)\alpha_1) \\ &\quad \oplus S^2(I_{-2}) \oplus V \otimes \Lambda^2(I_{-2}) \oplus M, \end{aligned}$$

where M is a direct sum irreducible highest weight representations (of level > 4) of $A_2^{(2)}$. Therefore we obtain

$$H_3(L_-^3)_{-4} \cong \begin{cases} V(-4\alpha_3 - (a+1)\alpha_2) & \text{if } a=1, b=2 \text{ or } b=1, a=3 \\ V(-4\alpha_3 - (p+1)\alpha_1) & \text{if } bp=4a, p=3 \text{ or } bp=8a, p=1 \\ S^2(I_{-2}) & \text{if } b=1 \\ =0 & \text{otherwise.} \end{cases}$$

Now we shall determine I_{-5}. We start with the exact sequence $0 \longrightarrow N_-^{(3)} \longrightarrow L_-^{(4)} \longrightarrow L_-^{(3)} \longrightarrow 0$ and the corresponding spectral sequence $\{E_{p,q}^r\}$ converging to $H^*(L_-^{(4)})$ such that $E_{p,q}^2 \cong H_p(L_-^{(3)}) \otimes \Lambda^q(I_{-3})$. We will compute $H_3(L_-^{(4)})_{-5}$ from this spectral sequence. It is easy to show that d_2: $E_{2,0}^2 \longrightarrow E_{0,1}^2$ is an isomorphism and that $E_{2,0}^\infty = 0$. Consider the sequence,

$$0 \longrightarrow E_{3,0}^2 \xrightarrow{d_2} E_{1,1}^2 \longrightarrow 0.$$

From the earlier equation we get $E_{3,0}^2 \cong H_3(L_-^{(3)})$ and $E_{1,1}^2 \cong H_1(L_-^{(3)}) \otimes I_{-3} \cong V \otimes I_{-3}$ and the level differs in $E_{3,0}^2$ and $E_{1,1}^2$. Hence $d_2 = 0$. So $H_3(L_-^{(3)}) = 0$ and hence we get $H_2(L_-^{(4)}) \cong I_{-4}$. We have $E_{1,1}^\infty = E_{1,1}^3 = E_{1,1}^2/\Im d_2 \cong V \otimes I_{-3}/\Im d_2$ is a direct summand of $H_2(L_-^{(4)})$. So $E_{0,2}^\infty = 0$ which implies that $E_{0,2}^4 = 0$. Thus the homomorphism d_3 is surjective in the sequence

$$0 \longrightarrow E_{3,0}^3 \xrightarrow{d_3} E_{0,2}^3 \longrightarrow 0.$$

Since $E_{0,2}^3$ is a submodule of $E_{0,2}^2 \cong \Lambda^2(I_{-3})$, by comparing levels of $E_{3,0}^3$ and $E_{0,2}^3$, we see that $d_3 = 0$. Thus $E_{3,0}^\infty = E_{3,0}^4 = E_{3,0}^3 = E_{3,0}^2$. So we have

$$(E_{3,0}^\infty)_{-5} = (E_{3,0}^2)_{-5} \cong H_3(L_-^{(3)})_{-5} = \begin{cases} V(-5\alpha_3 - 2\alpha_2 - \alpha_1) & \text{if } b=2 \\ V(-5\alpha_3 - 5\alpha_1) & \text{if } a=2, p=4 \\ 0 & \text{otherwise.} \end{cases}$$

Similarly we have $(E_{2,1}^\infty)_{-5} = (E_{1,2}^\infty)_{-5} = (E_{0,3}^\infty)_{-5} = 0$. Therefore we get

$$H_3(L_-^{(4)})_{-5} = \begin{cases} V(-5\alpha_3 - 2\alpha_2 - \alpha_1) & \text{if } b=2 \\ V(-5\alpha_3 - 5\alpha_1) & \text{if } a=2, p=4 \\ 0, & \text{otherwise} \end{cases}$$

$I_{-5} \cong (V \otimes I_{-4})/H_3(L_-^{(4)})_{-5}$, where $H_3(L_-^{(4)})_{-5}$ is given above. From the above equations, we get the structure of the components of the maximal ideal I (upto level 5) in the extended-hyperbolic Kac-Moody algebra $EHA_2^{(2)}$. Thus we have proved the following theorem.

Theorem 31. *With the usual notations, let* $L = \bigoplus_{n\in\mathbb{Z}} L_n$ *be the realization of* $EHA_2^{(2)}$ *associated with the GCM* $\begin{pmatrix} 2 & -4 & -p \\ -1 & 2 & -a \\ -bp/4a & -b & 2 \end{pmatrix}$ *with* $ab > 4$. *Then we have the following:*

(1) $I_{-2} \cong V(-(b+1)\alpha_3 - \alpha_2) \oplus V(-(1+bp/4a)\alpha_3 - \alpha_1)$
(2) $I_{-3} \cong (V \otimes I_{-2})/H_3(L_-^{(2)})_{-3}$, *where*

$$H_3(L_-^{(2)})_{-3} = \begin{cases} V(-3\alpha_3 - 3\alpha_2) & \text{if } b=1, a=2 \\ V(-3\alpha_3 - 3\alpha_1) & \text{if } b=2a, p=2 \\ 0 & \text{otherwise} \end{cases}$$

(3) $I_{-4} \cong (V \otimes I_{-3})/H_3(L_-^{(3)})_{-4}$, *where*

$$H_3(L_-^{3})_{-4} \cong \begin{cases} V(-4\alpha_3 - (a+1)\alpha_2) & \text{if } a=1, b=2 \text{ or } b=1, a=3 \\ V(-4\alpha_3 - (p+1)\alpha_1) & \text{if } bp=4a, p=3 \text{ or } bp=8a, p=1 \\ S^2(I_{-2}) & \text{if } b=1 \\ =0 & \text{otherwise} \end{cases}$$

(4) $I_{-5} \cong (V \otimes I_{-4})/H_3(L_-^{(4)})_{-5}$, *where*

$$H_3(L_-^{(4)})_{-5} = \begin{cases} V(-5\alpha_3 - 2\alpha_2 - \alpha_1) & \text{if } b=2 \\ V(-5\alpha_3 - 5\alpha_1) & \text{if } a=2, p=4 \\ 0 & \text{otherwise.} \end{cases}$$

For complete details about finding root multiplicities for $EHA_2^{(2)}$, and the proof (along with meanings of the notations) of the following theorem, one can refer Sthanumoorthy and Uma Maheswari [60].

Theorem 32. *Let* $L = \oplus_{n\in\mathbb{Z}} L_n$ *be the realization of the extended-hyperbolic Kac-Moody algebra* $EHA_2^{(2)}$ *associated with the GCM* $\begin{pmatrix} 2 & -4 & -p \\ -1 & 2 & -a \\ -\frac{bp}{4a} & -b & 2 \end{pmatrix}$ *with* $ab > 4$. *Then the multiplicities of roots upto level 3 are given as follows:*

(1) $\dim\,(L_{-1})_\alpha = P(\frac{1}{2a+p} - \frac{a(\alpha,\alpha)}{b(2a+p)})$,
(2) $\dim\,(L_{-2})_\alpha = \dim(G_{-2})_\alpha - \dim(I_{-2})_\alpha$,
(3) $\dim\,(L_{-3})_\alpha = \dim(G_{-3})_\alpha - \dim(I_{-3})_\alpha$,

where

$$\dim\,(G_{-2})_\alpha = \begin{cases} B(\alpha), & \text{if } (\alpha|2) \nmid \alpha \\ B(\alpha) - 1/2P\left(\frac{b-a(\alpha,\alpha)}{4(2ab+bp)}\right), & \text{if } (\alpha/2)|\alpha \end{cases}$$

$$\dim (G_{-3})_\alpha = \begin{cases} B(\alpha), & \textit{if } (\alpha|3) \nmid \alpha \\ B(\alpha) - (1/3)P\left(\frac{b-a(\alpha,\alpha)}{9(2ab+bp)}\right), & \textit{if } (\alpha/3)|\alpha \end{cases}$$

$$\dim (I_{-2})_\alpha = \begin{cases} D(4_{m+1}) & \textit{if } b = 1 \textit{ and } 4_{m+1} \textit{ as given above} \\ \mathcal{P}_{(b+1)\lambda-\alpha_2,\alpha} & \textit{if } b > 1 \end{cases}$$

$$\dim (I_{-3})_\alpha = \begin{cases} \Sigma c_m R(3m+1) + d_m S(3m) + e_m S(3m) + f_m R(3m+1) \textit{ for } b = 1 \\ \Sigma c_m \mathcal{P}_{(b+2)\lambda-\alpha_2-m\delta,\alpha} + d_m \mathcal{P}_{(b+2)\lambda-m\delta,\alpha} + e_m \mathcal{P}_{(\frac{bp}{4a}+2)\lambda-m\delta,\alpha} \\ +f_m \mathcal{P}_{(\frac{bp}{4a}+2)\lambda-\alpha_1-m\delta,\alpha} \textit{ for } b > 1. \end{cases}$$

Remark 64. For filtered Lie algebras associated with graded Lie algebras of Cartan type, one can refer [238].

Exercises

2.1 Let $A = \begin{pmatrix} 2 & -1 & 0 \\ -1 & 2 & -1 \\ 0 & -2 & 2 \end{pmatrix}$ be a Cartan matrix. Construct a Lie algebra, whose Cartan matrix is A and state the bracket relations.

2.2 In the usual notation, derive the Weyl group of $C_3^{(1)}$.

2.3 Construct a Cartan matrix of $A_2^{(1)}$ from that of A_2, using the formula of finding GCM from that of finite-dimensional Cartan matrix.

2.4 **(i)** Using the theory of realization of affine algebras, construct an untwisted affine KM algebra $A_3^{(1)}$ from that of A_3.
(ii) Using the theory of realization of affine algebras, construct a twisted affine KM algebra $D_3^{(2)}$ from that of D_3.

2.5 Prove that in the usual notation, the root $a\alpha_1 + 2\alpha_2$ is a special imaginary root for $\mathcal{G}(A)$, where $A = \begin{pmatrix} 2 & -a \\ -b & 2 \end{pmatrix}$.

2.6 Show that an indecomposable GCM is of affine type if and only if it is degenerate and all its principal minors are nonnegative.

2.7 In the usual notation, if $(\mathfrak{h}, \Pi, \Pi^\vee)$ satisfies linearly independency of the sets Π and $\Pi^\vee$ and $\langle \alpha_i^\vee, \alpha_j \rangle = a_{ij}$ for all $i, j \in I$, then prove that $\dim \mathfrak{h} \geq 2n - l$.

2.8 Let $A = \begin{pmatrix} 2 & -2 \\ -5 & 2 \end{pmatrix}$ be a GCM. Then in the usual notation, prove that $\Gamma_1(A) = \{\gamma \in K_A\colon \gamma \pm \alpha_1 \notin \Delta(A)\} \neq \emptyset$.

2.9 Let

$$A = \begin{pmatrix} 2 & -1 \\ -5 & 2 \end{pmatrix}$$

be the GCM associated to the hyperbolic Kac-Moody algebra. Then prove that $\Delta^{pim} \backslash \Delta^{sim} \neq 0$.

2.10 Prove that for any root $\alpha \in \Delta$ (set of all root system), $supp(\alpha)$ is connected.

2.11 Let $(\mathfrak{h}, \Pi, \Pi^\vee)$ be a minimal realization of A. Then prove that there exists a minimal realization $(\mathfrak{h}_J, \Pi_J, \Pi_J^\vee)$ of A_J, where $\mathfrak{h}_J \subset \mathfrak{h}$, $\Pi_J \subset \Pi$, $\Pi_J^\vee \subset \Pi^\vee$.

2.12 Let A be an indecomposable GCM of finite or affine type. Then prove that A is symmetrizable.

2.13 Let A be a symmetrizable GCM of finite, affine, or hyperbolic type. If $\alpha \in Q$ (the root lattice of A) and $|\alpha|^2 \le \min_i |\alpha_i|^2$, then show that either α or $-\alpha$ lies in Q_+.

2.14 With the usual notations, let $L = \oplus_{n \in Z} L_n$ be the realization of $EHA_1^{(1)}$ associated with the GCM, $A = \begin{pmatrix} 2 & -2 & 0 \\ -2 & 2 & -a \\ 0 & -b & 2 \end{pmatrix}$ with $ab > 4$. Prove the following,

(i) $I_{-2} \cong V(-(b+1)\alpha_3 - \alpha_2)$

(ii) $I_{-3} \cong V \otimes I_{-2}$

(iii) $I_{-4} \cong \begin{cases} V \otimes I_{-3}/(S^2(I_{-2}) \oplus V(-4\alpha_3 - 3\alpha_2 - \alpha_1)) & \text{if } b = 1 \\ V \otimes I_{-3} & \text{if } b > 1 \end{cases}$

(iv) $I_{-5} \cong \begin{cases} V \otimes I_{-4}/V \otimes \Lambda^2(I_{-2}) & \text{if } b = 1 \\ V \otimes I_{-4} & \text{if } b > 1. \end{cases}$

Here $S^2(I_{-2})$ denotes the symmetric algebra and $\Lambda(I_{-2})$ denotes the exterior algebra of I_{-2}.

2.15 Let $L = \oplus_{n \in Z} L_n$ be the realization of $EHA_1^{(1)}$ associated with the GCM, $A = \begin{pmatrix} 2 & -2 & 0 \\ -2 & 2 & -a \\ 0 & -b & 2 \end{pmatrix}$ with $ab > 4$. In the usual notation, prove that the root multiplicities upto level are 3 as follows:

(i) $\dim(L_{-1})_\alpha = P\left(\frac{1}{a} - \frac{(\alpha,\alpha)}{2b}\right)$

(ii) $\dim(L_{-2})_\alpha = \dim(G_{-2})_\alpha - \dim(I_{-2})_\alpha$

(iii) $\dim(L_{-3})_\alpha = \dim(G_{-3})_\alpha - \dim(I_{-3})_\alpha$.

2.16 Let $L = \oplus_{n \in Z} L_n$ be the realization of $EHA_1^{(1)}$ associated with the GCM, $A = \begin{pmatrix} 2 & -2 & 0 \\ -2 & 2 & -a \\ 0 & -b & 2 \end{pmatrix}$ with $b = 1, a \ge 5$. In the usual notation, then prove that

$$\dim(L_{-1})_\alpha = P((1/a) - (1/2)(\alpha, \alpha)),$$

$$\dim(L_{-2})_\alpha = \begin{cases} B(\alpha) - (1/2)B(\tau) - E((4/a) - 1 - (1/2)(\alpha,\alpha)) \\ \quad \text{if there exists a } \tau \text{ such that } \tau | \alpha \\ B(\alpha) - E((4/a) - 1 - (1/2)(\alpha,\alpha)), \quad \text{otherwise} \end{cases}$$

$$\dim(L_{-3})_\alpha = \begin{cases} B(\alpha) - (1/3)B(\tau) - \dim(I_{-3})_\alpha \\ \quad \text{if there exists a } \tau \text{ such that } \tau | \alpha \\ B(\alpha) - \dim(I_{-3})_\alpha, \quad \text{otherwise.} \end{cases}$$

Generalized Kac-Moody algebras

3

Chapter Outline

3.1 Borcherds-Cartan matrix (BCM), Generalized Generalized Cartan matrix (GGCM), Borcherds Kac-Moody (BKM) algebras, and Generalized Kac-Moody (GKM) algebras 136
3.2 Dynkin diagrams of GKM algebras 139
3.3 Root systems and Weyl groups of GKM algebras 144
3.4 Special imaginary roots in GKM algebras and their complete classifications 150
3.5 Strictly imaginary roots in GKM algebras and their complete classifications 164
3.6 Purely imaginary roots in GKM algebras and their complete classifications 169
3.7 Representations of GKM algebras 180
3.8 Homology modules and root multiplicities in GKM algebras 187

Abstract

Borcherds initiated the study of generalized Kac-Moody (GKM) algebras [27, 95, 96]. For automorphism of GKM algebras, one can refer Fuchs et al. [97]. In Kac-Moody algebras all simple roots are real whereas in the GKM algebras, one can have imaginary simple roots with norms > 0 and multiplicities > 1 [37]. The concept of special imaginary roots was extended from Kac-Moody algebras to GKM algebras in Li et al. [98] and therein they found out a special imaginary root for a particular GKM algebra. Barwald and Gebert [99] presented the results of some calculations of the multiplicities of simple imaginary roots for a particular Borcherds superalgebra and its maximal Kac-Moody subalgebra.

In this chapter, starting from the definition of generalized Kac-Moody algebras along with number of examples, Dynkin diagrams and real and imaginary roots for different classes of GKM algebras with examples are given. Also the extension of the concept of purely imaginary roots from Kac-Moody algebras to GKM algebras and complete classification [100] of GKM algebras possessing purely imaginary property were given here. Moreover, we have also studied here some properties of roots of certain classes of

Introduction to Finite and Infinite Dimensional Lie (Super)algebras. http://dx.doi.org/10.1016/B978-0-12-804675-3.00003-0

GKM algebras other than Kac-Moody algebras whose Generalized Generalized Cartan matrices (GGCMs) (of order ≤3) are extensions of Cartan matrices (of order ≤2) of finite, affine, and hyperbolic types [101] and also found out special imaginary roots of all these classes [102]. Complete classifications [82, 103] of different classes of imaginary roots, namely, special and strictly are given with number of examples wherever necessary, including Monstrous Lie algebras. In general, for characterization of GKM algebras and Monster Lie algebras, one can refer Borcherds [104, 105].

Kang [72, 73, 106] derived a closed form root multiplicity formula for all roots in GKM algebras. In particular, Kang and Kim [37] discussed many interesting applications of their dimension formula to the various classes of graded Lie algebras such as free Lie algebras, Kac-Moody algebras, and GKM algebras. This formula was also derived in Naito [107], Jurisich [108, 109], and Jurisich et al. [110]. A homological theory for the graded Lie algebras was developed by Kang [65], and a closed form root multiplicity formula for all symmetrizable GKM algebras was obtained in Kang [62]. For a generalization of Borcherds algebra, denominator formula, and the recursive dimension formula, one can refer Miyamoto [111].

In this chapter, integrable representation of GKM algebras, in particular, fundamental representation of Borcherds algebras with one simple root is explained with an example. Moreover, the theorems regarding homology modules and root multiplicities for some classes of GKM algebras, in particular, GKM algebras with one simple imaginary root are proved in [112]. Homology modules and root multiplicities and representations of GKM algebras are also discussed.

3.1 Borcherds-Cartan matrix (BCM), Generalized Generalized Cartan matrix (GGCM), Borcherds Kac-Moody (BKM) algebras, and Generalized Kac-Moody (GKM) algebras

Definition 106. Let I be a finite or countably infinite index set. A real Cartan matrix $A = (a_{ij})_{i,j \in I}$ is called a *Borcherds-Cartan matrix* (*BCM*) if

(i) $a_{ii} = 2$ or $a_{ii} \le 0$ for all $i \in I$.
(ii) $a_{ij} \le 0$ if $i \neq j$ and $a_{ij} \in \mathbb{Z}$ if $a_{ii} = 2$ and
(iii) $a_{ij} = 0 \Longleftrightarrow a_{ji} = 0$.

Remark 65.

(1) We define Ire and I^{im}, subsets of I such that $I^{\mathrm{re}} = \{i \in I \mid a_{ii} = 2\}, I^{\mathrm{im}} = \{i \in I \mid a_{ii} \le 0\}$.
(2) If I is a *finite index set*, then the corresponding real Cartan matrix defined above is called (also known as) a *Generalized Generalized Cartan matrix*, abbreviated as *GGCM*. If the above matrix A is symmetric, then it is enough to write conditions (i) and (ii) above.

Definition 107 (Generalized Kac-Moody algebra associated to GGCM [27, 107–109]). An auxiliary Lie algebra (GKM algebra) or derived GKM Lie algebra $\mathcal{G}'(A)$

associated to the GGCM $A = (a_{ij})_{i,j\in I}$ (defined above) is a Lie algebra, given by the generators e_i, f_i, and h_i for $i \in I$ satisfying the relations:

$$
\begin{aligned}
&[h_i, h_j] = 0 \text{ for all } i,j \in I,\\
&[h_i, e_j] = a_{ij}e_j,\ [h_i,f_j] = -a_{ij}f_j \text{ for all } i,j \in I,\\
&[e_i,f_j] = \delta_{ij}h_i \text{ for all } i,j \in I,\\
&(ade_i)^{(1-a_{ij})}(e_j) = (adf_i)^{(1-a_{ij})}(f_j) = 0 \text{ if } a_{ii} = 2 \text{ and } i \neq j,\\
&[e_i, e_j] = [f_i,f_j] = 0 \text{ if } a_{ij} = 0 \text{ for all } i,j \in I.
\end{aligned}
$$

We define the following Lie subalgebras of $\mathcal{G}'(A)$.

$$
\begin{aligned}
&\mathfrak{h}' := \sum_{i\in I} \mathbb{C}h_i, \text{ Cartan subalgebra,}\\
&\eta^+ := \{e_i\}_{i\in I}, \text{ generated by } e_i \text{ for } i \in I,\\
&\eta^- := \{f_i\}_{i\in I}, \text{ generated by } f_i \text{ for } i \in I.
\end{aligned}
$$

Then we have triangular decomposition of $\mathcal{G}'(A)$:

$$\mathcal{G}'(A) = \eta^- \oplus \mathfrak{h}' \oplus \eta^+.$$

Set $\deg h_i = (0,0,\dots,0) \in \mathbb{Z}^I$, $\deg e_i = -\deg f_i = (0,0,\dots,0,1,0,\dots,0) \in \mathbb{Z}^I$, where 1 is in the ith position. Then we have

$$\mathcal{G}'(A) = \bigoplus_{(n_1,n_2,\dots)\in\mathbb{Z}^I} \mathcal{G}'(n_1,n_2,\dots)$$

(Lie algebra grading) where $\mathcal{G}'(n_1,n_2,\dots)$ is the subspace of $\mathcal{G}'(A)$ of degree $(n_1,n_2,\dots) \in \mathbb{Z}^I$.

For $i \in I$, let d_i be the ith degree derivation of the Lie algebra $\mathcal{G}'(A)$ (that is d_i acts on the subspace $\mathcal{G}'(n_1,n_2,\dots)$ as multiplication by the scalar n_i). We extend the Lie algebra $\mathcal{G}'(A)$ by taking the semidirect product with the space $\partial_0 = \sum_{i\in I}\mathbb{C}d_i$: $\mathcal{G}^e(A) := \partial_0 \ltimes \mathcal{G}'(A)$. Note that $\mathfrak{h}^e := \partial_0 \oplus \mathfrak{h}'$ is an abelian subalgebra of $\mathcal{G}^e(A)$.

Now define $\alpha_i \in \mathfrak{h}^*$ by $\alpha_i(h_j) = a_{ij}, \alpha_i(d_j) = \delta_{ij}$ for $i,j \in I$. We fix a suitable subspace ∂ of ∂_0 such that the restrictions $\alpha_i|_{\mathfrak{h}}$ of $\alpha_i \in (\mathfrak{h}^e)^*$ for $i \in I$ to the subspace $\mathfrak{H} := \partial \oplus \mathfrak{h} \subset \mathfrak{h}^e$ are linearly independent. Then we put

$$\mathcal{G}(A) := \partial \ltimes \mathcal{G}'(A),$$

and call it, the GKM algebra $\mathcal{G}(A)$, associated to the matrix $A = (a_{ij})_{i,j\in I}$. In the usual notation, we write α_i for $\alpha_i|_{\mathfrak{h}} \in \mathfrak{h}^*$.

For $\alpha \in \mathfrak{h}^*$, we put

$$\mathcal{G}_\alpha := \{x \in \mathcal{G}(A) | [h, x] = \alpha(h)x \text{ for all } h \in \mathfrak{h}\}.$$

Here $\Delta := \{\alpha \in \mathfrak{h}^* \backslash \{0\} | \mathcal{G}_\alpha \neq 0\}$ (set of all roots), $\Delta = \Delta^+ \cup \Delta^-$, and $\Delta^- = -\Delta^+$ with $\Delta^+ := \Delta \cap (\sum_{i \in I} \mathbb{Z}_{\geq 0} \alpha_i)$ (set of all positive roots). The elements of the set $\Pi := \{\alpha_i \in \mathfrak{h}^* | i \in I\} \subset \Delta^+$, are called the simple roots of $\mathcal{G}(A)$.

Let $\underline{m} = \{m_i \in \mathbb{Z}_{>0} | i \in I\}$, be a collection of positive integers such that $m_i = 1$ for all $i \in I^{\text{re}}$. We have the root space decomposition of $\mathcal{G}(A)$ with respect to $\mathfrak{h}$:

$$\mathcal{G}(A) = \mathfrak{h} \oplus \bigoplus_{\alpha \in \Delta} \mathcal{G}_\alpha.$$

Furthermore, we have:

$$\eta^\pm = \bigoplus_{\alpha \in \Delta^\pm} \mathcal{G}_\alpha, \quad \mathcal{G}_0 = \mathfrak{h}, \quad \mathcal{G}_{\alpha_i} = \mathbb{C} e_i, \quad \mathcal{G}_{-\alpha_i} = \mathbb{C} f_i \ \ (i \in I).$$

Definition 108. A BCM or GGCM, A is called symmetrizable if there exists a diagonal matrix $D = \text{diag}(\varepsilon_1 \cdots \varepsilon_n)$ with $\varepsilon_i \in \mathbb{R}$, $\varepsilon_i > 0$ for all i such that DA is symmetric.

Definition 109 (BKM Lie algebra associated with symmetrizable Cartan matrix). The *Borcherds Kac-Moody algebra (BKM algebra or BKM Lie algebra) or Generalized Kac-Moody algebra (GKM algebra) (if the index set I is finite)* is $\mathcal{G} = \mathcal{G}(A, \underline{m})$ associated with a *symmetrizable BCM*, $A = (a_{ij})_{i,j \in I}$ of charge $\underline{m} = (m_i | i \in I)$ is the Lie algebra generated by the elements $h_i, d_i (i \in I), e_{ik}, f_{ik} (i \in I, k = 1, \ldots, m_i)$ such that

$$\begin{aligned}
\left[h_i, h_j\right] &= [d_i, d_j] = [h_i, d_j] = 0, \\
\left[h_i, e_{jl}\right] &= a_{ij} e_{jl}, [h, f_{jl}] = -a_{ij} f_{jl}, \\
\left[d_i, e_{jl}\right] &= \delta_{ij} e_{jl}, [d_i, f_{jl}] = -\delta_{ij} f_{jl}, \\
\left[e_{ik}, f_{jl}\right] &= \delta_{ij} \delta_{kl} h_i, \\
(ade_{ik})^{(1-a_{ij})}(e_{jl}) &= (adf_{ik})^{(1-a_{ij})}(f_{jl}) = 0 \text{ if } a_{ii} = 2 \quad \text{and} \quad i \neq j \quad \text{and} \\
\left[e_{ik}, e_{jl}\right] &= [f_{ik}, f_{jl}] = 0 \text{ if } a_{ij} = 0, \forall i, j \in I, k = 1, \ldots, m_i, \ l = 1, \ldots, m_j.
\end{aligned}$$

The subalgebra $\mathfrak{h} = (\bigoplus_{i \in I} \mathbb{C} h_i) \oplus (\bigoplus_{i \in I} \mathbb{C} d_i)$ is the Cartan subalgebra of $\mathcal{G}$.

Remark 66. In Definition 107 of GKM algebra, Cartan subalgebra is generated by $h_i, i \in I$ whereas in Definition 109, the Cartan subalgebra is generated by $h_i, d_i, i \in I$. For each $i \in I$, define a linear functional $\alpha_i \in \mathfrak{h}^\star$ by $\alpha_i(h_j) = a_{ji}, \alpha_i(d_j) = \delta_{ij} (i, j \in I)$. As usual, the linear functionals $(\alpha_i)(i \in I)$ are called the simple roots of $\mathcal{G}$. In Definition 109, as the Cartan subalgebra is generated by h_i and d_i and the relation between h_i, d_i, e_{ik}, f_{ik} are given. So Definition 109 is considered to be the main definition for GKM algebra or BKM algebra associated with a symmetrizable BCM.

Definition 110 (GKM algebra associated to a symmetrizable Cartan matrix with certain restrictions). If there is no d_i, the index set I is finite (Cartan subalgebra generated by h_i only) and charge m is not defined, then GKM algebra associated to the symmetrizable Cartan matrix can be defined as follows.

GKM algebra associated to a symmetrizable $GGCM = (a_{ij})_{i,j \in I}$ is the Lie algebra (over $\mathbb{C}$) generated by the above vector space $\mathfrak{h}$ and the elements e_i and f_i for all $i \in I$ such that

$$\begin{aligned}
[h, h'] &= 0 \text{ for } h, h' \in \mathfrak{h}, \\
[h, e_i] &= \langle \alpha_i, h \rangle e_i, \\
[h, f_i] &= -\langle \alpha_i, h \rangle f_i, \\
\left[e_i, f_j\right] &= \delta_{ij} \alpha_i^{\vee} \, (i, j \in I), \\
(ade_i)^{1-a_{ij}} e_j &= 0, (adf_i)^{1-a_{ij}} f_j = 0 \text{ if } a_{ii} = 2 \text{ and } j \neq i, \\
\left[e_i, e_j\right] &= 0 = [f_i, f_j] \text{ if } a_{ii}, a_{jj} \leq 0 \text{ and } a_{ij} = 0.
\end{aligned}$$

Remark 67. Definitions 109 and 110 of GKM algebras are generally considered in this chapter for all discussions. Second and third conditions of Definition 110 can be written as $[h_i, e_j] = a_{ij} e_j$ and $[h_i, f_j] = -a_{ij} f_j$ for $h_i \in \mathfrak{h}$, respectively.

Definition 111. A GGCM, "A" is called indecomposable if it cannot be reduced to a block diagonal form by shuffling its rows and columns. As in the case of Kac-Moody Lie algebras, we have the following results:

For an indecomposable BCM, "A," one of the following three possibilities holds. Let $A = (a_{ij})_{i,j=1}^{n}$ be an indecomposable GGCM. Then

(i) Finite: $\det A \neq 0$, there exists $u > 0$ such that $Au > 0$ $(Au \geq 0 \Rightarrow u > 0 \text{ or } u = 0)$.
(ii) Affine: $corank = 1$, there exists $u > 0$ such that $Au = 0$ $(Au \geq 0 \Rightarrow Au = 0)$.
(iii) Indefinite: $\det A \neq 0$, there exists $u > 0$ such that $Au < 0$ $(Au > 0, u > 0 \Rightarrow u = 0)$.

We call such a matrix as finite, affine, or indefinite type and write $A \in$ Fin, Aff, or Ind, respectively.

3.2 Dynkin diagrams of GKM algebras

Definition 112. Let $A = (a_{ij})_{i,j=1}^{n}$ be a GGCM. Then the corresponding *Dynkin diagram $S(A)$ of A*, defined as follows, has n vertices.

(i) Two vertices i and j are connected by $\max\{|a_{ij}|, |a_{ji}|\}$ number of lines if $a_{ij}a_{ji} \leq 4$ and there is an arrow pointing toward i if $|a_{ij}| > 1$.
(ii) If $a_{ij}a_{ji} > 4$, i and j are connected by a bold lined edge equipped with the ordered pair $(|a_{ij}|, |a_{ji}|)$ of integers.
(iii) If $a_{ii} = 2$, the ith vertex will be denoted by white circle ◯.
(iv) If $a_{ii} = 0$, the ith vertex will be denoted by crossed circle $\otimes$.

(v) If $a_{ii} = -k(k > 0)$, the ith vertex will be denoted by a white circle with $(-k)$ written within the parenthesis above the white circle.

Remark 68.

(i) As BCM is defined for real numbers, we restrict the entries of BCM to set of real numbers only.

(ii) In Sthanumoorthy and Lilly [82, 100], the crossed circle $\otimes$ was drawn as $\oplus$. Here in this book, the crossed circle is drawn as $\otimes$ as given in Wan [3] and Kac [8].

Example 23. Some examples of BCMs and the corresponding Dynkin diagrams.

Case 1: $n = 1$

$BCM = (a_{11})$. Let $a_{11} = -k$, where $k \in \mathbb{R}_+ \cup \{2\}$. Here $\mathbb{R}_+$ is the set of all real numbers including 0.

Dynkin diagrams corresponding to above BCM:

(i) For $k = 0$, the Dynkin diagram:

(ii) For $k > 0$, the Dynkin diagram:

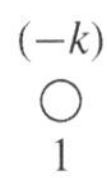

(iii) For $a_{11} = 2, (k = -2)$, the Dynkin diagram:

Case 2: $n = 2$

$$BCM = \begin{pmatrix} -k_1 & -a_1 \\ -b_1 & -k_2 \end{pmatrix}, \text{ where } k_1, k_2, a_1, b_1 \in \mathbb{R}_+ \cup \{-2\}.$$

Dynkin diagrams corresponding to above BCM:

(i) For $k_1 = k_2 = 0, a_1 b_1 = 1 (a_1 = b_1 = 1)$, Dynkin diagram:

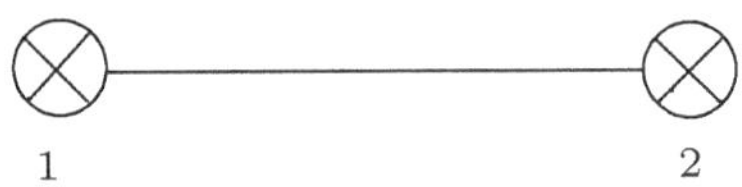

(ii) For $k_1 = 0, k_2 > 0, a_1 b_1 > 4$, Dynkin diagram:

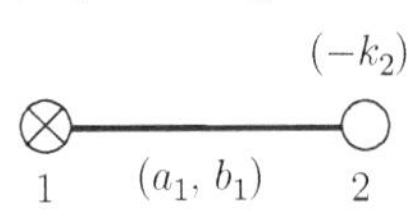

(iii) For $k_1 > 2, k_2 = -2, a_1 = 2, b_1 = 1$, Dynkin diagram

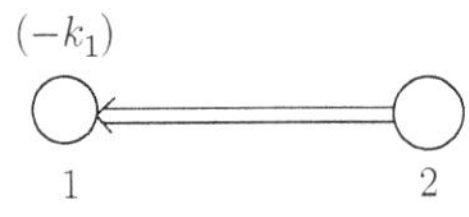

Case 3: $n = 3$

$$BCM = \begin{pmatrix} -k_1 & -a_1 & -a_2 \\ -b_1 & -k_2 & -p \\ -b_2 & -q & -k_3 \end{pmatrix}, \text{ where } k_i, a_i, b_i, p, q \in \mathbb{R}_+ \cup \{-2\} \text{ for all } i.$$

Dynkin diagrams corresponding to above BCM:

(i) For $k_i > 0$ for all i and $a_1b_1 = a_2b_2 = pq = 1$, Dynkin diagram:

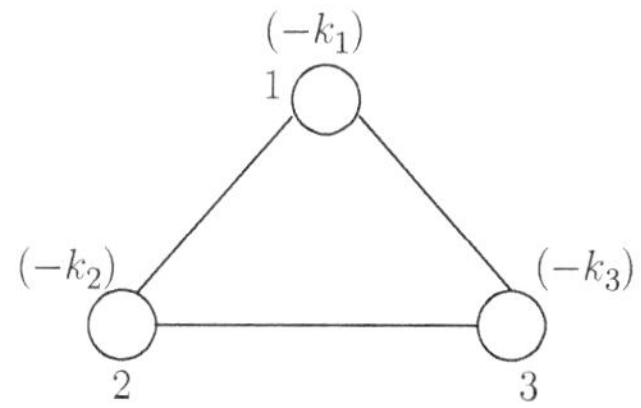

(ii) For $k_1 = k_2 = -2, k_3 = 0, a_1b_1 > 4, a_2b_2 = 1$ and $p = q = 2$, Dynkin diagram:

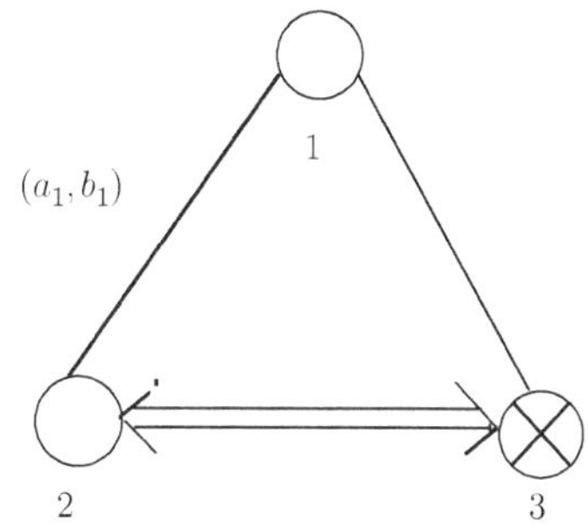

(iii) For $k_i = 0, a_ib_i > 4$ for all i and $p = 1, q = 4$, Dynkin diagram:

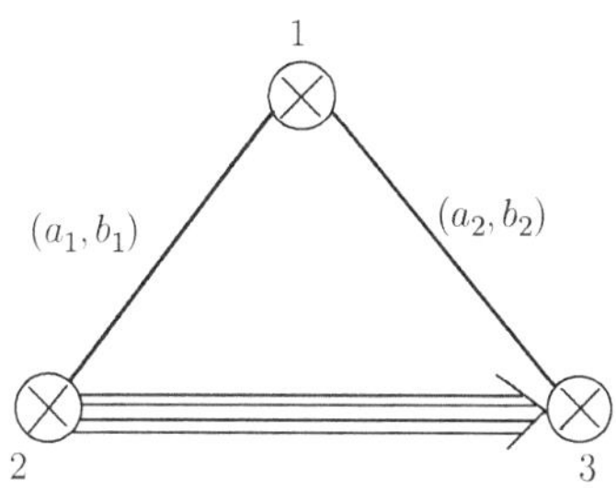

(iv) For $k_1 = -2, k_2 = 0, k_3 = -2, a_1b_1 = 1, a_2 = 0, p = 1$ and $q = 4$, Dynkin diagram:

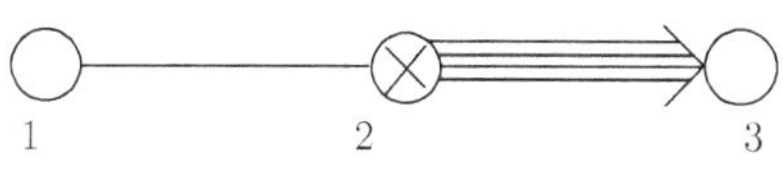

(v) $k_1 = 0, k_2 = k_3 = -2, a_1 = 0, a_2b_2 > 4, p = 1$ and $q = 4$, Dynkin diagram:

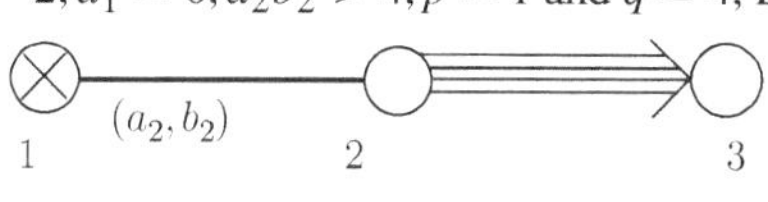

Case 4: $n = 4$

$$BCM = \begin{pmatrix} -k_1 & -a_1 & -a_2 & -a_3 \\ -b_1 & -k_2 & c_2 & -c_3 \\ -b_2 & -d_2 & -k_3 & -p \\ -b_3 & -d_3 & -q & -k_4 \end{pmatrix},$$

where $k_i, a_i, b_i, c_i, d_i, p, q \in \mathbb{R}_+ \cup \{-2\}$ for all i.

Dynkin diagrams corresponding to above BCM:

(i) For $k_i = 0$ and $a_ib_i = c_id_i = pq = 1$ for all i, Dynkin diagram:

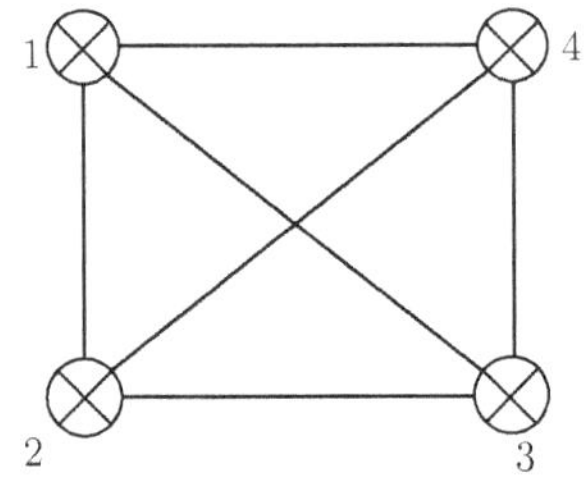

(ii) For $k_i > 0$ for all $i, a_3 = c_3 = 0, a_1b_1 = 1, a_2b_2 > 4, pq > 4$ and $c_2d_2 > 4$, Dynkin diagram:

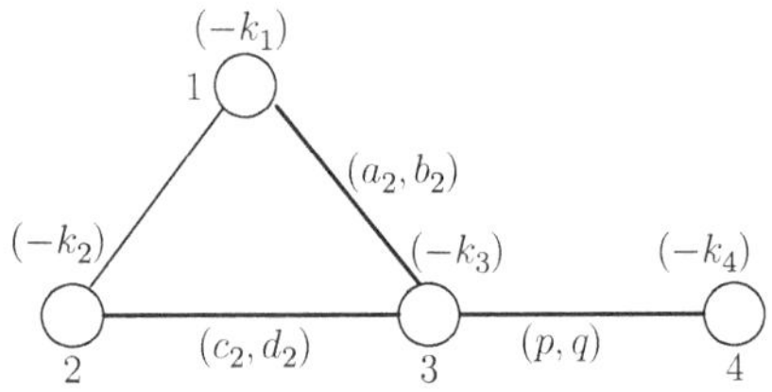

(iii) For $k_i = -2$ for all $i, a_2 = a_3 = c_3 = 0, a_1b_1 > 4, c_2 = d_2 = 2, p = 1$ and $q = 4$, Dynkin diagram:

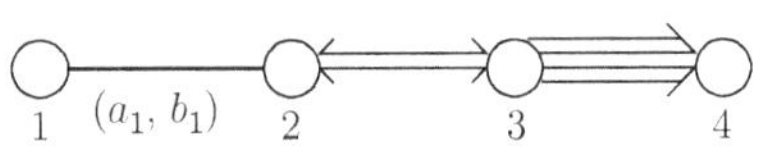

Case 5: $n = 5$

$$BCM = \begin{pmatrix} -k_1 & -a_1 & -a_2 & -a_3 & -a_4 \\ -b_1 & -k_2 & c_1 & -c_2 & -c_3 \\ -b_2 & -d_1 & -k_3 & -e_1 & -e_2 \\ -b_3 & -d_2 & -f_1 & -k_4 & -p \\ -b_4 & -d_3 & -f_2 & -q & -k_5 \end{pmatrix}$$ where $k_i, a_i, b_i, c_i, d_i, e_i, f_i, p, q \in \mathbb{R}_+ \cup \{-2\}$ for all i.

Dynkin diagrams corresponding to above BCM:

(i) For $k_1 = 0, k_2 = 0, k_4, k_5 > 0, k_3 = -2$ and $a_ib_i = c_id_i = e_if_i = 1$ for all i, $p = q = 2$, Dynkin diagram:

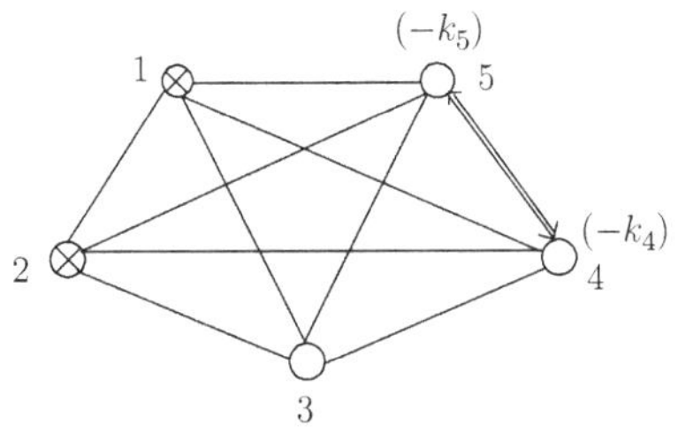

General case: In general, for any n, consider the following particular case with the BCM

$$\begin{pmatrix} -k_1 & -a_1 & 0 & & \cdots & 0 & 0 \\ -b_1 & -k_2 & -a_2 & & \cdots & 0 & 0 \\ 0 & -b_2 & -k_3 & -a_3 & \cdots & 0 & 0 \\ \vdots & \vdots & \vdots & \ddots & \cdots & \vdots & \vdots \\ 0 & 0 & 0 & \cdots & -b_{n-1} & -k_n & -a_n \\ 0 & 0 & 0 & \cdots & 0 & -b_n & -k_{n+1} \end{pmatrix},$$

where, $k_i, a_i, b_i \in \mathbb{R}_+ \cup \{-2\}$ for all i.

For $k_1 = -2, k_i > 0 (i = 2, \ldots, n+1), a_1 = b_1 = 2$ and $a_i = b_i = 1 (i = 2, \ldots, n+1)$, Dynkin diagram:

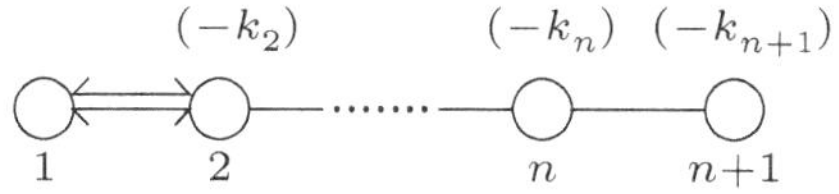

Definition 113. We say that a Dynkin diagram $S(A)$ of a GKM algebra is *completely connected*, if $a_{ij} \neq 0$ for all $i, j = 1, \ldots, n$.

Example 24. The example given in the above case 5 is the example for completely connected GKM algebra.

Definition 114. An indecomposable BCM (or GGCM), A is of *hyperbolic type*, if it is symmetrizable indefinite type and if every proper indecomposable principal submatrix of A is a BCM (or GGCM) of finite or affine type.

Remark 69. We consider BCM consisting of elements of $\mathbb{Z}$ only.

Definition 115. We define an indefinite nonhyperbolic GGCM, $A = (a_{ij})_{i,j=1}^n$ to be of *extended-hyperbolic type* if every proper, connected subdiagram of $S(A)$ is of finite, affine, or hyperbolic type **GGCM**.

We say that the GKM algebra associated with the GGCM, A (respectively $S(A)$) is of extended-hyperbolic type if A is of extended-hyperbolic type.

Remark 70. We note that all the extended-hyperbolic Kac-Moody algebras are also extended-hyperbolic GKM algebras. We give below two examples of extended-hyperbolic GKM algebras that are not extended-hyperbolic Kac-Moody algebras.

1. $GGCM = \begin{pmatrix} 0 & -1 & -2 \\ -1 & -k_1 & -2 \\ -3 & -2 & -k_2 \end{pmatrix}$ and the associated Dynkin diagram is

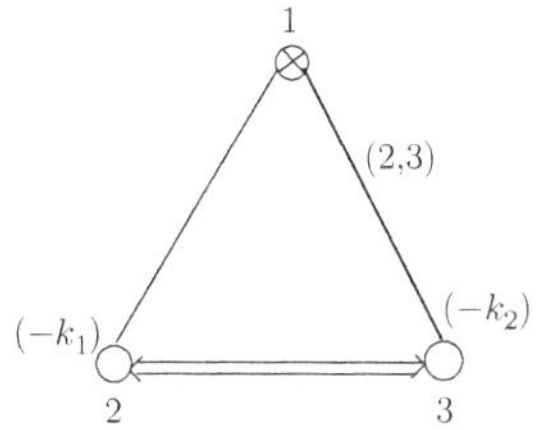

2. $GGCM = \begin{pmatrix} 0 & -2 & 0 \\ -2 & 0 & -4 \\ 0 & -5 & 0 \end{pmatrix}$ and the associated Dynkin diagram is

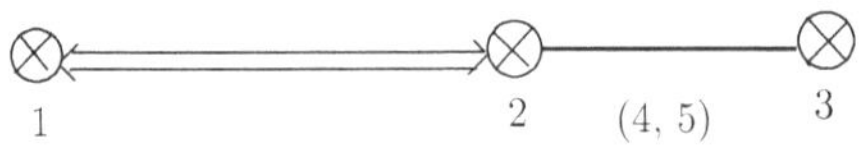

Definition 116. Let A be a BCM. As in the case of Kac-Moody algebras if BCM $A = (a_{ij})_{i,j\in I}$ is symmetrizable, then there exists a nondegenerate, symmetric, invariant *bilinear form* $(.,.)$ on $\mathcal{G}(A)$.

As the restriction of this to $\mathfrak{h}$ is also nondegenerate, there exists on $\mathfrak{h}^\star$, a nondegenerate symmetric, w-invariant bilinear form, also denoted by $(\ ,\ )$, through the linear isomorphism $\nu : \mathfrak{h} \to \mathfrak{h}^\star$.

In particular, $(\alpha_i, \alpha_j) = \varepsilon_i a_{ij}$, for all $1 \le i, j \le n$.

Let $\rho \in \mathfrak{h}^\star$ be a linear functional satisfying $\rho(h_i) = \frac{1}{2} a_{ii}$ for all i and $(\rho, \alpha_i) = \frac{1}{2}(\alpha_i, \alpha_i)$, for all i. α becomes real if $(\alpha, \alpha) > 0$ and imaginary if $(\alpha, \alpha) \le 0$.

Definition 117. Let A be a BCM, $(a_{ij})_{i,j\in I}$. Consider a triple, $(\mathfrak{h}, \Pi, \Pi^\vee)$ with $\Pi = \{\alpha_i\}_{i\in I}$ and $\Pi^\vee = \{\alpha_i^\vee\}_{i\in I}$ satisfying,

(1) $\mathfrak{h}$ is a finite dimensional (dim) (complex) vector space such that dim $\mathfrak{h} = 2n - \text{rank } A$.
(2) $\Pi = \{\alpha_i\}_{i\in I} \subset \mathfrak{h}^\star$ is linearly independent and $\Pi^\vee = \{\alpha_i^\vee\}_{i\in I} \subset \mathfrak{h}$ is linearly independent where $\mathfrak{h}^\star = \text{Hom}_{\mathbb{C}}(\mathfrak{h}, \mathbb{C})$.
(3) $\langle \alpha_j, \alpha_i^\vee \rangle = \alpha_j(\alpha_i^\vee) = a_{ij}$.

Here $\langle , \rangle$ denotes a duality pairing between $\mathfrak{h}$ and $\mathfrak{h}^*$. This triple $(\mathfrak{h}, \Pi, \Pi^\vee)$ is called a *realization of* A.

The elements of Π (respectively $\Pi^\vee$) are called the simple roots (respectively simple coroots) of $\mathcal{G}$.

3.3 Root systems and Weyl groups of GKM algebras

Let $Q = \bigoplus_{i\in I} \mathbb{Z}\alpha_i$. Then Q is the root lattice. Define $Q_+ = \bigoplus_{i\in I} \mathbb{Z}\alpha_{i \ge 0}$ and $Q_- = -Q_+$.

Define a partial ordering $\ge$ on $\mathfrak{h}^\star$ by $\lambda \ge \mu$ if $\lambda - \mu \in Q_+$. For $\alpha = \sum_{i\in I} k_i \alpha_i$, height of α is ht $\alpha = \sum_{i\in I} k_i$, and the support of α is supp $\alpha = \{i \in I / k_i \ne 0\}$.

The GKM algebra, $\mathcal{G} = \mathcal{G}(A, \underline{m})$ associated with a symmetrizable BCM, A of charge $\underline{m}$ has the following root space decomposition:

$\mathcal{G} = \bigoplus_{\alpha\in Q} \mathcal{G}_\alpha$, where $\mathcal{G}_\alpha = \{x \in \mathcal{G} |\ [h, x] = \alpha(h)x$ for all $h \in \mathfrak{h}\}$ is the root space associated to α. We have $\mathcal{G}_0 = \mathfrak{h} \in Q$. $\alpha \in Q$ is called a root if $\alpha \ne 0$, and $\mathcal{G}_\alpha \ne 0$.

Multiplicity of the root α = *multi* α = dim $\mathcal{G}_\alpha$. Moreover, *multi* α_i = *multi* $(-\alpha_i) = m_i$, for all $i \in I$. In general, *multi* α = *multi* $(-\alpha)$. $\alpha > 0$ $(\alpha < 0)$ is called positive (negative) root. Denote by Δ, the set of all roots, Δ_+, the set of all positive roots, and Δ_-, the set of all negative roots.

We have $\Delta = \Delta_+ \cup \Delta_-$. Let $\mathcal{G}_+ = \bigoplus_{\alpha>0} \mathcal{G}_\alpha$ and $\mathcal{G}_- = \bigoplus_{\alpha<0} \mathcal{G}_\alpha$. We have the triangular decomposition, $\mathcal{G} = \mathcal{G}_+ \oplus \mathcal{G}_0 \oplus \mathcal{G}_-$.

Remark 71. Note that, if $a_{ii} = 2$ for all i, then $\widehat{\mathcal{G}}(\widehat{A})$ is the same as the ordinary Kac-Moody algebra with symmetrizable Cartan matrix $\widehat{A}$. In general, $\widehat{\mathcal{G}}(\widehat{A})$ has almost all the properties that ordinary Kac-Moody algebras have, and the only major difference is that GKM algebras are allowed to have imaginary simple roots. In what follows, we will exclusively deal with the case of a Borcherds algebra with one imaginary simple root.

Definition 118. For each $i \in I^{\mathrm{re}}$, define simple reflection $r_i \in \mathfrak{h}^\star : r_i(\lambda) = \lambda - \langle \lambda, \alpha_i^\vee \rangle \alpha_i$ for $\lambda \in \mathfrak{h}^\star$.

The *Weyl group* W of $\mathcal{G}$ is the subgroup of $GL(\mathfrak{h}^\star)$ generated by r_i for all $i \in I^{\mathrm{re}}$. Here $\{W, r_i\}$ is called a Coxeter system. A simple root α_i is real if $a_{ii} = 2$ and imaginary if $a_{ii} \leq 0$. Simple imaginary roots have multiplicity >1. Denote by $\Pi^{\mathrm{re}} = \{\alpha_i \in \Pi \mid a_{ii} = 2\}$ (set of all real simple roots). Let W be the Weyl group of the GKM algebra. The set of all real roots is $\Delta^{\mathrm{re}} = W(\Pi^{\mathrm{re}})$. The set of all imaginary roots is $\Delta^{\mathrm{im}} = \Delta \backslash \Delta^{\mathrm{re}}$.

It can be proved that the set of all positive imaginary roots is $\Delta_+^{\mathrm{im}} = \bigcup_{w \in K} w(K)$, where $K = \{\alpha \in Q_+ \backslash \{0\} | \langle \alpha, \alpha_i^\vee \rangle \leq 0 \text{ if } a_{ii} = 2 \text{ and supp } \alpha \text{ is connected}\} \backslash \bigcup_j j\Pi^{\mathrm{im}}$.

Let $\Pi^{\mathrm{im}} = \{\alpha_i \in \Pi \mid a_{ii} \leq 0\}$ (set of all imaginary simple roots). In particular, the set Δ_+^{im} of positive imaginary roots is W-invariant.

Example 25 (Some examples of BCMs which appear as extension of generalized Cartan matrices [101]). We give below some examples of BCMs which appear as the extensions of generalized Cartan matrices which are symmetric or nonsymmetric. In the following, if GX denotes Cartan matrix of Kac-Moody algebra, GGX will denote Cartan matrix of GKM algebra which appears as an extension of GX.

We consider GKM algebras other than Kac-Moody algebras whose generalized Cartan matrices are extensions of Cartan matrices of finite, affine, and hyperbolic types. Let X be a Kac-Moody algebra of finite, affine, or hyperbolic type of order n.

We have

$$GGX = \begin{pmatrix} -\kappa & -a_1 & -a_2 & \cdots & -a_n \\ -b_1 & & & & \\ -b_2 & & & & \\ \vdots & & GX & & \\ -b_n & & & & \end{pmatrix},$$

where $a_i, b_i (i = 1, \ldots, n)$ are positive integers and $\kappa \in \mathbb{Z}_{\geq 0}$.

(a) BCMs which appear as the extensions of symmetric generalized Cartan matrices. Some Classes of GKM algebras:

Class 1. Extension of $GA_1 = GGA_1 = \begin{pmatrix} -k & -a \\ -a & 2 \end{pmatrix}$, where $GA_1 = (2)$ (finite).

Class 2. Extension of $GA_1^{(1)} = GGA_1^{(1)} = \begin{pmatrix} -k & -a & -b \\ -a & 2 & -2 \\ -b & -2 & 2 \end{pmatrix}$, where

$$GA_1^{(1)} = \begin{pmatrix} 2 & -2 \\ -2 & 2 \end{pmatrix} \text{ (affine).}$$

(b) BCMs which appear as the extensions of nonsymmetric generalized Cartan matrices

Class 1. Extension of nonsymmetric GG_2 is $GGG_2 = \begin{pmatrix} -k & -a & -b \\ -c & 2 & -1 \\ -d & -3 & 2 \end{pmatrix}$, where $GG_2 = \begin{pmatrix} 2 & -1 \\ -3 & 2 \end{pmatrix}$(finite).

Class 2. Extension of nonsymmetric $GA_2^{(2)}$ is $GGA_2^{(2)} = \begin{pmatrix} -k & -a & -b \\ -c & 2 & -4 \\ -d & -1 & 2 \end{pmatrix}$, where $GA_2^{(2)} = \begin{pmatrix} 2 & -4 \\ -1 & 2 \end{pmatrix}$(affine).

Class 3. Extension of GHA_1 is $GGHA_1 = \begin{pmatrix} -k & -a & -b \\ -c & 2 & -p \\ -d & -q & 2 \end{pmatrix}$, where $GHA_1 = \begin{pmatrix} 2 & -p \\ -q & 2 \end{pmatrix}$ and $pq > 4$ (hyperbolic).

Here HA_1 stands for the hyperbolic Kac-Moody algebra which is an extension of finite Lie algebra of type A_1 and GHA_1 stands for the generalized Cartan matrix which is an extension of the Cartan matrix of HA_1.

Computation of the set of all real and imaginary roots of different classes of GKM algebras

(i) Extension of finite type:

Consider $GGA_2 = \begin{pmatrix} -k & -a & -b \\ -c & 2 & -1 \\ -d & -1 & 2 \end{pmatrix}$

We have $\Pi = \{\alpha_1, \alpha_2, \alpha_3\}, \Pi^{\text{re}} = \{\alpha_2, \alpha_3\}, \Pi^{\text{im}} = \{\alpha_1\}$.

Weyl reflections of real simple roots are r_2 and r_3 and hence Weyl group is generated by r_2 and r_3. So $W = \{1, r_2, r_3, r_2r_3, r_3(r_2r_3), (r_2r_3)^2\}$ = Weyl group of A_2.

We have $\Delta^{\text{re}} = W(\Pi^{\text{re}}) = \{\pm\alpha_2, \pm\alpha_3, \pm(\alpha_1 + \alpha_2)\}$. Using $r_i(\lambda) = \lambda - \langle \lambda, \alpha_i \rangle \alpha_i$. It can be verified that

$$r_2(\alpha_1) = \alpha_1 + c\alpha_2,\ r_2(\alpha_2) = -\alpha_2,$$

$$r_2(\alpha_3) = \alpha_3 + \alpha_2,\ r_3(\alpha_1) = \alpha_1 + d\alpha_3,$$

$$r_3(\alpha_2) = \alpha_2 + \alpha_3,\ r_3(\alpha_3) = -\alpha_3,$$

$$r_2r_3(\alpha_2) = \alpha_3,\ r_2r_3(\alpha_3) = -\alpha_3 - \alpha_2 \quad \text{and}$$

$$r_3(r_2r_3)(\alpha_1) = \alpha_1 + (c+d)(\alpha_2 + \alpha_3).$$

$$\Delta^{\text{re}} = W(\Pi^{\text{re}}) = W\{\alpha_2, \alpha_3\} = \{1, r_2, r_3\}(\alpha_2, \alpha_3) = \pm\{\alpha_2, \alpha_3, \alpha_2 + \alpha_3\}.$$

We have

$$K = \{\alpha \in Q_+ \backslash \{0\} | \langle \alpha, \alpha_i^\vee \rangle \le 0 \text{ if } a_{ii} = 2 \text{ and supp } \alpha \text{ is connected}\} \Big\backslash \bigcup_{j \ge 2} j\Pi^{\text{im}}.$$
$$= \{\alpha = k_1\alpha_1 + k_2\alpha_2 + k_3\alpha_3 | \langle \alpha, \alpha_2^\vee \rangle \le 0, \langle \alpha, \alpha_3^\vee \rangle \le 0\} \Big\backslash \bigcup_{j \ge 2} j\alpha_1.$$

It can be proved that, $\alpha(\alpha_2^\vee) = -ck_1 + 2k_2 - k_3 \le 0 \Rightarrow k_2 \le \frac{1}{2}(ck_1 + k_3)$. Similarly $\alpha(\alpha_3^\vee) \le 0 \Rightarrow k_2 \ge 2k_3 - dk_1$. Hence $2k_3 - dk_1 \le k_2 \le \frac{1}{2}(ck_1 + k_3)$.
$k_2 = 0 \Rightarrow 2k_3 \le dk_1$ and $k_3 = 0 \Rightarrow 2k_2 \le k_1$. Also
$k_2 = k_3 = 0 \Rightarrow k_1 = 1$. Hence
$K = \{\sum_{i=1}^{3} k_i\alpha_i | k_i \in N, 2k_3 - dk_1 \le k_2 \le \frac{1}{2}(ck_1 + k_3)\}$.
$W = \{< 1, r_2, r_3 >\}$. Let $c = d = 1$. Then it can be proved that
$\Delta_+^{\text{im}} = \{k_1\alpha_1 + (2k_1 - k_3)\alpha_2 + p\alpha_3\}$
$\cup \{k_1\alpha_1 + (k_1 - k_2 + k_3)\alpha_2 + l\alpha_3\} \cup \{k_1\alpha_1 + k_2\alpha_2 + k_3\alpha_3\}$, where
$p = k_1 + k_2 - k_3$ or $2k_1 - k_2$, $l = k_3$ or $2k_1 - k_2$, $r = k_3$ or $k_1 + k_2 - k_3$.
Note that the sum of two distinct real roots is real if it is a root.

(ii) Extension of nonsymmetric affine type:

Nonsymmetric $GGA_2^{(2)} = \begin{pmatrix} -k & -a & -b \\ -c & 2 & -4 \\ -d & -1 & 2 \end{pmatrix}$ (extension of affine).

It can be easily verified that the above matrix $GGA_2^{(2)}$ is symmetrizable if and only if $4ad = bc$. It is nonsingular and indecomposable.

We have, $\Pi = \{\alpha_1, \alpha_2, \alpha_3\}$. It is clear that

$$(\alpha_1, \alpha_1) = -ck, \quad (\alpha_1, \alpha_2) = -ca, \quad (\alpha_1, \alpha_3) = -bc,$$
$$(\alpha_2, \alpha_2) = 2a, \quad (\alpha_2, \alpha_3) = -4a, \quad (\alpha_3, \alpha_3) = 8a.$$

Note that $\Pi^{\text{re}} = \{\alpha_2, \alpha_3\}$ and $\Pi^{\text{im}} = \{\alpha_1\}$. We have,

$$r_2(\alpha_1) = \alpha_1 + c\alpha_2, \quad r_2(\alpha_2) = -\alpha_2, \quad r_2(\alpha_3) = \alpha_3 + 4\alpha_2,$$
$$r_3(\alpha_1) = \alpha_1 + d\alpha_3, \quad r_3(\alpha_2) = \alpha_2 + \alpha_3 \quad \text{and} \quad r_3(\alpha_3) = -\alpha_3.$$

$W = \{1, r_2(r_3r_2)^j, r_3(r_2r_3)^j, (r_2r_3)^{j+1}, (r_3r_2)^{j+1} \mid j \text{ is a nonnegative integer}\}$.
Using Weyl reflections,

$$\Delta^{\text{re}} = \Delta_S^{\text{re}} \cup \Delta_L^{\text{re}},$$

where

$$\Delta_L^{\text{re}} = \pm\{k_2\alpha_2 + k_3\alpha_3 | k_2, k_3 \in \mathbb{Z}_+, k_2 - 2k_3 = \pm 2\} \quad \text{and}$$
$$\Delta_S^{\text{re}} = \pm\{k_2\alpha_2 + k_3\alpha_3 | k_2, k_3 \in \mathbb{Z}_+, k_2 - 2k_3 = \pm 1\}.$$

Below we write down explicitly the imaginary roots for a particular case. Let $a = 1, b = 2, c = 2, d = 1 (4ad = bc)$. Using the definition of K, we find

$$K = \{k_1\alpha_1 + k_2\alpha_2 + k_3\alpha_3 | k_1, k_2, k_3 \in \mathbb{N}, 2k_3 - k_1 \leq k_2 \leq k_1 + 2k_3, k_1 = 0 \\ \Rightarrow k_2 = 2k_3 = 0 \Rightarrow 2k_3 \leq k_1, k_3 = 0 \Rightarrow k_2 \leq k_1 \text{ and } k_2 = k_3 = 0 \Rightarrow k_1 = 1\}.$$

$\Delta_+^{\text{im}} = \cup_{w\in W} w(K)$. Here Weyl group is infinite.

Remark 72. For $GGA_2^{(2)}$, the real roots have two different lengths and the sum of two distinct real roots is either real or imaginary if it is a root.

(iii) Extension of nonsymmetric hyperbolic type:

Nonsymmetric $GGHA_1 = \begin{pmatrix} -k & -a & -b \\ -c & 2 & -p \\ -d & -q & 2 \end{pmatrix}$, where $pq > 4$ (extension of hyperbolic).

It can be easily verified that the above matrix $GGHA_1$ is symmetrizable if and only if $pad = qbc$. It is nonsingular and indecomposable.

We have $\Pi = \{\alpha_1, \alpha_2, \alpha_3\}$. It is clear that

$$(\alpha_1, \alpha_1) = -kdp, \quad (\alpha_1, \alpha_2) = -adp, \quad (\alpha_1, \alpha_3) = -dbp,$$
$$(\alpha_2, \alpha_2) = 2bq, \quad (\alpha_2, \alpha_3) = -bpq, \quad (\alpha_3, \alpha_3) = 2bp.$$

Note that $\Pi^{\text{re}} = \{\alpha_1, \alpha_2\}$ and $\Pi^{\text{im}} = \{\alpha_1\}$. We have,

$$r_2(\alpha_1) = \alpha_1 + c\alpha_2, \quad r_2(\alpha_2) = -\alpha_2, \quad r_2(\alpha_3) = \alpha_2 + p\alpha_3,$$
$$r_3(\alpha_1) = \alpha_1 + d\alpha_3, \quad r_3(\alpha_2) = \alpha_2 + q\alpha_3, \quad r_3(\alpha_3) = -\alpha_3.$$

$W = \{1, r_2(r_3, r_2)^j, r_3(r_2r_3)^j, (r_2r_3)^{j+1}, (r_3r_2)^{j+1} \mid j \text{ is a nonnegative integer}\}$.

Below we write down explicitly the real roots and imaginary roots for a particular case.

Let $a = 5, b = 1, c = 1, d = 1, p = 1, q = 5$, here $pad = qbc$

$$\Delta^{\text{re}} = \Delta_s^{\text{re}} \cup \Delta_{\text{L}}^{\text{re}}, \text{ where,}$$
$$\Delta_{\text{L}}^{\text{re}} = \pm\{k_2\alpha_2 + k_3\alpha_3 | k_2, k_3 \in \mathbb{Z}_+, 5k_2^2 - 5k_2k_3 + k_3^2 = 5\} \quad \text{and}$$
$$\Delta_{\text{S}}^{\text{re}} = \pm\{k_2\alpha_2 + k_3\alpha_3 | k_2, k_3 \in \mathbb{Z}_+, 5k_2^2 - 5k_2k_3 + k_3^2 = 1\}.$$

Using the definition of K, we find

$$K = \{k_1\alpha_1 + k_2\alpha_2 + k_3\alpha_3 | k_1, k_2, k_3 \in \mathbb{N}, \tfrac{-k_1+2k_3}{5} \leq k_2 \leq \tfrac{k_1+k_3}{2}, k_1 = 0 \Rightarrow 2k_3 \leq 5k_2 \\ \leq \tfrac{5k_3}{2}, k_2 = 0 \Rightarrow 2k_3 \leq k_1, k_3 = 0 \Rightarrow 2k_2 \leq k_1 \text{ and } k_2 = k_3 = 0 \Rightarrow k_1 = 1\}.$$

$\Delta_+^{\text{im}} = \cup_{w\in W} w(K)$. Here Weyl group is infinite.

Remark 73. If $q > p, \alpha_2$ is long and α_3 is short and if $q < p, \alpha_2$ is short and α_3 is long. Here all the properties of roots in $GGA_2^{(2)}$ are true.

Properties of real roots and imaginary roots of GKM algebras

We use the following properties of real roots of GKM algebras as given in the following two lemmas [142].

Lemma 5. *For $\alpha \in \Delta$ and $i \in I^{\mathrm{re}}$, one has the following:*

1. *The set $\{j \in \mathbb{Z};\ \alpha + j\alpha_i \in \Delta\}$ is a finite set. Let p be the minimum contained in this set, and let q be the maximum in this set. Then,*
   (i) $p + q = -\langle \alpha_i^\vee, \alpha \rangle$
   (ii) $\{j \in \mathbb{Z};\ \alpha + j\alpha_i \in \Delta\} = \{j \in \mathbb{Z};\ p \le j \le q\}$ *and*
   (iii) *The sequence $\{\text{multi } (\alpha + j\alpha_i)\}_{p \le j \le q}$ is bilaterally symmetric, and the left half of this sequence is monotone nondecreasing, namely,*
   (a) $j + k = p + q \Longrightarrow$ *multi* $(\alpha + j\alpha_i) =$ *multi* $(\alpha + k\alpha_i)$ *and*
   (b) $p \le j \le k \le \frac{1}{2}(p+q) \Longrightarrow$ *multi* $(\alpha + j\alpha_i) \le$ *multi* $(\alpha + k\alpha_i)$.
2. $(\alpha, \alpha_i) > 0 \Longrightarrow \alpha - \alpha_i \in \Delta_+$,
   $(\alpha, \alpha_i) < 0 \Longrightarrow \alpha + \alpha_i \in \Delta_+$.
3. $\alpha + \alpha_i \notin \Delta \Longrightarrow (\alpha, \alpha_i) \ge 0$,
   $\alpha - \alpha_i \notin \Delta \Longrightarrow (\alpha, \alpha_i) \le 0$.
4. *If α is a root, then supp α is connected.*
5. *For $\alpha \in \Delta$, $i \in I^{\mathrm{re}}$, if $\alpha + \alpha_i \notin \Delta$ and $\alpha - \alpha_i \in \Delta$, then $(\alpha, \alpha_i) > 0$.*

Lemma 6.

1. *Δ_+^{im} is invariant under the action of the Weyl group W.*
2. *If $\alpha \in \Delta_+^{\mathrm{im}}$, then there exists $w \in W$ satisfying $w\alpha \in -C^\vee$.*
3. *For $\alpha \in \Delta_+$ with $\alpha \in \Delta_+^{\mathrm{im}} \Longleftrightarrow (\alpha, \alpha) \le 0$.*

We state the following results, proved in Wakimoto [142].

Results.

(1) For $i \in I^{\mathrm{re}}$, one has the following:
   (i) $r_i(\alpha_i) = -\alpha_i$.
   (ii) If $i \in I^{\mathrm{re}}$, then $r_i(\Delta_+ \setminus \{\alpha_i\}) = \Delta_+ \setminus \{\alpha_i\}$.
   Let $C^\vee = \{h \in \mathfrak{h}'_{\mathbb{R}};\ \langle \alpha_i^\vee, h \rangle \ge 0\ (i \in I^{\mathrm{re}})\}$, $\mathfrak{h}'_{\mathbb{R}} = \sum_{i=1}^n \mathbb{R}\alpha_i^\vee = \sum_{i=1}^n \mathbb{R}\alpha_i$.
(2) If $\alpha \in Q_+ \cap (-C^\vee)$, then $(\alpha, \alpha) \le 0$.
(3) For $i \in I^{\mathrm{im}}$, take $\alpha \in \Delta_+ \setminus \{\alpha_i\}$, so that supp $(\alpha + \alpha_i)$ is connected. Then, for all $j \in \mathbb{Z}_{\ge 0}$, one has $\alpha + j\alpha_i \in \Delta_+$.
(4) For $i \in I^{\mathrm{im}}$ and $\alpha \in \Delta_+ \setminus \{\alpha_i\}$, if supp $(\alpha + \alpha_i)$ is connected, then $(\alpha, \alpha_i) < 0$.

Proposition 19.

(i) *Let A be an indecomposable, symmetrizable* GGCM. *Let $\alpha, \beta \in \Delta_+^{\mathrm{im}}$ be such that* supp *$(\alpha + \beta)$ is connected. Then $\alpha + \beta \in \Delta_+^{\mathrm{im}}$.*
(ii) *If $\alpha \in \Delta_+^{\mathrm{im}}$, $\alpha = \sum_{i=1}^n k_i\alpha_i$, then at least one of the k_i's are >0.*

Proof.

(i) Let $X^\vee = \bigcup_{w \in W} w(C^\vee)$ be the dual Tit's convex cone. Since $X^\vee$ is a convex cone, we can assume that $\alpha + \beta \in -C^\vee$ for $\alpha, \beta \in \Delta_+^{\mathrm{im}}$. This implies that $\langle \alpha_i^\vee, \alpha + \beta \rangle \le 0$ if $a_{ii} = 2$ and we are also given that supp $(\alpha + \beta)$ is connected. Hence $\alpha + \beta \in K \subset \Delta_+^{\mathrm{im}}$.
(ii) This is clear because if $\alpha \ne 0$, then $k_i \ne 0$ for at least one $i \in$ supp α.

□

3.4 Special imaginary roots in GKM algebras and their complete classifications

Below we generalize the notion of special imaginary roots from KM algebras to GKM algebras with a finite index set I [82].

Definition 119. Let α be an imaginary root of $\mathcal{G}(A)$ which is a symmetrizable GKM algebra. We call α a *special imaginary root*, if α satisfies the following conditions:

(i) $(\alpha, \alpha) \neq 0$;
(ii) $r_\alpha(\Delta) = \Delta, r_\alpha(\Delta^{\mathrm{re}}) = \Delta^{\mathrm{re}}, r_\alpha(\Delta^{\mathrm{im}}) = \Delta^{\mathrm{im}}$ and
(iii) r_α preserves root multiplicities.

It is clear that, if α is a special imaginary root, then $r_\alpha \in -W$ or an element of $-W$ multiplied by a diagram automorphism.

Lemma 7. *If α is a special imaginary root and β is a real simple root of a symmetrizable GKM algebra, then $(\beta, \alpha) = 0$.*

Proof. Suppose $(\beta, \alpha) \neq 0$. Then $r_\alpha(\beta) = \beta - \frac{2(\beta,\alpha)}{(\alpha,\alpha)}\alpha$ is not a real root. This contradicts the definition of special imaginary roots. □

Complete classification of special imaginary roots of generalized Kac-Moody algebras (GKM algebras or BKM algebras with finite index set)

For the purpose of finding a complete classification of GKM algebras possessing special imaginary roots, we broadly divide the GKM algebras into following four classes:

Class 1. GKM algebras all whose simple roots are real.
Class 2. GKM algebras all whose simple roots are imaginary.
Class 3. GKM algebras with three and more simple imaginary roots.
Class 4. GKM algebras with one or two simple imaginary roots.

In classes 3 and 4, if we delete rows and columns containing the negative diagonal elements of GGCMs, remaining matrices will be generalized Cartan matrices of Kac-Moody algebras.

Below we discuss these four classes separately.

Class 1. GKM algebras all whose simple roots are real:

These are Kac-Moody algebras for which a complete classification of Kac-Moody algebras possessing special imaginary roots was given by Bennett [54].

Class 2. GKM algebras all whose simple roots are imaginary:

These are GKM algebras whose GGCMs do not appear as extensions of generalized Cartan matrices of Kac-Moody algebras. So if we leave the corresponding rows and columns containing negative diagonal elements of a GGCM, we will not get generalized Cartan matrix of a Kac-Moody algebra. Hence Weyl group in each case is empty because these algebras do not contain simple real roots. So there is no special imaginary root.

Class 3. GKM algebras with three and more simple imaginary roots:

Here three and more negative elements will occur anywhere along the diagonals of the GGCM. In this case leaving the rows and columns containing these negative diagonal elements, we will get generalized Cartan matrices of the Kac-Moody algebras. Without loss of generality, we assume in the following proposition that the first r diagonal elements are negative.

Proposition 20 ([82]). *Let A be the symmetrizable*

$$GGX = \begin{pmatrix} -k_1 & -a_1 & -a_2 & \cdots & -a_{r-1} & -a_r & \cdots & -a_{n+r-1} \\ -b_1 & -k_2 & -c_2 & \cdots & -c_{r-1} & -c_r & \cdots & -c_{n+r-1} \\ -b_2 & -d_2 & -k_3 & \cdots & \cdots & \cdots & \cdots & \cdots \\ \vdots & \vdots & & \ddots & & & & \\ -b_{r-1} & -d_{r-1} & \cdots & \cdots & -k_r & \cdots & \cdots & \cdots \\ -b_r & -d_r & & & \vdots & & & \\ \vdots & \vdots & & & & GX & & \\ -b_{n+r-1} & -d_{n+r-1} & & & & & & \end{pmatrix},$$

where $k_i(1 \leq i \leq r,\ r \geq 3) \in \mathbb{Z}_{\geq 0}$, a_i, b_i, c_i, d_i are positive integers, GX is GCM of finite, affine, or indefinite type of order n. Let $\alpha = \sum_{i=1}^{n+r} l_i\alpha_i (\in K \subseteq \Delta_+^{\text{im}}$, for $l_i \in \mathbb{N})$ be any positive imaginary root of $\mathfrak{g}(A)$. Then α is not a special imaginary root of $\mathfrak{g}(A)$.

Proof. For the case of three and more simple imaginary roots, without loss of generality, assume that $\alpha_i(1 \leq i \leq r)$ are simple imaginary roots and α_i $(r+1 \leq i \leq n+r)$ are simple real roots. So $\Pi^{\text{re}} = \{\alpha_{r+1}, \ldots, \alpha_{n+r}\}$, $\Pi^{\text{im}} = \{\alpha_1, \ldots, \alpha_r\}$, and $\Pi = \{\alpha_1, \ldots, \alpha_r, \alpha_{r+1}, \ldots, \alpha_{n+r}\}$. Suppose, $\alpha = \sum_{i=1}^{n+r} l_i\alpha_i \in K \subseteq \Delta_+^{\text{im}}$, $(l_i \in \mathbb{N})$ is a special imaginary root of $\mathfrak{g}(A)$. We know that $\langle \alpha, \alpha_i^\vee \rangle \leq 0$ for $i = r+1$ to $n+r$ and $(\alpha, \alpha) < 0$. Let $(\alpha, \alpha) = a$. Using the definition of reflection on $\mathfrak{h}^*$, namely,

$$r_\alpha(\lambda) = \lambda - \frac{2(\lambda, \alpha)}{(\alpha, \alpha)}\alpha$$

for $\lambda \in \mathfrak{h}^*$ and α is an imaginary root, we have

$$r_\alpha(\alpha_j) = \alpha_j - p_j\alpha, \quad \text{where } p_j = \frac{2(\alpha_j, \alpha)}{a} \text{ for } 1 \leq j \leq r$$

$$r_\alpha(\alpha_i) = \alpha_i - p_i\alpha, \quad \text{where } p_i = \frac{2(\alpha_i, \alpha)}{a} \text{ for } r+1 \leq i \leq n+r. \tag{3.1}$$

Using Lemma 7, from Equation (3.1), we get

$$p_i = 0, \quad r+1 \le i \le n+r.$$

Solving these n equations in $n+r$ unknowns l_i, we get l_i, $(r+1 \le i \le n+r)$ in terms of $l_1, \ldots, l_r$.

So for $r+1 \le i \le r+n$, $l_i = 0$ for affine GX, because each l_i is a multiple of determinant of GX, which is zero and $l_i < 0$ for indefinite GX, because each l_i is a multiple of determinant of GX, which is less than zero.

Both these cases are not possible, because l_i must be positive. Hence α is not a special imaginary root for affine and indefinite cases. For finite case, let $l_i = \sum_{j=1}^{r} m_{ji} l_j$, $(r+1 \le i \le n+r)$ where m_{ji} are some positive integers. So

$$\alpha = \sum_{i=1}^{r} l_i \alpha_i + \sum_{i=r+1}^{r+n} \left(\sum_{j=1}^{r} m_{ji} l_j \right) \alpha_i.$$

Hence α is a special imaginary root if α satisfies

$$r_\alpha(\alpha_i) = -t.\, w_0(\alpha_i) \quad (1 \le i \le r),$$

where t is a permutation on n symbols and w_0 is the longest element of the Weyl group of GX. This is not possible, because α will not satisfy these r equations simultaneously and hence α has no solution. □

Example 26. Consider the symmetrizable GGCM,

$$A = \begin{pmatrix} -k_1 & -a_1 & -a_2 & -a_3 \\ -b_1 & -k_2 & -c_2 & -c_3 \\ -b_2 & -d_2 & -k_2 & -e_3 \\ -b_3 & -d_3 & -f_3 & 2 \end{pmatrix},$$

where $a_i, b_i, c_i, d_i, e_i, f_i, k_1$ are as in Proposition 20. Here α_1, α_2, and α_3 are simple imaginary roots and α_4 is a simple real root. Let $\alpha = l_1\alpha_1 + l_2\alpha_2 + l_3\alpha_3 + l_4\alpha_4 \in K \subseteq \Delta_+^{\mathrm{im}}$, $l_i \in \mathbb{N}$ be a special imaginary root of $\mathfrak{g}(A)$.

We know that $\langle \alpha, \alpha_4^\vee \rangle \le 0$ and $(\alpha, \alpha) < 0$. Let $(\alpha, \alpha) = a$. Using the definition of reflection of roots, we have

$$r_\alpha(\alpha_i) = \alpha_i - p_i\alpha \quad \text{for} \quad p_i = \frac{2(\alpha_i, \alpha)}{a} \text{ for all } i \tag{3.2}$$

Using Lemma 7, from Equation (3.2), we obtain $p_4 = 0$, and we get

$$l_4 = \frac{1}{2}(b_3 l_1 + d_3 l_2 + f_3 l_3).$$

So,

$$\alpha = l_1\alpha_1 + l_2\alpha_2 + l_3\alpha_3 + \frac{1}{2}(b_3l_1 + d_3l_2 + f_3l_3)\,\alpha_4.$$

This α will be a special imaginary root if it satisfies $r_\alpha(\alpha_1) = -r_4(\alpha_1), r_\alpha(\alpha_2) = -r_4(\alpha_2)$, and $r_\alpha(\alpha_3) = -r_4(\alpha_3)$. Hence

$$\begin{aligned}(1 - p_1l_1)\alpha_1 - p_1l_2\alpha_2 - p_1l_3\alpha_3 - p_1l_4\alpha_4 &= -(\alpha_1 + b_3\alpha_4),\\ -p_2l_1\alpha_1 + (1 - p_2l_2)\alpha_2 - p_2l_3\alpha_3 - p_2l_4\alpha_4 &= -(\alpha_2 + d_3\alpha_4), \text{ and}\\ -p_3l_1\alpha_1 - p_3l_2\alpha_2 + (1 - p_3l_3)\alpha_3 - p_3l_4\alpha_4 &= -(\alpha_3 + f_3\alpha_4).\end{aligned}$$

As α will not satisfy these three equations simultaneously, we will not get any special imaginary root.

Class 4. GKM algebras with one or two simple imaginary roots:
In class 4, there are two cases.

Case 1 in Class 4: GKM algebras with one simple imaginary root.
Case 2 in Class 4: GKM algebras with two simple imaginary roots.

First, we discuss the case 1 in class 4 in the following.

Case 1 in Class 4: GKM algebras with one simple imaginary root:

One negative element will occur anywhere along the diagonal. In this case, leaving the row and column containing this negative diagonal element, we will get the generalized Cartan matrix of a Kac-Moody algebra. For $n \leq 3$, we have completely classified and found out special imaginary roots for GKM algebras in Sthanumoorthy and Lilly [82, 102, 103].

Proposition 21. *Let A be the symmetrizable*

$$GGX = \begin{pmatrix} -k_1 & -a_1 & -a_2 & \cdots & -a_n \\ -b_1 & & & & \\ -b_2 & & & & \\ \vdots & & & GX & \\ -b_n & & & & \end{pmatrix},$$

where $k_1 \in \mathbb{Z}_{\geq 0}$ and a_i, b_i are positive integers. Let GX be a GCM of finite, affine, or indefinite type of order n and

$$\alpha = \sum_{i=1}^{n+1} l_i\alpha_i \in K \subseteq \Delta_+^{\text{im}}, \quad l_i \in \mathbb{N} \quad \textit{for all } i$$

be any positive imaginary root of $\mathfrak{g}(A)$. Further

(i) *If GX is a GCM of finite type, then α is a special imaginary root of $\mathfrak{g}(A)$ and*
(ii) *If GX is a GCM of affine or indefinite type, then α is not a special imaginary root of $\mathfrak{g}(A)$.*

Proof.

(i) Let GX be a GCM of finite type. For the case of one simple imaginary root, without loss of generality, assume that α_1 is the simple imaginary root and α_i, $i = 2$ to $n+1$ are simple real roots. Suppose $\alpha = \sum_{i=1}^{n+1} l_i\alpha_i (\alpha \in K \subseteq \Delta_+^{\text{im}}, l_i \in \mathbb{N})$, be a special imaginary root of $\mathfrak{g}(A)$. We know that $\langle \alpha, \alpha_i^\vee \rangle \leq 0$, $i = 2$ to $n+1$ and $(\alpha, \alpha) < 0$. Let $(\alpha, \alpha) = a$ (say). Using the definition of imaginary roots, we have

$$r_\alpha(\alpha_1) = \alpha_1 - p_1\alpha \quad \text{for } p_1 = \frac{2(\alpha_1, \alpha)}{a}$$

and

$$r_\alpha(\alpha_i) = \alpha_i - p_i\alpha \text{ for } i = 2 \text{ to } n+1 \text{ and } p_i = \frac{2(\alpha_i, \alpha)}{a}. \tag{3.3}$$

Using Lemma 7, from Equation (3.3), we obtain

$$p_i = 0 \text{ for } i = 2 \text{ to } n+1.$$

Solving these n equations in $n+1$ unknowns l_i, we get $l_i(2 \leq i \leq n+1)$ in terms of l_1. Let $l_i = m_i l_1$ for $2 \leq i \leq n+1$ and m_i are some positive integers. So,

$$\alpha = l_1\alpha_1 + l_1\left(\sum_{i=2}^{n+1} m_i\alpha_i\right).$$

This α will be a special imaginary root if it satisfies

$$r_\alpha(\alpha_1) = \begin{cases} -w_0(\alpha_1), & \text{if GX is of type B, C, F, G, E}_7 \text{ or E}_8 \\ -tw_0(\alpha_1), & \text{if GX is of type A, D or E}_6. \end{cases}$$

Here w_0 is the longest element of the Weyl group of GX and t is some permutation on n symbols. Hence under these conditions, we get special imaginary roots.

(ii) Let GX be a GCM of affine or indefinite type. The proof is exactly similar to affine and indefinite cases in Proposition 20 with $r = 1$.

□

Example 27. If $X = A_3$, then

$$GX = \begin{pmatrix} 2 & -1 & 0 \\ -1 & 2 & -1 \\ 0 & -1 & 2 \end{pmatrix} \text{ and } GGX = \begin{pmatrix} -k & -a_1 & -a_2 & -a_3 \\ -b_1 & 2 & -1 & 0 \\ -b_2 & -1 & 2 & -1 \\ -b_3 & 0 & -1 & 2 \end{pmatrix}.$$

It can be easily verified that when $A = GGA_3, DA$ will be symmetric (that is, A is symmetrizable) if we take

$$D = \begin{pmatrix} b_1\epsilon & 0 & 0 & 0 \\ 0 & a_1\epsilon & 0 & 0 \\ 0 & 0 & a_1\epsilon & 0 \\ 0 & 0 & 0 & a_1\epsilon \end{pmatrix}.$$

where $\epsilon \in \mathbb{Q}+$ such that $a_1b_2 = a_2b_1$ and $a_3b_1 = a_1b_3$. Also this matrix GGA_3 is nonsingular and indecomposable.

It is clear that $\Pi^{\text{re}} = \{\alpha_2, \alpha_3, \alpha_4\}, \Pi^{\text{im}} = \{\alpha_1\}$ and $\Pi = \{\alpha_1, \alpha_2, \alpha_3, \alpha_4\}$. We have

$$r_2(\alpha_1) = \alpha_1 + b_1\alpha_2; \quad r_2(\alpha_2) = -\alpha_2; \quad r_2(\alpha_3) = \alpha_2 + \alpha_3; \quad r_2(\alpha_4) = \alpha_4;$$
$$r_3(\alpha_1) = \alpha_1 + b_2\alpha_3; \quad r_3(\alpha_2) = \alpha_2 + \alpha_3; \quad r_3(\alpha_3) = -\alpha_3; \quad r_3(\alpha_4) = \alpha_3 + \alpha_4;$$
$$r_4(\alpha_1) = \alpha_1 + b_3\alpha_4; \quad r_4(\alpha_2) = \alpha_2; \quad r_4(\alpha_3) = \alpha_3 + \alpha_4; \quad \text{and } r_4(\alpha_4) = -\alpha_4.$$

$$\begin{aligned} W = \{&1,\ r_2,\ r_3r_4,\ r_2r_3,\ r_2r_4,\ r_3r_4,\ r_3r_2,\ r_4r_3,\ r_2r_3r_4,\ r_2r_3r_2, \\ &r_2r_4r_3,\ r_3r_2r_4,\ r_3r_4r_3,\ r_4r_3r_2,\ r_2r_3r_2r_4,\ r_2r_3r_4r_3,\ r_2r_4r_3r_2, \\ &r_3r_2r_4r_3,\ r_3r_4r_3r_2,\ r_2r_3r_2r_4r_3,\ r_3r_2r_4r_3r_2,\ r_4r_2r_3r_2r_4,\ r_4r_2r_3r_2r_4r_3\}, \end{aligned}$$

which is same as Weyl group of A_3 and the order of W is 24. By the definition, $K = \{\alpha \in Q_+ \backslash \{0\} | \langle \alpha, \alpha_i \rangle \leq 0 \text{ if } a_{ii} = 2 \text{ and supp } (\alpha) \text{ is connected } \} \backslash \cup_{j \geq 2} j\Pi^{\text{im}}$. So,

$$\begin{aligned} K = \{&k_1\alpha_1 + k_2\alpha_2 + k_3\alpha_3 + k_4\alpha_4 | k_1, k_2, k_3, k_4 \in \mathbb{N} \text{ satisfying } 2k_2 \leq b_1k_1 + k_3, \\ &2k_3 \leq b_2k_1 + k_2 + k_4 \text{ and } 2k_4 \leq b_3k_1 + k_3 \text{ with} \\ &k_2 = 0 \Rightarrow 2k_3 \leq b_2k_1 + k_4, k_3 = 0 \Rightarrow 2k_2 \leq b_1k_1 \text{ and } 2k_4 \leq b_3k_1, \\ &k_4 = 0 \Rightarrow 2k_3 \leq b_2k_1 + k_2, k_2 = k_4 = 0 \Rightarrow 2k_3 \leq b_2k_1, \\ &k_2 = k_3 = k_4 = 0 \Rightarrow k_1 = 1\}. \end{aligned}$$

As $\Delta_+^{\text{im}} = \cup_{w \in W} w(K)$, one can find out the set of all Δ_+^{im}, positive imaginary roots.

Let

$$k_2 = k_4 = \frac{3b_1}{2}k_1 \text{ and } k_3 = 2b_1k_1.$$

Then

$$\alpha = \frac{k_1}{2}(2\alpha_1 + 3b_1\alpha_2 + 4b_1\alpha_3 + 3b_1\alpha_4), k_1 \in \mathbb{N}.$$

We have

$$(\alpha, \alpha) = \frac{-k_1^2 b_1\epsilon}{4}(k + 5a_1b_1) < 0.$$

Using the definition of reflection of imaginary root, we have

$$r_\alpha(\alpha_1) = -(\alpha_1 + 3b_1\alpha_2 + 4b_1\alpha_3 + 3b_1\alpha_4).$$

Also, we have

$$-tr_4r_2r_3r_2r_4r_3(\alpha_1) = -(\alpha_1 + 3b_1\alpha_2 + 4b_1\alpha_3 + 3b_1\alpha_4).$$

Therefore

$$r_\alpha(\alpha_1) = -tr_4r_2r_3r_2r_4r_3(\alpha_1).$$

Similarly, we can verify that,

$$\begin{aligned} r_\alpha(\alpha_2) &= -tr_4r_2r_3r_2r_4r_3(\alpha_2),\\ r_\alpha(\alpha_3) &= -tr_4r_2r_3r_2r_4r_3(\alpha_3), \text{ and}\\ r_\alpha(\alpha_4) &= -tr_4r_2r_3r_2r_4r_3(\alpha_4). \end{aligned}$$

Since, $\det A = -4(k + a_1b_1) \neq 0$, we have $\dim \mathfrak{h}^\star = 4$ and $\Pi = \{\alpha_1, \alpha_2, \alpha_3, \alpha_4\}$ is a basis of $\mathfrak{h}^\star$. Thus, $r_\alpha = -tr_4r_2r_3r_2r_4r_3 \in -tW$ as a reflection on $\mathfrak{h}^\star$. Conversely, let

$$\alpha = k_1\alpha_1 + k_2\alpha_2 + k_3\alpha_3 + k_4\alpha_4 \in K \subseteq \Delta_+^{\text{im}}.$$

We know that $k_1, k_2, k_3, k_4 \in \mathbb{N}$ with $2k_2 \leq b_1k_1 + k_3, 2k_3 \leq b_1k_1 + k_2 + k_4$ and $2k_4 = b_1k_1 + k_3$.

Then

$$(\alpha, \alpha) = A\epsilon < 0,$$

where

$$\begin{aligned} A = &-kk_1^2b_1 - 2a_1b_1k_1k_2 - 2a_1b_1k_1k_3 - 2a_1b_1k_1k_4\\ &- 2a_1k_2k_3 - 2a_1k_3k_4 + 2a_1k_2^2 + 2a_1k_3^2 + 2a_1k_4^2. \end{aligned}$$

Using the definition of reflection of roots, we have

$$r_\alpha(\alpha_1) = \alpha_1 + \frac{2b_1}{A}(kk_1 + a_1k_2 + a_1k_3 + a_1k_4)\alpha, \tag{3.4}$$

$$r_\alpha(\alpha_2) = \alpha_2 - \frac{2a_1}{A}(2k_2 - b_1k_1 - k_3)\alpha, \tag{3.5}$$

$$r_\alpha(\alpha_3) = \alpha_3 - \frac{2a_1}{A}(2k_3 - b_1k_1 - k_2 - k_4)\alpha, \tag{3.6}$$

$$r_\alpha(\alpha_4) = \alpha_4 - \frac{2a_1}{A}(2k_4 - b_1k_1 - k_3)\alpha. \tag{3.7}$$

We claim that if $r_\alpha \in -tW$, then $r_\alpha = -tr_4r_2r_3r_2r_4r_3$. If this is not true, for example,

$$r_\alpha = -tr_2, \text{ then } r_\alpha(\alpha_2) = \alpha_4.$$

From Equation (3.4), we have $2a_1k_1(2k_2-b_1k_1-k_3)\alpha_1+(2a_1k_2(2k_2-b_1k_1-k_3)-A)\alpha_2+2a_1k_3(2k_2-b_1k_1-k_3)\alpha_3+(A+2a_1k_4(2k_2-b_1k_1-k_3))\alpha_4 = 0$. This implies each coefficient of $\alpha_i = 0$. Hence $a_1k_1(2k_2-b_1k_1-k_3) = 0$, $2a_1k_2(2k_2-b_1k_1-k_3)-A = 0$, $a_1k_3(2k_2 - b_1k_1 - k_3) = 0$, and $A + 2a_1k_4(2k_2 - b_1k_1 - k_3) = 0$.

Since $k_1, k_3 \neq 0, 2k_2 - b_1k_1 - k_3 = 0$ and hence $A = 0$, which is a contradiction. Similarly, we can prove that for each element in $-tW$ except $-tr_4r_2r_3r_2r_4r_3$, we get a contradiction.

Suppose, $r_\alpha = -tr_4r_2r_3r_2r_4r_3$. Then $r_\alpha(\alpha_2) = \alpha_2; r_\alpha(\alpha_3) = \alpha_3$ and $r_\alpha(\alpha_4) = \alpha_4$. From Equations (3.4)–(3.7), we obtain $(2k_2 - b_1k_1 - k_3)\alpha = (2k_3 - b_1k_1 - k_2 - k_4)\alpha = (2k_4 - b_1k_1 - k_3)\alpha = 0$. Hence $k_2 = k_4 = \frac{3b_1}{2}k_1$ and $k_3 = 2b_1k_1$. So we have proved the following: Let A be the symmetrizable GGA_3, t be the permutation $(2,4)$ and $\alpha = k_1\alpha_1 + k_2\alpha_2 + k_3\alpha_3 + k_4\alpha_4 \in K \subseteq \Delta_+^{\text{im}}$, with $k_1, k_2, k_3, k_4 \in \mathbb{N}$ and $b_1 = b_2 = b_3$. Then $r_\alpha \in -tW$ if and only if $k_2 = k_4 = \frac{3b_1}{2}k_1$ and $k_3 = 2b_1k_1 \in \mathbb{N}$.

Hence, $\alpha = \frac{k_1}{2}(2\alpha_1 + 3b_1\alpha_2 + 4b_1\alpha_3 + 3b_1\alpha_4)$ is a special imaginary root of $\mathcal{G}(A)$, for every $k_1 \in \mathbb{N}$, where A is a symmetrizable GGA_3 as defined with $b_1 = b_2 = b_3$. If this condition $b_1 = b_2 = b_3$ is not satisfied, we will not get special imaginary roots.

Remark 74. As the proof for the general case in (i) was discussed above, we give below the following example, $GX = B_3$ for finite type.

Example 28. Consider the symmetrizable GGCM,

$$A = \begin{pmatrix} -k_1 & -a_1 & -a_2 & -a_3 \\ -b_1 & 2 & -1 & 0 \\ -b_2 & -1 & 2 & -1 \\ -b_3 & 0 & -2 & 2 \end{pmatrix},$$

where a_i, b_i, k_1 $(i = 1, 2, 3)$ are as in Proposition 21.

It can be easily verified that A is symmetrizable if and only if $a_1b_2 = a_2b_1$ and $a_2b_3 = 2a_3b_2$. It is clear that $\Pi^{\text{re}} = \{\alpha_2, \alpha_3, \alpha_4\}$, $\Pi^{\text{im}} = \{\alpha_1\}$ and $\Pi = \{\alpha_1, \alpha_2, \alpha_3, \alpha_4\}$. Let $\alpha = l_1\alpha_1 + l_2\alpha_2 + l_3\alpha_3 + l_4\alpha_4 (\alpha \in K \subseteq \Delta_+^{\text{im}}, l_i \in \mathbb{N})$ be a special imaginary root of $\mathfrak{g}(A)$.

We know that $\langle \alpha, \alpha_i^\vee \rangle \leq 0$, $(2 \leq i \leq 4)$ and $(\alpha, \alpha) < 0$. Let $(\alpha, \alpha) = a$. Using the definition of reflection of roots, we have

$$r_\alpha(\alpha_1) = \alpha_1 - p_1\alpha \quad \text{for} \quad p_1 = \frac{2(\alpha_1, \alpha)}{a} \quad \text{and}$$
$$r_\alpha(\alpha_i) = \alpha_i - p_i\alpha \quad (2 \leq i \leq 4) \text{ for } p_i = \frac{2(\alpha_i, \alpha)}{a}. \tag{3.8}$$

Using Lemma 7, from Equation (3.8), we obtain $p_i = 0$ for $2 \leq i \leq 4$ and we get

$$2l_2 = (2b_1+2b_2+b_3)l_1, \quad l_3 = (b_1+2b_2+b_3)l_1, \quad \text{and} \quad 2l_4 = (2b_1+4b_2+3b_3)l_1.$$

So

$$\alpha = \frac{l_1}{2}\left(2\alpha_1 + (2b_1 + 2b_2 + b_3)\alpha_2 + 2(b_1 + 2b_2 + b_3)\alpha_3 + (2b_1 + 4b_2 + 3b_3)\alpha_4\right).$$

This α also satisfies

$$r_\alpha(\alpha_1) = -r_4 r_3 (r_4 r_2 r_3)^2 r_2(\alpha_1),$$

because $r_4 r_3 (r_4 r_2 r_3)^2 r_2$ is the longest element of the Weyl group of B_3. Hence

$$\alpha = \frac{l_1}{2}\left(2\alpha_1 + (2b_1 + 2b_2 + b_3)\alpha_2 + 2(b_1 + 2b_2 + b_3)\alpha_3 + (2b_1 + 4b_2 + 3b_3)\alpha_4\right)$$

is a special imaginary root of $\mathfrak{g}(A)$.

Theorem 33 ([102]). *There exists no special imaginary root for $\mathcal{G}(A)$, where A is a symmetrizable $GGA_2^{(1)}$.*

Proof. For an extension of affine GCM with $X = A_2^{(1)}$, we have

$$GX = \begin{pmatrix} 2 & -1 & -1 \\ -1 & 2 & -1 \\ -1 & -1 & 2 \end{pmatrix} \quad \text{and} \quad GGX = \begin{pmatrix} -k & -a_1 & -a_2 & -a_3 \\ -b_1 & 2 & -1 & -1 \\ -b_2 & -1 & 2 & -1 \\ -b_3 & -1 & -1 & 2 \end{pmatrix}.$$

It can be easily verified that when $A = GGA_2^{(1)}, DA$ will be symmetric (that is, A is symmetrizable) if we take D as

$$\begin{pmatrix} b_1\epsilon & 0 & 0 & 0 \\ 0 & a_1\epsilon & 0 & 0 \\ 0 & 0 & a_1\epsilon & 0 \\ 0 & 0 & 0 & a_1\epsilon \end{pmatrix},$$

where $\epsilon \in \mathbb{Q}_+$ such that $a_1b_2 = a_2b_1$ and $a_2b_3 = a_3b_2$. Also this matrix $GGA_2^{(1)}$ is nonsingular and indecomposable. It is clear that

$$\Pi^{\mathrm{re}} = \{\alpha_2, \alpha_3, \alpha_4\}, \quad \Pi^{\mathrm{im}} = \{\alpha_1\}, \quad \text{and} \quad \Pi = \{\alpha_1, \alpha_2, \alpha_3, \alpha_4\}.$$

We have

$$r_2(\alpha_1) = \alpha_1 + b_1\alpha_2; \quad r_2(\alpha_2) = -\alpha_2; \quad r_2(\alpha_3) = \alpha_2 + \alpha_3; \quad r_2(\alpha_4) = \alpha_4 + \alpha_2;$$

$r_3(\alpha_1) = \alpha_1 + b_2\alpha_3;\quad r_3(\alpha_2) = \alpha_2 + \alpha_3;\quad r_3(\alpha_3) = -\alpha_3;\quad r_3(\alpha_4) = \alpha_3 + \alpha_4;$
$r_4(\alpha_1) = \alpha_1 + b_2\alpha_4;\quad r_4(\alpha_2) = \alpha_2 + \alpha_4;\quad r_4(\alpha_3) = \alpha_3 + \alpha_4;$ and $r_4(\alpha_4) = -\alpha_4.$

Hence, W is the group generated by the reflections r_2, r_3, and r_4. Using the definition of K, we find

$$\begin{aligned} K = \{&k_1\alpha_1 + k_2\alpha_2 + k_3\alpha_3 + k_4\alpha_4 | k_1, k_2, k_3, k_4 \in \mathbb{N} \text{ satisfying } 2k_2 \leq b_1k_1 + k_3 + k_4, \\ &2k_3 \leq b_2k_1 + k_2 + k_4 \text{ and } 2k_4 \leq b_3k_1 + k_2 + k_3 \text{ with} \\ &k_2 = 0 \Rightarrow 2k_3 \leq b_2k_1 + k_4 \text{ and } 2k_4 \leq b_3k_1 + k_3, \\ &k_3 = 0 \Rightarrow 2k_2 \leq b_1k_1 + k_4 \text{ and } 2k_4 \leq b_3k_1 + k_2, \\ &k_4 = 0 \Rightarrow 2k_2 \leq b_1k_1 + k_3 \text{ and } 2k_3 \leq b_2k_1 + k_2, \\ &k_2 = k_3 = 0 \Rightarrow 2k_4 \leq b_3k_1, k_2 = k_4 = 0 \Rightarrow 2k_3 \leq b_2k_1, \\ &k_2 = k_4 = 0 \Rightarrow 2k_4 \leq b_1k_1, k_2 = k_3 = k_4 = 0 \Rightarrow k_1 = 1\} \end{aligned}$$

and

$$\Delta_+^{\text{im}} = \cup_{w \in W} w(K).$$

Let $\alpha = k_1\alpha_1 + k_2\alpha_2 + k_3\alpha_3 + k_4\alpha_4 \in K \subseteq \Delta_+^{\text{im}}$ be a special imaginary root of $\mathcal{G}(A)$. Using the above computation of K, we can conclude that $k_1, k_2, k_3, k_4 \in \mathbb{N}$ with $2k_2 \leq b_1k_1 + k_3 + k_4, 2k_3 \leq b_2k_1 + k_2 + k_4$ and $2k_4 \leq b_3k_1 + k_2 + k_3$. Then

$$(\alpha, \alpha) = A\epsilon < 0,$$

where

$$\begin{aligned} A = {}& 2a_1k_2^2 + 2a_1k_3^2 + 2a_1k_4^2 - b_1kk_1^2 - 2a_1b_1k_1k_2 \\ & - 2a_1b_2k_1k_3 - 2a_1b_3k_1k_4 - 2a_1k_2k_3 \\ & - 2a_1k_2k_4 - 2a_1k_3k_4. \end{aligned}$$

Using the definition of reflection of imaginary roots, we have

$$r_\alpha(\alpha_1) = \alpha_1 + \frac{2b_1}{A}(kk_1 + a_1k_2 + a_2k_3 + a_3k_4)\alpha, \tag{3.9}$$

$$r_\alpha(\alpha_2) = \alpha_2 - \frac{2a_1}{A}(2k_2 - b_1k_1 - k_3 - k_4)\alpha, \tag{3.10}$$

$$r_\alpha(\alpha_3) = \alpha_3 - \frac{2a_1}{A}(2k_3 - b_2k_1 - k_2 - k_4)\alpha, \text{ and} \tag{3.11}$$

$$r_\alpha(\alpha_4) = \alpha_4 - \frac{2a_1}{A}(2k_4 - b_3k_1 - k_2 - k_3)\alpha. \tag{3.12}$$

Then, from definition of special imaginary roots, we get

$$r_\alpha(\alpha_2) = \alpha_2; \quad r_\alpha(\alpha_3) = \alpha_3; \quad \text{and} \quad r_\alpha(\alpha_4) = \alpha_4.$$

From Equations (3.9)–(3.12), we obtain

$$\begin{aligned}(2k_2 - b_1k_1 - k_3 - k_4)\alpha &= (2k_3 - b_2k_1 - k_2 - k_4)\alpha \\ &= (2k_4 - b_3k_1 - k_2 - k_3)\alpha = 0.\end{aligned}$$

Therefore $\left(\sum_{i=1}^{3} b_i\right) k_1 = 0$. This is not possible as both $b_i (i = 1 \text{ to } 3)$ and k_1 are nonzero.

So we have proved the following, if A is a symmetrizable $GGA_2^{(1)}$ and $\alpha = k_1\alpha_1 + k_2\alpha_2 + k_3\alpha_3 + k_4\alpha_4 \in K \subseteq \Delta_+^{\text{im}}, k_1, k_2, k_3, k_4 \in \mathbb{N}$. Then $r_\alpha \notin -W$.

Hence, there exists no special imaginary root for $\mathcal{G}(A)$, where A is a symmetrizable $GGA_2^{(1)}$. □

Theorem 34. *Let A be the symmetrizable $GGHA_2^{(2)}$ and $\alpha = k_1\alpha_1 + k_2\alpha_2 + k_3\alpha_3 + k_4\alpha_4 \in K \subseteq \Delta_+^{\text{im}}$, $k_1, k_2, k_3, k_4 \in \mathbb{N}$. Then, $r_\alpha \notin -W$. Hence there exists no special imaginary root for $\mathfrak{g}(A)$, where A is a symmetrizable $GGHA_2^{(2)}$.*

Proof. Let $\alpha = k_1\alpha_1 + k_2\alpha_2 + k_3\alpha_3 + k_4\alpha_4 \in K \subseteq \Delta_+^{\text{im}}$ be a special imaginary root of $\mathfrak{g}(A)$. We know that $k_1, k_2, k_3, k_4 \in \mathbb{N}$, $2k_2 \leq b_1k_1 + k_3$, $2k_3 \leq b_2k_1 + k_2 + k_4$, and $2k_4 \leq b_3k_1 + 3k_3$. Then, $(\alpha, \alpha) = A\epsilon < 0$, where

$$\begin{aligned}A = {} & 2a_1k_2^2 + 2a_1k_3^2 + 8a_1k_4^2 - b_1kk_1^2 - 2a_1b_1k_1k_2 \\ & - 2a_2b_1k_1k_3 - 2a_3b_1k_1k_4 - 2a_1k_2k_3 - 8a_1k_3k_4.\end{aligned}$$

Using the definition of special imaginary roots, we have

$$r_\alpha(\alpha_1) = \alpha_1 + \frac{2b_1}{A}(kk_1 + a_1k_2 + a_2k_3 + a_3k_4)\alpha, \tag{3.13}$$

$$r_\alpha(\alpha_2) = \alpha_2 - \frac{2a_1}{A}(2k_2 - b_1k_1 - k_3)\alpha, \tag{3.14}$$

$$r_\alpha(\alpha_3) = \alpha_3 - \frac{2a_1}{A}(2k_3 - b_2k_1 - k_2 - 4k_4)\alpha, \text{and} \tag{3.15}$$

$$r_\alpha(\alpha_4) = \alpha_4 - \frac{8a_1}{A}(2k_4 - b_3k_1 - k_3)\alpha. \tag{3.16}$$

Then, from the definition of special imaginary roots, we get

$$r_\alpha(\alpha_2) = \alpha_2; \quad r_\alpha(\alpha_3) = \alpha_3; \quad \text{and} \quad r_\alpha(\alpha_4) = \alpha_4. \tag{3.17}$$

From Equations (3.13)–(3.16), we obtain

$$\begin{aligned}(2k_2 - b_1k_1 - k_3)\alpha &= (2k_3 - b_2k_1 - k_2 - 4k_4)\alpha \\ &= (2k_4 - b_3k_1 - k_3)\alpha = 0.\end{aligned}$$

Therefore, $(b_2 + 2b_3)k_1 + k_2 = 0$, which is not true.

Hence, there exists no special imaginary root for $\mathfrak{g}(A)$, where A is a symmetrizable $GGHA_2^{(2)}$. □

Remark 75. Similarly, the following result can also be proved: There exists no special imaginary root for $\mathcal{G}(A)$, where A is a symmetrizable $GGHA_1^{(1)}$.

Now, we discuss case 2 in Class 4 in the following.

Case 2 in Class 4: GKM algebras with two simple imaginary roots:

Two negative elements will occur anywhere along the diagonal of GGCMs. In this case, leaving the rows and columns containing these negative diagonal elements, we will get the generalized Cartan matrix of the Kac-Moody algebra. This may be finite, affine, or indefinite type.

Proposition 22 ([82]). *Let A be the symmetrizable*

$$GGX = \begin{pmatrix} -k_1 & -a_1 & -a_2 & \cdots & -a_{n+1} \\ -b_1 & -k_2 & -c_2 & \cdots & -c_{n+1} \\ -b_2 & -d_2 & & & \\ \vdots & \vdots & & GX & \\ -b_{n+1} & -d_{n+1} & & & \end{pmatrix},$$

where k_1, $k_2 \in \mathbb{Z}_{\geq 0}$, a_i, b_i, c_i, d_i are positive integers, GX is GCM of finite, affine, or indefinite type of order n and $\alpha = \sum_{i=1}^{n+1} l_i\alpha_i (\alpha \in K \subseteq \Delta_+^{\text{im}}, l_i \in \mathbb{N})$ be any positive imaginary root of $\mathfrak{g}(A)$. Further

(i) *If GX is GCM of finite type, then α is a special imaginary root of some $\mathfrak{g}(A)$ and*

(ii) *If GX is GCM of affine or indefinite type, then α is not a special imaginary root of $\mathfrak{g}(A)$.*

Proof.

(i) Let GX be a GCM of finite type. For the case of two simple imaginary roots and n simple real roots, without loss of generality, assume that α_1 and α_2 are the simple imaginary roots and α_i for $i = 3$ to $n+2$ are simple real roots.

Suppose

$$\alpha = \sum_{i=1}^{n+2} l_i\alpha_i \quad (\alpha \in K \subseteq \Delta_+^{\text{im}}, l_i \in \mathbb{N})$$

be a special imaginary root of $\mathfrak{g}(A)$. We know that $\langle \alpha, \alpha_i^\vee \rangle \leq 0$, $i = 3$ to $n+2$ and $(\alpha, \alpha) < 0$. Let $(\alpha, \alpha) = a$. Using the definition of imaginary roots, we have

$$\begin{aligned} r_\alpha(\alpha_j) &= \alpha_j - p_j\alpha \quad \text{for } p_j = \frac{2(\alpha_j, \alpha)}{a}, j = 1, 2 \text{ and} \\ r_\alpha(\alpha_i) &= \alpha_i - p_i\alpha \quad \text{for } p_i = \frac{2(\alpha_i, \alpha)}{a}, i = 3 \text{ to } n+2. \end{aligned} \tag{3.18}$$

Using Lemma 7, from Equation (3.18), we obtain

$$p_i = 0 \text{ for } i = 3 \text{ to } n+2.$$

Solving these n equations in $n+2$ unknowns l_i, we get l_i for $3 \leq i \leq n+2$ in terms of l_1 and l_2. Let $l_i = r_i l_1 + s_i l_2$, $3 \leq i \leq n+2$ and r_i, s_i are some positive integers. So, $\alpha = l_1\alpha_1 + l_2\alpha_2 + l_1\left(\sum_{i=3}^{n+2} p_i\alpha_i\right) + l_2\left(\sum_{i=3}^{n+2} q_i\alpha_i\right)$.

This α will be a special imaginary root, if it satisfies

$$r_\alpha(\alpha_i) = \begin{cases} -tw_0(\alpha_i), & i = 1,2 \text{ if GX is of type B, C, F, G, } E_7 \text{ or } E_8 \\ -tt_1w_0(\alpha_i), & i = 1,2 \text{ if GX is of type A, D or } E_6, \end{cases}$$

where t is the permutation $(1,2)$ and t_1 is another permutation on n symbols and w_0 is the longest element of the Weyl group of GX. Hence under the above condition, we get special imaginary roots.

(ii) Let GX be a GCM of affine or indefinite type. The proof is exactly similar to affine and indefinite cases in Proposition 20 with $r = 2$.

□

Example 29. Consider the symmetrizable GGCM,

$$A = \begin{pmatrix} -k_1 & -a_1 & -a_2 \\ -b_1 & -k_2 & -c_2 \\ -b_2 & -d_2 & 2 \end{pmatrix},$$

where a_i, b_i, c_2, d_2 are as in Proposition 22. It can be easily seen that A is symmetrizable if and only if $a_1c_2b_2 = b_1d_2a_2$.

Here α_1 and α_2 are simple imaginary roots and α_3 is a simple real root.

Let $\alpha = l_1\alpha_1 + l_2\alpha_2 + l_3\alpha_3$ $(\alpha \in K \subseteq \Delta_+^{\text{im}}, l_i \in \mathbb{N})$ be a special imaginary root of $\mathfrak{g}(A)$. We know that $\langle \alpha, \alpha_3^\vee \rangle \leq 0$. Let $(\alpha, \alpha) = a < 0$(say). Using the definition of reflection of roots, we have

$$\begin{aligned} r_\alpha(\alpha_1) &= \alpha_1 - p_1\alpha \quad \text{for } p_1 = \frac{2(\alpha_1,\alpha)}{a}, \\ r_\alpha(\alpha_2) &= \alpha_2 - p_2\alpha \quad \text{for } p_2 = \frac{2(\alpha_2,\alpha)}{a}, \text{ and} \\ r_\alpha(\alpha_3) &= \alpha_3 - p_3\alpha \quad \text{for } p_3 = \frac{2(\alpha_3,\alpha)}{a}. \end{aligned} \tag{3.19}$$

Then using Lemma 7, from Equation (3.19), we obtain $p_3 = 0$ and we get $l_3 = \frac{b_2l_1+d_2l_2}{2}$. So

$$\alpha = l_1\alpha_1 + l_2\alpha_2 + \frac{b_2l_1 + d_2l_2}{2}\alpha_3.$$

This α will be a special imaginary root if it satisfies $r_\alpha(\alpha_1) = -tr_3(\alpha_1)$ and $r_\alpha(\alpha_2) = -tr_3(\alpha_2)$, where t is the permutation $(1, 2)$. So,

$$(1 - p_1 l_1)\alpha_1 - p_1 l_2 \alpha_2 - p_1 l_3 \alpha_3 = -(\alpha_2 + b_2 \alpha_3) \text{ and}$$
$$-p_2 l_1 \alpha_1 + (1 - p_2 l_2)\alpha_2 - p_2 l_3 \alpha_3 = -(\alpha_1 + d_2 \alpha_3).$$

Solving the above equations, $l_2 = l_1$ and $b_2 = d_2$. Hence $\alpha = l_1(\alpha_1 + \alpha_2 + b_2\alpha_3)$ is a special imaginary root of $\mathfrak{g}(A)$ with $d_2 = b_2$.

Complete classification of special imaginary roots of BKM algebras possessing infinite number of simple imaginary roots

For GKM algebras corresponding to BCMs of infinite order, we have the following cases:

(i) All simple roots are imaginary: These are GKM algebras whose BCMs do not appear as extensions of generalized Cartan matrices of Kac-Moody algebras. So if we leave the corresponding rows and columns containing negative diagonal elements of BCMs, we will not get generalized Cartan matrices of Kac-Moody algebras. Hence Weyl group in each case is empty because these algebras do not contain simple real roots. So there is no special imaginary root.

(ii) Finite (nonzero) number of simple real roots and infinite number of simple imaginary roots: Here infinite number of negative elements will occur anywhere along the diagonal of BCMs. In this case, leaving the rows and columns containing these negative diagonal elements, we will get the generalized Cartan matrices of the Kac-Moody algebras. This may be finite, affine, or indefinite type. Using the definition of special imaginary roots, one can easily conclude that if α is a special imaginary root, the equations corresponding to Weyl reflections r_α cannot be solved. So GKM algebras with infinite number of simple imaginary roots and finite (nonzero) number of real simple roots do not possess special imaginary roots.

For example, consider the Monster Lie algebra. This algebra has one real simple root α_1 and infinite number of simple imaginary roots α_i for $2 \leq i \leq \infty$. Using the definition of special imaginary roots, one can easily conclude that if α is a special imaginary root, the equations corresponding to Weyl reflections r_α cannot be solved. The coefficients l_i of α_i $(2 \leq i \leq \infty)$ in the equations corresponding to Weyl reflections r_α of different simple imaginary roots α_i cannot satisfy simultaneously all the equations. *Hence Monster Lie algebra does not possess special imaginary roots.*

(iii) Infinite number of real simple roots and finite number of imaginary simple roots: As the number of real simple roots is infinite, the order of the Weyl group is infinite. So if α is a special imaginary root, one cannot verify simultaneously infinite number of equations for α to satisfy and hence we conclude that there exists no special imaginary root.

(iv) All simple roots are real: Here the Weyl group has infinite number of elements and there is no imaginary simple root. Hence we conclude that there exists no special imaginary root.

3.5 Strictly imaginary roots in GKM algebras and their complete classifications

In this section, we consider all GKM algebras for the purpose of finding out strictly imaginary roots [82, 100, 103].

Definition 120. An imaginary root γ in a BKM algebra (the index set I may be finite or infinite) is said to be *strictly imaginary*, if for every $\alpha \in \Delta^{\text{re}}$, either $\alpha + \gamma$ or $\alpha - \gamma$ is a root. Let Δ^{sim}, Δ_{+}^{sim}, and Δ_{-}^{sim} denote the set of all strictly imaginary roots, the set of all positive strictly imaginary roots, and the set of all negative strictly imaginary roots, respectively.

Similar to the definition of SIM property for generalized Cartan matrix(GCM), we defined SIM property for GGCM.

Definition 121. We say that a GGCM, A satisfies the *strictly imaginary property* (SIM property), if $\Delta_{+}^{\text{sim}}(A) = \Delta_{+}^{\text{im}}(A)$. We say that the GKM algebra $\mathfrak{g}(A)$ has the SIM property, if A satisfies the SIM property.

For the sake of finding out complete classification of GKM algebras possessing SIM property, we broadly divide the GKM algebras with finite number of simple roots into following three classes

Class 1. GKM algebras all whose simple roots are real.
Class 2. GKM algebras all whose simple roots are imaginary.
Class 3. GKM algebras with finite (nonzero) number of imaginary simple roots and finite (nonzero) number of real simple roots.

Below we discuss, these three classes separately.

Class 1. GKM algebras all whose simple roots are real:

These are Kac-Moody algebras for which complete classification of SIM property was given by Casperson [57] (Chapter 2).

Class 2. GKM algebras all whose simple roots are imaginary:

These are GKM algebras whose GGCMs do not appear as extensions of generalized Cartan matrices of Kac-Moody algebras. So all the diagonal elements are negative and there is no real simple root. Hence all the roots are imaginary and also strictly imaginary.

Class 3. GKM algebras with finite (nonzero) number of real simple roots and finite (nonzero) number of imaginary simple roots:

Here the negative diagonal elements will occur anywhere along the diagonals of the GGCM. In this case, leaving the rows and columns containing these negative diagonal elements, we will get generalized Cartan matrices of the Kac-Moody algebras.

We prove the following theorem [82, 100].

Theorem 35.

(a) *If $\alpha \in \Delta_{+}$ and $\langle \alpha_i^{\vee}, \alpha \rangle < 0$ for all $i \in I^{\text{re}}$, then $\alpha \in \Delta_{+}^{\text{sim}}$.*

(b) *If $\alpha \in \Delta_{+}^{\text{im}}$, $r_{\gamma}(\alpha) \neq \alpha$ for all $\gamma \in \Delta^{\text{re}}$, then $\alpha \in \Delta_{+}^{\text{sim}}$.*

(c) *If* $\alpha \in \Delta_+^{\text{sim}}$ *and* $\langle \alpha_i^\vee, \alpha \rangle \leq 0$ *for all* $i \in I^{\text{re}}$, *then* $\alpha + \beta \in \Delta_+$ *for all* $\beta \in \Delta_+$.
(d) *If* $\alpha \in \Delta_+^{\text{sim}}$, $\beta \in \Delta_+^{\text{im}}$, *then* $\alpha + \beta \in \Delta_+^{\text{im}}$.
(e) Δ_+^{sim} *is a semigroup.*

Proof.

(a) Using Lemma 5, $\alpha + \beta \in \Delta_+$, for every positive real root β and therefore $\alpha \in \Delta_+^{\text{sim}}$.

(b) Again using Lemma 5, $\langle \alpha_i^\vee, r_\gamma(\alpha) \rangle < 0$ for all $i \in I^{\text{re}}$ and $\gamma \in \Delta_+^{\text{re}}$. By (a), $r_\gamma(\alpha) \in \Delta_+^{\text{sim}}$ and so $\alpha \in \Delta_+^{\text{sim}}$, by Lemma 6.

(c) If $\alpha_i \in \Pi^{\text{re}}$, then $\langle \alpha_i^\vee, \alpha \rangle \leq 0$ and either $\alpha + \alpha_i$ or $\alpha - \alpha_i$ is in Δ_+. This forces $\alpha + \alpha_i \in \Delta_+$ by Lemma 5. Thus if $\beta \in \Pi^{\text{re}}$, $\alpha + \beta \in \Delta_+$. Now suppose $\beta \in K$. So, $\alpha + \beta \in Q_+ \cap -C^\vee$. We know that $\alpha + \alpha_i \in \Delta_+$, for all $i \in I^{\text{re}}$. Hence supp $\alpha \cup \{i\}$ is connected for all $i \in I^{\text{re}}$ and in particular supp $\alpha \cup$ supp β is connected. Thus $\alpha + \beta \in K \subseteq \Delta_+$, proving our claim for $\beta \in K$. We now argue by induction on the height of β, ht β. We have just shown that if $\beta \in \Pi^{\text{re}} \cup K$, then $\alpha + \beta \in \Delta_+$. If $\beta \notin \Pi^{\text{re}} \cup K$, there exists an $\alpha_i \in \Pi^{\text{re}}$ such that $r_i(\beta) < \beta$ and $r_i(\beta) \in \Delta_+$.

By induction, $\alpha + r_i(\beta) \in \Delta_+$. As $\Delta_+ \setminus \{\alpha_i\}$, $i \in I^{\text{re}}$ is invariant under the action of r_i, it follows that $r_i(\alpha + r_i(\beta)) = r_i(\alpha) + \beta \in \Delta_+$. Then, we have

$$\begin{aligned} \alpha + r_i(\beta) &= \alpha + \beta - p\alpha_i \\ &\leq \alpha + \beta \\ &\leq r_i(\alpha) + \beta = \alpha + \beta + q\alpha_i \quad \text{for some } p, q \geq 0. \end{aligned}$$

By Lemma 5, $\alpha + \beta$ must be a root as it is "squeezed" between the roots $\alpha + r_i(\beta)$ and $r_i(\alpha) + \beta$, proving the claim.

(d) As both Δ_+^{im} and Δ_+^{sim} are W-invariant, we may assume that $\alpha \in K \cap \Delta_+^{\text{im}} \setminus \Pi^{\text{im}}$. As β is an imaginary root different from simple imaginary root, 2β is also an imaginary root which is different from simple imaginary. By Lemma 5, $\alpha + 2\beta \in \Delta_+$. Applying Lemma 5 once more, we get $2(\alpha + \beta) \in \Delta_+$, proving that $\alpha + \beta \in \Delta_+^{\text{im}}$.

(e) Suppose $\alpha, \beta \in \Delta_+^{\text{sim}}$, are positive strictly imaginary roots. From (*d*) above, we know that $\alpha + \beta \in \Delta_+^{\text{im}}$. So to complete the proof, we must show that for any real root γ either $\alpha + \beta + \gamma$ or $\alpha + \beta - \gamma$ is a root. Without loss of generality, we may assume that $\beta \in K \cap \Delta_+^{\text{sim}}$, and that $\gamma \in \Delta_+^{\text{re}}$. There are two easy cases:

(i) If $\alpha + \gamma \in \Delta_+$, then $\beta + \alpha + \gamma \in \Delta_+$ (by (*c*)),

(ii) If $\langle \alpha + \beta, \gamma^\vee \rangle \neq 0$, then $r_\gamma(\alpha + \beta) \neq \alpha + \beta$. So either $\alpha + \beta - \gamma$ or $\alpha + \beta + \gamma$ is in Δ_+ (by Results stated after Lemma 6).

Otherwise $\beta + \gamma \notin \Delta$, $\langle \alpha + \beta, \gamma^\vee \rangle = 0$, and we proceed as follows:

As β is strictly imaginary and $\beta + \gamma$ is not a root, $\beta - \gamma$ is a root, and therefore $\langle \beta, \gamma^\vee \rangle > 0$. However, we are assuming that $\langle \alpha + \beta, \gamma^\vee \rangle = 0$, so we must have, $\langle \alpha, \gamma^\vee \rangle < 0$ or $\langle 2\alpha, \gamma^\vee \rangle \leq -2$. Thus $2\alpha + 2\gamma \in \Delta_+$, which proves that $\alpha + \gamma \in \Delta_+^{\text{im}}$. As $\beta \in \Delta_+^{\text{sim}}$, by (d), $\beta + \gamma + \alpha$ is a root. This completes the proof that $\beta + \alpha \in \Delta_+^{\text{sim}}$.

From the property (*d*) above of strictly imaginary roots, we have, $\Delta_+^{\text{sim}} + \Delta_+^{\text{sim}} \subset \Delta_+^{\text{im}}$. In general it can easily be established that $\Delta_+^{\text{im}} + \Delta_+^{\text{im}} \not\subset \Delta_+^{\text{im}}$. Here '+' between the sets denotes the usual addition of roots from the sets. □

Without loss of generality, we assume that the first l diagonal elements are negative in the matrix A of the following theorem [82]:

Theorem 36. *Let A be the symmetrizable*

$$GGX = \begin{pmatrix} -k_1 & -a_1 & -a_2 & \cdots & -a_{l-1} & -a_l & \cdots & -a_{l+n-1} \\ -b_1 & -k_2 & -c_2 & \cdots & -c_{l-1} & -c_l & \cdots & -c_{l+n-1} \\ -b_2 & -d_2 & -k_3 & \cdots & \cdots & \cdots & \cdots & \cdots \\ \vdots & \vdots & \vdots & \ddots & & & & \\ -b_{l-1} & -d_{l-1} & \cdots & \cdots & -k_l & -w_l & \cdots & -w_{l+n-1} \\ -b_l & -d_l & \cdots & \cdots & -x_l & & & \\ \vdots & \vdots & \vdots & \vdots & \vdots & & GX & \\ -b_{l+n-1} & -d_{l+n-1} & \cdots & \cdots & -x_{l+n-1} & & & \end{pmatrix},$$

where $k_i (1 \le i \le l) \in \mathbb{Z}_{\ge 0}$, $a_i, b_i, \ldots, w_i, z_i$ are positive integers, GX is GCM of finite, affine, or indefinite type of order n. The matrix A has l imaginary simple roots and n real simple roots. Then the following results are true:

(1) *If GX is a GCM of finite type and all $b_i, d_i, \ldots, x_i, w_i > 1$ for $l \le i \le l+n-1$, then $\mathfrak{g}(A)$ satisfies SIM property.*

(2) *If GX is a GCM of untwisted affine type and all $b_i, d_i, \ldots, x_i > 1$, for $l \le i \le l+n-1$, then $\mathfrak{g}(A)$ satisfies SIM property.*

(3) *If GX is a GCM of twisted affine type, then $\mathfrak{g}(A)$ does not satisfy SIM property.*

(4) *If GX is a GCM of indefinite type, then $\mathfrak{g}(A)$ does not always satisfy SIM property.*

Proof. We have $\Pi^{\text{re}} = \{\alpha_{l+1}, \ldots, \alpha_{l+n}\}$, the real simple roots and $\Pi^{\text{im}} = \{\alpha_1, \ldots, \alpha_l\}$, the imaginary simple roots are so that $\Pi = \{\alpha_1, \ldots, \alpha_l, \alpha_{l+1}, \ldots, \alpha_{l+n}\}$.

(1) If $\alpha \in \Delta_+$, then $\langle \alpha, \alpha_i^\vee \rangle < 0$ for all $i \in I^{\text{re}}$, for all $b_i, d_i, x_i > 1$ and $l \le i \le l+n-1$. By Theorem 35, $\alpha \in \Delta_+^{\text{sim}}$. Hence $\Delta^{\text{im}} \backslash \Delta^{\text{sim}} = \emptyset$ if all $b_i, d_i, x_i > 1$ for $l \le i \le l+n-1$. If some of the above elements b_i, d_i or x_i are equal to 1, then $\Delta^{\text{im}} \backslash \Delta^{\text{sim}}$ is nonempty and hence there are imaginary roots which are not strictly imaginary.

(2) If GX is a GCM of untwisted affine type, it satisfies SIM property (by Proposition 17). For all other imaginary roots of GKM algebras associated with A, the proof is exactly same as in (1). Hence $\Delta^{\text{im}} \backslash \Delta^{\text{sim}} = \emptyset$ if all $b_i, d_i, x_i > 1$ for $l \le i \le l+n-1$. If some of the above elements b_i, d_i, or x_i is equal to 1, then $\Delta^{\text{im}} \backslash \Delta^{\text{sim}}$ is nonempty and hence there are imaginary roots which are not strictly imaginary.

(3) If GX is a GCM of twisted affine type, it does not satisfy SIM property (by Theorem 19). That is, there are imaginary roots which are not strictly imaginary. Hence $\Delta^{\text{im}} \backslash \Delta^{\text{sim}} \neq \emptyset$.

(4) If GX is a GCM of indefinite type, it satisfies the SIM property if
 (i) it satisfies the condition NC1 and
 (ii) it has no principal submatrix contained in the list given in Theorem 19.

For all other imaginary roots of GKM algebras associated with A, the proof is exactly same as (1). Hence $\Delta^{\text{im}} \backslash \Delta^{\text{sim}} = \emptyset$ if $b_i, d_i, x_i > 1$ and GX satisfies SIM property. If some of the above elements b_i, d_i, or x_i is equal to 1, then $\Delta^{\text{im}} \backslash \Delta^{\text{sim}}$ is nonempty and hence there are imaginary roots which are not strictly imaginary.

□

Remark 76. We give below some examples to illustrate the above theorem. The examples are streamlined in such a way that they explain the theorem in full, when $l = 2$ and GX is of order 2.

Example 30. Consider the matrix $A = \text{GGX}$, where the corresponding GX is

$$\begin{pmatrix} 2 & -p \\ -q & 2 \end{pmatrix}.$$

We have,

$$A = \begin{pmatrix} -k_1 & -a_1 & -a_2 & -a_3 \\ -b_1 & -k_2 & -c_2 & -c_3 \\ -b_2 & -d_2 & 2 & -p \\ -b_3 & -d_3 & -q & 2 \end{pmatrix} \text{ with } k_1, k_2 \in \mathbb{Z}_{\geq 0} \text{ and all } a_i, b_i, c_i, d_i, p, q \in \mathbb{Z}_{>0}.$$

We get the following three cases: Case(i): $pq \leq 3$; Case (ii): $pq = 4$; and Case (iii): $pq > 4$.

We discuss below the following three cases separately.

Case (i): $pq \leq 3$. So GX is of finite type.

Here,

$$\Delta_+^{\text{im}} \backslash \Delta_+^{\text{sim}} = \begin{cases} \{2\alpha_1 + \alpha_3\} \text{ if } b_2 = 1 \text{ and all } b_3, d_2, d_3 > 1 \\ \{2\alpha_2 + \alpha_3\} \text{ if } d_2 = 1 \text{ and all } b_2, d_3, b_3 > 1 \\ \{2\alpha_1 + \alpha_4\} \text{ if } b_3 = 1 \text{ and all } b_2, d_2, d_3 > 1 \\ \{2\alpha_2 + \alpha_4\} \text{ if } d_3 = 1 \text{ and } b_2, b_3, d_2 > 1 \end{cases}$$

$$\Delta_+^{\text{im}} \backslash \Delta_+^{\text{sim}} = \begin{cases} \{2\alpha_1 + \alpha_3, 2\alpha_2 + \alpha_3\} \text{ if } b_2 = d_2 = 1 \text{ and } b_3, d_3 > 1 \\ \{2\alpha_1 + \alpha_3, 2\alpha_1 + \alpha_4\} \text{ if } b_2 = b_3 = 1 \text{ and } d_2, d_3 > 1 \\ \{2\alpha_1 + \alpha_3, 2\alpha_2 + \alpha_4\} \text{ if } b_2 = d_3 = 1 \text{ and } b_3, d_2 > 1 \\ \{2\alpha_2 + \alpha_3, 2\alpha_2 + \alpha_4\} \text{ if } d_2 = b_3 = 1 \text{ and } b_2, d_3 > 1 \\ \{2\alpha_2 + \alpha_3, 2\alpha_2 + \alpha_4\} \text{ if } d_2 = d_3 = 1 \text{ and } b_2, b_3 > 1 \\ \{2\alpha_1 + \alpha_4, 2\alpha_2 + \alpha_4\} \text{ if } b_3 = d_3 = 1 \text{ and } b_2, d_2 > 1 \\ \{2\alpha_1 + \alpha_3, 2\alpha_1 + \alpha_4, 2\alpha_2 + \alpha_3\} \text{ if } b_2 = b_3 = d_2 = 1 \text{ and } d_3 > 1 \\ \{2\alpha_1 + \alpha_3, 2\alpha_2 + \alpha_3, 2\alpha_2 + \alpha_4\} \text{ if } b_2 = d_2 = d_3 = 1 \text{ and } b_3 > 1 \\ \{2\alpha_1 + \alpha_3, 2\alpha_1 + \alpha_4, 2\alpha_2 + \alpha_4\} \text{ if } b_2 = b_3 = d_3 = 1 \text{ and } d_2 > 1 \\ \{2\alpha_2 + \alpha_3, 2\alpha_1 + \alpha_4, 2\alpha_2 + \alpha_4\} \text{ if } d_2 = b_3 = d_3 = 1 \text{ and } b_2 > 1 \\ \{2\alpha_1 + \alpha_3, 2\alpha_2 + \alpha_3, 2\alpha_1 + \alpha_4, 2\alpha_2 + \alpha_4\} \text{ if } b_2 = d_2 = b_3 = d_3 = 1, \end{cases}$$

which is nonempty. But by Theorem 35, one can conclude that $\Delta_+^{\text{im}} \backslash \Delta_+^{\text{sim}} = \emptyset$ only if all $b_2, b_3, d_2, d_3 > 1$. Hence $\mathfrak{g}(A)$ satisfies SIM property if all $b_2, b_3, d_2, d_3 > 1$.

Case (ii): $pq = 4$. So GX is of affine type.

We get the following three separate cases. Either $p = q = 2$ (or) $p = 1$ and $q = 4$ (or) $q = 1$ and $p = 4$. Let $p = q = 2$. Then GX is of untwisted affine type. Here, the set $\Delta_+^{\text{im}} \backslash \Delta_+^{\text{sim}}$ is same as in Case (i) above. Hence $\mathfrak{g}(A)$, here satisfies SIM property if all $b_2, b_3, d_2, d_3 > 1$.

Let $p = 1$ and $q = 4$. Then GX is twisted affine type. Here, the set $\Delta_+^{\mathrm{im}} \backslash \Delta_+^{\mathrm{sim}}$, in addition to the imaginary roots found in Case (i) above, contains more imaginary roots of the form $\{(2n+1)(\alpha_3 + 2\alpha_4) | n \geq 0\}$, corresponding to the algebra of the matrix $\begin{pmatrix} 2 & -1 \\ -4 & 2 \end{pmatrix}$. So this set $\Delta_+^{\mathrm{im}} \backslash \Delta_+^{\mathrm{sim}}$ is nonempty. Hence $\mathfrak{g}(A)$ does not satisfy SIM property. Let $q = 1$ and $p = 4$. Then GX is of twisted affine type. Here, the set $\Delta_+^{\mathrm{im}} \backslash \Delta_+^{\mathrm{sim}}$, in addition to the imaginary roots found in Case (i) above, contains more imaginary roots of the form $\{(2n+1)(2\alpha_3 + \alpha_4) | n \geq 0\}$, corresponding to the algebra of the matrix $\begin{pmatrix} 2 & -4 \\ -1 & 2 \end{pmatrix}$. So this set $\Delta_+^{\mathrm{im}} \backslash \Delta_+^{\mathrm{sim}}$ is nonempty. Hence $\mathfrak{g}(A)$ does not satisfy SIM property.

Case (iii): $pq > 4$: Then GX is of hyperbolic type.

We have two separate cases: Either $p = 1$ and $q > 4$ (or) $q = 1$ and $p > 4$. Let $p = 1$ and $q > 4$. In this case, the set $\Delta_+^{\mathrm{im}} \backslash \Delta_+^{\mathrm{sim}}$, in addition to the imaginary roots found in Case (i), contains one more imaginary root $\alpha_3 + 2\alpha_4$, corresponding to the algebra of the matrix $\begin{pmatrix} 2 & -1 \\ -q & 2 \end{pmatrix}$. So this set $\Delta_+^{\mathrm{im}} \backslash \Delta_+^{\mathrm{sim}}$ is nonempty. Hence $\mathfrak{g}(A)$ does not satisfy SIM property. Let $q = 1$ and $p > 4$. In this case, the set $\Delta_+^{\mathrm{im}} \backslash \Delta_+^{\mathrm{sim}}$, in addition to the imaginary roots found in Case (i), contains one more imaginary root $2\alpha_3 + \alpha_4$, corresponding to the algebra of the matrix $\begin{pmatrix} 2 & -p \\ -1 & 2 \end{pmatrix}$. So this set $\Delta_+^{\mathrm{im}} \backslash \Delta_+^{\mathrm{sim}}$ is nonempty. Hence $\mathfrak{g}(A)$ does not satisfy SIM property.

Complete classification of strictly imaginary property of BKM algebras with infinite index set which includes the Monster Lie algebras also [82]

In this section, we discuss strictly imaginary roots of GKM algebras associated with BCMs with countably infinite number of diagonal elements. We have the following three cases:

(i) **All simple roots are imaginary:** Since all diagonal elements are negative here, these algebras satisfy SIM property.

(ii) **One simple real root and infinite number of simple imaginary roots:** In this case, if the corresponding BCM contains -1 in a column which has the corresponding diagonal element 2, then these GKM algebras do not satisfy SIM property. For example, consider the Monster Lie algebra. This algebra has one real simple root α_1 and infinite number of simple imaginary roots α_i for $i > 1$. Since $2\alpha_2 + \alpha_1 + \alpha_1 = 2(\alpha_1 + \alpha_2)$ and $2\alpha_2 + \alpha_1 - \alpha_1 = 2\alpha_2$ are not roots of the Monster Lie algebra, $2\alpha_2 + \alpha_1$ is not a strictly imaginary root. But using Results stated after Lemma 6, one can easily conclude that all other imaginary roots are strictly imaginary. So $\Delta_+^{\mathrm{im}} \backslash \Delta_+^{\mathrm{sim}} = \{2\alpha_2 + \alpha_1\}$ and hence $\{2\alpha_2 + \alpha_1\}$ is the only imaginary root which is not strictly imaginary. *Hence we conclude that Monster Lie algebra does not satisfy SIM property.* Other algebras in this case satisfy SIM property.

(iii) **Two or more simple real roots and infinite number of simple imaginary roots:** By the argument as in (ii) above, one can conclude that some algebras satisfy SIM property and some others do not satisfy SIM property.

3.6 Purely imaginary roots in GKM algebras and their complete classifications

Whenever GKM algebras are considered, index set I is taken to be the finite index set and the corresponding matrix is called GGCM and whenever BKM algebra is used index set I may be taken as finite or infinite and the corresponding matrix is called BCM. So for BKM algebra, infinite index set is also included. We give a complete classification of BKM algebras or GKM algebras (finite index set) with the property that all of their imaginary roots are purely imaginary and then compare the properties of strictly imaginary roots and purely imaginary roots of BKM algebras and using the classification of BKM algebras possessing purely imaginary property, we find, among these classes, the algebras all whose purely imaginary roots are strictly imaginary [100]. Below we give an example.

Example 31. In the following diagram,

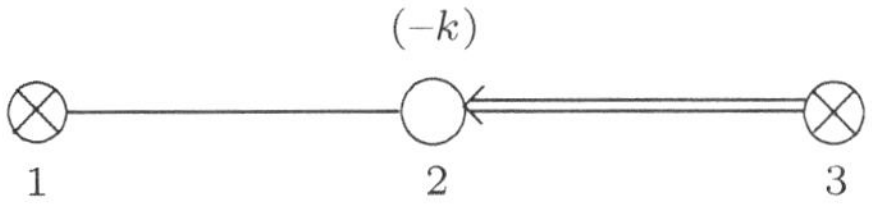

we can easily check that α_1 and α_3 are imaginary roots whereas $\alpha_1 + \alpha_3$ is not a root because supp $(\alpha_1 + \alpha_3)$ is not connected.

Just like Kac-Moody algebras, we can extend the notion of purely imaginary roots from Kac-Moody algebras to GKM algebras. Note that, if α and β are imaginary roots and if $\alpha + \beta$ is a root, then $\alpha + \beta$ must be imaginary. Hence, one can try to collect all those imaginary roots for a given GKM algebra having the property that, when added with another imaginary root gives an imaginary root.

Definition 122. Let $\alpha \in \Delta_+^{\text{im}}$. We say that α is *purely imaginary* if for any $\beta \in \Delta_+^{\text{im}}, \alpha + \beta \in \Delta_+^{\text{im}}$. Similarly we say that a negative root $\gamma \in \Delta_-^{\text{im}}$ is purely imaginary if $-\gamma$ is a purely imaginary root. Denote by

$$\Delta_+^{\text{pim}}(A) = \Delta_+^{\text{pim}} = \{\alpha \in \Delta_+^{\text{im}} \mid \alpha \text{ is purely imaginary}\}$$

and

$$\Delta_-^{\text{pim}}(A) = \Delta_-^{\text{pim}} = \{\alpha \in \Delta_-^{\text{im}} \mid \alpha \text{ is purely imaginary}\}.$$

Then the set of all purely imaginary roots is $\Delta^{\text{pim}} = \Delta_+^{\text{pim}} \cup \Delta_-^{\text{pim}}$. The symmetry of the root system means that we need only to prove the results for positive imaginary roots.

Remark 77. From Theorem 35 (d), we see that every positive strictly imaginary root is a purely imaginary root. One can see from the following example that the converse is not true.

Example 32. Consider the GKM algebra associated with the $GCM = \begin{pmatrix} -k_1 & -1 & 0 \\ -1 & 2 & -1 \\ 0 & -1 & 2 \end{pmatrix}$. Here $\alpha = 2\alpha_1 + \alpha_2 + \alpha_3$ is a purely imaginary root, whereas α is not strictly imaginary, since neither $\alpha + \alpha_2 + \alpha_3$ nor $\alpha - (\alpha_2 + \alpha_3)$ is a root. $\alpha = \alpha_1$ is a purely imaginary root, whereas α is not strictly imaginary, since neither $\alpha + \alpha_3$ nor $\alpha - \alpha_3$ is a root. $\alpha = 2\alpha_1 + \alpha_2$ is a purely imaginary root, whereas α is not strictly imaginary, since neither $\alpha + \alpha_2$ nor $\alpha - \alpha_2$ is a root.

Remark 78. In general, the set $\Delta^{\text{sim}} \setminus \Pi^{\text{im}}$ is not closed under the operation of taking rational multiples.

In Example 32, $4\alpha_1 + 2\alpha_2$ is a strictly imaginary root but $2\alpha_1 + \alpha_2$ is an imaginary root but not strictly imaginary. Also in general, the set Δ_+^{im} is not a semigroup.

Definition 123. We say that a GGCM, A satisfies the purely imaginary property if $\Delta_+^{\text{pim}}(A) = \Delta_+^{\text{im}}(A)$. We say that the GKM algebra $\mathcal{G}(A)$ has the purely imaginary property if A satisfies this property.

Also we prove the following propositions.

Proposition 23.

(1) Δ_+^{pim} *is a semigroup.*

(2) $W \cdot \Delta_+^{\text{pim}} = \Delta_+^{\text{pim}}$.

(3) *Let* $A = (a_{ij})_{i,j=1}^n$ *be an indecomposable GGCM. If* $\mathcal{G}(A)$ *is an affine or a symmetrizable hyperbolic GKM algebra, then* $\Delta_+^{\text{im}}(A) = \Delta_+^{\text{pim}}(A)$.

Proof.

(1) Let $\alpha, \beta \in \Delta_+^{\text{pim}}$. We want to show that $\alpha + \beta \in \Delta_+^{\text{pim}}$. Let $\delta \in \Delta_+^{\text{im}}$. Since $\beta \in \Delta_+^{\text{pim}}$, $\beta + \delta \in \Delta_+^{\text{im}}$. So $\alpha + \beta + \delta = \alpha + (\beta + \delta) \in \Delta_+^{\text{pim}}$ (since $\alpha \in \Delta_+^{\text{pim}}$).

(2) If suffices to show that $w(\alpha) + \beta \in \Delta_+^{\text{im}}$ for every $\beta \in \Delta_+^{\text{im}}$, $\alpha \in \Delta_+^{\text{pim}}$ and $w \in W$. Since $W \cdot K = \Delta_+^{\text{im}}$, $\beta = w_1(\gamma)$ for some $w_1 \in W$ and $\gamma \in K$. $w(\alpha) + \beta = w(\alpha + w^{-1}w_1(\gamma)) \in \Delta_+^{\text{im}}$, since $\alpha \in \Delta_+^{\text{pim}}$ and Δ_+^{im} is W-invariant. Thus $w(\alpha) + \beta \in \Delta_+^{\text{im}}$ for any $\beta \in \Delta_+^{\text{im}}$.

(3) Let A be of affine type. Then $\Delta_+^{\text{im}}(A) = \{k\delta \mid k = 0, 1, 2, \ldots\}$ where $\delta = \sum_{i=1}^n a_i\alpha_i$, and a_i's are coefficients occurring in $S(A)$. Any $\alpha, \beta \in \Delta_+^{\text{im}}$ can be written as $\alpha = n_1\delta$, $\beta = m_1\delta$ for some $n_1, m_1 \in \mathbb{Z}_+$. So $\alpha + \beta = (m_1 + n_1)\delta \in \Delta_+^{\text{im}}(A)$. Thus $\alpha \in \Delta_+^{\text{pim}}$ for all $\alpha \in \Delta_+^{\text{im}}$.

Now let A be a symmetrizable hyperbolic GGCM. For $\alpha, \beta \in \Delta_+^{\text{im}}$, $(\alpha+\beta, \alpha+\beta) = (\alpha, \alpha) + (\beta, \beta) + 2(\alpha, \beta) \leq 0$. Hence $\alpha + \beta \in \Delta_+^{\text{im}}$. This implies that for all $\alpha \in \Delta_+^{\text{im}}(A)$, $\alpha \in \Delta_+^{\text{pim}}(A)$.

We use the following proposition for the classification of GKM algebras with the purely imaginary property. □

Proposition 24.

(i) *Let* $A = (a_{ij})_{i,j=1}^n$ *be an indecomposable symmetrizable GGCM. Suppose* $\alpha \in \Delta_+^{\text{im}}$ *is such that supp* α *contains at least* $n - 1$ *vertices, then* $\alpha \in \Delta_+^{\text{pim}}$.

(ii) *Let $A = (a_{ij})_{i,j=1}^n$ be an indecomposable symmetrizable GGCM. Then $\Delta_+^{\mathrm{pim}} = \Delta_+^{\mathrm{im}}$.*

(iii) *If $\alpha \in \Delta_+^{\mathrm{im}} \backslash \Pi^{\mathrm{im}}, \alpha = \sum_{i=1}^n k_i\alpha_i$, then at least two of the k_i's are >0.*

(iv) *Let $A = (a_{ij})_{i,j=1}^n$ be an indecomposable symmetrizable GGCM. Suppose $\alpha \in \Delta_+^{\mathrm{im}} \backslash \Pi^{\mathrm{im}}$ is such that supp α contains at least $n-2$ vertices, then $\alpha \in \Delta_+^{\mathrm{pim}}$.*

(v) *Let $A = (a_{ij})_{i,j=1}^n$ be an indecomposable symmetrizable GGCM with $n \leq 4$. Then,*

$$\Delta_+^{\mathrm{pim}}(A) = \Delta_+^{\mathrm{im}}(A) \setminus \Pi^{\mathrm{im}}.$$

(vi) *Let $A = (a_{ij})_{i,j=1}^n$ be an indecomposable symmetrizable GGCM such that $a_{ij} \neq 0 \; \forall \; i, j \in \mathbb{N}$ and $i \neq j$. Then $\Delta_+^{\mathrm{pim}} = \Delta_+^{\mathrm{im}}$.*

Proof.

(i) It suffices to show that supp $(\alpha + \beta)$ is connected $\forall \; \beta \in \Delta_+^{\mathrm{im}}$. Since $\beta \in \Delta_+^{\mathrm{im}}$, supp β contains at least one vertex. If supp $\alpha \cap$ supp $\beta = \phi$, then supp $(\alpha+\beta)$ contains n vertices and hence connected. If supp $\alpha \cap$ supp $\beta \neq \phi$, then obviously supp $(\alpha+\beta)$ is connected.

(ii) For any $\alpha \in \Delta_+^{\mathrm{im}}$, supp α contains at least $n-1$ vertices and hence by the result (i), $\alpha \in \Delta_+^{\mathrm{pim}}$.

(iii) As $\alpha \neq 0, k_i \neq 0$ for at least one $i \in$ supp α. If $k_i \neq 0$ for exactly one i and $k_j = 0 \; \forall \; j \neq i$, $j \in$ supp α, then $\alpha = k_i\alpha_i$, which implies α is either simple real or simple imaginary root if $k_i = 1$ and if $k_i > 1, \alpha$ is not a root. This implies that at least two of the k_i's are >0 because, $\alpha \in \Delta_+^{\mathrm{im}} \backslash \Pi^{\mathrm{im}}$.

(iv) Because of (i), it suffices to show that supp $(\alpha + \beta)$ is connected $\forall \; \beta \in \Delta_+^{\mathrm{im}} \backslash \Pi^{\mathrm{im}}$. Since $\beta \in \Delta_+^{\mathrm{im}} \backslash \Pi^{\mathrm{im}}$, supp β contains at least two vertices. If supp$\alpha \cap$ supp $\beta = \phi$, then supp $(\alpha+\beta)$ contains n vertices and hence connected. If supp$\alpha \cap$ supp $\beta \neq \phi$, then obviously supp $(\alpha+\beta)$ is connected.

(v) For any $\alpha \in \Delta_+^{\mathrm{im}} \backslash \Pi^{\mathrm{im}}$, supp α contains at least $n-2$ vertices and hence by the above result (iv), $\alpha \in \Delta_+^{\mathrm{pim}}$.

(vi) Since $a_{ij} \neq 0 \; \forall \; i, j \in \mathbb{N}$. $S(A)$ is such that every vertex is connected to every other vertex of $S(A)$. Hence, for any $\alpha, \beta \in \Delta_+^{\mathrm{im}}$, supp $(\alpha+\beta)$ is connected. Hence by (4) under Properties of Kac-Moody algebras possessing purely imaginary roots in subsection 2.8.3 (Chapter 2). $\alpha \in \Delta_+^{\mathrm{pim}}$. This implies that $\Delta_+^{\mathrm{pim}} = \Delta_+^{\mathrm{im}}$.

□

Remark 79. From (iv), we understand that for an indecomposable GGCM $A = (a_{ij})_{i,j=1}^n$ with $n \leq 2$, all imaginary roots are purely imaginary roots.

Example 33. Let $n = 3$ and

$$A = \begin{pmatrix} -2 & -1 & -1 \\ -1 & 2 & -2 \\ -1 & -2 & 2 \end{pmatrix}.$$

Then $\alpha = \alpha_1, \alpha_2 + \alpha_3$, and $\alpha_1 + \alpha_2 + \alpha_3$ (such that supp α contains at least one vertex) are purely imaginary roots.

Now we prove Theorems 37 and 38. In Theorem 37, we give a necessary and sufficient condition for any $\alpha \in \Delta_+^{\text{im}}$ to be purely imaginary, which further leads to the classification of GKM algebras with the purely imaginary property.

Theorem 37. *Let A be an indecomposable symmetrizable GGCM and let* $\alpha \in \Delta_+^{\text{im}}(A)$. *Then* $\alpha \in \Delta_+^{\text{pim}}$ *if and only if every connected subdiagram of* $S(A)$, *which is of affine or indefinite type, is connected to at least one of the vertices of supp* α.

Proof. Let $\alpha \in \Delta_+^{\text{pim}}(A)$. Suppose there exists a connected subdiagram $S'(A)$ of $S(A)$, which is of nonfinite type and which is not connected to any of the vertices of supp α. Let $\beta \in \Delta_+^{\text{im}}(A)$ be such that supp $\beta \subset S'(A)$ (such a choice is possible because $S'(A)$ is of nonfinite type). Then, supp $(\alpha + \beta)$ is not connected, which is a contradiction, since $\alpha \in \Delta_+^{\text{pim}}(A)$. Thus, any connected subdiagram of $S(A)$, of nonfinite type, must be connected to at least one of the vertices of supp α.

Now, assume that every connected subdiagram of $S(A)$ which is of affine or indefinite type is connected to at least one of the vertices of supp α. To show that $\alpha \in \Delta_+^{\text{pim}}(A)$, it suffices to show that supp $(\alpha + \beta)$ is connected for all $\beta \in \Delta_+^{\text{im}}(A)$. Now for any $\beta \in \Delta_+^{\text{im}}(A)$, supp β is of nonfinite type and hence by hypothesis, supp β is connected to at least one of the vertices of supp α, which implies that supp $(\alpha + \beta)$ is connected. Hence the theorem is proved. □

The following theorem gives a complete classification of all those symmetrizable GKM algebras with the purely imaginary property.

Theorem 38. *An indecomposable symmetrizable GGCM* $A = (a_{ij})_{i,j=1}^n$ *satisfies the purely imaginary property if and only if for any two subdiagrams* $S_1(A)$ *and* $S_2(A)$ *of* $S(A)$, *each one being a connected subdiagram of nonfinite type, there exists a vertex* $i \in S_1(A)$ *and* $j \in S_2(A)$ *such that i and j are connected.*

Proof. The proof follows immediately from the previous theorem: For, $\Delta_+^{\text{im}}(A) \neq \Delta_+^{\text{pim}}(A)$ implies that there exists an imaginary root $\alpha \in \Delta_+^{\text{im}}, \alpha \notin \Delta_+^{\text{pim}}$. By Theorem 37, there exists a connected subdiagram $S_1(A)$, which is not connected to supp α. Take $S_2(A) =$ supp α. Then each of $S_1(A)$ and $S_2(A)$ are nonfinite connected subdiagrams of $S(A)$, such that no vertex $i \in S_1(A)$ is connected to any $j \in S_2(A)$.

Conversely, suppose that there exist two connected subdiagrams $S_1(A)$ and $S_2(A)$ of $S(A)$ both of nonfinite type, such that $S_1(A)$ is disconnected from $S_2(A)$. Take $\alpha \in \Delta_+^{\text{im}}$ such that supp $\alpha \subset S_1(A)$ and $\beta \in \Delta_+^{\text{im}}$ such that supp $\beta \subset S_2(A)$. Then, supp $(\alpha + \beta)$ is not connected because $S_1(A)$ is not connected to $S_2(A)$. This implies that $\alpha + \beta$ is not a root, which implies $\Delta_+^{\text{im}}(A) \neq \Delta_+^{\text{pim}}(A)$. □

From Proposition 24 (v) and (vi), we obtain the following remarks.

Remark 80. For all indecomposable, symmetrizable GGCM, $A = (a_{ij})_{i,j=1}^n$, we have the following:

(1) When $n \leq 2$, $\Delta_+^{\text{im}}(A) = \Delta_+^{\text{pim}}(A)$
(2) When $n \geq 3$, there exist algebras which satisfy the purely imaginary property and there also exist algebras which do not satisfy the purely imaginary property.

For $n = 3$, Theorems 37 and 38 give algebras possessing the purely imaginary property. The following example gives an algebra not possessing purely imaginary property.

Example 34. We give an example of a GKM algebra with the purely imaginary property. Consider the Dynkin diagram associated with the GGCM

$$\begin{pmatrix} 0 & -1 & -2 \\ -1 & -k_1 & -2 \\ -3 & -2 & -k_2 \end{pmatrix}.$$

This is a completely connected GGCM and hence by Proposition 24(vi), the corresponding GKM algebra satisfies the purely imaginary property.

Remark 81. Among the extended-hyperbolic GKM algebras there exist algebras which satisfy the purely imaginary property and there also exist algebras which do not satisfy the purely imaginary property. Hence this class of extended-hyperbolic GKM algebras has the property which is not similar to the property of the extended-hyperbolic Kac-Moody algebras. In the class of indefinite nonhyperbolic type GKM algebras which do not belong to the class of extended-hyperbolic, we have also examples and counter examples of algebras possessing the purely imaginary property similar to Kac-Moody algebras. These examples for Kac-Moody algebras were given in Chapter 2.

The proof of the following theorem is clear from definition of extended-hyperbolic GGCM and from Theorem 38.

Theorem 39. *Every indecomposable, symmetrizable extended-hyperbolic GKM algebra $\mathcal{G}(A)$ that has no mutually orthogonal imaginary simple roots satisfy the purely imaginary property.*

Example 35. An example of extended-hyperbolic GGCM, A which satisfies purely imaginary property:

The GGCM is

$$\begin{pmatrix} 2 & -1 & -1 \\ -1 & 0 & -1 \\ -1 & -1 & 2 \end{pmatrix},$$

and the associated Dynkin diagram is

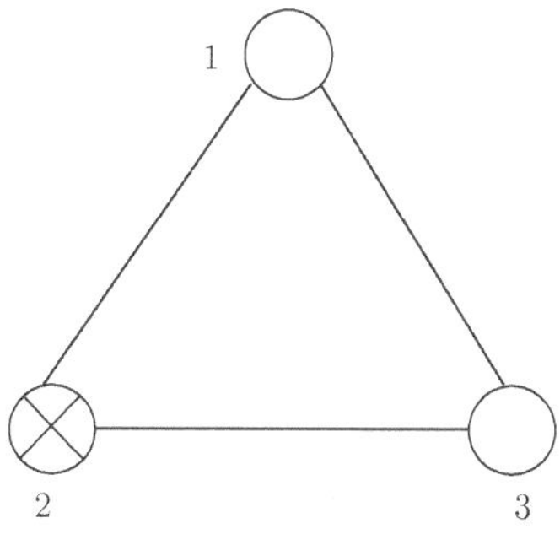

Example 36. An example of a symmetric, extended-hyperbolic GGCM, A which does not satisfy the purely imaginary property.

The GGCM is

$$\begin{pmatrix} 0 & -1 & 0 \\ -1 & 0 & -2 \\ 0 & -2 & 2 \end{pmatrix},$$

and the associated Dynkin diagram is

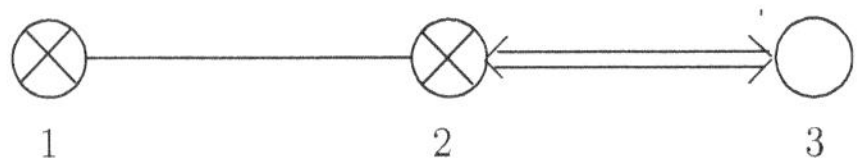

But Example 35 gives an example of a nonsymmetric, extended-hyperbolic GGCM, A whose Dynkin diagram is completely connected but satisfies the purely imaginary property.

The following example gives an indefinite nonhyperbolic GKM algebra which is not an extended-hyperbolic GKM algebra but has the purely imaginary property.

The GGCM is

$$\begin{pmatrix} 0 & -2 & 0 \\ -2 & 2 & -2 \\ 0 & -2 & 0 \end{pmatrix},$$

and the associated Dynkin diagram is

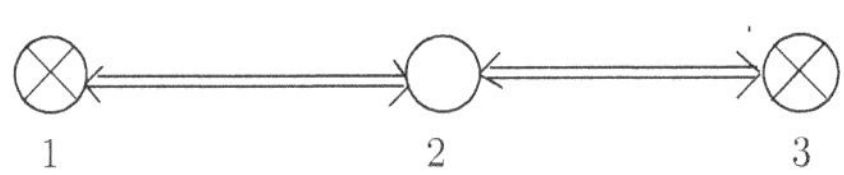

But Example 37 gives an indefinite nonhyperbolic GKM algebra which is an extended-hyperbolic GKM algebra that does not satisfy the purely imaginary property.

3.6.1 Complete classification of GKM algebras possessing purely imaginary property

In Theorems 37 and 38, we give a necessary and sufficient condition for an imaginary root to be purely imaginary and thereby giving a complete classification of those GKM algebras with the purely imaginary property, which includes, in particular, the following:

1. Every GKM algebra $\mathcal{G}(A)$ of finite type ($\Delta_+^{\mathrm{im}}(A) = \Delta_+^{\mathrm{pim}}(A) = \phi$).
2. Every affine GKM algebra.
3. Every symmetrizable hyperbolic GKM algebra.
4. Every GKM algebra whose symmetrizable GGCM is $A = (a_{ij})_{i,j=1}^n$ for $n \leq 2$.
5. Every GKM algebra whose symmetrizable GGCM is $A = (a_{ij})_{i,j=1}^n$ for $n \leq 3$ that has no mutually orthogonal imaginary simple roots.
6. Every symmetrizable extended-hyperbolic, indefinite type GKM algebra that has no mutually orthogonal imaginary simple roots.
7. Every symmetrizable GKM algebra whose Dynkin diagram is completely connected.

8. Every GKM algebra whose indecomposable GGCM is an extension of finite GGCM of any order.
9. Every GKM algebra whose indecomposable GGCM is an extension of affine GGCM of any order.
10. Every GKM algebra whose indecomposable GGCM is an extension of hyperbolic GCM of order 2.

From the complete classification, one can understand that the GKM algebras which do not possess the purely imaginary property are in the class of indefinite GKM algebras other than the hyperbolic GKM algebras and can also identify the algebras possessing and not possessing this property.

Remark 82. From Theorem 35(d) and (e), it is clear that,

(i) If $\alpha \in \Delta_+^{\mathrm{sim}}, \beta \in \Delta_+^{\mathrm{im}}$, then $\alpha + \beta \in \Delta_+^{\mathrm{im}}$.
(ii) Δ_+^{sim} is a semigroup.

From the property (i) of the strictly imaginary roots, it is clear that every positive strictly imaginary root is a purely imaginary root. One can see from the following example that the converse is not true.

Example 37. Let

$$A = \begin{pmatrix} -k_1 & -a & 0 \\ -1 & 2 & -1 \\ 0 & -1 & 2 \end{pmatrix} \text{ with } k_1 > 0 \text{ and } a > 0.$$

Clearly A is a symmetrizable GGCM. $\beta_1 = 2\alpha_1+\alpha_2+\alpha_3$ is not a strictly imaginary root because neither $\beta_1 + (\alpha_2 + \alpha_3)$ nor $\beta_1 - (\alpha_2 + \alpha_3)$ is a root. Similarly $\beta_2 = 2\alpha_1 + \alpha_2$ is not a strictly imaginary root because neither $\beta_2 + \alpha_2$ nor $\beta_2 - \alpha_2$ is a root. Also $\beta_3 = \alpha_1$ is not a strictly imaginary root because neither $\beta_3 + \alpha_3$ nor $\beta_3 - \alpha_3$ is a root.

The set of all GKM algebras possessing purely imaginary property but not possessing strictly imaginary property ($\Delta^{\mathrm{pim}} \backslash \Delta^{\mathrm{sim}}$ for different classes of GKM algebras)

It is clear that the set of all strictly imaginary roots of GKM algebra is a subset of set of all purely imaginary roots, that is, $\Delta^{\mathrm{pim}} \backslash \Delta^{\mathrm{sim}}$ may be a null set or nonempty set depending upon the algebra we take. So for the above-mentioned classes of GKM algebras possessing purely imaginary property, we verify whether the set $\Delta^{\mathrm{pim}} \backslash \Delta^{\mathrm{sim}}$ is empty or not.

1. For every GKM algebra of finite type, since Δ_+^{im} is empty, there exists no purely imaginary root and no strictly imaginary root.
2. For affine GKM algebra, $\Delta^{\mathrm{pim}} \backslash \Delta^{\mathrm{sim}}$ may be empty or nonempty.
   (i) For every GKM algebra associated with the GGCM of untwisted affine type and for GKM algebra associated with the GGCM (0), all purely imaginary roots are strictly imaginary roots.
   (ii) For GKM algebras associated with the GGCM of twisted affine type, $\Delta^{\mathrm{pim}} \backslash \Delta^{\mathrm{sim}}$ is not empty, that is, there are purely imaginary roots which are not strictly imaginary roots. This property is similar to KM algebras in Chapter 2.

For example, consider the GKM algebra associated with the GGCM,

$$\begin{pmatrix} 2 & -1 \\ -4 & 2 \end{pmatrix}.$$

For this GCM, it was already proved in Casperson [57] that

$$\Delta^{\mathrm{pim}}\backslash\Delta^{\mathrm{sim}} = \{(2n+1)(\alpha_1 + 2\alpha_2) \mid n \geq 0\}$$

is nonempty.

3. Every symmetrizable hyperbolic GKM algebra is either
 (a) a hyperbolic Kac-Moody algebra (or)
 (b) a hyperbolic GKM algebra which is not a Kac-Moody algebra.
 These two cases are discussed below.
 (a) Hyperbolic Kac-Moody algebra.
  (i) For the algebra associated with the GCM $\begin{pmatrix} 2 & -a \\ -b & 2 \end{pmatrix}$ with $ab > 4$ and either a or b is equal to 1, there are purely imaginary roots which are not strictly imaginary. This was explained in Chapter 2.
  (ii) For the algebra associated with the GCM $\begin{pmatrix} 2 & -a \\ -b & 2 \end{pmatrix}$ with $ab > 4$ and neither a nor b is equal to 1, all purely imaginary roots are strictly imaginary roots. This was also explained in Chapter 2.
 (b) Hyperbolic GKM algebra which is not a hyperbolic Kac-Moody algebra.
  (i) Consider the GKM algebra associated with the GGCM, $\begin{pmatrix} 0 & -a \\ -b & 2 \end{pmatrix}$. If $b > 1$, it is clear that all purely imaginary roots are strictly imaginary roots. If $b = 1, \Delta^{\mathrm{pim}}\backslash\Delta^{\mathrm{sim}} = \{2\alpha_1 + \alpha_2\}$ is nonempty. So there are purely imaginary roots which are not strictly imaginary.
  (ii) Consider the GKM algebra associated with the GGCM, $\begin{pmatrix} 0 & -a \\ -b & 0 \end{pmatrix}$. Here all purely imaginary roots are strictly imaginary roots.

4. For every GKM algebra whose symmetrizable GGCM is $A = (a_{ij})_{i,j=1}^{n}$ for $n < 2$, we have the following case only.
 We have $n = 1$ and the GKM algebra associated with the GGCM is $(-k_1)$ with $k_1 > 0$. Here all purely imaginary roots are strictly imaginary roots.

5. For every GKM algebra whose symmetrizable GGCM is $A = (a_{ij})_{i,j=1}^{n}$ for $n \geq 2$, we have the following cases:
 (a) GKM algebras with all simple roots are imaginary: here all purely imaginary are strictly imaginary.
 (b) GKM algebras with all simple roots are real and imaginary (at least there exist one simple imaginary root):
  (i) For $n = 2$, every GKM algebra whose symmetrizable GGCM is $A = (a_{ij})_{i,j=1}^{2} = \begin{pmatrix} -k_1 & -a \\ -b & 2 \end{pmatrix}$.
  If $b > 1, \Delta^{\mathrm{pim}}\backslash\Delta^{\mathrm{sim}}$ is empty. Hence all purely imaginary roots are strictly imaginary roots. If $b = 1, \Delta^{\mathrm{pim}}\backslash\Delta^{\mathrm{sim}} = \{2\alpha_1 + \alpha_2\}$ is nonempty. So, there are purely imaginary roots which are not strictly imaginary.

(ii) For $n = 3$, every GKM algebra whose symmetrizable GGCM is

$$A = \begin{pmatrix} -k_1 & -a_1 & -a_2 \\ -b_1 & -k_2 & -p \\ -b_2 & -q & 2 \end{pmatrix} \text{ with } k_1,\ k_2 \geq 0.$$

If both b_2 and q are > 1, all purely imaginary roots are strictly imaginary roots. We have

$$\Delta^{\text{pim}} \backslash \Delta^{\text{sim}} = \begin{cases} \{2\alpha_1 + \alpha_3\} & \text{if } b_2 = 1 \quad \text{and } q > 1 \\ \{2\alpha_2 + \alpha_3\} & \text{if } q = 1 \quad \text{and } b_2 > 1 \\ \{2\alpha_1 + \alpha_3, 2\alpha_2 + \alpha_3\} & \text{if } b_2 = 1 \quad \text{and } q = 1 \end{cases}$$

So if either b_2 or q equal to 0 or 1, there are purely imaginary roots which are not strictly imaginary.

(iii) Consider the GKM algebra associated with the GGCM,

$$\begin{pmatrix} -k_1 & -a_1 & -a_2 \\ -b_1 & 2 & -p \\ -b_2 & -q & 2 \end{pmatrix} \text{ with } pq \leq 3 \text{ and } q = 1.$$

If both b_1 and $b_2 > 1$, all purely imaginary roots are strictly imaginary roots. Here

$$\Delta^{\text{pim}} \backslash \Delta^{\text{sim}} = \begin{cases} \{2\alpha_1 + \alpha_2 + \alpha_3, 2\alpha_1 + \alpha_2, \alpha_1\} & \text{if } b_1 = 1 \quad \text{and } b_2 = 0, \\ \{2\alpha_1 + \alpha_2 + \alpha_3, 2\alpha_1 + \alpha_3, \alpha_1\} & \text{if } b_1 = 0 \quad \text{and } b_2 = 1, \\ \{2\alpha_1 + \alpha_2, 2\alpha_1 + \alpha_3\} & \text{if } b_1 = 1 \quad \text{and } b_2 = 1. \end{cases}$$

So if either b_1 or b_2 equal to 0 or 1, there are purely imaginary roots which are not strictly imaginary.

(iv) Consider the GKM algebra associated with the GGCM,

$$\begin{pmatrix} -k_1 & -a_1 & -a_2 \\ -b_1 & 2 & -p \\ -b_2 & -q & 2 \end{pmatrix} \text{ with } pq = 4 \text{ and } p = q.$$

If both b_1 and $b_2 > 1$, all purely imaginary roots are strictly imaginary. We have

$$\Delta^{\text{pim}} \backslash \Delta^{\text{sim}} = \begin{cases} \{2\alpha_1 + \alpha_2, 2\alpha_1 + \alpha_3\}, & \text{if } b_1 = 1 \quad \text{and } b_2 = 1. \\ \{2\alpha_1 + \alpha_2\}, & \text{if } b_1 = 1, \quad \text{and } b_2 > 1. \\ \{2\alpha_1 + \alpha_3\}, & \text{if } b_2 = 1 \quad \text{and } b_1 > 1. \end{cases}$$

So if either b_1 or b_2 equal to 0 or 1, there are purely imaginary roots which are not strictly imaginary.

(v) Consider the GKM algebra associated with the GGCM,

$$\begin{pmatrix} -k_1 & -a_1 & -a_2 \\ -b_1 & 2 & -p \\ -b_2 & -q & 2 \end{pmatrix} \text{ with } pq = 4 \text{ and } p \neq q.$$

If b_1, b_2, p, and q are all >1, all purely imaginary roots are strictly imaginary. We have,

$$\Delta^{\text{pim}} \backslash \Delta^{\text{sim}} = \begin{cases} \{(2n+1)(\alpha_2+2\alpha_3)|n \geq 0\} \cup \{2\alpha_1+\alpha_2\}, \text{ if } b_1 = p = 1 \text{ and } b_2 > 1. \\ \{(2n+1)(2\alpha_2+\alpha_3)|n \geq 0\} \cup \{2\alpha_1+\alpha_2\}, \text{ if } b_1 = q = 1 \text{ and } b_2 > 1. \\ \{(2n+1)(\alpha_2+2\alpha_3)|n \geq 0\} \cup \{2\alpha_1+\alpha_3\}, \text{ if } b_2 = p = 1 \text{ and } b_1 > 1. \\ \{(2n+1)(2\alpha_2+\alpha_3)|n \geq 0\} \cup \{2\alpha_1+\alpha_3\}, \text{ if } b_2 = q = 1 \text{ and } b_1 > 1. \\ \{(2n+1)(\alpha_2+2\alpha_3)|n \geq 0\} \cup \{2\alpha_1+\alpha_2, 2\alpha_1+\alpha_3\}, \text{ if } b_1 = b_2 = p = 1. \\ \{(2n+1)(2\alpha_2+\alpha_3)|n \geq 0\} \cup \{2\alpha_1+\alpha_2, 2\alpha_1+\alpha_3\}, \text{ if } b_1 = b_2 = q = 1. \\ \{(2n+1)(\alpha_2+2\alpha_3)|n \geq 0\}, \text{ if } p = 1 \text{ and } b_1, b_2 > 1. \\ \{(2n+1)(2\alpha_2+\alpha_3)|n > 0\}, \text{ if } q = 1 \text{ and } b_1, b_2 > 1. \end{cases}$$

So if either b_1 or b_2 equal to 0 or 1 or p or q equal to 1, there are purely imaginary roots which are not strictly imaginary roots.

(vi) Consider the GKM algebra, associated with the GGCM,

$$\begin{pmatrix} -k_1 & -a_1 & -a_2 \\ -b_1 & 2 & -p \\ -b_2 & -q & 2 \end{pmatrix} \text{ with } pq > 4.$$

If b_1, b_2, p, and q are all >1, all purely imaginary roots are strictly imaginary roots. We have,

$$\Delta^{\text{pim}} \backslash \Delta^{\text{sim}} = \begin{cases} \{\alpha_2+2\alpha_3\}, \text{ if } p = 1 \text{ and } b_1, b_2 > 1. \\ \{2\alpha_2+\alpha_3\}, \text{ if } q = 1 \text{ and } b_1, b_2 > 1. \\ \{2\alpha_1+\alpha_2\}, \text{ if } b_1 = 1 \text{ and } b_2, p, q > 1. \\ \{2\alpha_1+\alpha_3\}, \text{ if } b_2 = 1 \text{ and } b_1, p, q > 1. \\ \{2\alpha_1+\alpha_2, \alpha_2+2\alpha_3\}, \text{ if } b_1 = p = 1 \text{ and } b_2 > 1. \\ \{2\alpha_1+\alpha_2, 2\alpha_2+\alpha_3\}, \text{ if } b_1 = q = 1 \text{ and } b_2 > 1, p > 4. \\ \{2\alpha_1+\alpha_3, \alpha_2+2\alpha_3\}, \text{ if } b_2 = p = 1 \text{ and } b_1 > 1, q > 4. \\ \{2\alpha_1+\alpha_3, 2\alpha_2+\alpha_3\}, \text{ if } b_2 = q = 1 \text{ and } b_1 > 1, p > 4. \\ \{2\alpha_1+\alpha_2, 2\alpha_1+\alpha_3, \alpha_2+2\alpha_3\}, \text{ if } b_1 = b_2 = p = 1 \text{ and } q > 4. \\ \{2\alpha_1+\alpha_2, 2\alpha_1+\alpha_3, 2\alpha_2+\alpha_3\}, \text{ if } b_1 = b_2 = q = 1 \text{ and } p > 4. \end{cases}$$

If either b_1 or b_2 equal to 0 or 1 or p or q equal to 1, there are purely imaginary roots which are not strictly imaginary.

6. Every symmetrizable extended-hyperbolic indefinite type GKM algebra is a subclass of GKM algebras discussed in case 5.

7. For every symmetrizable GKM algebra whose Dynkin diagram is completely connected, we have the following two cases:

(i) Consider the GKM algebra associated with the GGCM of finite order with all simple roots are imaginary, then all purely imaginary roots are strictly imaginary roots.

(ii) Consider the GKM algebra associated with the GGCM $A = \begin{pmatrix} -k_1 & -a_1 & -a_2 \\ -b_1 & -k_2 & -p \\ -b_2 & -q & 2 \end{pmatrix}$.

Here if b_2 and $q > 1$, all purely imaginary roots are strictly imaginary roots. We have,

$$\Delta^{\text{pim}}\backslash\Delta^{\text{sim}} = \begin{cases} \{2\alpha_1+\alpha_3\}, & \text{if } b_2 = 1 \quad \text{and } q > 1. \\ \{2\alpha_2+\alpha_3\}, & \text{if } q = 1, \quad \text{and } b_2 > 1. \\ \{2\alpha_1+\alpha_3, 2\alpha_2+\alpha_3\}, & \text{if } b_2 = 1 \quad \text{and } q = 1. \end{cases}$$

If either b_2 or q or both equal to 1, there are purely imaginary roots which are not strictly imaginary.

8. Every symmetrizable GKM algebra whose indecomposable GGCM is an extension of finite GGCM of any order. Consider the GKM algebra associated with the GGCM, $A = \begin{pmatrix} -k_1 & -a_1 & \cdots & -a_n \\ -b_1 & & & \\ \vdots & & GX_n & \\ -b_n & & & \end{pmatrix}$, where GX_n is a $n \times n$ finite Cartan matrix.

Here, if $b_i \neq 1$ for all i, all purely imaginary roots are strictly imaginary roots.

We have, $\Delta^{\text{pim}}\backslash\Delta^{\text{sim}} = \{2\alpha_1+\alpha_2, 2\alpha_1+\alpha_3, \ldots, 2\alpha_1+\alpha_{n+1}\}$, if $b_i = 1$ for all i. If one or all b_i equal to 1, there are purely imaginary roots which are not strictly imaginary.

9. For every GKM algebra whose indecomposable GGCM is an extension of affine GGCM of any order, we have the following cases (i) and (ii):

(i) Consider the GKM algebra associated with the GGCM,

$$A = \begin{pmatrix} -k_1 & -a_1 & \cdots & -a_{n+1} \\ -b_1 & & & \\ \vdots & & GX_n^{(k)} & \\ -b_{n+1} & & & \end{pmatrix}, \text{ with } k_1 \geq 0 \text{ and } b_i > 0 (1 \leq i \leq n+1)$$

where $GX_n^{(1)}$ is the Cartan matrix of the untwisted affine Lie algebra. If $b_i > 1$ for all i, all purely imaginary roots are strictly imaginary roots. We have, $\Delta^{\text{pim}}\backslash\Delta^{\text{sim}} = \{2\alpha_1 + \alpha_2, 2\alpha_1+\alpha_3, \ldots, 2\alpha_1+\alpha_{n+2}\}$ if $b_i = 1$ for all i. If one or all b_i equal to 1, there are purely imaginary roots which are not strictly imaginary roots.

(ii) Consider the GKM algebra associated with the GGCM,

$$A = \begin{pmatrix} -k_1 & -a_1 & \cdots & -a_{n+1} \\ -b_1 & & & \\ \vdots & & GX_n^{(k)} & \\ -b_{n+1} & & & \end{pmatrix}, \text{ with } k_1 \geq 0 \text{ and } b_i > 0 (1 \leq i \leq$$

$n+1$), where $GX_n^{(k)}, k = 2$ or 3 is the Cartan matrix of the twisted affine Lie algebra. Here this GKM algebra does not satisfy strictly imaginary property. Always the set $\Delta^{\text{pim}}\backslash\Delta^{\text{sim}}$ is nonempty.

10. Every GKM algebra whose indecomposable GGCM is an extension of hyperbolic GCM of order 2. This was already discussed in the case 5(vi).

Remark 83. We have noticed from the above discussion that the GKM algebras all whose purely imaginary roots are strictly imaginary ($\Delta^{\text{pim}}\backslash\Delta^{\text{sim}}$ is empty) are the GKM algebras discussed in the cases 2(i), 3*a*(ii), and 7(i). For all other cases, for some particular algebras, all purely imaginary roots are strictly imaginary and for some other particular cases, there are purely imaginary roots which are not strictly imaginary.

3.7 Representations of GKM algebras

Definition 124 (Integrable representations of GKM algebras). As in the cases of generalized Cartan matrices and the corresponding Kac-Moody algebras, we can define the representations for GKM algebras. Let A be a symmetrizable BKM matrix and $\mathcal{G}(A)$ be the corresponding GKM algebra. Let π be a representation of $\mathcal{G}(A)$ in a complex vector space V. We denote by xv, the element $\pi(x)v$, for $x \in \mathcal{G}$ and $v \in V$. Let $\mathfrak{h}$ be the Cartan subalgebra of $\mathcal{G}(A)$. For $\lambda \in \mathfrak{h}^*$, let

$$V_\lambda = \{v \in V \mid hv = \lambda(h)v \ \ (h \in \mathfrak{h})\}.$$

If $V_\lambda \neq 0$, λ is called the *weight* of this representation and V_λ is called the *weight space* of weight λ. The dimension of V_λ is called multiplicity of λ ($multi_V(\lambda)$) in this representation (π, V).

The set of all weights of a representation (π, V) is denoted by $P(V)$, which is a subset of $\mathfrak{h}^*$.

Definition 125. If in a representation (π, V) of $\mathcal{G}$, each V_λ is of finite dimension and $V = \bigoplus_{\lambda \in P(V)} V_\lambda$, then the representation π is called $\mathfrak{h}$-diagonalizable or the vector space V is called $\mathfrak{h}$-diagonalizable $\mathcal{G}$-module.

Definition 126. If V is a $\mathfrak{h}$-diagonalizable $\mathcal{G}$-module, one can define,

$$ch\ V = \sum_{\lambda \in P(V)} (\dim V_\lambda) \cdot e^\lambda$$

called the character of V. The value of the character, $ch\ V$ at $h \in \mathfrak{h}$,

$$(ch\ V)(h) = \sum_{\lambda \in P(V)} (\dim V_\lambda) \cdot e^{\lambda(h)}$$

This series may converge or diverge. If this converges then, the value of the character is finitely determined at $h \in \mathfrak{h}$. If $V = \bigoplus_{\lambda \in P(V)} V_\lambda$ is the eigen space decomposition of V with respect to the action of π, we get,

$$\pi(h)|_{V_\lambda} = \lambda(h) \cdot Id_{V_\lambda}$$

for each V_λ, and hence $\pi(h)$ is a scalar operator. So we get

$$e^{\pi(h)}|_{V_\lambda} = e^{\lambda(h)} \cdot Id_{V_\lambda}.$$

Hence the character of V can be written as

$$(ch\ V)(h) = \sum_{\lambda \in P(V)} tr_V(e^{\pi(h)}).$$

Here e^{λ} for $\lambda \in \mathfrak{h}^*$ are the basic elements of the group algebra $\mathbb{C}[\mathfrak{h}^*]$ with the multiplication given by $e^{\lambda_1} \cdot e^{\lambda_2} = e^{\lambda_1+\lambda_2}$.

Definition 127. Let V be an $\mathfrak{h}$-diagonalizable and e_i and f_i be the Chevalley generators of V. If for each $i \in I^{\mathrm{re}}$, $\pi(e_i)$, and $\pi(f_i)$ are locally nilpotent on V, then V is said to be *integrable*. It can be verified that if V is an $\mathfrak{h}$-diagonalizable integrable $\mathcal{G}$- module, then $ch(V)$ is invariant under the action of Weyl group.

Definition 128. Let V be a $\mathfrak{h}$-diagonalizable $\mathcal{G}$-module, $P(V)$ be the set of weights and $\Lambda \in P(V)$. If $v_\Lambda \neq 0$ of V_Λ satisfies the conditions,

(1) $xv_\Lambda = 0$ for all $x \in \eta_+$ and
(2) $U(\mathcal{G}) \cdot v_\Lambda = V$,

then V is called *highest weight $\mathcal{G}$-module* and v_Λ is called highest weight vector of V and corresponding Λ is called *highest weight*. In fact, $x \cdot v_\Lambda = 0$ for all $x \in \eta_+$ $\Rightarrow e_i \cdot v_\Lambda = 0$ for all $i \in I$.

By PBW theorem of finite dimensional Lie algebras, we know that

$$U(\mathcal{G}) = U(\eta_-) \cdot U(\mathfrak{h}) \cdot U(\eta_+).$$

Hence the highest weight vector v_Λ satisfies

$$U(\eta_-) \cdot v_\Lambda = V.$$

Properties of highest weight $\mathcal{G}$-module V with highest weight Λ

(i) $\lambda = \Lambda - \sum_{i=1}^{n} m_i\alpha_i$ for $\lambda \in P(V)$ and $m_i \in Z_{\geq 0}$.
(ii) V_λ is an one-dimensional vector space, that is, $V_\lambda = \mathbb{C}v_\lambda$.

In fact, if V is a highest weight $\mathcal{G}$-module with highest weight λ, then

$$e^{-\Lambda} ch\, V = \sum_{m_1,\ldots,m_n \geq 0} (\dim V_{\lambda - \sum_i m_i\alpha_i}) e^{-\sum_{i=1}^{n} m_i\alpha_i}.$$

Here, $e^{-\Lambda} ch\, V$ is an element of ring of formal power series $\mathbb{C}[[e^{-\alpha_1}, \ldots, e^{-\alpha_n}]]$.

Definition 129. Let $\mathcal{G}$ be the GKM algebra with BKM matrix A. Let $P_+ := \{\lambda \in \mathfrak{h}^* | \langle \lambda, \alpha_i^\vee \rangle \geq 0$ for all $i \in I^{\mathrm{re}}\}$ and $P_{(+)} = \{\lambda \in P_+ | \langle \lambda, \alpha_i^\vee \rangle \geq 0$ for all $i \in I\}$, called *the dominant integral forms*. Then we define a subset $S(\Lambda)$ of $\mathfrak{h}^*$ to be the set of all sums of distinct, pairwise perpendicular, imaginary simple roots perpendicular to Λ, where an imaginary simple root $\alpha_i \in \Pi^{\mathrm{im}}$ is said to be perpendicular to $\lambda \in \mathfrak{h}^*$ if $\lambda(h_i) = 0$.

Remark 84. For a Kac-Moody algebra $\mathcal{G}(A)$, denote the integrable highest weight module by L(Λ). The following lemma for Kac-Moody algebra was proved in Kac [8].

Lemma 8. *The $\mathcal{G}(A)$-module $L(\Lambda)$ is integrable if and only if $\Lambda \in P_+$.*

The above lemma is true for GKM algebras also.

Definition 130. The map $U(\eta_-) \to V$ sending $x \mapsto xv_\Lambda$ is a surjective map. If it is also one-to-one, the map is linear isomorphism. Here V is called a *Verma module* with highest weight Λ, denoted by $M(\Lambda)$.

Definition 131 (Existence of Verma module). Consider an one-dimensional vector space $\mathbb{C} \cdot 1_\Lambda$, where 1_Λ is just a symbol. Let

$$h \cdot 1_\Lambda = \Lambda(h)1_\Lambda(h \in \mathfrak{h})$$
$$x \cdot 1_\Lambda = 0(x \in \eta_+)$$
$$U(\Lambda) = U(g) \otimes_{\mathfrak{h}\oplus\eta_+} \mathbb{C} \cdot 1_\Lambda$$

Now it is clear that the one-dimensional vector space $\mathbb{C} \cdot 1_\Lambda$ becomes an $(\mathfrak{h} \oplus \eta_+)$-module. Then $U(\Lambda)$ becomes an $U(\mathcal{G})$-module. Hence $U(\Lambda)$ is a $\mathcal{G}$-module using the left multiplication of elements of $U(\mathcal{G})$.

Moreover, the map $U(\eta_-) \to U(\mathcal{G}) \otimes_{\mathfrak{h}\oplus\eta_+} \mathbb{C} \cdot 1_\Lambda$ sending $x \mapsto x \otimes 1_\Lambda$ has a zero kernel. So the above $U(\Lambda)$ is a Verma module. The Verma module constructed above is unique because of the following universal property.

Universal property of the Verma module

Let V be a highest weight $\mathcal{G}$-module with highest weight λ. Then the following properties are true:

(i) There exists a uniquely determined nonzero homomorphism of $\mathcal{G}$-modules from $M(\Lambda) \to V$.

(ii) There exists a uniquely determined invariant subspace $U(\Lambda)$ of $M(\Lambda)$ such that $M(\Lambda)/U(\Lambda) \simeq V$.

Definition 132 (Character of Verma module). For the Verma module $M(\Lambda)$, it can be proved that

$$ch\, M(\Lambda) = e^\Lambda \prod_{\alpha\in\Delta_+} \frac{1}{(1 - e^{-\alpha})^{multi(\alpha)}}.$$

It can also be written as

$$ch\, M(\Lambda) = \frac{e^\Lambda}{R}$$

where $R = \prod_{\alpha\in\Delta_+} (1 - e^{-\alpha})^{multi(\alpha)}$.

Weyl-Kac-Borcherds character formula for GKM algebras

Theorem 40. *Let $\mathcal{G}(A)$ be the GKM algebra associated to a symmetric Kac-Moody algebra A, $\Delta \in P^+$ a dominant integral form, and $L(\Lambda)$ the irreducible highest weight*

$\mathcal{G}(A)$-module with highest weight Λ. Let $\rho \in \mathfrak{h}^*$ be an element such that $\rho(h_i) = \frac{1}{2}a_{ii}$ for all $i \in I$. Then we have

$$e(\rho) \prod_{\alpha \in \Delta^+} (1 - e^{(-\alpha)})^{dim g_\alpha} ch\, L(\Delta) = \sum_{(w,\beta)} (-1)^{l(w)+ht(\beta)} e(w(\Delta + \rho - \beta)).$$

When $\Lambda = 0$ in the theorem above, we have $ch\, L(0) = ch\, \mathbb{C} = e(0) = 1$, where $\mathbb{C}$ is the one-dimensional trivial $\mathcal{G}(A)$-module. Thus we obtain the following denominator identity.

Corollary 3. *Under the notation above, we have*

$$\prod_{\alpha \in \Delta^+} (1 - e^{(-\alpha)})^{dim g_\alpha} = \sum_{(w,\beta) \in W \times S} (-1)^{l(w)+ht(\beta)} e(w(\rho - \beta) - \rho).$$

Fundamental representation of Borcherds algebra with one imaginary simple root [114]

Here we consider only GKM algebras with finite index set (BKM algebras with finite index set) with *real symmetric matrix* ($a_{ij} = a_{ji}$ for all i, j). The Cartan subalgebra (abelian subalgebra of GKM algebra $\hat{\mathcal{G}}(\hat{A})$), $\mathfrak{h}(A)$ has a basis h_i ($i = 1, 2, \ldots, n$). It has a triangular decomposition,

$$\hat{\mathcal{G}}(\hat{A}) = \hat{\eta}_- \oplus \hat{\mathfrak{h}} \oplus \hat{\eta}_+.$$

We have $\hat{\eta_-} = \eta_- \backslash \tau_-$ and $\hat{\eta_+} = \eta_+ \backslash \tau_+$. Here η_- is free algebra generated by f_i, the ideal τ_- is generated by f_{ij}, η_+ is free algebra generated by e_i and the ideal τ_+ is generated by e_{ij}.

Now let us consider GKM algebra $\hat{\mathcal{G}}(\hat{A})$ with one simple imaginary root. In this case, deleting the row and the column corresponding to the imaginary root in $\hat{A}(GGX)$, the imaginary root disappears and we get the corresponding Kac-Moody algebra $\mathcal{G}(A)$ associated to the generalized Cartan matrix $A(GX)$. We have the triangular decomposition

$$\mathcal{G}(A) = \eta_- \oplus \mathfrak{h} \oplus \eta_+$$

which implies decomposition of the universal enveloping algebra:

$$U(\mathcal{G}(A)) = U(\eta_-) \oplus U(\mathfrak{h}) \oplus U(\eta_+).$$

An irreducible $\mathcal{G}(A)$-module V_λ is called *integrable highest-weight module* if there exists a dominant integral weight $\lambda \in \mathfrak{h}^*$ and a nonzero vector $\omega \in V_\lambda$ such that

$$\begin{aligned} h(\omega) &= \lambda(h)\omega \text{ for } h \in \mathfrak{h}, \\ \eta_+(\omega) &= 0, \\ U(\eta_-)(\omega) &= V_\lambda. \end{aligned}$$

We denote by $T(V_\lambda)$, the tensor algebra over V_λ,

$$T(V_\lambda) := \bigoplus_{n=0}^{\infty} V_\lambda^n \equiv \mathbb{C}\cdot 1 \oplus V_\lambda \oplus (V_\lambda \otimes V_\lambda) \oplus (V_\lambda \otimes V_\lambda \otimes V_\lambda) \oplus \cdots$$

Using our notations, we write the following theorem. For the proof one can refer Gebert and Teschner [114].

Theorem 41. *Let $\hat{A} = (a_{ij}), 0 \le i,j \le n$, be a symmetric integer matrix satisfying the following properties: (i) $a_{00} \le 0, a_{ii} = 2$ for $1 \le i \le n$, (ii) $a_{ij} \le 0$ if $i \ne j$. Let F_λ be the integrable highest weight module over the Kac-Moody algebra $\mathcal{G}(A)$ associated to the Cartan matrix, $A = (a_{ij}), 1 \le i,j \le n$, with highest weight λ defined by $\lambda(h_i) := -a_{0i}, 1 \le i \le n$, and the highest weight vector ω. Then the tensor algebra $\Im(V_\lambda)$ over F_λ is a $\hat{\mathcal{G}}(\hat{A})$-module isomorphic to the highest weight module $L(\wedge), \wedge(h_i) = \delta_{i0}, 0 \le i \le n$ of $\hat{\mathcal{G}}(\hat{A})$.*

Remark 85. We illustrate the above theorem by means of the following example. In this example, we use the notation V_λ instead of F_λ and $T(V_\lambda)$ instead of $\Im(V_\lambda)$ as used in the above theorem.

Example 38. Let $\hat{A} = \begin{pmatrix} -2 & -1 & 0 \\ -1 & 2 & -1 \\ 0 & -1 & 2 \end{pmatrix}$ be the generalized Cartan matrix of the corresponding generalized Kac-Moody Lie algebra, $\hat{\mathcal{G}}(\hat{A})$ and $A = \begin{pmatrix} 2 & -1 \\ -1 & 2 \end{pmatrix}$ be the Cartan matrix of the corresponding underlying Lie algebra $\mathcal{G}(A)$. Let us consider generators of Kac-Moody algebra $\hat{\mathcal{G}}(\hat{A})$ as $h_0, h_1, h_2, e_0, e_1, e_2, f_0, f_1, f_2$.

Let V_λ be the integrable weight module over the Kac-Moody algebra $\mathcal{G}(A)$ associated with the Cartan matrix A with the highest weight λ defined by $\lambda(h_1) = -a_{01} = 1,\ \lambda(h_2) = -a_{02} = 0$.

The Kac-Moody generators $e_i, h_i, f_i, i = 1,2$ act trivially on the "vacuum" vector 1 and as highest weight representation of V_λ. We extend this action to the tensor algebra $T(V_\lambda)$ by Leibnitz' rule.

The generator h_0 acts as diagonal:

$$\begin{aligned} h_0(1) &:= 1, \\ h_0(\omega) &:= (1 - a_{00})\omega = 3\omega, \\ h_0(f_1\varphi) &:= f_1\varphi + f_1 h_0(\varphi) \text{ for } \varphi \in V_\lambda, \\ h_0(f_2\varphi) &:= f_2 h_0(\varphi) \text{ for } \varphi \in V_\lambda, \\ h_0(\Phi \otimes \Psi) &:= h_0(\Phi) \otimes \Psi + \Phi \otimes h_0(\Psi) - \Phi \otimes \Psi \text{ for } \Phi, \Psi \in T(V_\lambda). \end{aligned}$$

Moreover, $f_0(\Psi) := \omega \otimes \Psi$ for $\Psi \in T(V_\lambda)$. For e_0, define $e_0(1) := 0$. Let $V_\lambda^n = \mathcal{U}(n_-)(\omega \otimes V_\lambda^{n-1})$ for $n = 0, 1, 2$. We have to prove, inductively,

$$e_0(f_i(\Psi)) := f_i(e_0(\Psi)),$$
$$e_0(\omega \otimes \Psi) := h_0(\Psi) + \omega \otimes e_0(\Psi),$$

for $\Psi \in V_\lambda^n, n = 0, 1, 2$. Now we will prove $T(V_\lambda)$ has a $\widehat{\mathcal{G}}(\widehat{A})$-module structure. Here h_0 and the $h_i's$ act diagonally on the tensor algebra as per the definition. Moreover $h's$ commute with each other. Hence all commutation relations involving only Kac-Moody generators e_i, h_i, f_i are true. Now one can check the following relations:

$$(e_0f_0 - f_0e_0)(\Psi) = e_0(\omega \otimes \Psi) - \omega \otimes e_0(\Psi) = h_0(\Psi),$$
$$(e_0f_i - f_ie_0)(\Psi) = 0,$$
$$(e_if_0 - f_0e_i)(\Psi) = e_i(\omega \otimes \Psi) - \omega \otimes e_i(\Psi) = 0.$$

Using matrix entries of A

$$(h_0f_0 - f_0h_0)(\Psi) = h_0(\omega \otimes \Psi) - \omega \otimes h_0(\Psi) = -a_{00}f_0(\Psi) = 2f_0(\Psi),$$
$$(h_1f_0 - f_0h_1)(\Psi) = -a_{01}f_0(\Psi) = f_0(\Psi),$$
$$(h_2f_0 - f_0h_2)(\Psi) = -a_{02}f_0(\Psi) = 0,$$
$$(h_0f_i - f_ih_0)(1) = -f_i(1) = 0 = -a_{0i}f_i(1), \text{ for } i = 1, 2,$$
$$(h_0f_i - f_ih_0)(\varphi) = -a_{0i}f_i\varphi, \quad i = 1, 2.$$

Hence

$$(h_0f_1 - f_1h_0)(\varphi) = -a_{01}f_1\phi = f_1\varphi,$$
$$(h_0f_2 - f_2h_0)(\varphi) = -a_{02}f_2\phi = 0,$$
$$(h_0f_i - f_ih_0)(\Phi \otimes \Psi) = -a_{0i}f_i(\Phi \otimes \Psi) \text{ for } i = 1, 2.$$

So,

$$(h_0f_1 - f_1h_0)(\Phi \otimes \Psi) = f_1(\Phi \otimes \Psi),$$
$$(h_0f_2 - f_2h_0)(\Phi \otimes \Psi) = 0,$$
$$(h_0e_i - e_ih_0)(1) = 0 = a_{0i}e_i(1) \text{ for } i = 1, 2.$$

So,

$$(h_0e_i - e_ih_0)(\omega) = 0 = a_{0i}e_i(\omega) \text{ for } i = 1, 2,$$
$$(h_0e_i - e_ih_0)(f_k\varphi) = a_{0i}e_i(f_k\varphi) \text{ for } i = 1, 2,$$
$$(h_0e_1 - e_1h_0)(f_k\varphi) = -e_1(f_k\varphi),$$
$$(h_0e_2 - e_2h_0)(f_k\varphi) = 0,$$
$$(h_0e_i - e_ih_0)(\Phi \otimes \Psi) = a_{0i}e_i(\Phi \otimes \Psi) \text{ for } i = 1, 2.$$

So,

$$
\begin{aligned}
(h_0e_1 - e_1h_0)(\Phi \otimes \Psi) &= a_{01}e_1(\Phi \otimes \Psi) = -e_1(\Phi \otimes \Psi),\\
(h_0e_2 - e_2h_0)(\Phi \otimes \Psi) &= a_{02}e_2(\Phi \otimes \Psi) = 0,\\
(h_ie_0 - e_0h_i)(1) &= 0 = a_{i0}e_0(1) \text{ for } i = 1, 2,\\
(h_ie_0 - e_0h_i)(\omega) &= a_{i0}e_0(\omega) \text{ for } i = 1, 2,\\
(h_ie_0 - e_0h_i)(f_k\phi) &= 0 = a_{i0}e_0(f_k\phi),\\
(h_ie_0 - e_0h_i)(\omega \otimes \Psi) &= a_{i0}e_0(\omega \otimes \Psi) \text{ for } i = 1, 2.
\end{aligned}
$$

So,

$$
\begin{aligned}
(h_1e_0 - e_0h_1)(\omega \otimes \Psi) &= -e_0(\omega \otimes \Psi),\\
(h_2e_0 - e_0h_2)(\omega \otimes \Psi) &= 0,\\
(h_ie_0 - e_0h_i)(f_k\varphi \otimes \Psi) &= a_{i0}e_0(f_k\varphi \otimes \Psi) \text{ for } i = 1, 2,\\
(h_1e_0 - e_0h_1)(f_k\varphi \otimes \Psi) &= -e_0(f_k\varphi \otimes \Psi),\\
(h_2e_0 - e_0h_2)(f_k\varphi \otimes \Psi) &= 0,\\
(h_0e_0 - e_0h_0)(1) &= 0 = a_{00}e_0(1),\\
(h_0e_0 - e_0h_0)(\omega) &= a_{00}e_0(\omega) = -2e_0(\omega),\\
(h_0e_0 - e_0h_0)(f_k\varphi) &= a_{00}e_0(f_k\varphi) = 0,\\
(h_0e_0 - e_0h_0)(\omega \otimes \Psi) &= -2e_0(\omega \otimes \Psi) \text{ by induction},\\
(h_0e_0 - e_0h_0)(f_k\varphi \otimes \Psi) &= (e_0h_0 - h_0e_0)(\varphi \otimes f_k(\Psi)) + f_k((h_0e_0 - e_0h_0)(\omega \otimes \Psi)).\\
&= a_{00}(-e_0(\phi \otimes f_k(\Psi)) + f_k(e_0(\varphi \otimes \Psi))) \text{ by induction}\\
&= a_{00} - e_0(f_k\phi \otimes \Psi) = -2e_0(f_k\varphi \otimes \Psi),
\end{aligned}
$$

for all $\varphi \in V_\lambda$ and $\Phi, \Psi \in T(V_\lambda)$.

Now proceeding as in the proof of the above theorem (proof in Gebert and Teschner [114]), to prove $T(V_\lambda) \simeq L(\lambda)$ as a $\widehat{\mathcal{G}}(\widehat{A})$-module, using linearity, first define a map $\nu : \mathfrak{U}(\eta_-)v_\wedge \to T(V_\lambda)$ by

$$\nu(f_{i_1}f_{i_2}v_\wedge) := f_{i_1}f_{i_2}(1),$$

where $i_1, i_2 \in \{0, 1, 2\}$.

Using Serre relations and Lemma 8, one can prove that ν will induce a well defined map $\nu' : \mathfrak{U}(\widehat{\eta}_-)v_\wedge \to T(V_\lambda)$, which is $\widehat{\mathcal{G}}(\widehat{A})$-module homomorphism. Because of $f_i^{1+\Lambda(h_i)}(1) = f_i(1) = 0$, one can conclude that ν' will induce $\nu'' : L(\lambda) \to T(V_\lambda)$, which is an isomorphism.

3.8 Homology modules and root multiplicities in GKM algebras

In Section 3.7, we have defined representations of GKM algebras. Let $\mathcal{G}$ be a GKM algebra and V be a $\mathcal{G}$-module.

Let $b_+ = \mathfrak{h} + \mathcal{G}_+$ be the Borel subalgebra of $\mathcal{G}$ and $\mathbb{C}_\Lambda$ be the one-dimensional b_+-module defined by $\mathcal{G}_+ \cdot 1 = 0, h \cdot 1 = \Lambda(h)1$ for all $h \in \mathfrak{h}$. The induced module $M(\Lambda) = \mu(\mathcal{G}) \otimes_{\mu(b_+)} \mathbb{C}_\Lambda$ becomes the *Verma module* with highest weight Λ. Every highest weight $\mathcal{G}$-module with highest weight Λ is a quotient of $M(\Lambda)$.

The Verma module $M(\Lambda)$ contains a unique maximal submodule $J(\Lambda)$ and irreducible quotient $M(\Lambda)/J(\Lambda)$ will be denoted by $V(\Lambda)$.

Remark 86. Let $V(\Lambda)$ be the irreducible highest weight module over $\mathcal{G}$ with highest weight Λ. We denote by T, the set of all imaginary simple roots counted with multiplicities and for $F \subset T$, we write $F \perp \Lambda$, if $\Lambda(h_i) = 0$ for all $\alpha_i \in F$.

Definition 133 (Different classes of root systems, root lattices, Weyl group W_J of $J \subseteq I^{re}$). Let $\Delta, \Delta^+, \Delta^-, Q, Q^+$ and Q^- be respectively, the root system, set of all positive roots, set of all negative roots, root lattice, positive root lattice, and negative root lattice of the GKM algebra $\mathcal{G}$ with realization $(\mathfrak{h}, \Pi, \Pi^\vee)$. Let J be a finite subset of Ire. We denote by $\Delta_J = \Delta \cap (\sum_{j\in J} \mathbb{Z}\alpha_j), \Delta_J^\pm = \Delta_J \cap \Delta^\pm$, and $\Delta^\pm(J) = \Delta^\pm \backslash \Delta_J^\pm$. We also denote by $Q_J = Q \cap (\sum_{j\in J} \mathbb{Z}\alpha_j), Q_J^\pm = Q_J \cap Q^\pm$, and $Q^\pm(J) = Q^\pm \backslash Q_J^\pm$. Define $\mathcal{G}_0^{(J)} = \mathfrak{h} \bigoplus \left(\bigoplus_{\alpha\in\Delta_J} \mathcal{G}_\alpha\right)$, and $\mathcal{G}_\pm^{(J)} = \bigoplus_{\alpha\in\Delta^\pm(J)} \mathcal{G}_\alpha$. Then we have the triangular decomposition:

$$\mathcal{G} = \mathcal{G}_-^{(J)} \bigoplus \mathcal{G}_0^{(J)} \bigoplus \mathcal{G}_+^{(J)},$$

where $\mathcal{G}_0^{(J)}$ is the Kac-Moody algebra (with an extended Cartan subalgebra) associated with the generalized Cartan matrix $A_J = (a_{ij})_{i,j\in J}$, and $\mathcal{G}_-^{(J)}$ (resp. $\mathcal{G}_+^{(J)}$) is a direct sum of irreducible highest weight (resp. lowest weight) modules over $\mathcal{G}_0^{(J)}$.

Let $W_J = \langle r_j | j \in J \rangle$ be the subgroup of W generated by the simple reflections $r_j (j \in J)$, and let $W(J) = \{w \in W | w\Delta^- \cap \Delta^+ \subset \Delta^+(J)\}$. Then W_J is the Weyl group of the Kac-Moody algebra $\mathcal{G}_0^{(J)}$ and $W(J)$ is the set of right coset representatives of W_J in W. Moreover, the elements in $W(J)$ can be determined by the following lemma. For the actual computation of the homology modules, the following lemma [37] is very important.

Lemma 9. *Suppose $w = w'r_j$ and $l(w) = l(w') + 1$. Then $w \in W(J)$ if and only if $w' \in W(J)$ and $w'(\alpha_j) \in \Delta^+(J)$.*

As in the case of symmetrizable Kac-Moody algebras, we view $\mathcal{G}_-^{(J)}$ as $Q_-^{(J)}$-graded Lie algebra. We omit the proof of the following proposition [37] in which Kostant's formula was generalized to GKM algebras to determine the homology modules $H_k^{(J)}$.

Proposition 25. *In the usual notations,*

$$H_k^{(J)} = \bigoplus_{\substack{w \in W(J) \\ F \subset T \\ l(w)+|F|=k}} V_J(w(\rho - s(F)) - \rho),$$

where $V_J(\mu)$ denotes the irreducible highest weight module over $\mathcal{G}_0^{(J)}$ with highest weight μ, and F runs over all the finite subsets of T such that any two elements in F are mutually perpendicular. Here, we denote by $|F|$, the number of elements in F and $s(F)$, the sum of elements in F. Define the homology space $H^{(J)}$ of $\mathcal{G}_-^{(J)}$ to be

$$H^{(J)} = \sum_{k=1}^{\infty} (-1)^{k+1} H_k^{(J)} = \sum_{\substack{w \in W(J) \\ F \subset T \\ l(w)+|F|=k}} (-1)^{l(w)+|F|+1} V_J(w(\rho - s(F)) - \rho).$$

We have
$$P(H^{(J)}) = \{\alpha \in Q^-(J) \mid \dim H_\alpha^{(J)} \neq 0\} = \{\tau_1, \tau_2, \tau_3, \ldots\},$$

and $d(i) = \dim H_{\tau_i}^{(J)}$ for $i = 1, 2, 3, \ldots$. For $\tau \in Q^-(J)$, we denote by $T^{(J)}(\tau)$, the set of all partitions of τ into a sum of $\tau_i^{\prime}s$, that is,

$$T^{(J)}(\tau) = \left\{ n = (n_i)_{i \geq 1} \mid n_i \in \mathbb{Z}_{\geq 0}, \sum n_i \tau_i = \tau \right\}.$$

For $n \in T^{(J)}(\tau)$, we will use the notation $|n| = \sum n_i$ and $n! = \Pi n_i!$. Now, for $\tau \in Q^-(J)$, we define a function,

$$W^{(J)}(\tau) = \sum_{n \in T^{(J)}(\tau)} \frac{(|n|-1)!}{n!} \Pi d(i)^{n_i}.$$

We will call $W^{(J)}(\tau)$, the Witt partition function. Then, using the formal power series, $\log(1-t) = -\sum_{k=1}^{\infty} \frac{t^k}{k}$ and the Möbius inversion formula, that is, the fact that for complex valued functions f and g defined on $\mathbb{Z}_{\geq 0}$,

$$g(n) = \sum_{d|n} f(d) \Leftrightarrow f(n) = \sum_{d|n} \mu\left(\frac{n}{d}\right) g(d),$$

we obtain a closed form root multiplicity formula for all symmetrizable GKM algebras. We use the following theorem, the proof of which is given in Kang [62] and Kang and Kim [37].

Theorem 42. *Let $\alpha \in \Delta^-(J)$ be a root of a symmetrizable GKM algebra $\mathcal{G}$. Then we have*

$$\begin{aligned}\dim \mathcal{G}_\alpha &= \sum_{d|\alpha} \frac{1}{d}\mu(d) W^{(J)}\left(\frac{\alpha}{d}\right)\\ &= \sum_{d|\alpha} \frac{1}{d}\mu(d) \sum_{n \in T^{(J)}(\frac{\alpha}{d})} \frac{(|n|-1)!}{n!} \prod d(i)^{n_i},\end{aligned}$$

where μ is the classical Möbius function defined as follows:

$$\mu(n) = \begin{cases} 1, & \text{for } n = 1,\\ (-1)^k, & \text{for } n = p_1 \cdots p_k (p_1, \ldots, p_k : \text{ distinct primes}),\\ 0, & \text{if it is not square free.}\end{cases}$$

Here n is a natural number. Suppose the BCM $A = (a_{ij})_{i,j \in I}$ of charge $\underline{m} = (m_i \in \mathbb{Z}_{>0} | i \in I)$ satisfies (i) the set Ire is finite, (ii) $a_{ij} \neq 0$ for all $i, j \in I^{\text{im}}$. If we take $J = I^{\text{re}}$, the set of all real indices, then the algebra $\mathcal{G}_0^{(J)} = \mathfrak{h} \bigoplus (\bigoplus_{\alpha \in \Delta_J} \mathcal{G}_\alpha)$ is the Kac-Moody algebra (with an extended Cartan subalgebra) associated with the generalized Cartan matrix $A_J = (a_{ij})_{i,j \in J}$. Here the GKM algebra $\mathcal{G} = \mathcal{G}_-^{(J)} \bigoplus \mathcal{G}_0^{(J)} \bigoplus \mathcal{G}_+^{(J)}$ can be realized as the minimal graded Lie algebra with the local part $V \bigoplus \mathcal{G}_0^{(J)} \bigoplus V^*$, where $V = \bigoplus_{i \in I^{\text{im}}} V_J(-\alpha_i)^{\oplus m_i}$, V^* is the contragredient module of V, and $m_i's$ are the multiplicities of the imaginary simple roots $\alpha_i's$. In our computation we also use the following proposition [108, 109, 115].

Proposition 26. *Suppose that a BCM $A = (a_{ij})_{i,j \in I}$ of charge $\underline{m} = (m_i \in \mathbb{Z}_{>0} | i \in I)$ satisfies: (i) the set Ire is finite, (ii) $a_{ij} \neq 0$ for all $i, j \in I^{\text{im}}$. Let $J = I^{\text{re}}$. Consider the corresponding triangular decomposition of the GKM algebra:*

$$\mathcal{G} = \mathcal{G}(A, \underline{m}) = \mathcal{G}_-^{(J)} \bigoplus \mathcal{G}_0^{(J)} \bigoplus \mathcal{G}_+^{(J)}.$$

Then the algebra $\mathcal{G}_-^{(J)} \bigoplus_{\alpha \in \Delta^-(J)} \mathcal{G}_\alpha$ (resp. $\mathcal{G}_+^{(J)} = \bigoplus_{\alpha \in \Delta^+(J)} \mathcal{G}_\alpha$) is isomorphic to the free Lie algebra generated by the space $V = \bigoplus_{i \in I^{\text{im}}} V_J(-\alpha_i)^{\bigoplus m_i}$ (resp. $V^* = \bigoplus_{i \in I^{\text{im}}} V_J^*(-\alpha_i)^{\oplus m_i}$), where $V_J(\mu)$ (resp. $V_J^*(\mu)$) denotes the irreducible highest weight (resp. lowest weight) module over the Kac-Moody algebra $\mathcal{G}_0^{(J)}$ with highest weight μ (respectively lowest weight $-\mu$).

Proposition 26 shows that, under the above assumptions, the GKM algebra $\mathcal{G} = \mathcal{G}_-^{(J)} \bigoplus \mathcal{G}_0^{(J)} \bigoplus \mathcal{G}_+^{(J)}$ is isomorphic to the maximal graded Lie algebra with local part $V \bigoplus \mathcal{G}_0^{(J)} \bigoplus V^*$.

Root multiplicities of some GKM algebras with one simple imaginary root

Below, we compute the root multiplicities of GKM algebras associated with BCMs which appear as extensions of finite A_3.

Proposition 27. *Let $\mathcal{G} = \mathcal{G}(A, \underline{m})$ be the GKM algebra associated with the BCM,*

$$A = \begin{pmatrix} -k & -1 & 0 & 0 \\ -1 & 2 & -1 & 0 \\ 0 & -1 & 2 & -1 \\ 0 & 0 & -1 & 2 \end{pmatrix}$$

of charge $\underline{m} = (r, 1, 1, 1)$ with $k,\ r \in \mathbb{Z}_{>0}$.

Then, for the root $\alpha = -k_1\alpha_1 - k_2\alpha_2 - k_3\alpha_3 - k_4\alpha_4$ with $k_1,\ k_2,\ k_3,\ k_4 \in \mathbb{Z}_{\geq 0}$, we have

$$\dim \mathcal{G}_\alpha = \frac{1}{k_1} \sum_{d|(k_1,k_2,k_3,k_4)} \mu(d) \begin{pmatrix} \frac{k_1}{d} \\ \frac{k_4}{d} \end{pmatrix} \begin{pmatrix} \frac{k_1-k_4}{d} \\ \frac{k_2-k_4}{d} \end{pmatrix} \begin{pmatrix} \frac{k_2-k_4}{d} \\ \frac{k_3-k_4}{d} \end{pmatrix} r^{\frac{k_1}{d}}. \tag{3.20}$$

Proof. Consider the GKM algebra, $\mathcal{G} = \mathcal{G}(A, \underline{m})$ associated with the above BCM of charge $\underline{m} = (r, 1, 1, 1)$ with $k,\ r \in \mathbb{Z}_{>0}$. This matrix is an extension of A_3.

Let $I = \{1, 2, 3, 4\}$ be the index set for the simple roots of $\mathcal{G}$. Then, α_1 is the imaginary simple root with multiplicity $r \geq 1$ and α_2, α_3, and α_4 are the real simple roots. Thus we have, $T = \{\alpha_1, \ldots, \alpha_1\}$ (counted r times), and since

$$(\alpha_1, \alpha_1) = -k < 0,$$

F can be either empty or $\{\alpha_1\}$. If we take $J = \{2, 3, 4\}$, then

$$\mathcal{G}_0^{(J)} = \mathcal{G}_0 \bigoplus \mathbb{C}h_1,$$

where

$$\mathcal{G}_0 = \langle e_2, f_2, h_2, e_3, f_3, h_3, e_4, f_4, h_4 \rangle \cong \mathrm{sl}(4, \mathbb{C}) \text{ and } W(J) = \{1\}.$$

By Proposition 25, we obtain

$$H_1^{(J)} = V_J(-\alpha_1) \bigoplus \cdots \bigoplus V_J(-\alpha_1) \text{ (} r \text{ copies)}$$
$$H_k^{(J)} = 0 \text{ for } k \geq 2.$$

This implies

$$H^{(J)} = V_J(-\alpha_1) \bigoplus \cdots \bigoplus V_J(-\alpha_1) \;\; (r \text{ copies}),$$

where $V_J(-\alpha_1)$ is the four-dimensional irreducible representation of sl(4, $\mathbb{C}$). By identifying $-k_1\alpha_1 - k_2\alpha_2 - k_3\alpha_3 - k_4\alpha_4 \in Q^-$ with $(k_1, k_2, k_3, k_4) \in \mathbb{Z}_{\geq 0} \times \mathbb{Z}_{\geq 0} \times \mathbb{Z}_{\geq 0} \times \mathbb{Z}_{\geq 0}$, we have

$$P(H^{(J)}) = \{(1,0,0,0), (1,1,0,0), (1,1,1,0), (1,1,1,1)\},$$

where

$$\dim H^{(J)}_{(1,0,0,0)} = \dim H^{(J)}_{(1,1,0,0)} = \dim H^{(J)}_{(1,1,1,0)} = \dim H^{(J)}_{(1,1,1,1)} = r.$$

For $k_1,\ k_2,\ k_3,\ k_4 \in \mathbb{Z}_{\geq 0}$, the only partition of (k_1, k_2, k_3, k_4) into a sum of (1,0,0,0), (1,1,0,0), (1,1,1,0) and (1,1,1,1) is

$$\begin{aligned}(k_1, k_2, k_3, k_4) &= (k_1 - k_2)(1,0,0,0) + (k_2 - k_3)(1,1,0,0) \\ &\quad + (k_3 - k_4)(1,1,1,0) + k_4(1,1,1,1).\end{aligned}$$

Thus, the Witt partition function becomes

$$W^{(J)}(k_1, k_2, k_3, k_4) = \frac{1}{k_1}\begin{pmatrix} k_1 \\ k_4 \end{pmatrix}\begin{pmatrix} k_1 - k_4 \\ k_2 - k_4 \end{pmatrix}\begin{pmatrix} k_2 - k_4 \\ k_3 - k_4 \end{pmatrix} r^{k_1}.$$

□

Remark 87. The above algebra $\mathcal{G}$ is isomorphic to the maximal graded Lie algebra with local part $H^{(J)} \bigoplus (sl(4, \mathbb{C}) + \mathfrak{h}) \bigoplus H^{(J)^*}$. Some particular cases:

1. If $k_4 > k_1$ or $k_3 > k_2$ or $k_2 > k_1$, then $\dim \mathcal{G}_{(k_1,k_2,k_3,k_4)} = 0$.

2. If $k_1 = k_2 = k_3 = k_4$, the formula (3.20) yields

$$\dim \mathcal{G}_{(k_1,k_1,k_1,k_1)} = \frac{1}{k_1}\sum_{d|k_1} \mu(d) r^{\frac{k_1}{d}}.$$

Hence, the subalgebra

$$F = \bigoplus_{k_1 \geq 1} \mathcal{G}_{(k_1,k_1,k_1,k_1)}$$

is the free Lie algebra on the r-dimensional space $\mathcal{G}_{(1,1,1,1)}$.

3. Similarly, we have

$$\dim \mathcal{G}_{(k_1,0,0,0)} = \frac{1}{k_1}\sum_{d|k_1} \mu(d) r^{\frac{k_1}{d}}$$

and the subalgebra

$$F_0 = \bigoplus_{k_1 \geq 1} \mathcal{G}_{(k_1,0,0,0)}$$

is the free Lie algebra on the r-dimensional space $\mathcal{G}_{(1,0,0,0)}$.

For example, for the root $(5,4,3,2)$, by the formula (3.20), we have

$$\dim \mathcal{G}_{(5,4,3,2)} = \frac{1}{5}\begin{pmatrix} 5 \\ 2 \end{pmatrix}\begin{pmatrix} 3 \\ 2 \end{pmatrix}\begin{pmatrix} 2 \\ 1 \end{pmatrix} r^5 = 12r^5.$$

In particular, when $r = 3$

$$\dim \mathcal{G}_{(5,4,3,2)} = 2916.$$

Similarly, for the root $(8,6,4,2)$, by the formula (3.20), we have

$$\dim \mathcal{G}_{(8,6,4,2)} = 315r^8 - 3r^4.$$

For the sake of completeness, we state the following Remark 89 (refer [29, 56, 116]).

Remark 88. Let $V(\Lambda_0)$ be the basic representation of the affine Kac-Moody algebra $A_n^{(1)}$, and let λ be a weight of $V(\Lambda_0)$. Then, we have

$$\dim(V(\Lambda_0)_\lambda) = p^{(n)}\left(1 - \frac{(\lambda,\lambda)}{2}\right),$$

where the function $p^{(n)}(m)$ are defined by

$$\sum_{m=0}^{\infty} p^{(n)}(m)q^m = \frac{1}{\phi(q)^n} = \frac{1}{\prod_{j\geq 1}(1-q^j)^n}.$$

The above formula can be restated as $\dim(V(\Lambda_0))_\lambda = p^{(n)}(m)$, where $\lambda = \Lambda_0 - m\delta$, Λ_0 is the highest weight, δ is the null root and $m \in \mathbb{Z}_+$.

Next we compute the root multiplicities of GKM algebras associated with BCMs which appear as extensions of affine $A_2^{(1)}$.

Proposition 28. *Let $\mathcal{G} = \mathcal{G}(A, \underline{m})$ be the GKM algebra associated with the BCM*

$$A = \begin{pmatrix} -k & -1 & 0 & 0 \\ -1 & 2 & -1 & -1 \\ 0 & -1 & 2 & -1 \\ 0 & -1 & -1 & 2 \end{pmatrix}$$

of charge $\underline{m} = (r, 1, 1, 1)$ *with* $k,\ r \in \mathbb{Z}_{>0}$*. Then, for the root* $\alpha = -k_1\alpha_1 - k_2\alpha_2 - k_3\alpha_3 - k_4\alpha_4$ *with* $k_1,\ k_2,\ k_3,\ k_4 \in \mathbb{Z}_{\geq 0}$*, we have*

$$\dim \mathcal{G}_\alpha = \sum_{d|\alpha} \frac{1}{d}\mu(d) \sum_{n \in T^{(J)}(\frac{\alpha}{d})} \frac{(|n|-1)!}{n!} r^{|n|} \prod \left(p^{(2)}\left(\frac{-k-(\tau_i,\tau_i)}{2}\right)\right)^{n_i} \tag{3.21}$$

Proof. Consider the GKM algebra $\mathcal{G} = \mathcal{G}(A, \underline{m})$ associated with the BCM of charge $\underline{m} = (r, 1, 1, 1)$ with $k,\ r \in \mathbb{Z}_{>0}$. This matrix is an extension of $A_2^{(1)}$. As in the above, we have the following results:

Let $I = \{1, 2, 3, 4\}$ be the index set for the simple roots of $\mathcal{G}$. Then, α_1 is the imaginary simple root with multiplicity $r \geq 1$ and α_2, α_3, and α_4 are the real simple roots. Thus we have

$$T = \{\alpha_1 \cdots \alpha_1\} \text{ (counted } r \text{ times)}$$

and since $(\alpha_1, \alpha_1) = -k < 0$, F can be either empty or $\{\alpha_1\}$.

If we take $J = \{2, 3, 4\}$, then

$$\mathcal{G}_0^{(J)} = \mathcal{G}_0 \bigoplus \mathbb{C}h_1,$$

where

$$\mathcal{G}_0 = \langle e_2, f_2, h_2, e_3, f_3, h_3, e_4, f_4, h_4 \rangle \cong A_2^{(1)} \text{ and } W(J) = \{1\}.$$

By Proposition 25, we obtain

$$H_1^{(J)} = V_J(-\alpha_1) \bigoplus \cdots \bigoplus V_J(-\alpha_1) \ \ (r \ \text{ copies})$$
$$H_k^{(J)} = 0 \text{ for } \ k \geq 2.$$

This implies

$$H^{(J)} = V_J(-\alpha_1) \bigoplus \cdots \bigoplus V_J(-\alpha_1) \ \ (r \text{ copies}),$$

where $V_J(-\alpha_1)$ is the standard representation of $A_2^{(1)}$ with highest weight $-\alpha_1$. The algebra $\mathcal{G}$ is isomorphic to the maximal graded Lie algebra with local part $H^{(J)} \bigoplus (A_2^{(1)} + \mathfrak{h}) \bigoplus H^{(J)*}$. Using the Remark 89, for $A_n^{(1)}$, $\Lambda_0 = -\alpha_1$, which implies $m = \frac{-k-(\lambda,\lambda)}{2}$.

By identifying $-j\alpha_1 - l\alpha_2 - m\alpha_3 - n\alpha_4 \in Q^-$ with $(j, l, m, n) \in \mathbb{Z}_{\geq 0} \times \mathbb{Z}_{\geq 0} \times \mathbb{Z}_{\geq 0} \times \mathbb{Z}_{\geq 0}$, weights of $V_J(-\alpha_1)$ are $(1, 0, 0, 0)$, $(1, 1, 0, 0)$, $(1, 1, 0, 1)$, etc. For a weight $\lambda = (1, l, m, n)$, by the above remark, we get

$$\dim(V_J(-\alpha_1))_\lambda = p^{(2)}\left(\frac{-k-(\lambda,\lambda)}{2}\right).$$

Using $\frac{-k-(\lambda,\lambda)}{2} = l + lm + ln + mn - l^2 - m^2 - n^2$, we get

$$\begin{aligned}\dim(V_J(-\alpha_1))_\lambda &= p^{(2)}\left(\frac{-k-(\lambda,\lambda)}{2}\right)\\ &= p^{(2)}(l + lm + ln + mn - l^2 - m^2 - n^2),\end{aligned}$$

where the function $p^{(2)}(m)$ are defined by

$$\sum_{m=0}^{\infty} p^{(2)}(m)q^m = \frac{1}{\phi(q)^2} = \frac{1}{\prod_{j\geq 1}(1-q^j)^2}.$$

So,

$$\dim H_\lambda^{(J)} = rp^{(2)}\left(\frac{-k-(\lambda,\lambda)}{2}\right).$$

We have,

$$P(H^{(J)}) = \{\tau_i | i \geq 1\},$$

where $\tau_1 = (1,0,0,0)$, $\tau_2 = (1,1,0,0)$, $\tau_3 = (1,1,0,1)$, $\tau_4 = (1,1,1,0)$, etc. Every root of $\mathcal{G}$ is of the form (j,l,m,n) for $j \geq 1$ and $l,\ m,\ n \geq 0$. Thus, the Witt partition function becomes

$$W^{(J)}(\tau) = \sum_{n\in T^{(J)}(\tau)} \frac{(|n|-1)!}{n!} r^{|n|} \prod \left(p^{(2)}\left(\frac{-k-(\tau_i,\tau_i)}{2}\right)\right)^{n_i}.$$

Therefore, using Theorem 42, we obtain:

$$\dim \mathcal{G}_\alpha = \sum_{d|\alpha} \frac{1}{d}\mu(d) \sum_{n\in T^{(J)}(\frac{\alpha}{d})} \frac{(|n|-1)!}{n!} r^{|n|} \prod \left(p^{(2)}\left(\frac{-k-(\tau_i,\tau_i)}{2}\right)\right)^{n_i}.$$

□

Table 3.1 **Weights of $H^{(J)}$ and their multiplicities**

Weight	Multiplicity	Weight	Multiplicity	Weight	Multiplicity
(1,0,0,0)	r	(1,2,1,1)	$2r$	(1,3,2,1)	r
(1,1,0,0)	r	(1,2,1,2)	$2r$	(1,3,2,2)	$5r$
(1,1,1,0)	r	(1,2,2,1)	$2r$	(1,3,2,3)	$5r$
(1,1,0,1)	r	(1,2,2,2)	$5r$	(1,3,2,4)	r
(1,1,1,1)	$2r$	(1,2,2,3)	$2r$	(1,3,3,2)	$5r$
(1,1,1,2)	r	(1,2,3,2)	$2r$	(1,3,3,3)	$10r$
(1,1,2,1)	r	(1,2,3,3)	$2r$	(1,3,3,4)	$5r$
(1,1,2,2)	r	(1,3,1,2)	r	(1,3,4,2)	r

Example 39. Consider the root $\alpha = (2,2,2,2)$. $T^{(J)}(2,2,2,2)$ corresponds to the partition of (2,2,2,2) into two parts. Hence, the partitions of the root $(2,2,2,2)$ into weights of $H^{(J)}$ are given in the following table:

$(1,0,0,0)$, $(1,2,2,2)$
$(1,1,0,0)$, $(1,1,2,2)$
$(1,1,1,0)$, $(1,1,1,2)$
$(1,1,0,1)$, $(1,1,2,1)$
$(1,1,1,1)$, $(1,1,1,1)$

Hence, by the formula (3.21), for the root $(2,2,2,2)$, we have

$$\dim \mathcal{G}_{(2,2,2,2)} = 10r^2 - r.$$

Below, we compute the root multiplicities of GKM algebras associated with BCMs which appear as extensions of hyperbolic HA_1.

Proposition 29. *Let $\mathcal{G} = \mathcal{G}(A, \underline{m})$ be the GKM algebra associated with the BCM,*

$$A = \begin{pmatrix} -k & 0 & -1 \\ 0 & 2 & -4 \\ -1 & -4 & 2 \end{pmatrix}$$

of charge $\underline{m} = (r, 1, 1)$ with $k,\ r \in \mathbb{Z}_{>0}$. Then, for the root $\alpha = -k_1\alpha_1 - k_2\alpha_2 - k_3\alpha_3$ with $k_1,\ k_2,\ k_3 \in \mathbb{Z}_{\geq 0}$, we have

$$\dim \mathcal{G}_\alpha = \sum_{d|\alpha} \frac{1}{d}\mu(d) \sum_{n \in T^{(J)}(\frac{\alpha}{d})} \frac{(|n|-1)!}{n!} r^{|n|} \prod \left(\dim \left(V_J(-\alpha_1)\right)_{\tau_i}\right)^{n_i}. \qquad (3.22)$$

Proof. Consider the GKM algebra $\mathcal{G} = \mathcal{G}(A, \underline{m})$ associated with the above BCM of charge $\underline{m} = (r, 1, 1)$ with $k,\ r \in \mathbb{Z}_{>0}$. Here also we have to repeat the following for the sake of completeness.

Let $I = \{1,2,3\}$ be the index set for the simple roots of $\mathcal{G}$. Then, α_1 is the imaginary simple root with multiplicity $r \geq 1$ and α_2 and α_3 are the real simple roots. Thus we have

$$T = \{\alpha_1, \ldots, \alpha_1\} \text{ (counted } r \text{ times)}$$

and since $(\alpha_1, \alpha_1) = -k < 0$, F can be either empty or $\{\alpha_1\}$.

If we take $J = \{2,3\}$, then

$$\mathcal{G}_0^{(J)} = \mathcal{G}_0 \bigoplus \mathbb{C}h_1,$$

where

$$\mathcal{G}_0 = \langle e_2, f_2, h_2, e_3, f_3, h_3 \rangle \cong HA_1 \text{ and } W(J) = \{I\}.$$

By Proposition 25, we obtain

$$H_1^{(J)} = V_J(-\alpha_1) \bigoplus \cdots \bigoplus V_J(-\alpha_1) \quad (r \text{ copies})$$

$$H_k^{(J)} = 0 \text{ for } k \geq 2.$$

This implies

$$H^{(J)} = V_J(-\alpha_1) \bigoplus \cdots \bigoplus V_J(-\alpha_1) \quad (r \text{ copies}),$$

where $V_J(-\alpha_1)$ is an irreducible highest weight representation with highest weight $-\alpha_1$. The algebra $\mathcal{G}$ is isomorphic to the maximal graded Lie algebra with local part $H^{(J)} \bigoplus (HA_1 + \mathfrak{h}) \bigoplus H^{(J)*}$.

By identifying $-l\alpha_1 - m\alpha_2 - n\alpha_3 \in Q^-$ with $(l,m,n) \in \mathbb{Z}_{\geq 0} \times \mathbb{Z}_{\geq 0} \times \mathbb{Z}_{\geq 0}$, we have

$$P(H^{(J)}) = \{\tau_i | i \geq 1\},$$

where $\tau_1 = (1,0,0)$, $\tau_2 = (1,0,1)$, $\tau_3 = (1,1,1)$, $\tau_4 = (1,1,2)$, $\tau_5 = (1,1,3)$, etc.

From Table 3.2, we can obtain $\dim(V_J(-\alpha_1))_{\tau_i}$.

Every root of $\mathcal{G}$ is of the form (l,m,n) for $l \geq 1$ and $m,n \geq 0$. Here

$$d(i) = \dim(H^{(J)})_{\tau_i} = r \ \dim V_J(-\alpha_1)_{\tau_i}.$$

Thus, the Witt partition function becomes

$$W^{(J)}(\tau) = \sum_{n \in T^{(J)}(\tau)} \frac{(|n|-1)!}{n!} r^{|n|} \prod (\dim V_J(-\alpha_1)_{\tau_i})^{n_i}.$$

Therefore, using Theorem 42, we obtain:

$$\dim \mathcal{G}_\alpha = \sum_{d|\alpha} \frac{1}{d} \mu(d) \sum_{n \in T^{(J)}(\frac{\alpha}{d})} \frac{(|n|-1)!}{n!} r^{|n|} \prod \left(\dim (V_J(-\alpha_1))_{\tau_i}\right)^{n_i}.$$

Table 3.2 Weights of $H^{(J)}$ and their multiplicities

Weight	Multiplicity	Weight	Multiplicity	Weight	Multiplicity
(1,0,0)	r	(1,2,3)	$3r$	(1,3,4)	$10r$
(1,0,1)	r	(1,2,4)	$4r$	(1,3,5)	$14r$
(1,1,1)	r	(1,2,5)	$4r$	(1,3,6)	$16r$
(1,1,2)	r	(1,2,6)	$3r$	(1,3,7)	$16r$
(1,1,3)	r	(1,2,7)	$2r$	(1,4,1)	r
(1,1,4)	r	(1,3,1)	r	(1,4,2)	$4r$
(1,2,1)	r	(1,3,2)	$3r$	(1,4,3)	$10r$
(1,2,2)	$2r$	(1,3,3)	$6r$	(1,4,4)	$20r$

Now to find $\dim (V_J(-\alpha_1))_{\tau_i}$, one can use the Table 5.1 in [117]. One can also refer, [112]. □

Example 40. Consider the root $\alpha = (3,3,3)$. $T^{(J)}(3,3,3)$ corresponds to the partition of (3,3,3) into three parts. Hence, the partitions of the root $(3,3,3)$ into weights of $H^{(J)}$ are given in the following table:

$(1,0,0),\ (1,0,0),\ (1,3,3)$
$(1,0,0),\ (1,0,1),\ (1,3,2)$
$(1,0,0),\ (1,1,1),\ (1,2,2)$
$(1,0,0),\ (1,1,2),\ (1,2,1)$
$(1,0,1),\ (1,0,1),\ (1,3,1)$
$(1,0,1),\ (1,1,1),\ (1,2,1)$
$(1,1,1),\ (1,1,1),\ (1,1,1)$

Hence, by the formula (3.22), for the root $(3,3,3)$, we have,

$$\dim \mathcal{G}_{(3,3,3)} = \frac{64r^3 - r}{3}.$$

In particular, when $r = 3$

$$\dim \mathcal{G}_{(3,3,3)} = 575.$$

Similarly, for the root $(4,4,4)$, we have,

$$\dim \mathcal{G}_{(4,4,4)} = \frac{655r^4 - 7r^2}{4}.$$

In particular, when $r = 4$

$$\dim \mathcal{G}_{(4,4,4)} = 41{,}892.$$

Root multiplicities of some GKM algebras having more than one simple imaginary root

Proposition 30. *Let* $\mathcal{G} = \mathcal{G}(A, \underline{m})$ *be the GKM algebra associated with the BCM*

$$A = \begin{pmatrix} -k_1 & -a_1 & \cdots & -a_{n-1} \\ -b_1 & -k_2 & \cdots & \\ \vdots & & \ddots & \\ -b_{n-1} & & \cdots & -k_n \end{pmatrix}$$

of charge $\underline{m} = (r_1, \ldots, r_n)$ *with* $r_i,\ k_i\ (\ i = 1, \ldots, n) \in \mathbb{Z}_{>0}$.

Then, for the root $\alpha = -m_1\alpha_1 - m_2\alpha_2 - \cdots - m_n\alpha_n$, *with* $m_i\ (i = 1, \ldots, n) \in \mathbb{Z}_{\geq 0}$, *we have*

$$\dim \mathcal{G}_\alpha = \frac{1}{\sum_{i=1}^n m_i} \sum_{d|\ (m_1,\ldots,m_n)} \mu(d) \begin{pmatrix} \frac{\sum_{i=1}^n m_i}{d} \\ \frac{\sum_{i=1}^{n-1} m_i}{d} \end{pmatrix} \begin{pmatrix} \frac{\sum_{i=1}^{n-1} m_i}{d} \\ \frac{\sum_{i=1}^{n-2} m_i}{d} \end{pmatrix} \cdots \begin{pmatrix} \frac{m_1+m_2}{d} \\ \frac{m_1}{d} \end{pmatrix} \prod_{i=1}^n r_i^{\frac{m_i}{d}}. \tag{3.23}$$

Proof. Consider the GKM algebra $\mathcal{G} = \mathcal{G}(A, \underline{m})$ associated with the above BCM of charge $\underline{m} = (r_1, r_2, \ldots, r_n)$ with $k_i,\ r_i,\ a_i,\ b_i \in \mathbb{Z}_{>0}$ for all i. Let $I = \{1, \ldots, n\}$ be the index set for the simple roots of $\mathcal{G}$. Then $\alpha_i,\ i = 1, \ldots, n$ are imaginary simple roots with multiplicities $r_i \geq 1,\ i = 1, \ldots, n$. Thus we have

$$T = \bigcup_{i=1}^n \{\alpha_i, \ldots, \alpha_i\} \text{ (counted } r_i \text{ times)},$$

and since

$$(\alpha_i, \alpha_i) < 0 \text{ and } (\alpha_i, \alpha_j) < 0 \text{ for all } i \text{ and } j,$$

F can be either $\emptyset$ or $\{\alpha_i\}$ for $i = 1, \ldots, n$. Since there is no real simple root, we should take $J = \emptyset$ and $W(J) = \{I\}$. By Proposition 25, we obtain

$$\begin{aligned} H_1^{(J)} &= C_{-\alpha_1} \oplus \cdots \oplus C_{-\alpha_1} \ \ (r_1 \text{ copies}) \\ &\quad \oplus C_{-\alpha_2} \oplus \cdots \oplus C_{-\alpha_2} \ \ (r_2 \text{ copies}) \\ &\quad \vdots \\ &\quad \oplus C_{-\alpha_n} \oplus \cdots \oplus C_{-\alpha_n} \ \ (r_n \text{ copies}) \\ H_k^{(J)} &= 0 \quad \text{for } k \geq 2. \end{aligned}$$

This implies

$$\begin{aligned} H^{(J)} &= C_{-\alpha_1} \oplus \cdots \oplus C_{-\alpha_1} \quad (r_1 \text{ copies}) \\ &\oplus\ C_{-\alpha_2} \oplus \cdots \oplus C_{-\alpha_2} \quad (r_2 \text{ copies}) \\ &\vdots \\ &\oplus\ C_{-\alpha_n} \oplus \cdots \oplus C_{-\alpha_n} \quad (r_n \text{ copies}). \end{aligned}$$

By identifying $-m_1\alpha_1 - m_2\alpha_2 - \cdots - m_n\alpha_n \in Q^-$ with $(m_1, \ldots, m_n) \in \mathbb{Z}_{\geq 0} \times \mathbb{Z}_{\geq 0} \times \cdots \times \mathbb{Z}_{\geq 0}$, we have

$$P(H^{(J)}) = \{(1, 0, \ldots, 0), (0, 1, 0, \ldots, 0), \ldots, (0, 0, \ldots, 1)\},$$

where

$$\dim H^{(J)}_{(i,0,\ldots,0)} = r_i \ \text{ for } i = 1, \ldots, n.$$

Thus, the Witt partition function becomes

$$W^{(J)}(m_1, \ldots, m_n) = \frac{1}{\sum_{i=1}^n m_i} \binom{\sum_{i=1}^n m_i}{\sum_{i=1}^{n-1} m_i} \binom{\sum_{i=1}^{n-1} m_i}{\sum_{i=1}^{n-2} m_i} \cdots \binom{m_1 + m_2}{m_1} \prod_{i=1}^n r_i^{m_i}.$$

□

Remark 89. The above algebra $\mathcal{G}$ is isomorphic to the maximal graded Lie algebra with local part $H^{(J)} \bigoplus \mathfrak{h}) \bigoplus H^{(J)*}$. For example, let $n = 4$. For the root $\alpha = (4, 4, 4, 4)$, by the formula (3.23), we have

$$\begin{aligned} \dim \mathcal{G}_{(4,4,4,4)} &= \frac{1}{16} \sum_{d|(4,4,4,4)} \mu(d) \binom{\frac{16}{d}}{\frac{12}{d}} \binom{\frac{12}{d}}{\frac{8}{d}} \binom{\frac{8}{d}}{\frac{4}{d}} r_1^{\frac{4}{d}} r_2^{\frac{4}{d}} r_3^{\frac{4}{d}} r_4^{\frac{4}{d}} \\ &= \frac{1}{16} \left(\binom{16}{12} \binom{12}{8} \binom{8}{4} r_1^4 r_2^4 r_3^4 r_4^4 \right. \\ &\quad \left. - \binom{8}{6} \binom{6}{4} \binom{4}{2} r_1^2 r_2^2 r_3^2 r_4^2 \right) \\ &= \frac{1}{4} (455\, r_1^4 r_2^4 r_3^4 r_4^4 - 7\, r_1^2 r_2^2 r_3^2 r_4^2). \end{aligned}$$

In particular, when $r_1 = 2,\ r_2 = 3,\ r_3 = 4,\ r_4 = 5$

$$\dim \mathcal{G}_{(4,4,4,4)} = 47{,}174{,}374{,}800.$$

Remark 90.

(1) For finding the root multiplicities of GKM algebras which appear as extensions of finite, affine, or indefinite type, one needs the corresponding weight multiplicities of the modules of the underlying Kac-Moody algebras. This problem was done here for some particular cases. For GKM algebras which are not algebras of above types, one can use directly the formula of Kang [72] and Kim and Shin [117], but each case has its own importance. Generally one can compute root multiplicities of GKM algebras whose BCMs appear as extensions of generalized Cartan matrices of any order and can also compute root multiplicities of any GKM algebra with any number of simple imaginary roots and in general for any GKM algebra.

(2) **(i)** On new developments of BKM algebras related to automorphic forms, one can refer Ray [118].

(ii) For connections between Siegel modular forms, sporadic simple groups, BKM algebras, and Mathieu moonshine, one can refer Govindarajan [119].

Exercises

3.1 Compute the set of all real and imaginary roots of following GKM Lie algebras whose generalized Cartan matrices are given below:

(i) $GGA_1 = \begin{pmatrix} -k & -a \\ -c & 2 \end{pmatrix}$

(ii) $GGB_2 = \begin{pmatrix} -k & -a & -b \\ -c & 2 & -1 \\ -d & -2 & 2 \end{pmatrix}$.

3.2 Let $\mathcal{G} = \mathcal{G}(A, \underline{m})$ be the GKM algebra associated with the BCM,

$$A = \begin{pmatrix} 2 & -1 & -1 \\ -1 & -k_1 & -a \\ -1 & -a & -k_2 \end{pmatrix}$$

of charge $\underline{m} = (1, r_1, r_2)$ with $k_1,\ k_2,\ r_1,\ r_2 \in \mathbb{Z}_{>0}$.

Prove that, for the root $\alpha = -n_1\alpha_1 - n_2\alpha_2 - n_3\alpha_3$ with $n_1,\ n_2,\ n_3 \in \mathbb{Z}_{\geq 0}$, we have

$$\dim \mathcal{G}_\alpha = \frac{1}{n_2 + n_3} \sum_{d|(n_1,n_2,n_3)} \mu(d) \begin{pmatrix} \frac{n_2+n_3}{d} \\ \frac{n_3}{d} \end{pmatrix} \left(\sum_{t=0}^{\min\{n_2, n_1-n_3\}} \begin{pmatrix} \frac{n_3}{d} \\ \frac{n_1-t}{d} \end{pmatrix} \begin{pmatrix} \frac{n_2}{d} \\ \frac{t}{d} \end{pmatrix} \right) r_1^{\frac{n_2}{d}} r_2^{\frac{n_3}{d}}.$$

3.3 Let $\mathcal{G} = \mathcal{G}(A, \underline{m})$ be the GKM algebra associated with the BCM,

$$A = \begin{pmatrix} -k_1 & -a \\ -b & -k_2 \end{pmatrix}$$

of charge $\underline{m} = (r_1, r_2)$ with $k_1,\ k_2,\ r_1,\ r_2,\ a,\ b \in \mathbb{Z}_{>0}$. Then prove that, for the root $\alpha = -m_1\alpha_1 - m_2\alpha_2$ with $m_1,\ m_2 \in \mathbb{Z}_{\geq 0}$, we have

$$\dim \mathcal{G}_\alpha = \frac{1}{m_1 + m_2} \sum_{d|(m_1,m_2)} \mu(d) \begin{pmatrix} \frac{m_1+m_2}{d} \\ \frac{m_1}{d} \end{pmatrix} r_1^{\frac{m_1}{d}} r_2^{\frac{m_2}{d}}.$$

3.4 Consider the nonsymmetric $GGB_2 = \begin{pmatrix} -k & -a & -b \\ -c & 2 & -1 \\ -d & -2 & 2 \end{pmatrix}$. Prove that in the usual notation

$$\Delta^{\rm re} = \{\alpha_2, \alpha_3, -\alpha_2, \alpha_2 + \alpha_3, \alpha_2 + 2\alpha_3, -\alpha_3, -\alpha_2 + 2\alpha_3, -(\alpha_2 + \alpha_3)\}.$$
$$\Delta_{\rm S}^{\rm re} = \pm\{\alpha_3, \alpha_2 + \alpha_3\}, \Delta_{\rm L}^{\rm re} = \pm\{\alpha_2, \alpha_2 + 2\alpha_3\}.$$

3.5 Verify the results of Lemma 5 and Proposition 19 for GGA_1, GGA_2, $GGA_1^{(1)}$, and $GGHA_1$.

3.6 Prove that $\alpha = \frac{k_1}{2}(2\alpha_1 + c\alpha_2)$ is a special imaginary root of $\mathcal{G}(A)$, for every $k_1 \in \mathbb{N}$, where A is a symmetrizable $GGA_1 = \begin{pmatrix} -k_1 & -a \\ -a & 2 \end{pmatrix}$.

3.7 Prove that $\alpha = k_1(\alpha_1 + c\alpha_2 + c\alpha_3)$ is a special imaginary root of $\mathcal{G}(A)$, for every $k_1 \in \mathbb{N}$, where A is a symmetrizable $GGA_2 = \begin{pmatrix} -k_1 & -a & -b \\ -a & 2 & -1 \\ -b & -1 & 2 \end{pmatrix}$.

3.8 Prove that $\alpha = k_1(2\alpha_1 + (2c+d)\alpha_2 + 2(c+d)\alpha_3)$is a special imaginary root of $\mathcal{G}(A)$, for every $k_1 \in \mathbb{N}$, where A is a symmetrizable $GGB_2 = \begin{pmatrix} -k_1 & -a & -b \\ -c & 2 & -1 \\ -d & -2 & 2 \end{pmatrix}$.

3.9 Prove that $\alpha = k_1(\alpha_1 + (2c+d)\alpha_2 + (3c+2d)\alpha_3)$ is a special imaginary root of $\mathcal{G}(A)$, for every $k_1 \in \mathbb{N}$, where A is a symmetrizable $GGG_2 = \begin{pmatrix} -k_1 & -a & -b \\ -c & 2 & -1 \\ -d & -3 & 2 \end{pmatrix}$.

3.10 Prove that, there exists no special imaginary root for $\mathcal{G}(A)$, where A is a symmetrizable $GGA_1^{(1)} = \begin{pmatrix} -k_1 & -a & -b \\ -a & 2 & -2 \\ -b & -2 & 2 \end{pmatrix}$ or $GGA_2^{(2)} = \begin{pmatrix} -k_1 & -a & -b \\ -a & 2 & -1 \\ -b & -4 & 2 \end{pmatrix}$ or $GGHA_1 = \begin{pmatrix} -k_1 & -a & -b \\ -a & 2 & -p \\ -b & -q & 2 \end{pmatrix}$ such that $pq > 4$.

3.11 Let $\begin{pmatrix} -k_1 & -1 & 0 \\ -1 & 2 & -1 \\ 0 & -2 & 2 \end{pmatrix}$ be a GGCM. Show that $\alpha = 3\alpha_1 + 2\alpha_2 + \alpha_3$ is a purely imaginary root.

3.12 Let $GGCM = \begin{pmatrix} -k_1 & -a_1 & -a_2 \\ -b_1 & -k_2 & -p \\ -b_2 & -q & -k_3 \end{pmatrix}$ with $k_1, k_2, k_3 \geq 0$. Then prove that all purely imaginary roots are strictly imaginary roots.

3.13 Let $A = \begin{pmatrix} 0 & -1 \\ -1 & 2 \end{pmatrix}$ be a symmetric GGCM of hyperbolic type. Find whether $\alpha = 2\alpha_1 + \alpha_2$ is a purely imaginary root or strictly imaginary root.

3.14 Find dimension formula for the extended hyperbolic Kac Moody algebra associated with the BCM $A = \begin{pmatrix} 0 & -1 & 0 \\ -1 & 2 & -1 \\ 0 & -2 & 2 \end{pmatrix}$.

3.15 Using Proposition 29, find dimension of $\mathcal{G}_\alpha$ for the root $\alpha = (9, 5, 4)$.

3.16 Using Proposition 30, find dimension of $\mathcal{G}_\alpha$ for the root $\alpha = (7, 6, 4)$.

Lie superalgebras

4

Chapter Outline

4.1 Basic concepts in Lie superalgebras with examples 204
4.2 Coloring matrices, θ-colored Lie superalgebras, and examples 212
4.3 Subsuperalgebras, ideals of Lie superalgebras, abelian Lie superalgebras, solvable, and nilpotent Lie superalgebras 215
4.4 General linear Lie superalgebras 216
4.5 Simple and semisimple Lie superalgebras and bilinear forms 216
4.6 Representations of Lie superalgebras 218
4.7 Different classes of classical Lie superalgebras 220
4.8 Universal enveloping algebras of Lie superalgebras and θ-colored Lie superalgebras 224
4.9 Cartan subalgebras and root systems of Lie superalgebras 230
4.10 Killing forms on Lie superalgebras 237
4.11 Dynkin diagrams of Lie superalgebras 239
4.12 Lie superalgebras over an algebraically closed field of characteristic zero 241
4.13 Classification of non-classical Lie superalgebras 245
4.14 Lie superalgebras of vector fields 248

Abstract

In Mathematics, graded Lie algebras were known in the context of deformation theory. Graded Lie algebras also appear in Physics in the context of "Supersymmetrices" relating to particles of differing statistics. The first basic example of graded Lie algebras was provided by Nijenhuis [120] and then by Frolicher and Nijenhuis [121]. In a paper of "Berezin and Kats [122]," Lie superalgebras appear as Lie algebras of certain generalized groups, called Lie supergroups. A satisfactory theory similar to Lie's theory, has been developed on the connection between Lie supergroups and Lie superalgebras [123, 124]. Kac [34] gave a construction of the theory of "Lie superalgebras" (as the Physicists call them, "$\mathbb{Z}_2$-graded Lie algebras") on the lines of Lie algebras. A detailed exposition of the theory of Lie superalgebras can also be seen in Scheunert [35]. Lie superalgebras also appear in Cohomology theories. Gerstenhaber [125] described cohomology structure of an associative ring.

Introduction to Finite and Infinite Dimensional Lie (Super)algebras. http://dx.doi.org/10.1016/B978-0-12-804675-3.00004-2

In this chapter, we describe Lie superalgebras. Starting from the basic concepts in Lie superalgebras and coloring matrix, θ-colored Lie superalgebras are also explained with examples. Ideals, subsuperalgebras, solvable, and nilpotent Lie superalgebras are given with examples. General linear Lie superalgebras, simple and semisimple Lie superalgebras with their dimensions and bilinear forms, representations of Lie superalgebras are also discussed. Basic concepts in representation theory, different classes of Classical Lie superalgebras, universal enveloping algebras of Lie superalgebras and θ-colored Lie superalgebras, Cartan subalgebras, and root systems of Lie superalgebras are also given. Killing form, Dynkin diagrams of Lie superalgebras are also discussed with examples. Classification of Lie superalgebras over an algebraically closed field of characteristic zero and that of non-classical Lie superalgebras are also given.

4.1 Basic concepts in Lie superalgebras with examples

Lie superalgebras can be defined in one of the following two ways:

(a) Starting from graded vector spaces, one can define $\mathbb{Z}_2$-graded vector spaces and finally arrive at Lie superalgebras (or)
(b) One can start with graded algebras then superalgebras and finally arrive at Lie superalgebras.

Initially, we discuss the above two ways of constructing Lie superalgebras here. First we define Lie superalgebras using $\mathbb{Z}_2$-graded vector spaces.

Definition 134. Let Γ be the ring of integers. A vector space V over the field K is said to be *Γ-graded* if it is equipped with a Γ-gradation, namely,

$$V = \bigoplus_{\gamma \in \Gamma} V_\gamma,$$

where $(V_\gamma)_{\gamma\in\Gamma}$ is the family of subspaces of V.

Remark 91. Every element of $y \in V$ has a unique decomposition of the form,

$$y = \Sigma_{\gamma\in\Gamma} y_\gamma;\; y_\gamma \in V_\gamma$$

(only finitely many y_γ are different from zero). The element y_γ is called the homogeneous component of y of *degree* γ or each element of V_γ is homogenous of degree γ for $\gamma \in \Gamma$. Hence an element of V is called *homogeneous of degree* $\gamma, \gamma \in \Gamma$, if it is an element of V_γ. If $x \in V_\gamma$, we write $|x| = \deg(x) = \gamma$.

Definition 135. Suppose a vector space V and two of its vector subspaces $V_{\bar{0}}$ and $V_{\bar{1}}$ are given. If V is the direct sum of these two subspaces, namely, $V = V_{\bar{0}} \oplus V_{\bar{1}}$, then V (with this decomposition) is called a *$\mathbb{Z}_2$-graded vector space*. $V_{\bar{0}}$ and $V_{\bar{1}}$ are called the *even part and odd part* of V, respectively. Here $\mathbb{Z}_2$ is the residue class ring mod 2 with elements $\overline{0}$ and $\overline{1}$. The elements of $V_{\overline{0}}$ (resp.$V_{\overline{1}}$) are also called even (resp. odd). A subspace U of V is called Γ-graded if

$$U = \bigoplus_{\gamma \in \Gamma} (U \cap V_\gamma).$$

If $V = \bigoplus_{j \in Z} V_j$ is a Z-graded vector space, then one can reduce a Z_2-gradation induced from that of Z-gradation given by

$$V_{\overline{0}} = \bigoplus_{j \in Z} V_{2j}; \quad V_{\overline{1}} = \bigoplus_{j \in Z} V_{2j+1}.$$

Definition 136. Let V and W be two Γ-graded vector spaces such that $V = \bigoplus_{\gamma \in \Gamma} V_\gamma$ and $W = \bigoplus_{\gamma \in \Gamma} W_\gamma$. We define a linear mapping $f : V \longrightarrow W$ satisfying

$$f(V_\alpha) \subset W_{\alpha+\gamma} \quad \text{for all } \alpha \in \Gamma.$$

Then f is said to be *homogenous of degree γ for $\gamma \in \Gamma$*. If f is homogenous of degree zero, then the mapping f becomes a homomorphism of two Γ-graded vector spaces from V to W. Similarly, an isomorphism or an automorphism of Γ-graded vector spaces can be defined.

Definition 137. If U and V are $\mathbb{Z}_2$-graded vector spaces, then a bilinear mapping, $b : V \times V \to U$ is called *supersymmetric*, if

$$b(y,x) = (-1)^{\xi\eta}\, b(x,y), \text{for all } x \in V_\xi, y \in V_\eta;\ \xi,\ \eta \in \mathbb{Z}_2.$$

The bilinear mapping $b : V \times V \to U$ is called *skew supersymmetric*, if

$$b(y,x) = -(-1)^{\xi\eta}\, b(x,y), \text{ for all } x \in V_\xi,\ y \in V_\eta;\ \xi,\ \eta \in \mathbb{Z}_2.$$

Definition 138. A $\mathbb{Z}_2$-graded vector space $\mathcal{G} = \mathcal{G}_{\overline{0}} \oplus \mathcal{G}_{\overline{1}}$, possessing the operation called the *bracket product*,

$$[\,.,\ .] : \mathcal{G} \times \mathcal{G} \ni (x,y) \overset{\text{bilinear map}}{\longrightarrow} [x,y] \in \mathcal{G},$$

is called a *Lie superalgebra* if it satisfies the following conditions:

(i) $[\mathcal{G}_i, \mathcal{G}_j] \subset \mathcal{G}_{i+j}$, $(i,j \in \mathbb{Z}_2)$,
(ii) $[y,x] = -(-1)^{|x||y|}[x,y]$,
(iii) $[x,[y,z]] - (-1)^{|x||y|}[y,[x,z]] = [[x,y],z]$,

for all $x, y, z \in \mathcal{G}$. Here $|x|$ denotes the degree of x.

Remark 92. The definition of the Lie superalgebra defined above is as per (a) in the introduction of this chapter.

Below we define graded algebras, superalgebras, and finally Lie superalgebras.

Definition 139. If A is an algebra and M is an abelian group, then an M-grading of A is a decomposition of A into a direct sum of subspaces $A = \oplus_{\alpha \in M} A_\alpha$ for which $A_\alpha A_\beta \subseteq A_{\alpha+\beta}$. *An algebra* A, equipped with an M-grading, is called M-graded algebra. In other words, an algebra is γ-graded if the corresponding vector space is γ-graded.

Definition 140. A *subspace* B of an M-graded algebra A is called M-graded if $B = \oplus_{\alpha \in M}(B \cap A_\alpha)$. A *subalgebra* (*or ideal*) of an M-graded algebra is an M-graded subalgebra (or ideal). Let A and A' be two M-graded algebras. A *homomorphism* $\Phi : A \to A'$ of M-graded algebras preserves grading in the sense that $\Phi(A_\alpha) \subseteq A'_{\Phi(\alpha)}$, where Φ is an *automorphism* of M.

Example 41.

(i) Any algebra A over a field K can be considered as M-graded (M is an additive commutative semigroup), if $A_0 = A, A_m = \{0\}$ for $m \neq 0$.

(ii) Let $\mathcal{G}$ be a complex Lie algebra of rank r, possessing a Cartan decomposition,

$$\mathcal{G} = \mathcal{G}_0 \oplus \sum_{\alpha \in \Phi} \mathcal{G}_\alpha,$$

where Φ is the root system of $\mathcal{G}$ with respect to its Cartan subalgebra $\mathcal{G}_0 = \mathfrak{h}$. From the above decomposition, it is clear that this Lie algebra $\mathcal{G}$ is $\mathbb{Z}^r$-graded with $\mathcal{G}_\alpha \mathcal{G}_\beta \subset \mathcal{G}_{\alpha+\beta}$ for $\alpha, \beta \in \Phi$.

(iii) Let V be a vector space over field K and $G(V)$ be the Grassman Lie algebra. Then $G(V)$ has a Z-grading,

$$G = \oplus_{k=0}^{\infty} G_k,$$

where G_k is the linear span of k-vectors of the form $v_1 \wedge v_2 \wedge \cdots \wedge v_k$ $(v_i \in V)$. $G(V)$ can also be considered as Z_2-grading with $G(V) = G_{\overline{0}} \oplus G_{\overline{1}}$, where $G_{\overline{0}}$ and $G_{\overline{1}}$ are linear span of k-vectors with k even or k odd, respectively.

Definition 141. Let $A = A_{\overline{0}} \oplus A_{\overline{1}}$ be a Z_2-graded algebra, then A is called a superalgebra. The elements of $A_{\overline{0}}$ are called *even* and those of $A_{\overline{1}}$ are called *odd.* In particular, if $A = A_{\overline{0}} \oplus A_{\overline{1}}$ is a superalgebra, then $A_\alpha . A_\beta \subset A_{\alpha+\beta}$ for $\alpha, \beta \in \mathbb{Z}_2 = \{\overline{0}, \overline{1}\}$. If "$a$" is an element of either $A_{\overline{0}}$ or $A_{\overline{1}}$, then a is *homogeneous of degree* $\overline{0}$ or $\overline{1}$, respectively. A superalgebra A is said to be Z-graded if for a given family, $(A_j)_{j \in Z}$ of Z_2-*graded subspaces* of A, we have $A = \bigoplus_{j \in Z} A_j$ and $A_i A_j \subset A_{i+j}$ for all $i, j \in Z$. This Z-gradation is said to be consistent with Z_2-gradation of A if $A_{\overline{0}} = \bigoplus_{j \in Z} A_{Z_j}$ and $A = \bigoplus_{j \in Z} A_{Z_{j+1}}$.

Definition 142. We can define a bracket [,] in a superalgebra A, in a natural way by $[a, b] = ab - (-1)^{(dega)(degb)} ba$. If the bracket satisfies $[a, b] = 0$, for all $a, b \in A$, then the superalgebra is called *commutative superalgebra.*

Associative superalgebra: If the bracket satisfies $[[a,b],c] = [a,[b,c]]$, then the superalgebra is called *associative superalgebra*.

Remark 93. For *an associative superalgebra A*, we have

$$[a,bc] = [a,b]c + (-1)^{(dega)(degb)}b[a,c].$$

Definition 143. A *Lie superalgebra* is a superalgebra $\mathcal{G} = \mathcal{G}_{\bar{0}} \oplus \mathcal{G}_{\bar{1}}$ with a bilinear operation [,], satisfying the following relations

$$[a,b] = (-1)^{(dega)(degb)}[b,a] \quad \text{(anti commutativity or graded skew-symmetry)}$$

and

$$(-1)^{(degc)(dega)}[a,[b,c]]+(-1)^{(dega)(degb)}[b,[c,a]]+(-1)^{(degb)(degc)}[c,[a,b]] = 0$$

(graded Jacobi identity or Jacobi identity).

The set of all even elements of a Lie superalgebra $\mathcal{G} = \mathcal{G}_{\bar{0}} \oplus \mathcal{G}_{\bar{1}}$ is a Lie algebra called the even subspace. We have $[\mathcal{G}_{\bar{0}}, \mathcal{G}_{\bar{1}}] \subset \mathcal{G}_{\bar{1}}$.

Using the above Jacobi identity, one can get,

$$[[a,b],c] = [a,[b,c]] - [b,[a,c]]$$

with $a \in \mathcal{G}_{\bar{0}}$, $b \in \mathcal{G}_{\bar{1}}$ and $c \in \mathcal{G}_{\bar{1}}$. Hence the commutator of $\mathcal{G}$ makes $\mathcal{G}_{\bar{1}}$ into a $\mathcal{G}_{\bar{0}}$-module. Moreover, the restriction of commutator to $\mathcal{G}_{\bar{1}}$ defines a bilinear symmetric mapping $\phi : \mathcal{G}_{\bar{1}} \times \mathcal{G}_{\bar{1}} \to \mathcal{G}_{\bar{0}}$. Hence a superalgebra $\mathcal{G} = \mathcal{G}_{\bar{0}} \oplus \mathcal{G}_{\bar{1}}$ is a Lie algebra such that

- **(i)** $\mathcal{G}_{\bar{0}}$ is an ordinary Lie algebra.
- **(ii)** The multiplication on the left by elements of $\mathcal{G}_{\bar{0}}$ determines a structure of a $\mathcal{G}_{\bar{0}}$-module on $\mathcal{G}_{\bar{1}}$.
- **(iii)** Multiplication by elements of $\mathcal{G}_{\bar{1}}$ determines a homomorphism of $\mathcal{G}_{\bar{0}}$ modules.
- **(iv)** The bilinear mapping $[,] : \mathcal{G}_{\bar{1}} \times \mathcal{G}_{\bar{1}} \to \mathcal{G}_{\bar{0}}$ is symmetric and $\mathcal{G}_{\bar{0}}$-invariant.
- **(v)** For $b_1, b_2, b_3 \in \mathcal{G}_{\bar{1}}$, one has

$$[[b_1,b_2],b_3] + [[b_2,b_3],b_1] + [[b_3,b_1],b_2] = 0.$$

Remark 94. The definition of the Lie superalgebra defined above in Definition 143 is as per (b) in the introduction of this chapter.

Definition 144. Any Z_2-graded vector space, $\mathcal{G} = \mathcal{G}_{\bar{0}} \oplus \mathcal{G}_{\bar{1}}$, is a Lie superalgebra with $[g_1, g_2] = 0$ for $g_1, g_2 \in \mathcal{G}$. This Lie superalgebra is called *supercommutative Lie superalgebra*.

Example 42.

(1) Let M be an abelian group and $V = \bigoplus_{\alpha \in M} V_\alpha$, an M-graded space. The *associative algebra, namely, End V*, can be equipped with the induced M-grading, End $V = \oplus_{\alpha \in M} End_\alpha V$, where $End_\alpha V = \{a \in End V | a(V_s) \subseteq V_{s+\alpha}\}$. In particular, for $M = \mathbb{Z}_2$, we obtain the *associative superalgebra, End* $V = End_{\bar{0}} V \oplus End_{\bar{1}} V$.

In fact, if $G = G_{\bar{0}} \oplus G_{\bar{1}}$ is an associative superalgebra, then we can introduce a commutator on G by $[a, b] = ab - (-1)^{\alpha\beta} ba$, where $a \in G_\alpha$ and $b \in G_\beta$, $\alpha, \beta \in \{\bar{0}, \bar{1}\}$. Then G is a Lie superalgebra.

In other words, if A is an associative superalgebra, then the bracket,

$$[a, b] = ab - (-1)^{(dega)(degb)} ba$$

turns A into a Lie superalgebra. As *EndV* is associative, *EndV* equipped with $\mathbb{Z}_2$-grading, denoted as $(EndV)_L$ is an associative superalgebra, which becomes a *Lie superalgebra arising from this gradation.*

(2) *Lie superalgebra arising from* $gl(V)$: Let $V = V_{\bar{0}} \oplus V_{\bar{1}}$ be a $\mathbb{Z}_2$-graded vector space. Consider the associative algebra $gl(V)$ of endomorphisms of V. It has a natural $\mathbb{Z}_2$-grading:

$$gl(V)_{\bar{0}} = \{f \in gl(V) : f(V_{\bar{n}}) \subseteq V_{\bar{n}}, \bar{n} \in \mathbb{Z}_2\}$$

and

$$gl(V)_{\bar{1}} = \{f \in gl(V) : f(V_{\bar{n}}) \subseteq V_{\overline{n+1}}, \bar{n} \in \mathbb{Z}_2\}.$$

The Lie bracket here is defined as follows:

$$[x, y] = \begin{cases} xy - yx, & \text{if} \quad x \text{ or } y \in gl(V)_{\bar{0}} \\ xy + yx, & \text{if} \quad x, y \in gl(V)_{\bar{1}}. \end{cases}$$

Let $v_1, v_2, \ldots, v_n$ be a basis of V with $v_1, v_2, \ldots, v_m \in V_{\bar{0}}$ and $v_{m+1}, v_{m+2}, \ldots, v_n \in V_{\bar{1}}$. Then any $x \in gl(V)$ can be written as $x = \begin{pmatrix} x_{0,0} & x_{0,1} \\ x_{1,0} & x_{1,1} \end{pmatrix}$ with respect to the above homogeneous basis. Now the supertrace of x is $str(x) = tr(x_{0,0}) - tr(x_{1,1})$.

(3) Any Lie algebra $G = G_{\bar{0}} \oplus G_{\bar{1}}$ is a Lie superalgebra with $G_{\bar{0}} = G$ and $G_{\bar{1}} = 0$. That is, a Lie algebra is a Lie superalgebra with trivial odd part.

(4) If $\mathcal{G}$ is a Lie algebra and V is a $\mathcal{G}$-module then taking $\mathcal{G}_{\bar{0}} = \mathcal{G}; \mathcal{G}_{\bar{1}} = V$, we get a Lie superalgebra $\mathcal{G} = \mathcal{G}_{\bar{0}} \oplus \mathcal{G}_{\bar{1}}$. Here, we define zero multiplication in $\mathcal{G}_{\bar{1}}$ and the multiplication of other elements by those in $\mathcal{G}$ by the action of $\mathcal{G}$ on V.

(5) The most obvious example of a Lie superalgebra is the set of all linear maps on a $\mathbb{Z}_2$-graded vector space.

(6) *Lie superalgebra arising from Grassmann superalgebra* $\wedge(n)$ [34]: First we define Grassman algebra $\wedge(n)$ of order n. The real (resp. complex) Grassman algebra is the algebra over $\mathbb{R}$ (resp. $\mathbb{C}$) generated by unit element 1 and n quantities θ_i (Grassman variables) satisfying the following anti-commutation relations:

$$\{\theta_i, \theta_j\} = 0.$$

This algebra has 2^n generators $1, \theta_i, \theta_i\theta_j, \theta_i\theta_j\theta_k, \ldots, \theta_1, \ldots, \theta_n$.

Denote by $\wedge(n)$, the *Grassmann algebra* in n variables $\xi_1, \ldots, \xi_n$. If we set $deg\xi_i = \bar{1}, i = 1, \ldots, n$, $\wedge(n)$ becomes $\mathbb{Z}_2$-graded Grassmann algebra, called *Grassmann superalgebra.* This Grassmann superalgebra satisfies commutative and associative properties. Moreover, $\wedge(m) \otimes \wedge(n) = \wedge(m+n)$. We can generalize this $\wedge(n)$ to $\wedge(m,n) = k[x_1, \ldots, x_m] \otimes \wedge(n)$. Here $k[x_1, \ldots, x_m]$ *is a superalgebra with trivial* $\mathbb{Z}_2$*-grading polynomial algebra.*

Let $G(V)$ be the Grassman algebra defined from vector space V over the field K with Z_2-grading $G(V) = G_{\overline{0}} \oplus G_{\overline{1}}$. If this Z_2 grading satisfies $[x,y] = xy - (-1)^{\alpha\beta}yx$, defined on homogenous elements $x \in G_\alpha$ and $y \in G_\beta$ for $\{\overline{0}, \overline{1}\} = \{\alpha, \beta\}$. Then $G(V)$ is a commutative or abelian Lie superalgebra.

Remark 95.

(1) If $\mathcal{G}$ is a Lie superalgebra and $\wedge(n)$ is a Grassmann superalgebra, then $\mathcal{G} \otimes \wedge(n)$ is also a Lie superalgebra.

Superalgebras on differential forms

Definition 145 (Superalgebra $\Omega(n)$). [34] Let $\Lambda(n)$ be the Grassmann superalgebra on $\xi_1, \ldots, \xi_n$. We denote by $\Omega(n)$ the associative superalgebra over $\Lambda(n)$ with the generators $d\xi_1, \ldots, d\xi_n$ and the defining relations $d\xi_i \circ d\xi_j = d\xi_j \circ d\xi_i$ and $deg\ (d\xi_i) = \overline{0}, i,j = 1, \ldots, n$. It is clear that $\Omega(n)$ is commutative (in the sense of bracket); that is, in particular $\xi_i(d\xi_j) = (d\xi_j)\xi_i$. Every element $\psi \in \Omega(n)$ can be written uniquely as a sum of elements of the form $\psi_k = \sum_{i_1 \leq \cdots \leq i_k} a_{i_1,\ldots,i_k}\ d\xi_{i_1}, \circ \cdots \circ d\xi_{i_k}$ where $a_{i_1,\ldots,i_k} \in \wedge(n)$. We define on $\Omega(n)$, the differential d as the derivation of $deg\overline{1}$ satisfying $d(\xi_i) = d\xi_i$, $d(d\xi_i) = 0, i = 1, \ldots, n$. This derivation exists and is unique. $\Omega(n)$ becomes a superalgebra called the superalgebra on differential forms with commutating differentials.

The differential d defined above has the following properties:

(a) $d(\phi \circ \psi) = d\phi \circ \psi + (-1)^{deg\phi}\phi \circ d\psi, \phi, \psi \in \Omega(n)$.
(b) $df = \sum_i \partial f/\partial \xi_i, f \in \wedge(n)$.
(c) $d^2 = 0$.
(d) Every derivation D of $\wedge(n)$ extends uniquely to a derivation $\tilde{D}$ of $\Omega(n)$ for which $[\tilde{D}, \mathrm{d}] = 0$.
(e) Every automorphism of $\wedge(n)$ extends uniquely to an automorphism of $\Omega(n)$ commuting with d.

For proof of the above properties, one can refer [34].

Definition 146 (*Superalgebra* $\Theta(n)$). [34] We denote by $\Theta(n)$, the associative superalgebra over $\wedge(n)$, with the generators $\theta\xi_1, \ldots, \theta\xi_n$ and defining relations,

$$\theta\xi_i \wedge \theta\xi_j = -\theta\xi_j \wedge \theta\xi_i, \quad deg\theta\xi_i = \overline{1}, i,j = 1, \ldots, n.$$

It is clear that $\Theta(n)$ is commutative (in the sense of bracket).

We have $\xi_i(\theta\xi_j) = -(\theta\xi_j)\xi_i$. Every element $w \in \Theta(n)$ can be written uniquely as a sum of elements of the form,

$$w_k = \sum_{i_1,\ldots,i_k} a_{i_1,\ldots,i_k}\theta\xi_{i_1} \wedge \cdots \wedge \theta\xi_{i_k}.$$

We define a differential θ on $\Theta(n)$ as the derivation of degree $\overline{0}$ for which $\theta(\xi_i) = \theta\xi_i$ and $\theta(\theta(\xi_i)) = 0, i = 1, 2, \ldots, n$.

This differential θ satisfies the following properties:

(1) $\theta(w_1 \wedge w_2) = \theta(w_1) \wedge w_2 + w_1 \wedge \theta(w_2)$.
(2) $\theta(f) = \sum_i (\theta\xi_i)(\partial f/\partial \xi_i), f \in \Lambda(n)$.
For more details, refer [34].

Consistently graded Lie superalgebras: Consistently graded Lie superalgebras can be defined in one of the following two ways. Starting from the definition of Z-graded Lie algebra, we can define consistently Z-graded Lie superalgebra as follows.

Definition 147 (Z-graded and consistently Z-graded Lie superalgebras). A Lie superalgebra $\mathcal{G}$ is called Z-graded if $\mathcal{G} = \bigoplus_{n\in Z} \mathcal{G}_n$ with $[\mathcal{G}_n, \mathcal{G}_m] \subset \mathcal{G}_{n+m}$ for all $n, m \in Z$, where $(\mathcal{G}_n)_{n\in Z}$ is a set of Z_2-graded subspaces of $\mathcal{G}$. This gradation $(\mathcal{G}_n)_{n\in Z}$ is called consistent with the Z_2-gradation of $\mathcal{G}$ if

$$\mathcal{G}_{\bar{0}} = \bigoplus_{n\in Z} \mathcal{G}_{2n} \; ; \; \mathcal{G}_{\bar{1}} = \bigoplus_{n\in Z} \mathcal{G}_{2n+1}.$$

Here the algebra $\mathcal{G}$ is called *consistently Z-graded.*

A Lie superalgebra with a Z-grading satisfying the above said equation, is called a *consistently graded Lie superalgebra.*

Remark 96. One can deduce from the definition (145), $[\mathcal{G}_0, \mathcal{G}_n] \subset \mathcal{G}_n$ for all $n \in Z$. Hence $\mathcal{G}_0$ is the graded subalgebra of $\mathcal{G}$ and adjoint representation of $\mathcal{G}$ will give a natural graded representation of $\mathcal{G}_0$ in $\mathcal{G}_n$.

Definition 148. (Z graded algebra and consistently Z-graded Lie superalgebras) Starting from the definition of Z-graded algebra, one can define Lie superalgebra and consistently graded Lie superalgebra as shown below. A *Z-graded Lie algebra* $\mathcal{G}$, is a direct sum, $\mathcal{G} = \oplus_{\alpha\in Z}\mathcal{G}_\alpha$ with a bilinear mapping $[\ ,\ ] : \mathcal{G} \times \mathcal{G} \to \mathcal{G}$ such that

(a) $[\mathcal{G}_\alpha, \mathcal{G}_\beta] \subseteq \mathcal{G}_{\alpha+\beta}$,
(b) $[a, b] = -(-1)^{\alpha\beta}[b, a]$,
(c) $(-1)^{\alpha\gamma}[a, [b, c]] + (-1)^{\alpha\beta}[b, [c, a]] + (-1)^{\beta\gamma}[c, [b, a]] = 0$

for all $a \in \mathcal{G}_\alpha$, $b \in \mathcal{G}_\beta$, $c \in \mathcal{G}_\gamma$ are true.

Now any Z-graded Lie algebra can be made into Lie superalgebra by setting

$$\mathcal{G}_0 = \bigoplus_{\alpha\in Z} \mathcal{G}_{2\alpha}, \quad \mathcal{G}_1 = \bigoplus_{\alpha\in Z} \mathcal{G}_{2\alpha+1}.$$

Any Lie superalgebra with Z-grading satisfying the above condition is called a *consistently graded Lie superalgebra.*

Remark 97. Using the definition of skew symmetric bilinear mapping of $V \times V$ into U (U and V are two Z_2-graded vector spaces) one can prove that the product mapping in a Lie superalgebra is skew symmetric.

Definition 149. [142] Let $V = V_{\bar{0}} \oplus V_{\bar{1}}$ be a Z_2-graded vector space with a nondegenerate skew-symmetric bilinear form $\langle ., . \rangle$. Then the vector space, $\mathcal{G} = V \oplus Cz$, spanned by V and another element z is called *Heisenberg superalgebra* if it satisfies,

$$[v_1, v_2] = < v_1, v_2 > z \; for \; v_1, v_2 \in V$$

and

$$[g, z] = 0 \; for \; all \; g \in \mathcal{G}.$$

Definition 150. [34] Let $\hat{\mathcal{G}} = \hat{\mathcal{G}}_{\bar{0}} \oplus \hat{\mathcal{G}}_{\bar{1}}$ be a Z_2-graded space. Let it be decomposed into a direct sum of Z-graded subspaces, $\hat{\mathcal{G}} = \mathcal{G}_{-1} \oplus \mathcal{G}_0 \oplus \mathcal{G}_1$. If $|i+j| \leq 1$, one can define a bilinear operation $\mathcal{G}_i \times \mathcal{G}_j \to \mathcal{G}_{i+j} : (x, y) \mapsto [x, y]$ satisfying the axiom of anticommutativity and the Jacobi identity for Lie superalgebras, provided that all the commutators in this identity are defined. Then $\mathcal{G}$ is called a *local Lie superalgebra*. To a Z-graded Lie superalgebra, $\mathcal{G} = \oplus_{i \in Z} \mathcal{G}_i$, there corresponds a *local Lie superalgebra*, $\mathcal{G}_{-1} \oplus \mathcal{G}_0 \oplus \mathcal{G}_1$, which we call the local part of $\mathcal{G}$.

Definition 151. (Filtration) A sequence of embedded Z_2-graded subspaces in a superalgebra $\mathcal{G} : \mathcal{G} = \mathcal{G}_{-1} \supset \mathcal{G}_0 \supset \mathcal{G}_1 \supset ....$ is called a filtration if $\mathcal{G}_i \mathcal{G}_j \subseteq \mathcal{G}_{i+j}$ and $\cap \mathcal{G}_i = 0, i, j \in \mathbb{Z}$.

A Lie algebra $\mathcal{G}$ with a filtration is called transitive if for any $a \in \mathcal{G}_i \backslash \mathcal{G}_{i+1}, i \geq 0$, there is an element $b \in \mathcal{G}$ for which $[a, b] \notin \mathcal{G}_i$. This condition can also be written as follows:

$$\mathcal{G}_i = \{a \in \mathcal{G}_{i-1} \mid [a, \mathcal{G}] \subseteq \mathcal{G}_{i-1}\}, i > 0.$$

Remark 98. For any Z-graded Lie superalgebra, one can define transitive gradation and bitransitive gradation as given below. Let $\mathcal{G} = \bigoplus_{n \in Z} \mathcal{G}_n$ be a Z-graded Lie superalgebra. Then

(i) the above gradation is called *transitive*, if $\{A \in \mathcal{G}_n | [A, \mathcal{G}_{-1}] = \{0\}\} = \{0\}$ for all $n \geq 0$ and
(ii) the above gradation is called *bitransitive* if in addition to the above condition, one should have $\{A \in \mathcal{G}_n | [A, \mathcal{G}_1] = \{0\}\} = \{0\}$ for all $n \leq 0$.

Definition 152. [34, 35] Let $\mathcal{G}$ be a superalgebra. A derivation D of degree s ($s \in Z_2$) of $\mathcal{G}$ is an endomorphism, $D \in End_s \mathcal{G}$ such that

$$D(ab) = D(a)b + (-1)^{s.dega} aD(b).$$

The space of all derivations of degree s is denoted by $der_s \mathcal{G} \subset End_s \mathcal{G}$. Let $der\mathcal{G} = der_{\bar{0}}\mathcal{G} \oplus der_{\bar{1}}\mathcal{G}$. Here $der_{\bar{0}}(\mathcal{G})$ is the set of all even derivations. $der_{\bar{1}}(\mathcal{G})$ is the set of all odd derivations and $der\mathcal{G}$ is called superalgebra of derivations of $\mathcal{G}$, which is a subalgebra of $(End\mathcal{G})_L$. Every element of $der\mathcal{G}$ is called *superderivation* of $\mathcal{G}$.

Alternatively, one can define superderivations, starting from superalgebra. Suppose that the vector space V is a superalgebra, $V = V_{\bar{0}} \oplus V_{\bar{1}}$. Let $D_\alpha(V), \alpha \in Z_2$, be the subspace of all $M \in l_\alpha(V)$ such that,

$$M(xy) = (Mx)y + (-1)^{\alpha\psi} x(My) \quad \text{for all } x \in V_\psi, y \in V; \psi \in Z_2.$$

Here $l_\alpha(V) = Hom(V)_\alpha = \{M \in Hom(V) | M(V_\beta) \subset V_{\alpha+\beta}, \beta \in Z_2\}$ for all $\alpha \in Z_2$. The algebra Hom(V) denotes the set of all k-linear mappings of V into V. If $\alpha = \bar{0}(resp. \alpha = \bar{1})$, the subspace $D_\alpha(V)$ consists of the even derivations (resp. of the odd derivations) of the superalgebra V. If we define,

$$D(V) = D_{\bar{0}}(V) \oplus D_{\bar{1}}(V),$$

then it is easy to see that $D(V)$ is a graded superalgebra of $l(V)$. Hence $D(V)$ becomes a Lie superalgebra. The elements of $D(V)$ are called superderivations of V.

Definition 153. For a Lie superalgebra $\mathcal{G}$, from the Jacobi identity one can directly verify that $ad\ a : b \to [a, b]$ is a derivation of $\mathcal{G}$. These derivations are called *inner*. Infact, $[D, ad\ a] = ad\ D\ a$ for $D \in der\mathcal{G}$, hence the set of all *inner derivations* is an ideal of $der\mathcal{G}$.

Definition 154. Let $V = V_{\bar{0}} + V_{\bar{1}}$ be a $\mathbb{Z}_2$-graded space with $\dim V_{\bar{0}} = m, \dim V_{\bar{1}} = n$. If we regard the same decomposition $V = V_{\bar{0}} \oplus V_{\bar{1}}$ as a $\mathbb{Z}$-grading of V, then it corresponds to a $\mathbb{Z}$-grading of $l(V)$, which is compatible with the $\mathbb{Z}_2$-grading, $l(V) = l_{-1} \oplus l(V)_{\bar{0}} \oplus l_1$. Let $e_1, \ldots, e_m, e_{m+1}, \ldots, e_{m+n}$ be a *basis of* V, formed from bases of $V_{\bar{0}}$ and $V_{\bar{1}}$. It is natural to call such a basis homogeneous.

In this basis, the *matrix of an operator* a from $l(V)$ can be written in the form $\begin{pmatrix} \alpha & \beta \\ \gamma & \delta \end{pmatrix}$, where α is an $(m \times m)$ matrix, δ is an $n \times n$ matrix, β is an $m \times n$ matrix and γ is an $n \times m$ matrix.

The matrices of even elements have the form $\begin{pmatrix} \alpha & 0 \\ 0 & \delta \end{pmatrix}$ and those of odd ones of the form $\begin{pmatrix} 0 & \beta \\ \gamma & 0 \end{pmatrix}$. Here l_1 consists of the matrices of the form $\begin{pmatrix} 0 & \beta \\ 0 & 0 \end{pmatrix}$ and l_{-1} consists of the matrices of the form $\begin{pmatrix} 0 & 0 \\ \gamma & 0 \end{pmatrix}$. Hence it is clear that the $l_{\bar{0}}$-modules l_1 and l_{-1} are contragradient and the $l_{\bar{0}}$-module l_1 is isomorphic to $gl_m \otimes gl_n$. On $l(V)$, we define *the supertrace, a linear function,* $str : l(V) \to k$ by $str(a) = tr\alpha - tr\delta$. One can easily observe that the supertrace of the matrix of an operator $a \in l(V)$ *does not depend on the choice of a homogeneous basis.* The bilinear form $(a, b) = str(ab)$ on $l(V)$ is (i) *consistent*, (ii) *supersymmetric*, and (iii) *invariant*.

Remark 99. In Kang and Kim [37], they define θ-colored Lie superalgebra using the Borcherds Cartan matrix and a bimultiplicative form using a coloring matrix. BCM matrix was already defined in Chapter 3. Below we define Borcherds Cartan matrix restricted to a coloring matrix.

4.2 Coloring matrices, θ-colored Lie superalgebras, and examples

For Definitions 155–157, one can refer [115].

Definition 155. Let $C = (c_{ij})_{i,j \in I}$ be a complex matrix satisfying $c_{ij}c_{ji} = 1$ for all $i,j \in I$. Therefore, we have $c_{ii} = \pm 1$ for all $i \in I$. We call $i \in I$, an even index, if $c_{ii} = 1$ and an odd index, if $c_{ii} = -1$. We denote by I^{even} (respectively I^{odd}) the set of all *even (respectively odd) indices.* A Borcherds-Cartan matrix $A = (a_{ij})_{i,j \in I}$ is *restricted (or colored)* with respect to C if it satisfies the following conditions: If $a_{ii} = 2$ and $c_{ii} = -1$, then a_{ij} are even integers for all $j \in I$. In this case, the matrix C is called a *coloring matrix* of A.

For this Borcherds-Cartan matrix A restricted to coloring matrix C, one can define root lattice in the usual sense.

Let $\mathfrak{h} = (\oplus_{i\in I}\mathbb{C}h_i) \oplus (\oplus_{i\in I}\mathbb{C}d_i)$ be a complex vector space with a basis $\{h_i, d_i; i \in I\}$. Then for each $i \in I$, define a linear functional $\alpha_i \in \mathfrak{h}^\star$ by,

$$\alpha_i(h_j) = a_{ji}, \text{ and } \alpha_i(d_j) = \delta_{ij} \text{ for all } j \in I. \tag{4.1}$$

Now, one can define the *root lattice* $Q = \sum_{i=1}^{n} \mathbb{Z}\alpha_i$ and the *positive root lattice* $Q_+ = \sum_{i=1}^{n} \mathbb{Z}_{\geq 0}\alpha_i$ Also $Q_- = Q\backslash Q_+$.

Example 43. Restricted to the coloring matrix, C= $\begin{pmatrix} -1 & c_1 & c_2 & c_3 \\ c_1^{-1} & 1 & c_4 & c_5 \\ c_2^{-1} & c_4^{-1} & 1 & c_6 \\ c_3^{-1} & c_5^{-1} & c_6^{-1} & -1 \end{pmatrix}$, a Borcherds Cartan matrix is $A = \begin{pmatrix} -k & -a & -b & -c \\ -a & 2 & -e & -d \\ -b & -e & 2 & -f \\ -c & -d & -f & 2 \end{pmatrix}$, where c, d, f are all even integers with $\psi = \{1, 4\}$.

Remark 100. If A is symmetrizable, then there exists a symmetric bilinear form $(.|.)$ on $\mathfrak{h}^\star$ satisfying $(\alpha_i|\alpha_j) = \epsilon_i a_{ij} = \epsilon_j a_{ji}$ for all $i, j \in I$.

Definition 156. The coloring matrix, $C = (c_{ij})_{i,j\in I}$ defines, a *bimultiplicative form* $\theta : Q \times Q \to \mathbb{C}^\times$ by

$$\begin{aligned} \theta(\alpha_i, \alpha_j) &= c_{ij} \text{ for all } i, j \in I, \\ \theta(\alpha + \beta, \gamma) &= \theta(\alpha, \gamma)\theta(\beta, \gamma), \\ \theta(\alpha, \beta + \gamma) &= \theta(\alpha, \beta)\theta(\alpha, \gamma) \end{aligned}$$

for all $\alpha, \beta, \gamma \in Q$.

Note that θ satisfies $\theta(\alpha, \beta)\theta(\beta, \alpha) = 1$ for all $\alpha, \beta \in Q$, since $c_{ij}c_{ji} = 1$ for all $i, j \in I$. In particular, $\theta(\alpha, \alpha) = \pm 1$, for all $\alpha \in Q$. We say that $\alpha \in Q$, is *even*, if $\theta(\alpha, \alpha) = 1$ and *odd*, if $\theta(\alpha, \alpha) = -1$.

Definition 157. A *θ-colored Lie superalgebra* is a Q-graded vector space $\mathcal{G} = \oplus_{\alpha\in Q}\mathcal{G}_\alpha$, together with a bilinear product $[\ ,\ ] : \mathcal{G} \times \mathcal{G} \to \mathcal{G}$ satisfying

$$\begin{aligned} &[\mathcal{G}_\alpha, \mathcal{G}_\beta] \subset \mathcal{G}_{\alpha+\beta}, \\ &[x, y] = -\theta(\alpha, \beta)[y, x], \\ &[x, [y, z]] = [[x, y], z] + \theta(\alpha, \beta)[y, [x, z]], \end{aligned}$$

for all $\alpha, \beta \in Q$ and $x \in \mathcal{G}_\alpha$, $y \in \mathcal{G}_\beta$, $z \in \mathcal{G}$. In a θ-colored Lie superalgebra $\mathcal{G} = \oplus_{\alpha\in Q}\mathcal{G}_\alpha$, for $x \in \mathcal{G}_\alpha$, we have $[x, x] = 0$, if α is *even* and $[x, [x, x]] = 0$, if α is *odd*.

Remark 101. In Example 43, bimultiplicative form is defined on the root lattice $Q = \sum_{i=1}^{n} \mathbb{Z}\alpha_i$ with values in the above coloring matrix. θ-colored Lie superalgebra is a Q-graded vector space, $L = \oplus_{\alpha\in Q}L_\alpha$, satisfying the required relations.

In the above definition of θ-colored Lie superalgebra, the bimultiplicative form θ is defined on root lattice Q with non-zero values in coloring matrix C. Instead of the root lattice Q and the coloring matrix C, if the bimultiplicative form θ is defined on Z_2 with values in $\{1, -1\}$, then one can get ordinary θ-colored Lie superalgebra without a coloring matrix. To distinguish between the above defined θ-colored Lie superalgebra and θ-colored Lie superalgebra without coloring matrix, we give below independently the definition of ϵ-colored Lie superalgebra [38] (without coloring matrix C).

Definition 158. Let K be a commutative ring with unity 1 and invertible elements 2,3 (here $2, 3 \in K^*$, the multiplicative group of invertible elements in K). Let A be an abelian additive group and let $\mathcal{G} = \bigoplus_{a \in A} \mathcal{G}_a$, where each $\mathcal{G}_a$ is a graded Lie algebra. Consider a bilinear alternating form $\epsilon : A \times A \to K^*$, satisfying

$$\epsilon(a+b, c) = \epsilon(a, c)\epsilon(b, c) \text{ and } \epsilon(a, b+c) = \epsilon(a, b)\epsilon(a, c)$$

for $a, b, c \in A$ and $\epsilon(a, b) = \epsilon(b, a)^{-1}$.

If K is a ring without zero divisors then $\epsilon(a, a) = \pm 1$ and hence

$$A = A_+ \cup A_-, \text{ where } A_\pm = \{a \in A | \epsilon(a, a) = \pm 1\}.$$

(It is obvious that A_+ is a subgroup in A of index ≤ 2.)

The A-graded algebra, $\mathcal{G} = \bigoplus_{a \in A} \mathcal{G}_a$, is called an ϵ-colored Lie superalgebra if for any $g, h, k \in \mathcal{G}$, we have

$$[g, h] = -\epsilon(a, b)[h, g]$$

and

$$[[g, h], k] = [g, [h, k]] - \epsilon(a, b)[h, [g, k]].$$

In particular, if $A = Z_2$, and $\epsilon : Z_2 \times Z_2 \to \{1, -1\}$ with $\epsilon(0, 0) = \epsilon(0, 1) = \epsilon(1, 0) = 1$, $\epsilon(1, 1) = -1$, then $\mathcal{G}$ becomes an ordinary Lie superalgebra.

Remark 102. If A=0, we get an ordinary Lie algebra.

One can obtain a Colored Lie superalgebra by introducing a color commutator on a A-graded algebra $A = \oplus_{a \in A} A_a$ by setting

$$[x, y] = xy - \epsilon(x, y)yx.$$

Here we use $\epsilon(x, y)$ in place of $\epsilon(a, b)$, if x is homogeneous of degree a and y is homogeneous of degree b.

Example 44. Consider the extended-hyperbolic Borcherds-Cartan matrix $A = \begin{pmatrix} -k & -a & -4 \\ -a & 2 & -1 \\ -4 & -2 & 2 \end{pmatrix}$ and the corresponding coloring matrix $C = \begin{pmatrix} -1 & c_1 & c_2 \\ c_1^{-1} & 1 & c_3 \\ c_2^{-1} & c_3^{-1} & -1 \end{pmatrix}$

with $c_1, c_2, c_3 \in C^X$. Then the corresponding extended-hyperbolic θ-colored Borcherds superalgebra is $\mathfrak{g} = \mathfrak{g}(A, \underline{m}, C)$, which is defined in Chapter 5. One can directly verify the axioms required for θ-colored Lie superalgebra.

4.3 Subsuperalgebras, ideals of Lie superalgebras, abelian Lie superalgebras, solvable, and nilpotent Lie superalgebras

Definition 159. Let $\mathcal{G}$ be a Lie superalgebra. If a Z_2-graded vector space H of $\mathcal{G}$ satisfies $[H, H] \subset H$, then H is called a *Lie sub-superalgebra* of $\mathcal{G}$. A left ideal in a Lie superalgebra $\mathcal{G}$ is a subspace A of $\mathcal{G}$ such that $[\mathcal{G}, A] \subseteq A$. The left ideals are not assumed to be Z_2-graded subspaces. On the other hand, if a Z_2-graded vector subspace H of $\mathcal{G}$ satisfies $[H, \mathcal{G}] \subset H$, then H is called an *ideal* of $\mathcal{G}$.

Definition 160. An *abelian Lie superalgebra* is a Lie superalgebra in which the commutator is zero for all the elements.

Remark 103. Any abelian Lie algebra is a trivial example of abelian Lie superalgebra with odd parts being zero.

Definition 161. A Lie superalgebra $\mathcal{G}$ is said to be *solvable*, if for the derived series

$$[\mathcal{G}, \mathcal{G}] = \mathcal{G}^{(1)}, [\mathcal{G}^{(1)}, \mathcal{G}^{(1)}] = \mathcal{G}^{(2)}, \ldots, [\mathcal{G}^{(i-1)}, \mathcal{G}^{(i-1)}] = \mathcal{G}^{(i)},$$

there exists an integer r such that $\mathcal{G}^{(r)} = 0$.

Remark 104. Let $\mathcal{G} = \mathcal{G}_{\bar{0}} \oplus \mathcal{G}_{\bar{1}}$ be a superalgebra with series of ideals satisfying

$$\mathcal{G} = H_0 \supset H_1 \supset \cdots \supset H_{l-1} \supset H_1 = \{0\}$$

such that the quotient H_i/H_{i+1} is abelian for $i = 0, 1, \ldots, l-1$. Then $\mathcal{G}$ is a solvable Lie superalgebra.

For proof of the following two theorems, one can see [35].

Theorem 43. *The Lie superalgebra $\mathcal{G} = \mathcal{G}_{\bar{0}} \oplus \mathcal{G}_{\bar{1}}$ is solvable if and only if $\mathcal{G}_{\bar{0}}$ is solvable.*

Remark 105. The above theorem is also true, in particular for Lie superalgebra of characteristic zero.

Theorem 44. *Finite dimensional Lie superalgebra, $\mathcal{G} = \mathcal{G}_{\bar{0}} \oplus \mathcal{G}_{\bar{1}}$ over a field K of characteristic zero is solvable if and only if $\mathcal{G}_{\bar{0}}$ is a solvable Lie algebra.*

Definition 162. A Lie superalgebra $\mathcal{G}$ is called *nilpotent* if for (lower central series)

$$[\mathcal{G}, \mathcal{G}] = \mathcal{G}^{[1]}, [\mathcal{G}, \mathcal{G}^{[1]}] = \mathcal{G}^{[2]}, \ldots, [\mathcal{G}, \mathcal{G}^{[i-1]}] = \mathcal{G}^{[i]},$$

there exists an integer r such that $\mathcal{G}^{r+1} = 0$.

Remark 106. The set of all strictly upper triangular matrices of any finite order is an example of nilpotent Lie algebra. Trivially as every Lie algebra is a Lie superalgebra with odd parts being zero, the set of all strictly upper triangular matrices of finite order is a trivial example of nilpotent Lie superalgebra. Similarly, set of all upper triangular matrices is a trivial example of solvable Lie superalgebra.

4.4 General linear Lie superalgebras

Definition 163. A linear map $\phi : \mathcal{G} \longrightarrow \mathcal{G}'$ between two Lie superalgebras, $\mathcal{G} = \mathcal{G}_{\overline{0}} \oplus \mathcal{G}_{\overline{1}}$ and $\mathcal{G}' = \mathcal{G}'_{\overline{0}} \oplus \mathcal{G}'_{\overline{1}}$ satisfying $\phi(\mathcal{G}_i) \subset \mathcal{G}'_i$ for $i = \overline{0}$ *or* $\overline{1}$ and $\phi([x, y]) = [\phi(x), \phi(y)]$ is called a *homomorphism of Lie superalgebras*. If in addition ϕ is also bijective it is called an *isomorphism of Lie superalgebras*.

Definition 164. Consider an associative superalgebra, $S = S_{\overline{0}} \oplus S_{\overline{1}}$. We can define a multiplication [,] on S by

$$[A, B] = AB - (-1)^{\alpha\beta}BA,$$

for all $A \in S_\alpha, B \in S_\beta, \alpha, \beta \in Z_2$. With this multiplication, S becomes a *Lie superalgebra* denoted by $\tilde{S}$(associated with S).

Definition 165 (The general linear Lie superalgebra $pl(V)$). Let $V = V_{\overline{0}} \oplus V_{\overline{1}}$ be a $\mathbb{Z}_2$-graded vector space and $Hom(V)$ be the set of all K-linear mappings of V into itself. In $Hom(V)$, define a Z_2-gradation by

$$Hom(V)_\alpha = \{A \in Hom(V) \mid A(V_\beta) \subset V_{\alpha+\beta}, \beta \in \mathbb{Z}_2\}, \forall\, \alpha \in \mathbb{Z}_2.$$

Then $Hom(V)$ becomes an associative superalgebra. For each $\alpha \in Z_2$, $Hom(V)_\alpha$ consists of linear mappings of V into itself which are homogeneous of degree α. The Lie superalgebra associated with the associative superalgebra, $Hom(V)$ (by introducing the bracket [,] as defined above) is denoted as $pl(V)$ called *the general linear Lie superalgebra of* V. Just like the $gl(V)$ in ordinary Lie algebra, $pl(V)$ plays the same role in the Lie superalgebra.

Remark. In particular, any Lie superalgebra $\mathcal{G}$ is a sub-superalgebra of $gl(\mathcal{G})$ via the adjoint action "ad," where $\mathcal{G}$ is considered as a $\mathbb{Z}_2$-graded vector space. Adjoint representation of Lie superalgebras will be defined in Section 4.6.

4.5 Simple and semisimple Lie superalgebras and bilinear forms

Definition 166. If a Lie superalgebra $\mathcal{G}$ has no graded ideals different from $\{0\}$ and $\mathcal{G}$ and $[\mathcal{G}, \mathcal{G}] \neq \{0\}$, then $\mathcal{G}$ is called *simple*. In other words, the Lie superalgebra $\mathcal{G}$ is called simple, if it is not abelian and the only Z_2-graded ideals of $\mathcal{G}$ are $\{0\}$ and $\mathcal{G}$. A non-abelian Lie superalgebra $\mathcal{G} = \mathcal{G}_{\bar{0}} \oplus \mathcal{G}_{\bar{1}}$ is simple, if it does not contain any non-trivial ideal.

The following results can be directly proved. (For proof, one can refer [35].)

Theorem 45. *Let $\mathcal{G}$ be a Lie superalgebra such that $\mathcal{G}_0 \neq 0 \neq \mathcal{G}_1$. Then $\mathcal{G}$ is simple if and only if the following conditions are true.*

(a) *If A is a non-zero $\mathcal{G}_0$-submodule of $\mathcal{G}_1$ such that $[\mathcal{G}_1, [\mathcal{G}_1, A]] \subseteq A$, then $[\mathcal{G}_1, A] = \mathcal{G}_0$.*
(b) *$\mathcal{G}_1$ is a faithful $\mathcal{G}_0$-module under the adjoint action.*
(c) $[\mathcal{G}_0, \mathcal{G}_1] = \mathcal{G}_1$.

Remark 107.

(1) If $\mathcal{G}$ is a Lie superalgebra, all of whose finite dimensional representations are completely reducible, then $\mathcal{G}$ is direct sum of simple Lie algebras.
(2) The only left ideals in a simple Lie superalgebra $\mathcal{G}$ are $\{0\}$ and $\mathcal{G}$.
For more details on simple Lie superalgebras, one can also refer [127].

Definition 167. Let $\mathcal{G}$ be a Lie superalgebra. For a *C-valued bilinear form* $\langle .,. \rangle$ on $\mathcal{G}$, one can define the following notions:

(a) *super-invariance:* $\langle [x, y], z \rangle = \langle x, [y, z] \rangle$.
(b) *super-symmetry:* $\langle y, x \rangle = (-1)^{|x||y|} \langle x, y \rangle$.
(c) *even:* $\langle \mathcal{G}_{\bar{0}}, \mathcal{G}_{\bar{1}} \rangle = 0$.
(d) *odd:* $\langle \mathcal{G}_{\bar{0}}, \mathcal{G}_{\bar{0}} \rangle = \langle \mathcal{G}_{\bar{1}}, \mathcal{G}_{\bar{1}} \rangle = \{0\}$.

If the above defined bilinear form on a Lie superalgebra satisfies super-invariance, super-symmetry, even(odd) property, then one can define the Radical of the bilinear form as,

$$Rad\langle .,. \rangle = \{x \in \mathcal{G} | < x, y >= 0 \text{ for any } y \in \mathcal{G}\}.$$

This is an ideal of $\mathcal{G}$.
Some results on invariant bilinear forms: Let $\mathcal{G}$ be a simple Lie super algebra.

(1) Any invariant bilinear form on $\mathcal{G}$ is either nondegenerate or equal to zero.
(2) Any invariant bilinear form on $\mathcal{G}$ is supersymmetric, that is, $\langle a, b \rangle = (-1)^{\bar{a}\bar{b}} \langle b, a \rangle$ for $a, b \in \mathcal{G}$.
(3) Any two non-zero invariant bilinear forms on $\mathcal{G}$ are proportional.
(4) The invariant bilinear forms on $\mathcal{G}$ are either all odd or all even.

Remark 108. Just like semisimple Lie algebras, there are infinite number of families of semisimple Lie superalgebras and finite number of exceptional Lie superalgebras.

Definition 168. A Lie superalgebra $\mathcal{G}$ is called semisimple if it does not contain any non-trivial solvable ideal.
Some results on simple and semisimple Lie superalgebras:

(1) A necessary condition for a Lie superalgebra, $\mathcal{G} = \mathcal{G}_{\bar{0}} \oplus \mathcal{G}_{\bar{1}}$ to be simple, is that the representation of $\mathcal{G}_{\bar{0}}$ on $\mathcal{G}_{\bar{1}}$ is faithful and $[\mathcal{G}_{\bar{1}}, \mathcal{G}_{\bar{1}}] = \mathcal{G}_{\bar{0}}$.
(2) For a Lie superalgebra $\mathcal{G} = \mathcal{G}_{\bar{0}} \oplus \mathcal{G}_{\bar{1}}$, the representation of $\mathcal{G}_{\bar{0}}$ on $\mathcal{G}_{\bar{1}}$ is irreducible then $\mathcal{G}$ is simple.
(3) Any semisimple Lie algebra can be written as direct sum of simple Lie superalgebras, whereas any semisimple Lie superalgebra can not be written as direct sum of simple Lie superalgebras.
(4) If $\mathcal{G}$ is a Lie superalgebra and I is the maximal solvable ideal, then $\mathcal{G}/I$ is a semisimple Lie superalgebra.

For proof of the following proposition, one can refer to [34].

Proposition 31. *Let* $\mathcal{G} = \mathcal{G}_{-1} \oplus \mathcal{G} \oplus \mathcal{G}_1$, *be a local Lie superalgebra. Then there is a maximal and minimal Z-graded Lie superalgebra whose local parts are isomorphic to* $\mathcal{G}$.

4.6 Representations of Lie superalgebras

Definition 169. Let $V = V_{\bar{0}} \oplus V_{\bar{1}}$ be a $\mathbb{Z}_2$-graded linear space. A *linear representation* ρ of a Lie superalgebra, $\mathcal{G} = \mathcal{G}_{\bar{0}} \oplus \mathcal{G}_{\bar{1}}$ in V is a homomorphism, $\rho : \mathcal{G} \to l(V)$. We say that V is a $\mathcal{G}$-module. Instead of $\rho(g)(v)$, one can write $g(v)$ for $g \in \mathcal{G}, v \in V$. By definition, $\mathcal{G}_i(V_j) \subseteq V_{i+j}, i, j \in \mathbb{Z}_2$ and $[g_1, g_2](v) = g_1(g_2(v)) - (-1)^{(deg\ g_1)(deg\ g_2)} g_2(g_1(v))$. The map $ad : \mathcal{G} \to l(\mathcal{G})$ for which $(adg)(a) = [g, a]$ is a linear representation of $\mathcal{G}$. It is called the *adjoint representation*. Moreover, $str([a, b]) = 0$ for any $a, b \in l(v)$.

Definition 170. [35] Let $\mathcal{G} = \mathcal{G}_{\bar{0}} \oplus \mathcal{G}_{\bar{1}}$ be a Lie superalgebra. Consider the adjoint representation 'ad'. Then it induces a *representation of the Lie algebra $\mathcal{G}_{\bar{0}}$ in the odd subspace $\mathcal{G}_{\bar{1}}$*. This representation is called the adjoint representation of $\mathcal{G}_{\bar{0}}$ in $\mathcal{G}_{\bar{1}}$ and is denoted by ad'. As for a Lie superalgebra, $\mathcal{G} = \mathcal{G}_{\bar{0}} \oplus \mathcal{G}_{\bar{1}}$, $\mathcal{G}_{\bar{0}}$ is a Lie algebra, the restriction of ad to $\mathcal{G}_{\bar{0}}$ is a representation of $\mathcal{G}_{\bar{0}}$ in the vector space $\mathcal{G}$.
Here the adjoint action "ad" means that for each element x of $\mathcal{G}$, ad(x) is an element of $gl(\mathcal{G})$ defined by

$$\mathrm{ad}(x)(y) := [x, y] \quad (y \in \mathcal{G}).$$

Definition 171. [35] (Graded Jacobi identity and adjoint representation). Let $\mathcal{G}$ be a Lie superalgebra, $\mathcal{G} = \mathcal{G}_{\bar{0}} \oplus \mathcal{G}_{\bar{1}}$. For $a \in \mathcal{G}$, we define a linear mapping,

$$ad\ a : \mathcal{G} \longrightarrow \mathcal{G}$$

by

$$(ad\ a)(b) = [a, b]\ for\ all\ b \in \mathcal{G}.$$

Taking the graded skew symmetry as $[a, b] = ab - (-1)^{\alpha\beta} ba$, the *graded Jacobi identity* is equivalent to each of the following requirements:

(1) ad is a homomorphism of the superalgebra $\mathcal{G}$ into the Lie superalgebra $l(\mathcal{G}) = pl(\mathcal{G})$. This homomorphism is called the adjoint representation of $\mathcal{G}$.
(2) "ad a" is a superderivation of $\mathcal{G}$, for all $a \in \mathcal{G}$.

From these results, we see that "ad" is a homomorphism of the Lie superalgebra $\mathcal{G}$ into the Lie superalgebra $D(\mathcal{G})$, the set of derivations of $\mathcal{G}$. The superderivations of $\mathcal{G}$ which are of the form $ad\ a$, $a \in \mathcal{G}$, are called *inner*.

Definition 172. If $\mathcal{G}$ is a Lie superalgebra and V is a graded $\mathcal{G}$-module, then $x \in V$ is called an invariant with respect to given representation of $\mathcal{G}$ in V, if it satisfies $A(x) = 0$ for all $A \in \mathcal{G}$. We denote by $V^{\mathcal{G}}$, the set of all $\mathcal{G}$-invariant elements of V. An element of V is $\mathcal{G}$-invariant, if and only if, all its homogenous components are $\mathcal{G}$-invariant. So $V^{\mathcal{G}}$ is the graded vector space of V.

Remark 109. If $\mathcal{G}$ is a Lie superalgebra and V is a graded $\mathcal{G}$-module and if x is a homogenous element of V, then $H = \{A \in \mathcal{G} | A(x) = 0\}$ is a graded subalgebra of $\mathcal{G}$ and x is H-invariant. Hence we call H, the graded subalgebra of $\mathcal{G}$ consisting of elements of $\mathcal{G}$ which leave x invariant.

Definition 173. Let $\mathcal{G}$ be a Lie superalgebra and W be a Z_2-graded vector space. The set Hom(W) has a natural Z_2-gradation which makes it an associative superalgebra. An even linear mapping $\gamma : \mathcal{G} \to Hom(W)$ is called a graded representation of $\mathcal{G}$ in W, if

$$\gamma([X,Y]) = \gamma(X)\gamma(Y) - (-1)^{\alpha\beta}\gamma(Y)\gamma(X),$$

for all $X \in \mathcal{G}_\alpha, Y \in \mathcal{G}_\beta; \alpha, \beta \in Z_2$. Naturally, if a Z_2-graded vector space is equipped with *Z_2-graded representation* of $\mathcal{G}$, then it is called a (left) *Z_2-graded $\mathcal{G}$-module.*

Definition 174. Let $\mathcal{G} = \bigoplus_{n\in Z} \mathcal{G}_n$ be a Z-graded Lie superalgebra. The gradation, $\bigoplus_{n\in Z} \mathcal{G}_n$ is *irreducible* if the graded representation of $\mathcal{G}_0$ in $\mathcal{G}_{-1}$ is irreducible.

Let $\mathcal{G} = \mathcal{G}_{\bar{0}} \oplus \mathcal{G}_{\bar{1}}$ be a basic Lie superalgebra with Cartan subalgebra $\mathfrak{h}$ and $\mathfrak{h}^*$, the dual of $\mathfrak{h}$. Let $\mathcal{G} = \eta^+ \oplus \mathfrak{h} \oplus \eta^-$ be a Borel decomposition, where η^+ and η^- are subalgebras such that $[\mathfrak{h}, \eta^+] \subset \eta^+$ and $[\mathfrak{h}, \eta^-] \subset \eta^-$ with dim $\eta^+ =$ dim η^-. Here η^+(resp. η^-) is spanned by the positive(resp.negative) root generators of $\mathcal{G}$.

Definition 175. A representation $\pi : \mathcal{G} \to EndV$ with representation space V is called a *highest weight representation* with highest weight $\Lambda \in \mathfrak{h}^*$ if (in the usual notations) there exists a non-zero vector $v_\Lambda \in V$ such that

$$\eta^+.v_\Lambda = 0$$
$$h \cdot v_\Lambda = \Lambda(h)v_\Lambda \ (h \in \mathfrak{h}).$$

The $\mathcal{G}$-module V is called a highest weight module, denoted by $V(\Lambda)$ and the vector $v_\Lambda \in V$, a highest weight vector.

Definition 176. Let $\mathcal{G}$ be a basic Lie algebra and K be a subalgebra of $\mathcal{G}$. Denote by $U(\mathcal{G})$ and $U(K)$, the corresponding universal enveloping algebras. From a K-module V, it is possible to construct a $\mathcal{G}$-module in the following way. The vector space V is naturally extended to a $U(K)$-module. One considers the factor space $U(\mathcal{G}) \otimes_{U(K)} V$ consisting of elements of $U(\mathcal{G}) \otimes V$ such that the elements $h \otimes v$ and $1 \otimes h(v)$ have been identified for $h \in K$ and $v \in V$. This space acquires the structure of a $\mathcal{G}$-module by setting $g(u \otimes v) = gu \otimes v$ for $u \in U(\mathcal{G}), g \in \mathcal{G}$, and $v \in V$. The $\mathcal{G}$-module $U(\mathcal{G}) \otimes_{U(K)} V$ is called the *induced module* from the K-module V and denoted by $Ind_K^{\mathcal{G}} V$.

Definition 177. Let $\mathcal{G} = \mathcal{G}_{\bar{0}} \oplus \mathcal{G}_{\bar{1}}$ be a basic Lie superalgebra with Cartan subalgebra $\mathfrak{h}$. Consider $V(\Lambda)$, a highest weight representation of $\mathcal{G}$ with highest weight Λ. The weight decomposition of V with respect to $\mathfrak{h}$ is

$$V(\Lambda) = \oplus_\lambda V_\lambda \ where \ V_\lambda = \{v \in V | h(v) = \lambda(h)v, h \in \mathfrak{h}\}.$$

Let e^λ be the formal exponential function on $\mathfrak{h}^*$(dual of $\mathfrak{h}$) such that $e^\lambda(\mu) = \delta_{\lambda,\mu}$ for two elements $\lambda, \mu \in \mathfrak{h}^*$, which satisfies $e^\lambda e^\mu = e^{\lambda+\mu}$. The *character and supercharacter* of $V(\lambda)$ are defined by

$$ch(V(\lambda)) = \sum_{\lambda}(\dim V_\lambda)e^{\lambda} \text{ and}$$
$$sch(V(\lambda)) = \sum_{\lambda}(-1)^{deg\lambda}(\dim V_\lambda)e^{\lambda}$$

respectively.

4.7 Different classes of classical Lie superalgebras

Definition 178. A finite dimensional Lie superalgebra, $\mathcal{G} = \mathcal{G}_{\overline{0}} \oplus \mathcal{G}_{\overline{1}}$, is called *classical*, if it is simple and the representation of $\mathcal{G}_{\overline{0}}$ on $\mathcal{G}_{\overline{1}}$ is completely reducible.

Below we state some results on classical Lie superalgebras.

Theorem 46. [129]

(1) *A simple Lie superalgebra $\mathcal{G} = \mathcal{G}_{\overline{0}} \oplus \mathcal{G}_{\overline{1}}$ is classical, if and only if, its even part $\mathcal{G}_{\overline{0}}$ is a reductive Lie algebra.*

(2) *If $\mathcal{G} = \mathcal{G}_{\overline{0}} \oplus \mathcal{G}_{\overline{1}}$ is a classical Lie superalgebra, then the representation of $\mathcal{G}_{\overline{0}}$ on $\mathcal{G}_{\overline{1}}$ is one of the following two types:*

Type I:The representation of $\mathcal{G}_{\overline{0}}$ on $\mathcal{G}_{\overline{1}}$ is the direct sum of two irreducible representations of $\mathcal{G}_{\overline{0}}$. Here we have, $\mathcal{G}_{\overline{1}} = \mathcal{G}_{-1} \oplus \mathcal{G}_1$ with

$$[\mathcal{G}_{-1}, \mathcal{G}_1] = \mathcal{G}_{\overline{0}} \text{ and } [\mathcal{G}_1, \mathcal{G}_1] = [\mathcal{G}_{-1}, \mathcal{G}_{-1}] = 0.$$

Here the superalgebra $\mathcal{G}$ is of Type I.

Type II:The representation of $\mathcal{G}_{\overline{0}}$ on $\mathcal{G}_{\overline{1}}$ is irreducible. Superalgebra here is said to be of type II.

Remark 110. The tables showing Types I and II, classical Lie superalgebras will be given after the definitions of all types of classical Lie superalgebras.

Classical Lie superalgebras [34]:

(1) *Lie superalgebra $A(m,n)$*: We have the following definition of l(V) or l(m, n). Let $V = V_{\overline{0}} \oplus V_{\overline{1}}$ be a Z_2-graded space with $\dim V_{\overline{0}} = m$ and $\dim V_{\overline{1}} = n$. One can consider, the associative algebra $EndV = \oplus_{i \in \mathbb{Z}_2} End_i V$, where

$$End_i V = \{a \in EndV | aV_s \subseteq V_{i+s}\}, i, s \in Z_2 = \{\overline{0}, \overline{1}\}.$$

Now $EndV$ becomes an associative superalgebra. Here the bracket $[a,b] = ab - (-1)^{(dega)(degb)}ba$. This makes End V into a Lie superalgebra, denoted by l(V) or l(m,n). On l(V), one can define $str : l(V) \to k$ with the property $str([a,b]) = 0$ for $a, b \in l(V)$. Define

$$sl(m,n) = \{a \in l(m,n) | str\ a = 0\}.$$

This is an ideal of l(m,n) of co-dimension 1. The $\mathbb{Z}$-grading, $sl(m,n) = \mathcal{G}_{-1} \oplus sl(m,n)_{\overline{0}} \oplus \mathcal{G}_1$, in a homogeneous basis of V is as follows: $sl(m,n)_{\overline{0}}$ is the set of matrices of the form $\begin{pmatrix} \alpha & 0 \\ 0 & \delta \end{pmatrix}$, where $tr\alpha = tr\delta$. $\mathcal{G}_1$ and $\mathcal{G}_{-1}$ are the set of matrices of the form $\begin{pmatrix} 0 & \beta \\ 0 & 0 \end{pmatrix}$ and

$\begin{pmatrix} 0 & 0 \\ \gamma & 0 \end{pmatrix}$, respectively. Now $sl(m,n)$ contains the one-dimensional ideal consisting of the scalar matrices λI_{2n}. Clearly, sl(n,n) contains the one dimensional ideal consisting of the scalar matrices λI_{2n}.

The Lie superalgebra $sl(1,1)$ is *three dimensional* and nilpotent. Let

$$A(m,n) = sl(m+1, n+1) \; for \; m \neq n, m, n \geq 0$$

and

$$A(n,n) = sl(n+1, n+1)/\langle I_{2n+2} \rangle, n > 0.$$

Now, the Z-grading of $sl(m+1, n+1)$ will induce a Z-grading of $A(m,n) = \mathcal{G}_{-1} \oplus \mathcal{G}_0 \oplus \mathcal{G}_1$.

(2) *The Lie superalgebras $B(m,n), D(m,n)$ and $C(n)$*: Let $V = V_{\bar{0}} \oplus V_{\bar{1}}$ be a $\mathbb{Z}_2$-graded space with $\dim V_{\bar{0}} = m$ and $\dim V_{\bar{1}} = n$. We define a nondegenerate *consistent supersymmetric bilinear form* F on V, such that $V_{\bar{0}}$ and $V_{\bar{1}}$ are orthogonal and the restriction of F to $V_{\bar{0}}$ is symmetric and to $V_{\bar{1}}$ is skew symmetric (in particular $n = 2r$ even). In $l(m,n)$, we define the subalgebra $osp(m,n) = osp(m,n)_{\bar{0}} \oplus osp(m,n)_{\bar{1}}$, where $osp(m,n)_s = \{a \in l(m,n)_s | F(a(x), y) = -(-1)^{s(\deg x)} F(x, a(y))\}, s \in \mathbb{Z}_2$. Here $osp(m,n)$ becomes *an orthogonal-symplectic superalgebra.* If $n = 0$, $osp(m,0)$ becomes an orthogonal Lie algebra and if $m = 0$, $osp(0,n)$ becomes a symplectic Lie algebra. Below we find the *explicit matrix form of the elements of osp(m,n)*.

(i) $m = 2l+1$. Choosing a basis, the matrix of the form F can be written as

$$\left[\begin{array}{ccc|cc} 0 & I_l & 0 & & \\ I_l & 0 & 0 & & \\ 0 & 0 & 1 & & \\ \hline & & & 0 & I_r \\ & & & -I_r & 0 \end{array}\right].$$

Hence matrix in $osp(m,n)$ is of the form

$$\left[\begin{array}{ccc|cc} a & b & u & x & x_1 \\ c & -a^T & v & y & y_1 \\ -v^T & -u^T & 0 & z & z_1 \\ \hline y_1^T & x_1^T & z_1^T & d & e \\ -y^T & -x^T & -z^T & f & -d^T \end{array}\right].$$

Here a is any $(l \times l)$-matrix; b and c are skew-symmetric $(l \times l)$-matrices; d is any $(r \times r)$-matrix; e and f are symmetric $(r \times r)$-matrices; u and v are $(l \times 1)$-matrices; x, y are $(l \times r)$-matrices and z is $(r \times 1)$-matrix. We have, $osp(m,n)_{\bar{0}}$ is a Lie algebra of type $B_l \oplus C_r$. $osp(m,n)_{\bar{0}}$-module $osp(m,n)_{\bar{1}}$ is isomorphic to $so_m \otimes sp_n$.

(ii) $m = 2l$. Choosing a basis, the matrix of F can be written as

$$\left[\begin{array}{cc|cc} 0 & I_l & & \\ I_l & 0 & & \\ \hline & & 0 & I_r \\ & & -I_r & 0 \end{array}\right].$$

A matrix in $osp(m,n)$ is of the form $\left[\begin{array}{cc|cc} a & b & x & x_1 \\ c & -a^T & y & y_1 \\ \hline y_1^T & x_1^T & d & e \\ -y^T & -x^T & f & -d^T \end{array}\right]$ (this is of the same form as that of $osp(m,n)$ with $m = 2n+1$ and the middle row and column deleted). In particular, $osp(m,n)_{\bar{0}}$ for $l \geq 2$ is a Lie algebra of type $D_l \oplus C_r$. Moreover $osp(m,n)_{\bar{0}}$-module $osp(m,n)_{\bar{1}}$ is isomorphic to $so_m \otimes sp_n$. Using Cartan's notation, we write

B(m, n) = osp(2m + 1, 2n) for **m ≥ 0, n > 0**

D(m, n) = osp(2m , 2n) for **m ≥ 2, n > 0**

and

C(n) = osp(2, 2n-2) for **n > 2**.

Now, consider the Lie superalgebra **C(n)**. It has subalgebra $C(n)_{\bar{0}}$ consisting of the matrices of the form $\left(\begin{array}{cc|cc} \alpha & 0 & & \\ 0 & -\alpha & & \\ \hline & & a & b \\ & & c & -a^T \end{array}\right)$ where a, b, c are $(n-1) \times (n-1)$ matrices with b and c being symmetric and $\alpha \in k$.

Moreover, $C(n)$ has the consistent $\mathbb{Z}$-grading, namely $C(n) = \mathcal{G}_{-1} \oplus C(n)_{\bar{0}} \oplus \mathcal{G}_1$. Here $\mathcal{G}_{-1}$ and $\mathcal{G}_1$ are the matrices of the form

$$\left(\begin{array}{cc|cc} & & 0 & 0 \\ & & y & y_1 \\ \hline y_1^T & 0 & & \\ -y^T & 0 & & \end{array}\right), \left(\begin{array}{cc|cc} & & x & x_1 \\ & & 0 & 0 \\ \hline 0 & x_1^T & & \\ 0 & -x^T & & \end{array}\right)$$ respectively.

The representations of $C(n)_{\bar{0}}$ on $\mathcal{G}_{-1}$ and $\mathcal{G}_1$ are contragradient, and the $C(n)_{\bar{0}}$-module $\mathcal{G}_1$ is isomorphic to osp_{2n-2}.

(3) *The Lie superalgebra* $P(n)$, $n \geq 2$: This is a subalgebra of $sl(n+1, n+1)$ consisting of the matrices of the form, $\left(\begin{array}{c|c} a & b \\ \hline c & -a^T \end{array}\right)$, where $tr(a) = 0$, b, a symmetric matrix and c, a skew symmetric matrix.

(4) *The Lie superalgebra* $Q(n)$, $n \geq 2$: Let $\widetilde{Q}(n)$ denote a subalgebra of $sl(n+1, n+1)$, consisting of the matrices of the form $\left(\begin{array}{c|c} a & b \\ \hline b & a \end{array}\right)$, where $tr(b) = 0$. The Lie superalgebra, $\widetilde{Q}(n)$ has a one-dimensional center $C = \langle I_{2n+2} \rangle$. Let $Q(n) = \widetilde{Q}(n)/C$.

(5) *The Lie superalgebras* $F(4), G(3)$ *and* $D(2,1;\alpha)$: The proofs of the following results are in [34].

  (a) There is a one and only one 40-dimensional classical Lie superalgebra $F(4)$ for which $F(4)_{\bar{0}}$ is a Lie algebra of type $B_3 \oplus A_1$ and its representation on $F(4)_{\bar{1}}$ is $spin_7 \otimes sl_2$.

  (b) There is one and only one *31-dimensional classical Lie superalgebra* $G(3)$ for which $G(3)_{\bar{0}}$ is a Lie algebra of type $G_2 \oplus A_1$ and its representation on $G(3)_{\bar{1}}$ is $G_2 \otimes sl_2$.

  (c) There is a one-parameter family of 17-dimensional *Lie superalgebras* $D(2,1,\alpha)$, $(\alpha \in k^* \backslash \{0,-1\})$, consisting of all simple Lie superalgebras for which $D(2,1,\alpha)_{\bar{0}}$ is a Lie algebra of type $A_1 \oplus A_1 \oplus A_1$ and its representation on $D(2,1,\alpha)_{\bar{1}}$ is $sl_2 \otimes sl_2 \otimes sl_2$.

Definition 179. [34] (Construction of contragradient Lie superalgebras). Let $A = (a_{ij})_{i,j=1}^{r}$ be an $(r \times r)$-matrix with elements from a field K and τ be a subset of $I = \{1, 2, \ldots, r\}$. Consider vector spaces, $\mathcal{G}_{-1}, \mathcal{G}_0$ and $\mathcal{G}_1$(over K) with bases $\{f_i\}, \{h_i\}$ and $\{e_i\}, i \in I$ respectively. One can easily verify the following relations which determine the structure of a local Lie superalgebra $\hat{\mathcal{G}}(A,\tau)$ on the space $\mathcal{G}_{-1} \oplus \mathcal{G}_0 \oplus \mathcal{G}_1$:

$$
\begin{aligned}
&[e_i, f_j] = \delta_{ij} h_i, \quad [h_i, h_j] = 0, \\
&[h_i, e_j] = a_{ij} e_j, \quad [h_i, f_j] = -a_{ij} f_j, \\
&deg h_i = \bar{0}, \quad deg f_i = deg e_i = \bar{0} \, for \, i \notin \tau, \\
&deg f_i = deg e_i = \bar{1} \, for \, i \in \tau.
\end{aligned}
$$

There exists a minimal Z-graded Lie superalgebra $\mathcal{G}(A,\tau)$ with local part $\hat{\mathcal{G}}(A,\tau)$. We call $\mathcal{G}(A,\tau)$, *the contragradient Lie superalgebra*, where A is its Cartan matrix and r is its rank.

Classification of finite dimensional contragradient Lie superalgebras

We state the following propositions 32, 33, and theorem 47, which are in [34].

Proposition 32. *Let* $\mathcal{G}(A,\tau)$ *be a simple finite-dimensional contragradient Lie superalgebra and* C *be its center. Then,* $\mathcal{G}(A,\tau)/C$ *is simple, if and only if, for any* $i, j \in I$ *there exists a sequence* $i_1, \ldots, i_t \in I$ *for which* $a_{ii_1}\ a_{i_1 i_2}, \ldots, a_{i_t j} \neq 0$.

Proposition 33. *Let* $\mathcal{G}$ *be one of* $A(m,n), B(m,n), C(n), D(m,n), D(2,1;\alpha), F(4)$ *or* $G(3)$ *and let* $\mathcal{G} \simeq \mathcal{G}(A,\tau)/C$, *where* C *is the center. Then* $C \neq 0$ *only for* $A(n,n)$ *and in this case dim* $C = 1$.

Theorem 47. *Let* $\mathcal{G}(A,\tau)$ *be a finite-dimensional contragradient Lie superalgebra. Let the Cartan matrix,* $A = (a_{ij})_{i,j=1}^{r}$ *satisfy Proposition 32. Let* C *be its center. Then* $\mathcal{G}' = \mathcal{G}(A,\tau)/C$ *is classical and* (A,τ) *is equivalent either to one of the pairs listed in Proposition 33 or to* (A,ϕ), *where* A *is the Cartan matrix of a simple Lie algebra.*

4.7.1 Construction of contragradient Lie superalgebras from simple classical Lie superalgebras

Let $\mathcal{G}$ be the superalgebra of the form, $sl(m+1, n+1), B(m,n), C(n), D(m,n), D(2,1,\alpha)$, $F(4)$ *or* $G(3)$. If H is a Cartan subalgebra, Π is its set of simple roots and e_i and f_i are the corresponding non-zero vectors in $\mathcal{G}_{\alpha_i}$ and $\mathcal{G}_{-\alpha_i}$ respectively, for each $\alpha_i \in \Pi$, then clearly, $[e_i, f_i] = h_i$ will form a basis of H. Let $dege_i = -degf_i = 1, degh_i = 0$. Now define a Z-grading on $\mathcal{G}$. As $\mathcal{G}$ is simple modulo its center, $\mathcal{G}$ is the minimal $\mathbb{Z}$-graded Lie superalgebra with the local part $\mathcal{G}_{-1} \oplus \mathcal{G}_0 \oplus \mathcal{G}_1$. Thus one can introduce contragradient Lie superalgebras in $\mathcal{G}$. Cartan matrix of contragradient Lie superalgebra is $A = (\alpha_j(h_i))$, with $\tau = \{i \in I | \alpha_i \in \Delta_1\}$.

We now list all the resulting pairs (A, τ), up to equivalence:
Properties of classical Lie superalgebras [34]:

(a) All the Lie superalgebras $A(m,n), B(m,n), C(n), D(m,n), D(2,1;\alpha), F(4), G(3), P(n), Q(n)$ are classical.

(b) For the Lie superalgebras $B(m,n), D(m,n), D(2,1;\alpha), F(4), G(3)$ and $Q(n)$ the $\mathcal{G}_{\bar{0}}$-module $\mathcal{G}_{\bar{1}}$ is irreducible and isomorphic to the modules in the following list [34]:

$\mathcal{G}$	$\mathcal{G}_{\bar{0}}$	$\mathcal{G}_{\bar{0}} \vert \mathcal{G}_{\bar{1}}$
B(m,n)	$B_m \oplus C_n$	$so_{2m+1} \otimes sp_{2n}$
D(m,n)	$D_m \oplus C_n$	$so_{2m} \otimes sp_{2n}$
$D(2,1;\alpha)$	$A_1 \oplus A_1 \oplus A_1$	$sl_2 \otimes sl_2 \otimes sl_2$
F(4)	$B_3 \oplus A_1$	$spin_7 \otimes sl_2$
G(3)	$G_2 \oplus A_1$	$G_2 \otimes sl_2$
Q(n)	A_n	$adsl_{n+1}$

(c) The Lie superalgebras $A(m,n)$, $C(n)$, and $P(n)$ admit a unique consistent Z-grading of the form $\mathcal{G}_{-1} \oplus \mathcal{G}_0 \oplus \mathcal{G}_1$. Moreover, the $\mathcal{G}_0$-modules $\mathcal{G}_1$ and $\mathcal{G}_{-1}$ are irreducible. $A(m,n)$ and $C(n)$ are contragradient and they are isomorphic to the modules in the following list [34]:

$\mathcal{G}$	$\mathcal{G}_0$	$\mathcal{G}_0 \vert \mathcal{G}_{-1}$
A(m,n)	$A_m \oplus A_n \oplus k$	$sl_{m+1} \otimes sl_{n+1} \otimes k$
A(n,n)	$A_m \oplus A_n$	$sl_{n+1} \otimes sl_{n+1}$
C(n)	$C_{n-1} \oplus k$	osp_{2n-2}
P(n)	A_n	$\Lambda^2 sl^*_{n+1}$

For $C(n)$, $\mathcal{G}_{\bar{0}} | \mathcal{G}_{\bar{1}} = osp^*_{2n-1}$. For $p(n)$, $\mathcal{G}_{\bar{0}} | \mathcal{G}_{\bar{1}} = S^2 sl_{n+1}$.

4.8 Universal enveloping algebras of Lie superalgebras and θ-colored Lie superalgebras

Definition 180. Let $\mathcal{G} = \mathcal{G}_{\bar{0}} \oplus \mathcal{G}_{\bar{1}}$ be a Lie superalgebra. Considering $\mathcal{G}$ as a vector space over $K = R$ or C, one can construct $T(\mathcal{G})$, the tensor algebra of $\mathcal{G}$. The definition of the universal enveloping superalgebra $U(\mathcal{G})$ is similar to the definition in the algebraic case. Using the Z_2-gradation of $\mathcal{G}$, one can get Z_2-gradation of $T(\mathcal{G})$, so that $\mathcal{G} \longrightarrow T(\mathcal{G})$ is an even linear mapping. Hence, we get $T(\mathcal{G})$ as a superalgebra. In $T(\mathcal{G})$, consider the two sided ideal I generated by elements of the form

$$X \otimes Y - (-1)^{\alpha\beta} Y \otimes X - [X, Y],$$

where

$$X \in \mathcal{G}_\alpha, Y \in \mathcal{G}_\beta; \alpha, \beta \in Z_2.$$

These elements are homogeneous of degree $\alpha + \beta$. So I is a graded ideal. Hence, one can define

$$U(\mathcal{G}) = T(\mathcal{G})/I.$$

So $U(\mathcal{G})$ is an associative superalgebra, called *universal enveloping algebra* of $\mathcal{G}$. From the compositions of $\mathcal{G} \longrightarrow T(\mathcal{G})$ and $T(\mathcal{G}) \longrightarrow T(\mathcal{G})/I = U(\mathcal{G})$, one can obtain the canonical even linear mapping,

$$\sigma : \mathcal{G} \longrightarrow U(\mathcal{G})$$

satisfying,

$$\sigma([X, Y]) = \sigma(X)\sigma(Y) - (-1)^{\alpha\beta}\sigma(Y)\sigma(X), \text{ for all } X \in \mathcal{G}_\alpha, Y \in \mathcal{G}_\beta; \alpha, \beta \in Z_2.$$

Clearly, every element of $U(\mathcal{G})$ is a linear combination of the products of the form

$$\sigma(X_1) \cdots \sigma(X_n), \textit{ where } X_i \in \mathcal{G}_{\alpha_i}, \textit{ for } 1 \leq i \leq n \text{ and } \alpha_i \in Z_2.$$

Remark 111. If $n = 0$, one can define this product to be 1. Moreover, the product

$$\sigma(X_1) \cdots \sigma(X_n)$$

is a homogeneous element of $U(\mathcal{G})$ of degree $\alpha_1 + \cdots + \alpha_n$.

The universal property of $U(\mathcal{G})$: If $\mathcal{A}$ is an associative algebra with unit element and ϕ is a linear mapping of $\mathcal{G}$ into $\mathcal{A}$, such that,

$$\phi([X, Y]) = \phi(X)\phi(Y) - (-1)^{\alpha\beta}\phi(Y)\phi(X), \text{ for all } X \in \mathcal{G}_\alpha; Y \in \mathcal{G}_\beta; \alpha, \beta \in Z_2,$$

then there exists a unique homomorphism $\overline{\phi} : U(\mathcal{G}) \to \mathcal{A}$ such that $\phi = \overline{\phi} \circ \sigma$, with $\overline{\phi}(1) = 1$. We have,

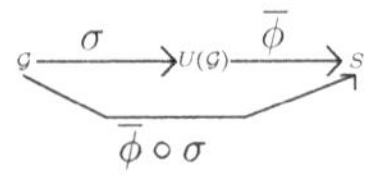

Homomorphisms between Lie superalgebras and their universal enveloping algebras:

(1) In particular, if $\mathcal{A}$ is a superalgebra and g is a homomorphism of degree 0, then $\overline{\phi}$ is also of degree 0.

(2) If $\mathcal{G} \overset{\sigma}{\to} U(\mathcal{G})$ and $\mathcal{G}' \overset{\sigma'}{\to} U(\mathcal{G}')$ are the canonical mappings and $\psi : \mathcal{G} \to \mathcal{G}'$ is a homomorphism, then there exists a unique homomorphism, $\overline{\psi} : U(\mathcal{G}) \to U(\mathcal{G}')$ such that $\sigma' \circ \psi = \overline{\psi} \circ \sigma$ and $\overline{\psi}(1) = 1$. We have,

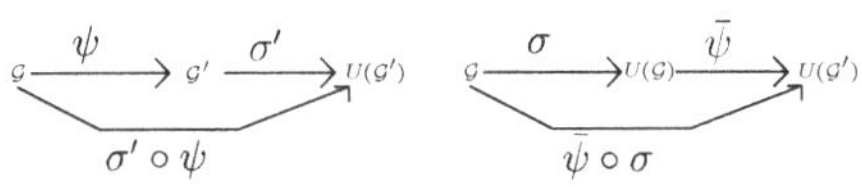

This implies

$$\mathcal{G} \overset{\sigma'\circ\psi}{\underset{\overline{\psi}\circ\sigma}{\longrightarrow}} U(\mathcal{G}').$$

If $U(\mathcal{G})$ is an enveloping algebra of $\mathcal{G}$ and $\sigma : \mathcal{G} \to U(\mathcal{G})$ is the canonical mapping, then one can identify $\mathcal{G}$ with a graded subspace of $U(\mathcal{G})$ using the above mapping σ. Here σ is an injection of $\mathcal{G}$ into $U(\mathcal{G})$. If ρ is a graded representation of $\mathcal{G}$ in some graded vector space W, then using the universal property of $U(\mathcal{G})$, one can get a unique homomorphism of associative superalgebras, $\overline{\rho} : U(\mathcal{G}) \to Hom(\mathcal{G})$, which is an extension of ρ such that $\overline{\rho}(X) = \rho(X)$, for all $X \in \mathcal{G}$ and $\overline{\rho}(1) = id$. Moreover, $\overline{\rho}(U(\mathcal{G})_\alpha)W_\beta \subset W_{\alpha+\beta}$ for $\alpha, \beta \in Z_2$. So, $\overline{\rho}$ is the graded representation of the associative superalgebra $U(\mathcal{G})$ in the graded vector space W or W is another Z_2-graded left $U(\mathcal{G})$-module. Conversely, if W is a Z_2 graded $U(\mathcal{G})$-module and $\omega : U(\mathcal{G}) \to Hom(W)$ is the corresponding homomorphism of associative superalgebras, then the restriction $\rho(\omega)$ to $\mathcal{G}$ is a graded representation of $\mathcal{G}$ in W. Hence a *graded* $\mathcal{G}$*-module* is same as *left graded* U($\mathcal{G}$)*-module*.

Definition 181. If $\mathcal{G}$ and $\mathcal{G}'$ are two Lie superalgebras and M and M' are the corresponding $\mathcal{G}$-modules, then one can consider M as a graded $U(\mathcal{G})$-module and M' as a graded $U(\mathcal{G}')$-module. Now $M \otimes M'$ has a natural structure of a graded $U(\mathcal{G})\overline{\otimes}U(\mathcal{G}')$-module. The Z_2-gradation in $M \otimes M'$ can be given by

$$(M \otimes M')_\alpha = \bigoplus_{\beta+\gamma=\alpha} (M_\beta \otimes M'_\gamma),$$

if $\alpha \in Z_2$. Moreover,

$$(P \otimes P')(q \otimes q') = (-1)^{\xi'\eta}(Pq) \otimes (P'q').$$

Here $P \in U(\mathcal{G})$, $P' \in U(\mathcal{G}')_{\xi'}$, $q \in M_\eta$, $q' \in M'$; ξ', $\eta \in Z_2$. Now $U(\mathcal{G}) \overline{\otimes} U(\mathcal{G}')$ is canonically isomorphic to universal enveloping algebra of $\mathcal{G} \times \mathcal{G}'$. We have $U(\mathcal{G})\overline{\otimes}U(\mathcal{G}') \cong U(\mathcal{G} \otimes \mathcal{G}') \cong$ the enveloping algebra of $\mathcal{G} \times \mathcal{G}'$. Moreover, $M \otimes M'$ has a natural graded $\mathcal{G} \times \mathcal{G}'$-module structure, namely

$$(x, x')(q \otimes q') = (xq) \otimes q' + (-1)^{\alpha'\eta} q \otimes (x'q'),$$

for all $x \in \mathcal{G}, x' \in \mathcal{G}'_\alpha, q \in M_\eta, q' \in M'; \alpha', \eta \in Z_2$. Now $\mathcal{G} \times \mathcal{G}'$-module $M \otimes M'$ is called the tensor product of the graded $\mathcal{G}$-module M with the graded $\mathcal{G}'$-module M'.

Example 45. Let A be an associative superalgebra with odd subspace of A being equal to zero. Then there exists a unique homomorphism ϵ of superalgebras ϵ : $U(\mathcal{G}) \to A$ such that $\epsilon \circ \sigma = 0$, $\epsilon(1) = 1$. One can identify canonically $U(\mathcal{G}) \otimes A$ and $A \otimes U(\mathcal{G})$ with $U(\mathcal{G})$ and get

$$(\epsilon \otimes id_U) \circ c = (id_U \otimes \epsilon) \circ c = id_U.$$

This homomorphism ϵ is called the *counit* of $U(\mathcal{G})$.

Now considering A as a vector space, it has a trivial structure of a graded $\mathcal{G}$-module. Consider Z_2 gradation of A defined by,

$$A_{\bar{0}} = A, \ A_{\bar{1}} = \{0\}.$$

Now the representation of $\mathcal{G}$ in A is equal to zero. If ϵ is the counit of U($\mathcal{G}$), then

$$P_A a = \epsilon(P) a \, for \, all \, P \in U(\mathcal{G}), \ a \in A.$$

Homomorphism of $\mathcal{G}$-modules to homomorphism of $U(\mathcal{G})$-modules [35]: A homomorphism of a graded $\mathcal{G}$-module M into a graded $\mathcal{G}$-module N is an even linear mapping,

$$g : M \to N$$

such that

$$g(ay) = ag(y) \, for \, all \, a \in \mathcal{G}, y \in M.$$

Hence,

$$g(Xy) = Xg(y) \, for \, all \, X \in U(\mathcal{G}), y \in M.$$

So, g is a homomorphism of graded $U(\mathcal{G})$-modules.

Definition 182 (Representations of product superalgebras). Let $\mathcal{G}$ be a Lie superalgebra and U($\mathcal{G}$) be the corresponding universal enveloping algebra. We can construct ϕ, an even linear mapping $\phi : U(\mathcal{G}) \to U(\mathcal{G})$ such that

$$\phi(XY) = (-1)^{\xi\eta}\phi(Y)\phi(X)$$
$$\text{for all } X \in U(\mathcal{G})_\xi,\ Y \in U(\mathcal{G})_\eta;\ \xi, \eta \in Z_2$$
$$\phi(\sigma(A)) = -\sigma(A), \text{for all } A \in \mathcal{G} \text{ and } \phi(1) = 1 \text{ with } \phi^2 = id.$$

Here σ is the canonical mapping from $\mathcal{G}$ to $U(\mathcal{G})$.

If $\mathcal{G}$ and $\mathcal{G}'$ are two Lie superalgebras with graded modules W and W', respectively, then we can consider W as a graded $U(\mathcal{G})$ module and W' as a graded $U(\mathcal{G}')$ module. The space of all k-linear mappings of W into W' is denoted by $Hom(W, W')$. It has a natural structure of a graded $U(\mathcal{G})\overline{\otimes}U(\mathcal{G}')$-module. A Z_2-gradation can be given as

$$Hom(W, W')_\alpha = \{h \in Hom(W, W')|h(W_\beta) \subset W'_{\alpha+\beta}, \beta \in Z_2\}$$

for all $\alpha \in Z_2$. Moreover, the module structure is

$$(X \otimes X')f = (-1)^{\xi(\xi'+\gamma)}X_{W'} \circ f \circ (\theta X)_W$$

for all $X \in U(\mathcal{G})_\xi, X' \in U(\mathcal{G}')'_\xi, f \in Hom(W, W')_\gamma; \xi, \xi', \gamma \in Z_2$.

The corresponding representation of $\mathcal{G} \times \mathcal{G}'$ can be given by

$$(a, a')f = a'_{W'} \circ f - (-1)^{\alpha\gamma} f \circ a_W$$

for all $a \in \mathcal{G}_\alpha, a' \in \mathcal{G}', f \in Hom(W, W')_\gamma; \alpha, \gamma \in Z_2$.

Remark 112.

(1) Every vector space has a trivial structure of a graded $\mathcal{G}$-module. In fact, if a Z_2-gradation of a vector space V is defined by $V_{\overline{0}} = V$ and $V_{\overline{1}} = 1$, then the representation of $\mathcal{G}$ in V is trivial.

(2) Using the general definitions of graded algebraic structure, a homomorphism of $\mathcal{G}$-graded module V into a $\mathcal{G}$-graded module W can be defined by an even linear mapping $f : V \to W$ such that $f(Ay) = Af(y)$ for all $A \in \mathcal{G}, y \in V$.

(3) The representation of H in $\mathcal{G} = \oplus_\lambda \mathcal{G}_\lambda(H)$ will be completely reducible. If this is the case, then we have, for all $\lambda \in H^*$,

$$\mathcal{G}_\lambda(H) = \{x \in \mathcal{G}|[h, x] = \lambda(h)x \, for\ all\ h \in \mathcal{G}\}.$$

Note that this remark applies if the Lie algebra in $\mathcal{G}_{\overline{1}}$ is completely reducible.

(4) Just like the definition of universal enveloping algebra of the Lie superalgebra, one can define universal enveloping algebra of θ-colored Lie superalgebra [37].

Definition 183. Let $\mathcal{G}$ be a θ-colored Lie superalgebra and $T(\mathcal{G})$ be the corresponding tensor algebra of $\mathcal{G}$. Let J be the ideal of $T(\mathcal{G})$ generated by the elements $[x, y] - x \otimes y + \theta(\alpha, \beta) y \otimes x$, $(x \in \mathcal{G}_\alpha, y \in \mathcal{G}_\beta)$. Now $T(\mathcal{G})/J$ is the universal enveloping algebra of $\mathcal{G}$.

Let $\mathfrak{h} = (\oplus_{i \in I}\mathbb{C}h_i) \oplus (\oplus_{i \in I}\mathbb{C}d_i)$ be a complex vector space with a basis $\{h_i, d_i; i \in I\}$. For each $i \in I$, define a linear functional $\alpha_i \in \mathfrak{h}^\star$ by

$$\alpha_i(h_j) = a_{ji},\ \alpha_i(d_j) = \delta_{ij} \text{ for all } j \in I,$$

where $\mathfrak{h}^* = Hom_C(\mathfrak{h}, \mathbb{C})$.

Definition 184. Consider a Z_2-graded vector space $W = W_{\bar{0}} \oplus W_{\bar{1}}$. In the usual notation, $T(W)$ is the tensor algebra of $W.T(W)$ becomes a Z-graded algebra,

$$T(W) = \bigoplus_{m \in Z} T_m(W).$$

Here $T_m(W)$ is the vector space W of tensors of order m if $m \geq 0$ and $T_n(W) = \{0\}$ if $n \leq -1$. Using the Z_2-gradation of W, one can give Z_2-gradation of $T(W)$. Now all the $T_m(W)$ are Z_2-graded subspaces of $T(W)$. This makes $T(W)$ as an associative Z-graded superalgebra.

Let $\tilde{I}$ be the two sided ideal of $T(W)$, generated by the tensors of the form,

$$X \otimes Y - (-1)^{\alpha\beta} Y \otimes X,$$
$$\text{with } X \in W_\alpha, Y \in W_\beta; \alpha, \beta \in Z_2.$$

These tensors, $X \otimes Y$ are homogeneous with respect to both the Z_2-gradation and the Z-gradation of $T(W)$. So

$$\tilde{U}(W) = T(W)/\tilde{I}$$

is an associative Z-graded superalgebra, called the *supersymmetric algebra* of the Z_2-graded vector space W.

The multiplication in $\tilde{U}(W)$ is supercommutative. The above Z_2-graded vector space W becomes an abelian Lie superalgebra if we define the multiplication $[W, W] = \{0\}$ and in this case $\tilde{U}(W)$ is an enveloping algebra of this abelian superalgebra.

Definition 185. Let $\mathcal{G}$ be a Lie superalgebra and $\sigma : \mathcal{G} \to U(\mathcal{G})$ be its canonical form. For $n \geq 0$, let $U^n(\mathcal{G})$ be the subspace of the $U(\mathcal{G})$, generated by the products of the form $\sigma(A_1)\sigma(A_2)\cdots\sigma(A_m)$ with $0 \leq m \leq n$ and $A_1, A_2, \ldots, A_m \in \mathcal{G}$. Let $U^n(\mathcal{G})$ be the Z_2-graded subspaces of $U(\mathcal{G})$. The family $(U^n(\mathcal{G}))_{n \in Z}$ is called the *canonical filtration* of the enveloping algebra, $U(\mathcal{G})$ if it satisfies the usual conditions of filtration.

Theorem 48 (PBW theorem (Poincare, Birkhoff, Witt)). *In above notation, let*

$$G_n(\mathcal{G}) = U^n(\mathcal{G})/U^{n-1}(\mathcal{G})$$

and

$$G(\mathcal{G}) = \bigoplus_{n \in Z} G_n(\mathcal{G}).$$

Then the canonical homomorphism, $\tilde{\phi} : \tilde{U}(\mathcal{G}) \to G(\mathcal{G})$ *is an isomorphism of Z-graded superalgebras.*

Another form of PBW theorem: If $\mathcal{G}$ *is a Lie superalgebra,* σ *is the canonical form from* $\mathcal{G}$ *into* $U(\mathcal{G})$ *and* $(E_i)_{i\in I}$ *is a basis of* $\mathcal{G}$ *such that elements of* E_i *are homogenous, then the products*

$$\sigma(E_{i_1})\sigma(E_{i_2})\cdots\sigma(E_{i_r})$$

form a basis of $U(\mathcal{G})$ *where* $\{i_1 i_2 \cdots i_r\} \subseteq I$ *satisfying* $r \geq 0$, $i_1 \leq i_2 \leq \cdots \leq i_r$ *with* $i_p < i_{p+1}$ *if* E_{i_p} *and* $E_{i_{p+1}}$ *are odd.*

Remark 113. The canonical form $\sigma : \mathcal{G} \to U(\mathcal{G})$ is injective.

Definition 186 (PBW basis). [129] Let $b_1, \ldots, b_B$ $(B = \dim \mathcal{G}_{\bar{0}})$ be a basis of the even part $\mathcal{G}_{\bar{0}}$ and $f_1, f_2, \ldots, f_F$ $(F = \dim \mathcal{G}_{\bar{1}})$ be a basis of the odd part $\mathcal{G}_{\bar{1}}$. The elements

$$b_1^{i_1} \cdots b_B^{i_B} f_1^{j_1} \cdots f_F^{j_F} \text{ with } i_1, \ldots, i_B \geq 0 \text{ and } j_1, \ldots, j_F \in \{0, 1\}$$

form a basis of the universal enveloping superalgebra $U(\mathcal{G})$, called the Poincare-Birkhoff-Witt (PBW) basis.

4.9 Cartan subalgebras and root systems of Lie superalgebras

Definition 187. [34] (The root space decomposition and Cartan subalgebras of the classical Lie superalgebras). Let $\mathcal{G} = \mathcal{G}_{\bar{0}} \oplus \mathcal{G}_{\bar{1}}$ be a Lie superalgebra. Then a Cartan subalgebra H of $\mathcal{G}_{\bar{0}}$ is also a Cartan subalgebra of $\mathcal{G}$. Moreover, every inner automorphism of $\mathcal{G}_{\bar{0}}$ extends to an inner automorphism of $\mathcal{G}$. Hence Cartan subalgebras of Lie superalgebra are conjugate like Cartan subalgebras of a Lie algebra. Moreover, Cartan subalgebras of classical Lie superalgebras are diagonalizable. We have the following root space decomposition:

$$\mathcal{G} = \bigoplus_{\alpha \in H^*} \mathcal{G}_\alpha, \text{ where } \mathcal{G}_\alpha = \{a \in \mathcal{G} | [h, a] = \alpha(h)a \text{ for } h \in H\}.$$

The set $\Delta = \{\alpha \in H^* | \mathcal{G}_\alpha \neq 0\}$ is called the root system. Let $\Delta_{\bar{0}}$ be the root system of $\mathcal{G}_{\bar{0}}$ and $\Delta_{\bar{1}}$ be the system of weights of the representation of $\mathcal{G}_{\bar{0}}$ on $\mathcal{G}_{\bar{1}}$, then $\Delta = \Delta_{\bar{0}} \cup \Delta_{\bar{1}}$. Here $\Delta_{\bar{0}}$ is called the system of even roots and $\Delta_{\bar{1}}$ is called the odd roots.

Remark 114. [129]

1) In the case of the superalgebra Q(n), a root can be both even and odd.
2) Except for A(1, 1), P(n) and Q(n), using the invariant bilinear form defined on the superalgebra $\mathcal{G}$, one can define a bilinear form on the root space $\mathfrak{h}^*$ by $(\alpha_i, \alpha_j) = (H_i, H_j)$ where the H_i form a basis of $\mathfrak{h}$.
3) $\mathcal{G}_{(\alpha=0)} = \mathfrak{h}$ except for Q(n).

4) dim $\mathcal{G}_\alpha = 1$ when $\alpha \neq 0$ except for A(1,1), P(2), P(3) and Q(n).
5) Except for $A(1,1), P(n)$ and $Q(n)$, one has
   (i) $[\mathcal{G}_\alpha, \mathcal{G}_\beta] \neq 0$ if and only if $\alpha, \beta, \alpha+\beta \in \Delta$.
   (ii) $(\mathcal{G}_\alpha, \mathcal{G}_\beta) = 0$ for $\alpha + \beta \notin 0$.
   (iii) If $\alpha \in \Delta$(resp. $\Delta_{\bar{0}}, \Delta_{\bar{1}}$), then $-\alpha \in \Delta$ (resp. $\Delta_{\bar{0}}, \Delta_{\bar{1}}$).
   (iv) $\alpha \in \Delta$ implies $2\alpha \in \Delta$ if and only if $\alpha \in \Delta_{\bar{1}}$ and $\alpha.\alpha \neq 0$
   (v) $\Delta_{\bar{0}}$ and $\Delta_{\bar{1}}$ are invariant under the action of the Weyl group of $\mathcal{G}_{\bar{0}}$.

Definition 188. Let $\mathcal{G} = \mathcal{G}_{\bar{0}} \oplus \mathcal{G}_{\bar{1}}$ be a Lie superalgebra with root system $\Delta = \Delta_{\bar{0}} \cup \Delta_{\bar{1}}$. Here $\Delta_{\bar{0}}$ is the set of even roots and $\Delta_{\bar{1}}$ is the set of odd roots. The *Weyl group* $W(\mathcal{G})$ of $\mathcal{G}$ is generated by the Weyl reflections w with respect to the even roots. We have

$$w_\alpha(\beta) = \beta - 2\frac{(\alpha|\beta)}{(\alpha|\alpha)}\alpha, \text{ where } \alpha \in \Delta_{\bar{0}} \text{ and } \beta \in \Delta.$$

Definition 189 (The system of roots and simple roots for classical Lie superalgebra). Let Δ be the root system of a classical Lie superalgebra $\mathcal{G}$. A system of roots, $\Pi = \{\alpha_1, \ldots, \alpha_r\} \subseteq \Delta$ is said to be simple if

(1) there are vectors $e_i \in \mathcal{G}_{\alpha_i}, f_i \in \mathcal{G}_{-\alpha_i}$, such that $[e_i, f_j] = \delta_{ij}h_i \in H$,
(2) the vectors e_i and f_i, $i = 1, \ldots, r$ generate $\mathcal{G}$, and
(3) Π is minimal with these properties.

Definition 190. [129] A classical Lie superalgebra is called *basic* if there exists a non-degenerate invariant bilinear form on $\mathcal{G}$. A classical Lie superalgebra which is not basic is called *strange*.

Definition 191. [129] Let $\mathcal{G} = \mathcal{G}_{\bar{0}} \oplus \mathcal{G}_{\bar{1}}$ be a basic Lie superalgebra with Cartan subalgebra $\mathfrak{h}$ and root system $\Delta = \Delta_{\bar{0}} \cup \Delta_{\bar{1}}$. Then one can get a *Borel decomposition* of $\mathcal{G}$ as $\mathcal{G} = \eta^+ \oplus \mathfrak{h} \oplus \eta^-$, where η^+ and η^- are subalgebras such that $[\mathfrak{h}, \eta^+] \subset \eta^+$ and $[\mathfrak{h}, \eta^-] \subset \eta^-$ with dim $\eta^+ =$ dim η^-.

Definition 192. [129] (Simple root system and Borel subalgebra of basic Lie superalgebra). If one can consider a root decomposition of $\mathcal{G}$ as $\mathcal{G} = \mathfrak{h} \oplus_\alpha \mathcal{G}_\alpha$, then α is called *positive,* if $\mathcal{G}_\alpha \cap \eta^+ \neq 0$ and called *negative,* if $\mathcal{G}_\alpha \cap \eta^- \neq 0$. Moreover, a root is *simple,* if it is not decomposed into a sum of positive roots. The set of all simple roots, denoted by Δ^0, is called simple root system. $B = \mathfrak{h} \oplus \eta^+$ is called a *Borel subalgebra* of $\mathcal{G}$. Here one can verify that Borel subalgebra is solvable, but not maximal solvable.

Definition 193. [129] (Distinguished Z-gradation, distinguished simple root system and distinguished Cartan matrix). Let $\mathcal{G}$ be a basic Lie superalgebra. Then there exists a particular simple root system for which the number of odd roots is equal to one. For a classical basic Lie superalgebra, $\mathcal{G} = \mathcal{G}_{\bar{0}} \oplus \mathcal{G}_{\bar{1}}$, there exists a consistent Z-gradation, $\mathcal{G} = \oplus_{i \in Z}\mathcal{G}_i$, called *distinguished Z-gradation* satisfying the following rules:

(i) For superalgebras of type I, $\mathcal{G}_i = 0$ for $|i| > 1$ and $\mathcal{G}_{\bar{0}} = \mathcal{G}_0; \mathcal{G}_{\bar{1}} = \mathcal{G}_1 \oplus \mathcal{G}_{-1}$.
(ii) For superalgebras of type II, $\mathcal{G}_i = 0$ for $|i| > 2$ and $\mathcal{G}_{\bar{0}} = \mathcal{G}_{-2} \oplus \mathcal{G}_0 \oplus \mathcal{G}_2$, $\mathcal{G}_{\bar{1}} = \mathcal{G}_1 \oplus \mathcal{G}_{-1}$.
For a classical basic Lie superalgebra $\mathcal{G}$, let $\mathcal{G} = \oplus \mathcal{G}_i$ be the distinguished Z-gradation of $\mathcal{G}$.

Then there exists a Z-gradation consistent (with the Z_2-gradation) satisfying

$$\mathcal{G}_{\bar{0}} = \sum_{i \in Z} \mathcal{G}_{2i} \text{ and } \mathcal{G}_{\bar{1}} = \sum_{i \in Z} \mathcal{G}_{2i+1}.$$

For each basic Lie superalgebra, there exists a particular simple root system for which number of odd roots is equal to one. Consider the distinguished Z-gradation of $\mathcal{G} = \oplus_{i \in Z} \mathcal{G}_i$. The even simple roots are given by the simple root system of the Lie subalgebra $\mathcal{G}_0$ and the odd simple root is the lowest weight of the representation $\mathcal{G}_1$ of $\mathcal{G}_0$. Such a simple root system is called *distinguished simple root system.* Two different distinguished simple root systems are related by means of Weyl reflections.

All equivalent distinguished simple root systems lead to the same Cartan matrix as defined above. Such a Cartan matrix is called the *distinguished Cartan matrix*.

Below we describe for $A(m,n)$, $B(m,n)$, $C(n)$, $D(m,n)$, $D(2,1;\alpha)$, $F(4)$ and $G(3)$, the systems of even roots $\Delta_{\bar{0}}$ and of odd roots $\Delta_{\bar{1}}$ and all systems of simple roots, upto W-equivalence. In all the examples, the Cartan subalgebra H is a subspace of the space D of diagonal matrices. The roots are expressed in terms of the standard basis ϵ_i of D^*.

Rank and Root systems of classical Lie superalgebras [129]:

(1) *The basic Lie superalgebra* $A(m-1,n-1) = sl(m,n)$: Let $\mathcal{G} = \mathcal{G}_{\bar{0}} \oplus \mathcal{G}_{\bar{1}}$. Here $\mathcal{G}_{\bar{0}} = sl(m) \oplus sl(n) \oplus U(1)$ and $\mathcal{G}_0 | \mathcal{G}_{-1} = sl_m \otimes sl_n \otimes k$, type I.

The rank and dimension of $A(m-1,n-1)$ are $m+n-1$ and $(m+n)^2 - 1$ respectively. The root system of $A(m-1,n-1)$ can be given as follows: In the usual notation,

$$\Delta = \{\epsilon_i - \epsilon_j, \delta_k - \delta_l, \epsilon_i - \delta_k, \delta_k - \epsilon_i\}$$

where

$$\Delta_{\bar{0}} = \overline{\Delta_{\bar{0}}} = \{\epsilon_i - \epsilon_j, \delta_k - \delta_l\}, \Delta_{\bar{1}} = \overline{\Delta_{\bar{1}}} = \{\epsilon_i - \delta_k, \delta_k - \epsilon_i\}.$$

Dimension of $\Delta_{\bar{0}}$ and $\Delta_{\bar{1}}$ are $m^2 + n^2 - m - n + 1$ and $2mn$ respectively. The distinguished simple root system (see Table 4.1) is

$$\{\alpha_1 = \delta_1 - \delta_2, \ldots, \alpha_{n-1} = \delta_{n-1} - \delta_n, \alpha_n = \delta_n - \epsilon_1, \alpha_{n+1} = \epsilon_1 - \epsilon_2, \ldots, \alpha_{m+n-1} = \epsilon_{m-1} - \epsilon_m\}.$$

Table 4.1 Distinguished set of simple roots for basic Lie superalgebras [129]

Lie superalgebra	Distinguished set of simple roots
$A(m-1,n-1)$	$\delta_1-\delta_2,\dots,\delta_{n-1}-\delta_n,\delta_n-\epsilon_1,\epsilon_1-\epsilon_2,\dots,\epsilon_{m-1}-\epsilon_m$
$B(m,n)$	$\delta_1-\delta_2,\dots,\delta_{n-1}-\delta_n,\delta_n-\epsilon_1,\epsilon_1-\epsilon_2,\dots,\epsilon_{m-1}-\epsilon_m,\epsilon_m$
$B(0,n)$	$\delta_1-\delta_2,\dots,\delta_{n-1}-\delta_n,\delta_n$
$C(n)$	$\epsilon-\delta_1,\delta_1-\delta_2,\dots,\delta_{n-1}-\delta_n,2\delta_n$
$D(m,n)$	$\delta_1-\delta_2,\dots,\delta_{n-1}-\delta_n,\delta_n-\epsilon_1,\epsilon_1-\epsilon_2,$ $\dots,\epsilon_{m-1}-\epsilon_m,\epsilon_{m-1}+\epsilon_m$
F(4)	$\frac{1}{2}(\delta-\epsilon_1-\epsilon_2-\epsilon_3),\epsilon_3,\epsilon_2-\epsilon_3,\epsilon_1-\epsilon_2$
G(3)	$\delta+\epsilon_3,\epsilon_1,\epsilon_2-\epsilon_1$
$D(2,1;\alpha)$	$\epsilon_1-\epsilon_2-\epsilon_3,2\epsilon_2,2\epsilon_3$

The set of distinguished positive roots for $1\le i<j\le m$ and $1\le k<l\le n$ is

$$\{\delta_k-\delta_l=\alpha_k+\cdots+\alpha_{l-1},\epsilon_i-\epsilon_j=\alpha_{n+i}+\cdots+\alpha_{n+j-1},\delta_k-\epsilon_i=\alpha_k+\cdots+\alpha_{n+i-1}\}.$$

The corresponding distinguished Cartan matrix is

$$\left(\begin{array}{ccccc|c|ccccc}
2 & -1 & 0 & \cdots & 0 & \cdots & \cdots & & & \cdots & 0\\
-1 & \ddots & \ddots & \ddots & & & & & & & \vdots\\
0 & \ddots & \ddots & \ddots & 0 & & & & & & \vdots\\
\vdots & \ddots & \ddots & \ddots & -1 & \ddots & & & & & \vdots\\
0 & & 0 & -1 & 2 & -1 & \ddots & & & & \vdots\\
\hline
\vdots & & & \ddots & -1 & 0 & -1 & \ddots & & \ddots & \\
\hline
\vdots & & & & \ddots & -1 & 2 & -1 & & & 0\\
 & & & & & \ddots & -1 & \ddots & \ddots & \ddots & \vdots\\
 & & & & & & 0 & \ddots & & \ddots & 0\\
 & & & & & & & \ddots & \ddots & \ddots & -1\\
0 & \cdots & & & \cdots & \cdots & 0 & \cdots & 0 & -1 & 2
\end{array}\right).$$

The highest distinguished root of the corresponding superalgebra is

$$-\alpha_0=\alpha_1+\cdots+\alpha_{m+n-1}=\delta_1-\epsilon_m.$$

(2) *The basic Lie superalgebra* $A(n-1,n-1)$: The structure is $\mathcal{G}_{\bar{0}}=sl(n)\oplus sl(n)$ and $\mathcal{G}_0|\mathcal{G}_{-1}=sl_n\otimes sl_n$ type I. The rank and dimension of $A(n-1,n-1)$ are $2n-2$ and $4n^2-2$ respectively. The root system is

$$\Delta=\{\epsilon_i-\epsilon_j,\delta_i-\delta_j,\epsilon_i-\delta_j,\delta_j-\epsilon_i\}$$

for $1 \leq i \neq j \leq n$. We have

$$\Delta_{\bar{0}} = \overline{\Delta_{\bar{0}}} = \{\epsilon_i - \epsilon_j, \delta_i - \delta_j\}, \Delta_{\bar{1}} = \overline{\Delta_{\bar{1}}} = \{\epsilon_i - \delta_j, \delta_j - \epsilon_i\}$$

where $\sum_{i=1}^{n} \epsilon_i = \sum_{i=1}^{n} \delta_i$. Here $\dim \Delta_{\bar{0}} = \dim \overline{\Delta_{\bar{0}}} = 2n^2 - 2n$ and $\dim \Delta_{\bar{1}} = \dim \overline{\Delta_{\bar{1}}} = 2n^2$. The distinguished simple root system is

$$\{\alpha_1 = \delta_1 - \delta_2, \ldots, \alpha_{n-1} = \delta_{n-1} - \delta_n, \alpha_n = \delta_n - \epsilon_1, \alpha_{n+1} = \epsilon_1 - \epsilon_2, \ldots, \alpha_{2n-1} = \epsilon_{n-1} - \epsilon_n\}.$$

Number of simple roots = 2n−1($\neq$ rank). The set of distinguished positive roots is

$$\{\delta_k - \delta_l = \alpha_k + \cdots + \alpha_{l-1},\ \epsilon_i - \epsilon_j = \alpha_{n+i} + \cdots + \alpha_{n+j-1},\ \delta_k - \epsilon_i = \alpha_k + \cdots + \alpha_{n+i-1}\}$$

for $1 \leq i < j \leq n$ and $1 \leq k < l \leq n$. The corresponding distinguished Cartan matrix is

$$\left(\begin{array}{ccccc|c|ccccc}
2 & -1 & 0 & \cdots & 0 & \cdots & \cdots & & & \cdots & 0 \\
-1 & \ddots & \ddots & \ddots & & & & & & & \vdots \\
0 & \ddots & \ddots & \ddots & 0 & & & & & & \vdots \\
\vdots & \ddots & \ddots & \ddots & -1 & \ddots & & & & & \vdots \\
0 & & 0 & -1 & 2 & -1 & \ddots & & & & \vdots \\
\hline
\vdots & & & & -1 & 0 & -1 & \ddots & & & \vdots \\
\hline
\vdots & & & & \ddots & -1 & 2 & -1 & 0 & & 0 \\
 & & & & & \ddots & -1 & \ddots & \ddots & \ddots & \vdots \\
 & & & & & & 0 & \ddots & & \ddots & 0 \\
 & & & & & & & \ddots & \ddots & \ddots & -1 \\
0 & \cdots & & & \cdots & \cdots & 0 & \cdots & 0 & -1 & 2
\end{array}\right).$$

The highest distinguished root of the corresponding superalgebra is

$$-\alpha_0 = \alpha_1 + \cdots + \alpha_{2n-1} = \delta_1 - \epsilon_m.$$

Factor group $Out(\mathcal{G}) = Aut(\mathcal{G})/Int(\mathcal{G})$:

$$Out(\mathcal{G}) = \mathbb{Z}_2 \text{ for } A(m,n) \text{ with } m \neq n \neq 0 \text{ and } A(0, 2n-1)$$
$$Out(\mathcal{G}) = \mathbb{Z}_4 \quad \text{for } A(0, 2n).$$

(3) *The basic Lie superalgebra $B(m,n)$*: The structure is $\mathcal{G}_{\bar{0}} = so(2m+1) \oplus sp(2n)$ and $\mathcal{G}_{\bar{1}} = (2m+1, 2n)$, type II. The rank is $m+n$ and the dimension is $2(m+n)^2 + m + 3n$. The root system $(1 \le i \ne j \le m$ and $1 \le k \ne l \le n)$:

$$\Delta = \{\pm\epsilon_i \pm \epsilon_j, \pm\epsilon_i, \pm\delta_k \pm \delta_l, \pm 2\delta_k, \pm\epsilon_i \pm \delta_k, \pm\delta_k\},$$
$$\Delta_{\bar{0}} = \{\pm\epsilon_i \pm \epsilon_j, \pm\epsilon_i, \pm\delta_k \pm \delta_l, \pm 2\delta_k\}, \text{ and } \Delta_{\bar{1}} = \{\pm\epsilon_i \pm \delta_k, \pm\delta_k\}.$$

The dimensions of $\Delta_{\bar{0}}$ and $\Delta_{\bar{1}}$ are $\dim \Delta_{\bar{0}} = 2m^2 + 2n^2, \dim \Delta_{\bar{1}} = 4mn + 2n$. The distinguished simple root system (see Table 4.1) is

$$\{\alpha_1 = \delta_1 - \delta_2, \ldots, \alpha_{n-1} = \delta_{n-1} - \delta_n, \alpha_n = \delta_n - \epsilon_1,$$
$$\alpha_{n+1} = \epsilon_1 - \epsilon_2, \ldots, \alpha_{n+m-1} = \epsilon_{m-1} - \epsilon_m, \alpha_{n+m} = \epsilon_m\}.$$

The corresponding distinguished Cartan matrix is

$$\left(\begin{array}{ccccc|c|ccccc}
2 & -1 & 0 & \cdots & 0 & \cdots & \cdots & & & \cdots & 0 \\
-1 & \ddots & \ddots & \ddots & & & & & & & \vdots \\
0 & \ddots & \ddots & \ddots & 0 & & & & & & \vdots \\
\vdots & \ddots & \ddots & \ddots & -1 & \ddots & & & & & \vdots \\
0 & & 0 & -1 & 2 & -1 & \ddots & & & & \vdots \\
\hline
\vdots & & & & -1 & 0 & -1 & \ddots & & & \vdots \\
\hline
\vdots & & & & \ddots & -1 & 2 & -1 & 0 & & 0 \\
 & & & & & \ddots & -1 & \ddots & \ddots & \ddots & \vdots \\
 & & & & & & 0 & \ddots & & \ddots & 0 \\
 & & & & & & & \ddots & \ddots & 2 & -1 \\
0 & \cdots & & & \cdots & \cdots & 0 & \cdots & 0 & -2 & 2
\end{array}\right).$$

(4) *The basic Lie superalgebra $C(n+1) = osp(2|2n)$*: The structure is $\mathcal{G}_{\bar{0}} = so(2) \oplus sp(2n)$ and $\mathcal{G}_{\bar{1}} = (2n) \oplus (2n)$, type I. The rank and dimension are $n+1$ and $2n^2 + 5n + 1$ respectively. The root system $(1 \le k \ne l \le n)$:

$$\Delta = \{\pm\delta_k \pm \delta_l, \pm 2\delta_k, \pm\epsilon \pm \delta_k\}$$
$$\Delta_{\bar{0}} = \{\pm\delta_k \pm \delta_l, \pm 2\delta_k\}, \text{ and } \Delta_{\bar{1}} = \{\pm\epsilon \pm \delta_k\}.$$

$\dim \Delta_{\bar{0}} = 2n^2, \dim \Delta_{\bar{1}} = 4n$. The distinguished simple root system (see Table 4.1) is

$$\{\alpha_1 = \epsilon - \delta_1, \alpha_2 = \delta_1 - \delta_2, \ldots, \alpha_n = \delta_{n-1} - \delta_n, \alpha_{n+1} = 2\delta_n\}.$$

The corresponding distinguished Cartan matrix is

$$\left(\begin{array}{c|ccccc}
0 & -1 & 0 & \cdots & \cdots & 0 \\
\hline
-1 & 2 & -1 & 0 & & 0 \\
0 & -1 & \ddots & \ddots & \ddots & \vdots \\
0 & 0 & \ddots & \ddots & -1 & 0 \\
\vdots & & \ddots & -1 & 2 & -2 \\
0 & \cdots & \cdots & 0 & -1 & 2
\end{array}\right).$$

(5) *The basic Lie superalgebra* $D(m,n) = osp(2m|2n)$: The structure is $\mathcal{G}_{\bar{0}} = so(2m) \oplus sp(2n)$ and $\mathcal{G}_{\bar{1}} = (2m, 2n)$, type II. The rank and dimension are $m+n$ and $2(m+n)^2 - m + n$, respectively. The root system ($1 \le i \ne j \le m$ and $1 \le k \ne l \le n$) :

$$\Delta = \{\pm\epsilon_i \pm \epsilon_j, \pm\delta_k \pm \delta_l, \pm 2\delta_k, \pm\epsilon_i \pm \delta_k\}$$
$$\Delta_{\bar{0}} = \{\pm\epsilon_i \pm \epsilon_j, \pm\delta_k \pm \delta_l, \pm 2\delta_k\},\ \Delta_{\bar{1}} = \{\pm\epsilon_i \pm \delta_k\}.$$

The dimensions are $\dim \Delta_{\bar{0}} = 2m^2 + 2n^2 - 2m$, $\dim \Delta_{\bar{1}} = 4mn$. The distinguished simple root system (Table 4.1) is

$$\{\alpha_1 = \delta_1 - \delta_2, \ldots, \alpha_{n-1} = \delta_{n-1} - \delta_n, \alpha_n = \delta_n - \epsilon_1,$$
$$\alpha_{n+1} = \epsilon_1 - \epsilon_2, \ldots, \alpha_{n+m-1} = \epsilon_{m-1} - \epsilon_m, \alpha_{n+m} = \epsilon_{m-1} + \epsilon_m\}.$$

The corresponding distinguished Cartan matrix is

$$\left(\begin{array}{ccccc|c|ccccc}
2 & -1 & 0 & \cdots & 0 & \cdots & \cdots & & & \cdots & 0 \\
-1 & \ddots & \ddots & \ddots & & & & & & & \vdots \\
0 & \ddots & \ddots & \ddots & 0 & & & & & & \vdots \\
\vdots & \ddots & \ddots & \ddots & -1 & \ddots & & & & & \vdots \\
0 & & 0 & -1 & 2 & -1 & \ddots & & & & \vdots \\
\hline
\vdots & & & & -1 & 0 & 1 & \ddots & & & \vdots \\
\hline
\vdots & & & & \ddots & -1 & 2 & -1 & 0 & & 0 \\
 & & & & & \ddots & -1 & \ddots & \ddots & \ddots & \vdots \\
 & & & & & & 0 & \ddots & & -1 & -1 \\
 & & & & & & & \ddots & -1 & 2 & 0 \\
0 & \cdots & & & \cdots & \cdots & 0 & \cdots & -1 & 0 & 2
\end{array}\right).$$

(6) *The basic Lie superalgebra* $F(4)$: The structure of $F(4)$ is $\mathcal{G}_{\bar{0}} = sl(2) \oplus so(7)$ and $\mathcal{G}_{\bar{1}} = (2, 8)$, type II. The rank and dimension is 4 and 40. The root system ($1 \le i \ne j \le 3$) :

$$\Delta = \{\pm\delta, \pm\epsilon_i \pm \epsilon_j, \pm\epsilon_i, \frac{1}{2}(\pm\epsilon_1 \pm \epsilon_2 \pm \epsilon_3 \pm \delta)\}$$
$$\Delta_{\bar{0}} = \{\pm\delta, \pm\epsilon_i \pm \epsilon_j, \pm\epsilon_i\},\ \Delta_{\bar{1}} = \{\frac{1}{2}(\pm\epsilon_1 \pm \epsilon_2 \pm \epsilon_3 \pm \delta)\}.$$

The dimensions are $\dim \Delta_{\bar{0}} = 20, \dim \Delta_{\bar{1}} = 16$. The distinguished simple root system is

$$\{\alpha_1 = \frac{1}{2}(\delta - \epsilon_1 - \epsilon_2 - \epsilon_3), \alpha_2 = \epsilon_3, \alpha_3 = \epsilon_2 - \epsilon_3, \alpha_4 = \epsilon_1 - \epsilon_2\}.$$

The corresponding distinguished Cartan matrix is $\begin{pmatrix} 0 & -1 & 0 & 0 \\ -1 & 2 & -2 & 0 \\ 0 & -1 & 2 & -1 \\ 0 & 0 & -1 & 2 \end{pmatrix}$.

(7) *The basic Lie superalgebra* $G(3)$: The structure of $F(4)$ is $\mathcal{G}_{\bar{0}} = sl(2) \oplus G_2$ and $\mathcal{G}_{\bar{1}} = (2, 7)$, type II (see Table 4.3). The rank and dimension are 3 and 31 respectively. The root system $(1 \le i \ne j \le 3)$:

$$\Delta = \{\pm 2\delta, \pm\epsilon_i, \epsilon_i - \epsilon_j, \pm\delta, \pm\epsilon_i \pm \delta\},$$
$$\Delta_{\bar{0}} = \{\pm 2\delta, \pm\epsilon_i, \epsilon_i - \epsilon_j\}, \ \Delta_{\bar{1}} = \{\pm\delta, \pm\epsilon_i \pm \delta\},$$

where $\epsilon_1 + \epsilon_2 + \epsilon_3 = 0$. The dimensions are $\dim \Delta_{\bar{0}} = 14, \dim \Delta_{\bar{1}} = 14$. The distinguished simple root system (Table 4.1) is

$$\{\alpha_1 = \delta + \epsilon_3, \alpha_2 = \epsilon_1, \alpha_3 = \epsilon_2 - \epsilon_1\}.$$

The corresponding distinguished Cartan matrix is $\begin{pmatrix} 0 & -1 & 0 \\ -1 & 2 & -3 \\ 0 & -1 & 2 \end{pmatrix}$.

(8) *The basic Lie superalgebra* $D(2, 1; \alpha)$: The structure of $D(2, 1; \alpha)$ is $\mathcal{G}_{\bar{0}} = sl(2) \oplus sl(2) \oplus sl(2)$ and $\mathcal{G}_{\bar{1}} = (2, 2, 2)$, type II. The rank and dimension are 3 and 17 respectively. The root system $(1 \le i \le 3)$:

$$\Delta = \{\pm 2\epsilon_i, \pm\epsilon_1 \pm \epsilon_2 \pm \epsilon_3\}$$
$$\Delta_{\bar{0}} = \{\pm 2\epsilon_i\}, \ \Delta_{\bar{1}} = \{\pm\epsilon_1 \pm \epsilon_2 \pm \epsilon_3\}.$$

The dimensions are $\dim \Delta_{\bar{0}} = 6, \dim \Delta_{\bar{1}} = 8$. The distinguished simple root system (see Table 4.1) is

$$\{\alpha_1 = \epsilon_1 - \epsilon_2 - \epsilon_3, \alpha_2 = 2\epsilon_2, \alpha_3 = 2\epsilon_3\}.$$

The corresponding distinguished Cartan matrix is $\begin{pmatrix} 0 & -1 & \alpha \\ -1 & 2 & 0 \\ -1 & 0 & 2 \end{pmatrix}$.

4.10 Killing forms on Lie superalgebras

Definition 194. Let $\mathcal{G}$ be a Lie superalgebra and ϕ be a representation of $\mathcal{G}$ on $\mathbb{R}$. Then a *bilinear form* b_ϕ *associated to* ϕ on $\mathcal{G}$ is a bilinear form on $\mathcal{G} \times \mathcal{G}$ into $\mathbb{R}$ such that

$$b_{\phi}(x, y) = str(\phi(x), \phi(y)), \text{ for all } x, y \in \mathcal{G}.$$

Definition 195. The *Killing form* on a superalgebra $\mathcal{G}$ is the *bilinear form on* $\mathcal{G}$ defined by

$$\kappa(x, y) = str((adx)(ady)) \text{ for x, y } \in \mathcal{G}.$$

Supertrace was already defined in definition 153.

Properties of Killing form on a Lie superalgebra: From the properties of the supertrace, we obtain the corresponding properties of the Killing form.

Here for a complex super square matrix M of the form $M = \begin{pmatrix} A & B \\ C & D \end{pmatrix}$, the supertrace of M is defined by

$$str(M) = tr(A) - (-1)^{degM} tr(D) = \begin{cases} tr(A) - tr(D) \text{ if } M \text{ is even} \\ tr(A) + tr(D) \text{ if } M \text{ is odd} \end{cases}.$$

The Killing form on a Lie superalgebra $\mathcal{G}_{\bar{0}} \oplus \mathcal{G}_{\bar{1}}$ has the following properties:

$$\kappa(x, y) = 0 \text{ for } x \in \mathcal{G}_{\bar{0}}, y \in \mathcal{G}_{\bar{1}} \text{ (consistency)}$$
$$\kappa(x, y) = (-1)^{(degx)(degy)} \kappa(y, x) \text{ (supersymmetric)}$$
$$\kappa([x, y], z) = \kappa(x, [y, z]) \text{ (invariant)}$$

Hence, Killing form of a Lie superalgebra $\mathcal{G}$ is consistent, supersymmetric and invariant. Moreover, Killing form κ on a Lie superalgebra satisfies

$$\kappa(\phi(x), \phi(y)) = \kappa(x, y) \text{ for all } \phi \in Aut(\mathcal{G}) \text{ and } x, y \in \mathcal{G}.$$

Remark 115.

(1) In general, a bilinear form is called an inner product on $\mathcal{G}$, if it is consistent, supersymmetric and invariant.
(2) The bilinear form associated to the adjoint representation of $\mathcal{G}$ is the Killing form on $\mathcal{G}$.

For the proofs of the following properties of the Killing form, one can refer to [35, 129] or [34].

(1) Let $\mathcal{G}$ be a Lie superalgebra such that $[\mathcal{G}, \mathcal{G}] = \mathcal{G}$. Then every invariant bilinear form on $\mathcal{G}$ is supersymmetric.
(2) Let $\mathcal{G}$ be a simple Lie superalgebra whose Killing form is non-degenerate. Then $\mathcal{G}$ is classical. Moreover, the Killing form on a Lie superalgebra $\mathcal{G}$ is non-degenerate if and only if $\mathcal{G}$ is the direct product (direct sum) of classical simple Lie superalgebras with nondegenerate Killing forms.

(3) Let $\mathcal{G}$ be a classical simple Lie superalgebra such that the center of $\mathcal{G}_{\bar{0}}$ is non-trivial. Then the Killing form of $\mathcal{G}$ is non-degenerate.

(4) Let $\mathcal{G}$ be a Lie superalgebra and let $\mathcal{G}'$ be a graded ideal in $\mathcal{G}$. If ϕ(resp.ϕ') is the Killing form of $\mathcal{G}$(resp .$\mathcal{G}'$), then the restriction of ϕ to $\mathcal{G}'$ is equal to ϕ'.

(5) Let $\mathcal{G}$ be a Lie superalgebra. If the Killing form on $\mathcal{G} = \mathcal{G}_{\bar{0}} \oplus \mathcal{G}_{\bar{1}}$ is non-degenerate, then its restriction to $\mathcal{G}_{\bar{0}}$ is non-degenerate and its restriction to $\mathcal{G}_{\bar{1}}$ gives a non-degenerate bilinear skew-symmetric form that is invariant under the representation of $\mathcal{G}_{\bar{0}}$ on $\mathcal{G}_{\bar{1}}$.

(6) Let $\mathcal{G}$ be a Lie superalgebra and K be a non-degenerate Killing form on $\mathcal{G}$. Then $\mathcal{G}$ splits into an orthogonal direct sum of Lie superalgebras with respect to a non-degenerate Killing form.

(7) Let $\mathcal{G}$ be a Lie superalgebra whose Killing form is non-degenerate. Any derivation of $\mathcal{G}$ is inner, that is of the form *ad a* for some $a \in \mathcal{G}$.

(8) Let $\mathcal{G} = \bigoplus \mathcal{G}_i$, be a Z-graded Lie superalgebra with non-degenerate Killing form. Then
  (a) $(\mathcal{G}_i, \mathcal{G}_j) = 0 \; for \; i \neq -j$
  (b) $(a, \mathcal{G}_i) \neq 0 \; for \; a \in \mathcal{G}_{-i}, a \neq 0.$
  (c) The representations of $\mathcal{G}_0$ on $\mathcal{G}_i$ and on $\mathcal{G}_{-i}$ are contragradient.
  (d) There is an element $z \in \mathcal{G}_0$ for which $[z, g] = sg \; for \; g \in \mathcal{G}_s$

(9) If $\mathcal{G}$ is a Lie superalgebra with a non-singular Killing form, then $\mathcal{G}$ is a direct sum of simple Lie superalgebras with non-singular Killing form.

4.11 Dynkin diagrams of Lie superalgebras [34, 129]

Let $\mathcal{G} = \mathcal{G}_{\bar{0}} \oplus \mathcal{G}_{\bar{1}}$, be a basic Lie superalgebra with Cartan subalgebra, H and root system, $\Delta = \Delta_{\bar{0}} \cup \Delta_{\bar{1}}$. Then $\mathcal{G}$ admits a Borel decomposition, $\mathcal{G} = \eta^+ \oplus H \oplus \eta^-$, where η^+ and η^- are subalgebras such that $[H, \eta^+] \subset \eta^+$ and $[H, \eta^-] \subset \eta^-$ with $\dim \eta^+ = \dim \eta^-$.

If $\mathcal{G} = H \oplus_\alpha \mathcal{G}_\alpha$ is the root decomposition of $\mathcal{G}$, a root α is called positive, if $\mathcal{G}_\alpha \cap \eta^+ \neq 0$ and negative, if $\mathcal{G}_\alpha \cap \eta^- \neq 0$. A root is called simple if it cannot be decomposed into a sum of positive roots. The set of all simple roots is called a simple root system of $\mathcal{G}$ and is denoted here by $\Delta_{\bar{0}}$.

Let $\mathcal{G}$ be a basic Lie superalgebra of rank r and dimension n. Let $\Delta_{\bar{0}} = \{\alpha_1, \alpha_2, \ldots, \alpha_r\}$ be a simple root system of $\mathcal{G}$ with associated Cartan matrix A and $A' = (a'_{ij})$ be the corresponding symmetric Cartan matrix defined by $a'_{ij} = \alpha_i.\alpha_j = B(H_{\alpha_i}, H_{\alpha'_j})$ with $\alpha_i\alpha_j = \alpha_i(H_{\alpha_j}) = \alpha_j(H_{\alpha_i})$. We can associate to the above simple root system $\{\alpha_1, \alpha_2, \ldots, \alpha_r\}$, *the Dynkin diagram by the following rules*:

(1) Let A be the Cartan matrix: The circles ○, ⊗ and • are called, respectively, white, gray(crossed), and black. Contragradient Lie superalgebras of rank r are depicted by a diagram consisting of r white, gray, or black circles; the ith circle is white if $i \notin \tau$ and gray or black if $i \in \tau$ and $a_{ii} = 0 \; or \; 2$ respectively. The ith and the jth circles are not joined if $a_{ij} = a_{ji} = 0$. Otherwise, they are joined as shown below (i) to (iv). Total number of nodes in each diagram is denoted by r and the index of non-white nodes is denoted by s.
  (i) **(a)** We draw a white circle corresponding to each simple even root.
  (b) We draw a black circle to each odd simple root of non-zero length ($a_{ii} \neq 0$).
  (c) We draw a gray circle(crossed circle) for each simple odd root of zero length ($a_{ii} = 0$).
  (ii) Let $\eta_{ij} = max\{|a_{ij}|, |a_{ji}|\}$. The ith and jth circles will be joined by η_{ij} lines.

(iii) If $\eta_{ij} > 1$ and $|a_{ij}| \neq |a_{ji}|$, we add an arrow on the lines connecting ith and jth circles. This arrow will point from j to i if $|a_{ij}| > 1$.

(iv) Let $\mathcal{G} = D(2, 1 : \alpha)$. Here $\eta_{ij} = 1$ if $a_{ij} \neq 0$ and $\eta_{ij} = 0$ if $a_{ij} = 0$. There is no arrow on the Dynkin diagram.

(2) Let A' be the symmetric Cartan matrix.

(i) **(a)** We draw a white circle corresponding to each simple even root.

(b) We draw a black circle to each odd simple root of non-zero length ($a'_{ii} \neq 0$).

(c) We draw a gray circle (crossed circle) for each simple odd root of zero length ($a'_{ii} = 0$).

(ii) The ith and jth lines are joined by η_{ij} lines. Here

(a) $\eta_{ij} = \dfrac{2|a'_{ij}|}{min(|a'_{ii}|,|a'_{jj}|)}$ if $a'_{ii}, a'_{jj} \neq 0$

(b) $\eta_{ij} = \dfrac{2|a'_{ij}|}{min(|a'_{ii}|,2)}$ if $a'_{ii} = 0,\ a'_{jj} \neq 0.$

(c) $\eta_{ij} = |a'_{ij}|$ if $a'_{ii},\ a'_{jj} = 0.$

(iii) **(a)** If $\eta_{ij} > 1$, we add an arrow on the lines connecting ith and jth circles pointing from i to j if $a'_{ii}.a'_{jj} \neq 0$ and $|a'_{ii}| > |a'_{jj}|$ or if $a'_{ii} = 0, a'_{jj} \neq 0, |a'_{jj}| > 2$. This arrow will point from j to i if $|a_{ij}| > 1$ and pointing towards j to i if $a'_{ii} = 0, a'_{jj} \neq 0$, $|a'_{jj}| > 2$.

(iv) Let $\mathcal{G} = D(2, 1 : \alpha)$. Here $\eta_{ij} = 1$ if $a'_{ij} \neq 0$ and $\eta_{ij} = 0$ if $a_{ij} = 0$. In this case, there is no arrow on the Dynkin diagram.

Below we draw the distinguished Dynkin diagrams of Lie superalgebras associated with a distinguished simple root systems.

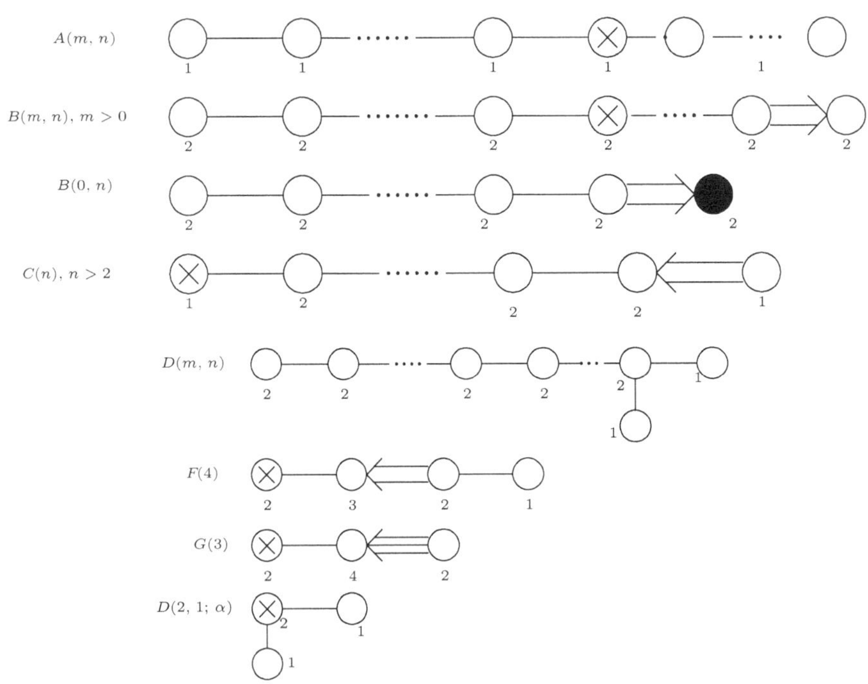

Definition 196 (Dynkin diagrams for low rank Lie superalgebras). The diagrams arising from Lie superalgebras of rank ≤ 2 are given below.

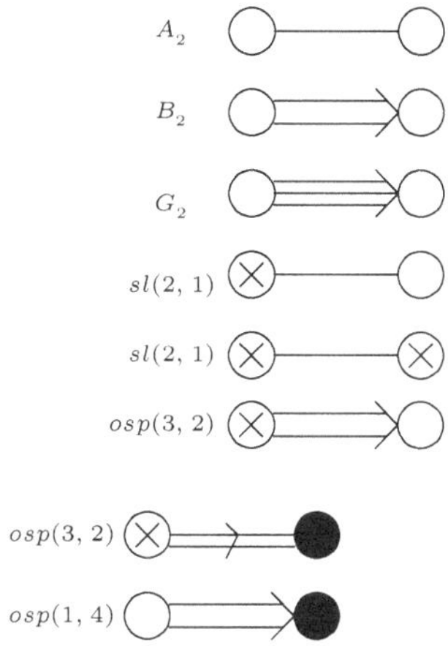

The nodes ◯, ⊗ and • are called white, gray(crossed), and black, respectively. The white, gray, and black nodes correspond to an even root, an odd isotropic root, and an odd nonisotropic root respectively.

Main features in the theory of finite dimensional Lie superalgebras which are different from those of Lie algebras:

(1) Let $\mathcal{G}$ be a finite dimensional Lie superalgebra. Then $\mathcal{G}$ contains a unique maximal solvable ideal R. But note that Levi's theorem on $\mathcal{G}$ being a semi direct sum of R and $\mathcal{G}/R$ is not true in general, for Lie superalgebras. The Lie superalgebra $\mathcal{G}/R$ is semisimple. Therefore, the theory of finite-dimensional Lie superalgebras is reduced in a certain sense to the theories of semisimple and of solvable Lie superalgebras.

(2) The main fact in the theory of solvable Lie algebras is Lie's theorem which asserts that every finite-dimensional irreducible representation of a solvable Lie algebra over $\mathbb{C}$ is one-dimensional. For Lie superalgebras, this is not true in general.

(3) A semisimple Lie algebra is a direct sum of simple ones whereas this is not true for Lie superalgebras. But using the construction, one can describe finite dimensional Lie superalgebras in terms of simple ones.

(4) When we study the fundamental problem of classifying finite-dimensional simple Lie superalgebras over an algebraically closed field of characteristic 0, the principle difficulty lies in the fact that the Killing form may be degenerate which can not happen in the case of simple Lie algebras. Therefore the classical technique of Killing-Cartan is not applicable here.

4.12 Lie superalgebras over an algebraically closed field of characteristic zero

Classification of classical Lie superalgebras

The classical Lie superalgebras that are not Lie algebras have the following three classes namely,

(a) Four series $A(m, n), B(m, n), c(n)$ and $D(m, n)$
(b) Three *exceptional* Lie superalgebras, namely

  (i) 40-dimensional $F(4)$,
  (ii) A family of 17-dimensional exceptional-Lie superalgebras $D(2, 1; \alpha)$,
  (iii) 31-dimensional $G(3)$,

(c) Two strange series $P(n)\&Q(n)$

Finite dimensional Lie superalgebras over an algebraically closed field of characteristic 0 can be divided into two main classes, namely,

(1) Classical Lie superalgebras(classical LSA).
(2) Non-classical Lie superalgebras(non-classical LSA). Further, in classical Lie superalgebras, we have the following two subdivisions:

  (i) classical Lie superalgebras with zero Killing form,
  (ii) classical Lie superalgebras with non-zero Killing form.

The above classification of simple finite dimensional Lie superalgebras over an algebraically closed field of characteristic 0 can also be described as follows:

Lie superalgebras with characteristic zero
$\swarrow \searrow$
classical LSA   nonclassical LSA
$\swarrow \searrow$
zero Killing form   nonzero Killing form

In the above classification (1), we have the following classical Lie superalgebras:

(i) (a) classical Lie superalgebra of zero Killing form, namely, $A(n, n), D(n + 1, n), P(n), Q(n)$ and $D(2, 1; \alpha)$,
  (b) classical Lie superalgebra of non-degenerate Killing form namely $A(m, n) m \neq n, B(m, n), C(n), D(m, n) (m - n \neq 1), F(4)$ and $G(3)$.

We state below two propositions [56].

Proposition 34. *Let $\mathcal{G} = \mathcal{G}_{\bar{0}} \oplus \mathcal{G}_{\bar{1}}$ be one of the Lie superalgebras $A(m, n), B(m, n), C(n), D(m, n), D(2, 1; \alpha), F(4), G(3), P(n)$ or $Q(n)$. Then the $\mathcal{G}_{\bar{0}}$-module $S^2\mathcal{G}_{\bar{1}}$ contains $\mathcal{G}_{\bar{0}}$ with multiplicity 1.*

Proposition 35. *Let $\mathcal{G} = \mathcal{G}_{\bar{0}} \oplus \mathcal{G}_{\bar{1}}$ be a simple Lie superalgebra for which the representation of $\mathcal{G}_{\bar{0}}$ on $\mathcal{G}_{\bar{1}}$ is the same as for one of the Lie superalgebras $A(m, n), B(m, n), C(n), D(m, n)($ where $(m, n) \neq (2, 1)), F(4), G(3), P(n)$ or $Q(n)$. Then $\mathcal{G}$ is isomorphic to this algebra.*

Remark 116. Kac classifies all finite dimensional *simple Lie superalgebras*. The remaining algebras which are not classical are often called *simple superalgebras of Cartan type*. As in the case of semisimple Lie algebras there are number of infinite families and finite number of exceptions for Lie superalgebras. For Lie superalgebras, the difference is the existence of infinite families of Lie superalgebras $D(2, 1; \alpha)$ depending on the continuous parameter α.

Classification of simple Lie superalgebras with non-degenerate Killing form

We state the following Theorems 49–52 [34].

Theorem 49. *Let $\mathcal{G} = \mathcal{G}_{\bar{0}} \oplus \mathcal{G}_{\bar{1}}$ be a simple Lie superalgebra with non-degenerate Killing form for which the representation of $\mathcal{G}_{\bar{0}}$ on $\mathcal{G}_{\bar{1}}$ is irreducible and $\mathcal{G}_{\bar{1}} \neq 0$. Then $\mathcal{G}$ is isomorphic to $B(m,n), D(m,n)$ with $m-n \neq 1$, $F(4)$, or $G(3)$.*

Theorem 50. *Let $\mathcal{G} = \mathcal{G}_{-1} \oplus \mathcal{G}_0 \oplus \mathcal{G}_1$ be a simple Lie superalgebra with a consistent Z-grading for which the representations of $\mathcal{G}_0$ on $\mathcal{G}_1$ and $\mathcal{G}_{-1}$ are faithful and irreducible and the Killing form is nondegenerate. Then $\mathcal{G}$ is isomorphic (even as a Z-graded superalgebra) to one of $A(m,n), m \neq n$, or $C(n)$.*

Theorem 51. *A classical Lie superalgebra $\mathcal{G} = \mathcal{G}_{\bar{0}} \oplus \mathcal{G}_{\bar{1}}$ with non-zero Killing form is isomorphic to one of $A(n,n), D(n+1,n), P(n), Q(n)$ or $D(2,1;\alpha)$.*

Theorem 52. *A simple finite-dimensional Lie superalgebra $\mathcal{G} = \mathcal{G}_{\bar{0}} \oplus \mathcal{G}_{\bar{1}}$ with non-degenerate Killing form is isomorphic to one of the simple Lie algebras or to one of the following classical Lie superalgebras:*

$$A(m,n) \text{ with } m \neq n,\ B(m,n),\ C(n),\ D(m,n) \text{ with } m-n \neq 1, F(4), \text{ or } G(3).$$

Definition 197. A Lie superalgebra is called *basic classical* if the algebra is simple, finite dimensional, the even part is reductive and carries an even, non-degenerate symmetric invariant bilinear form. They have been classified by Kac [34]. The basic classical Lie superalgebras and their corresponding decompositions of even parts are given in Table 4.2.

Classification of classical Lie superalgebras with zero Killing form

Proposition 36. *Let $\mathcal{G} = \mathcal{G}_{-1} \oplus \mathcal{G}_0 \oplus \mathcal{G}_1$ be a simple Lie superalgebra with a consistent Z-grading for which the representations of $\mathcal{G}_0$ on $\mathcal{G}_1$ and $\mathcal{G}_{-1}$ are faithful and irreducible and the Killing form is zero. Then $\mathcal{G}$ is isomorphic (even as a Z-graded superalgebra) to one of $A(n,n)$ or $P(n)$.*

Remark 117. One can refer [129], for the following details regarding the classification of Lie superalgebras (see Table 4.3).

Table 4.2 **List of basic classical Lie superalgebras and the decomposition of the even part [34]**

A(m,n)	$A_m + A_n + \mathbb{C}$	$m \geq 0, n \geq 0, m+n \geq 1$
B(m,n)	$B_m + C_n$	$m \geq 0, n \geq 1$
C(n)	$C_{n-1} + \mathbb{C}$	$n \geq 3$
D(m,n)	$D_m + C_n$	$m \geq 2, n \geq 1$
D(2,1,a)	$D_2 + A_1$	$a \neq 0, -1$
F(4)	$B_3 + A_1$	
G(3)	$G_2 + A_1$	

Table 4.3 **Classical Lie superalgebras [129]**

	Type I	**Type II**
BASIC (non-degenerate Killing form)	$C(n+1), n \geq 1$ A(m,n) $m > n \geq 0,$	B(m,n), $m \geq 0, n \geq 1$ D(m, n), $m \geq 2, n \geq 1, m \neq n+1$ F(4) G(3)
BASIC (zero Killing form)	A(n, n) $n \geq 1$	D(n+1,n), $n \geq 1$ $D(2,1;\alpha)$, $\alpha \notin \{0,-1\}$
STRANGE	P(n), $n \geq 2$	Q(n), $n \geq 2$

Some results on finite dimensional classical simple Lie superalgebras [34, 130]

General Theorem: A classical Lie superalgebra is isomorphic either to one of the simple Lie algebras $A_n, B_n, C_n, D_n, E_6, E_7, E_8, F_4, G_2$ or to one of $A(m,n), B(m,n), C(n),$ $D(m,n), D(2,1:\alpha), F(4), G(3), P(n), Q(n)$.

From the above results on finite dimensional classical simple Lie superalgebras, one can arrive at the following table. Below we give the tables on the classification of Lie superalgebras and their dimensions.

A finite dimensional classical simple Lie superalgebra $\mathcal{G}$ is either a simple Lie algebra or isomorphic to one of the following Lie superalgebras:

A(m,n)	= sl(m+1,n+1) with $m > n \geq 0,$
A(n,n)	= psl(n+1, n+1) with $n \geq 1$, $(A(1,0) \cong C(2))$
B(m,n)	= osp(2m+1, 2n) with $m \geq 0$, n>0
C(n)	= osp(2, 2n-2) with n > 2,
D(m, n)	= osp(2m, 2n) with $m \geq 2, n \geq 1,$
$D(2,1;\alpha)$	$= \Gamma(1,-1,-\alpha,\alpha)$ $\alpha \neq 0, -1 (D(2,1) \cong D(2,1,1).)$
p(n),	$n \geq 2;$
q(n),	$n \geq 2,$
G(3),	a simple algebra of dimension 31
F(4),	a simple algebra of dimension 40.

Moreover, there exists infinite families of Lie superalgebras $D(2,1;\alpha)$ depending upon the continuous parameter α.

List of simple Lie superalgebras (over an algebraically closed field of characteristic zero) and their dimensions [34]:

Lie superalgebras	Dimensions
$A(m,n)$	$(m+n+2)^2-1-\delta_{m,n}$
$B(m,n)$	$2(m+n)^2+m+3n$
$C(n)$	$2n^2+n-2$
$D(m,n)$	$2(m+n)^2-m+n$
$D(2,1;\alpha)$	17
$F(4)$	40
$G(3)$	31
$P(n)$	$2(n+1)^2-1$
$Q(n)$	$2(n+1)^2-2$

Table 4.4 **Z_2-gradation of the classical Lie superalgebras [129]**

Superalgebras $\mathcal{G}$	**Even part $\mathcal{G}_{\overline{0}}$**	**Odd part $\mathcal{G}_{\overline{1}}$**
A(n,n)	$A_n \oplus A_n$	$(n,\overline{n}) \oplus (\overline{n},n)$
C(n+1)	$C_n \oplus U(1)$	$(2n) \oplus (2n)$
B(m,n)	$B_m \oplus C_n$	(2m+1, 2n)
D(m,n)	$D_m \oplus C_n$	(2m, 2n)
F(4)	$A_1 \oplus B_3$	(2,8)
G(3)	$A_1 \oplus G_2$	(2,7)
$D(2,1;\alpha)$	$A_1 \oplus A_1 \oplus A_1$	(2,2,2)
P(n)	A_n	$[2] \oplus [1^{n-1}]$
Q(n)	A_n	$ad(A_n)$

Remark 118.

(i) The Z_2 gradation of classical Lie superalgebras are in Table 4.4.
(ii) If $\mathcal{G} = \mathcal{G}_{\overline{0}} \oplus \mathcal{G}_{\overline{1}}$ is a classical simple Lie superalgebra (with $\mathcal{G}_{\overline{1}} \neq 0$), then all finite dimensional representations of $\mathcal{G}$ are completely reducible if and only if $\mathcal{G} \cong osp(1,2n)$ for some $n \geq 1$.

Remark 119. Here $[.] \oplus [.]$ denotes the description of direct sum of the corresponding Lie superalgebras in terms of Young diagram (Appendix). Moreover, for the notation $(m,\overline{n}) \oplus (\overline{m},n)$, one can refer Appendix, Frappet et al. [129].

4.13 Classification of non-classical Lie superalgebras

Non-classical Lie superalgebras of Cartan type are $W(n), S(n)$ and $H(n)$ [similar to the series of simple infinite dimensional Lie algebras of Cartan type]. For the following discussion on Cartan superalgebra $W(n)$ and their properties, special Lie superalgebras $S(n), \tilde{S}(n)$, Hamiltonian Lie superalgebras $H(n)$ and $\tilde{H}(n)$ and their properties, one can refer [34].

Definition 198 (Cartan Lie superalgebra W(n)). Let $\wedge(n)$ be the Grassmann superalgebra with the generators $\xi_1, \ldots, \xi_n$. We denote *der* $\wedge(n)$ by W(n). We can prove that every derivation $D \in W(n)$ can be written in the form

$$D = \sum_i P_i \frac{\partial}{\partial \xi_i}.$$

Here $P_i \in \wedge(n)$ and $\frac{\partial}{\partial \xi_i}$ is the derivation defined by $\frac{\partial}{\partial \xi_i}(\xi_j) = \delta_{ij}$.

Let $deg \xi_i = 1, i = 1, \ldots, n$. Hence *a consistent z-grading* of $\wedge(n)$, induces a Z-grading of $W(n) = \bigoplus_{k \geq -1} W(n)_k$. Here

$$\begin{aligned} W(n)_k &= \{D \in W(n) | D(\wedge(n)_s) \subset \wedge(n)_{s+k}\} \\ &= \{\sum P_i \frac{\partial}{\partial \xi_i} | deg P_i = k+1, i = 1, \ldots, n\} \end{aligned}$$

In particular, $W(n)_{-1} = \langle \frac{\partial}{\partial \xi_i}, \ldots, \frac{\partial}{\partial \xi_n} \rangle$. Hence $[\frac{\partial}{\partial \xi_i}, \frac{\partial}{\partial \xi_j}] = 0$, this implies

$$\frac{\partial}{\partial \xi_i} \frac{\partial}{\partial \xi_j} = -\frac{\partial}{\partial \xi_j} \cdot \frac{\partial}{\partial \xi_i}$$

$W(n)$ is a Lie superalgebra, called Cartan Lie superalgebra. This is a Lie superalgebra arising from differential geometry.

Proposition 37.

(1) *$W(n) = \oplus_{i=-1}^{n-1} W(n)_i$ is transitive.*
(2) *The $W(n)_0$-module $W(n)_{-1}$ is isomorphic to gl_n.*
(3) *$W(n)_k = W(n)_k^l, k \geq 1$.*
(4) *$W(n)$ is simple for $n \geq 2$.*
(5) *If $G = \oplus_{i \geq -1} G_i$ is a transitive Z-graded Lie superalgebra for which the G_0-modules are $W(n)_{\pm 1}$, then $G \simeq W(n)$.*

Definition 199 (Special and Hamiltonian Lie superalgebras). *Special Lie superalgebras $S(n)$ and $\widetilde{S}(n)$.* One can define a volume form which is a differential form $\Theta(n)$, $w = f\theta\xi_1 \wedge \cdots \wedge \theta\xi_n, f \in \wedge(n)_{\bar{0}}, f(0) \neq 0$. To a volume form ω, there corresponds in $W(n)$, the subalgebra,

$$S(w) = \{D \in W(n) | Dw = 0\}.$$

Among these subalgebras, we consider two subalgebras as follows: $S(n) = S(\theta\xi_1 \wedge \cdots \wedge \theta\xi_n)$ and $\widetilde{S}(n) = S((1 + \xi_1 \cdots \xi_n)\theta\xi_1 \wedge \cdots \theta\xi_n)$ for $n = 2k$. If $\sum P_i(\frac{\partial}{\partial \xi_i})$ belongs to this $S(w)$, then one can write $\sum \frac{\partial}{\partial \xi_i}(fP_i) = 0$. Hence one can prove that $S(w)$ is the linear span of elements like $f^{-1} \frac{\partial a}{\partial \xi_i} \frac{\partial}{\partial \xi_j} + f^{-1} \frac{\partial a}{\partial \xi_j} \frac{\partial}{\partial \xi_i}, a \in \wedge(n)$. One can induce a filtration on $S(w)$ from $W(n)$. $S(n)$ has a even $\mathbb{Z}$-grading. $S(w)$ and $S(n)$ satisfy following properties:

(1) $GrS(w) \simeq S(n) = \oplus_{k=-1}^{n-2} S(n)_k$.
(2) The semisimple part of $S(w)_{\bar{0}}$ is isomorphic to sl_n.
(3) The $S(n)_{\bar{0}}$-module $S(n)_{-1}$ is isomorphic to sl_n.

Definition 200 (Hamiltonian forms and the Lie superalgebras $H(n)$&$\widetilde{H}(n)$). A Hamiltonian form is a closed differential form in $\Omega(n)$ of the form $\omega = \sum_{i,j=1}^{n} \omega_{ij} d\xi_i \circ$

$d\xi_j$, $\omega_{ij} \in \wedge(n)$, $\omega_{ij} = \omega_{ji}$, $det(\omega_{ij}(0)) \neq 0$. To a Hamiltonian form ω, there corresponds a subalgebra of $W(n)$:

$$\widetilde{H}(w) = \{D \in W(n) | Dw = 0\}$$

We set, $H(w) = [\widetilde{H}(w), \widetilde{H}(w)], \widetilde{H}(n) = \widetilde{H}((d\xi_1)^2 + \cdots + (d\xi_n)^2)$

and $H(n) = [\widetilde{H}(n), \widetilde{H}(n)]$.

Theorem 53 (For proof, one can refer [34]). *A simple finite dimensional Lie superalgebra over an algebraically closed field of characteristic zero is isomorphic either to one of the simple Lie algebras or to one of the Lie superalgebras* $A(m,n), B(m,n), C(n), D(m,n), D(2,1;\alpha), F(4), G(3)$, $P(n), Q(n), W(n), S(n)$, $\tilde{S}(n)$, *or* $H(n)$.

List of classification of classical Lie superalgebras was already given. Below we give the tables on the classification of non-classical Lie superalgebras and their dimensions.

Lie superalgebras	Dimensions
$W(n)$	$n \cdot 2^n$
$S(n)$	$(n-1)2^n + 1$
$\tilde{S}(n)$	$(n-1)2^n + 1$
$H(n)$	$2^n - 2$

List of non-classical Lie superalgebras (over an algebraically closed field of characteristic zero) and their dimensions:

Definition 201. The Lie superalgebras $W(n)$ for $n \geq 3$, $S(n)$ for $n \geq 4$, $\tilde{S}(n)$ for even $n \geq 4$ and $H(n)$ for $n \geq 5$ are called Cartan Lie superalgebras (for other values of n they are either not simple or isomorphic to classical Lie superalgebras: $W(2) \simeq A(1,0) \simeq C(2), S(3) \simeq P(2), H(4) \simeq A(1,1)$).

Remark 120. Z-gradations of the classical Lie superalgebras are given below in Table 4.5.

Table 4.5 **Z-gradation of the classical Lie superalgebras [129]**

superalgebras $\mathcal{G}$	**even part** $\mathcal{G}_{\overline{0}}$	$\mathcal{G}_1 \oplus \mathcal{G}_{-1}$	$\mathcal{G}_2 \oplus \mathcal{G}_{-2}$
A(m,n)	$A_m \oplus A_n \oplus U(1)$	$(m, \overline{n}) \oplus (\overline{m}, n)$	O
A(n,n)	$A_n \oplus A_n$	$(n, \overline{n}) \oplus (\overline{n}, n)$	O
C(n+1)	$C_n \oplus U(1)$	$(2n)_+ \oplus (2n)_-$	O
B(m,n)	$B_m \oplus A_{n-1} \oplus U(1)$	$(2m+1, n) \oplus (2m+1, n)$	$[2] \oplus [2^{n-1}]$
D(m,n)	$D_m \oplus A_{n-1} \oplus U(1)$	$(2m, n) \oplus (2m, \overline{n})$	$[2] \oplus [2^{n-1}]$
F(4)	$B_3 \oplus U(1)$	$8_+ \oplus 8_-$	$1_+ \oplus 1_-$
G(3)	$G_2 \oplus U(1)$	$7_+ \oplus 7_-$	$1_+ \oplus 1_-$
$D(2,1;\alpha)$	$A_1 \oplus A_1 \oplus U(1)$	$(2,2)_+ \oplus (2,2)_-$	$1_+ \oplus 1_-$

Classification of the simple Lie superalgebras can be described as follows [129]

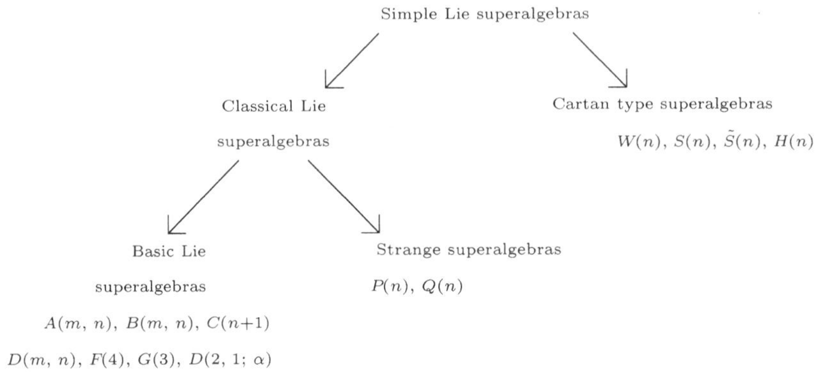

Remark 121.

(1) The following authors independently obtained classification *results on classical Lie superalgebras* under the following restrictions:
  (i) A. Paris and V. Rittenberg [131]: $\mathcal{G}_{\bar{0}}$ *is simple and the Killing form is non-degenerate.*
  (ii) D.Z. Djokovic [132], J. Pure and Appl. Algebra: $\mathbf{G}_{\bar{0}}$ *is simple.*
  (iii) P.G.O. Freund and I. Kaplansky [133]: $\mathbf{G}_{\bar{0}}$ *is reductive and* $\mathbf{G}$ *admits bilinear invariant form.*
  (iv) W. Nahm, V. Rittenberg and M. Scheunert [134]: *Killing form is non degenerate.*

(2) For details regarding, toroidal Lie superalgebra and integrable representations for toroidal Lie superalgebras one can refer to [85, 86].

(3) **(i)** For details regarding representations of the affine superalgebra germ $q(n)^2$, one can refer [135].
  (ii) For details regarding classification of finite growth general Kac-Moody superalgebras, one can refer [136].

(4) **(i)** For survey of some recent developments in the highest weight representation theory of the general linear Lie superalgebra $gl(n|m)(C)$, one can refer to [137].
  (ii) For development of a general theory of tilting modules for graded Lie superalgebras, extending work of Soergel for graded Lie algebras, one can refer [138].
  (iii) For computation of the characters of the finite dimensional irreducible representations of the Lie superalgebra $gl(m|n)$, one can refer to [139].

4.14 Lie superalgebras of vector fields

For supermanifolds, one can refer [140].

Definition 202 (Supermanifold). Let $U \subset R^p$ be an open set, $A(U)$ be an algebra and $\Theta(U)$ be the algebra of infinitely differentiable k-valued(real or complex) functions defined on U. Let $h : A(U) \to \Theta(U)$ be an isomorphism and $x_i \in \Theta(U)$ be such that $y_i = h^{-1}(x_i) \in A(U)$. Hence the elements y_i are topological generators in A(U). Let $\phi(x) = \phi(x_1, x_2, \ldots, x_p) \in \Theta(U)$. Now one can define $\phi(y) = \phi(y_1, y_2, \ldots, y_p)$ by $\phi(y) = h^{-1}(\phi) \in A(U)$. Hence the elements of $A(U)$ will be infinitely differentiable functions with generators y_i. Now assume a commutative algebra T with $\Lambda_T^q[\xi]$ being the exterior algebra over T with generators $\xi_1, \xi_2, \ldots, \xi_p$. Taking $\mathcal{U}_{p,q}(U) = \Lambda_{A(U)}^q[\xi]$, one can get an algebra $\mathcal{U}_{p,q}(U)$ with p even commuting generators y_i and odd q anti-commuting generators ξ_j. One can introduce Z_2-grading g on $\mathcal{U}_{p,q}(U)$ by $g(y_i) = 0$ and $g(\xi_i) = 1$. Any element of $\mathcal{U}_{p,q}(U)$ can be written as, $f(y, \xi) = \sum f^{i_1,\ldots,i_k}(y)\xi_{i_1}\cdots\xi_{i_k}$.

Let M be a smooth p-dimensional manifold. We assume the following conditions:

(1) For every coordinate neighborhood $U \subset M$ there exists an algebra $\mathcal{U}_{p,q}(U)$
(2) There exists a homomorphism $h_U : \mathcal{U}_{p,q}(U) \to \Lambda_{\Theta(U)}^q[\xi]$ such that $h_U(y_i) = x_i$ and $h_U(\xi_i) = \xi_i$.
(3) For $U \subset V$, there exists an epimorphism $\rho_V^U : \mathcal{U}_{p,q}(U) \to \mathcal{U}_{p,q}(V)$ preserving the grading and $(h_U(f))(x, \xi) = (h_V(\rho_V^U f))(x, \xi)$ for $f \in \mathcal{U}_{p,q}(U)$ and $x \in V$.
(4) The collection $\{\mathcal{U}_{p,q}(U), \rho_V^U\}$ of algebras and mappings is a fiber bundle, denoted by Θ_M^q. This bundle Θ_M^0 is isomorphic to the bundle of smooth functions on M.

Now the pair (M, Θ_M^q) is called *supermanifold.*

Definition 203. Any morphism in the category of pairs (M, Θ^q) is said to be morphism of supermanifolds.

Remark 122. In particular, an element of the algebra $\mathcal{U}_{p,q}(U) = \Gamma(U, \Theta_M^q)$ is a local cross section of the bundle Θ_M^q. This algebra is similar to the algebra of smooth functions defined in a coordinate neighborhood of an ordinary manifold. Moreover, the algebra of global cross sections $\Gamma(U, \Theta_M^q)$ is similar to the algebra of all smooth functions on an ordinary manifold. A supermanifold $(M', \Theta_{M'}^{q'})$ is a subsupermanifold of the supermanifold (M, Θ^q) if for every coordinate neighborhood $U \subset M$ which is small enough there exists an epimorphism

$$t_U : \Theta_{p,q}(U) \to \mathcal{U}_{p',q'}(U'), \quad U' = U \cap M'$$

such that $t_V\rho_V^U = \rho_{V'}^{\prime U'} t_U$, where $V \subset U, V' = V \cap M'$ and $\rho_V^U(\rho_{V'}^{\prime U'})$ is a restriction mapping of the bundle $\Theta_M^q(\Theta_{M'}^{q'})$.

A complex supermanifold can be defined using the sheaf of germs of holomorphic functions on an open set $U \subset \mathbb{C}^p$. For complex supermanifold and Lie superalgebras of vector fields, one can refer [239].

Definition 204 (Complex supermanifold). Let $\phi : U \to \wedge(\xi_1, \ldots, \xi_n)$ be an holomorphic function in a domain $U \subset \mathbb{C}^p$. Let z_0 be a point in U. Then ϕ can expanded in terms of power series $x_1, x_2, \ldots, x_p, \xi_1, \ldots, \xi_q$, where x_i are any local coordinates in a neighborhood of z_0 such that $z_0 = (0, 0, \ldots, 0)$. Hence one can assume that ϕ is an analytic function with even coordinates $x_1, \ldots, x_p$ and odd coordinates $\xi_1, \ldots, \xi_q$ such that $\phi = \phi(x_1, x_2, \ldots, x_p, \xi_1, \ldots, \xi_q)$. Assume that F is the sheaf of germs of ordinary holomorphic functions in U. Let $\wedge_F(\xi_1, \ldots, \xi_q) = F \otimes \wedge(\xi_1, \ldots, \xi_q)$ be the sheaf of holomorphic functions with values in $\wedge(\xi_1, \ldots, \xi_q)$.

A twister space (M, Θ) over the field $\mathbb{C}$ locally isomorphic to a space of the form $(U, \Lambda_F(\xi_1, \ldots, \xi_q))$, where U is an open set in $\mathbb{C}^p$ is a *complex supermanifold*(supermanifold). The sheaf Θ is called a structure sheaf.

Remark 123. Let M be a topological space. A supermanifold can be defined by an atlas in M. Let $x_1, \dots, x_p$ be the even coordinates and $\xi_1, \dots, \xi_q$ be the odd coordinates in the intersection with another chart where the $\tilde{x}_i, \tilde{\xi}_j$ are defined. We have the expressions of the form $\tilde{x}_i = \phi_i(x_1, \dots, x_p, \xi_1, \dots, \xi_q)$ and $\tilde{\xi}_i = \psi_i(x_1, \dots, x_p, \xi_1, \dots, \xi_q)$, where ϕ_i and ψ_j are analytic functions.

Example 46. Let M be an ordinary complex manifold of dimension n and $\Theta = \Omega$ be the sheaf of germs of holomorphic exterior forms on M. If $x_1, \dots, x_n$ is the local coordinates and $\mathcal{F}$ is the sheaf of germs of holomorphic forms on M, then locally $\Omega = \lambda_{\mathcal{F}}(dx_1, \dots, dx_n)$. Hence we obtain a supermanifold of dimension $n|n$ with even local coordinates x_i and odd local coordinates $\xi_j = dx_j$. If $\tilde{x}_1, \dots, \tilde{x}_n$ is another local coordinate system on M and $\tilde{\xi} = d\tilde{x}_j$, then the transition functions will be of the form

$$\tilde{x}_i = \phi_i(x, \dots, x_n),\ \tilde{\xi}_j = \sum_k \frac{\partial \phi_j}{\partial x_k} \xi_k.$$

Hence ϕ_i do not depend on ξ_j and ψ_j are linear in ξ_k. In general, if E is the holomorphic vector bundle over a complex manifold M and $\mathcal{E}$ is the sheaf of germs of its holomorphic cross sections, then assume $\Theta = \Lambda_{\mathcal{F}}\mathcal{E}$. We obtain a supermanifold (M, Θ). These supermanifolds are called split supermanifolds. We have

$$\Theta = \bigoplus_{k \in Z} \Theta_k,$$

where $\Theta_k = \Lambda^k_{\mathcal{F}}\mathcal{E}$. Hence a structure sheaf Θ of a split supermanifold is Z-graded in the above sense.

Any supermanifold of dimension $p|1$ is split. Any differentiable (real) supermanifold is split as well.

Definition 205 (Superalgebras of vector fields). Let (M, Θ) be a supermanifold. Denote by $Der\Theta$, the sheaf of derivations of Θ, which is a sheaf of Lie superalgebras. The filtration

$$x_{11}x_{12} \neq 0,\ x_{21}x_{12} \neq 0,\ x_{11}x_{22} \neq 0,\ x_{21}x_{22} \neq 0$$

generates a filtration in $Der\Theta$. The Lie superalgebra $\theta = \Gamma(M, Der\Theta)$ of sections of the sheaf $Der\Theta$ is called superalgebra of vector fields on (M, Θ). If M is compact, then $dim\, \theta$ is finite.

If the supermanifold (M, Θ) is split, then the sheaf $Der\Theta$ is the sheaf of graded Lie superalgebras and θ is a graded Lie algebra satisfying

$$\theta = \bigoplus_{S=-1}^{q} \theta_S.$$

Example 47. Consider the above supermanifold (M, Θ). Let $\mathcal{G}$ be the Lie algebra of holomorphic vector fields on M. Each vector field $v \in \mathcal{G}$ defines a derivation of degree 0 of the sheaf Θ the so called Lie derivative-which we shall also denote by v. Thus, $\mathcal{G}$ can be embedded into θ_0 as a superalgebra. On the other hand, a field $v \in \mathcal{G}$ defines a derivation $\tilde{v} \in \theta_{-1}$, inner multiplication on v. These derivations exhaust θ_{-1}. In θ_0 there is also a grading derivation ϵ defined by the formula $\epsilon(\phi) = s\phi$ for $\phi \in \Theta_s$. Finally the exterior differential d is an element of the space θ_1. Hence the following relations hold:

$$[v, \tilde{w}] = [v, w], [\tilde{v}, \tilde{w}] = [d, v] = 0,$$
$$[d, \tilde{v}] = v, \ [d, d] = 0.$$

It follows that $\tilde{\theta} = \theta_{-1} \oplus (g \oplus \langle\epsilon\rangle) \oplus < d >$ is a superalgebra of θ.

Below we give some definitions and properties of some *Lie superalgebras(basic superalgebras) of formal vector fields*. For details on these algebras, one can refer [141].

(1) *The Lie superalgebra $W(m,n)$:* Consider the Grassman superalgebra $\Lambda(n)$ in "n" odd indeterminates $\xi_1, \xi_2, \ldots, \xi_n$. Let $x_1, x_2, \ldots, x_m$ be "m" even indeterminates. Let $\Lambda(m,n) = \mathbb{C}[x_1, \ldots, x_m] \otimes \Lambda(n)$. It can be verified that $\Lambda(m,n)$ is an associative commutative superalgebra. Let $W(m,n)$ be the Lie superalgebra of derivations of $\Lambda(m,n)$. Then

$$W(m,n) = \left\{ \sum_{i=1}^{m} f_i \frac{\partial}{\partial x_i} + \sum_{i=1}^{n} g_i \frac{\partial}{\partial \xi_i} | (f_i, g_i) \in \Lambda(m,n) \text{ with} \right.$$

$\frac{\partial}{\partial x_i}$ is the even derivation uniquely determined by $\{\frac{\partial}{\partial x_i}(x_j) = \delta_{ij}$, $\frac{\partial}{\partial x_i}(\xi_j) = 0$

and $\frac{\partial}{\partial \xi_i}$ is the odd derivation uniquely determined by $\frac{\partial}{\partial \xi_i}(x_j) = 0$, $\frac{\partial}{\partial \xi_i}(\xi_j) = \delta_{ij}\}$

$W(m,n)$ is called general vectorial superalgebra.

To each vector field $X = \sum_{i=1}^{m} f_i \frac{\partial}{\partial x_i} + \sum_{i=1}^{m} g_i \frac{\partial}{\partial \xi_i}$, there exists its divergence,

$$div\, X = \sum_{i=1}^{m} \frac{\partial f_i}{\partial x_i} + \sum_{i=1}^{n} (-1)^{p(g_i)} \frac{\partial g_i}{\partial \xi_i}.$$

(2) *The Lie superalgebras $S(m,n)$, $S'(m,n)$:* Let $S'(m,n)$ be the subspace of $W(m,n)$ consisting of vector fields D with zero divergence. Hence $S'(m,n)$ is also a subalgebra of $W(m,n)$. Let $S(m,n)$ be the derived algebra of $S'(m,n)$. $S(m,n)$ coincides with $S'(m,n)$ if $m \neq 1$ or $m = 0$ and $n \geq 3$. One can define a gradation on $W(m,n)$ by $deg x_i = a_i = deg \frac{\partial}{\partial x_i}$ and $deg\, \xi_j = b_j = -deg \frac{\partial}{\partial \xi_j}$, where $a_i \in \mathbb{N}$ and $b_i \in \mathbb{Z}$. This gradation is called the gradation of type $(a_1, \ldots, a_m|b_1, \ldots, b_n)$. Hence $W(m,n) = \oplus_{j=-h}^{\infty} W(m,n)_j$, $(h \in \mathbb{N})$ is a graded Lie superalgebra of finite depth. These gradations will induce gradations on $S(m,n), S'(m,n)$ respectively. In particular, the gradation of type $(1,1,\ldots,1|1,1,\ldots,1)$ is called principal gradation of $W(m,n)$. Moreover, in this gradation $W(m,n)$ is the Lie superalgebra of depth 1 satisfying $W(m,n)_0 \cong gl(m,n)$ and $W(m,n)_{-1} \cong \mathbb{C}^{m|n}$, standard representation of $gl(m,n)$. Using these gradations, one can get gradations on $S(m,n)$ and $S'(m,n)$, called principal gradations. The gradation of type $(1,1,\ldots,1|0,0,\ldots,0)$ is called the subprincipal gradation of $W(m,n)$, $S(m,n)$ and $S'(m,n)$. Here *$S(m,n)$ is called special vectorial Lie superalgebra.*

Taking $degx_i = deg\xi_i = 1$, we will get principal gradation on $\Lambda(m,n)$ by $\Lambda(m,n) = \oplus_{j=0}^{\infty} \Lambda(m,n)_j$. Let $\Omega(m,n)$ be the superalgebra of differential forms over $\Lambda(m,n)$. Then one can show that $W(m,n)$ acts on $\Omega(m,n)$ via Lie derivatives.

(3) *The Lie superalgebras $H(2k,n)$ and $K(2k+1,n)$*: Let $p_1, p_2, \ldots, p_k, q_1, q_2, \ldots, q_k$ be 2k even indeterminates and $\xi_1, \xi_2, \ldots, \xi_n$ be n odd indeterminates. Let $\sigma = \sum_{i=1}^{k} dp_i dq_i + \sum_{i=1}^{n} d\xi_i d\xi_j$. Then $\sigma \in \Omega(2k,n)$. One can define the Hamiltonian Lie superalgebra as

$$H(2k,n) = \{D \in W(2k,n) | D\sigma = 0\}.$$

Now $\Lambda(2k,n) = \mathbb{C}[p_1, \ldots, p_k, q_1, q_2, \ldots, q_k] \otimes \Lambda(n)$. For $f, g \in \Lambda(2k,n)$, one can define Poisson bracket $[f,g] = \sum_{i=1}^{k} (\frac{\partial f}{\partial p_i}\frac{\partial g}{\partial q_i} - \frac{\partial f}{\partial q_i}\frac{\partial g}{\partial p_i}) - (-1)^{p(f)} \sum_{i=1}^{k} \frac{\partial f}{\partial \xi_i}\frac{\partial g}{\partial \xi_i}$. Denote by $\hat{H}(2k,n)$, the Lie superalgebra $\Lambda(2k,n)$ with the above Poisson bracket. One can define a surjective Lie superalgebra homomorphism from $\Lambda(2k,n)$ onto $H(2k,n)$ as

$$f \to \sum_{i=1}^{k} \left(\frac{\partial f}{\partial p_i}\frac{\partial}{\partial q_i} - \frac{\partial f}{\partial q_i}\frac{\partial}{\partial p_i} \right) - (-1)^{p(f)} \sum_{i=1}^{k} \frac{\partial f}{\partial \xi_i}\frac{\partial}{\partial \xi_i}.$$

Now $H(2k,n)$ can be identified with $\Lambda(2k,n)/\mathbb{C}1$ as a kernal of this map consists of constant functions. Hence *$H(2k,n)$ is a graded Lie superalgebra of depth* 1.

Let $t, p_1, p_2, \ldots, p_k, q_1, q_2, \ldots, q_k$ be $2k+1$ even indeterminates and $\xi_1, \xi_2, \ldots, \xi_n$ be the n odd indeterminates. Let $\sum = dt + \sum_{i=1}^{k} (p_i dq_i - q_i dp_i) + \sum_{i=1}^{n} \xi_i d\xi_i \in \Omega(2k+1,n)$ be the contact form. Now one can define the *contact superalgebra,*

$$K(2k+1,n) = \{D \in W(2k+1,n) | D\Sigma = f_D \Sigma\}, f_D \in \Lambda(2k+1,n)$$

with the contact bracket on the space $\Lambda(2k+1,n)$ by

$$[f,g] = (2-E)f\frac{\partial g}{\partial t} - \frac{\partial f}{\partial t}(2-E)g - \sum_{i=1}^{k} \left(\frac{\partial f}{\partial p_i}\frac{\partial g}{\partial q_i} - \frac{\partial f}{\partial q_i}\frac{\partial g}{\partial p_i} \right) + (-1)^{p(f)} \sum_{i=1}^{n} \frac{\partial f}{\partial \xi_i}\frac{\partial g}{\partial \xi_i}.$$

Here $E = \sum_{i=1}^{k} (p_i \frac{\partial}{\partial p_i} + q_i \frac{\partial}{\partial q_i}) + \sum_{i=1}^{n} \xi_i \frac{\partial}{\partial \xi_i}$ is the Euler operator. Hence $\Lambda(2k+1,n)$ becomes a Lie superalgebra and the map $\Lambda(2k+1,n) \to K(2k+1,n)$ given by $f \to (2-E)f\frac{\partial}{\partial t} + \frac{\partial f}{\partial t}E - \sum_{i=1}^{k} (\frac{\partial f}{\partial p_i}\frac{\partial}{\partial q_i} - \frac{\partial f}{\partial q_i}\frac{\partial}{\partial p_i}) + (-1)^{p(f)} \sum_{i=1}^{n} \frac{\partial f}{\partial \xi_i}\frac{\partial}{\partial \xi_i})$ is an isomorphism of Lie superalgebras.

The gradation of type $(2,1,\ldots,1|1,\ldots,1)$ is a gradation of depth 2, called the principal gradation of $K(2k+1,n)$.

(4) *The superalgebras $HO(n,n), SHO(n,n)$ and $SHO'(n,n)$:*

Let $\omega = \sum_{i=1}^{n} dx_i d\xi_i$ be a differential form on $\Omega(n,n)$. Then define the odd *Hamiltonian superalgebra* as

$$HO(n,n) = \{D \in W(n,n) | D\omega = 0\}.$$

Here the Lie superalgebra $HO(n,n)$ contains the *subalgebra of divergence zero vector fields*, namely,

$$SHO'(n,n) = \{D \in HO(n,n) | divD = 0\}.$$

Denote by $SHO(n,n), (n \geq 2)$, the derived algebra of $SHO'(n,n)$. Here $SHO(n,n)$ is an ideal of codimension 1 and it is simple if and only if $n \geq 3$. Moreover, we can obtain another realization of SHO(n,n) as follows:

One can define the Buttin bracket in $\Lambda(n,n)$ by

$$[f,g] = \sum_{i=1}^{k} \frac{\partial f}{\partial x_i}\frac{\partial g}{\partial \xi_i} + (-1)^{p(f)} \sum_{i=1}^{n} \frac{\partial f}{\partial \xi_i}\frac{\partial g}{\partial x_i}.$$

With this bracket, $\Lambda(n,n)$ becomes a Lie superalgebra. One can define a map $\Lambda(n,n) \to HO(n,n)$ by

$$f \to \sum_{i=1}^{n} \frac{\partial f}{\partial x_i}\frac{\partial}{\partial \xi_i} + (-1)^{p(f)} \sum_{i=1}^{n} \frac{\partial f}{\partial \xi_i}\frac{\partial}{\partial \xi_i}.$$

It is a surjective homomorphism of Lie superalgebras whose kernal consisting of constant functions. Hence one can identify $HO(n,n)$ with $\Lambda(n,n)/\mathbb{C}1$ with reversed parity. With this identification, one can write $SHO'(n,n) = \{f \in \Lambda(n,n)|\mathbb{C}1|\Delta(f) = 0,\}$ with $\Delta = \sum_{i=1}^{n} \frac{\partial^2}{\partial x_i \partial \xi_i}$. Here Δ is the odd Laplace operator. Moreover, $SHO(n,n)$ can be identified with the subspace consisting of the elements not containing the monomial $\xi_1\xi_2 \cdots \xi_n$.

The principal gradation of $W(n,n)$ will induce a principal gradation on $HO(n,n)$ and its subalgebras. *The Lie superalgebras $HO(n,n), SHO'(n,n)$, and $SHO(n,n)$ will become graded Lie superalgebras of depth* 1.

(5) *The Lie superalgebras $KO(n,n+1), SKO'(n,n+1;\beta)$ and $SKO(n,n+1;\beta)$:*

Assume that $x_1, x_2, ..x_n$ be n even indeterminates and $\xi_1, \xi_2, \ldots, \xi_n, \xi_{n+1} = \tau$ be odd n+1 odd indeterminates. Let $\Omega = d\tau + \sum_{i=1}^{n} (\xi_i dx_i + x_i d\xi_i) \in \Omega(n, n+1)$ be the odd contact form. The corresponding *odd contact superalgebra is* $KO(n,n+1)$, which is the subalgebra of $W(n,n+1)$ defined by

$$KO(n,n+1) = \{D \in W(n,n+1) \mid D\Omega = f_D \Omega\}$$

for some $f_D \in \mathbb{C}[x_1, x_2, \ldots, x_n, \xi_1, \xi_2, \ldots, \xi_n, \tau]$. The Lie superalgebra $KO(n,n+1)$ can be realized as follows: First we define odd contact bracket on the space $\Lambda(n,n+1)$ by

$$[f,g] = (2-E)f\frac{\partial g}{\partial \tau} + (-1)^{p(f)}\frac{\partial f}{\partial \tau}(2-E)g - \sum_{i=1}^{n}\left(\frac{\partial f}{\partial x_i}\frac{\partial g}{\partial \xi_i} + (-1)^{p(f)}\frac{\partial f}{\partial \xi_i}\frac{\partial g}{\partial x_i}\right).$$

Here $E = \sum_{i=1}^{n}(x_i \frac{\partial}{\partial x_i} + \xi_i \frac{\partial}{\partial \xi_i})$ is the Euler operator. Reversing parity, $\Lambda(n, n+1)$ with this bracket becomes a Lie superalgebra and the map $\Lambda(n, n+1) \to KO(n, n+1)$, given by $f \to (2-E)f\frac{\partial}{\partial \tau} - (-1)^{p(f)}\frac{\partial f}{\partial \tau}E - \sum_{i=1}^{n}(\frac{\partial f}{\partial x_i}\frac{\partial}{\partial \xi_i} + (-1)^{p(f)}\frac{\partial f}{\partial \xi_i}\frac{\partial}{\partial x_i})$, is an isomorphism of Lie superalgebras. Hence we may identify the Lie superalgebra $KO(n, n+1)$ with $\Lambda(n, n+1)$ with reversed parity.

Let $div_\beta = \Delta + (E - n\beta)\frac{\partial}{\partial x}$, for $\beta \in \mathbb{C}$. Here Δ is the odd Laplace operator. Let

$$SKO'(n, n+1; \beta) = \{f \in \Lambda(n, n+1) | div_\beta f = 0\}.$$

Denote by $SKO(n, n+1; \beta)$ the derived algebra of $SKO'(n, n+1; \beta)$. Hence the Lie algebra $SKO(n, n+1; \beta)$ is simple for $n \geq 2$ and coincides with $SKO'(n, n+1; \beta)$ if $\beta \neq 1$ or $\beta \neq \frac{n-2}{n}$. The Lie superalgebra $SKO(n, n+1; 1)$ consists of elements of $SKO'(n, n+1, 1)$ not containing the monomial $\tau\xi_1\xi_2\cdots\xi_n$. Similarly, the Lie superalgebra $SKO(n, n+1; \frac{n-2}{n})$ consists of elements of $SKO'(n, n+1, \frac{n-2}{n})$ not containing the monomial $\xi_1\xi_2\cdots\xi_n$. *One can note that $SKO(n, n+1; \frac{1}{n})$ is the subalgebra of $KO(n, n+1)$ consisting of divergence zero vector fields.*

Remark 124. For construction of exceptional simple Lie superalgebras of vector fields, one can refer [141].

Exercises

4.1 Prove that the graded vector space V, equipped with the inverted multiplication $(A, B) \to [B, A]$ is a Lie superalgebra.

4.2 If $\mathcal{G}$ is a superalgebra, then prove that a commutator algebra, $[\mathcal{G}, \mathcal{G}]$ of $\mathcal{G}$ is a graded ideal of $\mathcal{G}$.

4.3 Prove that the direct product of two Lie superalgebras is a superalgebra.

4.4 If $\mathcal{G}$ is a Lie superalgebra, then in the usual notation, prove that *Ad a* is a superderivation of $\mathcal{G}$ for all $a \in \mathcal{G}$.

4.5 Let $\mathcal{G}$ be a superalgebra and V be a graded $\mathcal{G}$-module. If x is the homogeneous element of V, then prove that $H = \{A \in \mathcal{G} | A_V(x) = 0\}(A_V(x)$ definition) is a graded subalgebra of $\mathcal{G}$ and x is H-invariant.

4.6 Let $\mathcal{G}$ be a simple Lie superalgebra. Prove that left or right graded ideal of $\mathcal{G}$ is a two sided ideal.

4.7 Prove that distinguished Cartan matrix and distinguished symmetric Cartan matrix of F(4) are $\begin{pmatrix} 0 & 1 & 0 & 0 \\ -1 & 2 & -2 & 0 \\ 0 & -1 & 2 & -1 \\ 0 & 0 & -1 & 2 \end{pmatrix}$ and $\begin{pmatrix} 0 & 1 & 0 & 0 \\ 1 & -2 & 2 & 0 \\ 0 & 2 & -4 & 2 \\ 0 & 0 & 2 & -4 \end{pmatrix}$ respectively associated to the distinguished simple root system $\Delta_0 = \{\alpha_1 = \frac{1}{2}(\delta - \epsilon_1 - \epsilon_2 - \epsilon_3), \alpha_2 = \epsilon_3, \alpha_3 = \epsilon_2 - \epsilon_3, \alpha_4 = \epsilon_1 - \epsilon_2\}$ with the scalar product $\epsilon_i.\epsilon_j = 2\delta_{ij}, \delta.\delta = 6, \epsilon_i.\delta = 0$.

4.8 Let $\mathcal{G}$ be a Lie superalgebra and $\wedge(n)$ a Grassmann superalgebra. Prove that $\mathcal{G} \otimes \wedge(n)$ is a Lie superalgebra.

4.9 If $\wedge(n) = \wedge_{\bar{0}}(n) \oplus \wedge_{\bar{1}}(n)$ is a Grassmann superalgebra, then find $der \wedge (n)$.

4.10 Prove that Lie superalgebra $\mathcal{G} = \mathcal{G}_{\bar{0}} \oplus \mathcal{G}_{\bar{1}}$ is solvable iff Lie algebra $\mathcal{G}_{\bar{0}}$ is solvable.

4.11 Prove that all the irreducible finite dimensional representations of a solvable Lie superalgebra $\mathcal{G} = \mathcal{G}_{\bar{0}} \oplus \mathcal{G}_{\bar{1}}$ are one-dimensional if and only if $[\mathcal{G}_{\bar{1}}, \mathcal{G}_{\bar{1}}] \subseteq [\mathcal{G}_{\bar{0}}, \mathcal{G}_{\bar{0}}]$.

4.12 Let $\mathcal{G}$ be a Lie superalgebra and A be a super commutative associative algebra over K. Set $\mathcal{G}_A = (\mathcal{G} \otimes A)_0 = (\mathcal{G}_0 \otimes A_0) \oplus (\mathcal{G}_1 \otimes A_1)$. Show that $\mathcal{G}_A$ becomes a Lie algebra, if we define

$$[x \otimes a, y \otimes b] = [x, y] \otimes ab$$

for $x, y \in \mathcal{G}, a, b \in A$.

4.13 If $\mathcal{G} = P(n)$, find a basis for each root space of $\mathcal{G}$, and show that if $\pm\alpha$ are both roots, then $[\mathcal{G}_\alpha, \mathcal{G}_{-\alpha}] \neq 0$.

4.14 Let $m = n + 1$ and for $A, B \in gl(m)$. Set $\langle A, B\rangle = \begin{vmatrix} A & B \\ B & A \end{vmatrix} \in \cap q(n)$. Show that the product in $\cap q(n)$ is given in this notation by

$$[< A, B >, < A', B' >] =< [A, A'] + [B, B'], [A, B'] + [B, A'] >$$

where $[B, B'] = BB' + B'B$.

4.15 Let $\mathcal{G}$ be a simple classical Lie superalgebra and $\alpha, \beta, \alpha + \beta$ are roots of $\mathcal{G}$. Prove that $[\mathcal{G}^\alpha, \mathcal{G}^\beta] = \mathcal{G}^{\alpha+\beta}$.

4.16 Let $\mathcal{G}$ be a Lie superalgebra whose Killing form is non-degenerate. Prove that any superderivation of $\mathcal{G}$ is inner.

4.17 If $\mathcal{G} = gl(m, n)$ and $A, B, C \in \mathcal{G}$, show that

(1) str(AB) = $(-1)^{\bar{A}\bar{B}}$str(BA),

(2) str([A, B]C) = str(A[B, C]),

(3) the bilinear form $\mathcal{G} \times \mathcal{G} \to K, (A, B) \mapsto str(AB)$ is nondegenerate.

4.18 If $\mathcal{G}$ is of type $A(m, n)$, show that $\kappa(x, y) = 2(m - n)str(xy)$ and if $\mathcal{G} = osp(m, n)$, show that $\kappa(x, y) = (m - n - 2)str(xy)$.

Borcherds Kac-Moody Lie superalgebras

5

Chapter Outline

5.1 BKM supermatrices and BKM Lie superalgebras 258
5.2 Dynkin diagrams of BKM Lie superalgebras and in particular Kac Moody Lie superalgebras and some examples 264
5.3 Domestic type and Alien type imaginary roots in BKM Lie superalgebras 272
5.4 Special imaginary roots in BKM Lie superalgebras and their complete classifications 281
5.5 Strictly and Purely imaginary roots in BKM Lie superalgebras and complete classification of BKM Lie superalgebras possessing purely imaginary roots 298
5.6 BKM Lie superalgebras possessing purely imaginary property but not strictly imaginary property 308
5.7 Complete classification of BKM Lie superalgebras possessing strictly imaginary property (SIM property) 315
5.8 Borcherds superalgebras and root supermultiplicities 325
5.9 Root supermultiplicities of Borcherds superalgebras which are extensions of Kac-Moody algebras and some combinatorial identities 351
5.10 Description of finite and infinite dimensional Lie algebras and Lie superalgebras and their different classes of root systems 368

Abstract

Wakimoto [142] introduced the definition of BKM supermatrix and BKM Lie superalgebras. A BKM supermatrix can be defined from a GKM matrix by introducing an additional subset corresponding to odd roots on the index set and hence we get a superalgebra structure. Hence BKM superalgebras can be considered as the Lie superalgebras associated to BKM supermatrices. In Kac-Moody algebras, all simple roots are real whereas just like GKM algebras, BKM Lie superalgebras can have imaginary simple roots. Moreover in the case of BKM superalgebras, simple roots (both real and imaginary) are basically of two types, namely, odd and even simple roots (both real and imaginary). Since an additional structure on the index set is introduced on a BKM supermatrix, the properties of imaginary roots in BKM Lie superalgebras will be different from those of BKM algebras. In BKM superalgebras, in addition to these

Introduction to Finite and Infinite Dimensional Lie (Super)algebras. http://dx.doi.org/10.1016/B978-0-12-804675-3.00005-4

classes of fundamental root systems of imaginary roots, there are alien and domestic type imaginary roots. The notion of Kac-Moody superalgebras was introduced by Kac [34] and therein the Weyl-Kac character formula for the irreducible highest weight modules with dominant integral highest weight which yields a denominator identity when applied to 1-dimensional representation was also derived. For canonical q-deformation of noncompact Lie (super)algebras, one can refer [143].

In this chapter, BKM superalgebras associated with BKM supermatrices, BKM superalgebras associated with symmetrizable supermatrices and Borcherds superalgebras associated with coloring matrices are separately defined. Basic concepts in BKM supermatrices, BKM Lie superalgebras, their Dynkin diagrams along with domestic and alien imaginary roots are explained with examples. In the context of these classes of basic root systems in BKM superalgebras, special imaginary roots, strictly imaginary roots and purely imaginary roots were studied. In this Chapter, all classes of BKM superalgebras which possess special imaginary roots and all those superalgebras which do not possess special imaginary roots were found out [144]. The concepts of strictly imaginary roots and purely imaginary roots of Borcherds Kac-Moody algebras (BKM algebras) were extended to BKM superalgebras and a complete classification of those BKM superalgebras with purely alien imaginary property and purely imaginary property were also given in this chapter [145]. Moreover, the properties of strictly imaginary roots and purely imaginary roots of BKM superalgebras were compared and the BKM superalgebras whose purely imaginary roots are also strictly imaginary were found out [145]. Complete classification of different classes of imaginary roots (special, strictly, purely) are also given.

Kim and Shin [117] derived a recursive dimension formula for all graded Lie algebras. Kang and Kim [115] constructed the dimension formula for graded Lie algebras. Moreover, Kang [126] derived a superdimension formula for the homogeneous subspaces of the graded Lie superalgebras, which enabled one to study the structure of the graded Lie superalgebras in a unified way. In Kang and Kwon [146], one can refer graded Lie superalgebras, supertrace formula and orbit Lie superalgebras. Using the Weyl-Kac-Borcherds formula and the denominator identity for the Borcherds superalgerbas, Kang and Kim [115] derived a dimension formula and combinatorial identities for the Borcherds superalgebras and found out the root multiplicities for Monstrous Lie superalgebras. In [147, 148], dimensional formulae, root supermultiplicities and corresponding combinatorial identities for the Borcherds superalgebras which are extensions of Kac-Moody algebras A_2, A_3 were computed. Moreover, a dimension formula for a Borcherds superalgebra which is an extension of a hyperbolic Kac-Moody algebra was computed. Particular cases of these Borcherds superalgebras were considered in [149] and therein dimension formulas were found out. Root supermultiplicities in Borcherds superalgebras along with superdimension formulae and corresponding combinatorial identities are also described in this chapter.

5.1 BKM supermatrices and BKM Lie superalgebras

In Chapter 3, BKM matrix (GKM matrix) and BKM algebras (GKM algebras) were already explained. We briefly explain below the fundamental concepts regarding BKM supermatrices and Borcherds Kac Moody Lie superalgebras (BKM Lie superalgebras) [142, 150].

Definition 206. Let $I = \{1, 2, \ldots, n\}$ be a countable (possibly infinite) index set, ψ be a subset of I and $A = (a_{ij})_{i,j \in I}$ be a real $n \times n$ matrix. The set (A, ψ) is called *Borcherds-Kac Moody supermatrix (BKM supermatrix)* if it satisfies the following conditions:

(1) $a_{ii} = 2$ or $a_{ii} \leq 0$ for all $i \in I$,
(2) $a_{ij} \leq 0$ if $i \neq j$ and $a_{ij} \in \mathbb{Z}$ if $a_{ii} = 2$,
(3) $a_{ij} = 0 \Leftrightarrow a_{ji} = 0$,
(4) If $i \in \psi$ and $a_{ii} = 2$ then $a_{ij} \in 2\mathbb{Z}$ for all j.

Remark 125. In addition to the conditions (1)–(3) required for Borcherds Cartan matrix (Chapter 3), condition (4) is required for BKM supermatrix. As in the case of BKM algebras, we define, the subsets Ire and I^{im} of I by

$$I^{\text{re}} = \{i \in I; a_{ii} = 2\} \text{ and } I^{\text{im}} = \{i \in I; a_{ii} \leq 0\}.$$

In the case of BKM supermatrix, the set ψ is non-empty. If ψ is an empty set then the BKM supermatrix coincides with the corresponding BKM matrix. Let

$$\psi^{\text{re}} := \{i \in \psi;\ a_{ii} = 2\} = I^{\text{re}} \cap \psi;\ \ \psi_0 := \{i \in \psi; a_{ii} = 0\};\ \ \psi_- := \{i \in \psi; a_{ii} < 0\}$$

and $\psi^{\text{im}} := \{i \in \psi; a_{ii} \leq 0\} = \psi_0 \cup \psi_-$.

Definition 207. Let I be an index set and (A, ψ) be an indecomposable BKM supermatrix, where $A = (a_{ij})_{i,j \in I}$ and $\psi \subseteq I$. Then one and only one of the following three possibilities holds for A.

(Fin): A is of *finite type* if $\det A \neq 0$; *there exists* $u > 0$ such that $Au > 0$ and $Au \geq 0 \Rightarrow u > 0$ or $u = 0$.
(Aff): A is of *affine type* if corank(A) =1; *there exists* $u > 0$ such that $Au = 0$ and $Au \geq 0 \Rightarrow Au = 0$.
(Ind): A is of *indefinite type* if $\det A \neq 0$; *there exists* $u > 0$ such that $Au < 0$ and $Au > 0,\ u > 0 \Rightarrow u = 0$.

Definition 208. We say that a BKM supermatrix (A, ψ) is of *hyperbolic type,* if it is indefinite type and every principal submatrix of A is either *finite or affine type* BKM supermatrix.

Definition 209. If a BKM supermatrix $A = (a_{ij})_{1 \leq i,j \leq n}$ decomposes as $A = DB$, where $D = (\varepsilon_i \delta_{ij})_{1 \leq i,j \leq n}$ is a diagonal matrix and $B = (b_{ij})_{1 \leq i,j \leq n}$ is a symmetric matrix, then A is said to be *symmetrizable*. Here ϵ_i (i =1,2, …,n) are non-zero integers and b_{ij} $(1 \leq i,j \leq n)$ are such that $a_{ij} \gtreqless 0 \Leftrightarrow b_{ij} \gtreqless 0$.

Remark 126. From the above definition, we have $a_{ij} = \varepsilon_i b_{ij}$, $a_{ji} = \varepsilon_j b_{ji}$ and since $b_{ij} = b_{ji}$, we have $\varepsilon_j a_{ij} = \varepsilon_i a_{ji}$ for each *i and j*.

Definition 210. For any BKM supermatrix (A, ψ), where $A = (a_{ij})_{1 \leq i,j \leq n}$, we have a triple $(\mathfrak{h}, \Pi, \Pi^{\vee})$, with $\Pi = \{\alpha_i; i \in I\}$ and $\Pi^{\vee} = \{\alpha_i^{\vee};\ i \in I\}$ satisfying the following relations:

(i) $\mathfrak{h}$ is a finite dimensional (complex) vector space such that $\dim \mathfrak{h} = 2n\text{-rank A}$,
(ii) $\Pi = \{\alpha_i\}_{i\in I} \subset \mathfrak{h}^*$ and $\Pi^\vee = \{\alpha_i^\vee\}_{i\in I} \subset \mathfrak{h}$ are linearly independent, where $\mathfrak{h}^* = \mathrm{Hom}_{\mathbb{C}}(\mathfrak{h}, \mathbb{C})$,
(iii) $\langle \alpha_j, \alpha_i^\vee \rangle = a_{ij}$, where $\langle , \rangle$ denotes a duality pairing between $\mathfrak{h}$ and $\mathfrak{h}^*$. This triple is called a *realization* of A. Call an element of Π (respectively $\Pi^\vee$) a *fundamental root* or a *simple root* (respectively *fundamental coroot* or a *simple coroot*).

Let $\Pi^{\mathrm{re}} = \{\alpha_i; i \in I^{\mathrm{re}}\}$ and $\Pi^{\mathrm{im}} = \{\alpha_i; i \in I^{\mathrm{im}}\}$. Here Π^{re} is called the *set of real simple roots* and Π^{im} is called the set of imaginary simple roots.

Remark 127. In addition to the above classification of fundamental roots, we have two more classes of fundamental roots namely,

$\Pi_{\mathrm{even}} := \{\alpha_i; i \in I\backslash\psi\}$, the set of all *even roots* and

$\Pi_{\mathrm{odd}} := \{\alpha_i; i \in \psi\}$, the set of all *odd roots*.

Example 48. Consider the BKM supermatrix (A, ψ) where $A = \begin{bmatrix} 0 & -1 & -2 & 0 & -2 \\ -2 & 2 & -2 & 0 & 0 \\ -2 & -2 & -k & -2 & -1 \\ 0 & 0 & -2 & 0 & -1 \\ -2 & 0 & -1 & -1 & 0 \end{bmatrix}$, with $k > 0$ and $\psi = \{2, 3, 5\}$.

Even and odd roots

In the above example

$\Pi^{\mathrm{re}}_{\mathrm{even}} = \emptyset$, $\Pi^{\mathrm{re}}_{\mathrm{odd}} = \{\alpha_2\}$, $\Pi^{\mathrm{im}}_{\mathrm{even}} = \{\alpha_1, \alpha_4\}$ and $\Pi^{\mathrm{im}}_{\mathrm{odd}} = \{\alpha_3, \alpha_5\}$.

Definition 211. We can define an inner product on $\mathfrak{h}$ as follows:

Let $\mathfrak{h}' = \sum_{i=1}^n \mathbb{C}\alpha_i^\vee$ and let $\mathfrak{h}''$ be the subspace of $\mathfrak{h}$ satisfying $\mathfrak{h} = \mathfrak{h}' \oplus \mathfrak{h}''$.

Now we can define the *symmetric bilinear form* $(\cdot|\cdot)$ on $\mathfrak{h}$ by

$$(\alpha_i^\vee | h) := \varepsilon_i \langle \alpha_i, h \rangle \text{ for all } h \in \mathfrak{h}, 1 \leq i \leq n \tag{5.1}$$

and $(h_1''|h_2'') := 0, \quad \forall\, h_1'', h_2'' \in \mathfrak{h}''$.

Remark 128. The bilinear form defined on $\mathfrak{h}$ for a BKM superalgebra is denoted by $(. | .)$, whereas the corresponding bilinear form for BKM algebra is denoted as $(., .)$. The bilinear form $(\cdot|\cdot)$ is non-degenerate, which induces an isomorphism $\nu : \mathfrak{h} \to \mathfrak{h}^*$.

For the sake of completeness, we state the following lemma. For proof, one can refer Wakimoto [142].

Lemma 10. *For* $1 \leq i, j \leq n$, *one has the following regarding the map* ν:

1. $\nu(\alpha_i^\vee) = \varepsilon_i \alpha_i$,
2. $(\alpha_i|\alpha_j) = b_{ij}$,
3. $(\alpha_i^\vee|\alpha_j^\vee) = \varepsilon_i \varepsilon_j b_{ij}$,
4. *If* $a_{ii} \neq 0$, *then* $\frac{2(\alpha_i|\lambda)}{(\alpha_i|\alpha_i)} = \frac{2\langle \alpha_i^\vee, \lambda \rangle}{a_{ii}}$, $\lambda \in \mathfrak{h}^*$.

   In particular if $a_{ii} = 2$, *then* $\frac{2(\alpha_i|\lambda)}{(\alpha_i|\alpha_i)} = \langle \alpha_i^\vee, \lambda \rangle (\lambda \in \mathfrak{h}^*)$.

Remark 129.

(1) In terms of inner product $(\cdot|\cdot)$, we have $\Pi^{\text{re}} = \{\alpha_i;\ (\alpha_i|\alpha_i) > 0\}$ and $\Pi^{\text{im}} = \{\alpha_i;\ (\alpha_i|\alpha_i) \leq 0\}$. The definitions for Π^{re} and Π^{im} are equivalent to the corresponding definitions for BKM algebra (refer Chapter 3).

(2) The definitions of graded vectorspaces, graded algebras, superalgebras, Lie superalgebras etc., were given in Chapter 4.

(3) Below we define BKM Lie superalgebra as defined in [142]. We give below two independent definitions of (i) BKM superalgebra associated with a BKM supermatrix, (ii) BKM superalgebra associated with a symmetrizable BKM supermatrix.

(i) BKM superalgebra associated with a BKM supermatrix

Definition 212. Let (A, ψ) be a BKM supermatrix. Denote by $\hat{\mathfrak{g}}(A, \psi)$, the Lie superalgebra, generated by the elements of $\mathfrak{h}$ and $e_1, e_2, \ldots, e_n, f_1, f_2, \ldots, f_n$ satisfying the following relations:

(i) $[h, h'] = 0$ for $h, h' \in \mathfrak{h}$,
(ii) $[h, e_i] = \alpha_i(h)e_i$ for $h \in \mathfrak{h}$, $1 \leq i \leq n$,
(iii) $[h, f_i] = -\alpha_i(h)f_i$ for $h \in \mathfrak{h}, 1 \leq i \leq n$,
(iv) $[e_i, f_j] = \delta_{ij}\alpha_i^\vee$ for $1 \leq i, j \leq n$.

Remark 130. If $\mathfrak{g}$ is a Lie superalgebra generated by Cartan subalgebra $\mathfrak{h}$ and generators $e_1, e_2, \ldots, e_n, f_1, f_2, \ldots, f_n$ satisfying the relations stated in Definition 212 and if there does not exist an ideal i of $\mathfrak{g}$ satisfying $i \cap \mathfrak{h} = \{0\}$, then $\mathfrak{g}$ is called a *simple* BKM Lie superalgebra.

$\hat{\mathfrak{g}}(A)$ has a natural decomposition of direct sum of root spaces as

$$\hat{\mathfrak{g}}(A) = \oplus_{\alpha \in \mathfrak{h}*} \hat{\mathfrak{g}}_\alpha,$$

where

$$\hat{\mathfrak{g}}_\alpha := \{X \in \hat{\mathfrak{g}}(A)\ ;\ [h, X] = \alpha(h)X \ \ (h \in \mathfrak{h})\}.$$

If $\alpha \neq 0$ and $\hat{\mathfrak{g}}_\alpha \neq \{0\}$, then α is called a root of $\hat{\mathfrak{g}}_\alpha$ and $\hat{\mathfrak{g}}_\alpha$ is called the root space of the root α. $\pm\alpha_i$ are roots of $\hat{\mathfrak{g}}(A)$ with $\hat{\mathfrak{g}}_{\alpha_i} = \mathbb{C}e_i$, $\hat{\mathfrak{g}}_{-\alpha_i} = \mathbb{C}f_i$.

A root α can be written as

$$\alpha = \sum_{i=1}^{n} m_i\alpha_i \ \ (m \in \mathbb{Z}),$$

m_i's are either all ≥ 0 or all ≤ 0. If m_i's are all ≥ 0 or all ≤ 0, then α is called a *positive root or a negative root* respectively. Also $\sum_{i=1}^{n} m_i$ is called the *height* of α and is denoted by $ht(\alpha)$. We denote by $\hat{\Delta}$, $\hat{\Delta}_+$, $\hat{\Delta}_-$, $\hat{\Delta}_{\bar{0}}$ and $\hat{\Delta}_{\bar{1}}$ the set of all roots, positive roots, negative roots, even roots and odd roots of $\hat{\mathfrak{g}}$ respectively.

Let

$$\hat{n}_+ := \sum_{\alpha \in \hat{\Delta}_+} \hat{\mathfrak{g}}_\alpha, \quad \hat{n}_- := \sum_{\alpha \in \hat{\Delta}_-} \hat{\mathfrak{g}}_\alpha.$$

Actually $\hat{n}_+$ is a free Lie superalgebra generated by $e_1, e_2, \ldots, e_n$ and $\hat{n}_-$ is a free Lie superalgebra generated by $f_1, f_2, \ldots, f_n$. We have a triangular decomposition of $\hat{\mathfrak{g}}(A)$,

$$\hat{\mathfrak{g}}(A) = \mathfrak{h} \oplus \hat{n}_+ \oplus \hat{n}_-.$$

Consider the elements

$$v_{ij}^+ := (ad\, e_i)^{1-a_{ij}} e_j,\ v_{ij}^- := (ad\, f_i)^{1-a_{ij}} f_j \text{ for } 1 \le i,j \le n.$$

Consider the ideal $\hat{\tau}$ of $\hat{\mathfrak{g}}(A)$ generated by $v_{ij}^{\pm}$, $[e_i, e_j]$ and $[f_i, f_j]$ for $1 \le i,j \le n$.

Let

$$\tilde{\mathfrak{g}}(A) := \hat{\mathfrak{g}}(A)/\hat{\tau}.$$

Remark 131. If A is symmetrizable BKM supermatrix, then $\tilde{\mathfrak{g}}(A)$ is a simple BKM Lie superalgebra. For the proof of this statement, one can refer [118, 142].

In general, if A is not a symmetrizable BKM supermatrix, consider the ideal $\tilde{\tau}$ of $\tilde{\mathfrak{g}}(A)$ satisfying the following conditions to get the associated simple Lie superalgebra $\mathfrak{g}(A)$:

(i) $\tilde{\tau}$ is an ideal of $\tilde{\mathfrak{g}}(A)$,
(ii) $\tilde{\tau} \subset \tilde{n}_+ \oplus \tilde{n}_-$, where $\tilde{n}_+$ and $\tilde{n}_-$ are free Lie superalgebras in $\tilde{\mathfrak{g}}(A)$ generated by $e_1, \ldots, e_n$ and $f_1, \ldots, f_n$ respectively,
(iii) $\tilde{\tau}$ is the maximum among those satisfying (i) and (ii).

Set $\mathfrak{g}(A) = \tilde{\mathfrak{g}}(A)/\tilde{\tau}$.

Definition 213. If (A, ψ) is a BKM supermartix, then $\mathfrak{g}(A)$ is called the *BKM Lie superalgebra or the BKM superalgebra* associated to (A, ψ).

(ii) BKM superalgebra associated with a symmetrizable BKM supermatrix

Remark 132. If A is symmetrizable, then one has $\tilde{\tau} = \{0\}$ and $\mathfrak{g}(A) = \tilde{\mathfrak{g}}(A)$. If A is symmetrizable, $\mathfrak{g}(A)$ is called a *BKM superalgebra associated with symmetrizable BKM supermatrix*. Below we define BKM Lie superalgebra associated with symmetrizable BKM supermatrix independently.

Definition 214 (Borcherds Kac-Moody Lie superalgebra $\tilde{\mathfrak{g}}(A, \psi)$ with symmetrizable Cartan matrix). The *Borcherds Kac-Moody Lie superalgebra* (abbreviated as BKM Lie superalgebra) or BKM superalgebra $\mathfrak{g}(A, \psi)$ associated to a symmetrizable BKM supermatrix $(A, \psi) = (a_{ij})_{1 \le i,j \le n}$ is the Lie superalgebra generated by the vector space $\mathfrak{h}$ and the elements $e_i, f_i (i \in I)$ satisfying the following relations:

(i) $[h, h'] = 0$ for $h, h' \in \mathfrak{h}$,
(ii) $[h, e_i] = \alpha_i(h) e_i$ for $h \in \mathfrak{h}$, $1 \le i \le n$,
(iii) $[h, f_i] = -\alpha_i(h) f_i$ for $h \in \mathfrak{h}$, $1 \le i \le n$,
(iv) $[e_i, f_j] = \delta_{ij} \alpha_i^\vee$ for $1 \le i,j \le n$,
(v) if $i \in I^{\mathrm{re}}$, $j \in I$, $i \ne j$, then $\left\{ \begin{array}{ll} (ade_i)^{1-a_{ij}} & e_j = 0 \\ (adf_i)^{1-a_{ij}} & f_j = 0 \end{array} \right\}$,

(vi) if $i,j \in I$, $i \neq j$, $a_{ij} = 0$, then $\left\{ \begin{array}{ll} [e_i, e_j] & = 0 \\ [f_i, f_j] & = 0 \end{array} \right\}$,

(vii) if $i \in \psi_0$ then $[e_i, e_i] = [f_i, f_i] = 0$ such that $\psi_0 = \{i \in \psi \mid a_{ii} = 0\}$.

Remark 133.

(i) In Definition 206, instead of $a_{ii} = 2$, one can also have $a_{ii} > 0$. In this case, in the above definition condition(v), can be replaced by

$$(ad(e_i))^{1-\frac{2a_{ij}}{a_{ii}}} e_j = 0, (ad(f_i))^{1-\frac{2a_{ij}}{a_{ii}}} f_j = 0 \ \ if \ \ a_{ii} > 0 \text{ and } i \neq j.$$

Whenever $a_{ii} = 2$, condition (v) follows.

(ii) Moreover, if A is a symmetric matrix, A is called the generalized symmetric Cartan matrix of the BKM Lie superalgebra. Also, whenever $a_{ii} > 0$, for all $i \in I$, then the associated BKM superalgebra is called *Kac Moody Lie superalgebra* associated with a symmetric generalized Cartan matrix.

(iii) For an affine Kac Moody Lie superalgebra, as $a_{ii} > 0$ for all $i \in I$, there are imaginary roots which are not simple.

(iv) If (A, ψ) is a BKM supermatrix then the associated algebra $\mathcal{G}(A)$ is called BKM Lie superalgebra or BKM superalgebra.

(v) If A is symmtrizable the associated superalgebra $\mathcal{G}(A)$ is called symmetrizable BKM Lie superalgebra or symmetrizable BKM superalgebra. In $\mathfrak{g}(A)$ with BKM supermatrix (A, ψ), we assume that the elements e_i and f_i $(i \in \psi)$, are odd elements and $e_j, f_j (j \notin \psi)$ and the elements of $\mathfrak{h}$ are even elements.

(vi) If ψ(set of all odd roots) is empty, then the associated matrix A is a BKM matrix or GKM matrix and the corresponding Lie algebra $\mathcal{G}(A)$ is called a BKM Lie algebra or BKM algebra or GKM algebra. We discussed these GKM algebras in Chapter 3.

(vii) In the GKM algebra, if the set consisting of simple imaginary roots is empty then GKM algebras will become Kac Moody algebras (KM algebras).

(viii) In the Cartan matrix all principle minors are positive definite. If all the principle minors of Cartan matrix are not necessarily positive definite then the corresponding Cartan matrix will become GKM matrix and corresponding KM algebras are discussed in Chapter 2.

We state the following theorem. For proof, one can refer [34, 142]. A symmetrizable BKM Lie superalgebra possesses a bilinear form given in the following theorem:

Theorem 54. *For a symmetrizable BKM supermatrix (A, ψ), there exists uniquely a non-degenerate super-symmetric super invariant even bilinear form $(\cdot|\cdot)$, whose restriction on $\mathfrak{h}$ coincides with that described in Lemma 10 and this bilinear form has the following properties:*

(i) *If $\alpha, \beta \in \Delta$ and $\alpha + \beta \neq 0$, then one has $(\mathfrak{g}_\alpha | \mathfrak{g}_\beta) = \{0\}$.*

(ii) *For $\alpha \in \Delta$, $(\cdot|\cdot)$ is a non-degenerate pairing between $\mathfrak{g}_\alpha$ and $\mathfrak{g}_{-\alpha}$.*

(iii) *For $\alpha \in \Delta$, one has $[x, y] = (x|y)\alpha$, for $x \in \mathfrak{g}_\alpha, y \in \mathfrak{g}_\beta$.*

We extend the definition of extended-hyperbolic BKM algebras (or extended-hyperbolic GKM algebras) to the extended-hyperbolic BKM superalgebras.

Definition 215. We define an indefinite non-hyperbolic BKM supermatrix (A, ψ), to be of extended-hyperbolic type, if every principal submatrix of A is of finite, affine or hyperbolic type BKM supermatrix.

We say that the BKM superalgebra associated with a BKM supermatrix (A, ψ) is of extended-hyperbolic type, if A is of extended-hyperbolic type.

5.2 Dynkin diagrams of BKM Lie superalgebras and in particular Kac Moody Lie superalgebras and some examples

Definition 216. To each BKM supermatrix $A = (a_{ij})_{1 \leq i,j \leq n}$, we associate a diagram $qDyn(A)$, called *quasi-Dynkin diagram* of A by the following rules:

1. Put 'n'' circles ○'s on a plane, call a circle ○, a vertex and name the vertices as $\alpha_1, \alpha_2, \ldots, \alpha_n$.
2. If $i \neq j$ and $a_{ij} < 0$, connect the two vertices α_i and α_j.
3. If $i \neq j$ and $a_{ij} = 0$, do not connect α_i and α_j.

Remark 134. $qDyn(A)$ is a diagram that only shows whether the vertices are connected or not.

Definition 217. It is clear that if the quasi-Dynkin diagram of A is connected, then A is *indecomposable* in the usual sense.

Below we define the Dynkin diagrams for all classes of BKM supermatrices completely [144].

Definition 218. Every BKM supermatrix (A, ψ), where $A = (a_{ij})_{i,j \in I}$, $\psi \subset I$, (the index set) is associated with a *Dynkin diagram*, $S(A)$ defined as follows:

$S(A)$ has n vertices and vertices i and j are connected by $\max\{|a_{ij}|, |a_{ji}|\}$ number of lines if $a_{ij}a_{ji} \leq 4$ and there is an arrow pointing towards i if $|a_{ij}| > 1$. If $a_{ij}.a_{ji} > 4$, then i and j are connected by a bold faced edge equipped with the ordered pair $(|a_{ij}|, |a_{ji}|)$. Also

(i) if $a_{ii} = 2$, $i \notin \psi$, the ith vertex will be denoted by a white circle,
(ii) if $a_{ii} = 2$, $i \in \psi$, the ith vertex will be denoted by a white circle with (od) written within parentheses and below the circle to denote the vertex corresponding to an odd root in this case,
(iii) if $a_{ii} = 0$, $i \notin \psi$, the ith vertex will be denoted by a crossed circle,
(iv) if $a_{ii} = 0$, $i \in \psi$, the ith vertex will be denoted by a crossed circle with (od) written within parentheses and below the circle to denote the vertex corresponding to an odd root in this case,
(v) if $a_{ii} = -k$, $k > 0$, $i \notin \psi$, the ith vertex will be denoted by a white circle with $(-k)$ written within parentheses and above the circle,
(vi) if $a_{ii} = -k$, $k > 0$, $i \in \psi$, the ith vertex will be denoted by a white circle with $(-k)$ written within parentheses and above the circle with (od) written within parentheses and below the circle to denote the vertex corresponding to an odd root in this case.

We give below some examples of Dynkin diagrams of Borcherds Kac-Moody supermatrices:

1. Let (A, ψ) be the BKM supermatrix with $A = (-k)$.

(i) Let $\psi = \{1\}$ and $k > 0$. Then the corresponding Dynkin diagram:

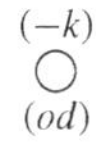

(ii) Let $\psi = \emptyset$ and $k > 0$. Then the corresponding Dynkin diagram:

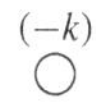

(iii) Let $\psi = \{1\}$ and $k = 0$. Then the corresponding Dynkin diagram:

2. Let (A, ψ) be the BKM supermatrix with $A = \begin{pmatrix} 2 & -a \\ -b & -k \end{pmatrix}$.

(i) Let $\psi = \{2\}$ with $a = b = 1$ and $k \in \mathbb{Z}_{>0}$. Then the Dynkin diagram:

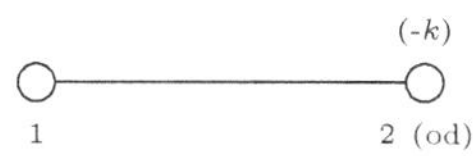

(ii) Let $\psi = \{2\}$ with $a = b = 1$ and $k = 0$. Then the corresponding Dynkin diagram:

(iii) Let $\psi = \{1, 2\}$ with $a = 2$ and $b = 1$ and $k \in \mathbb{Z}_{>0}$. Then the corresponding Dynkin diagram:

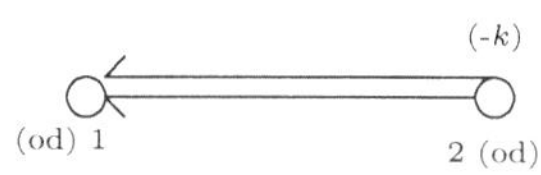

3. Let (A, ψ) be the BKM supermatrix with $A = \begin{pmatrix} -k & -1 & -1 \\ -2 & 2 & -2 \\ -1 & -1 & 2 \end{pmatrix}$.

(i) Let $\psi = \{1, 2\}$ and $k > 0$. Then the corresponding Dynkin diagram:

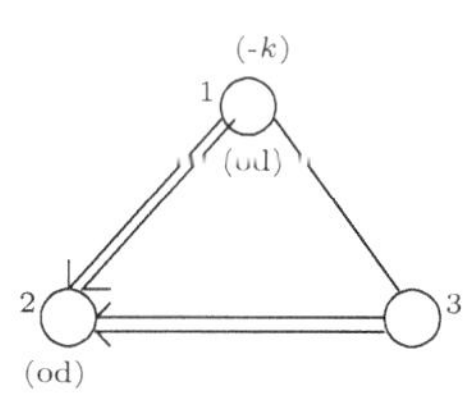

(ii) Let $\psi = \{2\}$ and $k = 0$. Then the corresponding Dynkin diagram:

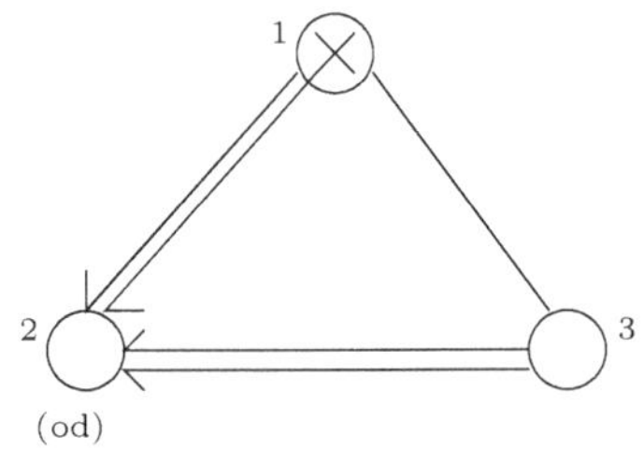

Similarly, for any given BKM supermatrix the corresponding Dynkin diagram can be drawn.

Let us denote the Dynkin diagram associated to an $n \times n$ *extended-hyperbolic BKM supermatrix (Super BKM extended hyperbolic matrix)* (A, ψ) by $SBEH\ A^{(n)}$, $n \geq 3$ and the corresponding super BKM hyperbolic matrix (BKM super hyperbolic matrix) by $SBHA^{(n-1)}$ for $n \geq 3$.

Below we give Dynkin diagrams of a few $SBHA^{(n-1)}$ and $SBEHA^{(n)}$ for $n \geq 3$.

1. $n = 3 : SBHA^{(n-1)} = SBHA^{(2)}$:

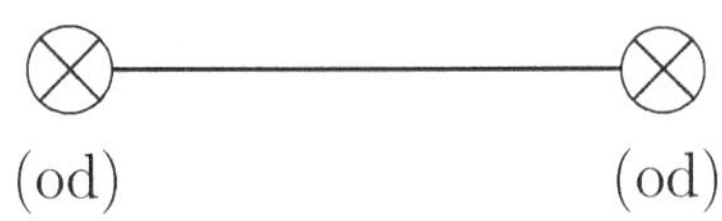

The corresponding $SBEHA^{(3)}$:

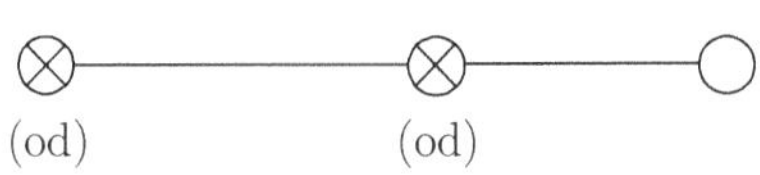

2. $n = 4$; $SBHA^{(n-1)} = SBHA^{(3)}$:

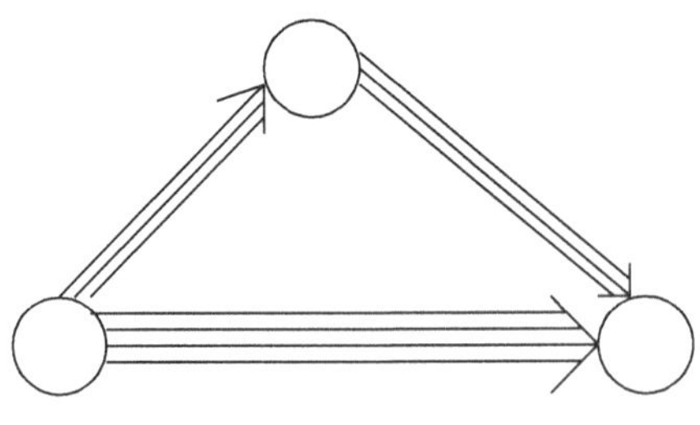

The corresponding $SBEHA^{(4)}$:

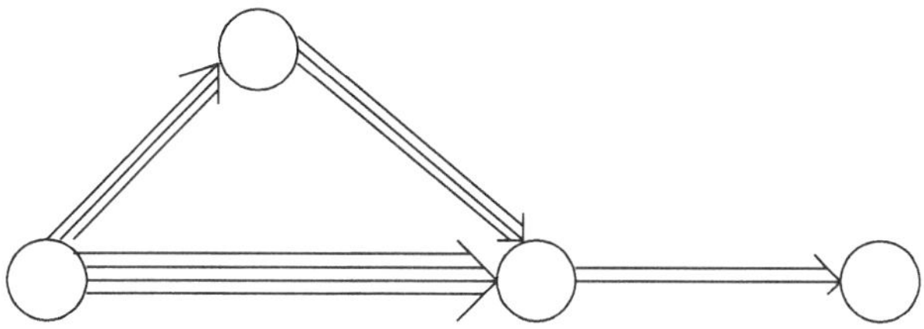

Definition 219. As in the cases of KM algebras, and GKM algebras, we say that a Dynkin diagram $S(A)$ is *completely connected*, if $a_{ij} \neq 0$ for all $i, j = 1, \ldots, n$.

Let us denote such Dynkin diagrams by $SBEHA_c^{(n)}$.

Example 49. Here, we give the complete set of BKM superalgebras $\mathfrak{g}(A)$ among the extended-hyperbolic types, whose Dynkin diagrams are *completely connected.*

(1) $SBEHA_c^{(3)}$: It consists of any indecomposable BKM supermatrix (A, ψ) where $A = (a_{ij})_{i,j=1}^3$ and any one or two or all $a_{ii} = 2$ or 0.

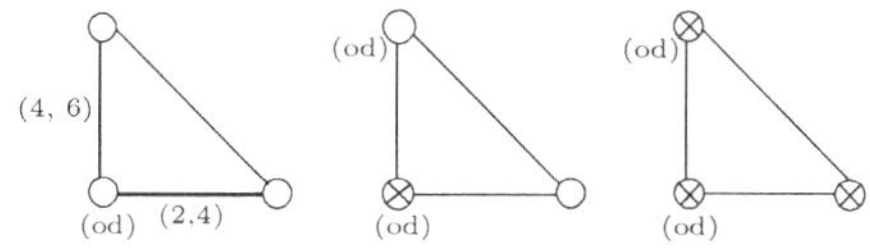

(2) $SBEHA_c^{(4)}$: It consists of any indecomposable BKM supermatrix (A, ψ), where $A = (a_{ij})_{i,j=1}^4$, of $SBEHA^{(4)}$ type such that $a_{ij} \neq 0 \; \forall \, i, j \in \{1, 2, 3, 4\}$.

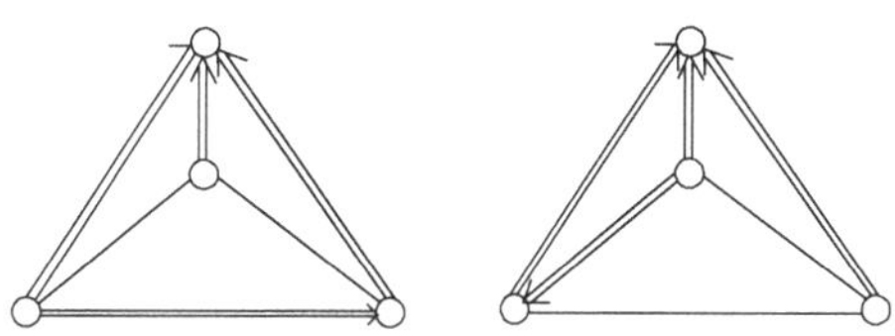

(3) $SBEHA_c^{(5)}$: There is exactly one diagram in this class given by

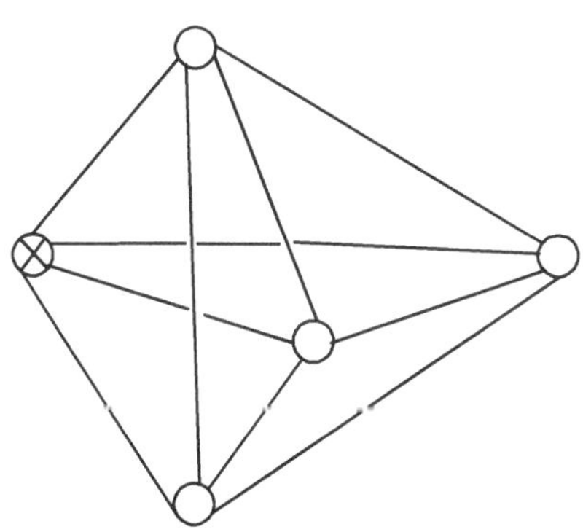

$SBEHA_c^{(n)}, n > 5$: There is no completely connected Dynkin diagram in $SBEHA_c^{(n)}$, $n > 5$.

Remark 135.

(1) Note that extended-hyperbolic BKM supermatrices (A, ψ) which are not extended-hyperbolic KM supermatrices are of order 3.
(2) We note that all the extended-hyperbolic BKM algebras are also extended-hyperbolic BKM superalgebras.

We give below two examples of extended-hyperbolic BKM superalgebras which are not extended-hyperbolic BKM algebras. In the following, (A, ψ) is in the usual notation.

1. $A = \begin{pmatrix} 0 & -1 & 0 \\ -1 & 0 & -1 \\ 0 & -1 & 0 \end{pmatrix}, \psi = \{2\}$

The corresponding Dynkin diagram:

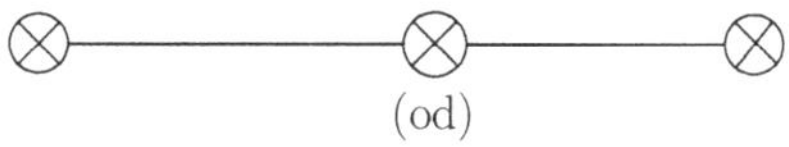

2. $A = \begin{pmatrix} 2 & -1 & -1 \\ -2 & 2 & -2 \\ -1 & -1 & 0 \end{pmatrix}, \psi = \{2, 3\}$

The corresponding Dynkin diagram:

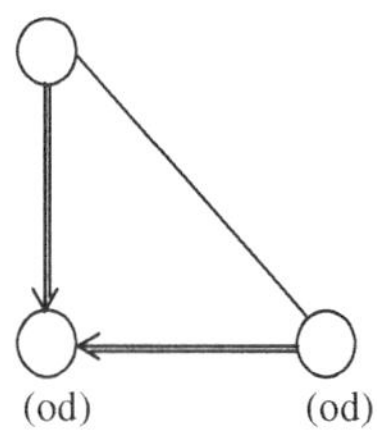

Root space decomposition in BKM Lie superalgebras

In this section, we give the basic definitions of the Weyl group and real and imaginary roots of Borcherds Kac Moody Lie superalgebras.

Remark 136. The following definitions are exactly similar to those for Kac-Moody algebras and generalized Kac-Moody algebras.

Definition 220. The BKM Lie superalgebra $\mathfrak{g}$ has the following natural *root space decomposition:*

For $\mathfrak{g} = \oplus_{\alpha \in Q} \mathfrak{g}_\alpha$, $\mathfrak{g}_\alpha = \{X \in \mathfrak{g}; [h, X] = \alpha(h)X, \ \forall\ h \in \mathfrak{h}\}$, is called the *root space* associated to α. An element $\alpha \in Q$ is called a *root*, if $\alpha \neq 0$ and $\mathfrak{g}_\alpha \neq 0$. The number mult$\alpha = \dim \mathfrak{g}_\alpha$ is called the *multiplicity* of the root α. A root α of $\mathfrak{g}(A)$ can be expressed as $\alpha = \sum_{i=1}^{n} m_i \alpha_i$, $\quad m_i \in \mathbb{Z}$ where m_i's are all ≥ 0 or all ≤ 0. Corresponding to whether m_i's are all ≥ 0 or all ≤ 0, α is called a *positive root or a negative root* respectively. Also $\sum_{i=1}^{n} m_i$ is called the *height* of α and is denoted by $ht(\alpha)$. We denote by $\Delta, \Delta_+, \Delta_-, \Delta_{\overline{0}}$ and $\Delta_{\overline{1}}$ the set of all roots, positive roots, negative roots, even roots and odd roots of $\mathfrak{g}$ respectively. Also note that $\mathfrak{g}_{\alpha_i} = \mathbb{C}e_i$ and $\mathfrak{g}_{-\alpha_i} = \mathbb{C}f_i$.

We have $Q = \sum_{i=1}^{n} \mathbb{Z}\alpha_i$ and $Q_+ = \sum_{i=1}^{n} \mathbb{Z}_{\geq 0}\alpha_i$. Here Q is called the *root lattice* and Q_+ is called the *positive root lattice.*

The root lattice Q becomes a partially *ordered set* by taking

$$\alpha \geq \beta \Leftrightarrow \alpha - \beta \in Q_+ \text{ for } \alpha, \beta \in Q.$$

Now, for $\alpha = \sum m_i\alpha_i \in Q$, support of α is $\text{supp}(\alpha) = \{i \in I; m_i \neq 0\}$. If $\text{supp}(\alpha)$ is a connected subset of the Quasi-Dynkin diagram of A, we say that $\text{supp}(\alpha)$ is connected.

Definition 221. For each $i \in I^{\text{re}}$, we define the *simple reflection* $r_i \in \mathfrak{h}^*$ by

$$r_i(\lambda) = \lambda - \langle \lambda, \alpha_i^\vee \rangle \alpha_i, \ \lambda \in \mathfrak{h}^*.$$

The *Weyl group* W of $\mathfrak{g}(A)$ is the subgroup of $GL(\mathfrak{h}^*)$ generated by the r_i's ($i \in I^{\text{re}}$). Note that $(W, \{r_i;\ i \in I^{\text{re}}\})$ is a Coxeter system.

We define the *reflection* r_α of $\mathfrak{h}^*$ with respect to α by

$$r_\alpha(\lambda) = \lambda - \langle \lambda, \alpha^\vee \rangle \alpha \ (\lambda \in \mathfrak{h}^*),$$

where $\alpha^\vee = w(\alpha_i^\vee) \in \mathfrak{h}$ is the dual real root of α. Note that

$$r_\alpha = w r_i w^{-1} \in W.$$

Hence the root α is *conjugate to the fundamental root* α_i under the action of the Weyl group. The set of all real roots is

$$\Delta^{\text{re}} := W(\Pi^{\text{re}}) \cup W(\{2\alpha_i; i \in \psi^{\text{re}}\}).$$

The set of all imaginary roots is

$$\Delta^{\text{im}} := \Delta \backslash \Delta^{\text{re}}.$$

Remark 137. It can be easily verified that the bilinear form $(\cdot|\cdot)$ on $\mathfrak{h}$ and $\mathfrak{h}^*$ is invariant under the action of the Weyl group and in particular, we have

$$(\alpha_i|\alpha_j) = \varepsilon_i a_{ij} \ \text{ for } 1 \leq i, j \leq n.$$

For proofs of the following fundamental results and Proposition 38, one can refer [142].

Fundamental results on real roots

For $\alpha \in \Delta^{\text{re}}$, one has the following:

1. $(\alpha|\alpha) > 0$,
2. $r_\alpha \in W$,
3. $\text{mult}(\alpha) = 1$
4. For $i \in I^{\text{re}}, r_i(\alpha_i) = -\alpha_i$.
5. For $i \in I^{re}\backslash\psi^{re}$, then $r_i(\Delta_+\backslash\{\alpha_i\}) = \Delta_+\backslash\{\alpha_i\}$
6. For $i \in \psi^{re}$, then $r_i(\Delta_+\backslash\{\alpha_i, 2\alpha_i\}) = \Delta_+\backslash\{\alpha_i, 2\alpha_i\}$.
7. If α is a root, then $\text{supp}(\alpha)$ is connected.

Now consider the following sets:

$K = \{\alpha \in Q_+ \text{ such that } \langle \alpha, \alpha_i^\vee \rangle \leq 0 (i \in I^{re}) \text{ and supp}(\alpha) \text{ is connected}\}.$
$\overset{\circ}{K} = \{\alpha \in K; |\text{supp}(\alpha)| \geq 2\}.$
$\mathfrak{h}'_{\mathbb{R}} = \sum_{i=1}^n \mathbb{R}\alpha_i^\vee = \sum_{i=1}^n \mathbb{R}\alpha_i.$
$C^\vee = \{h \in \mathfrak{h}'_{\mathbb{R}}; \langle \alpha_i^\vee, h \rangle \geq 0 (i \in I^{re})\}.$
$\overset{\circ}{C^\vee} = \{h \in \mathfrak{h}'_{\mathbb{R}}; \langle \alpha_i^\vee, h \rangle > 0 (i \in I^{re})\}.$

Definition 222. $C^\vee$ is called a fundamental domain or a "positive Weyl chamber." $\overset{\circ}{C^\vee}$ is the 'interior' of the fundamental domain. The set of differences $C^\vee \backslash \overset{\circ}{C^\vee}$ is called *the walls of the positive Weyl chamber.*

Proposition 38.

(1) *For $i \in I^{im}$, take $\alpha \in \Delta_+ \backslash \{\alpha_i\}$, so that supp $(\alpha + \alpha_i)$ is connected. Then, for all $j \in \mathbb{Z}_{\geq 0}$, one has $\alpha + j\alpha_i \in \Delta_+$.*

(2) **(i)** *For $i \in I^{im}$ and $\alpha \in \Delta_+ \backslash \{\alpha_i\}$, if supp$(\alpha + \alpha_i)$ is connected, then $(\alpha, \alpha_i) < 0$.*

(ii) *For $\alpha \in \Delta$ and $i \in I^{re}$, the set $\{j \in \mathbb{Z};\ \alpha + j\alpha_i \in \Delta\}$ is finite.*

(iii) **(a)** $(\alpha|\alpha_i) > 0 \Longrightarrow \alpha - \alpha_i \in \Delta_+,\ (\alpha|\alpha_i) < 0 \Longrightarrow \alpha + \alpha_i \in \Delta_+.$

(b) $\alpha + \alpha_i \notin \Delta \Longrightarrow (\alpha|\alpha_i) \geq 0,\ \ \alpha - \alpha_i \notin \Delta \Longrightarrow (\alpha|\alpha_i) \leq 0.$

(c) *For $\alpha \in \Delta$, $i \in I^{re}$, if $\alpha + \alpha_i \notin \Delta$ and $\alpha - \alpha_i \in \Delta$, then* $(\alpha, \alpha_i) > 0.$

In Remark 133, we have already defined Kac-Moody Lie superalgebra associated with a generalized symmetric Cartan matrix. Now we define independently the definition of Kac-Moody Lie superalgebras associated with a generalized symmetrizable Cartan matrix as follows:

Definition 223. *Kac-Moody Lie superalgebra associated with generalized symmetrizable Cartan matrix* $(A, \psi) = ((a_{ij})_{1 \leq i,j \leq n}, \psi)$ is the Lie superalgebra generated by the vector space $\mathfrak{h}$ and the elements $e_i, f_i (i \in I)$ satisfying the following relations:

(i) $[h, h'] = 0$ for $h, h' \in \mathfrak{h}$,
(ii) $[h, e_i] = \alpha_i(h)e_i$ for $h \in \mathfrak{h},\ 1 \leq i \leq n$,
(iii) $[h, f_i] = -\alpha_i(h)f_i$ for $h \in \mathfrak{h}, 1 \leq i \leq n$,
(iv) $[e_i, f_j] = \delta_{ij}\alpha_i^\vee$ for $1 \leq i, j \leq n$,
(v) if $i \in I^{re},\ j \in I,\ i \neq j$, then $\left\{ \begin{array}{ll} (ade_i)^{1 - \frac{2a_{ij}}{a_{ii}}} & e_j = 0 \\ (adf_i)^{1 - \frac{2a_{ij}}{a_{ii}}} & f_j = 0 \end{array} \right\}$,
(vi) if $i, j \in I,\ i \neq j,\ a_{ij} = 0$, then $\left\{ \begin{array}{l} [e_i, e_j] = 0 \\ [f_i, f_j] = 0 \end{array} \right\}$.

Types of Kac-Moody Lie superalgebras

Kac-Moody superalgebras can be divided into the following three classes

(i) *Finite dimensional Kac-Moody Lie superalgebras*: These are the finite dimensional Kac-Moody superalgebras, which are finite dimensional Lie superalgebras.

(ii) *Affine Kac-Moody superalgebras:* These are the set of all infinite dimensional Lie superalgebras and are of two types:

(a) *Untwisted affine Kac-Moody superalgebras* corresponding to identity automorphisms of corresponding simple Lie superalgebras.

(b) *Twisted affine Kac-Moody superalgebras* corresponding to outer automorphisms of order 2 or 4.

(iii) *Indefinite Kac-Moody superalgebras*: These are the set of all non-finite and non-affine Kac-Moody superalgebras.

Example 50.

(i) Finite dimensional Kac-Moody superalgebras: One can refer Chapter 4.

(ii) Affine Kac-Moody superalgebras:
The following gives list of affine Kac-Moody Lie superalgebras:
$B^{(1)}(0,n), B^{(1)}(0,1), A^{(2)}(0,2n-1), n \geq 3, A^{(2)}(0,3), C^{(2)}(n+1), n \geq 2, C^{(2)}(2), A^{(4)}$ $(0,2n), n \geq 2$ and $A^{(4)}(0,2)$.

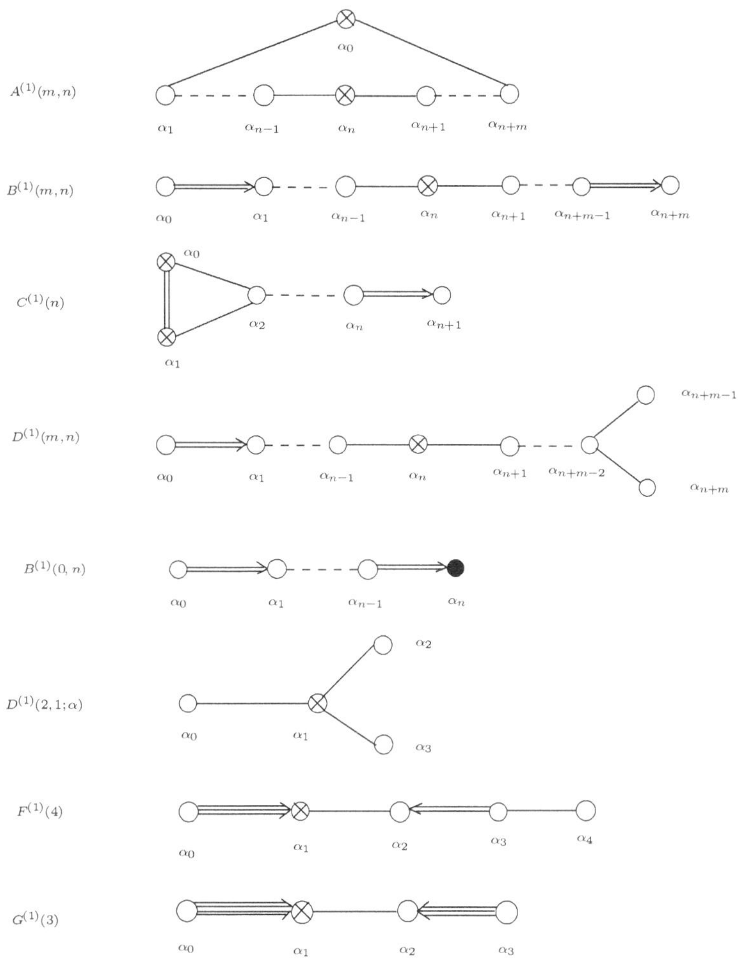

(iii) Indefinite Kac Moody superalgebras:

The set of all indefinite Kac-Moody superalgebras contains hyperbolic Kac Moody superalgebras and non-hyperbolic Kac Moody superalgebras. Here every leading principal submatrix of the generalized cartan matrix of the hyperbolic Kac Moody superalgebra decomposes into constituents of finite or affine type or equivalently, deletion of a vertex of the Dynkin Diagram gives finite or affine type Dynkin diagrams. Below we give one example for hyperbolic Kac moody superalgebras which is an extension of affine $C^{(1)}(n)$. Moreover we give another example of indefinite non-hyperbolic Kac Moody superalgebras which is an extension of $B^{(1)}(0,n)$. The above two algebras are denoted by $EC^{(1)}(n)$ and $EB^{(1)}(0,n)$ respectively. Similarly examples for any other hyperbolic Kac Moody superalgbras or indefinite Kac Moody superalgebras can also be given.

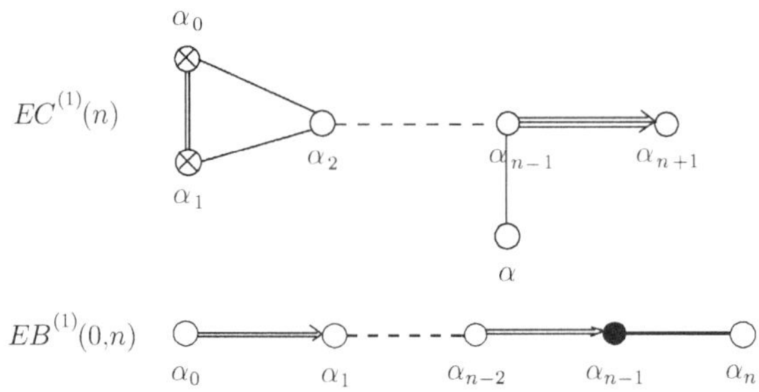

Remark 138. For affine Lie superalgebras, one can refer Eswara Rao [263].

5.3 Domestic type and Alien type imaginary roots in BKM Lie superalgebras

Definition 224. An imaginary root which is conjugate to a fundamental root under the action of the Weyl group is called *domestic-type imaginary* (domestic imaginary) *root*. We denote the set of domestic type imaginary root by $\Delta^{\text{dom}\cdot\text{im}}$.

Definition 225. An imaginary root which is not conjugate to a fundamental root under the action of the Weyl group is called *alien-type imaginary* (alien imaginary) *root*. We denote the set of all alien type imaginary roots by $\Delta^{\text{ali}\cdot\text{im}}$.

Since an imaginary root is either conjugate or not conjugate to a fundamental root under the action of Weyl group, each imaginary root is either domestic imaginary or alien imaginary. Hence $\Delta^{\text{im}} = \Delta^{\text{dom}\cdot\text{im}} \cup \Delta^{\text{ali}\cdot\text{im}}$. Also we have, $\Delta^{\text{dom}\cdot\text{im}} = W(\pm\Pi^{\text{im}}) \cup W(\{\pm 2\alpha_i; i \in \psi_-\})$ and $\Delta^{\text{ali}\cdot\text{im}} = \Delta^{\text{im}} \backslash \Delta^{\text{dom}\cdot\text{im}}$.

Below we write some basic results, the proof of which can be seen in [142].

Basic results on imaginary roots and Weyl groups

(1) **(i)** Δ_+^{im} is invariant under the action of the Weyl group.

(ii) If $\alpha \in \Delta_+^{\text{im}}$, then there exists $w \in W$ satisfying $w\alpha \in -C^\vee$.

(iii) For $\alpha \in \Delta_+, \alpha \in \Delta_+^{\text{im}} \Leftrightarrow (\alpha \mid \alpha) \leq 0$.

(iv) If $w \in W$, then $w(\Delta) = \Delta$.
(v) If $\alpha \in Q_+ \cap \{-C^\vee\}$, then $(\alpha|\alpha) \leq 0$.

(2) For a symmetrizable BKM supermatrix (A, ψ), if

$$\Delta^{\text{dom}\cdot\text{im}} := W(\pm\Pi^{\text{im}}) \cup W(\{\pm 2\alpha_i; i \in \psi_-\}),$$

$$\Delta^{\text{ali}\cdot\text{im}} := \Delta^{\text{im}} \backslash \Delta^{\text{dom}\cdot\text{im}},$$

then
(i) $\Delta_+^{\text{ali}\cdot\text{im}} = W(\overset{\circ}{K}) = \cup_{w\in W} w(\overset{\circ}{K})$.
(ii) $\Delta_+^{\text{im}} = W(\overset{\circ}{K}) \cup W(\Pi^{\text{im}}) \cup W(2\alpha_i; i \in \psi_-)$.

Examples for Domestic and Alien imaginary roots

Example 51. Consider the BKM superalgebra associated with the BKM supermatrix

$$A = \begin{pmatrix} -k & -1 \\ -1 & 2 \end{pmatrix}, k > 0,\ \psi = \{1\}.$$

Here $\psi_- = \{1\}$. Using $\Delta_+^{\text{dom}\cdot\text{im}} = W(\pm\Pi^{\text{im}}) \cup W(\{\pm 2\alpha_i; i \in \psi_-\})$, we get the set of all positive domestic imaginary roots,

$$\Delta_+^{\text{dom}\cdot\text{im}} = \{\alpha_1, \alpha_1 + \alpha_2, 2\alpha_1, 2(\alpha_1 + \alpha_2)\}.$$

Example 52. Consider the BKM superalgebra associated with the BKM supermatrix (A, ψ), where

$$A = \begin{pmatrix} 0 & -1 & -1 \\ -1 & 2 & -1 \\ -1 & -1 & 2 \end{pmatrix} \text{ with } \psi = \{1\}.$$

Consider $\alpha = 2\alpha_1 + \alpha_2 + \alpha_3 \in Q$. Using $\langle \alpha_j, \alpha_i^\vee \rangle = a_{ij}$, we get $\langle \alpha, \alpha_2^\vee \rangle = < 2\alpha_1 + \alpha_2 + \alpha_3, \alpha_2^\vee >= 2\langle \alpha_1, \alpha_2^\vee \rangle + \langle \alpha_2, \alpha_2^\vee \rangle + \langle \alpha_3, \alpha_2^\vee \rangle = 2a_{21} + a_{22} + a_{23} = -2 + 2 - 1 = -1 < 0$.

Similarly, we can verify that $\langle \alpha, \alpha_3^\vee \rangle < 0$ and $|supp(\alpha)| > 2$. By the definitions of K and $\overset{\circ}{K}$, it is clear that, $\alpha \in \overset{\circ}{K}$. Hence, using $\Delta_+^{ali.im} = \underset{w\in W}{\cup} w(\overset{\circ}{K})$, it is clear that α is an alien imaginary root as $1 \in W$.

Some examples of BKM Lie superalgebras which appear as extensions of KM algebras

We use the following notations: If *GX* stands for KM matrix of type *X* (where *X* is of finite, affine or hyperbolic type) then by $(SBGX, \psi)(= (A, \psi))$ we mean the super Borcherds Kac-Moody matrix (that is Borcherds Kac-Moody supermatrix, which is the same as the BKM supermatrix) which is an extension of *GX*, where ψ can be suitably chosen in each case. For finding out the real and imaginary roots, one has to choose ψ properly in each case.

In the following, let GX be an $n \times n$ Kac-Moody matrix or generalized Kac-Moody matrix and SBGX be an $(n + r) \times (n + r)$ BKM supermatrix, which is an extension of KM or GKM matrix by adding "r" rows and "r" columns to the above and the left of the rows and columns of GX as given below. Depending upon the definition of ψ, SBGX will become a BKM supermatrix or GKM matrix or KM matrix. We have

$$SBGX = \begin{pmatrix} -k_1 & -a_1 & -a_2 & \cdots & -a_{r-1} & -a_r & \cdots & -a_{n+r-1} \\ -b_1 & -k_2 & -c_2 & \cdots & -c_{r-1} & -c_r & \cdots & -c_{n+r-1} \\ -b_2 & -d_2 & -k_3 & \cdots & \cdots & \cdots & \cdots & \cdots \\ \vdots & \vdots & & \ddots & & & & \\ -b_{r-1} & -d_{r-1} & \cdots & \cdots & -k_r & \cdots & \cdots & \cdots \\ -b_r & -d_r & & & \vdots & & & \\ \vdots & \vdots & & & & GX & & \\ -b_{n+r-1} & -d_{n+r-1} & & & & & & \end{pmatrix}.$$

In this section, we consider the following classes of BKM Lie superalgebras. In the following, $a_1, a_2, a_3, b_1, b_2, b_3, p, q \in \mathbb{N}$, and $k_i \in \mathbb{N} \cup \{0\}$.

Class 1: Extension of finite KM matrix

(i) If $X = A_1$, then
$GX = [2]$ and $SBGX = \begin{pmatrix} -k & -a_1 \\ -b_1 & 2 \end{pmatrix}$.

(ii) If $X = B_2$, then
$$GX = \begin{pmatrix} 2 & -1 \\ -2 & 2 \end{pmatrix} \text{ and } SBGX = \begin{pmatrix} -k & -a_1 & -a_2 \\ -b_1 & 2 & -1 \\ -b_2 & -2 & 2 \end{pmatrix}.$$

(iii) If $X = C_3$, then
$$GX = \begin{pmatrix} 2 & -1 & 0 \\ -1 & 2 & -2 \\ 0 & -1 & 2 \end{pmatrix} \text{ and } SBGX = \begin{pmatrix} -k & -a_1 & -a_2 & -a_3 \\ -b_1 & 2 & -1 & 0 \\ -b_2 & -1 & 2 & -2 \\ -b_3 & 0 & -1 & 2 \end{pmatrix}.$$

Class 2: Extension of affine KM matrix

(iv) If $X = G_2^{(1)}$, then
$$GX = \begin{pmatrix} 2 & -1 & 0 \\ -1 & 2 & -1 \\ 0 & -3 & 2 \end{pmatrix} \text{ and } SBGX = \begin{pmatrix} -k & -a_1 & -a_2 & -a_3 \\ -b_1 & 2 & -1 & 0 \\ -b_2 & -1 & 2 & -1 \\ -b_3 & 0 & -3 & 2 \end{pmatrix}.$$

Class 3: Extension of Hyperbolic KM matrix

(v) If $X = HA_1^{(1)}$, then

$$GX = \begin{pmatrix} 2 & -1 & 0 \\ -1 & 2 & -2 \\ 0 & -2 & 2 \end{pmatrix} \text{ and } SBGX = \begin{pmatrix} -k & -a_1 & -a_2 & -a_3 \\ -b_1 & 2 & -1 & 0 \\ -b_2 & -1 & 2 & -2 \\ -b_3 & 0 & -2 & 2 \end{pmatrix}.$$

Some examples to find out the set of real roots, domestic type imaginary roots and alien type imaginary roots:

Class 1: Extension of a finite KM matrix

Example 53. Let $X = B_3$. Then

$$(SBGB_3, \psi) = \begin{pmatrix} -k & -a_1 & -a_2 & -a_3 \\ -b_1 & 2 & -1 & 0 \\ -b_2 & -1 & 2 & -1 \\ -b_3 & 0 & -2 & 2 \end{pmatrix} = (A, \psi).$$

Taking

$$D = \begin{pmatrix} 2b_1 & 0 & 0 & 0 \\ 0 & 2a_1 & 0 & 0 \\ 0 & 0 & 2a_1 & 0 \\ 0 & 0 & 0 & a_1 \end{pmatrix}$$

with $a_1b_2 = a_2b_1$ and $a_2b_3 = 2a_3b_2$, one can directly verify that $SBGB_3$ is symmetrizable. Also this matrix $SBGB_3$ is nonsingular and indecomposable. It is clear that

$$\Pi^{\text{re}} = \{\alpha_2, \alpha_3, \alpha_4\},\ \Pi^{\text{im}} = \{\alpha_1\} \quad \text{and } \Pi = \{\alpha_1, \alpha_2, \alpha_3, \alpha_4\}.$$

Let $\psi = \{1, 4\}$, if b_3 is an even integer. Hence we have $\psi^{\text{re}} = \{4\}$, if b_3 is an even integer. Also $\psi_0 = \{1\}$, if $k = 0$ and $\psi_- = \{1\}$, if $k > 0$. Hence $\Pi_{\text{even}} = \{\alpha_2, \alpha_3\}$, and $\Pi_{\text{odd}} = \{\alpha_1, \alpha_4\}$ for b_3 is an even integer. Using the definition of reflections,

$$r_2(\alpha_1) = \alpha_1 + b_1\alpha_2;\ r_2(\alpha_2) = -\alpha_2;\ r_2(\alpha_3) = \alpha_2 + \alpha_3;\ r_2(\alpha_4) = \alpha_4$$
$$r_3(\alpha_1) = \alpha_1 + b_2\alpha_3;\ r_3(\alpha_2) = \alpha_2 + \alpha_3;\ r_3(\alpha_3) = -\alpha_3;\ r_3(\alpha_4) = \alpha_3 + \alpha_4$$
$$r_4(\alpha_1) = \alpha_1 + b_3\alpha_4;\ r_4(\alpha_2) = \alpha_2;\ r_4(\alpha_3) = \alpha_3 + 2\alpha_4;\ r_4(\alpha_4) = -\alpha_4.$$

So we get the Weyl group W of the algebra corresponding to $SBGB_3$, which is same as that of GB_3, as

$$\begin{aligned} W = \{&1,\ r_2,\ r_3,\ r_4,\ r_3r_2,\ r_4r_2,\ r_2r_3,\ r_4r_3,\ r_3r_4,\ r_2r_3r_2, r_4r_3r_2,\ r_3r_4r_2, \\ &r_4r_2r_3, r_3r_4r_3,\ r_2r_3r_4,\ r_4r_3r_4, r_4r_2r_3r_2,\ r_3r_4r_3r_2,\ r_2r_3r_4r_2, \\ &r_4r_3r_4r_2,\ r_3r_4r_2r_3, r_2r_3r_4r_3,\ (r_4r_3)^2,\ r_4r_2r_3r_4, r_3r_4r_2r_3r_2, \\ &r_2r_3r_4r_3r_2, (r_4r_3)^2r_2,\ r_4r_2r_3r_4r_2,\ r_2r_3r_4r_2r_3, r_4r_3r_4r_2r_3,\ r_4r_2r_3r_4r_3, \end{aligned}$$

$r_3r_4r_2r_3r_4,\ r_2r_3r_4r_2r_3r_2,\ r_4r_3r_4r_2r_3r_2,\ r_4r_2r_3r_4r_3r_2, (r_3r_4r_2)^2,\ (r_4r_2r_3)^2,$
$r_3r_4r_2r_3r_4r_3,\ r_4r_3r_4r_2r_3r_4,\ (r_4r_2r_3)^2r_2,$
$r_3r_4r_2r_3r_4r_3r_2,\ r_4r_3r_4r_2r_3r_4r_2,\ r_3(r_4r_2r_3)^2,\ r_4r_3r_4r_2r_3r_4r_3,$
$r_3(r_4r_2r_3)^2r_2,\ r_4r_3r_4r_2r_3r_4r_3r_2,\ r_4r_3(r_4r_2r_3)^2,\ r_4r_3(r_4r_2r_3)^2r_2\}.$

The order of W is 48.
Actions of Weyl group elements on fundamental real roots are listed in Table 5.1.
Using the definition of bilinear form (.|.), we get

Table 5.1 The images of fundamental real roots under the action of Weyl group elements of $SBGB_3$

"Weyl group" element	Image of α_2 under the action of Weyl group element	Image of α_3 under the action of Weyl group element	Image of α_4 under the action of Weyl group element
1	α_2	α_3	α_4
r_2	$-\alpha_2$	$\alpha_2+\alpha_3$	α_4
r_3	$\alpha_2+\alpha_3$	$-\alpha_3$	$\alpha_3+\alpha_4$
r_4	α_2	$\alpha_3+2\alpha_4$	$-\alpha_4$
r_3r_2	$-(\alpha_2+\alpha_3)$	α_2	$\alpha_3+\alpha_4$
r_4r_2	$-\alpha_2$	$\alpha_2+\alpha_3+2\alpha_4$	$-\alpha_4$
r_2r_3	α_3	$-(\alpha_2+\alpha_3)$	$\alpha_2+\alpha_3+\alpha_4$
r_4r_3	$\alpha_2+\alpha_3+2\alpha_4$	$-(\alpha_3+2\alpha_4)$	$\alpha_3+\alpha_4$
r_3r_4	$\alpha_2+\alpha_3$	$\alpha_3+2\alpha_4$	$-(\alpha_3+\alpha_4)$
$r_2r_3r_2$	$-\alpha_3$	$-\alpha_2$	$\alpha_2+\alpha_3+\alpha_4$
$r_4r_3r_2$	$-(\alpha_2+\alpha_3+2\alpha_4)$	α_2	$\alpha_3+\alpha_4$
$r_3r_4r_2$	$-(\alpha_2+\alpha_3)$	$\alpha_2+2\alpha_3+2\alpha_4$	$-(\alpha_3+\alpha_4)$
$r_4r_2r_3$	$\alpha_3+2\alpha_4$	$-(\alpha_2+\alpha_3+2\alpha_4)$	$\alpha_2+\alpha_3+\alpha_4$
$r_2r_3r_4$	α_3	$\alpha_2+\alpha_3+2\alpha_4$	$-(\alpha_2+\alpha_3+\alpha_4)$
$r_4r_3r_4$	$\alpha_2+\alpha_3+2\alpha_4$	α_3	$-(\alpha_3+\alpha_4)$
$r_4r_2r_3r_2$	$-(\alpha_3+2\alpha_4)$	$-\alpha_2$	$\alpha_2+\alpha_3+\alpha_4$
$r_3r_4r_3r_2$	$-(\alpha_2+2\alpha_3+2\alpha_4)$	$\alpha_2+\alpha_3$	α_4
$r_3r_4r_3$	$\alpha_2+2\alpha_3+2\alpha_4$	$-(\alpha_3+2\alpha_4)$	α_4
$r_2r_3r_4r_2$	$-\alpha_3$	$\alpha_2+2\alpha_3+2\alpha_4$	$-(\alpha_2+\alpha_3+\alpha_4)$
$r_4r_3r_4r_2$	$-(\alpha_2+\alpha_3+2\alpha_4)$	$\alpha_2+2\alpha_3+2\alpha_4$	$-(\alpha_3+\alpha_4)$
$r_3r_4r_2r_3$	$\alpha_3+2\alpha_4$	$-(\alpha_2+2\alpha_3+2\alpha_4)$	$\alpha_2+\alpha_3+\alpha_4$
$r_2r_3r_4r_3$	$\alpha_2+2\alpha_3+2\alpha_4$	$-(\alpha_2+\alpha_3+2\alpha_4)$	α_4
$(r_4r_3)^2$	$\alpha_2+2\alpha_3+2\alpha_4$	$-\alpha_3$	$-\alpha_4$
$r_4r_2r_3r_4$	$\alpha_3+2\alpha_4$	$\alpha_2+\alpha_3$	$-(\alpha_2+\alpha_3+\alpha_4)$
$r_3r_4r_2r_3r_2$	$-(\alpha_3+2\alpha_4)$	$-(\alpha_2+\alpha_3)$	$\alpha_2+\alpha_3+\alpha_4$
$r_2r_3r_4r_3r_2$	$-(\alpha_2+2\alpha_3+2\alpha_4)$	α_3	α_4
$(r_4r_3)^2r_2$	$-(\alpha_2+2\alpha_3+2\alpha_4)$	$\alpha_2+\alpha_3+2\alpha_4$	$-\alpha_4$
$r_4r_2r_3r_4r_2$	$-(\alpha_3+2\alpha_4)$	$\alpha_2+2\alpha_3+2\alpha_4$	$-(\alpha_2+\alpha_3+\alpha_4)$
$r_2r_3r_4r_2r_3$	$\alpha_2+\alpha_3+2\alpha_4$	$-(\alpha_2+2\alpha_3+2\alpha_4)$	$\alpha_3+\alpha_4$
$r_4r_3r_4r_2r_3$	α_3	$-(\alpha_2+2\alpha_3+2\alpha_4)$	$\alpha_2+\alpha_3+\alpha_4$
$r_4r_2r_3r_4r_3$	$\alpha_2+2\alpha_3+2\alpha_4$	$-(\alpha_2+\alpha_3)$	$-\alpha_4$
$r_3r_4r_2r_3r_4$	$\alpha_3+2\alpha_4$	α_2	$-(\alpha_2+\alpha_3+\alpha_4)$

Table 5.1 The images of fundamental real roots under the action of Weyl group elements of $SBGB_3$—cont'd

"Weyl group" element	Image of α_2 under the action of Weyl group element	Image of α_3 under the action of Weyl group element	Image of α_4 under the action of Weyl group element
$r_2r_3r_4r_2r_3r_2$	$-(\alpha_2+\alpha_3+2\alpha_4)$	$-\alpha_3$	$\alpha_3+\alpha_4$
$r_4r_3r_4r_2r_3r_2$	$-\alpha_3$	$-(\alpha_2+\alpha_3+2\alpha_4)$	$\alpha_2+\alpha_3+\alpha_4$
$r_4r_2r_3r_4r_3r_2$	$-(\alpha_2+2\alpha_3+2\alpha_4)$	$\alpha_3+2\alpha_4$	$-\alpha_4$
$(r_3r_4r_2)^2$	$-(\alpha_3+2\alpha_4)$	$\alpha_2+\alpha_3+2\alpha_4$	$-(\alpha_2+\alpha_3+\alpha_4)$
$(r_4r_2r_3)^2$	$\alpha_2+\alpha_3$	$-(\alpha_2+2\alpha_3+2\alpha_4)$	$\alpha_3+\alpha_4$
$r_3r_4r_2r_3r_4r_3$	$\alpha_2+\alpha_3+2\alpha_4$	$-\alpha_2$	$-(\alpha_3+\alpha_4)$
$r_4r_3r_4r_2r_3r_4$	α_3	α_2	$-(\alpha_2+\alpha_3+\alpha_4)$
$(r_4r_2r_3)^2r_2$	$-(\alpha_2+\alpha_3)$	$-(\alpha_3+2\alpha_4)$	$\alpha_3+\alpha_4$
$r_3r_4r_2r_3r_4r_3r_2$	$-(\alpha_2+\alpha_3+2\alpha_4)$	$\alpha_3+2\alpha_4$	$-(\alpha_3+\alpha_4)$
$r_4r_3r_4r_2r_3r_4r_2$	$-\alpha_3$	$\alpha_2+\alpha_3$	$-(\alpha_2+\alpha_3+\alpha_4)$
$r_3(r_4r_2r_3)^2$	α_2	$-(\alpha_2+\alpha_3+2\alpha_4)$	α_4
$r_4r_3r_4r_2r_3r_4r_3$	$\alpha_2+\alpha_3$	$-\alpha_2$	$-(\alpha_3+\alpha_4)$
$r_3(r_4r_2r_3)^2r_2$	$-\alpha_2$	$-(\alpha_3+2\alpha_4)$	α_4
$r_4r_3r_4r_2r_3r_4r_3r_2$	$-(\alpha_2+\alpha_3)$	α_3	$-(\alpha_3+\alpha_4)$
$r_4r_3(r_4r_2r_3)^2$	α_2	$-(\alpha_2+\alpha_3)$	$-\alpha_4$
$r_4r_3(r_4r_2r_3)^2r_2$	$-\alpha_2$	$-\alpha_3$	$-\alpha_4$

$$(\alpha_1 \mid \alpha_1) = -2b_1k, \quad (\alpha_1 \mid \alpha_2) = -2b_1a_1, \quad (\alpha_1 \mid \alpha_3) = -2b_1a_2,$$
$$(\alpha_1 \mid \alpha_4) = -2b_1a_3, \quad (\alpha_2 \mid \alpha_2) = 4a_1, \quad (\alpha_2 \mid \alpha_3) = -2a_1,$$
$$(\alpha_2 \mid \alpha_4) = 0, \quad (\alpha_3 \mid \alpha_3) = 4a_1, \quad (\alpha_3 \mid \alpha_4) = -2a_1,$$
$$(\alpha_4 \mid \alpha_4) = 2a_1.$$

We have

$$\langle k_1\alpha_1 + k_2\alpha_2 + k_3\alpha_3 + k_4\alpha_4, \alpha_2^\vee \rangle = -b_1k_1 + 2k_2 - k_3,$$
$$\langle k_1\alpha_1 + k_2\alpha_2 + k_3\alpha_3 + k_4\alpha_4, \alpha_3^\vee \rangle = -b_2k_1 - k_2 + 2k_3 - k_4 \text{ and}$$
$$\langle k_1\alpha_1 + k_2\alpha_2 + k_3\alpha_3 + k_4\alpha_4, \alpha_4^\vee \rangle = -b_3k_1 - 2k_3 + 2k_4.$$

Using the definitions of K, $\mathring{K}$ and Δ^{re}, we get

$$K = \{k_1\alpha_1 + k_2\alpha_2 + k_3\alpha_3 + k_4\alpha_4; k_1 \in \mathbb{N},\ k_2,\ k_3,\ k_4 \in \mathbb{Z}_+ \text{ with}$$
$$2k_2 \le b_1k_1 + k_3,\ 2k_3 \le b_2k_1 + k_2 + k_4, 2k_4 \le b_3k_1 + 2k_3\}.$$

and

$$\mathring{K} = \{k_1\alpha_1 + k_2\alpha_2 + k_3\alpha_3 + k_4\alpha_4; k_1 \in \mathbb{N},\ k_2,\ k_3,\ k_4 \in \mathbb{Z}_+ \text{ with}$$
$$2k_2 \le b_1k_1 + k_3,\ 2k_3 \le b_2k_1 + k_2 + k_4,$$
$$2k_4 \le b_3k_1 + 2k_3\} \backslash \{k_1\alpha_1;\ k_1 \in \mathbb{N}\}.$$

Hence we find from definition of $\Delta^{\rm re}$ and the above tables of actions of Weyl group elements,

$$\Delta^{\mathrm{re}} = \pm\{\alpha_2, \alpha_3, \alpha_2+\alpha_3, \alpha_2+\alpha_3+2\alpha_4, \alpha_3+2\alpha_4, \alpha_2+2\alpha_3+2\alpha_4, \alpha_4, \alpha_3+\alpha_4, \alpha_2+\alpha_3+\alpha_4\} \cup 2\{\alpha_4, \alpha_3+\alpha_4, \alpha_2+\alpha_3+\alpha_4\}.$$

Similarly by applying the Weyl group elements on α_1 and the set $\mathring{K}$, we can find the following

$$\Delta^{\mathrm{dom\cdot im}} = \begin{cases} W(\pm\{\alpha_1, 2\alpha_1\}) & \text{for } k>0, \\ W(\pm\alpha_1) & \text{for } k=0, \end{cases}$$

where k is as in $SBGB_3$, and

$$\Delta_+^{\mathrm{ali\cdot im}} = \bigcup_{w\in W} w(\mathring{K}).$$

Class 2: Extension of an affine KM matrix

Example 54. Consider $X = G_2^{(1)}$. Then

$$(SBGG_2^{(1)}, \psi) = \begin{pmatrix} -k & -a_1 & -a_2 & -a_3 \\ -b_1 & 2 & -1 & 0 \\ -b_2 & -1 & 2 & -1 \\ -b_3 & 0 & -3 & 2 \end{pmatrix} = (A, \psi).$$

If

$$D = \begin{pmatrix} 3b_1 & 0 & 0 & 0 \\ 0 & 3a_1 & 0 & 0 \\ 0 & 0 & 3a_1 & 0 \\ 0 & 0 & 0 & a_1 \end{pmatrix}$$

with $a_1b_2 = a_2b_1$ and $a_2b_3 = 3b_2a_3$, the matrix $SBGG_2^{(1)}$ becomes symmetrizable. Also this matrix is non-singular and indecomposable. It is clear that $I^{\mathrm{re}} = \{2,3,4\}$, $I^{\mathrm{im}} = \{1\}$. $\Pi^{\mathrm{re}} = \{\alpha_2, \alpha_3, \alpha_4\}$, $\Pi^{\mathrm{im}} = \{\alpha_1\}$, $\Pi = \{\alpha_1, \alpha_2, \alpha_3, \alpha_4\}$.

We have $\psi = \{1\}$.
So $\psi^{\mathrm{re}} = \emptyset$.
If $k=0$, $\psi_0 = \{1\}$ and $\psi_- = \emptyset$.
If $k>0$, $\psi_0 = \emptyset$ and $\psi_- = \{1\}$.
Hence, for $k \geq 0$, $\psi^{\mathrm{im}} = \{1\}$.

We have $\Pi_{\mathrm{even}} = \{\alpha_2, \alpha_3, \alpha_4\}$ and $\Pi_{\mathrm{odd}} = \{\alpha_1\}$. Using the definitions of reflections of simple roots, we get the following results:

$$r_2(\alpha_1) = \alpha_1 + b_1\alpha_2; \quad r_2(\alpha_2) = -\alpha_2; \quad r_2(\alpha_3) = \alpha_3 + \alpha_2; \quad r_2(\alpha_4) = \alpha_4;$$
$$r_3(\alpha_1) = \alpha_1 + b_2\alpha_2; \quad r_3(\alpha_2) = \alpha_2 + \alpha_3; \quad r_3(\alpha_3) = -\alpha_3; \quad r_3(\alpha_4) = \alpha_4 + \alpha_3;$$
$$r_4(\alpha_1) = \alpha_1 + b_3\alpha_4; \quad r_4(\alpha_2) = \alpha_2; \quad r_4(\alpha_3) = \alpha_3 + 3\alpha_4; \quad r_4(\alpha_4) = -\alpha_4.$$

Hence we get $W = \textit{Infinite set}$ consisting of all products of the reflections r_2, r_3 and r_4 taken any number of times with any order.

Using the definition of the bilinear form $(.|.)$, we get the following results:

$$\begin{aligned}
&(\alpha_1 \mid \alpha_1) = -3kb_1; (\alpha_1 \mid \alpha_2) = -3a_1b_1; (\alpha_1 \mid \alpha_3) = -3a_2b_1;\\
&(\alpha_1 \mid \alpha_4) = -3a_3b_1; (\alpha_2 \mid \alpha_2) = 6a_1; (\alpha_2 \mid \alpha_3) = -3a_1;\\
&(\alpha_2 \mid \alpha_4) = 0; (\alpha_3 \mid \alpha_3) = 6a_1;\\
&(\alpha_3 \mid \alpha_4) = -3a_1; (\alpha_4 \mid \alpha_4) = 2a_1.
\end{aligned}$$

For $\alpha = \sum_{i=1}^{4} k_i\alpha_i \in Q$, we have

$$\begin{aligned}
&\langle k_1\alpha_1 + k_2\alpha_2 + k_3\alpha_3 + k_4\alpha_4, \alpha_2^\vee \rangle = -b_1k_1 + 2k_2 - k_3;\\
&\langle k_1\alpha_1 + k_2\alpha_2 + k_3\alpha_3 + k_4\alpha_4, \alpha_3^\vee \rangle = -b_2k_1 - k_2 + 2k_3 - k_4;\\
&\langle k_1\alpha_1 + k_2\alpha_2 + k_3\alpha_3 + k_4\alpha_4, \alpha_4^\vee \rangle = -b_3k_1 - 3k_3 + 2k_4.
\end{aligned}$$

Using the definitions of K and $\overset{\circ}{K}$, we get

$$\begin{aligned}
K = \{&k_1\alpha_1 + k_2\alpha_2 + k_3\alpha_3 + k_4\alpha_4; k_1 \in \mathbb{N},\ k_2, k_3, k_4 \in \mathbb{Z}_+ \text{ with}\\
&2k_2 \le b_1k_1 + k_3,\ 2k_3 \le b_2k_1 + k_2 + k_4 \text{ and}\\
&2k_4 \le b_3k_1 + 3k_3\},\\
\overset{\circ}{K} = \{&k_1\alpha_1 + k_2\alpha_2 + k_3\alpha_3 + k_4\alpha_4; k_1 \in \mathbb{N},\ k_2, k_3, k_4 \in \mathbb{Z}_+ \text{ with}\\
&2k_2 \le b_1k_1 + k_3,\ 2k_3 \le b_2k_1 + k_2 + k_4 \text{ and}\\
&2k_4 \le b_3k_1 + 3k_3\} \backslash \{k_1\alpha_1; k_1 \in \mathbb{N}\}.
\end{aligned}$$

Since the Weyl group is an infinite set, by fundamental results, we get

$$\Delta^{\text{re}} = W(\{\alpha_2, \alpha_3, \alpha_4\}).$$

$$\Delta^{\text{dom}\cdot\text{im}} = \begin{cases} W(\pm\{\alpha_1, 2\alpha_1\}) & \text{for } k \ge 0,\\ W(\pm\alpha_1) & \text{for } k = 0, \end{cases}$$

where k is as in $SBGG_2^{(1)}$.

Also we have

$$\Delta_+^{\text{ali}\cdot\text{im}} = \bigcup_{\omega \in W} \omega(\overset{\circ}{K})$$

Hence

$$\Delta_+^{\text{im}} = \Delta_+^{\text{dom}\cdot\text{im}} \bigcup \Delta_+^{\text{ali}\cdot\text{im}}.$$

Similarly one can find the root systems of the other cases of the extension of the affine KM matrices.

Class 3: Extension of a Hyperbolic KM matrix

Example 55. Consider $X = HA_2^{(2)}$. Then

$$(SBGHA_2^{(2)}, \psi) = \begin{pmatrix} -k & -a_1 & -a_2 & -a_3 \\ -b_1 & 2 & -1 & 0 \\ -b_2 & -1 & 2 & 4 \\ -b_3 & 0 & -1 & 2 \end{pmatrix} = (A, \psi).$$

If

$$D = \begin{pmatrix} b_1 & 0 & 0 & 0 \\ 0 & a_1 & 0 & 0 \\ 0 & 0 & a_1 & 0 \\ 0 & 0 & 0 & 4a_1 \end{pmatrix}$$

with $a_1 b_2 = a_2 b_1$ and $4a_2 b_3 = a_3 b_2$, then the matrix $SBGHA_2^{(2)}$ becomes symmetrizable. Also this matrix is non-singular and indecomposable. We have

$$\Pi^{\text{re}} = \{\alpha_2, \alpha_3, \alpha_4\}, \; \Pi^{\text{im}} = \{\alpha_1\} \text{ and } \Pi = \{\alpha_1, \alpha_2, \alpha_3, \alpha_4\}.$$

Let $\psi = \{1\}$. Then $\psi^{\text{re}} = \emptyset$. Also $\psi_- = \emptyset$; $\psi_0 = \{1\}$ for $k = 0$.
$\psi_0 = \emptyset$; $\psi_- = \{1\}$ for $k > 0$.
Hence $\psi^{\text{im}} = \{1\}$ for $k \geq 0$.
Also $\Pi_{\text{even}} = \{\alpha_2, \alpha_3, \alpha_4\}$, and $\Pi_{\text{odd}} = \{\alpha_1\}$.
Using the definition of reflections of simple real roots, we get

$$\begin{aligned} &r_2(\alpha_1) = \alpha_1 + b_1\alpha_2; \quad r_2(\alpha_2) = -\alpha_2; \quad r_2(\alpha_3) = \alpha_3 + \alpha_2; \quad r_2(\alpha_4) = \alpha_4; \\ &r_3(\alpha_1) = \alpha_1 + b_2\alpha_3; \quad r_3(\alpha_2) = \alpha_2 + \alpha_3; \quad r_3(\alpha_3) = -\alpha_3; \\ &r_3(\alpha_4) = \alpha_4 + 4\alpha_3; \quad r_4(\alpha_1) = \alpha_1 + b_3\alpha_4; \quad r_4(\alpha_2) = \alpha_2; \\ &r_4(\alpha_3) = \alpha_3 + \alpha_4; \quad \text{and} \quad r_4(\alpha_4) = -\alpha_4. \end{aligned}$$

Hence the Weyl group of the algebra corresponding to $SBGHA_2^{(2)}$, which is the same that of $GHA_2^{(2)}$, is $W =$ Infinite set consisting of all products of the reflections r_2, r_3 and r_4 taken any number of times with any order. Using the definitions of bilinear form $(.|.)$, we get the following:

$$\begin{aligned} &(\alpha_1 \mid \alpha_1) = -kb_1; \quad (\alpha_1 \mid \alpha_2) = -a_1b_1; \quad (\alpha_1 \mid \alpha_3) = -a_2b_1; \\ &(\alpha_1 \mid \alpha_4) = -a_3b_1; \quad (\alpha_2 \mid \alpha_2) = 2a_1; \\ &(\alpha_2 \mid \alpha_3) = -a_1; \quad (\alpha_2 \mid \alpha_4) = 0; \quad (\alpha_3 \mid \alpha_3) = 2a_1; \\ &(\alpha_3 \mid \alpha_4) = -4a_1; \quad (\alpha_4 \mid \alpha_4) = 8a_1. \end{aligned}$$

For $\alpha = \sum_{i=1}^{4} k_i\alpha_i \in Q$, we have

$$\langle k_1\alpha_1 + k_2\alpha_2 + k_3\alpha_3 + k_4\alpha_4, \alpha_2^\vee\rangle = -b_1k_1 + 2k_2 - k_3$$
$$\langle k_1\alpha_1 + k_2\alpha_2 + k_3\alpha_3 + k_4\alpha_4, \alpha_3^\vee\rangle = -b_2k_1 - k_2 + 2k_3 - 4k_4$$
$$\langle k_1\alpha_1 + k_2\alpha_2 + k_3\alpha_3 + k_4\alpha_4, \alpha_4^\vee\rangle = -b_3k_1 - k_3 + 2k_4.$$

Using the definitions of K and $\overset{\circ}{K}$, we have

$$K = \{k_1\alpha_1 + k_2\alpha_2 + k_3\alpha_3 + k_4\alpha_4; k_1 \in \mathbb{N},\ k_2, k_3, k_4 \in \mathbb{Z}_+ \text{ with}$$
$$2k_2 \le b_1k_1 + k_3;\ 2k_3 \le b_2k_1 + k_2 + 4k_4;\ 2k_4 \le b_3k_1 + k_3\} \text{ and}$$
$$\overset{\circ}{K} = \{k_1\alpha_1 + k_2\alpha_2 + k_3\alpha_3 + k_4\alpha_4; k_1 \in \mathbb{N},\ k_2, k_3, k_4 \in \mathbb{Z}_+ \text{ with}$$
$$2k_2 \le b_1k_1 + k_3;\ 2k_3 \le b_2k_1 + k_2 + 4k_4;\ 2k_4 \le b_3k_1 + k_3\}$$
$$\setminus \{k_1\alpha_1; k_1 \in \mathbb{N}\}.$$

Since the Weyl group is an infinite set, by the fundamental results, we have

$$\Delta^{\text{re}} = W\{\alpha_2, \alpha_3, \alpha_4\} \text{ and}$$

$$\Delta_+^{\text{dom}\cdot\text{im}} = \begin{cases} W(\pm\{\alpha_1, 2\alpha_1\}) & \text{for } k > 0 \\ W(\pm\alpha_1) & \text{for } k = 0, \end{cases}$$

Here, k is as in $SBGHA_2^{(2)}$. Also,

$$\Delta_+^{\text{ali}\cdot\text{im}} = \bigcup_{w \in W} w(\overset{\circ}{K}).$$

Hence

$$\Delta_+^{\text{im}} = \Delta_+^{\text{dom}\cdot\text{im}} \cup \Delta_+^{\text{ali}\cdot\text{im}}.$$

Similarly one can find the root systems for other superalgebras which are the extensions of hyperbolic types.

5.4 Special imaginary roots in BKM Lie superalgebras and their complete classifications

Definition 226. Let α be an imaginary root of a symmetrizable BKM superalgebra $\mathfrak{g}(A)$ associated with the BKM supermatrix (A, ψ). We call α, a special imaginary root, if α satisfies the following conditions:

(i) $(\alpha \mid \alpha) \neq 0$,
(ii) $r_\alpha(\Delta) = \Delta$, $r_\alpha(\Delta^{\text{re}}) = \Delta^{\text{re}}$, $r_\alpha(\Delta^{\text{dom}\cdot\text{im}}) = \Delta^{\text{dom}\cdot\text{im}}$, $r_\alpha(\Delta^{\text{ali}\cdot\text{im}}) = \Delta^{\text{ali}\cdot\text{im}}$ and
(iii) r_α preserves root multiplicities.

Let $\Delta^{sp.im}$ denote the set of all special imaginary roots in the corresponding BKM Lie superalgebra.

The proof of the following lemma is exactly similar to the same result which was proved for GKM algebras in Chapter 3.

Lemma 11. *If α is a special imaginary root and β is a real simple root of a symmetrizable BKM superalgebra, then $(\beta|\alpha) = 0$.*

Remark 139.

(1) As the Weyl groups of BKM superalgebras, the corresponding BKM algebras and the corresponding KM algebras are same, one can use the following results of Bennett [54] for KM algebras: Given a special imaginary root α of a Kac-Moody algebra $\mathfrak{g}(A)$, where A is the generalized Cartan matrix of the corresponding KM algebra, there exists a unique special imaginary root β of $\mathfrak{g}(A)$ such that
(i) $\alpha = w(\beta)$ *or* $-w(\beta)$ for some $w \in W$, and
(ii) β is of type 1 or type 2.
If α is a special imaginary root, then $r_\alpha \in -W$ or an element of $-W$ multiplied by a diagram automorphism. Moreover if $r_\alpha \in -W$ or an element of $-W$ multiplied by a diagram automorphism and α is an imaginary root, then α is special.

(2) Several propositions given below were proved in [144] for the existence and non-existence of special imaginary roots for BKM Lie superalgebras. We state these results below.

Existence of special imaginary roots

Proposition 39. *Let (A, ψ) be a BKM supermatrix, where $A =$ Symmetrizable*

$$SBGA_2 = \begin{pmatrix} -k & -a_1 & -a_2 \\ -b_1 & 2 & -1 \\ -b_2 & -1 & 2 \end{pmatrix},$$

t be the permutation $(2, 3)$ and $\alpha = k_1\alpha_1 + k_2\alpha_2 + k_3\alpha_3 \in \Delta_+^{\text{im}}$, $k_1, k_2, k_3 \in \mathbb{N}$. Let $b_1 = b_2$. Then, $r_\alpha \in -tW$ if and only if $k_2 = k_3 = b_1k_1 \in \mathbb{N}$. Hence $\alpha = k_1(\alpha_1 + b_1\alpha_2 + b_1\alpha_3)$ is a special imaginary root of $\mathfrak{g}(A)$, for every $k_1 \in \mathbb{N}$, with $b_1 = b_2$. If the condition $b_1 = b_2$ is not satisfied, we will not get special imaginary roots.

Proposition 40. *Let (A, ψ) be a BKM supermatrix, where $A =$ Symmetrizable $SBGB_2 = \begin{pmatrix} -k & -a_1 & -a_2 \\ -b_1 & 2 & -1 \\ -b_2 & -2 & 2 \end{pmatrix}$ and $\alpha = k_1\alpha_1 + k_2\alpha_2 + k_3\alpha_3 \in \Delta_+^{\text{im}}$, $k_1, k_2, k_3 \in \mathbb{N}$. Then, $r_\alpha \in -W$ if and only if $2k_2 = (2b_1 + b_2)k_1$ and $k_3 = (b_1 + b_2)k_1 \in \mathbb{N}$. Hence $\alpha = \frac{k_1}{2}\left(2\alpha_1 + (2b_1 + b_2)\alpha_2 + 2(b_1 + b_2)\alpha_3\right)$ is a special imaginary root of $\mathfrak{g}(A)$, for every $k_1 \in \mathbb{N}$.*

Proposition 41. *Let (A, ψ) be a BKM supermatrix, where A = Symmetrizable*

$$SBGA_3 = \begin{pmatrix} -k & -a_1 & -a_2 & -a_3 \\ -b_1 & 2 & -1 & 0 \\ -b_2 & -1 & 2 & -1 \\ -b_3 & 0 & -1 & 2 \end{pmatrix},$$

t be the permutation (2, 4) and $\alpha = k_1\alpha_1 + k_2\alpha_2 + k_3\alpha_3 + k_4\alpha_4 \in \Delta_+^{\text{im}}$, $k_1, k_2, k_3, k_4 \in \mathbb{N}$. *Let* $b_1 = b_2 = b_3$. *Then,* $r_\alpha \in -tW$ *if and only if* $k_2 = k_4 = \frac{3b_1}{2}k_1$ *and* $k_3 = 2b_1k_1 \in \mathbb{N}$. *Hence* $\alpha = \frac{k_1}{2}(2\alpha_1 + 3b_1\alpha_2 + 4b_1\alpha_3 + 3b_1\alpha_4)$ *is a special imaginary root of* $\mathfrak{g}(A)$, *for every* $k_1 \in \mathbb{N}$, *with* $b_1 = b_2 = b_3$. *If this condition* $b_1 = b_2 = b_3$ *is not satisfied, we will not get special imaginary roots.*

Proposition 42. *Let (A, ψ) be a BKM supermatrix, where A = symmetrizable*

$$SBGB_3 = \begin{pmatrix} -k & -a_1 & -a_2 & -a_3 \\ -b_1 & 2 & -1 & 0 \\ -b_2 & -1 & 2 & -1 \\ -b_3 & 0 & -2 & 2 \end{pmatrix}$$

and $\alpha = k_1\alpha_1 + k_2\alpha_2 + k_3\alpha_3 + k_4\alpha_4 \in \Delta_+^{\text{im}}$, $k_1, k_2, k_3, k_4 \in \mathbb{N}$. *Then,* $r_\alpha \in -W$ *if and only if* $2k_2 = (2b_1 + 2b_2 + b_3)k_1$, $k_3 = (b_1 + 2b_2 + b_3)k_1$ *and* $2k_4 = (2b_1 + 4b_2 + 3b_3)k_1 \in \mathbb{N}$. *Hence* $\alpha = \frac{k_1}{2}(2\alpha_1 + (2b_1 + 2b_2 + b_3)\alpha_2 + 2(b_1 + 2b_2 + b_3)\alpha_3 + (2b_1 + 4b_2 + 3b_3)\alpha_4)$, *for every* $k_1 \in \mathbb{N}$ *is a special imaginary root of* $\mathfrak{g}(A)$.

Non-existence of special imaginary roots

Proposition 43. *Let (A, ψ) be a BKM supermatrix, where A = symmetrizable*

$$SBGA_1^{(1)} = \begin{pmatrix} -k & -a_1 & -a_2 \\ -b_1 & 2 & -2 \\ -b_2 & -2 & 2 \end{pmatrix}$$

and $\alpha = k_1\alpha_1 + k_2\alpha_2 + k_3\alpha_3 \in \Delta_+^{\text{im}}$, $k_1, k_2, k_3 \in \mathbb{N}$. *Then,* $r_\alpha \notin -W$. *Hence there exists no special imaginary root for* $\mathfrak{g}(A)$.

Proposition 44. *Let (A, ψ) be a BKM supermatrix, where A = Symmetrizable*

$$SBGG_2^{(1)} = \begin{pmatrix} -k & -a_1 & -a_2 & -a_3 \\ -b_1 & 2 & -1 & 0 \\ -b_2 & -1 & 2 & -1 \\ -b_3 & 0 & -3 & 2 \end{pmatrix}$$

and $\alpha = k_1\alpha_1 + k_2\alpha_2 + k_3\alpha_3 + k_4\alpha_4 \in \Delta_+^{\mathrm{im}}$, $k_1, k_2, k_3, k_4 \in \mathbb{N}$. *Then,* $r_\alpha \notin -W$. *Hence there exists no special imaginary root for* $\mathfrak{g}(A)$.

Proposition 45. *Let* (A, ψ) *be a BKM supermatrix, where* $A =$ *symmetrizable*

$$SBGHA_1 = \begin{pmatrix} -k & -a_1 & -a_2 \\ -b_1 & 2 & -p \\ -b_2 & -q & 2 \end{pmatrix}$$

and $\alpha = k_1\alpha_1 + k_2\alpha_2 + k_3\alpha_3 \in \Delta_+^{\mathrm{im}}$, $k_1, k_2, k_3 \in \mathbb{N}$. *Then,* $r_\alpha \notin -W$. *Hence there exists no special imaginary root for* $\mathfrak{g}(A)$.

Proposition 46. *Let* (A, ψ) *be a BKM supermatrix, where* $A =$ *symmetrizable*

$$SBGHA_2^{(2)} = \begin{pmatrix} -k & -a_1 & -a_2 & -a_3 \\ -b_1 & 2 & -1 & 0 \\ -b_2 & -1 & 2 & -4 \\ -b_3 & 0 & -1 & 2 \end{pmatrix}$$

and $\alpha = k_1\alpha_1 + k_2\alpha_2 + k_3\alpha_3 + k_4\alpha_4 \in \Delta_+^{\mathrm{im}}$, $k_1, k_2, k_3, k_4 \in \mathbb{N}$. *Then,* $r_\alpha \notin -W$. *Hence there exists no special imaginary root for* $\mathfrak{g}(A)$.

Proposition 47. *Let* (A, ψ) *be a BKM supermatrix, where* $A =$ *symmetrizable*

$$SBGHA_1^{(1)} = \begin{pmatrix} -k & -a_1 & -a_2 & -a_3 \\ -b_1 & 2 & -1 & 0 \\ -b_2 & -1 & 2 & -2 \\ -b_3 & 0 & -2 & 2 \end{pmatrix}$$

and $\alpha = k_1\alpha_1 + k_2\alpha_2 + k_3\alpha_3 + k_4\alpha_4 \in \Delta_+^{\mathrm{im}}$, $k_1, k_2, k_3, k_4 \in \mathbb{N}$. *Then,* $r_\alpha \notin -W$. *Hence there exists no special imaginary root for* $\mathfrak{g}(A)$.

Remark 140. In [144], we found out all imaginary roots and special imaginary roots of the BKM superalgebras (Borcherds Kac-Moody Lie superalgebras other than BKM algebras) whose symmetrizable BKM supermatrices (of order ≤ 4) are extensions of KM matrices (of order ≤ 3) of finite, affine and hyperbolic types. We notice that all BKM superalgebras, whose BKM supermatrices are the extensions of KM matrices of finite-non-simply laced types, have special imaginary roots, but from among the BKM superalgebras whose BKM super matrices are extensions of KM matrices of simply laced finite type, some algebras possess special imaginary roots whereas some others do not. Also we found that all BKM superalgebras whose BKM supermatrices are extensions of KM matrices of affine and hyperbolic types do not possess special imaginary roots whereas the underlying hyperbolic KM algebras possess special imaginary roots. Moreover for BKM superalgebras, we found that only the alien-type imaginary roots are special imaginary roots whereas the domestic-type imaginary roots are not special imaginary roots. We also noticed that all alien-type imaginary roots are even imaginary roots.

Complete classification of BKM Lie superalgebras possessing special imaginary roots

In this section, we give a complete classification of BKM Lie superalgebras possessing special imaginary roots. Throughout this chapter, by "order" we mean the order of the simple roots, that is "number" of simple roots.

For the purpose of finding a complete classification of BKM Lie superalgebras possessing special imaginary roots, we divide the BKM Lie superalgebras into two categories.

Category 1: BKM Lie superalgebras *without any odd simple root* (GKM algebras only).
Category 2: BKM Lie superalgebras *with a non-empty set of odd roots.*

We discuss these two categories separately below:

Category 1: BKM Lie superalgebras without any odd simple root(GKM algebras only): For complete classification of GKM algebras(of finite order and infinite order) other than KM algebras possessing special imaginary roots, one can see Chapter 3. One can refer [54], for Kac-Moody algebras possessing special imaginary roots.

Category 2: BKM Lie superalgebras with a non-empty set of odd roots: We have the following two cases in category 2.

(a) BKM Lie superalgebras of finite order with a non-empty set of odd roots(which are not GKM algebras).
(b) BKM Lie superalgebras of infinite order with a non-empty set of odd roots(which are not GKM algebras).

We discuss these two cases separately.

Category 2(a). BKM Lie superalgebras of finite order with a non-empty set of odd roots(which are not GKM algebras): For the above case(a), we prove the following Theorem 55 and state Lemma 11 [152].

Theorem 55. *Let* $A = (a_{ij})_{i,j=1}^{n+r}$ *be the symmetrizable*

$$GGX = \begin{pmatrix} -k_1 & -a_1 & -a_2 & \cdots & -a_{r-1} & -a_r & \cdots & -a_{n+r-1} \\ -b_1 & -k_2 & -c_2 & \cdots & -c_{r-1} & -c_r & \cdots & -c_{n+r-1} \\ -b_2 & -d_2 & -k_3 & \cdots & \cdots & \cdots & \cdots & \cdots \\ \vdots & \vdots & & \ddots & & & & \\ -b_{r-1} & -d_{r-1} & \cdots & \cdots & -k_r & \cdots & \cdots & \cdots \\ -b_r & -d_r & & & \vdots & & & \\ \vdots & \vdots & & \vdots & & GX & & \\ -b_{n+r-1} & -d_{n+r-1} & & & \vdots & & & \end{pmatrix},$$

where $k_i(1 \leq i \leq r) \in \mathbb{Z}_{\geq 0}$, a_i, b_i, c_i, d_i *are positive integers. Here GX is a Generalized Cartan matrix(GCM) of finite, affine or indefinite type of order n and GGX is a Generalized Generalized Cartan supermatrix(GGCM supermatrix) or BKM supermatrix of finite, affine or indefinite type with r simple imaginary roots added to that of GX. Let* $\alpha = \sum_{i=1}^{n+r} l_i\alpha_i$ *(for* $l_i \in \mathbb{N}$*) be any positive imaginary root of* $\mathfrak{g}(A)$*. Then the following results are true:*

(i) *If* α *is a special imaginary root then* $\sum\limits_{i=1}^{n+r} l_i a_{ji} = 0$ *for each* $j \in I^{re}$ *and* $supp(\alpha) \geq 2$.

(ii) $\Delta^{sp.im.} \subset \Delta^{ali.im.}$. *Inclusion of odd roots has no effect on* $\Delta^{ali.im.}$ *and hence has no effect on* $\Delta^{sp.im.}$.

(iii) *If* α *is special imaginary root and* $j \in I^{re}$ *(*α_j *is a real simple root)then* $r_\alpha(\alpha_j) = \alpha_j$.

Proof. In the usual notation, let $I = \{1, 2, \dots, n + r\}$ with $I^{im} = \{1, 2, \dots, r\}$ and $I^{re} = \{r + 1, r + 2, \dots, n + r\}$. Here, $\Pi = \{\alpha_1, \alpha_2, \dots, \alpha_{n+r}\}$, the set of all simple roots with $\Pi^{im} = \{\alpha_1, \alpha_2, \dots, \alpha_r\}$, the set of all imaginary simple roots and $\Pi^{re} = \{\alpha_{r+1}, \alpha_{r+2}, \dots, \alpha_{n+r}\}$, the set of all real simple roots. In the usual notation, ψ, the index set of odd roots will take one of the following forms:

$$\psi = \begin{cases} \{i | i \in I^{im}\} \quad (or) \\ \{i | i \in I^{im}\} \cup \{j | a_{jk} \in 2\mathbb{Z}\ \forall\ k \ \ and \ \ j \in I^{re}\} \quad (or) \\ \{j | a_{jk} \in 2\mathbb{Z}\ \forall\ k \quad and \ \ j \in I^{re}\} \end{cases}$$

Let $\alpha = \sum_{i=1}^{n+r} l_i\alpha_i$ be a positive imaginary root for $l_i \in \mathbb{Z}\ \forall\ i$ with atleast one l_i being non-zero. In the usual notation, we have

$$\begin{aligned} \langle \alpha, \alpha_j^\vee \rangle &= \sum_{i=1}^{n+r} l_i a_{ji},\ for\ j \in I^{re}, \\ K &= \{\alpha \in Q_+ | \langle \alpha, \alpha_j^\vee \rangle \leq 0 (j \in I^{re})\ and\ supp(\alpha)\ is\ connected\} \\ &= \{\alpha \in Q_+ | \sum_{i=1}^{n+r} l_i a_{ji} \leq 0 (j \in I^{re})\ and\ supp(\alpha)\ is\ connected\} \\ with\ \overset{\circ}{K} &= \{\alpha \in K |\ |supp(\alpha)|\ \geq 2\} \\ &= \{\alpha \in K | \sum_{i=1}^{n+r} l_i a_{ji} \leq 0 (j \in I^{re})\} \backslash \{l_i\alpha_i | l_i \in \mathbb{Z}, (1 \leq i \leq r)\} \\ and\ \Delta^{ali.im} &= \bigcup_{w \in W} w(\overset{\circ}{K}). \end{aligned}$$

Moreover, $\langle \alpha_i, \alpha_j^\vee \rangle = a_{ji}\ with\ \ (\alpha_i | \alpha_j) = b_{ij} = \epsilon_i a_{ij}$.

(i) Let $\alpha = \sum_{i=1}^{n+r} l_i\alpha_i$ be a special imaginary root. We have

$$\begin{aligned} \langle \alpha_j, \alpha^\vee \rangle &= \frac{2(\alpha_j | \alpha)}{(\alpha | \alpha)} \qquad (5.2) \\ \Rightarrow \langle \alpha_j, \alpha^\vee \rangle &= \frac{2}{P} \sum_{i=1}^{n+r} l_i \epsilon_j a_{ji}\ for\ (\alpha | \alpha) = P < 0 \end{aligned}$$

As α being a special imaginary root and α_j being a simple real root (by Lemma 11), we have $\langle\alpha_j, \alpha^\vee\rangle = 0$. Hence, if $\alpha = \sum_{i=1}^{n+r} l_i\alpha_i$ is special imaginary root, then

$$\sum_{i=1}^{n+r} l_i a_{ji} = 0 \; as \; \epsilon_i, \epsilon_j > 0 \;\; and \; P < 0 \, for \, j \in I^{re}. \tag{5.3}$$

Moreover if $\alpha = \sum_{i=1}^{n+r} l_i\alpha_i \in \Delta_+^{im}$ is a special imaginary root and $supp(\alpha) < 2(that\ is, supp(\alpha) = 1)$, then $\alpha = l\alpha_i$ $(1 \leq i \leq r)$ $l = 1$ *or* 2 as per the definition of imaginary root. As $\alpha = l\alpha_i$ is a special imaginary root, by the above conclusion $0 = \sum l_i a_{ji} = l a_{ji} \Rightarrow a_{ji} = 0$ for all i as $l \neq 0$ which is a contradiction as $a_{ji}(1 \leq i \leq r)$'s are non-zero integers. Hence

$$supp(\alpha) \geq 2. \tag{5.4}$$

(ii) From (5.3) and (5.4) using the definition of $\overset{\circ}{K}$, it is clear that $\alpha \in \overset{\circ}{K}$ if α is a special imaginary root. So $\alpha \in \Delta^{ali.im}$.
Hence $\Delta^{sp.im} \subset \Delta^{ali.im}$. Moreover, it is clear from the above proof that whatever may be the index set ψ, the above conclusion is always true (that is the odd roots have no effect in the above conclusion).

(iii) If α is a special imaginary root for $j \in I^{re}$ (α_j is a real simple root), then

$$\begin{aligned} r_\alpha(\alpha_j) &= \alpha_j - \langle\alpha_j, \alpha^\vee\rangle\alpha \\ &= \alpha_j \;\; (\text{using } \langle\alpha_j, \alpha^\vee\rangle = 0 \text{ as } \alpha \text{ is a special imaginary root }). \end{aligned}$$

So, for a special imaginary root α, $r_\alpha(\alpha_j) = \alpha_j$ for all $j \in I^{re}$. □

Lemma 12. *Let* $A = (a_{ij})_{i,j=1}^{n+r}$ *be the symmetrizable*

$$GGX = \begin{pmatrix} -k_1 & -a_1 & -a_2 & \cdots & -a_{r-1} & -a_r & \cdots & -a_{n+r-1} \\ -b_1 & -k_2 & -c_2 & \cdots & -c_{r-1} & -c_r & \cdots & -c_{n+r-1} \\ -b_2 & -d_2 & -k_3 & \cdots & \cdots & \cdots & \cdots & \cdots \\ \vdots & \vdots & & \ddots & & & & \\ -b_{r-1} & -d_{r-1} & \cdots & \cdots & -k_r & \cdots & \cdots & \cdots \\ -b_r & -d_r & & & \vdots & & & \\ \vdots & \vdots & & & & GX & & \\ -b_{n+r-1} & -d_{n+r-1} & & & & & & \end{pmatrix},$$

where $k_i(1 \leq i \leq r) \in \mathbb{Z}_{\geq 0}$, a_i, b_i, c_i, d_i *are positive integers. Here GX is KM matrix of finite, affine or indefinite type of order n and GGX is a supermatrix of finite, affine or indefinite type with r simple imaginary roots added to that of GX. If* $\alpha = \sum_{i=1}^{n+r} l_i\alpha_i$ *is a special imaginary root, then we prove that* $\sum_{i=r+1}^{n+r} l_i(\alpha_i|\alpha) = 0$. *Hence* $(\alpha|\alpha) = \sum_{i=1}^{r} l_i(\alpha_i|\alpha)$.

Remark 141. In the case of one simple imaginary root, $(\alpha|\alpha) = l_1(\alpha_1|\alpha)$.

If the number of imaginary simple root is 2 ($r = 2$), by Lemma 11 we have, $(\alpha|\alpha) = l_1(\alpha_1|\alpha) + l_2(\alpha_2|\alpha)$.

For Category (2), case (a) of BKM Lie superalgebras possessing special imaginary roots, as in the case of GKM algebras (Chapter 3), we divide these BKM supermatrices of finite order into five classes as explained below.

(1) *BKM Lie superalgebras all whose simple roots are real*: There is no imaginary root for these BKM Lie superalgebras. So, as in the case of GKM algebras, this set of BKM Lie superalgebras does not possess special imaginary roots.

(2) *BKM Lie superalgebras all whose simple roots are imaginary*: As we do not have any simple real roots here, the Weyl group is empty and hence this set of BKM Lie superalgebras does not possess special imaginary roots.

(3) *BKM Lie superalgebras with one simple imaginary root*: As in the case of GKM algebras, we prove the following proposition("if part" of it is proved in Chapter 3 also).

Theorem 56. *Let A be the symmetrizable BKM supermatrix*

$$GGX = \begin{pmatrix} -k_1 & -a_1 & -a_2 & \cdots & -a_n \\ -b_1 & & & & \\ -b_2 & & & & \\ \vdots & & & GX & \\ -b_n & & & & \end{pmatrix},$$

where $k_1 \in \mathbb{Z}_{\geq 0}$, a_i, b_i are positive integers. Here GX is a Generalized Cartan matrix(GCM) of finite, affine or indefinite type of order n. Let

$$\alpha = \sum_{i=1}^{n+1} l_i\alpha_i \in K \subseteq \Delta_+^{\text{im}}, l_i \in \mathbb{N} \qquad for\ 1 \leq i \leq n+1$$

be any positive imaginary root of $\mathfrak{g}(A)$. Then

(i) *If GX is a GCM of finite type, then $\alpha = l_1\alpha_1 + l_1\left(\sum_{i=2}^{n+1} m_i\alpha_i\right)$ is a special imaginary root of $\mathfrak{g}(A)$ if and only if*

$$r_\alpha(\alpha_1) = \begin{cases} -w_0(\alpha_1), & if \quad GX \text{ is of type } B, C, F_4, G_2, E_7 \text{ or } E_8 \\ -tw_0(\alpha_1), & if \quad GX \text{ is of type } A, D \text{ or } E_6. \end{cases}$$

Here w_0 is the longest element of the Weyl group and t is the permutation on n symbols.

(ii) *If GX is a GCM of affine or indefinite type, then α is not a special imaginary root of $\mathfrak{g}(A)$.*

Proof.

(i) Let GX be of finite type of GCM. We have in the usual notations,

$$\Pi = \{\alpha_1, \alpha_2, \ldots, \alpha_{n+1}\},\ \Pi^{im} = \{\alpha_1\} \text{ and } \Pi^{re} = \{\alpha_2, \alpha_3, \ldots, \alpha_{n+1}\}$$
$$I = \{1, 2, \ldots, n+1\};\ I^{re} = \{2, 3, \ldots, n+1\} \text{ and } I^{im} = \{1\}.$$

The index set ψ of odd simple roots will be one of the following forms:

$$\psi = \{1\}\ (or)\ \{1\} \cup \{i | a_{ij} \in 2\mathbb{Z}\ for\ all\ j (i \in I^{re})\}\ (or)\ \{i | a_{ij} \in 2\mathbb{Z}\ for\ all\ j (i \in I^{re})\}.$$

For any value of ψ defined as above, from the proof of Theorem 55 it is clear that involvement of the odd root has no effect in the computation of special imaginary root. For the sake of completeness, we give the complete proof below.

Suppose

$$\alpha = \sum_{i=1}^{n+1} l_i \alpha_i \in K \subseteq \Delta_+^{\mathrm{im}}, l_i \in \mathbb{N}$$

is a special imaginary root of $\mathfrak{g}(A)$.

Then we know that $\langle \alpha, \alpha_i^\vee \rangle \leq 0$, $i = 2$ to $n+1$ and $(\alpha|\alpha) < 0$. Let $(\alpha|\alpha) = A$ (say).
Using the definition of imaginary roots, we have

$$r_\alpha(\alpha_i) = \alpha_i - p_i \alpha \text{ for } 1 \leq i \leq n+1, \text{ where } p_i = \frac{2(\alpha_i|\alpha)}{A}. \tag{5.5}$$

Using Lemma 11, from Equation (5.5), we obtain

$$p_i = 0 \text{ for } i = 2 \text{ to } n+1.$$

Solving these n equations in $n+1$ unknowns l_i, we get l_i $(2 \leq i \leq n+1)$ in terms of l_1. Let $l_i = m_i l_1$ for $2 \leq i \leq n+1$ where m_i are some positive integers.
So,

$$\alpha = l_1 \alpha_1 + l_1 \left(\sum_{i=2}^{n+1} m_i \alpha_i \right).$$

This α will be a special imaginary root if it satisfies

$$r_\alpha(\alpha_1) = \begin{cases} -w_0(\alpha_1), & \text{if} \quad \text{GX is of type B, C, F, G, } E_7 \text{or } E_8 \\ -tw_0(\alpha_1), & \text{if} \quad \text{GX is of type A, D or } E_6. \end{cases}$$

Here w_0 is the longest element of the Weyl group of GX and t is some permutation on n symbols. Hence under these conditions we get special imaginary roots.

Conversely, suppose $\alpha = l_1\alpha_1 + l_1 \sum_{i=2}^{n+1} m_i \alpha_i$ is an imaginary root. Let us prove that

$$r_\alpha(\alpha_1) = \begin{cases} -w_0(\alpha_1), & \text{if} \quad GX \textit{ is of type } B, C, F_4, G_2, E_7 \textit{ or } E_8 \\ -tw_0(\alpha_1), & \text{if} \quad GX \textit{ is of type } A, D \textit{ or } E_6. \end{cases}$$

We have,

$$\begin{aligned}(\alpha|\alpha) &= \left(\left(l_1\alpha_1 + l_1 \sum_{i-2}^{n+1} m_i \alpha_i \right) |\alpha \right) \\ &= l_1 \left\{ (\alpha_1|\alpha) + \sum_{i=2}^{n+1} m_i (\alpha_i|\alpha) \right\}.\end{aligned}$$

By definition of reflection of a root, we have,

$$r_\alpha(\alpha_1) = \alpha_1 - 2\frac{(\alpha_1|\alpha)}{(\alpha|\alpha)}\alpha$$

$$= \alpha_1 - 2\frac{(\alpha_1|\alpha)}{l_1\{(\alpha_1|\alpha) + \sum_{i=2}^{n+1} m_i(\alpha_i|\alpha)\}}\left(l_1\left(\alpha_1 + \sum_{i=2}^{n+1} m_i\alpha_i\right)\right).$$

Using Lemma 11, we get $\sum_{i=2}^{n+1} m_i(\alpha_i|\alpha) = 0$. Hence

$$r_\alpha(\alpha_1) = \alpha_1 - \frac{2(\alpha_1|\alpha)}{(\alpha_1|\alpha)}\left(\alpha_1 + \sum_{i=2}^{n+1} m_i\alpha_i\right)$$

$$= -\left(\alpha_1 + 2\sum_{i=2}^{n+1} m_i\alpha_i\right).$$

By the definition, w_0 being the longest element of the Weyl group, the root $-(\alpha_1 + 2\sum_{i=2}^{n+1} m_i\alpha_i)$ can be expressed as the image of w_0 on α_1 with certain restrictions on the entries of the corresponding matrix GGX. Hence under these restrictions on the underlying matrix we get $-(\alpha_1 + 2\sum_{i=2}^{n+1} m_i\alpha_i) = r_\alpha(\alpha_1) = w_0(\alpha_1)$.

(ii) Let GX be of affine or indefinite type. Hence $l_i = 0$ for affine GX, because each l_i is a multiple of determinant of GX, which is zero and $l_i < 0$ for indefinite GX, because each l_i is a multiple of determinant of GX, which is less than zero.

Both these cases are not possible, because l_i must be positive. Hence α is not a special imaginary root for affine and indefinite cases. □

Below we give one example for non-existence of special imaginary roots.

Example 56.

Let (A, ψ) be a symmetrizable supermatrix, where $A = SBGHA_1^{(1)}$ and $\alpha = l_1\alpha_1 + l_2\alpha_2 + l_3\alpha_3 + l_4\alpha_4 \in \Delta_+^{im}$, $l_1, l_2, l_3, l_4 \in \mathbb{N}$. Then $r_\alpha \notin -W$.

Proof: Consider the symmetrizable $SBGHA_1^{(1)} = \begin{pmatrix} -k & -a_1 & -a_2 & -a_3 \\ -b_1 & 2 & -1 & 0 \\ -b_2 & -1 & 2 & -2 \\ -b_3 & 0 & -2 & 2 \end{pmatrix}$.

This is symmetrizable if and only if $a_1b_2 = a_2b_1$, $a_1b_3 = a_3b_1$ and $a_2b_3 = a_3b_2$. In the usual notation, $\Pi^{re} = \{\alpha_2, \alpha_3, \alpha_4\}$, $\Pi^{im} = \{\alpha_1\}$ We define $\psi = \{1\}$ (*or*) $\{\{1, 4\}$ (*or*) $\{4\}$, *if b_3 is even integer*$\}$. Let $\alpha = l_1\alpha_1 + l_2\alpha_2 + l_3\alpha_3 + l_4\alpha_4 \in \Delta_+^{im}$ be a special imaginary root of $g(A)$. Since α is an imaginary root, $\langle\alpha, \alpha_i^\vee\rangle \le 0$, $(2 \le i \le 4)$ $\Rightarrow$ $2l_2 \le l_1b_1 + l_3$ $2l_3 \le l_1b_2 + l_2 + 2l_4$ and $2l_4 \le 2l_3 + l_1b_3$. Let $(\alpha|\alpha) = A$ where $A < 0$, we have

$$A = 2a_1l_2^2 + 2a_1l_3^2 - l_1^2kb_1 - 2a_1b_1l_1l_2 - 2a_1l_2l_3 - 2a_2b_1l_1l_3$$

Using the definition of reflection, we get the following results:

$$r_\alpha(\alpha_1) = \alpha_1 + \frac{2b_1}{A}(kl_1 + a_1l_2 + a_2l_3 + a_3l_4)\alpha \tag{5.6}$$

$$r_\alpha(\alpha_2) = \alpha_2 + \frac{2a_1}{A}(b_1l_1 - 2l_2 + l_3)\alpha \tag{5.7}$$

$$r_\alpha(\alpha_3) = \alpha_3 + \frac{2a_1}{A}(b_2l_1 + l_2 - 2l_3 + 2l_4)\alpha \tag{5.8}$$

$$r_\alpha(\alpha_4) = \alpha_4 + \frac{2b_1}{A}(b_3l_1 + 2l_3 - 2l_4)\alpha \tag{5.9}$$

As α is a special imaginary root and α_2, α_3 and α_4 are simple real roots, using Lemma 11, from Equations (5.7), (5.8), and (5.9) we have

$$\begin{aligned} b_1l_1 - 2l_2 + l_3 &= 0 \\ b_2l_1 + l_2 - 2l_3 + 2l_4 &= 0 \\ \textit{and}\ \ b_3l_1 + 2l_3 - 2l_4 &= 0. \end{aligned}$$

Solving these equations, we get $2l_2 = -l_1(b_2 + b_3)$, $-l_3 = l_1(b_1 + b_2 + b_3)$ and $2l_4 = l_1(3b_1 + 2b_2 + 2b_3)$. This is not possible as $l_1 \neq 0$ and l_i, b_i are positive integers.Hence α is not a special imaginary root for $SBGHA_1^{(1)}$.

(4) *BKM Lie superalgebras with two imaginary simple roots*: We prove the following theorem, a part of it was proved in Chapter 3 for GKM algebras.

Theorem 57. *Let A be the symmetrizable supermatrix*

$$GGX = \begin{pmatrix} -k_1 & -a_1 & -a_2 & \cdots & -a_{n+1} \\ -b_1 & -k_2 & -c_2 & \cdots & -c_{n+1} \\ -b_2 & -d_2 & & & \\ \vdots & \vdots & & GX & \\ -b_{n+1} & -d_{n+1} & & & \end{pmatrix},$$

where $k_1,\ k_2 \in \mathbb{Z}_{\geq 0}$, a_i, b_i, c_i, d_i are positive integers, GX is GCM of finite, affine or indefinite type of order n and GGX is a BKM supermatrix. If $\alpha = \sum_{i=1}^{n+2} l_i\alpha_i \in K \subseteq \Delta_+^{\mathrm{im}}, l_i \in \mathbb{N}$ be any positive imaginary root of $\mathfrak{g}(A)$. Then

(i) *If GX is of finite type, then $\alpha = l_1\alpha_1 + l_2\alpha_2 + \sum_{i=3}^{n+2}(l_1r_i + l_2s_i)\alpha_i$, where $r_i, s_i \in \mathbb{Z}$, is a special imaginary root of some $\mathfrak{g}(A)$ if and only if*

$$r_\alpha(\alpha_i) = \begin{cases} -tw_0(\alpha_i), & i = 1,2 \textit{ if GX is of type } B, C, F_4, G_2, E_7 \textit{ or } E_8 \\ -tt_1w_0(\alpha_i), & i = 1,2 \textit{ if GX is of type } A, D \textit{ or } E_6 \end{cases}$$

under different restrictions on the matrix entries of GGX for each case. Here w_0 is the longest Weyl group element of $\mathfrak{g}(A)$ and t is the permutation $(1,2)$ and t_1 is the another permutation on n symbols.

(ii) *If GX is of affine or indefinite type, then α is not a special imaginary root of $\mathfrak{g}(A)$.*

Proof.

(i) Let GX be of finite type. We have

$$\Pi = \{\alpha_1, \alpha_2, \ldots, \alpha_{n+2}\},\ \Pi^{re} = \{\alpha_3, \alpha_4, \ldots, \alpha_{n+2}\} \text{ and } \Pi^{im} = \{\alpha_1, \alpha_2\}.$$

We can take ψ as

$$\psi = \begin{cases} \{i|i \in I^{im}\} \text{ (or)} \\ \{i|i \in I^{im}\} \cup \{j|a_{jk} \in 2\mathbb{Z} \text{ for all } k\ (i \in I^{re})\}\ (or) \\ \{j|a_{jk} \in 2\mathbb{Z} \text{ for all } k\ (i \in I^{re})\}. \end{cases}$$

For any value ψ defined as above, using the proof of Theorem 55, it is clear that the involvement of the odd root has no effect in the computation of special imaginary root.

For the sake of completeness, we give below the proof.

(i) Let GX be of finite type. For the case of two simple imaginary roots, without loss of generality, assume that α_1 and α_2 are the simple imaginary roots and α_i for $i = 3$ to $n+2$ are simple real roots.

Let

$$\alpha = \sum_{i=1}^{n+2} l_i\alpha_i \in K \subseteq \Delta_+^{\text{im}}, \quad l_i \in \mathbb{N}$$

be a special imaginary root of $\mathfrak{g}(A)$. Then we can prove that α becomes

$$\alpha = l_1\left(\alpha_1 + \alpha_2 + \sum_{i=3}^{n+2}(r_i + s_i)\alpha_i\right).$$

As the proof here is similar to the proof which is given in Chapter 3 for Generalized Kac-Moody algebras for general case, we omit the proof. Conversely, suppose $\alpha = l_1(\alpha_1 + \alpha_2 + \sum_{i=3}^{n+2}(r_i + s_i)\alpha_i)$ is an imaginary root. We prove that $r_\alpha = -tw_0\ (or)\ -tt_1w_0$, where w_0 is the longest element of Weyl group. We have

$$(\alpha|\alpha) = l_1\left\{(\alpha_1|\alpha) + (\alpha_2|\alpha) + \sum_{i=3}^{n+2}(r_i + s_i)(\alpha_i|\alpha)\right\}.$$

Using the definition of reflection, we have

$$r_\alpha(\alpha_1) = \alpha_1 - 2\frac{(\alpha_1|\alpha)}{l_1\{(\alpha_1|\alpha) + (\alpha_2|\alpha) + \sum_{i=3}^{n+2}(r_i + s_i)(\alpha_i|\alpha)\}}\alpha.$$

Using Lemma 11, we get $\sum_{i=3}^{n+2}(r_i + s_i)(\alpha_i|\alpha) = 0$.

$$r_\alpha(\alpha_1) = \alpha_1 - \frac{2(\alpha_1|\alpha)}{(\alpha_1|\alpha) + (\alpha_2|\alpha)}\alpha.$$

Restricting the matrix entries of GGX by $k_1b_1 = k_2a_1$, we have

$$r_\alpha(\alpha_1) = -\left(\alpha_2 + 2\sum_{i=2}^{n+1} r_i\alpha_i + \sum_{i=3}^{n+2} s_i\alpha_i\right).$$

By the definition of w_0, the longest element of the Weyl group, $-(\alpha_2 + 2\sum_{i=2}^{n+1} r_i\alpha_i + 2\sum_{i=3}^{n+2} s_i\alpha_i)$ can be expressed as the image of α_1 under tt_1w_0. That is $-(\alpha_1 + 2\sum_{i=2}^{n+1} r_i\alpha_i + 2\sum_{i=3}^{n+2} s_i\alpha_i) = tt_1w_0(\alpha_1)$.

(ii) Let GX be of affine or indefinite type. The proof is exactly similar to affine and indefinite cases in Theorem 56 with $r = 2$. □

Remark 142. Below we give two examples to illustrate the above theorem.

Example 57. Let (A, ψ) be a symmetrizable supermatrix, where $A = SBGG_2$ and $\alpha = l_1\alpha_1 + l_2\alpha_2 + l_3\alpha_3 + l_4\alpha_4 \in \Delta_+^{im}$, $l_1, l_2, l_3, l_4 \in \mathbb{N}$. Then $r_\alpha = -tw_0\ (or)\ -tt_1w_0$, where w_0 is the longest element of the Weyl group and t is (1,2) permutation and t_1 is permutation on n symbols, if and only if $2(2b_2 + b_3) = l_3$ and $2(3b_2 + 2b_3) = l_4$. We have

$$SBGG_2 = \begin{pmatrix} -k_1 & -a_1 & -a_2 & -a_3 \\ -b_1 & -k_2 & -c_2 & -c_3 \\ -b_2 & -d_2 & 2 & -1 \\ -b_3 & -d_3 & -3 & 2 \end{pmatrix},$$

where $a_1b_2c_2 = a_2b_1d_2$, $c_2d_3 = 3d_2c_3$ and $a_2b_3 = 3a_3b_2$.
In the usual notation, $\Pi^{re} = \{\alpha_3, \alpha_4\}$ and $\Pi^{im} = \{\alpha_1, \alpha_2\}$. We define $\psi = \{1\}\ (or)\ \{2\}\ (or)\ \{1,2\}$. Let $\alpha = l_1\alpha_1 + l_2\alpha_2 + l_3\alpha_3 + l_4\alpha_4 \in \Delta_+^{im}$ be a special imaginary root and $l_1, l_2, l_3, l_4 \in \mathbb{N}$. Since α is an imaginary root, $\langle\alpha, \alpha_i^\vee\rangle \le 0$, $(3 \le i \le 4) \Rightarrow l_1b_2 + l_2d_2 + l_4 \le 2l_3$ and $l_1b_3 + l_2d_3 + 3l_3 \le 2l_4$.
Let $(\alpha|\alpha) = P < 0$ where

$$\begin{aligned} P = &-3b_1b_2k_1l_1^2 - 3b_2a_1k_2l_2 + 6b_1a_1l_3^2 + 2a_1b_1l_4^2 - 6a_1b_1b_2l_1l_2 - 6a_2b_1b_2l_1l_3 \\ &- 6a_3b_1b_2l_1l_4 - 6a_1c_2b_2l_2l_3 - 6a_1c_3b_2l_2l_4 - 6a_1b_1l_3l_4 \end{aligned}$$

Using the definition of reflection, we have the following results:

$$r_\alpha(\alpha_1) = \alpha_1 + \frac{6b_1b_2}{P}(l_1k_1 + l_2a_1 + l_3a_2 + l_4a_3)\alpha = \alpha_1 - p_1\alpha \tag{5.10}$$

$$r_\alpha(\alpha_2) = \alpha_2 + \frac{6a_1b_2}{P}(l_1b_1 + l_2k_2 + l_3c_2 + l_4c_3)\alpha = \alpha_2 - p_2\alpha \tag{5.11}$$

$$r_\alpha(\alpha_3) = \alpha_3 + \frac{6b_1a_2}{P}(l_1b_2 + l_2d_2 - 2l_3 + l_4)\alpha \tag{5.12}$$

$$r_\alpha(\alpha_4) = \alpha_4 + \frac{2b_1a_2}{P}(l_1b_3 + l_2d_3 + 3l_3 - 2l_4)\alpha \tag{5.13}$$

where $p_i = 2\frac{(\alpha_i|\alpha)}{(\alpha|\alpha)}$ *for* $i = 1, 2$.

By Lemma 11, as α is a special imaginary root and α_i $(i = 3,4)$ are simple real roots, we have, $l_1b_2 + l_2d_2 + l_4 = 2l_3$ and $l_1b_3 + l_2d_3 + 3l_3 = 2l_4$. Solving these two equations, we get

$$l_3 = l_1(2b_2 + b_3) + l_2(2d_2 + d_3);\ l_4 = l_1(3b_2 + 2b_3) + l_2(3d_2 + 2d_3).$$

Then

$$\alpha = l_1\alpha_1 + l_2\alpha_2 + (l_1(2b_2+b_3)+l_2(2d_2+d_3))\alpha_3 + (l_1(3b_2+2b_3)+l_2(3d_2+2d_3))\alpha_4.$$

We claim that $r_\alpha = -(r_3r_4)^3$. If this is not true, for example, let $r_\alpha = -r_4$. Then $-r_4(\alpha_3) = -\alpha_4 - 3\alpha_3$. Using this result in (5.12), we have

$$-(\alpha_4 + 3\alpha_3) = r_\alpha(\alpha_3) = \alpha_3 + \frac{6b_1a_2}{P}(l_1b_2 + l_2d_2 - 2l_3 + l_4)\alpha$$

which is the same as

$$\begin{aligned} &6b_1a_2(l_1b_2 + l_2d_2 + 2l_3 - l_4)l_1\alpha_1 + 6b_1a_2(l_1b_2 + l_2d_2 + 2l_3 - l_4)l_2\alpha_2 \\ &\quad + (4P + 6b_1a_2(l_1b_2 + l_2d_2 + 2l_3 - l_4))l_3\alpha_3 \\ &\quad + (P + 6b_1a_2(l_1b_2 + l_2d_2 + 2l_3 - l_4))l_4\alpha_4 = 0. \end{aligned}$$

Equating each coefficient to 0, we get

$$\begin{aligned} 6b_1a_2(l_1b_2 + l_2d_2 + 2l_3 - l_4)l_1 &= 0, \\ 6b_1a_2(l_1b_2 + l_2d_2 + 2l_3 - l_4)l_2 &= 0, \\ 4P + 6b_1a_2(l_1b_2 + l_2d_2 + 2l_3 - l_4)l_3 &= 0, \\ P + 6b_1a_2(l_1b_2 + l_2d_2 + 2l_3 - l_4)l_4 &= 0. \end{aligned}$$

Since l_1, l_2, l_3 *and* $l_4 \neq 0$, $l_1b_2 + l_2d_2 + 2l_3 - l_4 = 0$. Hence $P = 0$ which is a contradiction. One can directly verify that

$$r_\alpha(\alpha_3) = -(r_3r_4)^3(\alpha_3) = -\alpha_3.$$

Similarly, one can prove that $r_\alpha(\alpha_4) = -(r_3r_4)^3(\alpha_4) = -\alpha_4$. Hence $r_\alpha = -(r_3r_4)^3$. We have,

$$(r_3r_4)^3(\alpha_1) = \alpha_1 + 2(2b_2 + b_3)\alpha_3 + 2(3b_2 + 2b_3)\alpha_4 \tag{5.14}$$

$$(r_3r_4)^3(\alpha_2) = \alpha_2 + 2(2d_2 + d_3)\alpha_3 + 2(3d_2 + 2d_3)\alpha_4 \tag{5.15}$$

Taking $r_\alpha(\alpha_1) = -t(r_3r_4)^3(\alpha_1)$ and $r_\alpha(\alpha_2) = -t(r_3r_4)^3(\alpha_2)$ where t is (1,2) permutation and equating Equations (5.14) and (5.15) with Equations (5.10) and (5.11) respectively, we get

$$(1 - p_1 l_1)\alpha_1 - p_1(l_2\alpha_2 - (l_1(2b_2 + b_3) + l_2(2d_2 + d_3))$$
$$\alpha_3 - (l_1(3b_2 + 2b_3) + l_2(3d_2 + 2d_3))\alpha_4) \quad (5.16)$$
$$= -(\alpha_2 + 2(2d_2 + d_3)\alpha_3 + 2(3d_2 + 2d_3)\alpha_4) \quad (5.17)$$
$$(1 - p_2 l_2)\alpha_2 - p_2(l_1\alpha_1 - (l_1(2b_2 + b_3) + l_2(2d_2 + d_3))$$
$$\alpha_3 - (l_1(3b_2 + 2b_3) + l_2(3d_2 + 2d_3))\alpha_4) \quad (5.18)$$
$$= -(\alpha_1 + 2(2b_2 + b_3)\alpha_3 + 2(3b_2 + 2b_3)\alpha_4) \quad (5.19)$$

Using the permutation (1,2), equating Equations (5.17) and (5.19), we get $l_1 = l_2$, $b_2 = d_2$ *and* $b_3 = d_3$. This implies

$$\alpha = l_1(\alpha_1 + \alpha_2 + 2(2b_2 + b_3)\alpha_3 + 2(3b_2 + 2b_3)\alpha_4).$$

Conversely, suppose $\alpha = l_1\alpha_1 + l_2\alpha_2 + l_3\alpha_3 + l_4\alpha_4 \in \Delta_+^{im}$, with $2(2b_2 + b_3) = l_3$ and $2(3b_2 + 2b_3) = l_4$. Then

$$\alpha = l_1(\alpha_1 + \alpha_2 + 2(2b_2 + b_3)\alpha_3 + 2(3b_2 + 2b_3)\alpha_4).$$

We prove that $r_\alpha \in -tW\ (or)\ -tt_1W$. We have

$$(\alpha|\alpha) = -3b_1b_2\left(k_1 + \frac{a_1k_2}{b_1} + 2a_1 + 8a_2b_2 + 4a_2b_3 + 12a_3b_2 + 8a_3b_3\right).$$

Now if we restrict the matrix entries by $k_1 = \frac{a_1k_2}{b_1}$, then

$$(\alpha|\alpha) = -3b_1b_2(2k_1 + 2a_1 + 8a_2b_2 + 4a_2b_3 + 12a_3b_2 + 8a_3b_3).$$

Also,

$$(\alpha|\alpha) = -3b_1b_2(k_1 + 2a_1 + 4a_2b_2 + 2a_2b_3 + 6a_3b_2 + 4a_3b_3).$$

Using the definition of reflection, we have,

$$r_\alpha(\alpha_1) = -(\alpha_2 + 2(2b_2 + b_3)\alpha_3 + 2(3b_2 + 2b_3)\alpha_4) \quad (5.20)$$

But

$$-(r_3r_4)^3(\alpha_1) = -(\alpha_1 + 2(2b_2 + b_3)\alpha_3 + 2(3b_2 + 2b_3)\alpha_4).$$

Using $t = (1, 2)$ permutation, we have

$$-t(r_3r_4)^3(\alpha_1) = -(\alpha_2 + 2(2b_2 + b_3)\alpha_3 + 2(3b_2 + 2b_3)\alpha_4) \quad (5.21)$$

Therefore $r_\alpha(\alpha_1) = -t(r_3r_4)^3(\alpha_1)$. Here $(r_3r_4)^3$ is the longest Weyl group element of G_2. Similarly, we get

$$r_\alpha(\alpha_2) = -t(r_3r_4)^3(\alpha_2),$$
$$r_\alpha(\alpha_3) = -t(r_3r_4)^3(\alpha_3),$$
$$r_\alpha(\alpha_4) = -t(r_3r_4)^3(\alpha_4).$$

Hence α is special imaginary root.

Example 58. Let (A, ψ) be a symmetrizable supermatrix, where $A = SBGA_2^{(2)}$ and $\alpha = l_1\alpha_1 + l_2\alpha_2 + l_3\alpha_3 + l_4\alpha_4 \in \Delta_+^{im}$ $(l_1, l_2, l_3, l_4 \in \mathbb{N})$. We prove that $r_\alpha \neq -tw_0$ (or) $-tt_1w_0$, where w_0 is the longest element of the Weyl group, t is (1,2) permutation and t_1 is permutation on n symbols.

Proof. Consider symmetrizable $SBGA_2^{(2)} = \begin{pmatrix} -k_1 & -a_1 & -a_2 & -a_3 \\ -b_1 & -k_2 & -c_2 & c_3 \\ -b_2 & -d_2 & 2 & -4 \\ -b_3 & -d_3 & -1 & 2 \end{pmatrix}$, where $a_1b_2c_2 = a_2b_1d_2$, $4c_2d_3 = d_2c_3$ and $4a_2b_3 = a_3b_2$. In the usual notation, $\Pi^{re} = \{\alpha_3, \alpha_4\}$, $\Pi^{im} = \{\alpha_1, \alpha_2\}$. We define $\psi = \{1\}$ (or) $\{2\}$ (or) $\{1,2\}$ (or) ($\{1,3\}$ if b_2 and d_2 is even integer). Let $\alpha = l_1\alpha_1 + l_2\alpha_2 + l_3\alpha_3 + l_4\alpha_4 \in \Delta_+^{im}$ $l_1, l_2, l_3, l_4 \in \mathbb{N}$. So $r_\alpha = -w_0$ (or) $-tw_0$, where w_0 is the longest element of the Weyl group. Then α is a special imaginary root. Since α is an imaginary root, $\langle \alpha, \alpha_i^\vee \rangle \leq 0$, $(3 \leq i \leq 4)$. We have $2l_3 \leq l_1b_2 + l_2d_2 + 4l_4$ and $2l_4 \leq l_1b_3 + l_2d_3 + 2l_3$. Then $P = (\alpha|\alpha) < 0$, where

$$(\alpha|\alpha) = -l_1^2b_1b_2k_1 - 2l_1l_2b_1b_2a_1 - 2l_1l_3b_1b_2a_2 - 2l_1l_4b_1b_2a_3 - 2l_2l_3a_1b_2c_2 - l_2^2a_1b_2k_2 + 2l_3^2a_2b_1 - 2l_2l_4a_1b_2c_3 - 8l_3l_4b_1a_2 + 4l_4^2a_2b_1.$$

Using the definition of reflection, we have the following equations:

$$r_\alpha(\alpha_1) = \alpha_1 + \frac{2b_1b_2}{P}(l_1k_1 + l_2a_1 + l_3a_2 + l_4a_3)\alpha$$
$$r_\alpha(\alpha_2) = \alpha_2 + \frac{2a_1b_2}{P}(l_1b_1 + l_2k_2 + l_3c_2 + l_4c_3)\alpha$$
$$r_\alpha(\alpha_3) = \alpha_3 + \frac{2b_1a_2}{P}(l_1b_2 + l_2d_2 - 2l_3 + 4l_4)\alpha$$
$$r_\alpha(\alpha_4) = \alpha_4 + \frac{4b_1a_2}{P}(l_1b_3 + l_2d_3 + l_3 - 2l_4)\alpha.$$

Then from the definition of special imaginary roots, we can conclude that

$$r_\alpha(\alpha_3) = \alpha_3 \text{ and } r_\alpha(\alpha_4) = \alpha_4.$$

But the Weyl group is infinite, so we can not find the longest Weyl group. Hence, $r_\alpha \neq -w_0$, where w_0 is the longest element of the Weyl group. □

(5) *BKM Lie superalgebras with three or more simple imaginary roots*: These are BKM Lie superalgebras, whose BKM supermatrices are extended KM matrices of KM algebras with three or more imaginary roots. As in the case of GKM algebras, we have the following proposition.

Proposition 48. *Let A be the symmetrizable supermatrix*

$$GGX = \begin{pmatrix} -k_1 & -a_1 & -a_2 & \cdots & -a_{r-1} & -a_r & \cdots & -a_{n+r-1} \\ -b_1 & -k_2 & -c_2 & \cdots & -c_{r-1} & -c_r & \cdots & -c_{n+r-1} \\ -b_2 & -d_2 & -k_3 & \cdots & \cdots & \cdots & \cdots & \cdots \\ \vdots & \vdots & & \ddots & & & & \\ -b_{r-1} & -d_{r-1} & \cdots & \cdots & -k_r & \cdots & \cdots & \cdots \\ -b_r & -d_r & & & \vdots & & & \\ \vdots & \vdots & & & & GX & & \\ -b_{n+r-1} & -d_{n+r-1} & & & & & & \end{pmatrix},$$

where $k_i(1 \leq i \leq r,\ r \geq 3) \in \mathbb{Z}_{\geq 0}$, a_i, b_i, c_i, d_i are positive integers, GX is KM matrices of finite, affine or indefinite type of order n. Let $\alpha = \sum_{i=1}^{n+r} l_i\alpha_i (\in K \subseteq \Delta_+^{\text{im}}$, for $l_i \in \mathbb{N}$)be any positive imaginary root of $\mathfrak{g}(A)$. Then α is not a special imaginary root of $\mathfrak{g}(A)$.

For BKM Lie superalgebras with three or more imaginary roots, assume that simple imaginary roots are $\{\alpha_i \mid 1 \leq i \leq r;\ r \geq 3\}$ and simple real roots are $\{\alpha_j | r+1 \leq j \leq n+r\}$. We have

$$\Pi^{re} = \{\alpha_{r+1}, \alpha_{r+2}, \ldots, \alpha_{n+r}\},\ \Pi^{im} = \{\alpha_1, \alpha_2, \ldots, \alpha_r\} \text{ and } \Pi = \{\alpha_1, \alpha_2, \ldots, \alpha_{n+r}\}$$

where

$$I = \{1, 2, \ldots, n+r\}; I^{re} = \{r+1, r+2, \ldots, n+r\} \text{ and } I^{im} = \{1, 2, \ldots, r\}.$$

We can take ψ as

$$\psi = \{i : i \in I^{im}\} \text{ (or) } \{i; i \in I^{im}\} \cup \{j; a_{jk} \in 2\mathbb{Z}, \forall k (i \in I^{re})\} \text{ (or)}$$
$$\{j; a_{jk} \in 2\mathbb{Z}, \forall k (i \in I^{re})\}.$$

For any value ψ defined as above, the proof to show that α is not a special imaginary root runs on the same lines as the proof of Proposition 20, because by Theorem 55, the involvement of the odd root has no effect in the computation of special imaginary root.

Category 2(b): BKM Lie superalgebras of infinite order with non-empty set of odd roots(which are not GKM algebras): For BKM algebras of infinite order, with infinite number of imaginary simple roots, we have the following four cases:

(1) All simple roots are imaginary.
(2) Finite (non-zero) number of real simple roots and infinite number of imaginary simple roots. This case will cover Fake Monster Lie superalgebras also [151].
(3) Infinite number of real simple roots and finite number of imaginary simple roots.
(4) All simple roots are real.

For all the above cases, the BKM Lie superalgebras have no special imaginary root. As the involvement of odd root has no effect in the computation of special imaginary root, the conclusions of the above cases follow from Theorem 55 and from the conclusions of (1)–(4) in class (b) of category 1 in complete classification of GKM algebras possessing special imaginary roots in Chapter 3.

Remark 143. From the above classifications, we conclude that BKM Lie superalgebras possessing special imaginary roots are in the classes of BKM Lie superalgebras with one or two imaginary simple roots whose BKM supermatrices appear as extensions of finite dimensional Cartan matrices only.

5.5 Strictly and Purely imaginary roots in BKM Lie superalgebras and complete classification of BKM Lie superalgebras possessing purely imaginary roots

Below we define strictly domestic type imaginary roots, strictly alien type imaginary roots and strictly imaginary roots. Here the definition of strictly imaginary roots is exactly similar to the definition of strictly imaginary roots in GKM algebras. Similarly, we define the different classes of purely imaginary roots.

Definition 227. A domestic-type imaginary root γ in a BKM superalgebra is said to be *strictly domestic-type imaginary*, if for every $\alpha \in \Delta^{\text{re}}$, either $\alpha + \gamma$ or $\alpha - \gamma$ is a root. Let $\Delta^{\text{s·dom·im}}$, $\Delta_+^{\text{s·dom·im}}$, $\Delta_-^{\text{s·dom·im}}$ denote the set of all strictly domestic-type imaginary roots, positive strictly domestic-type imaginary roots and negative strictly domestic-type imaginary roots respectively.

Definition 228. An alien-type imaginary root γ in a BKM superalgebra is said to be *strictly alien-type imaginary*, if for every $\alpha \in \Delta^{\text{re}}$, either $\alpha + \gamma$ or $\alpha - \gamma$ is a root. Let $\Delta^{\text{s·ali·im}}$, $\Delta_+^{\text{s·ali·im}}$, $\Delta_-^{\text{s·ali·im}}$ denote the set of all strictly alien-type imaginary roots, positive strictly alien-type imaginary roots and negative strictly alien-type imaginary roots respectively.

Definition 229. An imaginary root γ in a BKM superalgebra is said to be *strictly imaginary*, if for every $\alpha \in \Delta^{\text{re}}$, either $\alpha + \gamma$ or $\alpha - \gamma$ is a root. The set of all strictly imaginary roots is denoted by Δ^{sim}. Let Δ^{sim}, Δ_+^{sim}, Δ_-^{sim} denote the set of all strictly imaginary roots, positive strictly imaginary roots and negative strictly imaginary roots respectively.

The following results are true for all $i \in I$ for KM algebras (Chapter 2). Instead of $i \in I$ for KM algebras one can easily verify these results for $i \in I^{\text{re}}$ for BKM Lie superalgebras. Hence the proof of the following theorem is not given.

Theorem 58.

(a) *If $\alpha \in \Delta_+$ and $\langle \alpha_i^\vee, \alpha \rangle < 0$ for all $i \in I^{\text{re}}$, then $\alpha \in \Delta_+^{\text{s.ali.im}}$.*
(b) *If $\alpha \in \Delta_+^{\text{ali·im}}$, $r_\gamma(\alpha) \neq \alpha$ for all $\gamma \in \Delta^{\text{re}}$, then $\alpha \in \Delta_+^{\text{s.ali.im}}$.*
(c) *If $\alpha \in \Delta_+^{\text{s.ali.im}}$ and $\langle \alpha_i^\vee, \alpha \rangle \leq 0$ for all $i \in I^{\text{re}}$, then $\alpha + \beta \in \Delta_+$ for all $\beta \in \Delta_+$.*
(d) *If $\alpha \in \Delta_+^{\text{s.ali.im}}$, $\beta \in \Delta_+^{\text{ali·im}}$, then $\alpha + \beta \in \Delta_+^{\text{ali·im}}$.*
(e) *$\Delta_+^{\text{s.ali.im}}$ is a semigroup.*

For the proof of the following theorem, one can refer [147].

Theorem 59.

(a) *If* $\alpha \in \Delta^{dom.im} \backslash \{\alpha_i\}$ *and* $supp(\alpha + \alpha_i)$ *is connected, then* $\alpha \in \Delta^{im}$.
(b) *If* $\alpha \in \Delta^{dom.im} \backslash \{\alpha_i\}$, $r_\gamma(\alpha) \neq \alpha$ *for all* $\gamma \in \Delta^{re}$, *then* $\alpha \in \Delta^{s.ali.im}$.

The following result is given for KM algebras in Chapter 2. The same result can be easily proved using Definition 228 of strictly imaginary roots for BKM superalgebras and hence the proof is not given here.

Theorem 60. *In the usual notation, if* $\gamma \in \overset{\circ}{K} \setminus \Delta_+^{\text{sim}}$, *for BKM superalgebras, then for some simple real root* $\alpha_i \in \Pi$, *neither* $\gamma + \alpha_i$ *nor* $\gamma - \alpha_i$ *is a root.*

Definition 230. A BKM supermatrix (A, ψ) is said to have *strictly domestic imaginary property*, if

$$\Delta_+^{\text{s.dom.im}}(A) = \Delta_+^{\text{dom.im}}(A).$$

Definition 231. A BKM supermatrix (A, ψ) is said to have *strictly alien imaginary property*, if

$$\Delta_+^{\text{s.ali.im}}(A) = \Delta_+^{\text{ali.im}}(A).$$

Definition 232. A BKM supermatrix (A, ψ) is said to have *strictly imaginary property* (SIM property), if

$$\Delta_+^{\text{sim}}(A) = \Delta_+^{\text{im}}(A).$$

From the property (*d*) of Theorem (58), we have

$$\Delta_+^{\text{sim}} + \Delta_+^{\text{im}} \subset \Delta_+^{\text{im}}.$$

We give the following example to prove

$$\Delta_+^{\text{im}} + \Delta_+^{\text{im}} \not\subset \Delta_+^{\text{im}}.$$

Example 59. Consider the BKM superalgebra associated with BKM supermatrix (A, ψ), where

$$A = \begin{bmatrix} 2 & -6 & 0 & 0 & 0 \\ -2 & 0 & -1 & 0 & 0 \\ 0 & -1 & 2 & -1 & 0 \\ 0 & 0 & -1 & 2 & -2 \\ 0 & 0 & 0 & -2 & 2 \end{bmatrix}, \quad \textit{with } \psi = \{1\}$$

we can easily check that $\alpha_1 + \alpha_2$ and $\alpha_4 + \alpha_5$ are imaginary roots whereas $\alpha_1 + \alpha_2 + \alpha_4 + \alpha_5$ is not a root because $\text{supp}(\alpha_1 + \alpha_2 + \alpha_4 + \alpha_5)$ is not connected.

Definition 233. Any $\alpha \in \Delta_+^{\text{dom}\cdot\text{im}}$, is called *purely domestic imaginary*(positive purely domestic imaginary root), if for any $\beta \in \Delta_+^{\text{dom}\cdot\text{im}}$, $\alpha + \beta \in \Delta_+^{\text{dom}\cdot\text{im}}$. Similarly negative purely domestic imaginary root can be defined. The set of all purely domestic imaginary roots is denoted by $\Delta^{dom.im}$. We use the notations,

$$\begin{aligned}\Delta_+^{\text{p}\cdot\text{dom}\cdot\text{im}}(A) &= \Delta_+^{\text{p}\cdot\text{dom}\cdot\text{im}} \\ &= \{\alpha \in \Delta_+^{\text{dom}\cdot\text{im}} \mid \alpha \text{ is a positive purely domestic imaginary}\}\end{aligned}$$

and

$$\begin{aligned}\Delta_-^{\text{p}\cdot\text{dom}\cdot\text{im}}(A) &= \Delta_-^{\text{p}\cdot\text{dom}\cdot\text{im}} \\ &= \{\alpha \in \Delta_-^{\text{dom}\cdot\text{im}} \mid \alpha \text{ is a negative purely domestic imaginary}\}.\end{aligned}$$

Hence,

$$\Delta^{\text{p}\cdot\text{dom}\cdot\text{im}} = \Delta_+^{\text{p}\cdot\text{dom}\cdot\text{im}} \cup \Delta_-^{\text{p}\cdot\text{dom}\cdot\text{im}}.$$

Definition 234. For $\alpha \in \Delta_+^{\text{ali}\cdot\text{im}}$, α is *purely alien imaginary root* (positive purely alien imaginary root), if for any $\beta \in \Delta_+^{\text{ali}\cdot\text{im}}$, $\alpha + \beta \in \Delta_+^{\text{ali}\cdot\text{im}}$. Similarly we say that a negative root $\gamma \in \Delta_-^{\text{ali}\cdot\text{im}}$ is purely alien imaginary if $-\gamma$ is a purely alien imaginary root. We use the following notations

$$\begin{aligned}\Delta_+^{\text{p}\cdot\text{ali}\cdot\text{im}}(A) &= \Delta_+^{\text{p}\cdot\text{ali}\cdot\text{im}} \\ &= \{\alpha \in \Delta_+^{\text{ali}\cdot\text{im}} \mid \alpha \text{ is a positive purely alien imaginary root}\}\end{aligned}$$

and

$$\begin{aligned}\Delta_-^{\text{p}\cdot\text{ali}\cdot\text{im}}(A) &= \Delta_-^{\text{p}\cdot\text{ali}\cdot\text{im}} \\ &= \{\alpha \in \Delta_-^{\text{ali}\cdot\text{im}} \mid \alpha \text{ is a negative purely alien imaginary root}\}.\end{aligned}$$

Hence

$$\Delta^{\text{p}\cdot\text{ali}\cdot\text{im}} = \Delta_+^{\text{p}\cdot\text{ali}\cdot\text{im}} \cup \Delta_-^{\text{p}\cdot\text{ali}\cdot\text{im}}.$$

Definition 235. Let $\alpha \in \Delta_+^{\text{im}}$, we say that α is *purely imaginary root* (positive purely imaginary root), if for any $\beta \in \Delta_+^{\text{im}}$, $\alpha + \beta \in \Delta_+^{\text{im}}$. Similarly, we say that a negative root $\gamma \in \Delta_-^{\text{im}}$ is purely imaginary if $-\gamma$ is a purely imaginary root. We use the following notations

$$\Delta_+^{\text{pim}}(A) = \Delta_+^{\text{pim}} = \{\alpha \in \Delta_+^{\text{im}} \mid \alpha \text{ is a positive purely imaginary root}\}$$

and

$$\Delta_-^{\text{pim}}(A) = \Delta_-^{\text{pim}} = \{\alpha \in \Delta_-^{\text{im}} \mid \alpha \text{ is a negative purely imaginary root}\}.$$

Hence

$$\Delta^{\text{pim}} = \Delta_{+}^{\text{pim}} \cup \Delta_{-}^{\text{pim}}.$$

The symmetry of the root system means that we need to prove any result for positive imaginary roots only.

The following proposition is immediate, the proof is given in [145].

Proposition 49. *Let* $A = (a_{ij})_{i,j=1}^{n}$ *be a BKM supermatrix, then*

$$\Delta_{+}^{\text{p.ali.im}}(A) = \Delta_{+}^{\text{pim}}(A) \text{ if } a_{ii} \neq 0 \text{ for all } i.$$

Remark 144. From the above definitions, $\Delta_{+}^{p.ali.im} \cup \Delta_{+}^{p.dom.im} \subset \Delta_{+}^{p.im}$. Let $A = (a_{ij})_{i,j=1}^{n}$ be a BKM supermatrix, then $\Delta_{+}^{\text{p.ali.im}}(A)$ is a proper subset of $\Delta_{+}^{\text{pim}}(A)$ if $a_{ii} \leq 0$ *for all i*.

The proof of the above remark is obvious for $a_{ii} < 0$ for some $i \in \psi$, and the proof for the case $a_{ii} = 0$ for some $i \in \psi$ follows from the following example.

Example 60. Consider the hyperbolic BKM superalgebra associated with the BKM supermatrix (A, ψ), where

$$A = \begin{pmatrix} 0 & -1 \\ -1 & 2 \end{pmatrix} \text{ with } \psi = \{1\}.$$

We have

$$\Delta_{+}^{\text{dom.im}} = \{\alpha_1, \alpha_1 + \alpha_2\}.$$

Here,

$$\alpha_1 + \alpha_1 + \alpha_2 = 2\alpha_1 + \alpha_2 \in \Delta_{+}^{\text{ali.im}} \subset \Delta_{+}^{\text{im}}.$$

Therefore, α_1 and $\alpha_1 + \alpha_2 \in \Delta_{+}^{\text{pim}}$. But α_1 and $\alpha_1 + \alpha_2$ are not at all purely alien imaginary roots.

Hence $\Delta_{+}^{\text{p.ali.im}}(A)$ is a proper subset of Δ_{+}^{pim}.

In general, each purely alien imaginary root need not be a strictly alien imaginary root and each purely imaginary root need not be a strictly imaginary root. We give the following example to prove this statement.

Example 61. Consider the BKM superalgebra associated with the BKM supermatrix (A, ψ), where

$$A = \begin{pmatrix} 0 & -1 & 0 \\ -1 & 2 & -1 \\ 0 & -1 & 2 \end{pmatrix} \text{ with } \psi = \{1\}.$$

Here, $\alpha = 2\alpha_1 + \alpha_2 + \alpha_3$ is a purely alien imaginary root and hence a purely imaginary root, whereas α is not strictly alien imaginary and hence not strictly imaginary, since neither $\alpha + \alpha_2 + \alpha_3$ nor $\alpha - (\alpha_2 + \alpha_3)$ is a root. Moreover, $\alpha = \alpha_1$ is a purely imaginary root, whereas α is not strictly imaginary, since neither $\alpha + \alpha_3$ nor $\alpha - \alpha_3$ is a root. $\alpha = 2\alpha_1 + \alpha_2$ is a purely alien imaginary root and hence a purely imaginary root, whereas α is not strictly alien imaginary and hence not a strictly imaginary, since neither $\alpha + \alpha_2$ nor $\alpha - \alpha_2$ is a root.

Remark 145. In general the sets Δ_+^{im}, $\Delta_+^{\text{ali}\cdot\text{im}}$ and $\Delta_+^{\text{dom}\cdot\text{im}}$ are not semigroups.

Example 62. Consider the BKM superalgebra associated with the BKM supermatrix (A, ψ) where

$$A = \begin{pmatrix} -k & -1 \\ -1 & 2 \end{pmatrix}, k > 0,\ \psi = \{1\}.$$

Here

$$\Delta_+^{\text{dom}\cdot\text{im}} = \{\alpha_1, \alpha_1 + \alpha_2, 2\alpha_1, 2(\alpha_1 + \alpha_2)\}.$$

We notice that $\alpha_1 + \alpha_1 + \alpha_2 = 2\alpha_1 + \alpha_2$ is not domestic imaginary root. Hence $\Delta_+^{\text{dom}\cdot\text{im}}$ is not a semigroup.

Also $\alpha_1 + 2\alpha_1 = 3\alpha_1$ is not a root. Hence Δ_+^{im} is not a semigroup.

Example 59 is an example to show that $\Delta_+^{\text{ali}\cdot\text{im}}$ is not a semigroup.

Definition 236. We say that a BKM supermatrix (A, ψ), satisfies the *purely imaginary property*, if

$$\Delta_+^{\text{pim}}(A) = \Delta_+^{\text{im}}(A).$$

We say that the BKM superalgebra $\mathfrak{g}(A)$ has the *purely imaginary property*, if (A, ψ) satisfies this property.

Definition 237. We say that a BKM supermatrix (A, ψ), satisfies the *purely domestic imaginary property*, if

$$\Delta_+^{\text{p}\cdot\text{ali}\cdot\text{im}}(A) = \Delta_+^{\text{dom}\cdot\text{im}}(A).$$

We say that the BKM superalgebra $\mathfrak{g}(A)$ has the *purely domestic imaginary property*, if (A, ψ) satisfies this property.

Definition 238. We say that a BKM supermatrix (A, ψ), satisfies the *purely alien imaginary property*, if

$$\Delta_+^{\text{p}\cdot\text{ali}\cdot\text{im}}(A) = \Delta_+^{\text{ali}\cdot\text{im}}(A).$$

We say that the BKM superalgebra $\mathfrak{g}(A)$ has the *purely alien imaginary property*, if (A, ψ) satisfies this property.

Proposition 50. *For BKM superalgebras, Δ_+^{pim} and $\Delta_+^{\text{p}\cdot ali\cdot im}$ are semigroups, whereas $\Delta_+^{\text{p}\cdot dom\cdot im}$ is not a semigroup and $\Delta_+^{\text{p}\cdot dom\cdot im}$ is an empty set.*

Remark 146. One can obviously notice that when two domestic imaginary roots are added, the result is not a root. Hence we conclude that $\Delta_+^{\text{p}\cdot\text{dom}\cdot\text{im}}$ is not a semigroup and $\Delta_+^{\text{p}\cdot\text{dom}\cdot\text{im}} = \emptyset$.

As the proofs of the following results from Proposition 51 to 53 are similar to the proofs of the corresponding results in KM algebras in Chapter 2 with suitable replacements for imaginary roots, the proofs of these are not given.

Proposition 51.

(a) $W \cdot \Delta_+^{\text{pim}} = \Delta_+^{\text{pim}}$.
(b) $W \cdot \Delta_+^{\text{p}\cdot ali\cdot im} = \Delta_+^{\text{p}\cdot ali\cdot im}$.

The following proposition shows the existence of BKM superalgebras with the purely imaginary property and purely alien imaginary property.

Proposition 52. Let (A, ψ) be an indecomposable BKM supermatrix. If $\mathfrak{g}(A)$ is an affine or a symmetrizable hyperbolic BKM superalgebra, then

(a) $\Delta_+^{\text{im}}(A) = \Delta_+^{\text{pim}}(A)$
and
(b) $\Delta_+^{\text{ali}\cdot\text{im}}(A) = \Delta_+^{\text{p}\cdot ali\cdot im}(A)$.

Remark 147. We notice that the above proposition is not true for the following hyperbolic BKM superalgebra associated to the BKM supermatrix (A, ψ) where

$$A = \begin{bmatrix} 0 & -a \\ -b & 0 \end{bmatrix} \text{ with } \psi = \{1, 2\}.$$

We notice that as $\mathring{K} = \emptyset$, we have $\Delta_+^{\text{ali}\cdot\text{im}} = \emptyset$. Also the domestic imaginary roots $\{\alpha_1, \alpha_2\}$ are not purely imaginary roots as $\alpha_1 + \alpha_2$ is not a root.

The following Lemmas 13 and 14, Corollaries 4, 5 and 6, and Proposition 53 are important for the classification of BKM superalgebras with the purely alien imaginary property.

Lemma 13. *Let (A, ψ) be an indecomposable, symmetrizable BKM supermatrix. Let α, $\beta \in \Delta_+^{\text{ali}\cdot\text{im}}$ be such that supp $(\alpha + \beta)$ is connected. Then, $\alpha + \beta \in \Delta_+^{\text{ali}\cdot im}$.*

Lemma 14.

(a) *If $\alpha \in \Delta_+^{\text{dom}\cdot im}$ with $\alpha = \sum_{i=1}^n k_i\alpha_i$, then atleast one of the k_is are > 0.*
(b) *If $\alpha \in \Delta_+^{\text{ali}\cdot\text{im}}$ with $\alpha = \sum_{i=1}^n k_i\alpha_i$, then atleast two of the k_is are > 0.*

Corollary 4. *Let (A, ψ) be an indecomposable symmetrizable $n \times n$ BKM supermatrix. Suppose $\alpha \in \Delta_+^{\text{ali}\cdot\text{im}}$ is such that supp α contains atleast $n - 2$ vertices, then $\alpha \in \Delta_+^{\text{p}\cdot\text{ali}\cdot\text{im}}$.*

Corollary 5. *Let (A, ψ) where $A = (a_{ij})_{i,j=1}^n$ be an indecomposable symmetrizable BKM supermatrix. Suppose $\alpha \in \Delta_+^{\text{im}}$ is such that supp α contains $n - 1$ vertices, then $\alpha \in \Delta_+^{\text{pim}}$.*

Corollary 6. *Let (A, ψ), where $A = (a_{ij})_{i,j=1}^n$ be an indecomposable symmetrizable $n \times n$ BKM supermatrix with $n \leq 4$, then $\Delta_+^{\text{p}\cdot\text{ali}\cdot\text{im}}(A) = \Delta_+^{\text{ali}\cdot\text{im}}(A)$.*

Proposition 53. *Let $A = (a_{ij})_{i,j=1}^{n}$ be an indecomposable symmetrizable BKM supermatrix such that $a_{ij} \neq 0$ for all $i, j \in \mathbb{N}$ and $i \neq j$. Then, $\Delta_+^{\text{p}\cdot\text{ali}\cdot\text{im}} = \Delta_+^{\text{ali}\cdot\text{im}}$.*

The following theorem gives a necessary and sufficient condition for any $\alpha \in \Delta_+^{\text{ali}\cdot\text{im}}$ to be purely alien imaginary, which further leads to the classification of BKM superalgebras with the purely alien imaginary property.

Theorem 61. *Let (A, ψ) be an indecomposable symmetrizable $n \times n$ BKM supermatrix and let $\alpha \in \Delta_+^{\text{ali}\cdot\text{im}}(A)$. Then, $\alpha \in \Delta_+^{\text{p}\cdot\text{ali}\cdot\text{im}}$ if and only if every connected subdiagram of $S(A)$, which is of affine or indefinite type, is connected to atleast one of the vertices of supp α.*

Proof. Let $\alpha \in \Delta_+^{\text{p}\cdot\text{ali}\cdot\text{im}}$. Suppose there exists a connected subdiagram $S'(A)$ of $S(A)$ which is not of finite type and which is not connected to any of the vertices of suppα. Let β be positive alien imaginary root whose support is connected in $S'(A)$ (such a choice is possible because $S'(A)$ is of non-finite type and hence possesses imaginary roots). Then supp$(\alpha + \beta)$ is not connected which is a contradiction since $\alpha \in \Delta_+^{\text{p}\cdot\text{ali}\cdot\text{im}}(A)$ implies $\alpha + \beta \in \Delta_+^{\text{ali}\cdot\text{im}}$. Thus, any connected subdiagram of $S(A)$, which is not of finite type must be connected to atleast one of the vertices of suppα.

Now, assume that every connected subdiagram of $S(A)$, which is of affine or indefinite type is connected to atleast one of the vertices of suppα. To show that $\alpha \in \Delta_+^{\text{p}\cdot\text{ali}\cdot\text{im}}$, it suffices to prove that supp$(\alpha + \beta)$ is connected for all $\beta \in \Delta_+^{\text{ali}\cdot\text{im}}$ For any $\beta \in \Delta_+^{\text{ali}\cdot\text{im}}$, supp$\beta$ is a connected subdiagram of non-finite type and hence by hypothesis, suppβ is connected to atleast one of the vertices of suppα, which implies that supp$(\alpha + \beta)$ is connected. □

The following theorem gives a complete classification of all those symmetrizable BKM superalgebras with the purely alien imaginary property.

Theorem 62. *Let (A, ψ) be an indecomposable symmetrizable $n \times n$ BKM supermatrix and let $A \in \Delta_+^{ali.im}(A)$. Then $A \in \Delta_+^{p.ali.im}(A)$ if and only if every connected subdiagram of $S(A)$, which is of affine or indefinite type is connected to atleast one of the vertices of suppα.*

Proof. The proof follows immediately from the previous theorem. For $\Delta_+^{\text{ali}\cdot\text{im}}(A) \neq \Delta_+^{\text{p}\cdot\text{ali}\cdot\text{im}}(A)$ implies that there exists an imaginary root $\alpha \in \Delta_+^{\text{ali}\cdot\text{im}}$, $\alpha \notin \Delta_+^{\text{p}\cdot\text{ali}\cdot\text{im}}$. By Theorem 61, there exists a connected subdiagram $S_1(A)$, which is not connected to suppα. Take $S_2(A)$ =suppα. Then each of $S_1(A)$ and $S_2(A)$ are of non-finite connected subdiagrams of $S(A)$, such that no vertex $i \in S_1(A)$ is connected to any $j \in S_2(A)$.

Conversely, suppose that there exists two connected subdiagrams $S_1(A)$ and $S_2(A)$ of $S(A)$, both of non-finite type such that $S_1(A)$ is disconnected from $S_2(A)$. Take $\alpha \in \Delta_+^{\text{im}}$ such that supp$\alpha \subset S_1(A)$ and $\beta \in \Delta_+^{\text{ali}\cdot\text{im}}$ such that supp$\beta \subset S_2(A)$. Then supp$(\alpha + \beta)$ is not connected because $S_1(A)$ is not connected to $S_2(A)$.

This implies that $\alpha + \beta$ is not a root, which implies

$$\Delta_+^{\text{ali}\cdot\text{im}}(A) \neq \Delta_+^{\text{p}\cdot\text{ali}\cdot\text{im}}(A).$$

□

From Corollary 6, we conclude the following:

Results: For all indecomposable, symmetrizable $n \times n$ BKM supermatrix, we have one of the following:

(i) when $n \leq 4$, $\Delta_+^{\text{ali}\cdot\text{im}}(A) = \Delta_+^{\text{p}\cdot\text{ali}\cdot\text{im}}(A)$ and

(ii) when $n \geq 5$ there exist BKM superalgebras whose Dynkin diagrams is not completely connected which satisfy the purely alien imaginary property and there also exist BKM superalgebras which do not satisfy the purely alien imaginary property.

For $n = 5$, Theorem 62 gives BKM superalgebras possessing the purely alien imaginary property and the following gives a BKM superalgebra not possessing the purely alien imaginary property.

Example 63. Consider the BKM supermatrix (A, ψ), where

$$A = \begin{bmatrix} 2 & -4 & 0 & 0 & 0 \\ -6 & -k & -1 & 0 & 0 \\ 0 & -1 & 0 & -1 & 0 \\ 0 & 0 & -2 & 2 & -2 \\ 0 & 0 & 0 & -2 & 2 \end{bmatrix}, \text{ with } \psi = \{1, 5\}.$$

Here $\alpha_1 + \alpha_2$ and $\alpha_4 + \alpha_5$ are alien imaginary roots, whereas $\alpha_1 + \alpha_2 + \alpha_4 + \alpha_5$ is not a root, since $\text{supp}(\alpha_1 + \alpha_2 + \alpha_4 + \alpha_5)$ is not connected. Example 59 is also an example for $n = 5$ for a BKM superalgebra not possessing purely alien imaginary property.

Example 64. We give below an example of a BKM superalgebra with the purely alien imaginary property.

Consider the BKM supermatrix (A, ψ), where

$$A = \begin{pmatrix} 0 & -1 & -1 & -1 & -1 \\ -1 & -3 & -1 & -1 & -1 \\ -1 & -1 & 2 & -1 & -1 \\ -1 & -1 & -1 & 2 & -1 \\ -2 & -2 & -2 & -2 & 2 \end{pmatrix}, \text{ with } \psi = \{1, 5\}.$$

This is a completely connected BKM supermatrix and hence by Results (ii) stated after Theorem 62, the corresponding BKM superalgebra satisfies the purely alien imaginary property.

Extended-hyperbolic BKM superalgebras satisfying the purely alien imaginary property

We can give a complete classification for Dynkin diagrams of indecomposable extended-hyperbolic type BKM superalgebras as in the case of KM algebras (Chapter 2).

All the symmetrizable BKM superalgebras associated with the $SBEHA_c^{(n)}$ for $n \leq 5$ satisfy purely alien imaginary property.

In the following theorem, we prove a more general statement namely, that any $SBEHA^{(n)}$, $n \geq 3$ satisfies the purely alien imaginary property.

The proof is similar to the proof of the corresponding theorem in KM algebras (Chapter 2).

Theorem 63. *Every indecomposable, symmetrizable extended-hyperbolic BKM superalgebra* $\mathfrak{g}(A)$ *satisfies the purely alien imaginary property.*

Example 65. Here we give the Dynkin diagram for the example of a symmetrizable, extended-hyperbolic BKM supermatrix whose Dynkin diagram is not connected but satisfies the purely imaginary property and purely alien-imaginary property.

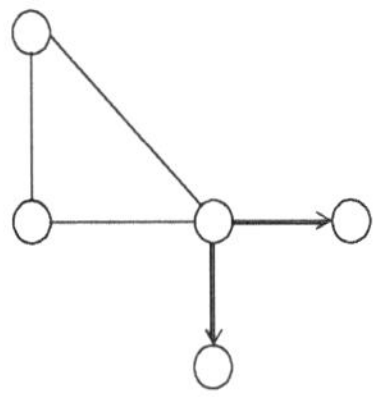

Example 66. The following example gives an indefinite BKM superalgebra not belonging to $SBEHA^{(n)}$ $(n = 3)$ type and which does not satisfy the purely imaginary property.

The BKM supermatrix is

$$A = \begin{pmatrix} -k_1 & -1 & 0 \\ -1 & 2 & -1 \\ 0 & -1 & -k_2 \end{pmatrix} \text{ with } \psi = \{1, 3\} \text{ and } k_1, k_2 > 0.$$

The corresponding Dynkin diagram:

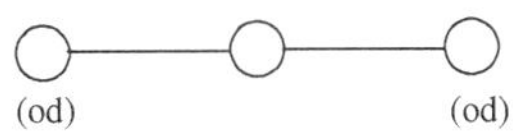

Example 67. The following example gives the Dynkin diagram of an indefinite non-hyperbolic superalgebra, not belonging to $SBEHA^{(n)}$, $n = 5$ but has the purely imaginary property and purely alien imaginary property.

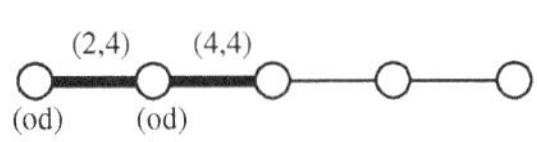

Remark 148. Here we notice that $\Pi^{im} = \phi$ and hence the set of all domestic imaginary roots is an empty set and we will have the alien imaginary roots. Also for any two alien imaginary roots α and β, we have, by the definition of K, $\langle \alpha, \alpha_i^\vee \rangle \leq 0$ and $\langle \beta, \alpha_i^\vee \rangle \leq 0$ for all $i \in \Pi^{re}$. As we obtain $\langle \alpha + \beta, \alpha_i^\vee \rangle \leq 0$ for all $i \in \Pi^{re}$, we get $\alpha + \beta$ is also an imaginary root. Hence the purely imaginary property and purely alien imaginary property are satisfied.

So, in the classes of indefinite non-hyperbolic type BKM superalgebras which do not belong to the class of $SBEHA^{(n)}$, we have examples and counter examples of algebras possessing the purely imaginary property and purely alien imaginary property.

Among the extended-hyperbolic BKM superalgebras there exist algebras which satisfy the purely imaginary property and there also exist algebras which do not satisfy the purely imaginary property. But all extended-hyperbolic BKM superalgebras satisfy purely alien imaginary property.

Example 68. This example gives a non-symmetric, extended-hyperbolic BKM supermatrix which does not satisfy a purely imaginary property but satisfies purely alien imaginary property.

The BKM supermatrix is

$\begin{pmatrix} 0 & -1 & 0 \\ -2 & 2 & -2 \\ 0 & -1 & 0 \end{pmatrix}$ with $\psi = \{1, 3\}$. The corresponding Dynkin diagram is

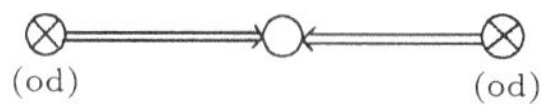

Here we notice that α_1 and α_2 are simple imaginary roots, but $\alpha_1 + \alpha_2$ is not at all a root, and hence the purely imaginary property is not satisfied.

The following example gives an indefinite non-hyperbolic BKM superalgebra which is not an extended-hyperbolic BKM superalgebra but has the purely alien imaginary property.

Example 69. Consider the BKM supermatrix $A = \begin{pmatrix} -k_1 & -1 & 0 \\ -1 & -k_2 & -1 \\ 0 & -1 & 2 \end{pmatrix}$ with $\psi = \{1, 2\}$ and $k_1, k_2 > 0$. The corresponding Dynkin diagram:

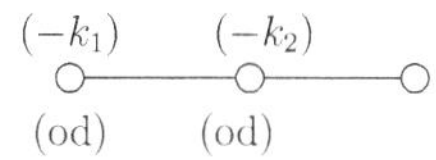

Complete classification of BKM Lie superalgebras possessing purely alien imaginary property: In Theorems 61 and 62, we have given a necessary and sufficient condition for an imaginary root to be purely alien imaginary and thereby giving the following complete classification of those BKM superalgebras with the *purely alien imaginary property:*

1. Every BKM superalgebra $\mathfrak{g}(A)$ of finite type (*here* $\Delta_+^{\text{ali}\cdot\text{im}}(A) = \Delta_+^{\text{p}\cdot\text{ali}\cdot\text{im}}(A) = \emptyset$).
2. Affine BKM superalgebras.
3. Symmetrizable hyperbolic BKM superalgebras.
4. Symmetrizable BKM supermatrix (A, ψ), where $A = (a_{ij})_{i,j=1}^{n}$, $n \leq 4$.
5. Symmetrizable extended-hyperbolic, indefinite type BKM superalgebras.
6. Symmetrizable BKM superalgebras whose Dynkin diagrams are completely connected.

Complete classification of BKM Lie superalgebras possessing purely imaginary property is given below:

1. Every BKM superalgebra $\mathfrak{g}(A)$ of finite type (*here* $\Delta_+^{im}(A) = \Delta_+^{\text{p}\cdot\text{im}}(A) = \emptyset$).
2. Every affine BKM superalgebra.
3. Every symmetrizable indecomposable hyperbolic BKM superalgebra.
4. Every BKM superalgebra whose indecomposable BKM supermatrix is an extension of finite BKM supermatrix of any order and has only even simple imaginary roots.

5. Every BKM superalgebra whose indecomposable BKM supermatrix is an extension of affine BKM supermatrix of any order and has only even simple imaginary roots.
6. Every BKM superalgebra whose indecomposable BKM supermatrix is an extension of hyperbolic KM matrix of order 2 and has only even simple imaginary roots.

Remark 149. From the above complete classification, one can understand that the BKM superalgebras which do not possess the purely alien imaginary property are in the class of indefinite BKM superalgebras other than the hyperbolic and extended-hyperbolic superalgebras and can also identify the superalgebras possessing and not possessing this property.

BKM superalgebras which do not possess the purely imaginary property are in the class of indefinite BKM superalgebras other than hyperbolic BKM superalgebras and can also identify the superalgebras possessing and not possessing this property.

5.6 BKM Lie superalgebras possessing purely imaginary property but not strictly imaginary property

Now we compare the strictly imaginary roots and purely imaginary roots of BKM superalgebras.

In [8], for KM algebras it was noticed that

(i) If $\alpha \in \Delta_+^{\text{sim}}$, $\beta \in \Delta_+^{\text{im}}$, then $\alpha + \beta \in \Delta_+^{\text{im}}$.
(ii) Δ_+^{sim} is a semigroup.

From above the property (i) of the strictly imaginary roots, it is clear that every positive strictly imaginary root is a purely imaginary root. One can see from the following example that the converse is not true.

Example 70. Let (A, ψ) be an indecomposable symmetrizable BKM supermatrix, where

$$A = \begin{pmatrix} 0 & -a & 0 \\ -1 & 2 & -1 \\ 0 & -1 & 2 \end{pmatrix} \quad \text{with } a > 0,\ \psi = \{1\}.$$

Clearly A is a symmetrizable BKM supermatrix $\beta_1 = 2\alpha_1 + \alpha_2 + \alpha_3$ is not a strictly imaginary root because neither $\beta_1 + (\alpha_2 + \alpha_3)$ nor $\beta_1 - (\alpha_2 + \alpha_3)$ is a root. Similarly, $\beta_2 = 2\alpha_1 + \alpha_2$ is not a strictly imaginary root because neither $\beta_2 + \alpha_2$ nor $\beta_2 - \alpha_2$ is a root. Also, $\beta_3 = \alpha_1$ is not a strictly imaginary root because neither $\beta_3 + \alpha_3$ nor $\beta_3 - \alpha_3$ is a root.

Remark 150. From [145], it is clear that the set of all strictly imaginary roots of any BKM superalgebra is a subset of set of all purely imaginary roots, that is, $\Delta^{\text{pim}} \backslash \Delta^{\text{sim}}$ may be an empty set or non-empty set depending upon the algebras. So, for the BKM superalgebras possessing purely imaginary property, we verify whether the set $\Delta^{\text{pim}} \backslash \Delta^{\text{sim}}$ is empty or not. In the case where $\Delta^{pim} \backslash \Delta^{sim} = \emptyset (\Delta^{pim} = \Delta^{sim})$ all purely imaginary roots are strictly imaginary as $\Delta^{sim} \subset \Delta^{pim}$ is always true. BKM superalgebras which satisfy the condition $\Delta^{pim} \backslash \Delta^{sim} = \emptyset$, will be in the class of BKM algebras possessing strictly imaginary property. Hence the condition $\Delta^{pim} = \Delta^{sim}$ is equivalent to $\Delta^{sim} = \Delta^{im}$, which is equivalent to SIM property.

It is clear that the set of all strictly imaginary roots of BKM superalgebra is a subset of set of all purely imaginary roots, that is, $\Delta^{\text{pim}} \backslash \Delta^{\text{sim}}$ may be an empty-set or non-empty set depending upon the algebras. So, for the BKM superalgebras possessing purely imaginary property, we verify whether the set $\Delta^{\text{pim}} \backslash \Delta^{\text{sim}}$ is empty or not.

In case $\psi = \phi$, every BKM superalgebra is a BKM algebra or GKM algebra for which the complete classification of strictly imaginary property was already known (Chapter 3). Hence for every BKM superalgebra which is not a BKM algebra($\psi \neq \phi$), we find below the set $\Delta^{pim} \backslash \Delta^{sim}$ for the following cases.

1. For every BKM superalgebra of finite type, since Δ_+^{im} is empty, there is neither a purely imaginary root, purely alien imaginary root nor a strictly imaginary root.
2. For affine BKM superalgebra, $\Delta^{\text{pim}} \backslash \Delta^{\text{sim}}$ may be empty or non-empty.

Every affine BKM superalgebra is either

Case (a): An affine BKM algebra ($\psi = \emptyset$), or
Case (b): An affine BKM superalgebra which is not an affine BKM algebra ($\psi \neq \emptyset$).

Case (a): Since $\Pi^{im} = \emptyset$, there is neither an purely imaginary root nor a purely alien imaginary root. There is also no strictly imaginary root.

Case (b): In [56], the Dynkin diagrams of Lie superalgebras were given in Tables 0–4. The numerical marks in Tables 1–4 [56] are the coefficients of the linear dependence between the corresponding column of the Cartan matrix. Moreover, for every node p_s, of the diagram $L^{(k)}$ with numerical a_s, there exists an automorphism of order $m = ka_s$ of the simple Lie superalgebra $\mathfrak{g}$ with the corresponding Z/mZ gradation $\mathfrak{g} = \oplus \mathfrak{g}_i$. Moreover, by the result of a proposition in [56], we have

$$\Delta^{sim} = k\Delta^{im} \quad \text{for k=1,2,3,4.}$$

So for every affine BKM superalgebra associated with the BKM supermatrix in Table 2–4 in [56],

$$\Delta^{pim} \backslash \Delta^{sim} = \Delta^{im} \backslash k\Delta^{im},$$

which is non-empty for $k = 2, 3, 4$.

For example consider the BKM superalgebra associated with BKM supermatrix (A, ψ), where

$$A = \begin{pmatrix} 2 & -2 \\ -2 & 2 \end{pmatrix}.$$

We have

$$\text{if } \psi = \{1, 2\}\ \Delta^{\text{pim}} \backslash \Delta^{\text{sim}} = \Delta^{\text{im}} \backslash 2\Delta^{\text{im}}.$$
$$\text{if } \psi = \{2\}\ \Delta^{\text{pim}} \backslash \Delta^{\text{sim}} = \Delta^{\text{im}} \backslash 4\Delta^{\text{im}}.$$

The following results are explained in [145].

Result. Consider a BKM superalgebra with n simple imaginary roots and m simple real roots associated with the BKM supermatrix (A, ψ), where

$$A = \begin{pmatrix} -k_1 & & & & \\ & -k_2 & & & \\ & & \ddots & & \\ & & & -k_n & \\ & & & & GX \end{pmatrix}$$

with $k_i \geq 0$ for all i and ψ is defined. Then we have, for $i \in \psi$ and $j \in I^{re}$.

(i) If $i \in \psi_-$ for each i (that is $k_i > 0$ for each i), then we have $\Delta^{pim} \backslash \Delta^{sim} = \emptyset$.
(ii) If $i \in \psi_0$ for each i (that is $k_i = 0$ for each i), then we have $\Delta^{pim} \backslash \Delta^{sim} = \emptyset$ if $a_{ji} > 1$ for $j \in I^{re} = \{n+1, n+2, \ldots, n+m\}$.
(iii) If $i \in \psi_-$ for some i (i.e., $k_i > 0$ for some i) and $i \in \psi_0$ for some i (i.e., $k_i = 0$ for some i) with $a_{ji} > 1$ for $j \in I^{re} = \{n+1, n+2, \ldots, n+m\}$ then we have $\Delta^{pim} \backslash \Delta^{sim} = \emptyset$.

The above results can be directly proved for each of the following cases.

(i) BKM superalgebras(which are not BKM algebras) with finite number of simple imaginary roots and finite number of simple real roots.
(ii) BKM superalgebras with finite number of simple imaginary roots and no simple real roots.
(iii) BKM superalgebras with finite number of simple imaginary roots and one simple real root (GX_m is single element, 2).
(iv) BKM superalgebras with finite number of simple imaginary roots and two simple real roots (GX_m is a 2×2 matrix).

Remark 151. In the following example for all cases $\psi \neq \emptyset$ is only considered, because the cases $\psi = \emptyset$ will be only for BKM algebras not for BKM superalgebras.

Example 71.

(1) Consider the hyperbolic BKM superalgebra associated with the BKM supermatrix (A, ψ) where

$$A = \begin{pmatrix} 0 & -a \\ -b & 2 \end{pmatrix}$$

(a) For $\psi = \{1\}$, $\psi = \{2\}$ and $b > 1$, all purely imaginary roots are strictly imaginary roots. Hence, $\Delta^{\text{pim}} \backslash \Delta^{\text{sim}} = \emptyset$.
(b) For $\psi = \{1\}$ and $b = 1$, we have $\Delta^{\text{pim}} \backslash \Delta^{\text{sim}} = \{2\alpha_1 + \alpha_2\}$.

(2) Consider the BKM superalgebra whose symmetrizable BKM supermatrix is (A, ψ), where $A = (a_{ij})_{i,j=1}^{n}$ for $n \leq 2$.
For $\psi \neq \emptyset$, we will have the following cases:
(a) For $n = 1$, the BKM superalgebra associated with the BKM supermatrix is $(-k_1)$ with $k_1 \geq 0$ and $\psi = \{1\}$, all purely imaginary roots are strictly imaginary roots.
(b) For $n = 2$, the BKM superalgebra associated with the BKM supermatrix (A, ψ) where

$$A = \begin{pmatrix} -k_1 & -a \\ -b & -k_2 \end{pmatrix} \text{ with } k_1,\ k_2 \geq 0.$$

For $\psi = \{1\}$ or $\{2\}$ or $\{1,2\}$, we have

$$\Delta^{\text{pim}} \backslash \Delta^{\text{sim}} = \emptyset.$$

(c) For $n = 2$, the BKM superalgebra associated with the BKM supermatrix (A, ψ) where

$$A = \begin{pmatrix} -k_1 & -a \\ -b & 2 \end{pmatrix} \text{ with } k_1 \geq 0$$

we have for $b > 1$, $\Delta^{\text{pim}} \backslash \Delta^{\text{sim}}$is empty , if $\psi = \{1\}$ or $\{2\}$ or $\{1,2\}$.
for $b = 1$, $\Delta^{\text{pim}} \backslash \Delta^{\text{sim}} = \{2\alpha_1 + \alpha_2\}$ if $k_1 = 0, \psi = \{1\}$ or $\{2\}$ or $\{1,2\} = \{2\alpha_1 + \alpha_2\}$ if $k_1 > 0, \psi = \{2\}$.

(3) Consider the BKM superalgebra whose symmetrizable BKM supermatrix (A, ψ) where $A = (a_{ij})_{i,j=1}^{n}$ for $n = 3$. For $\psi \neq \emptyset$, we have the following cases:

(a) Consider the BKM superalgebra associated with the BKM supermatrix (A, ψ) where

$$A = \begin{pmatrix} -k_1 & -a_1 & -a_2 \\ -b_1 & -k_2 & -p \\ -b_2 & -q & -k_3 \end{pmatrix} \text{ with } k_1,\ k_2,\ k_3 \geq 0$$

and $\psi = \{1\}$, or $\{2\}$, or $\{3\}$, or $\{1,2\}$, or $\{1,3\}$, or $\{2,3\}$, or $\{1,2,3\}$.
Here, all purely imaginary roots are strictly imaginary roots.

(b) Consider the BKM superalgebra associated with the BKM supermatrix (A, ψ) where

$$A = \begin{pmatrix} -k_1 & -a_1 & -a_2 \\ -b_1 & -k_2 & -p \\ -b_2 & -q & 2 \end{pmatrix} \text{ with } k_1,\ k_2 \geq 0.$$

(i) For $\psi = \{1\}, \text{or}\{2\}, \text{or}\{1,2\}$ and $k_1 = k_2 = 0$, if both b_2 and q are > 1, all purely imaginary roots are strictly imaginary roots. One can prove that $\Delta^{pim} \backslash \Delta^{sim} \neq \emptyset$. Hence, if either b_2 or q equal to 0 or 1, there are purely imaginary roots which are not strictly imaginary.

(ii) For $\psi = \{1\}, k_1 > 0, k_2 \geq 0$, if either b_2 or q equal to 0 or 1 and $b_2 > 1$, then there are purely imaginary roots which are not strictly imaginary. one can prove that $\Delta^{pim} \backslash \Delta^{sim} \neq \emptyset$. Hence there are purely imaginary roots which are not strictly imaginary.

(iii) For $\psi = \{2\}, k_1 \geq 0, k_2 > 0$, there are purely imaginary roots which are not strictly imaginary if either b_2 or q equal to 0 or 1 and $q > 1$. One can prove that $\Delta^{pim} \backslash \Delta^{sim} \neq \emptyset$. Hence there are purely imaginary roots which are not strictly imaginary.

(iv) For $\psi = \{1,2\}$ or $\{1,2,3\}$, $k_1, k_2 > 0$, then all purely imaginary roots are strictly imaginary roots.

(v) For $\psi = \{1,2\}$ or $\{1,2,3\}$, $k_1 = 0$, $k_2 > 0$ One can prove that $\Delta^{pim} \backslash \Delta^{sim} \neq \emptyset$. So there are purely imaginary roots which are not strictly imaginary roots.

(vi) For $\psi = \{1,2\}$ or $\{1,2,3\}$, $k_1 > 0$ and $k_2 = 0$. One can prove that $\Delta^{pim} \backslash \Delta^{sim}$ $\neq \emptyset$. So there are purely imaginary roots which are not strictly imaginary roots.

(c) Consider the BKM superalgebra associated with the BKM supermatrix (A, ψ) where

$$A = \begin{pmatrix} -k_1 & -a_1 & -a_2 \\ -b_1 & 2 & -p \\ -b_2 & -q & 2 \end{pmatrix} \text{ with } pq \leq 3 \text{ and } q = 1.$$

Let $\psi \neq \emptyset$.

(i) For $\psi = \{1\}$ or $\{2\}$, or $\{3\}$ or $\{1, 2, 3\}$, $k_1 = 0$. We have if both b_1 and $b_2 > 1$, all purely imaginary roots are strictly imaginary roots.
So, if either b_1 or b_2 equal to 0 or 1, there are purely imaginary roots which are not strictly imaginary.

(ii) For $\psi = \{1\}$ or $\{1, 2\}$, or $\{1, 2, 3\}$, $k_1 > 0$, then $\Delta^{\text{pim}} \backslash \Delta^{\text{sim}} = \emptyset$. That is all purely imaginary roots are strictly imaginary roots.

(d) Consider the BKM superalgebra associated with the BKM supermatrix (A, ψ) where

$$A = \begin{pmatrix} -k_1 & -a_1 & -a_2 \\ -b_1 & 2 & -p \\ -b_2 & -q & 2 \end{pmatrix} \text{ with } pq = 4 \text{ and } p = q.$$

If both b_1 and $b_2 > 1$, all purely imaginary roots are strictly imaginary roots.

(i) For $\psi = \{1\}$, $k_1 > 0$, then all purely imaginary roots are strictly imaginary roots.
(ii) $\psi = 1, k_1 = 0$,
(iii) $\psi = \{1, 2, 3\}, k_1 > 0$,
(iv) $\psi = \{1, 2, 3\}, k_1 = 0$,
for the above cases (ii)–(iv), either b_1 or b_2 are equal to 0 or 1, there are purely imaginary roots which are not strictly imaginary roots.

(e) Consider the BKM superalgebra associated with the BKM supermatrix (A, ψ) where

$$A = \begin{pmatrix} -k_1 & -a_1 & -a_2 \\ -b_1 & 2 & -p \\ -b_2 & -q & 2 \end{pmatrix} \text{ with } pq = 4 \text{ and } p \neq q.$$

If b_1, b_2, p and q are all > 1, all purely imaginary roots are strictly imaginary.
In the following cases we get $\Delta^{\text{pim}} \backslash \Delta^{\text{sim}} \neq \emptyset$,

(i) For $\psi = \{1\}$ and $k_1 = 0$, we have

$$\Delta^{\text{pim}} \backslash \Delta^{\text{sim}} = \begin{cases} \{(2n+1)(\alpha_2 + 2\alpha_3) | n \geq 0\} \cup \{2\alpha_1 + \alpha_2, \alpha_1\}, \\ \{(2n+1)(\alpha_2 + 2\alpha_3) | n \geq 0\} \cup \{2\alpha_1 + \alpha_2\}, \text{if } b_1 = p = 1 \text{ and } b_2 > 1. \end{cases}$$

So, if either b_1 or b_2 equal to 0 or 1 or p or q equal to 1, there are purely imaginary roots which are not strictly imaginary roots.

(ii) For $\psi = \{1\}$ and $k_1 > 0$, if either b_1 or b_2 equal to 0 or 1 or p or q equal to 1, there are purely imaginary roots which are not strictly imaginary roots.

(4) Consider symmetrizable extended-hyperbolic indefinite type BKM superalgebras. This is a subclass of BKM superalgebras discussed in 3(e).

(5) Consider the symmetrizable BKM superalgebra whose Dynkin diagram is completely connected.

(a) Consider the BKM superalgebra associated with the BKM supermatrix (A, ψ) where

$$A = \begin{pmatrix} -k_1 & -a_1 & \cdots & -a_n \\ -b_1 & -k_2 & \cdots & \\ \vdots & \vdots & \ddots & \\ -b_n & \cdots & & -k_{n+1} \end{pmatrix} \text{ with } k_i \geq 0 \text{ for all } \; i \text{ and for all } \psi.$$

Here, all purely imaginary roots are strictly imaginary roots.

(b) Consider the BKM superalgebra associated with the BKM supermatrix (A, ψ) where

$$A = \begin{pmatrix} -k_1 & -a_1 & -a_2 \\ -b_1 & -k_2 & -p \\ -b_2 & -q & 2 \end{pmatrix}.$$

(i) For $\psi = \{1\}$ or $\{2\}$ or $\{3\}$ and $k_1 = k_2 = 0$,
(ii) For $\psi = \{1\}$ and $k_1 > 0,\ k_2 = 0$,
(iii) For $\psi = \{1, 2\}$ and $k_1 > 0,\ k_2 > 0$, we have

$$\Delta^{\text{pim}} \backslash \Delta^{\text{sim}} = \emptyset$$

(iv) For $\psi = \{2\}$, $k_1 = 0,\ k_2 > 0$, if either b_1 or b_2 is equal to 0 or 1, then there are purely imaginary roots which are not strictly imaginary.

(6) Consider the symmetrizable BKM superalgebra whose indecomposable BKM supermatrix is an extension of finite BKM supermatrix of any order.
Consider the BKM superalgebra associated with the BKM supermatrix (A, ψ) with

$$A = \begin{pmatrix} -k_1 & -a_1 & \cdots & -a_n \\ -b_1 & & & \\ \vdots & & GX_n & \\ -b_n & & & \end{pmatrix},$$

where GX_n is a $n \times n$ finite Cartan matrix. Let $\psi \neq \emptyset$.

(a) For $\psi = \{1\}$, $k_1 = 0$, we have

$$\Delta^{\text{pim}} \backslash \Delta^{\text{sim}} = \{2\alpha_1 + \alpha_2,\ 2\alpha_1 + \alpha_3,\ \cdots,\ 2\alpha_1 + \alpha_{n+1}\}, \text{ if } b_i = 1 \text{ for all } i.$$

If one or all b_i equal to 1, there are purely imaginary roots which are not strictly imaginary.

(b) For $\psi = \{1\}$, $k_1 > 0$, then we have

$$\Delta^{\text{pim}} \backslash \Delta^{\text{sim}} = \emptyset.$$

(7) Consider the BKM superalgebra whose indecomposable BKM supermatrix (A, ψ) with

$$A = \begin{pmatrix} -k_1 & -a_1 & \cdots & -a_{n+1} \\ -b_1 & & & \\ \vdots & & GX_n^{(k)} & \\ -b_{n+1} & & & \end{pmatrix},$$

where $GX_n^{(k)}$, $k = 1$, 2 or 3 is the Cartan matrix of the affine Lie algebra. Let $\psi \neq \emptyset$

(a) For $\psi = \{1\}$ or $\{1, 2, \ldots, n+2\}$ and $k_1 = 0$ then, we have

$$\Delta^{\text{pim}} \backslash \Delta^{\text{sim}} = \{2\alpha_1 + \alpha_2, 2\alpha_1 + \alpha_3, \ldots, 2\alpha_1 + \alpha_{n+2}\} \text{ if } b_i = 1 \text{ for all } i.$$

If one or all b_i equal to 1, there are purely imaginary roots which are not strictly imaginary roots.

(b) For $\psi = \{1\}$ or $\{1, 2, \ldots, n+2\}$ and $k_1 > 0$ then, we have

$$\Delta^{\text{pim}} \backslash \Delta^{\text{sim}} = \emptyset.$$

(c) For $k = 2, 3$ or 4 then, we have $\Delta^{\text{pim}} \backslash \Delta^{\text{sim}}$ of the superalgebra corresponding to $GX_n^{(k)}$ is non-empty for all ψ.

(8) Consider the BKM superalgebra whose indecomposable BKM supermatrix is an extension of hyperbolic BKM supermatrix of order 2. This was already discussed in case 5(vi).

(9) Consider the BKM superalgebra an indefinite type other than hyperbolic type associated with the BKM supermatrix (A, ψ) where

$$A = \begin{pmatrix} -k_1 & -a_1 & -a_2 \\ -b_1 & 2 & -p \\ -b_2 & -q & 2 \end{pmatrix} \text{ with } k_1 \geq 0,\ pq > 4.$$

For $\psi \neq \emptyset$.

(a) If b_1, b_2, p and q are all > 1, all purely imaginary roots are strictly imaginary roots.

We have the following cases for ψ, and k_1.

(b) $\psi = \{1\}$, $k_1 = 0$,

(c) $\psi = \{1\}$, $k_1 > 0$,

(d) $\psi = \{1, 2, 3\}$ or $\{2, 3\}$ and $k_1 = 0$,

(e) $\psi = \{2, 3\}$ and $k_1 > 0$.

For the above cases (b), (c), (d) and (e), one can verify that, if either b_1 or b_2 equal to 0 or 1 or p or q equal to 1, then $\Delta^{pim} \backslash \Delta^{sim}$, is non empty that is there are purely imaginary roots which are not strictly imaginary.

(f) For $\psi = \{1, 2, 3\}$, $k_1 > 0$, then, we have

$$\Delta^{\text{pim}} \backslash \Delta^{\text{sim}} = \emptyset.$$

Remark 152. We have noticed from the above discussion that the BKM superalgebras all whose purely imaginary roots are strictly imaginary ($\Delta^{\text{pim}} \backslash \Delta^{\text{sim}}$ is empty) are the BKM algebras discussed in the cases 2(a), 3*a*, 3(b)(ii) and 7(a). For all other cases, for some particular algebras, all purely imaginary roots are strictly imaginary and for some other particular cases (mentioned there), there are purely imaginary roots which are not strictly imaginary.

5.7 Complete classification of BKM Lie superalgebras possessing strictly imaginary property (SIM property)

Casperson [57] found out the indefinite Kac-Moody algebras possessing strictly imaginary property and gave a complete classification of Kac-Moody algebras possessing strictly imaginary property. It was stated in Chapter 2. For more details regarding the following classification and results (Theorems 64, 65 and 66, Examples 72 and 73), one can refer [147].

As in the cases of special and purely imaginary roots we divide the classes of BKM superalgebras into two categories:

Category 1: BKM Lie superalgebras without odd roots (GKM algebras only): For this Category, complete classification of GKM algebras possessing strictly imaginary property was already given in Chapter 3.

Category 2: BKM Lie superalgebras with a non-empty set of odd roots: We divide this category 2 into the following two classes:

Category 2 Class (I): BKM Lie superalgebras of finite order Cartan matrices and with a non-empty set of odd roots,

Category 2 Class (II): BKM Lie superalgebras of infinite order Cartan matrices and with a non-empty set of odd roots.

We discuss below these two classes separately.

Category 2 Class (I): BKM Lie superalgebras with a non-empty set of odd roots: We classify these BKM superalgebras into three subclasses (i)–(iii).

(i) BKM superalgebras with all simple roots being real *with a non-empty set of odd roots*: These are BKM superalgebras which do not have any imaginary root. So this set of BKM superalgebras do not possess strictly imaginary property.

(ii) BKM superalgebras all whose simple roots are imaginary *with a non-empty set of odd roots*: These are BKM superalgebras whose supermatrices do not appear as the extensions of KM matrices. So all the diagonal elements are negative. Hence there is no real simple root and all the roots are imaginary and also strictly imaginary.

(iii) BKM superalgebras with finite (non-zero) number of real simple roots and finite (non-zero) number of imaginary simple roots *with a non-empty set of odd roots*.

Remark 153. Hereafter, we denote by GGX, a BKM supermatrix or BKM matrix.

Theorem 64. *Let $A = (a_{ij})_{i,j=1}^{n+r}$ (the symmetrizable GGX)*

$$
= \begin{pmatrix}
-k_1 & -a_1 & -a_2 & \cdots & -a_{r-1} & -a_r & \cdots & \cdots & -a_{n+r-1} \\
-b_1 & -k_2 & -c_2 & \cdots & -c_{r-1} & -c_r & \cdots & \cdots & -c_{n+r-1} \\
-b_2 & -d_2 & -k_3 & \cdots & \cdots & \cdots & \cdots & \cdots & \cdots \\
\vdots & \vdots & & & \ddots & & & & \\
-b_{r-1} & -d_{r-1} & \cdots & \cdots & \cdots & -k_r & x_{r+1} & \cdots & x_{n+r-1} \\
-b_r & -d_r & \vdots & \vdots & & w_r & & & \\
\vdots & \vdots & \vdots & \vdots & & \vdots & & GX & \\
-b_{n+r-1} & -d_{n+r-1} & & & & w_{n+r-1} & & &
\end{pmatrix}.
$$

Here $k_i(1 \leq i \leq r) \in \mathbb{Z}_{\geq 0}$, a_i, b_i, c_i, d_i are positive integers. Here GX is KM matrix of finite, affine or indefinite type of order $n \geq 1$ and GGX is a supermatrix of finite, affine or indefinite type with r simple imaginary roots added to that of GX. Then the following results are true for BKM superalgebras with odd roots.

(1) *GX is of finite type:*

(a) *If $\alpha = \sum_{i=1}^{n+r} l_i\alpha_i \in W(\overset{\circ}{K})$ with $\sum_{\substack{i=1\\ i\neq j}}^{n+r} l_i a_{ji} < 2l_j$ (for all $j \in I^{re}$) is true for all α, then the corresponding BKM superalgebra satisfies SIM property.*

(b) *If $\alpha \in W(\Pi^{im}\backslash\psi_0) \cup W(\psi_0) \cup W\{2\alpha_i | i \in \psi^-\}$ ($\psi_0 = \{i \in \psi^{im} | k_i = 0\}$ and $\psi^- = \{i \in \psi^{im} | k_i < 0\}$), with $b_i, d_i, \ldots, x_i, w_i > 1$ in the above GGX for $r+1 \leq i \leq n+r$ is true for all α, then the corresponding BKM superalgebra satisfies SIM property. Here in (a) and (b), $\psi = \psi_0 \cup \psi^-$ is the set of all odd roots and*

$$\Delta_+^{im} = \bigcup_{w\in W} w(\overset{\circ}{K}) \bigcup W(\Pi^{im}) \bigcup W\{2\alpha_i | \alpha_i \in \psi^-\}$$

is the set of all positive imaginary roots.

(2) *GX is of untwisted affine type:*

(a) *If $\alpha = \sum_{i=1}^{n+r} l_i\alpha_i \in W(\overset{\circ}{K})$ with $\sum_{\substack{i=1\\ i\neq j}}^{n+r} l_i a_{ji} < 2l_j$ (for all $j \in I^{re}$) is true for all α, then the corresponding BKM Lie superalgebra satisfies SIM property.*

(b) *If $\alpha \in W(\Pi^{im}\backslash\psi_0) \cup \psi_0 \cup W\{2\alpha_i | i \in \psi^-\}$ with $b_i, d_i, \ldots, x_i, w_i > 1$ in the above GGX for $r+1 \leq i \leq n+r$ is true for all α, then the corresponding BKM Lie superalgebra satisfies SIM property.*

(3) *If GX is of twisted affine type, then strictly imaginary property does not hold.*

(4) *If GX is of indefinite type, then strictly imaginary property does not always hold.*

Proof. In the usual notation, let $I = \{1, 2, \ldots, n + r\}$ with $I^{im} = \{1, 2, \ldots, r\}$ and $I = \{r + 1, r + 2, \ldots, n + r\}$. $\Pi = \{\alpha_1, \alpha_2, \ldots, \alpha_{n+r}\}$ is the set of all simple roots with $\Pi^{im} = \{\alpha_1, \alpha_2, \ldots, \alpha_r\}$ is the set of all simple imaginary roots and $\Pi^{re} = \{\alpha_{r+1}, \alpha_{r+2}, \ldots, \alpha_{n+r}\}$ is the set of all simple real roots.

In general $\langle\alpha_i, \alpha_j\rangle = a_{ij}, 1 \leq i, j \leq n + r$ and $\psi = \{i | i \in I^{im}\}\ (or)\ \{i | i \in I^{im}\} \cup \{j | a_{jk} \in 2\mathbb{Z}\ \forall\ k\ \ and\ \ j \in I^{re}\}\ \ (or)\ \ \{j | a_{jk} \in 2\mathbb{Z}\ \forall\ k\ \ and\ \ j \in I^{re}\}$.

(1) Let GX be of finite type GCM and α be an positive imaginary root. Then

$$\alpha \in \Delta_+^{im} = \bigcup_{w\in W} w(\overset{\circ}{K}) \bigcup W(\Pi^{im}) \bigcup W\{2\alpha_i | \alpha_i \in \psi^-\}.$$

Here $\overset{\circ}{K} = \{\alpha \in Q_+ | \langle\alpha, \alpha_i^\vee\rangle \leq 0 (i \in I^{re})\ and\ |supp(\alpha)| > 2\}$.

We discuss below Case (a), Case (b), and Case (c) separately.

Case (a): If $\alpha = \sum_{i=1}^{n+r} l_i\alpha_i \in W(\overset{\circ}{K})$, then we have

$$\langle\alpha, \alpha_j^\vee\rangle = \sum_{i=1}^{n+r} l_i a_{ji}.$$

By Theorem 58.(a), it is clear that if $\alpha \in \Delta_+^{im}$ and $\sum_{i=1}^{n+r} l_i a_{ji} < 0$, then strictly imaginary property holds.

Case (b): If $\alpha \in W(\Pi^{im} \backslash \psi_0) \cup \psi_0$, then α can be written as

$$\alpha = \prod_{i=r+1}^{r+n} r_i(\alpha_i), \text{ for all } i \in \Pi^{im}.$$

Here

$$\begin{aligned} r_{r+1}(\alpha_i) &= \alpha_i - a_{r+1,i}\alpha_{r+1} \\ r_{r+2}r_{r+1}(\alpha_i) &= \alpha_i - \sum_{j=r+1}^{r+2} a_{j,i}\alpha_j + \prod_{j=r+1}^{r+2} a_{j,i}a_{j+1,j}\alpha_{r+2} \\ r_{r+3}r_{r+2}r_{r+1}(\alpha_i) &= \alpha_i - \sum_{j=r+1}^{r+3} a_{j,i}\alpha_j + (a_{r+1,i}\sum_{j=r+2}^{r+3} a_{j,r+1}\alpha_j + a_{r+2,i}a_{r+3,r+2}\alpha_{r+3}) \\ &\quad + \prod_{j=r+1}^{r+3} a_{j,i}a_{j+1,j}\alpha_{r+3}. \end{aligned}$$

Finally,

$$\begin{aligned} &r_{r+n}r_{r+n-1}\cdots r_{r+3}r_{r+2}r_{r+1}(\alpha_i) \\ &= \alpha_i - \sum_{k=r+1}^{r+n} a_{ki}\alpha_k + \left(a_{r+1,i}\sum_{k=r+2}^{r+n} a_{k,r+1}\alpha_k + a_{r+2,i}\sum_{k=r+3}^{r+n} a_{k,r+2}\alpha_k \right. \\ &\quad \left. + \cdots + a_{r+n-1,i}a_{r+n,r+n-1}\alpha_{r+n} \right) + \cdots + \prod_{j=r+1}^{n+r} a_{j,i}a_{j+1,j}\alpha_{r+n}. \end{aligned}$$

We divide this case (b) into Case (b)(1), Case (b)(2) (Case (b)(2)(i), Case (b)(2)(ii)) and Case (b)(3) separately.

Case (b)(1): Let $\alpha = \alpha_i \in \Delta_+^{im}(n = 0)$. We have $(\alpha_i|\alpha_j) = \epsilon_i a_{ji}$ for $j \in I^{re}$. Since a_{ji}s are always negative integers and ϵ_is are always positive, by Theorem 59, it is clear that $\alpha_i \in \Delta^{sim}$.

Case (b)(2): Let $\alpha = r_{r+1}(\alpha_i)(n = 1)$.

Case (b)(2)(i): Let $\alpha = r_{r+1}(\alpha_i))(n = 1)$ with $r + 1 \neq j$.

If $r_j(\alpha_i - a_{r+1,i}\alpha_j) \neq \alpha_i - a_{r+1,i}\alpha_j$, by Theorem 59, $\alpha_i - a_{r+1,i}\alpha_j \in \Delta^{sim}$.

On the contrary, if $r_j(\alpha_i - a_{r+1,i}\alpha_j) = \alpha_i - a_{r+1,i}\alpha_j$. Then

$$\begin{aligned} &\alpha_i - a_{r+1,i}\alpha_i - a_{ji}\alpha_j + a_{r+1,i}a_{j,r+1}\alpha_j = \alpha_i - a_{r+1,i}\alpha_i. \\ &\quad \Rightarrow a_{ji} - a_{r+1,i}a_{j,r+1} = 0 \ (or) \ \alpha_j = 0. \end{aligned}$$

As α_j is a real simple root, $\alpha_j \neq 0$, $a_{ji} - a_{r+1,i}a_{j,r+1} = 0$. This is also not true, because $a_{ji}, a_{r+1,i}, a_{j,r+1}$ are negative integers. So by Theorem 59, $\alpha_i - a_{r+1,i}\alpha_i \in \Delta^{sim}$.

Case (b)(2)(ii): Let $\alpha = r_{r+1}(\alpha_i)(n = 1)$ with $r + 1 = j$. Then

$$r_j(\alpha_i - a_{ji}\alpha_i) = \alpha_i.$$

But $r_j(\alpha_i - a_{ji}\alpha_j) = \alpha_i - a_{ji}\alpha_j. \Rightarrow \alpha_i = \alpha_i - a_{ji}\alpha_j \Rightarrow\ a_{ji}\alpha_j = 0. \Rightarrow a_{ji} = 0$ or $\alpha_j = 0$. Here $\alpha_j \neq 0$ because α_j is a simple real root and $a_{ij} \neq 0$ for $j \in \psi^{re}$ with $a_{ji} \in 2\mathbb{Z}$ and $i \neq j$. So $r_j(\alpha_i - a_{ji}\alpha_j) \neq \alpha_i - a_{ji}\alpha_j \Rightarrow r_j(\alpha) \neq \alpha$ with $|a_{ji}| > 1$. Hence by Theorem 59, $\alpha \in \Delta^{sim}$ if $|a_{ji}| > 1$.
Case (b)(3): Let $\alpha = r_{r+n}r_{r+n-1}\cdots r_{r+3}r_{r+2}r_{r+1}(\alpha_i)(n > 1)$. We have

$$\begin{aligned} r_{r+n}r_{r+n-1}\cdots r_{r+3}r_{r+2}r_{r+1}(\alpha_i) = \alpha_i - &\sum_{k=r+1}^{r+n} a_{ki}\alpha_k + a_{r+1,i}\sum_{k=r+2}^{r+n} a_{k,r+1}\alpha_k \\ &+ a_{r+2,i}\sum_{k=r+3}^{r+n} a_{k,r+2}\alpha_{r+3} + \cdots \\ &+ a_{r+n-1,i}a_{r+n,r+n-1} \\ &+ \cdots + \prod_{j=r+1}^{r+n} a_{j,i}a_{j+1,j}\alpha_{n+r} \neq \alpha_i. \end{aligned}$$

As all the a_{ji}s are negative integers, by Theorem 59,

$$r_{r+n}r_{r+n-1}\cdots r_{r+3}r_{r+2}r_{r+1}(\alpha_i) \in \Delta^{sim}.$$

Case (c): If $\alpha \in W\{2\alpha_i | i \in \psi^-\}$, then α can be written as

$$\alpha = \prod_{i=r+1}^{r+n} r_i(2\alpha_i), \text{ for all } i \in \psi^-.$$

Here

$$\begin{aligned} r_{r+1}(2\alpha_i) &= 2\alpha_i - 2a_{r+1,i}\alpha_{r+1} \\ r_{r+2}r_{r+1}(2\alpha_i) &= 2\alpha_i - \sum_{j=r+1}^{r+2} 2a_{j,i}\alpha_j + \prod_{j=r+1}^{r+2} 2a_{j,i}a_{j+1,j}\alpha_{r+2} \\ r_{r+3}r_{r+2}r_{r+1}(2\alpha_i) &= 2\alpha_i - \sum_{j=r+1}^{r+3} 2a_{j,i}\alpha_j + (2a_{r+1,i}\sum_{j=r+2}^{r+3} a_{j,r+1}\alpha_j \\ &\quad + 2a_{r+2,i}a_{r+3,r+2}\alpha_{r+3}) \\ &\quad + \prod_{j=r+1}^{r+3} 2a_{j,i}a_{j+1,j}\alpha_{r+3}. \end{aligned}$$

Finally,

$$r_{r+n}r_{r+n-1}\cdots r_{r+3}r_{r+2}r_{r+1}(2\alpha_i) = 2\alpha_i - \sum_{k=r+1}^{r+n} 2a_{ki}\alpha_k + \Bigg(2a_{r+1,i}\sum_{k=r+2}^{r+n} a_{k,r+1}\alpha_k + 2a_{r+2,i}\sum_{k=r+3}^{r+n} a_{k,r+2}\alpha_k + \cdots + 2a_{r+n-1,i}a_{r+n,r+n-1}\alpha_{r+n}\Bigg) + \cdots + \prod_{j=r+1}^{n+r} 2a_{j,i}a_{j+1,j}\alpha_{r+n}.$$

We discuss below Case (c)(1), Case (c)(2) and Case (c)(3) separately in Case (c).
Case (c) (1): Let $\alpha = 2\alpha_i \in \Delta_+^{im}(n=0)$. We have $(2\alpha_i|\alpha_j) = 2\epsilon_i a_{ji}$ for $j \in I^{re}$. Since a_{ji}s are always negative integers and ϵ_is are always positive, by Theorem 59, it is clear that $2\alpha_i \in \Delta^{sim}$.
Case(c) (2): Let $\alpha = r_{r+1}(2\alpha_i)(n=1)$.
Case(c) (2)(i): Let $\alpha = r_{r+1}(2\alpha_i)(n=1)$ with $r+1 \neq j$.
If $r_j(2\alpha_i - 2a_{r+1,i}\alpha_j) \neq 2\alpha_i - 2a_{r+1,i}\alpha_j$, then $2\alpha_i - 2a_{r+1,i}\alpha_j \in \Delta^{sim}$.
We have, $r_j(2\alpha_i - 2a_{r+1,i}\alpha_j) = 2\alpha_i - 2a_{r+1,i}\alpha_i - 2a_{ji}\alpha_j + 2a_{r+1,i}a_{j,r+1}\alpha_j$.
But $2\alpha_i - 2a_{r+1,i}\alpha_i - 2a_{ji}\alpha_j + 2a_{r+1,i}a_{j,r+1}\alpha_j = 2\alpha_i - 2a_{r+1,i}\alpha_i$.

$$\Rightarrow 2a_{ji} - 2a_{r+1,i}a_{j,r+1} = 0 \ (or)\ 2\alpha_j = 0.$$

As α_j is a real simple root with $\alpha_j \neq 0$, $a_{ji} - a_{r+1,i}a_{j,r+1} = 0$. This is not true, because $a_{ji}, a_{r+1,i}, a_{j,r+1}$ are negative integers. So by Theorem 59, $\alpha_i - a_{r+1,i}\alpha_i \in \Delta^{sim}$.
Case(c)(2)(ii): Let $\alpha = r_{r+1}(2\alpha_i)(n=1)$ with $r+1 = j$. Then

$$r_j(2\alpha_i - 2a_{ji}\alpha_i) = 2\alpha_i.$$

But $r_j(2\alpha_i - 2a_{ji}\alpha_j) = 2\alpha_i - 2a_{ji}\alpha_j. \Rightarrow 2\alpha_i = 2\alpha_i - 2ae_{ji}\alpha_j \Rightarrow\ 2a_{ji}\alpha_j = 0$
$\Rightarrow a_{ji} = 0$ or $\alpha_j = 0$. Here $\alpha_j \neq 0$ because α_j is a simple real root and $a_{ij} \neq 0$ for $j \in \psi^{re}$ with $a_{ji} \in 2\mathbb{Z}$ and $i \neq j$. So $r_j(2\alpha_i - 2a_{ji}\alpha_j) \neq 2\alpha_i - 2a_{ji}\alpha_j$. $r_j(\alpha) \neq \alpha$ with $|a_{ji}| > 1$. By the Theorem 59, $\alpha \in \Delta^{sim}$ if $|a_{ji}| > 1$.
Case(c)(3): Let $\alpha = r_{r+n}r_{r+n-1}\cdots r_{r+3}r_{r+2}r_{r+1}(2\alpha_i)(n>1)$. We have

$$r_{r+n}r_{r+n-1}\cdots r_{r+3}r_{r+2}r_{r+1}(2\alpha_i) = 2\alpha_i - \sum_{k=r+1}^{r+n} a_{ki}2\alpha_k + 2a_{r+1,i}\sum_{k=r+2}^{r+n} a_{k,r+1}\alpha_k + 2a_{r+2,i}\sum_{k=r+3}^{r+n} a_{k,r+2}\alpha_{r+3} + \cdots + 2a_{r+n-1,i}a_{r+n,r+n-1} + \cdots + \prod_{j=r+1}^{r+n} 2a_{j,i}a_{j+1,j}\alpha_{n+r} \neq 2\alpha_i.$$

As all the a_{ji}s are negative integers, by Theorem 59, $r_{r+n}r_{r+n-1}\cdots r_{r+3}r_{r+2}r_{r+1}(2\alpha_i) \in \Delta^{sim}$.

(2) Let GX be of untwisted affine type. For KM algebras of untwisted affine type, SIM property holds as per Casperson [57]. For BKM algebras with odd roots which we get as extensions of KM algebras of untwisted affine type, the proof is exactly same as in case(1) and hence SIM property holds.

(3) Let GX be of twisted affine type. As per Casperson [57] mentioned above, SIM property does not hold for KM algebras and the same is true for BKM Lie superalgebras which appear as extension of KM algebras of twisted affine type. Hence SIM property does not hold.

(4) Let GX be of indefinite type. As far as indefinite BKM Lie superalgebras are concerned, extension of finite and untwisted affine type of KM algebras will hold SIM property, where as other algebras do not hold. The following example will illustrate the above theorem. □

Example 72 (Extension of a finite type). Let $A = (a_{ij})_{i,j=1}^{n+r}$ (the symmetrizable GGX)

$$= \begin{pmatrix} -k & -a_1 & -a_2 \\ -b_1 & 2 & -1 \\ -b_2 & -1 & 2 \end{pmatrix}.$$

This is a BKM supermatrix of indefinite type denoted by $SBGA_2$, which is an extension of finite type A_2. The Weyl group for corresponding BKM Lie superalgebra is $W = \{1, r_2, r_3, r_2r_3, r_3r_2, r_2r_3r_2\}$.

$$\Delta_+^{im} = \bigcup_{w\in W} w(\overset{\circ}{K}) \cup W(\Pi^{im}) \cup W\{2\alpha_i | i \in \psi^-\}.$$

Here

$$\bigcup_{w\in W} W(\overset{\circ}{K})$$

$$= \Big\{ p_1\alpha_1 + q_1\alpha_2 + r_1\alpha_3 \text{ where } (p_1, q_1, r_1) = (k_1, k_2, k_3) \; or \; (k_1, k_1b_1 - k_2 + k_3, k_3), \; or \; (k_1, k_2, k_1b_2 + k_2 - k_3) \; or \; (k_1, k_1(b_2 + b_1) - k_3, k_1b_2 + k_2 - k_3) \; or \; (k_1, k_1(b_1 + b_2) - k_3, k_1(b_1 + b_2) - k_2) \text{ with } 2k_3 - k_1b_2 \le k_2 \le \frac{k_1b_1 + k_3}{2} \Big\}$$

$$W(\Pi^{im}) = \{p_2\alpha_1 + q_2\alpha_2 + r_2\alpha_3 \text{ where } (p_2, q_2, r_2) = (1, 0, 0) \; or \; (1, b_1, 0) \; or \; (1, 0, b_2) \; or \; (1, (b_1 + b_2), 1) \; or \; (1, (b_1 + b_2), (b_1 + b_2)) \; or \; (1, b_1, (b_1 + b_2))\}$$

$$W\{2\alpha_i | i \in \psi^-\} = \{p_3\alpha_1 + q_3\alpha_2 + r_3\alpha_3 | (p_3, q_3, r_3) = (2, 0, 0) \; or \; (2, 2b_1, 0) \; or \; (2, 0, 2b_2) \; or \; (2, 2(b_1 + b_2), 2) \; or \; (2, 2(b_1 + b_2), 2(b_1 + b_2)) \; or \; (2, 2b_1, 2(b_1 + b_2))\}.$$

Case (a): Let $\alpha = p_1\alpha_1 + q_1\alpha_2 + r_1\alpha_3$, where the values of (p_1, q_1, r_1) are as in $W(\overset{\circ}{K})$ with $-k_1b_2 - k_2 < 2k_3$ and $-k_1b_1 - k_3 < 2k_2$. So (in general) if, $\sum_{\substack{i=1\\ i\neq j}}^{n+r} l_i a_{ji} < 2l_j$, then by Theorem 58, $\alpha \in \Delta^{sim}$ for all $\alpha \in W(\overset{\circ}{K})$.

Case (b): Let $\alpha = p_2\alpha_1 + q_2\alpha_2 + r_2\alpha_3$ where the values of (p_2, q_2, r_2) are as in $W(\Pi^{im})$. If $b_1, b_2 > 0$, then it is clear that $r_\gamma(\alpha) \neq \alpha$ for $\alpha = \alpha_1 + b_1\alpha_2, \alpha = \alpha_1 + b_2\alpha_3, \alpha = \alpha_1 + (b_1+b_2)\alpha_2 + \alpha_3, \alpha = \alpha_1 + (b_1+b_2)\alpha_2 + (b_1+b_2)\alpha_3, \alpha = \alpha_1 + b_1\alpha_2 + (b_1+b_2)\alpha_3$ and $\gamma = 2, 3$. Hence, by Theorem 59, SIM property holds. In general, if $b_1, b_2 > 1$ then $\alpha \in \Delta^{sim}$ for $\alpha \in W(\Pi^{im})$.

Case (c): Let $\alpha = p_3\alpha_1 + q_3\alpha_2 + r_3\alpha_3$ where the values of (p_3, q_3, r_3) are as in $W\{2\alpha_i | i \in \psi^-\}$. If $b_1, b_2 \in 2\mathbb{Z}$, then it is clear that $r_\gamma(\alpha) \neq \alpha$ for $\alpha = 2\alpha_1 + 2b_1\alpha_2$, $\alpha = 2\alpha_1 + 2b_2\alpha_3, \alpha = 2\alpha_1 + 2(b_1+b_2)\alpha_2 + 2\alpha_3, \alpha = 2\alpha_1 + 2(b_1+b_2)\alpha_2 + 2(b_1+b_2)\alpha_3$, $\alpha = 2\alpha_1 + 2b_1\alpha_2 + 2(b_1+b_2)\alpha_3$ and $\gamma = 2, 3$. Hence, by Theorem 59, SIM property holds.

In general, for $\alpha \in W\{2\alpha_i | i \in \psi_-\}$, SIM property holds if $b_1, b_2 > 1$.

Example 73. Extension of an untwisted affine type: Let $A = (a_{ij})_{i,j=1}^{n+r}$ (the symmetrizable GGX) $= \begin{pmatrix} -k & -a_1 & -a_2 \\ -b_1 & 2 & -2 \\ -b_2 & -2 & 2 \end{pmatrix}$. This is a BKM supermatrix of indefinite type denoted by $SBGA_1^{(1)}$, which is an extension of untwisted affine type $A_1^{(1)}$. The Weyl group of the corresponding BKM Lie superalgebra is

$$W = \{1, r_2(r_3r_2)^j, r_3(r_2r_3)^j, (r_2r_3)^{j+1}, (r_3r_2)^{j+1} | j \in \mathbb{Z}_+\}.$$

Then

$$\Delta_+^{im} = \bigcup_{w\in W} W(\overset{\circ}{K}) \cup W(\Pi^{im}) \cup W\{2\alpha_i | i \in \psi^-\}.$$

Here

$$\begin{aligned}
&\bigcup_{w\in W} W(\overset{\circ}{K}) \\
&\quad = \{\, p_1\alpha_1 + q_1\alpha_2 + r_1\alpha_3 \text{ where } (p_1, q_1, r_1) = (k_1, k_2, k_3) \; or \; (k_1, (k_1b_1 \\
&\qquad - k_2 + 2k_3), k_3) \; or \; (k_1, k_2, (k_1b_2 + 2k_2 - k_3)) \; or \; (k_1, (k_1(b_1 + 2b_2) \\
&\qquad + 3k_2 - 2k_3), k_3) \; or \; (k_1, (k_1b_1 - k_2 + 2k_3), (k_1(2b_1 + b_2) - 2k_2 \\
&\qquad + 3k_3)), \ldots \; with \; - k_1b_1 + 2k_2 - 2k_3 \leq 0 \; and \; - k_1b_2 - 2k_2 + 2k_3 \leq 0\,\} \\
&W(\Pi^{im}) \\
&\quad = \{p_2\alpha_1 + q_2\alpha_2 + r_2\alpha_3 \text{ where } (p_2, q_2, r_2) = (1, 0, 0) \; or \; (1, b_1, 0) \; or \; (1, 0, b_2) \\
&\qquad or \; (1, (b_1 + 2b_2), b_2) \; or \; (1, b_1, (b_2 + 2b_1)) \; or \; (1, (4b_1 + 2b_2), (b_2 + 2b_1)) \\
&\qquad or \; (1, (b_1 + 2b_2), 2(2b_2 + b_1)) \; or \; (1, (4b_1 + 2b_2), (4b_2 + 6b_1)) \; or
\end{aligned}$$

$$(1,(7b_1+6b_2),(4b_2+6b_1)) \text{ or } (1,(7b_1+6b_2),(7b_2+8b_1)) \text{ or } (1,1,(4b_1+6b_2)) \text{ or } \dots\}$$

$$W\{2\alpha_i | i \in \psi_-\}$$
$$= \{p_3\alpha_1 + q_3\alpha_2 + r_3\alpha_3 \text{ where } (p_3,q_3,r_3) = (2,0,0) \text{ or } (2,2b_1,0) \text{ or } (2,0,2b_2) \text{ or } (2,2(b_1+2b_2),2b_2) \text{ or } (2,2b_1,2(b_2+2b_1)) \text{ or } (2,2(4b_1+2b_2),2(b_2+2b_1)) \text{ or } (2,2(b_1+2b_2),4(2b_2+b_1)) \text{ or } (2,2(4b_1+2b_2),2(4b_2+6b_1)) \text{ or } (2,2(7b_1+6b_2),2(4b_2+6b_1)) \text{ or } (2,2(7b_1+6b_2),2(7b_2+8b_1)) \text{ or } (2,2,2(4b_1+6b_2)) \text{ or } \dots\}$$

Case(a): Let $\alpha = p_1\alpha_1 + q_1\alpha_2 + r_1\alpha_3$, where the values of (p_1,q_1,r_1) are as in $W(\overset{\circ}{K})$ with $-k_1b_1 + 2k_2 - 2k_3 \leq 0$ and $-k_1b_2 - 2k_2 + 2k_3 \leq 0$. So (in general) if, $\sum_{\substack{i=1\\ i\neq j}}^{n+r} l_i a_{ji} < 2l_j$, then by Theorem 58, $\alpha \in \Delta^{sim}$ for all $\alpha \in W(\overset{\circ}{K})$.

Case (b): Let $\alpha = p_2\alpha_1 + q_2\alpha_2 + r_2\alpha_3$ where the values of (p_2,q_2,r_2) are as in $W(\Pi^{im})$. If $b_1, b_2 > 0$, then it is clear that $r_\gamma(\alpha) \neq \alpha$ for $\alpha = \alpha_1, \alpha = \alpha_1 + b_1\alpha_2, \alpha = \alpha_1 + b_2\alpha_3$, $\alpha = \alpha_1 + (b_1+2b_2)\alpha_2 + \alpha_3, \alpha = \alpha_1 + (7b_1+6b_2)\alpha_2 + (8b_1+7b_2)\alpha_3$ and $\gamma = 2,3$. Hence, by Theorem 59, SIM property holds.

In general, if $b_1, b_2 > 1$ then $\alpha \in \Delta^{sim}$ for $\alpha \in W(\Pi^{im})$.

Case (c): Let $\alpha = p_3\alpha_1 + q_3\alpha_2 + r_3\alpha_3$ where the values of (p_3,q_3,r_3) are as in $W\{2\alpha_i | i \in \psi^-\}$. If $b_1, b_2 \in 2\mathbb{Z}$, then it is clear that $r_\gamma(\alpha) \neq \alpha$ for $\alpha = 2\alpha_1, \alpha = 2\alpha_1 + 2b_1\alpha_2, \alpha = 2\alpha_1 + 2b_2\alpha_3, \alpha = 2\alpha_1 + 2(b_1+2b_2)\alpha_2 + 2\alpha_3, \alpha = 2\alpha_1 + 2(7b_1+6b_2)\alpha_2 + 2(8b_1+7b_2)\alpha_3$ and $\gamma = 2,3$. Hence, by Theorem 59, SIM property holds if $b_1, b_2 > 1$.

Category 2: Class(II): BKM Lie superalgebras of infinite order and with a finite non-empty set of odd roots: We divide this class into three subclasses.

(i) All simple roots are imaginary (odd or even).
(ii) One simple real root (odd or even) and infinite number of imaginary roots (odd or even).
(iii) Finite number of simple real roots and infinite number of imaginary roots.

We discuss these cases below.

(i) *All simple roots are imaginary (odd or even)*: For this class, all the roots are imaginary. So these algebras satisfy strictly imaginary property.
(ii) One simple real root (odd or even) and infinite number of imaginary roots(odd or even)

Theorem 65. *Let $A = (a_{ij})_{i,j=1}^{\infty}$(the symmetrizable GGX)*

$$= \begin{pmatrix} 2 & -a_1 & -a_2 & \cdots & -a_{r-1} & -a_r & \cdots & \cdots \\ -b_1 & -k_1 & -c_2 & \cdots & -c_{r-1} & -c_r & \cdots & \cdots \\ -b_2 & -d_2 & -k_2 & \cdots & \cdots & \cdots & \cdots & \cdots \\ \vdots & \vdots & & \ddots & & & & \\ -b_r & -d_r & \cdots & \cdots & -k_r & \cdots & \cdots & \cdots \\ \vdots & \vdots & & & \vdots & \ddots & & \\ \vdots & \vdots & & & \vdots & & \ddots & \end{pmatrix}.$$

Here $k_i(i \geq 2) \in \mathbb{Z}_{\geq 0}$, a_i, b_i, c_i, d_i are positive integers and GGX is the BKM supermatrix with one real simple root and infinite number of imaginary roots. If $\alpha = \sum_{i=1}^{\infty} l_i\alpha_i$ with $\sum_{i=2}^{\infty} l_i a_{1i} + 2l_1 < 0$ is true for all α, then the corresponding BKM Lie superalgebra satisfies strictly imaginary property.

Proof. In the usual notation $I = \{1, 2, 3, \ldots\}$ with $I^{re} = \{1\}$ and $I^{im} = \{i \in I | i \geq 2\}$. $\Pi = \{\alpha_1, \alpha_2, \alpha_3, \ldots\}$, with $\Pi^{re} = \{\alpha_1\}$ and $\Pi^{im} = \{\alpha_i | i \in I^{im}\}$. We define $\psi \in \mathbb{N}$ and $\psi = \{1\}$ *(or)* $\psi = \{i \geq 2 | i \in I^{im}\}$ *(or)* $\psi = \{1\} \cup \{i \geq 2 | i \in I^{im}\}$. Let $\alpha = \sum_{i=1}^{\infty} l_i\alpha_i$ and $\beta = \alpha_1$. Then by Theorem 58.

$$\langle \alpha, \alpha_1^\vee \rangle = \langle \sum_{i=1}^{n+r} l_i\alpha_i, \alpha_1^\vee \rangle = \sum_{i=1}^{n+r} \langle l_i\alpha_i, \alpha_1^\vee \rangle = \sum_{i=1}^{n+r} l_i a_{1i}.$$

If $\sum_{i=1}^{n+r} l_i a_{1i} < 0$, which is same as $\sum_{i=2}^{\infty} l_i a_{1i} + 2l_1 < 0$ then strictly imaginary property holds. □

Remark 154. For BKM superalgebras which appear as extension of twisted affine type (Case 3) and extension of indefinite type (Case 4), examples for $\Delta^{sim} \backslash \Delta^{pim} = \Phi$ were given in [147](Section 4, Case 3, subcase 2).

BKM Lie superalgebras with finite number of real roots and infinite number of imaginary roots

Monster Lie superalgebras: We have seen above that in the case of BKM Lie superalgebras of infinite order with one simple real root (odd or even) and infinite number of imaginary roots (odd or even) for $\psi \subseteq \{i \geq 2 | i \in I^{im}\}$ SIM property holds only when $|a_{1i}|$ and $|a_{i1}|$ are all greater than one. As a counter example, we consider Monster Lie superalgebra with one simple real root and infinite number of simple imaginary roots(odd or even). Consider $\alpha = \sum_{n=1}^{\infty} l_i\alpha_i$ with $\sum_{i=2}^{\infty} l_i a_{1i} + 2l_1 < 0$ and $a_{1i} = 0$ for $2 \leq i \leq c(1) + 1$ (c(1): multiplicity of the root corresponding to -2). As $a_{1i} = 0$ $(2 \leq i \leq c(1) + 1)$, the corresponding BKM Lie superalgebra does not satisfy SIM property. We prove this below. Consider Monster Lie superalgebra which has the following supermatrix as defined below:

Let $I = \{-1\} \cup \{1, 2, 3, \ldots\}$ be an index set and consider the Borcherds-Cartan supermatrix $A = (-(i + j))_{i,j \in I}$ with charge $\underline{m} = (c(i) | i \in I)$, where $c(i)$ are the coefficients of the elliptic modular function

$$j(q) - 744 = q^{-1} + 196884q + 21493760q^2 + \cdots = \sum_{n=-1}^{\infty} c(n)q^n.$$

Here

$$A = (-(i+j))_{i,j\in I} = \begin{pmatrix} 2 & 0 & \cdots & \cdots & 0 & -1 & \cdots & \cdots \\ 0 & -2 & \cdots & \cdots & \cdots & -3 & \cdots & \cdots \\ \vdots & \vdots & & \ddots & & & & \\ 0 & \cdots & \cdots & \cdots & -2 & -3 & \cdots & \cdots \\ -1 & -3 & \cdots & \cdots & -3 & -4 & \cdots & \cdots \\ \vdots & \vdots & & & \vdots & & \ddots & \\ \vdots & \vdots & & & \vdots & & & \ddots \end{pmatrix},$$

is the BKM supermatrix and $I = \{-1\} \cup \{1, 2, 3, \ldots\}$. We define $\psi \subseteq \{i \in I | i \geq 2\}$. α_{-1} is the real root corresponding to the diagonal element 2 and $\alpha_i (2 \leq i \leq c(1)+1)$ are the imaginary roots corresponding to the diagonal element -2

We consider any α_i, *for* $2 \leq i \leq c(1)+1$, then

$$\langle \alpha_i, \alpha_{-1}^{\vee} \rangle = a_{-1i} = 0.$$

This implies α_i does not satisfy the strictly imaginary property for $2 \leq i \leq c(1)+1$. Hence strictly imaginary property does not hold for a Monster Lie superalgebra.

(iii) Finite number (atleast two) of simple real roots and infinite number of imaginary roots: We prove the following theorem for this case.

Theorem 66. *Let* $A = (a_{ij})_{i,j=1}^{\infty}$ *(the symmetrizable GGX)*

$$= \begin{pmatrix} 2 & -a'_1 & -a'_2 & \cdots & -a_r & \cdots & \cdots & \cdots & \cdots \\ -b'_1 & 2 & -c'_1 & \cdots & -c_r & \cdots & \cdots & \cdots & \cdots \\ \vdots & \vdots & \ddots & \vdots & \vdots & \vdots & \vdots & \vdots & \vdots \\ -b'_{r-1} & -d'_r & 2 & -a_1 & -a_2 & \cdots & -a_{r-1} & -a_r & \cdots \\ -b'_r & -d'_{r+1} & -b_1 & -k_1 & -c_2 & \cdots & -c_{r-1} & -c_r & \cdots \\ -b'_{r+1} & -d_{r+2} & -b_2 & -d_2 & -k_2 & \cdots & \cdots & \cdots & \cdots \\ \vdots & \vdots & \vdots & \vdots & \vdots & \ddots & \vdots & \vdots & \vdots \\ -d_{2r} & -d_{2r} & -b_r & -d_r & \cdots & \cdots & -k_r & \cdots & \cdots \\ \vdots & \vdots & \vdots & \vdots & \vdots & \vdots & \vdots & \ddots & \vdots \\ \vdots & \vdots & \vdots & \vdots & \vdots & \vdots & \vdots & \vdots & \ddots \end{pmatrix},$$

where $k_i \in \mathbb{Z}_{\geq 0}, a_i, b_i, c_i, d_i, a'_i, b'_i, c'_i, d'_i$ *are positive integers and GGX is the BKM supermatrix with* $r (r \geq 2)$ *real simple roots and infinite number of imaginary roots. If* $\alpha = \sum_{i=1}^{\infty} l_i \alpha_i$ *with* $\sum_{i=r+1}^{\infty} l_i a_{1i} + 2l_1 < 0$ *is true for all* α, *then the corresponding BKM Lie superalgebra satisfies strictly imaginary property.*

Proof. In the usual notations $I = \{1, 2, 3, \ldots\}$ with $I^{re} = \{1, 2, \ldots, r\} (r \geq 2)$ and $I^{im} = \{i | i \geq r+1\} (i \geq 3)$. Here, $\Pi = \{\alpha_1, \alpha_2, \alpha_3, \ldots\}$, the set of all simple roots with $\Pi^{re} = \{\alpha_1, \alpha_2, \ldots, \alpha_r\}$, the set of simple real root and $\Pi^{im} = \{\alpha_i | i \geq r+1\}$, the set of all simple imaginary roots. We define $\psi \in \mathbb{N}$ and $\psi = \{i | i \in I^{re}\}$ *(or)* $\{j | j \in I^{im}\}$ *(or)* $\{i | i \in I^{re}\} \cup \{j | j \in I^{im}\}$. Let $\alpha = \sum_{i=1}^{\infty} l_i \alpha_i$ and $\{\alpha_1, \alpha_2, \ldots, \alpha_r\} \in \Pi^{re}$.

$$\langle \alpha, \alpha_j^\vee \rangle = \langle \sum_{i=1}^{\infty} l_i \alpha_i, \alpha_j^\vee \rangle = \sum_{i=1}^{\infty} l_i a_{ji}, \ \forall j \in I^{re}.$$

As all $a_{ji}, i \neq j$ are negative integers and $a_{jj} = 2$, we have $\sum_{i=1}^{\infty} l_i a_{ji} < 0$. Hence $\langle \alpha_i, \alpha_j^\vee \rangle < 0$ and SIM property holds. □

Remark 155. As in the case of Monster Lie superalgebra with one simple real root and infinite number of imaginary simple roots with $a_{1i} = a_{i1} = 0$, we can consider BKM Lie superalgebras with two simple real roots and infinite number of imaginary simple roots with $a_{ji} = a_{ij} = 0$ for some i *and* for $j = 1, 2$. In this case, as $\langle \alpha_i, \alpha_j^\vee \rangle = a_{ji} = 0$ *for all* $j \in I^{re}$ and for some $i \in I^{im}$, (similar to Monster Lie superalgebra as in Theorem 66) SIM property does not hold. Hence we understand that for the infinite order case, the SIM property depends on the non-diagonal non-zero entries of the corresponding BKM supermatrix.

5.8 Borcherds superalgebras and root supermultiplicities

In this chapter, so far we discussed about nature of different classes of roots of BKM Lie superalgebras associated with BKM supermatrices. In this section, now we define Borcherds superalgebra associated with Borcherds Cartan matrix and a θ-colored Lie superalgebra (Chapter 4). In Chapter 4, using coloring matrix C, we get a bimultiplicative form θ on Q, the root lattice of Borcherds Cartan Matrix A. A θ-colored Lie superalgebra is a Q-graded vectorspace together with a bilinear product defined using θ. Basically a θ-colored Lie superalgebra $\mathfrak{g}$ will become an ordinary Lie superalgebra if $\mathfrak{g}$ is a Z_2-graded vectorspace with $\theta(0,0) = \theta(0,1) = \theta(1,0) = 1$ and $\theta(1,1) = -1$. (θ-colored Lie superalgebra, with a bimultiplicative form θ defined on the root lattice Q with values on elements of the coloring matrix C, is a Lie superalgebra if it is a $\mathbb{Z}_2$-graded vector space with $\theta(\alpha, \beta) = (-1)^{|x||y|}$, for all $x \in \mathfrak{g}_\alpha$, $y \in \mathfrak{g}_\beta$). Hence one can notice, from Definition 214 of the BKM Lie superalgebras and from the definition of the Borcherds superalgebras given below that, with the above identification for θ, every BKM Lie superalgebra associated with a BKM supermatrix A can be considered as a Borcherds superalgebra associated with the same Borcherds Cartan matrix A with the coloring matrix C.

In this section, we briefly give some basic concepts on the homology modules and the root supermultiplicities (or superdimensions of the weight spaces) of the Borcherds superalgebras.

Let I be a countable (possibly infinite) index set. Let the square matrix $A = (a_{ij})_{i,j \in I}$ be a Borcherds-Cartan matrix (Chapter 3). As in Chapter 3, we say that an index $i \in I$ is real if $a_{ii} = 2$ and imaginary if $a_{ii} \leq 0$. We denote by $I^{\text{re}} = \{i \in I; a_{ii} = 2\}$, $I^{\text{im}} = \{i \in I; a_{ii} \leq 0\}$. Let $\underline{m} = \{m_i \in \mathbb{Z}_{>0};\ i \in I\}$ be a collection of positive integers such that $m_i = 1$ for all $i \in I^{\text{re}}$ and $m_i \geq 1$ for all $i \in I^{\text{im}}$. We call $\underline{m}$ a *charge* of A.

Remark 156. Coloring matrix, Borcherds Cartan matrix restricted to a coloring matrix, Θ-colored Lie superalgebra are already defined in Chapter 4.

Definition 239. [115]. The *Borcherds superalgebra* $\mathfrak{g} = \mathfrak{g}(A, \underline{m}, C)$ associated with the symmetrizable Borcherds-Cartan matrix A of charge $\underline{m} = (m_i; i \in I)$ and the coloring matrix $C = (c_{ij})_{i,j \in I}$ is the θ-colored Lie superalgebra generated by the elements $h_i, d_i (i \in I)$, $e_{ik}, f_{ik} (i \in I, k = 1, 2, \ldots, m_i)$ with defining relations:

$$\begin{aligned}
&[h_i, h_j] = [h_i, d_j] = [d_i, d_j] = 0, \\
&[h_i, e_{jl}] = a_{ij} e_{jl}, \ [h_i, f_{jl}] = -a_{ij} f_{jl}, \\
&[d_i, e_{jl}] = \delta_{ij} e_{jl}, \ [d_i, f_{jl}] = -\delta_{ij} f_{jl}, \\
&[e_{ik}, f_{jl}] = \delta_{ij} \delta_{kl} h_i, \\
&(ade_{ik})^{1-a_{ij}} e_{jl} = (adf_{ik})^{1-a_{ij}} f_{jl} = 0 \text{ if } a_{ii} = 2 \text{ and } \ i \neq j, \\
&[e_{ik}, e_{jl}] = [f_{ik}, f_{jl}] = 0 \text{ if } \ a_{ij} = 0,
\end{aligned}$$

for $i, j \in I, k = 1, \ldots, m_i, \ l = 1, \ldots, m_j$.

The abelian subalgebra $\mathfrak{h} = (\oplus_{i \in I} \mathbb{C} h_i) \oplus (\oplus_{i \in I} \mathbb{C} d_i)$ is called the Cartan subalgebra of $\mathfrak{g}$ and the linear functionals $\alpha_i \in \mathfrak{h}^\star (i \in I)$ are called the simple roots of $\mathfrak{g}$. For each $i \in I^{\mathrm{re}}$, let $r_i \in GL(\mathfrak{h}^\star)$ be the simple reflection of $\mathfrak{h}^\star$ defined by

$$r_i(\lambda) = \lambda - \lambda(h_i)\alpha_i \quad (\lambda \in \mathfrak{h}^\star).$$

The subgroup W of $GL(\mathfrak{h}^\star)$ generated by the r_i's $(i \in I^{\mathrm{re}})$ is called the Weyl group of the Borcherds superalgebra $\mathfrak{g}$.

The Borcherds superalgebra $\mathfrak{g} = \mathfrak{g}(A, \underline{m}, C)$ has the root space decomposition $\mathfrak{g} = \oplus_{\alpha \in Q} \mathfrak{g}_\alpha$, where

$$\mathfrak{g}_\alpha = \{x \in \mathfrak{g} | \ [h, x] = \alpha(h)x, \text{ for all } \ h \in \mathfrak{h}\}.$$

Remark 157. The above definitions of Weyl group, root space decomposition and the root system, etc of the Borcherds superalgebras are exactly same as those defined for BKM Lie superalgebras in the beginning of this chapter, Section 5.1.

Example 74. In this example, we show how the nature of a root (even or odd) depends upon the choices of the diagonal elements of the coloring matrix. Consider the Borcherds superalgebra $\mathfrak{g} = \mathfrak{g}(A, \underline{m}, C)$ associated with the Borcherds-Cartan matrix

$$A = \begin{pmatrix} -k & -a \\ -b & 2 \end{pmatrix}$$

of charge $\underline{m}$ and with the coloring matrix C as defined each case below:

(i) $C = \begin{pmatrix} -1 & c \\ c^{-1} & 1 \end{pmatrix}$, with $c \in \mathbb{C}^\times$

(ii) $C = \begin{pmatrix} 1 & c \\ c^{-1} & -1 \end{pmatrix}$, with $c \in \mathbb{C}^\times$ and b is even.

(iii) $C = \begin{pmatrix} -1 & c \\ c^{-1} & -1 \end{pmatrix}$, with $c \in \mathbb{C}^\times$ and b is even.

We have, for (i), let $\alpha = k_1\alpha_1 + k_2\alpha_2$. In this case

$$\begin{aligned}\theta(\alpha,\alpha) &= \theta(k_1\alpha_1 + k_2\alpha_2, k_1\alpha_1 + k_2\alpha_2)\\ &= \theta(\alpha_1,\alpha_1)^{k_1^2}\theta(\alpha_1,\alpha_2)^{k_1k_2}\theta(\alpha_2,\alpha_2)^{k_2^2}\theta(\alpha_2,\alpha_1)^{k_1k_2}\\ &= (-1)^{k_1^2}(c)^{k_1k_2}(1)^{k_2^2}(c^{-1})^{k_1k_2}\\ &= (-1)^{k_1^2}\end{aligned}$$

then α is even (resp. odd) if k_1 is even (resp.odd).

We have, for (ii), let $\alpha = k_1\alpha_1 + k_2\alpha_2$. In this case

$$\begin{aligned}\theta(\alpha,\alpha) &= \theta(k_1\alpha_1 + k_2\alpha_2, k_1\alpha_1 + k_2\alpha_2)\\ &= \theta(\alpha_1,\alpha_1)^{k_1^2}\theta(\alpha_1,\alpha_2)^{k_1k_2}\theta(\alpha_2,\alpha_2)^{k_2^2}\theta(\alpha_2,\alpha_1)^{k_1k_2}\\ &= (1)^{k_1^2}(c)^{k_1k_2}(-1)^{k_2^2}(c^{-1})^{k_1k_2}\\ &= (-1)^{k_2^2}\end{aligned}$$

then α is even (resp. odd)if k_2 is even (resp.odd).

Similarly, we have, for (iii), let $\alpha = k_1\alpha_1 + k_2\alpha_2$. In this case, α is even if both k_1 and k_2 are even or odd and α is odd if k_1 is even and k_2 is odd or if k_1 is odd and k_2 is even.

Remark 158. As in the case of GKM algebra, we can define representations, modules, weight multiplicity, highest weight module, highest weight vector and character of highest weight module for Borcherds superalgebras.

Definition 240. A $\mathfrak{g}$-module V is called $\mathfrak{h}$-*diagonalizable*, if it admits a weight space decomposition $V = \bigoplus_{\mu\in\mathfrak{h}^*} V_\mu$, where

$$V_\mu = \{v \in V \mid h\cdot v = \mu(h)v \text{ for all } h \in \mathfrak{h}\}.$$

If $V_\mu \neq 0$, then μ is called a *weight* of V and dim V_μ is called the *multiplicity* of μ in V.

Definition 241. A $\mathfrak{h}$-diagonalizable $\mathfrak{g}$-module V is called a *highest weight module* with highest weight $\lambda \in \mathfrak{h}^*$, if there is a nonzero vector $v_\lambda \in V$ such that

(i) $e_{ik}\cdot v_\lambda = 0$, for all $i \in I$, $k = 1,\ldots,m_i$,
(ii) $h\cdot v_\lambda = \lambda(h)v_\lambda$ for all $h \in \mathfrak{h}$, and
(iii) $V = U(\mathfrak{g})\cdot v_\lambda$. The vector v_λ is called a *highest weight vector.*

Moreover, for a highest weight module V with highest weight λ, we have

(a) $V = U(\mathfrak{g}^-)\cdot v_\lambda$,
(b) $V = \oplus_{\mu\le\lambda} V_\mu$, $V_\lambda = \mathbb{C}v_\lambda$, and
(c) $\dim V_\mu < \infty$ for all $\mu \le \lambda$.

Definition 242. Let $P(V)$ denote the set of all weights of V. If all the weights spaces are finite dimensional, the *character* of V is defined by

$$\text{ch}V = \sum_{\mu\in\mathfrak{h}^*} (\dim V_\mu)e^\mu,$$

where e^{μ} are the basis elements of the group $\mathbb{C}[\mathfrak{h}^\star]$ with the multiplication given by $e^{\mu}e^{\nu} = e^{\mu+\nu}$ for $\mu,\ \nu \in \mathfrak{h}^\star$.

Definition 243. Let $b_+ = \mathfrak{h} \oplus \mathfrak{g}_+$ be the *Borel subalgebra* of $\mathfrak{g}$ and $\mathbb{C}_\lambda$ be the 1-dimensional b_+-module defined by $\mathfrak{g}_+ \cdot 1 = 0,\ h \cdot 1 = \lambda(h)1$ for all $h \in \mathfrak{h}$. The induced module $M(\lambda) = U(\mathfrak{g}) \otimes_{U(b_+)} \mathbb{C}_\lambda$ is called the *Verma module* over $\mathfrak{g}$ with highest weight λ. Every highest weight $\mathfrak{g}$-module with highest weight λ is a homomorphic image of $M(\lambda)$. The Verma module $M(\lambda)$ contains a unique maximal submodule $J(\lambda)$. Hence the quotient $V(\lambda) = M(\lambda)/J(\lambda)$ is irreducible.
Let P^+ be the set of all linear functionals $\lambda \in \mathfrak{h}^\star$ satisfying

$$\begin{cases} \lambda(h_i) \in \mathbb{Z}_{\geq 0} & \text{for all } i \in I^{\mathrm{re}} \\ \lambda(h_i) \in 2\mathbb{Z}_{\geq 0} & \text{for all } i \in I^{\mathrm{re}} \cap I^{\mathrm{odd}} \\ \lambda(h_i) \geq 0 & \text{for all } i \in I^{\mathrm{im}}. \end{cases}$$

The elements of P^+ are called the *dominant integral weights.*

Let $\rho \in \mathfrak{h}^\star$ be the $\mathbb{C}$-linear functional satisfying $\rho(h_i) = \frac{1}{2}a_{ii}$ for all $i \in I$.

Definition 244. Let T denote the set of all imaginary simple roots counted with multiplicities. For a subset $F \subset T$, define $F \perp \lambda$, *if* $\lambda(h_i) = 0$ *for all* $\alpha_i \in F$.

Definition 245. [115]: Let J be a finite subset of Ire and we denote by $\Delta_J = \Delta \cap (\sum_{j\in J} \mathbb{Z}\alpha_j)$, $\Delta_J^{\pm} = \Delta^{\pm} \cap \Delta_J$ and $\Delta^{\pm}(J) = \Delta^{\pm} \backslash \Delta_J^{\pm}$. Let

$$\mathfrak{g}_0^{(J)} = \mathfrak{h} \bigoplus \left(\bigoplus_{\alpha\in\Delta_J} \mathfrak{g}_\alpha \right)$$

and

$$\mathfrak{g}_\pm^{(J)} = \bigoplus_{\alpha\in\Delta^{\pm}(J)} \mathfrak{g}_\alpha.$$

Then $\mathfrak{g}_0^{(J)}$ is called the *restricted Kac-Moody superalgebra* (with an extended Cartan subalgebra) associated with the Cartan matrix $A_J = (a_{ij})_{i,j\in J}$ and the set of odd indices $J^{\mathrm{odd}} = J \cap I^{\mathrm{odd}} = \{j \in J;\ c_{jj} = -1\}$.

Then the *triangular decomposition* of $\mathfrak{g}$ is given by

$$\mathfrak{g} = \mathfrak{g}_-^{(J)} \bigoplus \mathfrak{g}_0^{(J)} \bigoplus \mathfrak{g}_+^{(J)}.$$

Let $W_J = \langle r_j | j \in J \rangle$ be the subgroup of W generated by the simple reflections r_j $(j \in J)$ and let $W(J) = \left\{ w \in W;\ \Delta_w \subset \Delta^+(J) \right\}$, where $\Delta_w = \{\alpha \in \Delta^+;\ w^{-1}\alpha < 0\}$.

The following lemma referred in [115], proved in [91], is useful in actual computation of the elements of $W(J)$.

Lemma 15. *Suppose* $w = w'r_j$ *and* $l(w) = l(w') + 1$. *Then* $w \in W(J)$ *if and only if* $w' \in W(J)$ *and* $w'(\alpha_j) \in \Delta^+(J)$.

Therefore, W_J is the *Weyl group* of the restricted Kac-Moody superalgebra $\mathfrak{g}_0^{(J)}$ and $W(J)$ is the set of right coset representatives of W_J in W. That is $W = W_J W(J)$.
Let $\Delta^{\pm}_{\bar{i},J} = \Delta_{\bar{i}} \cap \Delta_J (i = 0, 1)$ and $\Delta^{\pm}_{\bar{i}}(J) = \Delta^{\pm}_{\bar{i}} \backslash \Delta^{\pm}_{\bar{i},J} (i = 0, 1)$.

The following proposition gives a generalization of the denominator identity for Borcherds superalgebras. For proof, one can refer [115].

Proposition 54. *Let J be a finite subset of the set of all real indices Ire. Then*

$$\frac{\prod_{\alpha\in\Delta^-_{\bar{0}(J)}}(1-e^{\alpha})^{\mathfrak{g}_\alpha}}{\prod_{\alpha\in\Delta^-_{\bar{1}(J)}}(1+e^{\alpha})^{\mathfrak{g}_\alpha}} = \sum_{\substack{w\,\in\,W(J)\\ F\subset J}} (-1)^{l(w)+|F|}\, chV_J(w(\rho - s(F) - \rho)),$$

where $V_J(\mu)$ denotes the irreducible highest weight module over the restricted Kac-Moody superalgebra $\mathfrak{g}_0^{(J)}$ with highest weight μ and where F runs over all the finite subsets of T such that any two elements of F are mutually perpendicular. Here $l(w)$ denotes the length of w, $|F|$ the number of elements in F, and $s(F)$ the sum of the elements in F.

Definition 246. One can introduce another basis elements of the group algebra $\mathbb{C}[\mathfrak{h}^\star]$ by defining $E^{\alpha} = \theta(\alpha,\alpha)e^{\alpha}$. Now we define the *superdimension or supermultiplicity* Dim $\mathfrak{g}_\alpha$ of the root space $\mathfrak{g}_\alpha$ by

$$\text{Dim}\mathfrak{g}_\alpha = \theta(\alpha,\alpha)\dim\mathfrak{g}_\alpha.$$

The superdimension $\text{Dim}\mathfrak{g}_\alpha$ of the root space $\mathfrak{g}_\alpha$ associated with the root α can be called as the supermultiplicity of the root α.

Remark 159. Since $w(\rho - s(F)) - \rho$ is an element of Q_-, all the weights of the irreducible highest weight $\mathfrak{g}_0^{(J)}$-module, $V_J(w(\rho - s(F)) - \rho)$, are also elements of Q_-.

Definition 247. We can define the *superdimension* $\text{Dim}V_\mu$ of the weight space V_μ of the $\mathfrak{g}_0^{(J)}$-module $V_J(w(\rho - s(F)) - \rho)$. More generally, for an $\mathfrak{h}$-diagonalizable $\mathfrak{g}_0^{(J)}$-module $V = \oplus_{\mu\in\mathfrak{h}^\star} V_\mu$ such that $P(V) \subset Q$, we define the *superdimension* $\text{Dim}V_\mu$ of the *weight space* V_μ to be $\text{Dim}V_\mu = \theta(\mu,\mu)\dim V_\mu$.

For each $k \geq 1$, let $H_k^{(J)} = \bigoplus_{\substack{w\,\in\,W(J)\\ F\subset T\\ l(w)+|F|=k}} V_J(w(\rho - s(F)) - \rho)$. Define the homology space $H^{(J)}$ of $\mathfrak{g}_-^{(J)}$ to be

$$H^{(J)} = \sum_{k=1}^{\infty}(-1)^{k+1}H_k^{(J)} = H_1^{(J)} \ominus H_2^{(J)} \oplus H_3^{(J)} \ominus \cdots,$$

an alternating direct sum of the vector spaces.

For $\tau \in Q_-$, define the *superdimension* $\text{Dim}H_\tau^{(J)}$ *of the τ-weight space of $H^{(J)}$ to be*

$$\begin{aligned}
\mathrm{Dim}H_{\tau}^{(J)} &= \sum_{k=1}^{\infty}(-1)^{k+1}(\mathrm{Dim}H_{k}^{(J)})_{\tau} \\
&= \sum_{k=1}^{\infty}(-1)^{k+1} \sum_{\substack{w \in W(J) \\ F \subset T \\ l(w)+|F|=k}} \mathrm{Dim}V_J(w(\rho - s(F)) - \rho)_{\tau} \\
&= \sum_{\substack{w \in W(J) \\ F \subset T \\ l(w)+|F|\geq 1}} (-1)^{l(w)+|F|+1}\mathrm{Dim}V_J(w(\rho - s(F)) - \rho)_{\tau}.
\end{aligned}$$

Remark 160. Let $P(H^{(J)}) = \left\{\alpha \in Q_-(J) \mid \dim H_{\alpha}^{(J)} \neq 0\right\}$ and let $\{\tau_1, \tau_2, \tau_3, \ldots\}$ be an enumeration of the set $P(H^{(J)})$.

Remark 161.

(1) The elements of $P(H^{(J)})$ can be determined by applying the following proposition, proved in [8]. Let $\Lambda \in P^+$. Then $P(\Lambda) = W \cdot \{\lambda \in P^+ \mid \lambda$ is nondegenerate with respect to $\Lambda\}$.

(2) Let $D(i) = \mathrm{Dim}H_{\tau_i}^{(J)}$. For $\tau \in Q_-(J)$, define

$$T^{(J)}(\tau) = \left\{n = (n_i)_{i\geq 1} | n_i \in \mathbb{Z}_{\geq 0}, \sum n_i\tau_i = \tau, \text{set of all partitions of } \tau \text{into a sum of } \tau_i' \text{s}\right\}.$$

For $n \in T^{(J)}(\tau)$, use the notations $|n| = \sum n_i$ and $n! = \prod n_i!$.

Definition 248. Let $\tau \in Q_-(J)$. Then *Witt partition function* $W^{(J)}(\tau)$ is defined as

$$W^{(J)}(\tau) = \sum_{n\in T^{(J)}(\tau)} \frac{(|n|-1)!}{n!}\prod D(i)^{n_i}.$$

We state the following theorem, which was proved in [115].

Theorem 67. *Let J be a finite subset of Ire. Then, for $\alpha \in \Delta^-(J)$, we have*

$$\begin{aligned}
Dim\mathfrak{g}_{\alpha} &= \sum_{d|\alpha} \frac{1}{d}\mu(d)W^{(J)}\left(\frac{\alpha}{d}\right) \\
&= \sum_{d|\alpha} \frac{1}{d}\mu(d) \sum_{n\in T^{(J)}(\frac{\alpha}{d})} \frac{(|n|-1)!}{n!}\prod D(i)^{n_i},
\end{aligned}$$

where μ is the classical Möbius function, and, for a positive integer d, $d|\alpha$ denotes $\alpha = d\tau$ for some $\tau \in Q_-$, in which case $\tau = \frac{\alpha}{d}$.

Remark 162. (i) Closed form formula for the superdimension $\mathrm{Dim}\mathfrak{g}_{\alpha}$ of the root space $\mathfrak{g}_{\alpha}(\alpha \in \Delta^-(J))$ is given in the above theorem.

(ii) The classical Mobious function μ was already defined in Theorem 42 in Chapter 3.

Root supermultiplicities of some Borcherds superalgebras

In this section, we will apply Theorem 67 to compute the root supermultiplicities of some Borcherds superalgebras. For computing root supermultiplicities of other Borcherds superalgebras, one can refer [149].

Proposition 55. *Let $\mathfrak{g} = \mathfrak{g}(A, \underline{m}, C)$ be a Borcherds superalgebra associated with the data*

$$A = \begin{pmatrix} -k & -1 & 0 & 0 \\ -1 & 2 & -1 & 0 \\ 0 & -1 & 2 & -1 \\ 0 & 0 & -1 & 2 \end{pmatrix}$$

(Extension of A_3) of charge $\underline{m} = (r, 1, 1, 1)$ with $k,\ r \in \mathbb{Z}_{>0}$ and

$$C = \begin{pmatrix} -1 & c_1 & c_2 & c_3 \\ c_1^{-1} & 1 & c_4 & c_5 \\ c_2^{-1} & c_4^{-1} & 1 & c_6 \\ c_3^{-1} & c_5^{-1} & c_6^{-1} & 1 \end{pmatrix}$$

with c_1, c_2, c_3, c_4, c_5 and $c_6 \in C^X$. Then, for a root $\alpha = -l_1\alpha_1 - l_2\alpha_2 - l_3\alpha_3 - l_4\alpha_4$ with $l_1,\ l_2,\ l_3,\ l_4 \in \mathbb{Z}_{\geq 0}$, we have

$$Dim\mathfrak{g}_\alpha = \sum_{d|(l_1,l_2,l_3,l_4)} \frac{1}{d}\mu(d)W^{(J)}\left(\frac{l_1}{d}, \frac{l_2}{d}, \frac{l_3}{d}, \frac{l_4}{d}\right)$$

$$Dim\mathfrak{g}_\alpha = \sum_{d|(l_1,l_2,l_3,l_4)} \frac{1}{d}\mu(d) \sum_{l_4=0}^{\min(\frac{l_1}{d},\frac{l_2}{d},\frac{l_3}{d},)} \frac{(\frac{l_1}{d}-1)!}{(\frac{l_1}{d}-\frac{l_2}{d})!(\frac{l_2}{d}-\frac{l_3}{d})!(\frac{l_3}{d}-\frac{l_4}{d})!\frac{l_4}{d}!}(-r)^{\frac{l_1}{d}} \tag{5.22}$$

Proof. Consider the Borcherds superalgebra $\mathfrak{g} = \mathfrak{g}(A, \underline{m}, C)$ associated with the above Borcherds-Cartan matrix of charge $\underline{m} = (r, 1, 1, 1)$ with $k,\ r \in \mathbb{Z}_{>0}$. This matrix is an extension of A_3 and the corresponding coloring matrix is as in the Proposition.

Then $I = \{1, 2, 3, 4\}$ is the index set for the simple roots of $\mathfrak{g}$. Here, α_1 is the imaginary odd simple root with multiplicity $r \geq 1$ and α_2,α_3 and α_4 are the real even simple roots.

Also, for $\alpha = k_1\alpha_1 + k_2\alpha_2 + k_3\alpha_3 + k_4\alpha_4 \in Q$, we have $\theta(\alpha, \alpha) = (-1)^{k_1^2}$. Hence α is even (respy. odd) if k_1 is even (respy. odd). Thus we have

$T = \{\alpha_1, \ldots, \alpha_1\}$ (counted r times),

and since

$$(\alpha_1, \alpha_1) = -k < 0,$$

the subset $F \subset T$ can be either empty or $\{\alpha_1\}$. Let us consider the case $J = \{2, 3, 4\}$, then

$$\mathfrak{g}_0^{(J)} = \mathfrak{g}_0 \bigoplus \mathbb{C}h_1, \text{ where}$$
$$\mathfrak{g}_0 = \langle e_2, f_2, h_2, e_3, f_3, h_3, e_4, f_4, h_4 \rangle \cong \mathrm{sl}(4, \mathbb{C}) \text{ and } W(J) = \{1\}.$$

Hence, from Definition (247), we get

$$H_1^{(J)} = V_J(-\alpha_1) \bigoplus \cdots \bigoplus V_J(-\alpha_1) \ (r \text{ copies})$$
$$H_k^{(J)} = 0 \text{ for } \ k \geq 2.$$

This implies

$$H^{(J)} = V_J(-\alpha_1) \bigoplus \cdots \bigoplus V_J(-\alpha_1) \ \ (r \text{ copies}),$$

where $V_J(-\alpha_1)$ is the 4-dimensional irreducible representation of $\mathrm{sl}(4, \mathbb{C})$.
By identifying $-k_1\alpha_1 - k_2\alpha_2 - k_3\alpha_3 - k_4\alpha_4 \in Q^-$ with $(k_1, k_2, k_3, k_4) \in \mathbb{Z}_{\geq 0} \times \mathbb{Z}_{\geq 0} \times \mathbb{Z}_{\geq 0} \times \mathbb{Z}_{\geq 0}$, by the notation of $P(H^{(J)})$,

$$P(H^{(J)}) = \{(1,0,0,0), (1,1,0,0), (1,1,1,0), (1,1,1,1)\}, \text{ where}$$
$$\dim H^{(J)}_{(1,0,0,0)} = \dim H^{(J)}_{(1,1,0,0)} = \dim H^{(J)}_{(1,1,1,0)} = \dim H^{(J)}_{(1,1,1,1)} = r.$$

Hence, using the definition of super dimension, we get

$$DimH^{(J)}_{(1,0,0,0)} = DimH^{(J)}_{(1,1,0,0)} = DimH^{(J)}_{(1,1,1,0)} = DimH^{(J)}_{(1,1,1,1)} = -r.$$

For $k_1, \ k_2, \ k_3, \ k_4 \in \mathbb{Z}_{\geq 0}$, the only partition of (k_1, k_2, k_3, k_4) into a sum of (1,0,0,0), (1,1,0,0), (1,1,1,0) and $(1, 1, 1, 1)$ is

$$(k_1, k_2, k_3, k_4) = (k_1 - k_2)(1,0,0,0) + (k_2 - k_3)(1,1,0,0)$$
$$+ (k_3 - k_4)(1,1,1,0) + k_4(1,1,1,1).$$

Hence $T^{(J)}(\tau) = \{(k_1 - k_2, k_2 - k_3, k_3 - k_4, k_4); k_1, k_2, k_3, k_4 \in \mathbb{Z}_{\geq 0}$ with $k_4 = 0, 1, 2, \ldots, \min(k_1, k_2, k_3)\}$.

Thus, the Witt partition function becomes

$$W^{(J)}(k_1, k_2, k_3, k_4) = \sum_{k_4=0}^{\min(k_1,k_2,k_3)} \frac{(k_1 - 1)!}{(k_1 - k_2)!(k_2 - k_3)!(k_3 - k_4)!(k_4)!}(-r)^{k_1}.$$

Hence we obtain the following superdimension formula.

$$Dim\mathfrak{g}_{\alpha} = \sum_{d|(l_1,l_2,l_3,l_4)} \frac{1}{d}\mu(d)W^{(J)}\left(\frac{l_1}{d}, \frac{l_2}{d}, \frac{l_3}{d}, \frac{l_4}{d}\right)$$

$$Dim\mathfrak{g}_{\alpha} = \sum_{d|(l_1,l_2,l_3,l_4)} \frac{1}{d}\mu(d) \sum_{l_4=0}^{\min(\frac{l_1}{d},\frac{l_2}{d},\frac{l_3}{d},)} \frac{(\frac{l_1}{d}-1)!}{(\frac{l_1}{d}-\frac{l_2}{d})!(\frac{l_2}{d}-\frac{l_3}{d})!(\frac{l_3}{d}-\frac{l_4}{d})!\frac{l_4}{d}!}(-r)^{\frac{l_1}{d}} \tag{5.23}$$

Some particular cases:

1. If $l_1 = l_2 = l_3 = l_4$, the formula (5.23) yields

$$Dim\mathfrak{g}_{(l_1,l_1,l_1,l_1)} = \frac{1}{l_1}\sum_{d|l_1}\mu(d)r^{\frac{l_1}{d}}.$$

In particular for the root $\alpha = (6,6,6,6)$,

$$Dim\mathfrak{g}_{(6,6,6,6)} = \frac{1}{6}\sum_{d|6}\mu(d)(-r)^{\frac{6}{d}}$$
$$= \frac{1}{6}(r^6 + r^3 - r^2 - r).$$

For $r = 6$,
$$Dim\mathfrak{g}_{(6,6,6,6)} = 6^5 + 6^2 - 6 - 1$$
$$= 7805$$

2. When $l_2 = l_3 = l_4 = 0$, we have

$$Dim\mathfrak{g}_{(l_1,0,0,0)} = \frac{1}{l_1}\sum_{d|l_1}\mu(d)(-r)^{\frac{l_1}{d}}.$$

3. When $l_4 > l_3, l_3 > l_2 > l_1$, then

$$Dim\mathfrak{g}_{(l_1,l_2,l_3,l_4)} = 0$$

For example, for the root $(6,5,4,3,)$ by the formula 5.23 we have

$$Dim\mathfrak{g}_{(6,5,4,3)} = 120r^6.$$

If $r = 2$ then

$$Dim\mathfrak{g}_{(6,5,4,3)} = 7680.$$ □

Remark 163. For affine Kac-Moody algebras, we can find the dimension of $V_J(-\alpha_1)_\lambda$, using the following theorem which was proved in [56](also given in [29]).

Theorem 68. *Let $V(\Lambda_0)$ be the basic representation of the affine Kac-Moody algebra $A_n^{(1)}$, and let λ be a weight of $V(\Lambda_0)$. Then, we have*

$$\dim(V(\Lambda_0))_\lambda = p^{(n)}\left(1-\frac{(\lambda,\lambda)}{2}\right).$$

Here the functions $p^{(n)}(m)$ are defined by

$$\sum_{m=0}^{\infty} p^{(n)}(m)q^m = \frac{1}{\phi(q)^n} = \frac{1}{\prod_{j\geq 1}(1-q^j)^n}.$$

Remark 164. The above formula can be restated as $\dim(V(\Lambda_0))_\lambda = p^{(n)}(m)$, where $\lambda = \Lambda_0 - m\delta$, Λ_0 is the highest weight, δ is the null root and $m \in \mathbb{Z}_+$.

Proposition 56. *Let $\mathfrak{g} = \mathfrak{g}(A, \underline{m}, C)$ be the Borcherds superalgebra associated with the following data:*

$$A = \begin{pmatrix} -k & -1 & 0 & 0 \\ -1 & 2 & -1 & -1 \\ 0 & -1 & 2 & -1 \\ 0 & -1 & -1 & 2 \end{pmatrix}$$

(Extension of $A_2^{(1)}$) of charge $\underline{m} = (r, 1, 1, 1)$ with $k,\ r \in \mathbb{Z}_{>0}$ and with the θ-colored matrix

$$C = \begin{pmatrix} -1 & c_1 & c_2 & c_3 \\ c_1^{-1} & 1 & c_4 & c_5 \\ c_2^{-1} & c_4^{-1} & 1 & c_6 \\ c_3^{-1} & c_5^{-1} & c_6^{-1} & 1 \end{pmatrix}$$

with c_1, c_2, c_3, c_4, c_5 and $c_6 \in C^x$. Then for a root $\alpha = -k_1\alpha_1 - k_2\alpha_2 - k_3\alpha_3 - k_4\alpha_4 \in Q_-$ with $k_1, k_2, k_3, k_4 \in \mathbb{Z}_{\geq 0}$, we have

$$Dim\mathfrak{g}_\alpha = \sum_{d|\alpha} \frac{1}{d}\mu(d) \sum_{n\in T^{(J)}(\frac{\alpha}{d})} \frac{(|n|-1)!}{n!} \prod_i \left(DimH^{(J)}_{\frac{\alpha}{d}}\right)^{n_i} \tag{5.24}$$

Proof. Consider the Borcherds superalgebra $\mathfrak{g} = \mathfrak{g}(A, \underline{m}, C)$ associated with the above Borcherds-Cartan matrix of charge $\underline{m} = (r, 1, 1, 1)$ with $k,\ r \in \mathbb{Z}_{>0}$ and with the corresponding θ-colored matrix as stated above.

Here the matrix A is an extension of $A_2^{(1)}$. Let $I = \{1, 2, 3, 4\}$ be the index set for the simple roots of $\mathfrak{g}$. Here, α_1 is an odd imaginary simple root with multiplicity $r \geq 1$ and α_2,α_3 and α_4 are even real simple roots. Also for a root $\alpha = k_1\alpha_1 + k_2\alpha_2 + k_3\alpha_3 + k_4\alpha_4$, we have $\theta(\alpha, \alpha) = (-1)^{k_1^2}$. Hence α is even (resp. odd) if k_1 is even (resp. odd). Thus we have

$$T = \{\alpha_1 \cdots \alpha_1\} \ \text{(counted } r \text{ times)}$$

and since $(\alpha_1|\alpha_1) = -k < 0$ the subset $F \subset T$ is either empty or $\{\alpha_1\}$.

If we consider $J = \{2, 3, 4\}$, then

$$\mathfrak{g}_0^{(J)} = \mathfrak{g}_0 \bigoplus \mathbb{C}h_1, \text{ where}$$
$$\mathfrak{g}_0 = \langle e_2, f_2, h_2, e_3, f_3, h_3, e_4, f_4, h_4 \rangle \cong A_2^{(1)} \text{ and } W(J) = \{1\}.$$

By Definition 247, we obtain

$$H_1^{(J)} = V_J(-\alpha_1) \bigoplus \cdots \bigoplus V_J(-\alpha_1) \ \ (r \text{ copies})$$
$$H_k^{(J)} = 0 \ \text{ for } \ k \geq 2.$$

This implies

$$H^{(J)} = V_J(-\alpha_1) \bigoplus \cdots \bigoplus V_J(-\alpha_1) \ \ (r \text{ copies}),$$

where $V_J(-\alpha_1)$ is the basic representation of $A_2^{(1)}$ with highest weight $-\alpha_1$. Here, for $A_n^{(1)}$, $\lambda_0 = -\alpha_1$, which implies $m = \frac{-k-(\lambda,\lambda)}{2}$.

By identifying the root $\lambda = -j\alpha_1 - l\alpha_2 - m\alpha_3 - n\alpha_4 \in Q^-$ with $(j, l, m, n) \in \mathbb{Z}_{\geq 0} \times \mathbb{Z}_{\geq 0} \times \mathbb{Z}_{\geq 0} \times \mathbb{Z}_{\geq 0}$, weights of $V_J(-\alpha_1)$ are $(1, 0, 0, 0)$, $(1, 1, 0, 0)$, $(1, 1, 0, 1)$ etc.

For a weight $\lambda = (1, l, m, n)$, by the above remark, we get

$$\dim(V_J(-\alpha_1))_\lambda = p^{(2)}\left(\frac{-k-(\lambda,\lambda)}{2}\right).$$

Using $\frac{-k-(\lambda,\lambda)}{2} = l + lm + ln + mn - l^2 - m^2 - n^2$, we get

$$\begin{aligned}\dim(V_J(-\alpha_1))_\lambda &= p^{(2)}\left(\frac{-k-(\lambda,\lambda)}{2}\right)\\ &= p^{(2)}(l + lm + ln + mn - l^2 - m^2 - n^2),\end{aligned}$$

where the functions $p^{(2)}(m)$ are defined by

$$\sum_{m=0}^{\infty} p^{(2)}(m)q^m = \frac{1}{\phi(q)^2} = \frac{1}{\prod_{j\geq 1}(1-q^j)^2}.$$

So

$$\dim H_\lambda^{(J)} = rp^{(2)}\left(\frac{-k-(\lambda,\lambda)}{2}\right).$$

Hence the super dimension is given by

$$DimH_{\lambda}^{(J)} = \theta(\lambda,\lambda)rp^{(2)}\left(\frac{-k-(\lambda,\lambda)}{2}\right).$$

We have, by applying Remark 161,

$$P(H^{(J)}) = \{\tau_i | i \geq 1\}, \text{ where}$$

$\tau_1 = (1,0,0,0)$, $\tau_2 = (1,1,0,0)$, $\tau_3 = (1,1,0,1)$, $\tau_4 = (1,1,1,0)$ etc., as the Weyl group of $A_2^{(1)}$ is an infinite set. Every root of $\mathfrak{g}$ is of the form (j,l,m,n) for $j \geq 1$ and $l,\ m,\ n \geq 0$. Thus, the Witt partition function becomes

$$W^{(J)}(\tau) = \sum_{n \in T^{(J)}(\tau)} \frac{(|n|-1)!}{n!} \prod_i \left(DimH_{\tau}^{(J)}\right)^{n_i}$$

Therefore, using Theorem 67, we obtain:

$$Dim\mathfrak{g}_{\alpha} = \sum_{d|\alpha} \frac{1}{d}\mu(d) \sum_{n \in T^{(J)}(\frac{\alpha}{d})} \frac{(|n|-1)!}{n!} \prod_i \left(DimH_{\frac{\alpha}{d}}^{(J)}\right)^{n_i}$$

□

Remark 165. We notice the following regarding the weights of $H^{(J)}$ and their multiplicities for the above Borcherds superalgebra (Table 5.2)
Particular cases (i): Consider the root $\alpha = (2,3,3,4)$.

Then $T^{(J)}(2,3,3,4)$ corresponds the partition of (2,3,3,4) into two parts. Hence, the partitions of the root $(2,3,3,4)$ into weights of $H^{(J)}$ are given in the following table:

(1,1,1,1), (1,2,2,3)
(1,1,1,2), (1,2,2,2)
(1,1,2,2), (1,2,1,2)
(1,1,0,1), (1,2,3,3)
(1,0,0,0), (1,3,3,4)

Table 5.2 We notice the following regarding the weights of $H^{(J)}$ and their multiplicities

Weight	Supermultiplicity	Weight	Supermultiplicity	Weight	Supermultiplicity
(1,0,0,0)	-r	(1,2,1,1)	-2r	(1,3,2,1)	-r
(1,1,0,0)	-r	(1,2,1,2)	-2r	(1,3,2,2)	-5r
(1,1,1,0)	-r	(1,2,2,1)	-2r	(1,3,2,3)	-5r
(1,1,0,1)	-r	(1,2,2,2)	-5r	(1,3,2,4)	-r
(1,1,1,1)	-2r	(1,2,2,3)	-2r	(1,3,3,2)	-5r
(1,1,1,2)	-r	(1,2,3,2)	-2r	(1,3,3,3)	-10r
(1,1,2,1)	-r	(1,2,3,3)	-2r	(1,3,3,4)	-5r
(1,1,2,2)	-r	(1,3,1,2)	-r	(1,3,4,2)	-r

Hence, by the formula (5.24) for the root (2, 3, 3, 4), we have

$$Dim\mathfrak{g}_{(2,3,3,4)} = 18r^2.$$

In particular, when $r = 2$,

$$Dim\mathfrak{g}_{(2,3,3,4)} = 72.$$

(ii) Consider the root $\alpha = (3, 3, 3, 4)$.

$T^{(J)}(3, 3, 3, 4)$ corresponds the partition of the root (3, 3, 3, 4) into three parts. Hence, the partitions of the root (3, 3, 3, 4) into weights of $H^{(J)}$ are given in the following table:

(1,0,0,0), (1,0,0,0), (1,3,3,4)
(1,0,0,0), (1,1,0,1), (1,2,3,3)
(1,0,0,0), (1,1,1,1), (1,2,2,3)
(1,0,0,0), (1,1,1,2), (1,2,2,2)
(1,0,0,0), (1,1,2,2), (1,2,1,2)
(1,1,0,0), (1,1,1,2), (1,1,2,2)
(1,1,0,1), (1,1,1,2), (1,1,2,1)
(1,1,1,0), (1,1,1,2), (1,1,1,2)

Hence, by the formula (5.24), for the root (3, 3, 3, 4), we have

$$Dim\mathfrak{g}_{(3,3,3,4)} = -33r^3.$$

In particular, when r=3

$$Dim\mathfrak{g}_{(3,3,3,4)} = -891.$$

Proposition 57. *Let* $\mathfrak{g} = \mathfrak{g}(A, \underline{m}, C)$ *be the* extended -hyperbolic Borcherds superalgebra *associated with the extended-hyperbolic Borcherds Cartan matrix,*

$$A = \begin{pmatrix} 0 & -1 & -1 \\ -1 & 2 & -3 \\ -1 & -3 & 2 \end{pmatrix}$$

of charge $\underline{m} = (1, 1, 1)$ *with the coloring matrix,*

$$C = \begin{pmatrix} -1 & c_1 & c_2 \\ c_1^{-1} & 1 & c_3 \\ c_2^{-1} & c_3^{-1} & 1 \end{pmatrix}$$

with $c_1, c_2, c_3 \in \mathbb{C}^{\times}$. *Then, for the root* $\alpha = -l\alpha_1 - m\alpha_2 - n\alpha_3 \in Q_-$ $(l, m, n \in \mathbb{Z}_{\geq 0})$, *the supermultiplicity of the root* α *is given by*

$$Dim\mathfrak{g}_\alpha = \sum_{d|\alpha} \frac{1}{d}\mu(d) \sum_{n\in T^{(J)}(\frac{\alpha}{d})} \frac{(|n|-1)!}{n!} \prod \left(DimH^{(J)}_{\tau_i}\right)^{n_i}. \tag{5.25}$$

Proof. Consider the *extended-hyperbolic Borcherds superalgebra* $\mathfrak{g} = \mathfrak{g}(A, \underline{m}, C)$ associated with the extended-hyperbolic Borcherds-Cartan matrix,

$$A = \begin{pmatrix} 0 & -1 & -1 \\ -1 & 2 & -3 \\ -1 & -3 & 2 \end{pmatrix}$$

(which is an extension of HA_1) of charge $\underline{m} = (1, 1, 1)$ and with the coloring matrix,

$$C = \begin{pmatrix} -1 & c_1 & c_2 \\ c_1^{-1} & 1 & c_3 \\ c_2^{-1} & c_3^{-1} & 1 \end{pmatrix},$$

with $c_1, c_2, c_3 \in \mathbb{C}^\times$.

Then $I = \{1, 2, 3\}$ be the index set for the simple roots of $\mathfrak{g}$. Here, α_1 is an odd imaginary simple root and α_2 and α_3 are even real simple roots. Also, for a root $\alpha = l\alpha_1 + m\alpha_2 + n\alpha_3 \in Q$, we have $\theta(\alpha, \alpha) = (-1)^{l^2}$. Hence α is even (resp. odd) only if l is even (resp. odd).

Therefore, $T = \{\alpha_1\}$ and since $(\alpha_1|\alpha_1) = 0$, the subset $F \subset T$ can be either empty or $\{\alpha_1\}$. If we consider $J = \{2, 3\}$, then

$$\mathfrak{g}_0^{(J)} = \mathfrak{g}_0 \bigoplus \mathbb{C}h_1, \text{ where}$$
$$\mathfrak{g}_0 = \langle e_2, f_2, h_2, e_3, f_3, h_3 \rangle \cong HA_1.$$

Also we have $W(J) = \{1\}$.

By Definition 247, we obtain

$$H_1^{(J)} = V_J(-\alpha_1),$$
$$H_k^{(J)} = 0 \text{ for } k \geq 2.$$

This implies

$$H^{(J)} = V_J(-\alpha_1), \text{ where}$$

$V_J(-\alpha_1)$ is an irreducible highest weight representation with highest weight $-\alpha_1$.

By identifying a root $\tau = -l\alpha_1 - m\alpha_2 - n\alpha_3 \in Q_-$ with $(l, m, n) \in \mathbb{Z}_{\geq 0} \times \mathbb{Z}_{\geq 0} \times \mathbb{Z}_{\geq 0}$, since the Weyl group of HA_1 is an infinite set and by applying Proposition 67, we have,

$$P(H^{(J)}) = \{\tau_i | i \geq 1\},$$

Table 5.3 **Weights of $H^{(J)}$ and their supermultiplicities:**

Weight	Supermultiplicity	Weight	Supermultiplicity	Weight	Supermultiplicity
(1,0,0)	-1	(1,2,5)	-4	(1,4,3)	-18
(1,0,1)	-1	(1,2,6)	-2	(1,4,4)	-36
(1,1,0)	-1	(1,3,1)	-2	(1,4,5)	-54
(1,1,1)	-2	(1,3,2)	-6	(1,4,6)	-66
(1,1,2)	-2	(1,3,3)	-12	(1,4,7)	-66
(1,1,3)	-2	(1,3,4)	-18	(1,5,2)	-4
(1,1,4)	-1	(1,3,5)	-20	(1,5,3)	-20
(1,2,1)	-2	(1,3,6)	-18	(1,5,4)	-54
(1,2,2)	-4	(1,3,7)	-12	(1,5,5)	-106
(1,2,3)	-6	(1,4,1)	-1	(1,5,6)	-166
(1,2,4)	-6	(1,4,2)	-6	(1,5,7)	-214

where $\tau_1 = (1,0,0)$, $\tau_2 = (1,0,1)$, $\tau_3 = (1,1,0)$, $\tau_4 = (1,1,1)$, $\tau_5 = (1,1,2)$, $\tau_6 = (1,1,3)$, etc.

The dimension $\dim H^{(J)}_{\tau_i}$ can be obtained by using Table 6.4 of [117].

Weight multiplicities of $-\alpha_1(h_2) = -a_{21} = 1$ and $-\alpha_1(h_3) = -a_{31} = 1$.

Then the superdimension is given by

$$\mathrm{Dim}(H^{(J)})_{\tau_i} = \theta(\tau_i, \tau_i)\,\dim(H^{(J)})_{\tau_i}\,.$$

If we have, for any root τ,

$$T^{(J)}(\tau) = \{n = (n_i)_{i\geq 1};\ n_i \in \mathbb{Z}_{>0}\,,\ \sum n_i\tau_i = \tau\},$$

then the Witt partition function, becomes

$$W^{(J)}(\tau) = \sum_{n\in T^{(J)}(\tau)} \frac{(|n|-1)!}{n!} \prod (\mathrm{Dim}(H^{(J)})_{\tau_i})^{n_i}.$$

□

Remark 166. The weights of $H^{(J)}$ and their supermultiplicities are given in Table 5.3.

Example 75. Consider the root $\alpha = (2,4,5)$.

$T^{(J)}(2,4,5)$ corresponds the partition of $(2,4,5)$ into two parts. Hence, the partitions of the root $(2,4,5)$ into weights of $H^{(J)}$ (using the above table) are given in the following table:

(1,0,0), (1,4,5)
(1,0,1), (1,4,4)
(1,1,0), (1,3,5)
(1,1,1), (1,3,4)
(1,1,2), (1,3,3)
(1,1,3), (1,3,2)
(1,1,4), (1,3,1)
(1,2,1), (1,2,4)
(1,2,2), (1,2,3)

Hence, by Equation (5.25), for the root $(2,4,5)$, the supermultiplicity of the root is

$$\mathrm{Dim}\mathfrak{g}_{(2,4,5)} = 220.$$

Example 76. Consider the root $\alpha = (3,4,4)$. $T^{(J)}(3,4,5)$ corresponds the partition of $(3,4,4)$ into three parts. Hence, the partitions of the root $(3,4,4)$ into weights of $H^{(J)}$(using the above table) are given in the following tables:

(1,0,0), (1,0,0), (1,4,4)
(1,0,0), (1,0,1), (1,4,3)
(1,0,1), (1,0,1), (1,4,2)
(1,0,0), (1,1,0), (1,3,4)
(1,0,0), (1,1,1), (1,3,3)
(1,1,0), (1,0,1), (1,3,3)
(1,0,0), (1,1,2), (1,3,2)
(1,0,1), (1,1,1), (1,3,2)
(1,0,0), (1,1,3), (1,3,1)
(1,0,1), (1,1,2), (1,3,1)
(1,1,0), (1,1,0), (1,2,4)
(1,0,0), (1,2,1), (1,2,3)
(1,1,0), (1,1,1), (1,2,3)
(1,0,0), (1,2,2), (1,2,2)
(1,0,1), (1,2,1), (1,2,2)
(1,1,0), (1,1,2), (1,2,2)
(1,1,0), (1,1,3), (1,2,1)
(1,1,1), (1,1,2), (1,2,1)

Hence, by Equation (5.25), for the root $(3,4,5)$, the supermultiplicity of the root is

$$\mathrm{Dim}\mathfrak{g}_{(3,4,5)} = -384.$$

Proposition 58. *Let $\mathfrak{g} = \mathfrak{g}(A, \underline{m}, C)$ be the Borcherds superalgebra associated with the Borcherds-Cartan matrix*

$$A = \begin{pmatrix} -k & 0 & -1 \\ 0 & 2 & -4 \\ -1 & -4 & 2 \end{pmatrix}$$

(Extension of HA_1) of charge $\underline{m} = (r, 1, 1)$ with $k,\ r \in \mathbb{Z}_{>0}$ and with the coloring matrix

$$C = \begin{pmatrix} 1 & c_1 & c_2 \\ c_1^{-1} & -1 & c_3 \\ c_2^{-1} & c_3^{-1} & 1 \end{pmatrix}$$

with $c_1, c_2, c_3 \in C^X$. Then, for the root $\alpha = -k_1\alpha_1 - k_2\alpha_2 - k_3\alpha_3$ with $k_1,\ k_2,\ k_3 \in \mathbb{Z}_{\geq 0}$, we have

$$Dim\mathfrak{g}_\alpha = \sum_{d|\alpha} \frac{1}{d}\mu(d) \sum_{n \in T^{(J)}(\tau)} \frac{(|n|-1)!}{n!} \prod \left(Dim(H^{(J)})_{\tau_i}\right)^{n_i}. \tag{5.26}$$

Proof. Consider the Borcherds superalgebra $\mathfrak{g} = \mathfrak{g}(A, \underline{m}, C)$ associated with the above Borcherds-Cartan matrix of charge $\underline{m} = (r, 1, 1)$ with $k,\ r \in \mathbb{Z}_{>0}$ and the corresponding θ colored matrix (one of the possible choices of the diagonal elements).

Let $I = \{1, 2, 3\}$ be the index set for the simple roots of $\mathfrak{g}$. Here, α_1 is an odd imaginary simple root with multiplicity $r \geq 1$ and α_2 and α_3 are even real simple roots. Moreover, for root $\alpha = k_1\alpha_1 + k_2\alpha_2 + k_3\alpha_3$, we have $\theta(\alpha, \alpha) = (-1)^{k_2^2}$. Hence the root α is even (resp. odd) if k_2 is even (resp. odd).

Thus we have

$$T = \{\alpha_1, \ldots, \alpha_1\} \text{ (counted } r \text{ times).}$$

since $(\alpha_1, \alpha_1) = -k < 0$, the subset $F \subset T$ is either empty or $\{\alpha_1\}$.

If we take $J = \{2, 3\}$, then

$$\mathfrak{g}_0^{(J)} = \mathfrak{g}_0 \bigoplus \mathbb{C}h_1, \text{ where}$$

$$\mathfrak{g}_0 = \langle e_2, f_2, h_2, e_3, f_3, h_3 \rangle \cong HA_1 \text{ and } W(J) = \{I\}.$$

By Definition 247, we obtain

$$H_1^{(J)} = V_J(-\alpha_1) \bigoplus \cdots \bigoplus V_J(-\alpha_1) \ \ (r \text{ copies})$$

$$H_k^{(J)} = 0 \text{ for } k \geq 2.$$

This implies

$$H^{(J)} = V_J(-\alpha_1) \bigoplus \cdots \bigoplus V_J(-\alpha_1) \ \ (r \text{ copies}), \text{ where}$$

$V_J(-\alpha_1)$ is an irreducible highest weight representation with highest weight $-\alpha_1$.

One can identify a root $\tau = -l\alpha_1 - m\alpha_2 - n\alpha_3 \in Q^-$ with $(l, m, n) \in \mathbb{Z}_{\geq 0} \times \mathbb{Z}_{\geq 0} \times \mathbb{Z}_{\geq 0}$. As the Weyl group of HA_1 is an infinite set, by applying Remark 161, we get

$$P(H^{(J)}) = \{\tau_i | i \geq 1\},$$

where $\tau_1 = (1,0,0)$, $\tau_2 = (1,0,1)$, $\tau_3 = (1,1,1)$, $\tau_4 = (1,1,2)$, $\tau_5 = (1,1,3)$, etc.

$$T^{(J)}(\tau) = \{n = (n_i)_{i\geq 0};\ n_i \in \mathbb{Z}_{\geq 0};\ \sum_i n_i\tau_i = \tau\}.$$

The dimension of $H^{(J)}_{\tau_i}$ can be obtained from Table 5.1 in [117] as $\dim(H^{(J)})_{\tau_i} = r \dim V_J(-\alpha_i)_{\tau_i}$ and the super dimension is given by

$$\begin{aligned} Dim(H^{(J)})_{\tau_i} &= \theta(\tau,\tau) r \dim V_J(-\alpha_i)_{\tau_i} \\ &= -r \dim V_J(-\alpha_i)_{\tau_i}. \end{aligned}$$

Thus, the Witt partition function becomes

$$W^{(J)}(\tau) = \sum_{n\in T^{(J)}(\tau)} \frac{(|n|-1)!}{n!} r^{|n|} \prod \left(Dim(H^{(J)})_{\tau_i}\right)^{n_i}.$$

Therefore, using Theorem 67 we obtain the following superdimension formula:

$$Dim\mathfrak{g}_\alpha = \sum_{d|\alpha} \frac{1}{d}\mu(d) \sum_{n\in T^{(J)}(\frac{\alpha}{d})} \frac{(|n|-1)!}{n!} \prod_i \left(DimH^{(J)}_{\frac{\alpha}{d}}\right)^{n_i}$$

For the above proposition, we give below some examples to find out superdimension $Dim\mathfrak{g}_\alpha$.

We have the following regarding the weights of $H^{(J)}$ and their multiplicities:

Weight	super Multiplicity	Weight	super Multiplicity	Weight	super Multiplicity
$(1,0,0)$	r	$(1,2,3)$	$3r$	$(1,3,4)$	$-10r$
$(1,0,1)$	r	$(1,2,4)$	$4r$	$(1,3,5)$	$-14r$
$(1,1,1)$	$-r$	$(1,2,5)$	$4r$	$(1,3,6)$	$-16r$
$(1,1,2)$	$-r$	$(1,2,6)$	$3r$	$(1,3,7)$	$-16r$
$(1,1,3)$	$-r$	$(1,2,7)$	$2r$	$(1,4,1)$	r
$(1,1,4)$	$-r$	$(1,3,1)$	$-r$	$(1,4,2)$	$4r$
$(1,2,1)$	r	$(1,3,2)$	$-3r$	$(1,4,3)$	$10r$
$(1,2,2)$	$2r$	$(1,3,3)$	$-6r$	$(1,4,4)$	$20r$

One can easily note that the values of the super multiplicity differs for other choices of the coloring matrices. □

Particular cases (i) Consider the root $\alpha = (2,3,4)$.

$T^{(J)}(2,3,4)$ corresponds to the partition of (2,3,4) into three parts. Hence, the partitions of the root $(2,3,4)$ into weights of $H^{(J)}$ are given in the following table:

(1,0,0), (1,3,4)
(1,0,1), (1,3,3)
(1,1,1), (1,2,3)
(1,1,2), (1,2,2)
(1,1,3), (1,2,1)

Hence, by the formula (8.35.1), for the root $(2,3,4)$,we have

$$Dim\mathfrak{g}_{(2,3,4)} = -22\, r^2.$$

In particular, if $r = 3$, then

$$Dim\mathfrak{g}_{(2,3,4)} = -198.$$

(ii) Consider the root $\alpha = (3,4,4)$.

$T^{(J)}(3,4,4)$ corresponds to the partition of $(3,4,4)$ into four parts. Hence, the partitions of the root $(3,4,4)$ into weights of $H^{(J)}$ are given in the following tables:

(1,0,0), (1,0,0), (1,4,4)
(1,0,0), (1,0,1), (1,4,3)
(1,0,1), (1,0,1), (1,4,2)
(1,0,0), (1,1,1), (1,3,3)
(1,0,0), (1,1,2), (1,3,2)
(1,0,0), (1,1,3), (1,3,1)
(1,0,0), (1,2,1), (1,2,3)
(1,0,0), (1,2,2), (1,2,2)
(1,1,1), (1,1,1), (1,2,2)
(1,0,1), (1,2,1), (1,2,2)
(1,1,1), (1,1,2), (1,2,1)

Hence, by the formula (5.26) for the root $(3,4,4)$, we have

$$Dim\mathfrak{g}_{(3,4,4)} = 68\, r^3.$$

In particular, when $r = 1$

$$Dim\mathfrak{g}_{(3,4,4)} = 68.$$

Borcherds superalgebras having finite number of imaginary simple roots and having no real simple root:

Proposition 59. *Let $\mathfrak{g} = \mathfrak{g}(A, \underline{m}, C)$ be the Borcherds superalgebra associated with the Borcherds-Cartan matrix*

$$A = \begin{pmatrix} -k_1 & -a_1 & \cdots & -a_{n-1} \\ -b_1 & -k_2 & \cdots & \\ \vdots & & \ddots & \\ -b_{n-1} & & \cdots & -k_n \end{pmatrix}$$

of charge $\underline{m} = (r_1, \ldots, r_n)$ with $r_i,\ k_i\ (\ i = 1, \ldots, n) \in \mathbb{Z}_{>0}$ and the coloring matrix

$$C = \begin{pmatrix} -1 & & & & \\ & -1 & & c & \\ & & \ddots & & \\ & & & \ddots & \\ & c^{-1} & & \ddots & \\ & & & & -1 \end{pmatrix}, \text{ with } c, c^{-1} \in C^X$$

Then, for the root $\alpha = -k_1\alpha_1 - k_2\alpha_2 - \cdots - k_n\alpha_n \in Q_-$ with $k_i \in \mathbb{Z}_{\geq 0}$, for $i = 1, 2, \ldots, n$, we have

$$Dim\mathfrak{g}_\alpha = \frac{1}{\sum_{i=1}^n k_i} \sum_{d|(k_1,\ldots,k_n)} \mu(d) \begin{pmatrix} \frac{\sum_{i=1}^n k_i}{d} \\ \frac{\sum_{i=1}^{n-1} k_i}{d} \end{pmatrix} \begin{pmatrix} \frac{\sum_{i=1}^{n-1} k_i}{d} \\ \frac{\sum_{i=1}^{n-2} k_i}{d} \end{pmatrix} \cdots \begin{pmatrix} \frac{k_1+k_2}{d} \\ \frac{k_1}{d} \end{pmatrix} \prod_{i=1}^n (-r_i)^{\frac{k_i}{d}}. \tag{5.27}$$

Proof. Consider the Borcherds superalgebra $\mathfrak{g} = \mathfrak{g}(A, \underline{m}, C)$ associated with the above Borcherds-Cartan matrix of charge $\underline{m} = (r_1, r_2, \ldots, r_n)$ with $k_i,\ r_i,\ a_i,\ b_i \in \mathbb{Z}_{>0}$ for all i and with the coloring matrix as defined above (choosing a possible choice for the diagonal elements).

Let $I = \{1, \ldots, n\}$ be the index set for the simple roots of $\mathfrak{g}$. Then $\alpha_i,\ i = 1, \ldots, n$ are odd imaginary simple roots with multiplicities $r_i \geq 1,\ i = 1, \ldots, n$. Moreover, for a root $\alpha = l_1\alpha_1 + \cdots + l_n\alpha_n \in Q$, we have $\theta(\alpha, \alpha) = (-1)^{\sum_{i=1}^n l_i^2}$. Then α is even if $l_i (i = 1, \ldots, n)$ are even or $l_i (i = 1, 2, \ldots, n)$ are odd with n is even. Also α is odd if $l_i, i = 1, \ldots, n$ are odd and n is odd.

Then we have

$$T = \bigcup_{i=1}^n \{\alpha_i \cdots, \alpha_i\} \text{ (counted } r_i \text{ times)},$$

and since

$$(\alpha_i|\alpha_i) = -k_i < 0 \text{ and } (\alpha_i|\alpha_j) < 0 \text{ for all } i \text{ and } j,$$

the subset $F \subset T$ is either $\emptyset$ or $\{\alpha_i\}$ for $i = 1, \ldots, n$. Since there is no real simple root, we have to take $J = \emptyset$ and $W(J) = \{1\}$.

By Definition 247, we obtain

$$
\begin{aligned}
H_1^{(J)} &= C_{-\alpha_1} \oplus \cdots \oplus C_{-\alpha_1} \quad (r_1 \text{ copies}) \\
&\oplus\ C_{-\alpha_2} \oplus \cdots \oplus C_{-\alpha_2} \quad (r_2 \text{ copies}) \\
&\ \vdots \\
&\oplus\ C_{-\alpha_n} \oplus \cdots \oplus C_{-\alpha_n} \quad (r_n \text{ copies}) \\
H_k^{(J)} &= 0 \quad \text{for } k \geq 2.
\end{aligned}
$$

This implies

$$
\begin{aligned}
H_1^{(J)} &= C_{-\alpha_1} \oplus \cdots \oplus C_{-\alpha_1} \quad (r_1 \text{ copies}) \\
&\oplus\ C_{-\alpha_2} \oplus \cdots \oplus C_{-\alpha_2} \quad (r_2 \text{ copies}) \\
&\ \vdots \\
&\oplus\ C_{-\alpha_n} \oplus \cdots \oplus C_{-\alpha_n} \quad (r_n \text{ copies}).
\end{aligned}
$$

By identifying $-m_1\alpha_1 - m_2\alpha_2 - \ldots - m_n\alpha_n \in Q^-$ with $(m_1, \ldots, m_n) \in \mathbb{Z}_{\geq 0} \times \mathbb{Z}_{\geq 0} \times \cdots \times \mathbb{Z}_{\geq 0}$, we have

$$
\begin{aligned}
P(H^{(J)}) &= \{(1,0,\ldots,0), (0,1,0,\ldots,0), \ldots, (0,0,\ldots,1)\}, \text{ where} \\
\dim H_{(i,0,\ldots,0)}^{(J)} &= r_i \text{ for } i = 1, \ldots, n.
\end{aligned}
$$

Hence the super dimension is given by

$$
D(i) = DimH_{(i,0,\ldots,0)}^{(J)} = -r_i \text{ for } i = 1, \ldots, n.
$$

Therefore, the Witt partition function becomes

$$
W^{(J)}(m_1, \ldots, m_n) = \frac{1}{\sum_{i=1}^{n} m_i} \binom{\sum_{i=1}^{n} m_i}{\sum_{i=1}^{n-1} m_i} \binom{\sum_{i=1}^{n-1} m_i}{\sum_{i=1}^{n-2} m_i} \cdots \binom{m_1 + m_2}{m_1} \prod_{i=1}^{n} (-r_i)^{m_i}.
$$

□

Particular cases:

(i) Let $n = 4$.
For the root $\alpha = (1, 1, 1, 1)$, by the formula 5.27, we have

$$
\begin{aligned}
Dim\mathfrak{g}_{(1,1,1,1)} &= \frac{1}{4} \sum_{d|(1,1,1,1)} \mu(d) \binom{\frac{4}{d}}{\frac{3}{d}} \binom{\frac{3}{d}}{\frac{2}{d}} \binom{\frac{2}{d}}{\frac{1}{d}} (-r_1)^{\frac{1}{d}} (-r_2)^{\frac{1}{d}} (-r_3)^{\frac{1}{d}} (-r_4)^{\frac{1}{d}} \\
&= \frac{1}{4} \left(\binom{4}{3} \binom{3}{2} \binom{2}{1} \right) r_1 r_2 r_3 r_4 \\
&= 6 r_1 r_2 r_3 r_4.
\end{aligned}
$$

In particular, when $r_1 = 1,\ r_2 = 2,\ r_3 = 3,\ r_4 = 4$, we obtain

$$Dim\mathfrak{g}_{(1,1,1,1)} = 144.$$

(ii) Let $n = 5$. For the root $\alpha = (1,2,3,4,5)$, by the formula 5.27, we have

$$\begin{aligned}Dim\mathfrak{g}_{(1,2,3,4,5)} &= \frac{1}{15}\sum_{d|(1,2,3,4,5)}\mu(d)\binom{15}{10}\binom{10}{6}\binom{6}{3}\binom{2}{1}(-r_1)^1(-r_2)^2(-r_3)^3(-r_4)^4(-r_5)^5\\ &= -60060 r_1 r_2^2 r_3^3 r_4^4 r_5^5.\end{aligned}$$

In particular, when $r_1 = r_2 = r_3 = r_4 = r_5 = 1$

$$Dim\mathfrak{g}_{(1,2,3,4,5)} = -60060.$$

Remark 167. For the above case, if n=2, we get the following theorem.

Theorem 69. *Let $\mathfrak{g} = \mathfrak{g}(A, \underline{m}, C)$ be the Borcherds superalgebra associated with the data:*

$$A = \begin{pmatrix} -k_1 & -a \\ -b & -k_2 \end{pmatrix}$$

of charge $\underline{m} = (r_1, r_2)$ with $k_1,\ k_2,\ r_1,\ r_2 \in \mathbb{Z}_{>0}$ and with the coloring matrix

$$C = \begin{pmatrix} -1 & c \\ c^{-1} & -1 \end{pmatrix}$$

with $c \in C^X$. Then, for the root $\alpha = -m\alpha_1 - n\alpha_2$ with $m, n \in \mathbb{Z}_{\geq 0}$, we have

$$Dim\mathfrak{g}_\alpha = \frac{1}{m+n}\sum_{d|(m,n)}\mu(d)\binom{m+n}{\frac{m}{d}} - (r_1)^{\frac{m}{d}}(-r_2)^{\frac{n}{d}}. \tag{5.28}$$

Proof. Here we have $I = \{1,2\}$. Here α_1 and α_2 are odd imaginary simple roots with multiplicities $r_1,\ r_2 \geq 1$. Moreover, for $\alpha = m\alpha_1 + n\alpha_2 \in Q$, we have $\theta(\alpha,\alpha) = (-1)^{m^2+n^2}$. Hence α is even if both m and n are either odd or even and α is odd if either m is even and n is odd or m is odd and n is even.

Thus we have

$$T = \{\alpha_1, \ldots \alpha_1\}\ (\text{counted } r_1 \text{ times}) \bigcup \{\alpha_2, \ldots, \alpha_2\}\ (\text{counted } r_2 \text{ times}).$$

Since, $(\alpha_1,\alpha_1) = -bk_1 < 0$, $(\alpha_2,\alpha_2) = -ak_2 < 0$ and $(\alpha_1,\alpha_2) = -ab < 0$, F can be either $\emptyset$ or $\{\alpha_1\}$ or $\{\alpha_2\}$. Since there is no real simple root, we should take $J = \emptyset$ and $W(J) = \{1\}$.

By Definition 247, we obtain

$$H_1^{(\emptyset)} = C_{-\alpha_1} \oplus \cdots \oplus C_{-\alpha_1} \ (r_1 \quad \text{copies}) \\ \oplus \ C_{-\alpha_2} \oplus \cdots \oplus C_{-\alpha_2} \ (r_2 \quad \text{copies})$$

$$H_k^{(\emptyset)} = 0 \ \text{ for } k \geq 2.$$

This implies

$$H^{(\emptyset)} = C_{-\alpha_1} \oplus \cdots \oplus C_{-\alpha_1} \ (r_1 \quad \text{copies}) \\ \oplus \ C_{-\alpha_2} \oplus \cdots \oplus C_{-\alpha_2} \ (r_2 \quad \text{copies}).$$

By identifying a root $\tau = -m_1\alpha_1 - m_2\alpha_2 \in Q^-$ with $(m_1, m_2) \in \mathbb{Z}_{\geq 0} \times \mathbb{Z}_{\geq 0}$, we have

$$P(H^{(\emptyset)}) = \{(1,0),(0,1)\}, \text{ where}$$

$$\dim H^{(\emptyset)}_{(1,0)} = r_1 \quad \text{and} \ \dim H^{(\emptyset)}_{(0,1)} = r_2.$$

Hence the super dimension is given by

$$DimH^{(\emptyset)}_{(1,0)} = -r_1 \quad \text{and} \ DimH^{(\emptyset)}_{(0,1)} = -r_2.$$

Thus, the Witt partition function becomes

$$W^{(\phi)}(l_1,l_2) = \frac{(l_1+l_2-1)!}{l_1!l_2!}\left(DimH^{(\emptyset)}_{(1,0)}\right)^{l_1}\left(DimH^{(\emptyset)}_{(0,1)}\right)^{l_2}$$
$$= \frac{1}{l_1+l_2}\binom{l_1+l_2}{l_1}(-r_1)^{l_1}(-r_2)^{l_2}.$$

Therefore, using Theorem 67 we obtain the following superdimension formula:

$$Dim\mathfrak{g}_\alpha = \frac{1}{m+n}\sum_{d|(m,n)}\mu(d)\binom{m+n}{\frac{m}{d}} - (r_1)^{\frac{m}{d}}(-r_2)^{\frac{n}{d}}$$

□

Particular cases: For example, for the root $\alpha = (4,4)$, by the formula (5.28), we have

$$Dim\mathfrak{g}_{(4,4)} = \frac{1}{8}\sum_{d|(4,4)}\mu(d)\left(\begin{array}{c}\frac{8}{d}\\ \frac{8}{d}\end{array}\right)(-r_1)^{\frac{4}{d}}(-r_2)^{\frac{4}{d}}$$
$$= \frac{1}{8}\left(\begin{array}{c}8\\4\end{array}\right)(-r_1)^4(-r_2)^4 - \left(\begin{array}{c}1\\8\end{array}\right)\binom{4}{2}(-r_1)^2(-r_2)^2$$
$$= \frac{1}{8}(70r_1^4r_2^4 - 6r_1^2r_2^2).$$

In particular, when $r_1 = 4, r_2 = 2$, we obtain

$$Dim\mathfrak{g}_{(4,4)} = 35792.$$

Similarly, for the root $\alpha = (3, 5)$, by the formula 5.28, we have

$$\begin{aligned} Dim\mathfrak{g}_{(3,5)} &= \frac{1}{8} \sum_{d|(3,5)} \mu(d) \begin{pmatrix} \frac{8}{d} \\ \frac{5}{d} \end{pmatrix} (-r_1)^{\frac{3}{d}} (-r_2)^{\frac{5}{d}} \\ &= \frac{1}{8} \begin{pmatrix} 8 \\ 5 \end{pmatrix} (-r_1)^3 (-r_2)^5) \\ &= 7 r_1^3 r_2^5. \end{aligned}$$

In particular, when $r_1 = 2,\ r_2 = 1$

$$Dim\mathfrak{g}_{(3,5)} = 56.$$

Superdimensions of Borcherds superalgebras having one real simple root and two imaginary simple roots

Proposition 60. *Let $\mathfrak{g} = \mathfrak{g}(A, \underline{m}, C)$ be the Borcherds superalgebra associated with the Borcherds-Cartan matrix*

$$A = \begin{pmatrix} 2 & -1 & -1 \\ -1 & -k_1 & -a \\ -1 & -a & -k_2 \end{pmatrix}$$

of charge $\underline{m} = (1, r_1, r_2)$ with $k_1,\ k_2,\ r_1,\ r_2 \in \mathbb{Z}_{>0}$ and with the coloring matrix

$$C = \begin{pmatrix} 1 & c_1 & c_2 \\ c_1^{-1} & -1 & c_3 \\ c_2^{-1} & c_3^{-1} & -1 \end{pmatrix}$$

with $c_1, c_2, c_3 \in C^x$. Then, for a root $\alpha = -k_1\alpha_1 - k_2\alpha_2 - k_3\alpha_3$ with $k_1,\ k_2,\ k_3 \in \mathbb{Z}_{\geq 0}$, we have

$$Dim\mathfrak{g}_\alpha = \frac{1}{k_2 + k_3} \sum_{d|(k_1,k_2,k_3)} \mu(d) \begin{pmatrix} \frac{k_2+k_3}{d} \\ \frac{k_3}{d} \end{pmatrix} \left(\sum_{t=0}^{\min\left(\frac{k_2}{d}, \frac{k_1-k_3}{d}\right)} \begin{pmatrix} \frac{k_3}{d} & \frac{k_1-t}{d} \end{pmatrix} \begin{pmatrix} \frac{k_2}{d} \\ \frac{t}{d} \end{pmatrix} \right) (-r_2)^{\frac{k_2}{d}} (-r_3)^{\frac{k_3}{d}}.$$

Proof. Consider the Borcherds superalgebra $\mathfrak{g} = \mathfrak{g}(A, \underline{m}, C)$ associated with the above Borcherds-Cartan matrix of charge $\underline{m} = (1, r_1, r_2)$ with $k_1,\ k_2,\ r_1,\ r_2,\ a \in \mathbb{Z}_{>0}$ and with the corresponding coloring matrix (one of the possible choices of the diagonal elements) as in the Proposition.

Let $I = \{1, 2, 3\}$ be the index set for the simple roots of $\mathfrak{g}$. Then, α_1 is an even real simple root and α_2 and α_3 are odd imaginary simple roots with multiplicities $r_1,\ r_2 \geq 1$. Moreover, for a root $\alpha = l_1\alpha_1 + l_2\alpha_2 + l_3\alpha_3 \in Q$, we have $\theta(\alpha, \alpha) = (-1)^{l_2^2 + l_3^2}$.

Hence α is even if both l_2 and l_3 are either even or odd. Also α is odd if l_2 is even and l_3 is odd or l_2 is odd and l_3 is even.

Then we have

$$T = \{\alpha_2, \ldots, \alpha_2\} \text{ (counted } r_1 \text{ times)}$$
$$\bigcup \{\alpha_3, \ldots, \alpha_3\} \text{ (counted } r_2 \text{ times)}.$$

Since (α_2, α_2), (α_3, α_3), (α_2, α_3) are all < 0, the subset $F \subset T$ can be either $\emptyset$ or $\{\alpha_2\}$ or $\{\alpha_3\}$.

If we take $J = \{1\}$, then

$$\mathfrak{g}_0^{(J)} = \mathfrak{g}_0 \oplus \mathbb{C}h_2 \oplus \mathbb{C}h_3, \text{ where}$$
$$\mathfrak{g}_0 = \langle e_1, f_1, h_1 \rangle \simeq \mathrm{sl}(2, \mathbb{C}) \text{ and } W(J) = \{1\}.$$

By Definition 247, we obtain

$$H_1^{(J)} = V_J(-\alpha_2) \oplus \cdots \oplus V_j(-\alpha_2) \ (r_1 \text{ copies })$$
$$\oplus V_J(-\alpha_3) \oplus \cdots \oplus V_J(-\alpha_3) \ (r_2 \text{ copies})$$
$$H_k^{(J)} = 0 \quad \text{for } k \geq 2.$$

This implies

$$H^{(J)} = V_J(-\alpha_2) \oplus \cdots \oplus V_j(-\alpha_2) \ (r_1 \text{ copies })$$
$$\oplus V_J(-\alpha_3) \oplus \cdots \oplus V_J(-\alpha_3) \ (r_2 \text{ copies}),$$

where $V_J(-\alpha_2)$ is the 2-dimensional irreducible representation of $\mathrm{sl}(2, \mathbb{C})$ with highest weight $-\alpha_2$ and $V_J(-\alpha_3)$ is the 2-dimensional irreducible representation of $\mathrm{sl}(2, \mathbb{C})$ with highest weight $-\alpha_3$.

By identifying $-n_1\alpha_1 - n_2\alpha_2 - n_3\alpha_3 \in Q^-$ with $(n_1, n_2, n_3) \in \mathbb{Z}_{\geq 0} \times \mathbb{Z}_{\geq 0} \times \mathbb{Z}_{\geq 0}$, we have

$$P(H^{(J)}) = \{\tau_1 = (0, 1, 0), \tau_2 = (0, 0, 1), \tau_3 = (1, 1, 0), \tau_4 = (1, 0, 1)\}, \text{ where}$$
$$\dim H_{\tau_1}^{(J)} = \dim H_{\tau_3}^{(J)} = r_2 \text{ and } \dim H_{\tau_2}^{(J)} = \dim H_{\tau_4}^{(J)} = r_3.$$

Then, the super dimension is given by

$$DimH_{\tau_1}^{(J)} = DimH_{\tau_3}^{(J)} = -r_2 \text{ and } DimH_{\tau_2}^{(J)} = DimH_{\tau_4}^{(J)} = -r_3.$$

For $(n_1, n_2, n_3) \in \mathbb{Z}_{\geq 0}$, only the partition of (n_1, n_2, n_3) as a sum of τ_1, τ_2, τ_3 and τ_4 is

$$(n_1, n_2, n_3) = (n_2 - t)\tau_1 + (n_3 - n_1 + t)\tau_2 + t\tau_3 + (n_1 - t)\tau_4$$

with t ranges from 0 to $\min(n_2, n_1 - n_3)$. Therefore, the Witt partition function becomes

$$W^{(J)}(n_1, n_2, n_3) = \sum_{t=0}^{\min\{n_2, n_1-n_3\}} \frac{(n_2+n_3-1)!}{(n_2-t)!\,(n_3-n_1+t)!\,t!\,(n_1-t)!}(-r_2)^{n_2}(-r_3)^{n_3} \tag{5.29}$$

$$= \sum_{t=0}^{\min\{n_2, n_1-n_3\}} \frac{1}{n_2+n_3}\binom{n_2+n_3}{n_3}\binom{n_3}{n_1-t}\binom{n_2}{t}(-r_2)^{n_2}(-r_3)^{n_3}. \tag{5.30}$$

Particular cases (i): For example, for the root $\alpha = (5, 4, 3)$, by the formula (5.30), we have

$$\begin{aligned} Dim\mathfrak{g}_{(5,4,3)} &= \frac{1}{7}\sum_{d|(5,4,3)} \mu(d)\binom{\frac{7}{d}}{\frac{3}{d}}\left(\sum_{t=0}^{\min(\frac{4}{d},\frac{5-3}{d})}\binom{\frac{3}{d}}{\frac{5-t}{d}}\binom{\frac{4}{d}}{\frac{t}{d}}\right)\left((-r_2)^{\frac{4}{d}} - (r_2)^{\frac{3}{d}}\right) \\ &= \frac{1}{7}\binom{7}{3}\left(\sum_{t=0}^{\min(4,2)}\binom{3}{5-t}\binom{4}{t}\right)(-r_2)^4(-r_3)^3 \\ &= -30\, r_2^4 r_3^3. \end{aligned}$$

In particular, when $r_2 = 1,\ r_3 = 1$

$$Dim\mathfrak{g}_{(5,4,3)} = -30.$$

(ii) For the root $\alpha = (3, 3, 3)$, by the formula (5.30), we have

$$\begin{aligned} Dim\mathfrak{g}_{(3,3,3)} &= \frac{1}{6}\sum_{d|(3,3,3)} \mu(d)\binom{\frac{6}{d}}{\frac{3}{d}}\left(\sum_{t=0}^{\min(\frac{3}{d},\frac{3-3}{d})}\binom{\frac{3}{d}}{\frac{3-t}{d}}\binom{\frac{3}{d}}{\frac{t}{d}}\right)(-r_2)^{\frac{3}{d}}(-r_3)^{\frac{3}{d}} \\ &= \frac{1}{6}\left(\binom{3}{3}(-r_2)^3(-r_3)^3 - \binom{1}{1}(-r_2)^1(-r_3)^1\right) \\ &= \frac{1}{6}(20(r_2)^3(r_3)^3 - r_2 r_3). \end{aligned}$$

In particular, when $r_2 = 2,\ r_3 = 3$

$$Dim\mathfrak{g}_{(3,3,3)} = 719.$$

□

Remark 168. Hence, one can compute root (super) multiplicities of those types of Borcherds superalgebras whose colored Borcherds-cartan matrices appear as an extension of generalized Cartan matrices of any order and also that one can compute root (super) multiplicities of Borcherds superalgebras with finite number of simple imaginary roots. It is interesting to notice that the nature of the root whether a root is either even or odd (and hence the super dimension is positive or negative) depends on the choice of the coloring matrix. It is important to note that the root (super) multiplicities of Borcherds superalgebras associated with colored Borcherds-Cartan matrices, which appear as an extension of finite type,affine type or indefinite type can be computed only when the corresponding weight multiplicities of the modules of the underlying Kac-Moody algebras are known. For Borcherds superalgebras which are other than finite, affine and indefinite types, one can use the super dimension formula of Kang and Kim [115] to compute the root (super) multiplicities.

5.9 Root supermultiplicities of Borcherds superalgebras which are extensions of Kac-Moody algebras and some combinatorial identities

Theorems 70, 71, 72 and Examples 77, 78, 79 were proved in [148].

5.9.1 Superdimension formula and the corresponding combinatorial identity for the extended-hyperbolic Borcherds superalgebra which is an extension of A_2

Consider J as two different subsets of Π^{re}. Then we can get the dimension formula and the corresponding combinatorial identity for $\alpha = \tau = -4\alpha_1 - 2\alpha_2 - 3\alpha_3 \in Q_-$ with b=1 for the extended-hyperbolic Borcherds superalgebra $\mathfrak{g} = \mathfrak{g}(A, \underline{m}, C)$ associated with the extended-hyperbolic Borcherds-Cartan supermatrix $A = \begin{pmatrix} -k & -a & -b \\ -a & 2 & -1 \\ -b & -1 & 2 \end{pmatrix}$.

Consider the extended-hyperbolic Borcherds superalgebra $\mathfrak{g} = \mathfrak{g}(A, \underline{m}, C)$ associated with the extended-hyperbolic Borcherds-Cartan supermatrix $A = \begin{pmatrix} -k & -a & -b \\ -a & 2 & -1 \\ -b & -1 & 2 \end{pmatrix}$ and the corresponding coloring matrix be C= $\begin{pmatrix} -1 & c_1 & c_2 \\ c_1^{-1} & 1 & c_3 \\ c_2^{-1} & c_3^{-1} & 1 \end{pmatrix}$ with $c_1, c_2, c_3 \in C^X$.

Let $I = \{1, 2, 3\}$ be the index set with charge $\underline{m} = \{1, 1, 1\}$.

Let us consider the root $\alpha = k_1\alpha_1 + k_2\alpha_2 + k_3\alpha_3 \in Q$. Then we have $\theta(\alpha, \alpha) = (-1)^{k_1^2}$. Hence α is an even root(resp. odd root) if k_1 is even integer(resp. odd integer).

Also $T = \{\alpha_1\}$ and the subset $F \subset T$ is either empty or $\{\alpha_1\}$. Take $J \subset \Pi^{re}$ as $J = \{2\}$. By Lemma 15, this implies that $W(J) = \{1, r_3\}$. From Definition 247, the homological space can be written as

$$\begin{aligned}
H_1^{(J)} &= V_J(1(\rho - \alpha_1) - \rho) \oplus V_J(r_3(\rho -) - \rho) \\
&= V_J(-\alpha_1) \oplus V_J(-\alpha_3) \\
H_2^{(J)} &= V_J(r_3(\rho - \alpha_1) - \rho) = V_J(-\alpha_1 - (b+1)\alpha_3) \\
H_3^{(J)} &= 0 \, \textit{for all } k \geq 3 \\
\textit{and hence } H^{(J)} &= H_1^{(J)} \ominus H_2^{(J)} \\
&= V_J(-\alpha_1) \oplus V_J(-\alpha_3) \ominus V_J(-\alpha_1 - (b+1)\alpha_3)
\end{aligned}$$

with

$$\begin{aligned}
&DimH^{(J)}_{(1,0,0)} = -1 \neq 0; \;\; DimH^{(J)}_{(0,1,0)} = -1 \neq 0 \\
&DimH^{(J)}_{(0,0,1)} = 1 \neq 0; \;\; DimH^{(J)}_{(1,1,b+1)} = -1 \neq 0
\end{aligned}$$

So we have

$$P(H^{(J)}) = \{(1,0,0), (0,1,0), (0,0,1), (1,1,b+1)\}.$$

Let $\tau = \alpha = -p\alpha_1 - q\alpha_2 - t\alpha_3 \in Q_-$, with $(p, q, t) \in \mathbb{Z}_{\geq 0} \times \mathbb{Z}_{\geq 0} \times \mathbb{Z}_{\geq 0}$. Then by Remark 161 we get

$$\begin{aligned}
T^{(J)}(\tau) = \{(s_1, s_2, s_3, s_4) | s_1(1,0,0) + s_2(0,1,0) + s_3(0,0,1) \\
+ s_4(1,1,b+1) = (p,q,t)\}.
\end{aligned}$$

This implies

$$\begin{aligned}
s_1 + s_4 &= p \\
s_2 + s_4 &= q \\
s_3 + (b+1)s_4 &= t.
\end{aligned}$$

So, we have

$$s_1 = p - s_4; \; s_2 = q - s_4; \; s_3 = t - (b+1)s_4; \; s_4 = 0 \, \textit{to} \, min(p, q, [\frac{t}{b+1}])$$

Applying s_1, s_2, s_3, s_4 in Witt partition formula, we get

$$W^{(J)}(\tau) = \sum_{s_4=0}^{min(p,q,[\frac{t}{b+1}])} \frac{(p+q+t-(b+2)s_4 - 1)!(-1)^{p+q-s_4}}{(p-s_4)!(q-s_4)!(t-(b+1)s_4)!s_4!} \tag{5.31}$$

The superdimension of $\mathfrak{g}_\alpha$ is

$$Dim\mathfrak{g}_\alpha = \sum_{d|\alpha} \frac{1}{d}\mu(d) W^{(J)}\left(\frac{\alpha}{d}\right)$$
$$= \sum_{d|\alpha} \frac{1}{d}\mu(d) \sum_{n\in T^{(J)}(\frac{\alpha}{d})} \frac{(|n|-1)!}{n!} \prod D(i)^{n_i}.$$

Substituting the value of $W^{(J)}(\tau)$ from Equation (5.31) in the above super dimension formula, we get

$$Dim\mathfrak{g}_\alpha = \sum_{d|\alpha} \frac{1}{d}\mu(d) \sum_{s_4=0}^{min(\frac{p}{d},\frac{q}{d},[\frac{t}{d(b+1)}])} \frac{(\frac{p}{d}+\frac{q}{d}+\frac{r}{d}-(b+2)s_4-1)!(-1)^{\frac{p}{d}+\frac{q}{d}-s_4}}{(\frac{p}{d}-s_4)!(\frac{q}{d}-s_4)!(\frac{t}{d}-(b+1)s_4)!s_4!}.$$

On the other hand if we consider, $\{J\} = \{2,3\}$. Then $W(J) = \{1\}$ and $\mathfrak{g}_0^{(J)} =< e_1, f_1, h >\cong sl(2,\mathbb{C})$ and $T = \{\alpha_1\}$. $F \subset T$ is either empty or $\{\alpha_1\}$. Using Definition 247, we have $H^{(J)} = V_J(-\alpha_1)$. We consider $H^{(J)}$ as a four-dimensional irreducible representation of the Lie algebra $sl(2,\mathbb{C})$ and hence we obtain

$$P(H^{(J)}) = \{(1,0,1),(0,1,1),(1,1,1),(1,1,2)\}.$$

with

$$DimH^{(J)}_{(1,0,1)} = -1;\ DimH^{(J)}_{(0,1,1)} = -1$$
$$DimH^{(J)}_{(1,1,1)} = -1;\ DimH^{(J)}_{(1,1,2)} = -1$$

Let $\tau = -p\alpha_1 - q\alpha_2 - t\alpha_3 \in Q_-$, with $(p,q,t) \in \mathbb{Z}_{\geq 0}\times\mathbb{Z}_{\geq 0}\times\mathbb{Z}_{\geq 0}$. Then by Remark 161 we get

$$T^{(J)}(\tau) = \{(s_1,s_2,s_3,s_4)|(p-|\phi_1|, q-|\phi_2|, \phi_1, \phi_2), \textit{such that } p-|\phi_1| = 1,2,\ldots;$$
$$q-|\phi_2| = 0,1,2,\ldots,;\phi_1 \textit{ is partition of } q \textit{ with parts } \leq 2; \phi_2 \textit{ is partition of}$$
$$\textit{of } t \textit{ with parts } \leq 2\}.$$

Applying s_1, s_2, s_3, s_4 in Witt partition formula, we get

$$W^{(J)}(\tau) = \sum_{\phi\in T^{(J)}(\alpha)} \frac{(p-1)!(-1)^{p+q}}{(p-|\phi_1|)!(q-|\phi_2|)!\phi_1!\phi_2!} \tag{5.32}$$

Moreover the dimension of $\mathfrak{g}_\alpha$ is

$$
\begin{aligned}
Dim\mathfrak{g}_\alpha &= \sum_{d|\alpha} \frac{1}{d}\mu(d) W^{(J)}\left(\frac{\alpha}{d}\right) \\
&= \sum_{d|\alpha} \frac{1}{d}\mu(d) \sum_{n\in T^{(J)}\left(\frac{\alpha}{d}\right)} \frac{(|n|-1)!}{n!} \prod D(i)^{n_i}.
\end{aligned}
$$

Substituting the value of $W^{(J)}(\tau)$ in Equation (5.32), the above dimension formula becomes

$$
Dim\mathfrak{g}_\alpha = \sum_{d|\alpha} \frac{1}{d}\mu(d) \sum_{\phi\in T^{(J)}(\alpha)} \frac{(p-1)!(-1)^{p+q}}{(p-|\phi_1|)!(q-|\phi_2|)!\phi_1!\phi_2!}.
$$

Now let us consider Borcherds Cartan matrix $A = \begin{pmatrix} -k & -a & -1 \\ -a & 2 & -1 \\ -1 & -1 & 2 \end{pmatrix}$ (That is b=1 in the above Borcherds Cartan matrix). For this Borcherds Cartan matrix, we consider the root $\tau = \alpha = (4,2,3) \in Q_-$. Substituting $\tau = \alpha = (4,2,3) \in Q_-$ and $b = 1$ in Equation 5.31, we have

$$
\begin{aligned}
W^{(J)}(\tau) &= \sum_{s_4=0}^{min(p,q,[\frac{t}{b+1}])} \frac{(p+q+t-(b+2)s_4-1)!(-1)^{p+q-s_4}}{(p-s_4)!(q-s_4)!(t-(b+1)s_4)!s_4!} \\
&= \sum_{s_4=0}^{min(4,2,[\frac{3}{2}])} \frac{(4+2+3-3s_4-1)!(-1)^{6-s_4}}{(4-s_4)!(2-s_4)!(3-2s_4)!s_4!} \\
&= \frac{8!(-1)^6}{4!2!3!} + \frac{5!(-1)^5}{3!1!1!1!} = 120.
\end{aligned}
$$

Substituting the root (4,2,3) and $b = 1$ in Equation (5.32), we have

$$
\begin{aligned}
W^{(J)}(\tau) &= \sum_{\phi\in T^{(J)}(\tau)} \frac{(p+q-1)!(-1)^{p+q}}{(p-|\phi_1|)!(q-|\phi_2|)!\phi_1!\phi_2!} \\
&= \frac{5!}{2!1!1!1!} + \frac{5!}{2!2!} + \frac{5!}{3!} + \frac{5!}{3!2!} \\
&= 120.
\end{aligned}
$$

Hence the combinatorial identity

$$
\sum_{s_4=0}^{min(p,q,[\frac{t}{b+1}])} \frac{(p+q+t-(b+2)s_4-1)!(-1)^{p+q-s_4}}{(p-s_4)!(q-s_4)!(t-(b+1)s_4)!s_4!} =
$$

$$\sum_{\phi \in T^{(J)}(\tau)} \frac{(p+q-1)!(-1)^{p+q}}{(p-|\phi_1|)!(q-|\phi_2|)!\phi_1!\phi_2!}$$

holds for the particular root $\tau = \alpha = -4\alpha_1 - 2\alpha_2 - 3\alpha_3 \in Q_-$ for the Borcherds Cartan matrix $A = \begin{pmatrix} -k & -a & -1 \\ -a & 2 & -1 \\ -1 & -1 & 2 \end{pmatrix}$. Hence we have proved the following theorem.

Theorem 70. *For the extended-hyperbolic Borcherds superalgebra $\mathfrak{g} = \mathfrak{g}(A, \underline{m}, C)$ associated with the extended-hyperbolic Borcherds-Cartan supermatrix $A = \begin{pmatrix} -k & -a & -b \\ -a & 2 & -1 \\ -b & -1 & 2 \end{pmatrix}$ with charge $\underline{m} = \{1, 1, 1\}$, consider the root $\tau = \alpha = -p\alpha_1 - q\alpha_2 - t\alpha_3 \in Q_-$. Then the dimension of $\mathfrak{g}_\alpha$ can be written as*

$$Dim\mathfrak{g}_\alpha = \sum_{d|\alpha} \frac{1}{d}\mu(d) \sum_{s_4=0}^{min(\frac{p}{d},\frac{q}{d},[\frac{t}{d(b+1)}])} \frac{(\frac{p}{d}+\frac{q}{d}+\frac{t}{d}-(b+2)s_4-1)!(-1)^{\frac{p}{d}+\frac{q}{d}-s_4}}{(\frac{p}{d}-s_4)!(\frac{q}{d}-s_4)!(\frac{t}{d}-(b+1)s_4)!s_4!}$$

or

$$Dim\mathfrak{g}_\alpha = \sum_{d|\alpha} \frac{1}{d}\mu(d) \sum_{\phi \in T^{(J)}(\tau)} \frac{(p-1)!(-1)^{p+q}}{(p-|\phi_1|)!(q-|\phi_2|)!\phi_1!\phi_2!}.$$

Moreover, the following combinatorial identity holds for the particular root $\tau = \alpha = -4\alpha_1 - 2\alpha_2 - 3\alpha_3$ in the Borcherds superalgebra associated with Borcherds-Cartan supermatrix $A = \begin{pmatrix} -k & -a & -1 \\ -a & 2 & -1 \\ -1 & -1 & 2 \end{pmatrix}$:

$$\sum_{s_4=0}^{min(p,q,[\frac{t}{b+1}])} \frac{(p+q+t-(b+2)s_4-1)!(-1)^{p+q-s_4}}{(p-s_4)!(q-s_4)!(t-(b+1)s_4)!s_4!} = \sum_{\phi \in T^{(J)}(\tau)} \frac{(p+q-1)!(-1)^{p+q}}{(p-|\phi_1|)!(q-|\phi_2|)!\phi_1!\phi_2!} \tag{5.33}$$

where ϕ_1 is partition of "q" with parts ≤ 2; ϕ_2 is partition of "t" with parts ≤ 2.

Remark 169. If a=1, b=0, the above example is same as the example $A = \begin{pmatrix} -k & -a & -b \\ -a & 2 & -1 \\ -b & -1 & 2 \end{pmatrix}$ given in [149]. But in this present case, we consider two different cases of $J \subset \Pi^{re}$ and obtained the corresponding Dimension formulas. Although, the equality of above two values of $Dim\mathfrak{g}_\alpha$ is not proved here in general, the same is

proved for a particular root $\alpha = (4,2,3)$ and b=1 in the Borcherds-Cartan supermatrix
$A = \begin{pmatrix} -k & -a & -b \\ -a & 2 & -1 \\ -b & -1 & 2 \end{pmatrix}$.

Below, we find the dimension formulae and combinatorial identities for the Borcherds superalgebras which are extensions of A_2, A_3 (for a same set $J \in \Pi^{re}$) by solving $T^{(J)}(\tau)$ in two different methods.

5.9.2 Superdimension formula and the corresponding combinatorial identity for the extended-hyperbolic Borcherds superalgebra which is an extension of A_2

Consider the extended-hyperbolic Borcherds superalgebra $\mathfrak{g} = \mathfrak{g}(A, \underline{m}, C)$ associated with the extended-hyperbolic Borcherds-Cartan supermatrix $A = \begin{pmatrix} -k & -a & -b \\ -a & 2 & -1 \\ -b & -1 & 2 \end{pmatrix}$ and the corresponding coloring matrix be C= $\begin{pmatrix} -1 & c_1 & c_2 \\ c_1^{-1} & 1 & c_3 \\ c_2^{-1} & c_3^{-1} & 1 \end{pmatrix}$ with $c_1, c_2, c_3 \in C^X$.
Let $I = \{1,2,3\}$ be the index set for the simple roots of $\mathfrak{g}$. Here α_1 is the imaginary odd simple root with multiplicity $r \geq 1$ and α_2, α_3 are the real even simple roots. Let us consider the root $\alpha = k_1\alpha_1 + k_2\alpha_2 + k_3\alpha_3 \in Q$. We have $\theta(\alpha, \alpha) = (-1)^{k_1^2}$. Hence α is an even root(resp. odd root) if k_1 is even integer(resp. odd integer). Also $T = \{\alpha_1\}$ (counted r times) and the subset $F \subset T$ is either empty or $\{\alpha_1\}$. Take $J \subset \Pi^{re}$ as $J = \{2,3\}$. Then

$$\mathfrak{g}_0 =< e_2, f_2, h_2, e_3, f_3, h_3 > \cong sl(3, \mathbb{C}).$$

By Lemma 15, this implies that $W(J) = \{1\}$. From Definition 247, the homology space can be written as

$$\begin{aligned} H_1^{(J)} &= V_J(1(\rho - \alpha_1) - \rho) \\ &= V_J(-\alpha_1) \\ H_k^{(J)} &= 0 \ \forall k \geq 2 \\ \text{Therefore } H^{(J)} &= H_1^{(J)} \\ &= V_J(-\alpha_1) \oplus V_J(-\alpha_1) \oplus \cdots V_J(-\alpha_1) \ (\textit{counted r times}) \end{aligned}$$

with

$$\begin{aligned} DimH^{(J)}_{(1,0,0)} &= -r; \ DimH^{(J)}_{(1,1,0)} = -r \\ DimH^{(J)}_{(1,1,1)} &= -r; \ DimH^{(J)}_{(1,1,a+1)} = -r \end{aligned}$$

We take

$$P(H^{(J)}) = \{(1,0,0),(1,1,0),(1,1,1),(1,1,a+1)\}.$$

Let $\tau = \alpha = -p\alpha_1 - q\alpha_2 - t\alpha_3 \in Q_-$, with $(p,q,t) \in \mathbb{Z}_{\geq 0} \times \mathbb{Z}_{\geq 0} \times \mathbb{Z}_{\geq 0}$. Then by (2) under Remark 161, we get

$$\begin{aligned} T^{(J)}(\tau) = \{(s_1, s_2, s_3, s_4) | s_1(1,0,0) + s_2(1,1,0) + s_3(1,1,1) \\ + s_4(1,1,a+1) = (p,q,t)\}. \end{aligned}$$

This implies

$$s_1 + s_2 + s_3 + s_4 = p; \quad s_2 + s_3 + s_4 = q; \quad s_3 + (a+1)s_4 = t$$

Solving, we get

$$s_1 = p - q; \quad s_2 = q - t + as_4; \quad s_3 = t - (a+1)s_4; \; \textit{and}\; s_4 = 0 \; \textit{to min} \left(p, q, \left[\frac{t}{a+1}\right]\right)$$

Applying s_1, s_2, s_3, s_4 in Witt partition formula, we have

$$W^{(J)}(\tau) = \sum_{\substack{s_4=0 \\ (q-t+as_4)\geq 0}}^{min(p,q,[\frac{t}{a+1}])} \frac{(p-1)!(-r)^p}{(p-q)!(q-t+as_4)!(t-(a+1)s_4)!s_4!} \tag{5.34}$$

(The terms $(p-q) \geq 0$ and $(t-(a+1)s_4) \geq 0$ always.)
The superdimension of $\mathfrak{g}_\alpha$ is

$$\begin{aligned} Dim\mathfrak{g}_\alpha &= \sum_{d|\alpha} \frac{1}{d} \mu(d) W^{(J)}\left(\frac{\alpha}{d}\right) \\ &= \sum_{d|\alpha} \frac{1}{d} \mu(d) \sum_{n \in T^{(J)}(\frac{\alpha}{d})} \frac{(|n|-1)!}{n!} \prod D(i)^{n_i}, \end{aligned}$$

Substituting the value of $W^{(J)}(\tau)$ from Equation (5.34) in the above dimension formula, we have

$$Dim\mathfrak{g}_\alpha = \sum_{d|\alpha} \frac{\mu(d)}{d} \sum_{\substack{s_4=0 \\ (q-t+as_4/d)\geq 0}}^{min(\frac{p}{d},\frac{q}{d},[\frac{t}{d(a+1)}])} \frac{(\frac{p}{d}-1)!(-r)^{\frac{p}{d}}}{(\frac{p}{d}-\frac{q}{d})!(\frac{q}{d}-\frac{t}{d}+as_4)!(\frac{t}{d}-(a+1)s_4)!s_4!}$$

(The terms $(p-q) \geq 0$ and $(t-(a+1)s_4) \geq 0$ always.)

If we use the $P(H^{(J)})$ and substituting this partition in $T^{(J)}(\tau)$, we have

$$T^{(J)}(\tau) = \{(p-q), q-|\phi|, \phi\}$$

where ϕ is partition of "t" with parts $(1, a+1)$ of length "t." Applying $T^{(J)}(\tau)$ in Witt partition formula, we have

$$W^{(J)}(\tau) = \sum_{\phi \in T^{(J)}(\tau)} \frac{(p-1)!(-r)^p}{(p-q)!(q-|\phi|)!\phi!} \tag{5.35}$$

The superdimension of $\mathfrak{g}_\alpha$ is

$$\begin{aligned} Dim\mathfrak{g}_\alpha &= \sum_{d|\alpha} \frac{1}{d}\mu(d) W^{(J)}\left(\frac{\alpha}{d}\right) \\ &= \sum_{d|\alpha} \frac{1}{d}\mu(d) \sum_{n \in T^{(J)}\left(\frac{\alpha}{d}\right)} \frac{(|n|-1)!}{n!} \prod D(i)^{n_i}, \end{aligned}$$

Substituting the value of $W^{(J)}(\tau)$ from Equation (5.35) in the above dimension formula, we have

$$Dim\mathfrak{g}_\alpha = \sum_{d|\alpha} \frac{1}{d}\mu(d) \sum_{\phi \in T^{(J)}(\tau)} \frac{(p-1)!(-r)^p}{(p-q)!(q-|\phi|)!\phi!}$$

Consider Equation (5.35)

$$W^{(J)}(\tau) = \sum_{\phi \in T^{(J)}(\tau)} \frac{(p-1)!(-r)^p}{(p-q)!(q-|\phi|)!\phi!}$$

where ϕ is partitions of "t" in $\alpha = \tau = -p\alpha_1 - q\alpha_2 - t\alpha_3$ with parts (1, a+1) of length $\leq t$. That is, ϕ can be written as $(n_1, n_{(a+1)})$, where n_1 is number of 1s in the partition of "t" and $n_{(a+1)}$ is number of $(a+1)$s in the same partition of "t.' Then $\phi = \{(t,0), (t-(a+1), 1), \ldots, \textit{upto the term satisfying } t - n_1(a+1) > 0\}$.
Hence

$$\begin{aligned} W^{(J)}(\tau) = & \sum_{(n_1, n_{(a+1)}) \in T^{(J)}(\tau)} \frac{(p-1)!(-r)^p}{(p-q)!(q-n_1-n_{(a+1)})!n_1!n_{(a+1)}!} \\ = & \frac{(p-1)!(-r)^p}{(p-q)!(q-t-0)!t!0!} + \frac{(p-1)!(-r)^p}{(p-q)!(q-t+a+1-1)!(t-(a+1)1)!1!} \\ & + \frac{(p-1)!(-r)^p}{(p-q)!(q-t+2a+2-2)!(t-2(a+1))!2!} \end{aligned}$$

$$+\cdots \text{ upto the term satisfying } t-n_1(a+1)>0$$

$$= \frac{(p-1)!(-r)^p}{(p-q)!(q-t-0)!t!0!} + \frac{(p-1)!(-r)^p}{(p-q)!(q-t+a)!(t-(a+1))!1!}$$
$$+ \frac{(p-1)!(-r)^p}{(p-q)!(q-t+2a)!(t-2(a+1))!2!}$$
$$+\cdots \text{ upto the term satisfying } t-n_1(a+1)>0$$

Here we are considering the roots of type (p,q,t) such that $p \geq q \leq t$ and also $p \geq t$. Moreover, the sum ranges upto the term satisfying $t-n_1(a+1)>0$. So we can write the above summation as

$$W^{(J)}(\tau) = \sum_{\substack{s_4=0 \\ (q-t+as_4)\geq 0}}^{min(p,q,[\frac{t}{a+1}])} \frac{(p-1)!(-r)^p}{(p-q)!(q-t+as_4)!(t-(a+1)s_4)!s_4!}$$

which is same as Equation (5.34). (The terms $(p-q) \geq 0$ and $(t-(a+1)s_4) \geq 0$ always.) Hence, we get the following theorem.

Theorem 71. *For the extended-hyperbolic Borcherds superalgebra* $\mathfrak{g}=\mathfrak{g}(A,\underline{m},C)$ *associated with the extended-hyperbolic Borcherds-Cartan supermatrix* $A = \begin{pmatrix} -k & -a & -b \\ -a & 2 & -1 \\ -b & -1 & 2 \end{pmatrix}$ *with charge* $\underline{m}=\{r,1,1\}$, *consider the root* $\alpha = -p\alpha_1 - q\alpha_2 - t\alpha_3 \in Q_-$. *Then the superdimension of* $\mathfrak{g}_\alpha$ *is*

$$Dim\mathfrak{g}_\alpha = \sum_{\substack{s_4=0 \\ (q-t+as_4)\geq 0}}^{min(p,q,[\frac{t}{a+1}])} \frac{(p-1)!(-r)^p}{(p-q)!(q-t+as_4)!(t-(a+1)s_4)!s_4!}.$$

Moreover the following combinatorial identity holds:

$$\sum_{\substack{s_4=0 \\ (q-t+as_4)\geq 0}}^{min(p,q,[\frac{t}{a+1}])} \frac{(p-1)!(-r)^p}{(p-q)!(q-t+as_4)!(t-(a+1)s_4)!s_4!} = \sum_{\phi \in T^{(J)}(\tau)} \frac{(p-1)!(-r)^p}{(p-q)!(q-|\phi|)!\phi!} \tag{5.36}$$

Remark 170. In the above formula, $(p-q) \geq 0$ and $(t-(a+1)s_4) \geq 0$ always. The term in the LHS containing $(q-t+as_4)!$ in the denominator, if $(q-t+as_4)<0$, should be deleted.

Example 77. For the Borcherds-Cartan supermatrix $A = \begin{pmatrix} -k & -1 & -b \\ -1 & 2 & -1 \\ -b & -1 & 2 \end{pmatrix}$, consider a root $\alpha=\tau=(5,3,4)$ with r=2. Substituting $\alpha=\tau=(5,3,4), r=2, a=1$ in Equation (5.34), we have

$$W^{(J)}(\tau) = \sum_{\substack{s_4=0 \\ (q-t+as_4)\geq 0}}^{min(p,q,[\frac{t}{a+1}])} \frac{(p-1)!(-r)^p}{(p-q)!(q-t+as_4)!(t-(a+1)s_4)!s_4!}$$

$$= \sum_{s_4=1}^{min(5,3,[\frac{4}{2}])} \frac{(5-1)!(-2)^5}{(5-3)!(3-4+s_4)!(4-2s_4)!s_4!}$$

$$= \sum_{s_4=1}^{2} \frac{4!(-32)}{2!(-1+s_4)!(4-2s_4)!s_4!}$$

$$= \frac{4!(-32)}{2!0!2!} + \frac{4!(-32)}{2!1!0!2!}$$

$$= -192 - 192 = -384$$

Substituting $\alpha = \tau = (5,3,4), r = 2, a = 1$ in Equation (5.35), we have

$$W^{(J)}(\tau) = \sum_{\phi \in T^{(J)}(\alpha)} \frac{(p-1)!(-r)^p}{(p-q)!(q-|\phi|)!\phi!}$$

$$= \frac{4!(-32)}{2!2!0!1!} + \frac{4(-32)}{2!2!0!1!} = -192 - 192 = -384$$

Hence the equality (5.36) holds.

Example 78. For the Borcherds-Cartan supermatrix $A = \begin{pmatrix} -k & -2 & -b \\ -2 & 2 & -1 \\ -b & -1 & 2 \end{pmatrix}$, consider a root $\alpha = \tau = (4,2,3) \in Q_-$ with $r = 1$. Substituting $\alpha = \tau = (4,2,3), r = 1, a = 2$ in Equation (5.34), we have

$$W^{(J)}(\tau) = \sum_{\substack{s_4=0 \\ (q-t+as_4)\geq 0}}^{min(p,q,[\frac{r}{a+1}])} \frac{(p-1)!(-r)^p}{(p-q)!(q-t+as_4)!(t-(a+1)s_4)!s_4!}$$

$$= \sum_{s_4=1}^{min(4,2,[\frac{3}{3}])} \frac{(4-1)!(-1)^4}{(4-2)!(2-3+2s_4)!(3-3s_4)!s_4!}$$

$$= \frac{3!}{2!1!0!1!} = 3$$

Substituting $\alpha = \tau = (4,2,3), r = 1, a = 2$ in Equation (5.35), we have

$$W^{(J)}(\tau) = \sum_{\phi \in T^{(J)}(\tau)} \frac{(p-1)!(-r)^p}{(p-q)!(q-|\phi|)!\phi!}$$
$$= \frac{(4-1)!(-1)^4}{2!(3-2)!1!1!} = 3$$

Hence the equality (5.36) holds.

5.9.3 Superdimension formula and combinatorial identity for the Borcherds superalgebra which is an extension of A_3

Here we are finding the superdimension formula and combinatorial identity for the Borcherds superalgebra which is an extension of A_3 using the same $J \subset \Pi^{re}$ and solving $T^{(J)}(\tau)$ in two different ways.

Consider the extended-hyperbolic Borcherds superalgebra $\mathfrak{g} = \mathfrak{g}(A, \underline{m}, C)$ associated with the extended-hyperbolic Borcherds-Cartan supermatrix $A = \begin{pmatrix} -k & -a & -b & -c \\ -a & 2 & -1 & 0 \\ -b & -1 & 2 & -1 \\ -c & 0 & -1 & 2 \end{pmatrix}$ and the corresponding coloring matrix be C= $\begin{pmatrix} -1 & c_1 & c_2 & c_3 \\ c_1^{-1} & 1 & c_4 & c_5 \\ c_2^{-1} & c_4^{-1} & 1 & c_6 \\ c_3^{-1} & c_5^{-1} & c_6^{-1} & 1 \end{pmatrix}$ with $c_1, c_2, c_3, c_4 \in C^X$.
Let $I = \{1, 2, 3, 4\}$ be the index set with charge $\underline{m} = \{r, 1, 1, 1\}$.

Let us consider the root $\alpha = k_1\alpha_1 + k_2\alpha_2 + k_3\alpha_3 + k_4\alpha_4 \in Q$. We have $\theta(\alpha, \alpha) = (-1)^{k_1^2}$. Hence α is an even root(resp. odd root) if k_1 is even integer(resp. odd integer). Also $T = \{\alpha_1\}$ (counted "r" times) and the subset $F \subset T$ is either empty or $\{\alpha_1\}$. Take $J \subset \Pi^{re}$ as $J = \{2, 3, 4\}$. By Lemma 15, this implies that $W(J) = \{1\}$. Then

$$\mathfrak{g}_0 =< e_2, f_2, h_2, e_3, f_3, h_3, e_4, f_4, h_4 >\cong sl(4, \mathbb{C}).$$

From Definition 247, the homological space can be written as

$$H_1^{(J)} = V_J(1(\rho - \alpha_1) - \rho)$$
$$= V_J(-\alpha_1)$$
$$H_k^{(J)} = 0 \ \forall k \geq 2$$

$$\textit{Therefore } H^{(J)} = H_1^{(J)} = V_J(-\alpha_1) \oplus V_J(-\alpha_1) \oplus \cdots \oplus V_J(-\alpha_1)(\textit{counted 'r' times})$$

with

$$DimH^{(J)}_{(1,0,0,0)} = -r;\ DimH^{(J)}_{(1,1,0,0)} = -r;\ DimH^{(J)}_{(1,1,1,0)} = -r;$$
$$DimH^{(J)}_{(1,1,1,1)} = -r;\ DimH^{(J)}_{(1,1,1,a+1)} = -r$$

Hence we have

$$P(H^{(J)}) = \{(1,0,0,0),(1,1,0,0),(1,1,1,0),(1,1,1,1),(1,1,1,a+1)\}.$$

Let $\alpha = \tau = -p\alpha_1 - q\alpha_2 - u\alpha_3 - v\alpha_4 \in Q_-$, with $(p,q,u,v) \in \mathbb{Z}_{\geq 0} \times \mathbb{Z}_{\geq 0} \times \mathbb{Z}_{\geq 0} \times \mathbb{Z}_{\geq 0}$. Then by (2) under Remark 161, we get

$$\begin{aligned} T^{(J)}(\tau) = \{(s_1,s_2,s_3,s_4,s_5) | s_1(1,0,0,0) + s_2(1,1,0,0) + s_3(1,1,1,0) \\ + s_4(1,1,1,1) + s_5(1,1,1,a+1) = (p,q,u,v)\}. \end{aligned}$$

This implies

$$\begin{aligned} s_1 + s_2 + s_3 + s_4 + s_5 = p;\ s_2 + s_3 + s_4 + s_5 = q \\ s_3 + s_4 + s_5 = u;\ s_4 + (a+1)s_5 = v. \end{aligned}$$

We have

$$\begin{aligned} &s_1 = p - q;\ s_2 = q - u \\ &s_3 = u - v + s_5;\ s_4 = v - (a+1)s_5;\ s_5 = 0 \ to\ min(p,q,u,[\frac{v}{a+1}]). \end{aligned}$$

Applying s_1, s_2, s_3, s_4, s_5 in Witt partition formula, we have

$$W^{(J)}(\tau) = \sum_{\substack{s_5=0 \\ (u-v+s_5)\geq 0}}^{min(p,q,u,[\frac{v}{a+1}])} \frac{(p-1)!(-r)^p}{(p-q)!(q-r)!(u-v+s_5)!(v-(a+1)s_5)!s_5!}. \quad (5.37)$$

(The terms $(p-q) \geq 0, (q-r) \geq 0$ and $(v-(a+1)s_5) \geq 0$ always.)
The dimension $\mathfrak{g}_\alpha$ is

$$\begin{aligned} Dim\mathfrak{g}_\alpha &= \sum_{d|\alpha} \frac{1}{d}\mu(d) W^{(J)}\left(\frac{\alpha}{d}\right) \\ &= \sum_{d|\alpha} \frac{1}{d}\mu(d) \sum_{n \in T^{(J)}(\frac{\alpha}{d})} \frac{(|n|-1)!}{n!} \prod D(i)^{n_i}, \end{aligned}$$

Substituting the value of $W^{(J)}(\tau)$ from Equation (5.37) in the above dimension formula, we have

$$Dim\mathfrak{g}_\alpha = \sum_{\substack{s_5=0\\(u-v+s_5)\geq 0}}^{min(p,q,u,[\frac{v}{a+1}])} \frac{(p-1)!(-r)^p}{(p-q)!(q-r)!(u-v+s_5)!(v-(a+1)s_5)!s_5!}$$

(The terms $(p-q)\geq 0, (q-r)\geq 0$ and $(v-(a+1)s_5)\geq 0$ always.)

If we use the same $P(H^{(J)})$ and substituting in the partition, we have

$$T^{(J)}(\tau) = \{(p-q), q-u, u-|\phi|, \phi\}$$

where ϕ is partition of v with parts $\leq (1, a+1)$ and of length v. Applying $T^{(J)}(\tau)$ in Witt partition formula (Equation (2.2))

$$W^{(J)}(\tau) = \sum_{\phi\in T^{(J)}(\tau)} \frac{(p-1)!(-r)^p}{(p-q)!(q-u)!(u-|\phi|)!\phi!}. \tag{5.38}$$

The dimension $\mathfrak{g}_\alpha$ is

$$\begin{aligned} Dim\mathfrak{g}_\alpha &= \sum_{d|\alpha} \frac{1}{d}\mu(d) W^{(J)}\left(\frac{\alpha}{d}\right) \\ &= \sum_{d|\alpha} \frac{1}{d}\mu(d) \sum_{n\in T^{(J)}(\frac{\alpha}{d})} \frac{(|n|-1)!}{n!} \prod D(i)^{n_i}, \end{aligned}$$

Substituting the value of $W^{(J)}(\tau)$ from Equation (5.38) in the above dimension formula, we have

$$Dim\mathfrak{g}_\alpha = \sum_{d|\alpha} \frac{1}{d}\mu(d) W^{(J)} \sum_{\phi\in T^{(J)}(\tau)} \frac{(p-1)!(-r)^p}{(p-q)!(q-u)!(u-|\phi|)!\phi!}$$

Consider Equation (5.38),

$$W^{(J)}(\alpha) = \sum_{\phi\in T^{(J)}(\tau)} \frac{(p-1)!(-r)^p}{(p-q)!(q-u)!(u-|\phi|)!\phi!}$$

where ϕ is partitions of "v" in $\alpha = \tau = -p\alpha_1 - q\alpha_2 - u\alpha_3 - v\alpha_4$ with parts (1, a+1) of length $\leq v$. That is, ϕ can be written as $(n_1, n_{(a+1)})$, where n_1 is number of 1s in the partition of "v" and $n_{(a+1)}$ is number of $(a+1)$s in the same partition of "v." Then $\phi = \{(v,0), (v-(a+1), 1), (v-2(a+1), 2), \ldots$ *upto the term satisfying* $v - n_1(a+1) > 0\}$.

We have

$$\begin{aligned} W^{(J)}(\tau) &= \sum_{n_1, n_{(a+1)} \in T^{(J)}} (\tau) \frac{(p-1)!(-r)^p}{(p-q)!(q-u)!(u-n_1-n_{(a+1)})!n_1!n_{(a+1)}!} \\ &= \frac{(p-1)!(-r)^p}{(p-q)!(q-u)!(u-v-0)!t!0!} \\ &+ \frac{(p-1)!(-r)^p}{(p-q)!(q-u)!(u-v+a)!(v-(a+1))!1!} \\ &+ \frac{(p-1)!(-r)^p}{(p-q)!(q-u)!(u-v+2a)!(v-2(a+1))!2!} \\ &+ \cdots \textit{ upto the term satisfying } v - n_1(a+1) > 0 \end{aligned}$$

Here we are considering the roots of type (p, q, u, v) such that $p \geq q, u \leq v$ and also $p \geq v$. Moreover the sum ranges upto the term satisfying $v - n_1(a+1) > 0$. So we can write the above summation as

$$W^{(J)}(\tau) = \sum_{\substack{s_5=0 \\ (u-v+s_5)\geq 0}}^{min(p,q,u,[\frac{v}{a+1}])} \frac{(p-1)!(-r)^p}{(p-q)!(q-r)!(u-v+s_5)!(v-(a+1)s_5)!s_5!}$$

which is same as Equation (5.37). (The terms $(p-q) \geq 0, (q-r) \geq 0$ and $(v-(a+1)s_5) \geq 0$ always.) Hence we proved the following theorem.

Theorem 72. *For the extended-hyperbolic Borcherds superalgebra* $\mathfrak{g} = \mathfrak{g}(A, \underline{m}, C)$ *associated with the extended-hyperbolic Borcherds-Cartan supermatrix* $A = \begin{pmatrix} -k & -a & -b & -c \\ -a & 2 & -1 & 0 \\ -b & -1 & 2 & -1 \\ -c & 0 & -1 & 2 \end{pmatrix}$, *let us consider the root* $\alpha = \tau = -p\alpha_1 - q\alpha_2 - u\alpha_3 - v\alpha_4 \in Q_-$. *Then the dimension of* $\mathfrak{g}_\alpha$ *is*

$$Dim\mathfrak{g}_\alpha = \sum_{d|\alpha} \frac{1}{d}\mu(d)W^{(J)} \sum_{\substack{s_5=0 \\ (u-v+s_5)\geq 0}}^{min(p,q,u,[\frac{v}{a+1}])} \frac{(p-1)!(-r)^p}{(p-q)!(q-r)!(u-v+s_5)!(v-(a+1)s_5)!s_5!}.$$

Moreover the following combinatorial identity holds:

$$\begin{aligned} &\sum_{\substack{s_5=0 \\ (u-v+s_5)\geq 0}}^{min(p,q,u,[\frac{v}{a+1}])} \frac{(p-1)!(-r)^p}{(p-q)!(q-u)!(u-v+s_5)!(v-(a+1)s_5)!s_5!} \\ &\quad = \sum_{\phi \in T^{(J)}(\tau)} \frac{(p-1)!(-r)^p}{(p-q)!(q-u)!(u-|\phi|)!\phi!} \end{aligned} \tag{5.39}$$

Remark 171. In the above formula, $(p-q) \geq 0$, $(q-r) \geq 0$ and $(v-(a+1)s_5) \geq 0$ always. The term in the LHS containing $(u-v+s_5)!$ in the denominator, should be omitted if $(u-v+s_5) < 0$.

Example 79. For the Borcherds-Cartan supermatrix

$$A = \begin{pmatrix} -k & -1 & -b & -c \\ -1 & 2 & -1 & 0 \\ -b & -1 & 2 & -1 \\ -c & 0 & -1 & 2 \end{pmatrix},$$

consider the root $\alpha = \tau$=(5, 3, 2, 4) with $r = 2$ and $a = 1$. Using these values in Equation (5.37), we have

$$\begin{aligned}
&\sum_{\substack{s_5=0 \\ (u-v+s_5)\geq 0}}^{min(p,q,u,[\frac{v}{a+1}])} \frac{(p-1)!(-r)^p}{(p-q)!(q-u)!(u-v+s_5)!(v-(a+1)s_5)!s_5!} \\
&= \sum_{s_5=2}^{min(5,3,2,[\frac{4}{2}])} \frac{(5-1)!(-2)^5}{(5-3)!(3-2)!(2-4+2)!(4-2(2))!2!} \\
&= \frac{4!(-2)^5}{2!1!0!0!2!} \\
&= -192.
\end{aligned}$$

Now if we use $\alpha = (p,q,u,v) = (5,3,2,4), r = 2, a = 1$ in Equation (5.38), we have

$$\sum_{\phi \in T^{(J)}(\tau)} \frac{(p-1)!(-r)^p}{(p-q)!(q-u)!(u-|\phi|)!\phi!} = \frac{(5-1)!(-2)^5}{(5-3)!(3-2)!(2-|\phi|)!\phi!} = -192.$$

Hence the equality (5.39) holds.

Example 80. For the Borcherds-Cartan supermatrix

$$A = \begin{pmatrix} -k & -1 & -b & -c \\ -1 & 2 & -1 & 0 \\ -b & -1 & 2 & -1 \\ -c & 0 & -1 & 2 \end{pmatrix},$$

consider the root $\alpha = \tau$ =(7, 4, 3, 5) = (p,q,u,v) with r=1 and a=1. Applying in Equation (5.37), we have

$$\sum_{\substack{s_5=0\\(u-v+s_5)\geq 0}}^{min(p,q,u,[\frac{v}{a+1}])} \frac{(p-1)!(-r)^p}{(p-q)!(q-u)!(u-v+s_5)!(v-(a+1)s_5)!s_5!}$$

$$= \sum_{s_5=1}^{2} \frac{(7-1)!(-1)^7}{(7-4)!(4-3)!(3-5+2s_5)!(3-2s_5)!s_5!}$$

$$= \frac{6!(-1)}{3!0!1!1!} + \frac{6!(-1)}{3!1!2!(-1)!2!}$$

$$= -120$$

Consider the root $\tau = \alpha$ as $(7, 4, 3, 5)$ with r=1 and a=1. Applying in Equation (5.38), we have

$$\sum_{\phi \in T^{(J)}(\tau)} \frac{(p-1)!(-r)^p}{(p-q)!(q-u)!(u-|\phi|)!\phi!} = \frac{(7-1)!(-1)^7}{(7-4)!(4-3)!(3-2)!1!} = -120$$

Hence the equality (5.39) holds.

The following Theorems 73 and 74 were proved in [153].

Theorem 73. *For the extended-hyperbolic Borcherds superalgebra* $\mathfrak{g} = \mathfrak{g}(A, \underline{m}, C)$ *associated with the extended-hyperbolic Borcherds-Cartan supermatrix* $A = \begin{pmatrix} -k & -a & -b \\ -a & 2 & -1 \\ -b & -2 & 2 \end{pmatrix}$ *with charge* $\underline{m} = \{1, 1, 1\}$, *consider the root* $\alpha = -p\alpha_1 - q\alpha_2 - u\alpha_3 \in Q_-$. *Then the dimension of* $\mathfrak{g}_\alpha$ *is*

$$Dim\mathfrak{g}_\alpha = \sum_{d|\alpha} \frac{1}{d}\mu(d) \sum_{s_4=0}^{min(\frac{p}{d},\frac{q}{d},[\frac{t}{d(a+1)}])} \frac{(\frac{p}{d} - \frac{u}{d} + \frac{q}{d} + \frac{a}{d}s_4 - 1)!(-1)^{p/d-u/d+q/d+as_4/d}}{(\frac{p}{d} - \frac{q}{d})!(\frac{q}{d} - \frac{u}{d} + \frac{a}{d}s_4)!(\frac{u}{d} - \frac{(a+1)}{d}s_4)!(s_4/d)!}.$$

Moreover the following combinatorial identity holds:

$$\sum_{s_4=0}^{min(p,q,[\frac{u}{a+1}])} \frac{(p-u+q+as_4-1)!(-1)^{p-u+q+as_4}}{(p-q)!(q-u+as_4)!(u-(a+1)s_4)!s_4!} =$$

$$\sum_{\phi_1,\phi_2 \in T^{(J)}(\tau)} \frac{(p-u+\phi_1-1)!(-1)^{p-u+\phi_1}}{(p-q)!(\phi_1-u)!(u-(a+1)\phi_2)!\phi_2!}. \tag{5.40}$$

where ϕ_1 *is the partition of "u" with parts* $(1, a)$ *of length "*$u + as_4$*" and* ϕ_2 *is the partition of* s_4 *with parts upto* $min(p, q, [\frac{u}{a+1}])$.

Theorem 74. *For the extended-hyperbolic Borcherds superalgebra* $\mathfrak{g} = \mathfrak{g}(A, \underline{m}, C)$ *associated with the extended-hyperbolic Borcherds-Cartan supermatrix* $A = \begin{pmatrix} -k & -a & -b & -c \\ -a & 2 & -1 & 0 \\ -b & -1 & 2 & -1 \\ -c & 0 & -2 & 2 \end{pmatrix}$ *let us consider* $\alpha = \tau = -p\alpha_1 - q\alpha_2 - u\alpha_3 - v\alpha_4 \in Q_-$.

Then the dimension of $\mathfrak{g}_\alpha$ *is*

$$Dim\mathfrak{g}_\alpha =$$

$$\sum_{d|\alpha} \frac{1}{d}\mu(d) \sum_{\substack{s_5=0 \\ (u-v+s_5)\geq 0}}^{min(p/d,q/d,u/d,[\frac{v}{d(c+1)}])} \frac{(\frac{q}{d}-1)!(-1)^{q/d}}{(\frac{p}{d}-\frac{q}{d}+\frac{v}{d}+c\frac{s_5}{d})!(\frac{q}{d}-\frac{u}{d})!(\frac{u}{d}-\frac{v}{d}-c\frac{s_5}{d})!(\frac{q}{d}-\frac{p}{d}-\frac{s_5}{d})!\frac{s_5}{d}!}.$$

Moreover the following combinatorial identity holds:

$$\sum_{s_5=0}^{min(p,q,u,[\frac{v}{c+1}])} \frac{(q-1)!(-1)^q}{(p-q+v+cs_5)!(q-u)!(u-v-cs_5)!(q-p-s_5)!s_5!}$$

$$= \sum_{\phi_1,\phi_2 \in T^{(J)}(\tau)} \frac{(q-1)!(-1)^q}{(p-q+\phi_1)!(q-u)!(q-p-\phi_2)!(u-\phi_1)|\phi_2|!} \qquad (5.41)$$

where ϕ_1 *is partition of* v *with parts upto* $(1, c)$ *and of length* v *and* ϕ_2 *is the partition of* s_5 *with parts upto min(p,q,u,[v/c+1]) of length* s_5.

Remark 172.

(1) In Sections 5.9.2 and 5.9.3, the identities (5.34) and (5.39) hold for any root, because we have derived the identities by simply solving the $T^{(J)}(\tau)$ in two different ways.

(2) It is hoped that, in general, superdimensions of roots and the corresponding combinatorial identities for Borcherds superalgebras which are extensions of all finite dimensional Kac-Moody algebras and superdimensions for all other categories can also be found out.

Example 81. For the Borcherds-Cartan supermatrix

$$A = \begin{pmatrix} -k & -1 & -b & -c \\ -1 & 2 & -1 & 0 \\ -b & -1 & 2 & -1 \\ -c & 0 & -2 & 2 \end{pmatrix},$$

consider the root $\alpha = \tau = (7, 6, 4, 3) = (p, q, u, v)$ with c = 1. Applying in LHS of equation (5.41), we have

$$\sum_{s_5=0}^{min(p,q,u,[\frac{v}{c+1}])} \frac{(q-1)!(-1)^q}{(p-q+v+cs_5)!(q-u)!(u-v-cs_5)!(q-p-s_5)!s_5!}$$

$$= \sum_{s_5=0}^{1} \frac{(6-1)!(-1)^6}{(7-6+3+s_5)!(6-4)!(4-3-s_5)!(6-7-s_5)!s_5!}$$

$$= \frac{5!(1)}{4!2!1!0!} + \frac{5!(1)}{5!2!0!0!1!}$$

$$= \frac{5}{2} + \frac{1}{2} = 3$$

Consider the root $\tau = \alpha$ as $(7, 6, 4, 3)$ with c = 1. Applying in LHS of equation (5.41), we have

$$\sum_{\phi_1,\phi_2 \in T^{(J)}(\tau)} \frac{(q-1)!(-1)^q}{(p-q+\phi_1)!(q-u)!(q-p-\phi_2)!(u-\phi_1)|\phi_2|!}$$

$$= \frac{(6-1)!(-1)^6}{(7-6+3)!(6-4)!(4-3)!(6-7)!0!}$$

$$+ \frac{(6-1)!(-1)^6}{(7-6+3+1)!(6-4)!(4-3-1)!(6-7)!1!}$$

$$= \frac{5}{2} + \frac{1}{2} = 3$$

Hence the equality (5.41) holds.

5.10 Description of finite and infinite dimensional Lie algebras and Lie superalgebras and their different classes of root systems

Diagram 1 and Diagram 2: Finite and infinite dimensional Lie algebras and Lie superalgebras.

Diagram 3: Different classes of root systems in finite dimensional and infinite dimensional Lie algebras and Lie superalgebras.

Diagram 1:

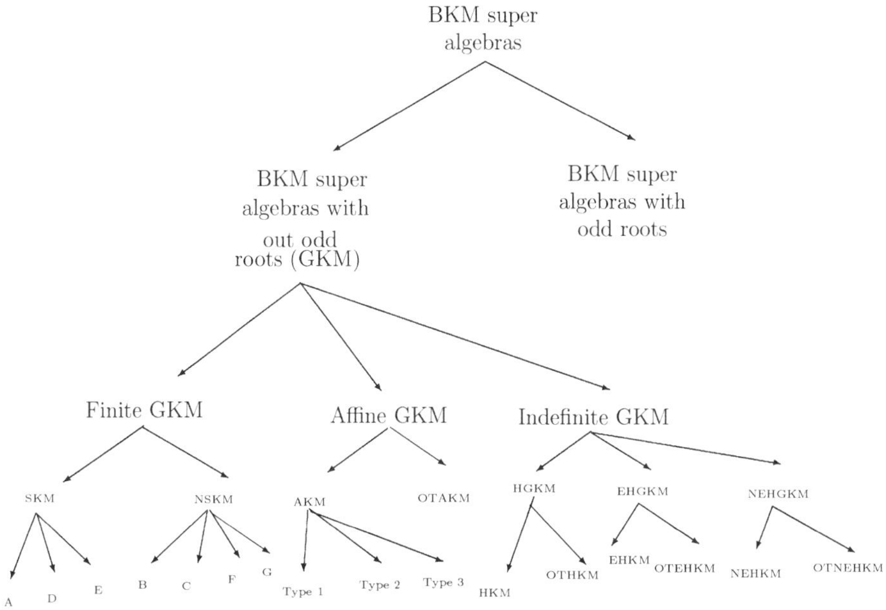

Here
SKM — Simply laced Kac-Moody algebras (simply laced finite dimensional Lie algebras)
NSKM — Non-simply laced Kac-Moody algebras (non-simply laced finite dimensional Lie algebras)
AKM — Affine Kac-Moody algebras
OTAKM — Other than Affine Kac-Moody algebras in affine GKM algebras
HGKM — Hyperbolic generalized Kac-Moody algebras
EHGKM — Extended hyperbolic generalized Kac-Moody algebras
NEHGKM — Non-extended hyperbolic generalized Kac-Moody algebras
HKM — Hyperbolic Kac-Moody algebras
OTHKM — Other than hyperbolic Kac-Moody algebras in hyperbolic GKM algebras
EHKM — Extended hyperbolic Kac-Moody algebras
OTEHKM — Other than extended hyperbolic Kac-Moody algebras in extended Hyperbolic GKM algebras
NEHKM — Non extended hyperbolic Kac-Moody algebras
OTNEHKM — Other than nonextended hyperbolic Kac-Moody algebras in nonextended Hyperbolic GKM algebras

Diagram 2:

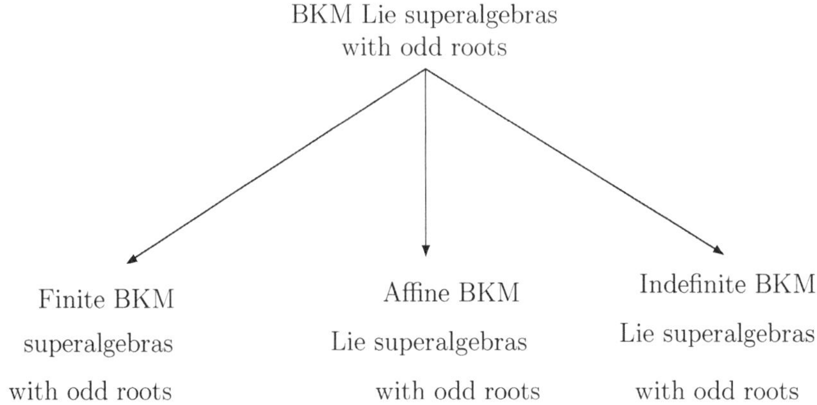

Remark 173.

(1) In the diagram 2, other branches can also be drawn.
(2) From the above diagram 1, one can easily understand that starting from a Borcherds Kac-Moody supermatrix(BKM supermatrix) or from the corresponding Borcherds Kac-Moody superalgebra(BKM superalgebra), one can arrive at the final branch on the left side, namely, ordinary Cartan matrix or corresponding finite dimensional Lie algebra by removing one by one all restrictions or conditions in the definition of the corresponding BKM supermatrix.

Diagram 3:

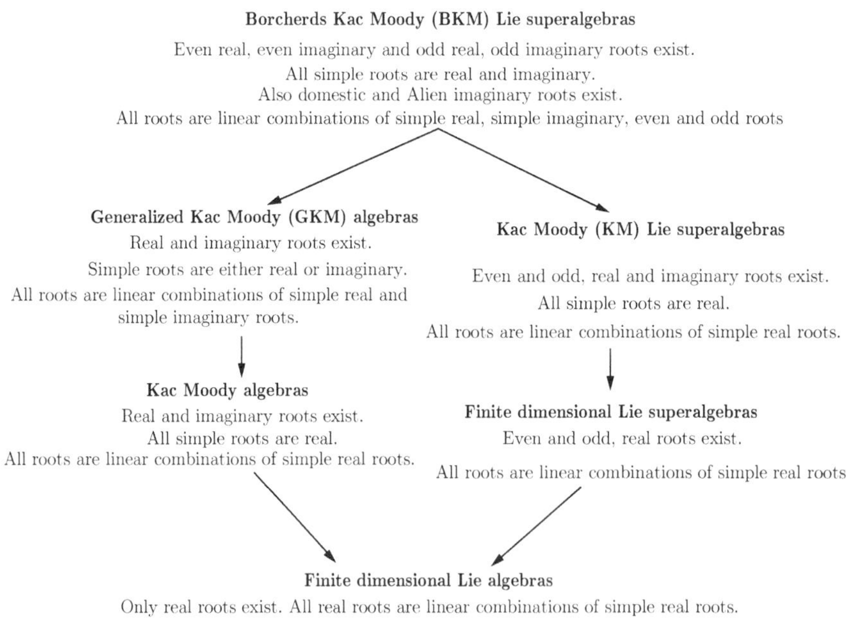

Remark: Existence of distinguished roots and distinguished simple roots in finite dimensional Lie superalgebras were explained in Chapter 4. Similarly, existence of special imaginary roots, strictly imaginary roots and purely imaginary roots in KM algebras, GKM algebras, and BKM Lie superalgebras were explained in Chapters 2 and 3 and is explained in this chapter.

Exercises

5.1 For the Borcherds Kac-Moody superalgebras associated Borcherds Kac-Moody supermatrices,

(i) $(SBGA_1, \psi) = \begin{pmatrix} -k & -a_1 \\ -b_1 & 2 \end{pmatrix} = (A, \psi)$

(ii) $(SBGA_1^{(1)}, \psi) = \begin{pmatrix} -k & -a_1 & -a_2 \\ -b_1 & 2 & -2 \\ -b_2 & -2 & 2 \end{pmatrix} = (A, \psi)$, and

(iii) $(SBGHA_1, \psi) = \begin{pmatrix} -k & -a_1 & -a_2 \\ -b_1 & 2 & -p \\ -b_2 & -q & 2 \end{pmatrix} = (A, \psi)$, where $pq > 4$, find the set of all Domestic and Alien imaginary roots in each case.

5.2 If $\mathfrak{g} = \mathfrak{g}(A, \underline{m}, C)$ is the extended-hyperbolic Borcherds superalgebra associated with Borcherds Cartan matrix

$$A = \begin{pmatrix} 0 & -1 & 0 \\ -1 & 2 & -1 \\ 0 & -1 & 2 \end{pmatrix}$$ of charge $\underline{m} = (1, 1, 1)$ and with the coloring matrix

$$C = \begin{pmatrix} -1 & c_1 & c_2 \\ c_1^{-1} & 1 & c_3 \\ c_2^{-1} & c_3^{-1} & 1 \end{pmatrix}$$ with $c_1, c_2, c_3 \in C^x$ then, for the root $\alpha = -k_1\alpha_1 - k_2\alpha_2 - k_3\alpha_3 \in Q_-$

with $k_1, k_2, k_3 \in \mathbb{Z}_{\geq 0}$, prove that the super dimension of the root space $\mathfrak{g}_\alpha$, corresponding to the root α is given by

$$\operatorname{Dim} \mathfrak{g}_\alpha = \frac{1}{k_1} \sum_{d|(k_1,k_2,k_3)} \mu(d) \begin{pmatrix} \frac{k_1-k_3}{d} \\ \frac{k_2-k_3}{2} \end{pmatrix} \begin{pmatrix} \frac{k_1}{d} \\ \frac{k_3}{2} \end{pmatrix} (-1)^{\frac{k_1}{d}}.$$

5.3 Let $\mathfrak{g} = \mathfrak{g}(A, \underline{m}, C)$ be an extended -hyperbolic Borcherds superalgebra associated with the data:

$$A = \begin{pmatrix} 0 & -1 & 0 \\ -1 & 2 & -2 \\ 0 & -2 & 2 \end{pmatrix}$$

$\underline{m} = (1, 1, 1)$ and

$$C = \begin{pmatrix} -1 & c_1 & c_2 \\ c_1^{-1} & 1 & c_3 \\ c_2^{-1} & c_3^{-1} & 1 \end{pmatrix}.$$

Prove that for a root $\alpha = l\alpha_1 + m\alpha_2 + n\alpha_3 \in Q$ with $l \geq 1$ and $m, n \geq 0$, we have

$$\text{Dim } \mathfrak{g}_\alpha = \sum_{d|\alpha} \frac{1}{d}\mu(d) \sum_{n \in T^{(J)}(\tau)} \frac{(|n|-1)!}{n!} \prod_i \left(DimH_{\tau_i}^{(J)}\right)^{n_i}.$$

5.4 Let $\mathfrak{g} = \mathfrak{g}(A, \underline{m}, C)$ be the extended -hyperbolic Borcherds superalgebra associated with the data

$$A = \begin{pmatrix} 0 & -1 & -1 \\ -1 & 2 & -3 \\ -1 & -3 & 2 \end{pmatrix}$$

$\underline{m} = (1, 1, 1)$ with the coloring matrix

$$C = \begin{pmatrix} -1 & c_1 & c_2 \\ c_1^{-1} & 1 & c_3 \\ c_2^{-1} & c_3^{-1} & 1 \end{pmatrix}$$

for $c_1, c_2, c_3 \in C^X$. Prove that, for the root $\alpha = -l\alpha_1 - m\alpha_2 - n\alpha_3 \in Q_-$ with $l, m, n \in \mathbb{Z}_{\geq 0}$, we have

$$Dim\mathfrak{g}_\alpha = \sum_{d|\alpha} \frac{1}{d}\mu(d) \sum_{n \in T^{(J)}(\frac{\alpha}{d})} \frac{(|n|-1)!}{n!} \prod \left(DimH_{\tau_i}^{(J)}\right)^{n_i}.$$

5.5 If the extended-hyperbolic Borcherds superalgebra $\mathfrak{g} = \mathfrak{g}(A, \underline{m}, C)$ associated with the extended-hyperbolic Borcherds-Cartan supermatrix is $A = \begin{pmatrix} -2 & -1 & -1 \\ -2 & 2 & -4 \\ -1 & -2 & 2 \end{pmatrix}$, then for the root $\tau = -l\alpha_1 - m\alpha_2 - n\alpha_3 \in Q^-$ such that $l, m, n \in \mathbb{Z}_{\geq 0}$, prove that $Dim\mathfrak{g}_\alpha = \theta(\lambda, \lambda)p(m - (m-n)^2)$.

5.6 Let (A, ψ) be a BKM supermatrix, where $A =$ Symmetrizable $SBGG_2$ and $\alpha = k_1\alpha_1 + k_2\alpha_2 + k_3\alpha_3 \in \Delta_+^{\text{im}}$, $k_1, k_2, k_3 \in \mathbb{N}$. Then, prove that $r_\alpha \in -W$ if and only if $k_2 = (2b_1 + b_2)k_1$ and $k_3 = (3b_1 + 2b_2)k_1 \in \mathbb{N}$. Also prove that $\alpha = k_1(\alpha_1 + (2b_1 + b_2)\alpha_2 + (3b_1 + 2b_2)\alpha_3)$ is a special imaginary root of $\mathfrak{g}(A)$, for every $k_1 \in \mathbb{N}$.

5.7 Let (A, ψ) be a BKM supermatrix, where $A =$ Symmetrizable $SBGA_1$ and $\alpha = k_1\alpha_1 + k_2\alpha_2 \in \Delta_+^{\text{im}}$, $k_1, k_2 \in \mathbb{N}$. Prove that $r_\alpha \in -W$ if and only if $2k_2 = b_1k_1 \in \mathbb{N}$. Also prove that $\alpha = \frac{k_1}{2}(2\alpha_1 + b_1\alpha_2)$ is a special imaginary root of $\mathfrak{g}(A)$, for every $k_1 \in \mathbb{N}$.

5.8 Let (A, ψ) be a BKM supermatrix, where $A =$ Symmetrizable $SBGA_2^{(2)}$ and $\alpha = k_1\alpha_1 + k_2\alpha_2 + k_3\alpha_3 \in \Delta_+^{\text{im}}$, $k_1, k_2, k_3 \in \mathbb{N}$. Prove that $r_\alpha \notin -W$.

5.9 Let (A, ψ) be a BKM supermatrix, where $A =$ Symmetrizable $SBGA_2^{(1)}$ and $\alpha = k_1\alpha_1 + k, 2\alpha_2 + k_3\alpha_3 + k_4\alpha_4 \in \Delta_+^{\text{im}}$, $k_1, k_2, k_3, k_4 \in \mathbb{N}$. Prove that $r_\alpha \notin -W$.

5.10 For the Borcherds-Cartan supermatrix $A = \begin{pmatrix} -k & -2 & -b \\ -2 & 2 & -1 \\ -b & -1 & 2 \end{pmatrix}$, prove that a root $\alpha = (7,5,6) \in Q_-$ with r = 1 satisfies the following combinatorial identity for extension of A_2,

$$\sum_{s_4=0}^{min(p,q,[\frac{t}{a+1}])} \frac{(p-1)!(-r)^p}{(p-q)!(q-t+as_4)!(t-(a+1)s_4)!s_4!} = \sum_{\phi \in T^{(J)}(\tau)} \frac{(p-1)!(-r)^p}{(p-q)!(q-|\phi|)!\phi!}.$$

5.11 For the Borcherds-Cartan super matrix $A = \begin{pmatrix} -k & -1 & -b \\ -1 & 2 & -1 \\ -b & -2 & 2 \end{pmatrix}$, consider a root $\alpha = \tau = (5,4,3)$ with $a = 1$. Prove that the following identity holds.

$$\sum_{s_4=0}^{min(p,q,[\frac{u}{a+1}])} \frac{(p-u+q+as_4-1)!(-1)^{p-u+q+as_4}}{(p-q)!(q-u+as_4)!(u-(a+1)s_4)!s_4!}$$
$$= \sum_{\phi_1,\phi_2 \in T^{(J)}(\tau)} \frac{(p-u+\phi_1-1)!(-1)^{p-u+\phi_1}}{(p-q)!(\phi_1-u)!(u-(a+1)\phi_2)!\phi_2!}.$$

Lie algebras of Lie groups, Kac-Moody groups, supergroups, and some specialized topics in finite- and infinite-dimensional Lie algebras

6

Chapter Outline

6.1 Lie groups and Lie algebras of Lie groups 376
6.2 Kac-Moody groups, supergroups, and some applications 386
6.3 Homogeneous spaces, corresponding Lie algebras, and spectra of some differential operators on homogeneous spaces 390
6.4 Spectral invariants of zeta function of the Laplace-Beltrami operator 394
6.5 Generalization of Macdonald's identities for some Kac-Moody algebras 402
6.6 Some special infinite dimensional Lie algebras 406
6.7 Hirota bilinear differential operators and soliton solutions for KdV equation 411
6.8 Principal vertex operator construction of basic representation and homogeneous vertex operator construction of the basic representation 415
6.9 Principal vertex operators and the corresponding Hirota bilinear equations for $A_2^{(1)}, A_2^{(2)}, C_2^{(1)}$, and $B_3^{(1)}$ 426
6.10 Fermionic Fock space, Clifford algebra, Bosonic Fock space, and Boson-Fermion correspondence 459
6.11 Remark on Quantum groups, String theory and Mathematical Physics with some references 460

Abstract

In this chapter, first we discuss about some fundamental concepts in Lie groups, Lie algebras of Lie groups, Kac-Moody groups, supergroups, etc. and then some fundamental applications of Lie algebras and Lie groups to differential geometry, to number theory, and finally to differential equations. As an application to differential geometry is concerned, we define some differential operators and give the spectra of these differential operators on quadric hypersurface and quaternionic projective space. As an application to number theory, we explain about zeta and eta functions and give some results on

Introduction to Finite and Infinite Dimensional Lie (Super)algebras. http://dx.doi.org/10.1016/B978-0-12-804675-3.00006-6

spectral invariants of the zeta function of the Laplace-Beltrami operator acting on 1-forms on the above spaces and 2-forms on $(4r - 1)$ dimensional sphere and eta function and spectral asymmetry of the operator $B = \pm({}^*d - d^*)$ acting on 2-forms on $(4r - 1)$ dimensional sphere. We also give some generalizations of Macdonald's identities for some Kac-Moody algebras. As an application to differential equation, we discuss about Hirota bilinear differential operators, principal and homogeneous vertex operators and their corresponding Hirota bilinear equations and super Hirota bilinear equations for some simply and non-simply laced Kac-Moody algebras. Basic properties of Fermionic Fock space, Clifford algebra, Bosonic Fock space and Boson-Fermion correspondence along with some references to Quantum groups, String theory and Mathematical Physics are also given in this chapter.

6.1 Lie groups and Lie algebras of Lie groups

Lie groups and their algebras arise most often in physics as symmetric groups of dynamical systems and their relations to conservation laws. Moreover, Lie groups and Lie algebras appear together in geometry of differential operators. For all fundamentals regarding Lie groups and Lie algebras, one can refer Lie [6], Belinfante et al. [154, 156], Belinfante and Kolman [155], Varadarajan [157], Helgason [42], and Knapp [158], Arfken [269], Chevalley [270]. For some applications of Lie algebras, one can also refer Baurle and De Kerf [159] and for Lie algebras and Lie groups, one can refer Bourbaki [160], Hausner and Schwartz [261], Kac [262]. For representations of groups, one can refer Boerner [161]. For multiplicity formula of certain semisimple Lie groups, one can refer Lepowsky [162]. For infinite dimensional groups and applications, one can refer Kac [201]. For two approaches to the study of the orbits, invariants of a linear reductive group operating on a finite dimensional vector space, one can refer [202].

Definition 249. Let H be a Hausdorff space and G be a topological group. For each $p \in H$, $g \in G$, one can associate a homeomorphism $p \mapsto g.p$ of H into itself such that

(1) $g_1g_2 \cdot p = g_1 \cdot (g_2 \cdot p)$ for $p \in H$ and $g_1, g_2 \in G$ and
(2) the mapping $(g, p) \to g \cdot p$ is a continuous mapping of the product space $G \times H$ onto H.

Then the group G is called *topological transformation group* of H. It is clear that $e \cdot p = p$ for all $p \in H$. If e is the only element which has this property, then group G is called *effective* and G is acting effectively on H.

Definition 250. A *Lie group* is a group G which is also an analytic manifold such that the mapping $(\sigma, \tau) \to \sigma\tau^{-1}$ of the product manifold $G \times G$ into G is analytic. A Lie group is also defined as a differentiable manifold G which is also a group such that, the group operations $G \times G \to G$ defined by $(g, h) \mapsto gh$ and $G \to G$ defined by $g \mapsto g^{-1}$ are differentiable.

Definition 251. A function $f(z_1, \ldots, z_n)$ of n-complex variables is called holomorphic, if at each point $(z_1^0, \ldots, z_n^0)$, there exists a power series,

$$\sum_{a_{m_1,\ldots,m_n}} (z_1 - z_1^0)^{m_1}, \ldots, (z_n - z_n^0)^{m_n},$$

which converges absolutely to $f(z_1, \ldots, z_n)$ in a neighborhood of the point.

Definition 252. A *complex Lie group* is a complex manifold with a group structure such that the product and inverse maps are holomorphic.

Definition 253. Let G be a Lie group. A submanifold H of G is called a *Lie subgroup* if H is a subgroup of the (abstract) group G and H is a topological group.

Example 82.

(1) Lower dimensional Lie groups:
  (i) $U(1) \simeq S^1 = \{e^{i\theta \mid 0 \leq \theta < 2\pi}\}$.
  (ii) Three parameter group, $SU(2) \simeq S^3$, where

$$u = \begin{pmatrix} a & b \\ -\bar{b} & \bar{a} \end{pmatrix} = \begin{pmatrix} x_1 + ix_2 & x_3 + ix_4 \\ -x_3 + ix_4 & x_1 - ix_2 \end{pmatrix} \in SU(2),$$

  for $a, b \in \mathbb{C}$, $x_i \in \mathbb{R}$, $i = 1, 2, 3, 4$ with the condition

$$\det u = |a|^2 + |b|^2 = \sum_{i=1}^{4} (x_i)^2 = 1.$$

(2) $GL(n, C)$: The group of complex $n \times n$ matrices of determinant $\neq 0$. That is, $GL(n, C) =$ {nonsingular $n \times n$ complex matrices}. Similarly, $GL(n, R)$: The group of all real $n \times n$ matrices of determinant $\neq 0$.

(3) $SL(n, C)$: The group of complex $n \times n$ matrices of determinant 1. That is,

$$SL(n, C) = \{g \in GL(n, C) | \det g = 1\}.$$

  Similarly, $SL(n, R)$: The group of real $n \times n$ matrices of determinant 1.

(4) **(i)** $U(p, q)$: [42] The group of matrices g in $GL(p + q, C)$ which leave invariant the Hermitian form,

$$-z_1\bar{z}_1 - \cdots - z_p\bar{z}_p + z_{p+1}\bar{z}_{p+1} + \cdots + z_{p+q}\bar{z}_{p+q}, \quad \text{that is } g^t I_{p,q}\bar{g} = I_{p,q}.$$

  The unitary group denoted by $U(n)$, where the set $U(0, n) = U(n, 0)$.
  (ii) Define $SU(p, q) = U(p, q) \cap SL(p + q, C)$ and $SU(n) = U(n) \cap SL(n, C)$. Let $S(U_p \times U_q)$, denote the set of matrices $\begin{pmatrix} g_1 & 0 \\ 0 & g_2 \end{pmatrix}$, where $g_1 \in U(p), g_2 \in U(q)$ and det $g_1 \cdot \det g_2 = 1$.

Remark 174. The set of all $n \times n$ unitary matrices is the unitary group $U(n)$ which is also a subgroup of $Gl(n, C)$. The set of all unitary matrices with determinant 1 is a special unitary group $SU(n)$ which is also a subgroup of $Gl(n, C)$.

(5) **(i)** Complex orthogonal group $O(n, C)$ is a subgroup of $Gl(n, C)$ consisting of all complex orthogonal matrices. So, an $n \times n$ complex matrix A is in $O(n, C)$ if and only if $A^T A = I$. Hence, $O(n, C)$ is a matrix of Lie group such that $|A| = \pm 1$.
  (ii) $SO(n, C)$ is the group of matrices g in $SL(n, C)$, which leave invariant the quadratic form, $z_1^2 + \cdots + z_n^2$, hence $g^t g = I_n$. We have

$$SO(n, C) = \{g \in SL(n, C) | \, g^t g = I\}.$$

  The set of all $n \times n$ real orthogonal matrices is the orthogonal group $O(n)$, it is a subgroup of $Gl(n, C)$ [42].

(iii) $SO(n)$, the special orthogonal group is a subgroup of $O(n)$ consisting of all orthogonal matrices with determinant 1. Hence $SO(n)$ is $\{\sigma \in SL(n,\mathbb{C})|\sigma = \overline{\sigma} = \sigma^*\}$, where $SL(n,\mathbb{C})$ is the group of all complex $n \times n$ matrices of determinant 1 and $\overline{\sigma}$ is the complex conjugate of σ and σ^* is the matrix $(\sigma^t)^{-1}$, the inverse of the transpose of σ. Moreover $SO(n)$ is a subgroup of $O(n)$, and so subgroup of $GL(n,C)$.

(6) The group $Sp(n+1,\mathbb{C})$ [183] is the group of all matrices in $GL(2n+2,\mathbb{C})$ which leave invariant the bilinear form, $\sum_{r=1}^{n+1}(X_rY_{r+n+1} - X_{n+r+1}Y_r)$. If I_n denotes the unit matrix of degree n, then the matrix of the coefficients of this bilinear form is $J = \begin{pmatrix} 0 & I_n \\ -I_n & 0 \end{pmatrix}$. Consequently, $Sp(n+1,\mathbb{C}) = \{A \in GL(2n+2,\mathbb{C}); A^tJA = J\}$. So

$$Sp(n+1,C) = \left\{g \in SL(2n+2,C) \mid g^tJg = J \textit{ for } J = \begin{pmatrix} 0 & I_n \\ -I_n & 0 \end{pmatrix}\right\}.$$

Remark 175. As $U(2n+2)$ is the group of all unitary matrices in $Sp(n+1,\mathbb{C})$, the symplectic group $G = Sp(n+1) = Sp(n+1,\mathbb{C}) \cap U(2n+2)$.

(7) $Sp(p,q)$ [42] is the group of matrices g in $Sp(p+q,\mathbb{C})$ which leave invariant the Hermitian form

$$Z^tK_{p,q}\overline{Z}, \text{ hence, } g^tK_{p,q}\overline{g} = K_{p,q}.$$

Example 83. Consider the set $G = \mathbb{R}\times\mathbb{R}\times S^1$, where S^1 is unit circle in the complex plane. We can define a product in G, $(G \times G) \to G$ by

$$(x_1, y_1, u_1)\cdot(x_2, y_2, u_2) = (x_1 + x_2,\ y_1 + y_2,\ \mathrm{e}^{\mathrm{i}x_1y_2}u_1u_2).$$

Here $x_1, x_2, y_1, y_2 \in \mathbb{R}$ and $u_1, u_2 \in S^1$. This operation makes G into a group.

$$((x_1, y_1, u_1)\cdot(x_2, y_2, u_2))\cdot(x_3, y_3, u_3) = (x_1 + x_2 + x_3,\ y_1 + y_2 + y_3,\ \mathrm{e}^{\mathrm{i}(x_1y_2 + x_1y_3 + x_2y_3)}u_1u_2u_3).$$

Here $x_1, x_2, y_1, y_2, x_3, y_3 \in \mathbb{R}$ and $u_1, u_2, u_3 \in S^1$. Clearly (0,0,1) is unit element and inverse element of (x, y, u) is $(-x, -y, \mathrm{e}^{\mathrm{i}xy}u^{-1})$. The operations $G \times G \to G$: $(g, h) \mapsto gh$ and $g \to g^{-1}$ are clearly smooth mappings where $g = (x_1, y_1, u_1)$ and $h = (x_2, y_2, u_2)$. Hence G is a Lie group, which is not a matrix Lie group.

Some relations between subgroups of complex general linear group GL(n, C)

$$SL(n,\mathbb{R}) = SL(n,\mathbb{C}) \cap GL(n,\mathbb{R}).$$

$$SO(n,\mathbb{C}) = O(n,\mathbb{C}) \cap SL(n,\mathbb{C}).$$

$$SU(n) = U(n) \cap SL(n,\mathbb{C}).$$

$$O(n,\mathbb{R}) = O(n,\mathbb{C}) \cap GL(n,\mathbb{R})$$

$$SO(n,\mathbb{R}) = O(n,\mathbb{R}) \cap SL(n,\mathbb{C}).$$

Remark 176. For discrete subgroups of Lie groups, one can refer Raghunathan [163].

Definition 254. A matrix Lie group G is called *compact* if

(i) any sequence $\{A_n\}$ of matrices in G will converge to a matrix $A \in G$ and
(ii) for all, $A = (A_{ij})(1 \leq i,j \leq n) \in G$, $|A_{ij}| \leq C$ where C is a constant.

Example 84.

(a) Compact groups:
  (i) Special orthogonal group, $SO(n) = \{g \in SL(n, \mathbb{C}) \mid g^t g = I \text{ and has real entries}\}$.
  (ii) Unitary group, $U(n) = \{g \in GL(n, \mathbb{C}) \mid \bar{g}^t g = I\}$.
  (iii) Special unitary group, $SU(n) = \{g \in U(n) \mid \det g = 1\}$.
  (iv) Symplectic group, $Sp(n) = \left\{g \in U(2n) \mid g^t J g = J \text{ for } J = \begin{pmatrix} 0 & I_n \\ -I_n & 0 \end{pmatrix}\right\}$.
  ($\dim SO(n) = \dim O(n)$ and $\dim SU(n) = \dim U(n) - 1$.)
(b) Noncompact groups:
  (i) $O(n, \mathbb{C})$ and $SO(n, \mathbb{C})$ are noncompact groups.
  (ii) $Sp(n, \mathbb{R}), Sp(n, \mathbb{C})$, and Heisenberg group are noncompact groups. Similarly $\mathbb{R}^n$, $\mathbb{R}^*$, and $\mathbb{C}^*$ are noncompact.

Remark 177. One-dimensional connected Lie group must be abelian. There are only two cases:

(1) The additive group of the set of all real numbers R, which is simply connected but not compact.
(2) $U(1)$ group, which is a unit circle in complex plane with group multiplication being multiplication of complex numbers. Here,

$$U(1) = \{z \in C \mid |z| = 1\} \simeq S^1 \simeq R/Z \simeq SO(2).$$

It is compact and connected, but not simply connected. In general, one can prove that any connected abelian group is a Cartesian product of these two kinds of groups.

Definition 255. A *representation of a Lie group* G is a homomorphism of G into a matrix group. A *faithful* representation is a representation which is a *isomorphism into*. Moreover if a group has a faithful representation, then it is isomorphic to a subgroup of matrix group.

Remark 178. Instead of matrices, one can use $gl(V)$, the set of all linear transformations of a vector space and we say that representation of G is on V.

Example 85. $\rho : \mathbb{C}^* \to gl(2, \mathbb{R})$, defined by $\rho(x + iy) = \begin{pmatrix} x & -y \\ y & x \end{pmatrix}$ is a faithful representation for the Lie group $\mathbb{C}^*$ of nonzero complex numbers.

Finite dimensional representations of compact groups and Lie groups

A compact group is a topological group, whose underlying topology is compact Hausdorff. A finite dimensional representation of a compact group G on finite dimensional complex vector space V is a continuous homomorphism ϕ of G into $GL(V)$, the set of all nonsingular linear transformations on V. In addition, if G is a Lie group and ϕ is a smooth map, the differential of this smooth map at the identity will give a representation of the real Lie algebra $\mathcal{G}$ of G on the vector space V.

Example 86. For (i) $G = U(n)$ or $SU(n)$ and (ii) $G = O(n)$ or $SO(n)$, assume $V = C^n$. Define G to act on C^n, by matrix multiplication by,

$$\phi(g)\begin{pmatrix} z_1 \\ \vdots \\ z_n \end{pmatrix} = g\begin{pmatrix} z_1 \\ \vdots \\ z_n \end{pmatrix}.$$

In both the cases, φ *is called standard representation of* (*i*) $G = U(n)$ *or* $SU(n)$ *and* (*ii*) $G = O(n)$ *or* $SO(n)$ *respectively*.

Definition 256. Let G be a Lie group. Let $\rho \in G$. Consider the left translation $L_\rho : G \to G$, defined by $L_\rho(g) = \rho g, g \in G$. This is the left multiplication by ρ which is a smooth map as the product map of $G \times G$ to G is smooth. Then L_ρ is an *analytic diffeomorphism.*

Definition 257. Let G be a Lie group. For any element $g \in G$, define the left translation L_g such that $L_g(H) = gh$. This L_g is a diffeomorphism with inverse L_g^{-1}. A vector field $X \in C^\infty(M, T(M))$ is called left invariant if $L'_g(X) = X$ for all $g \in G$. If $L'_g(X) = X$ and $L'_g(Y) = y$, then $L'_g[X, Y] = [L'_g(X), L'_g(Y)] = [X, Y]$. Hence set of all left invariant vector fields forms Lie algebra of vector field. Here L'_g denotes the corresponding diffeomorphism on the Lie algebra of vector fields on G.

Definition 258.

(i) Let G be a Lie group. A vector field Z on G is called *left invariant* if $dL_\rho Z = Z$ for all $\rho \in G$, where dL_ρ is defined on the set of all vector fields on G.

(ii) If $f : G \to R$ is a smooth function and $g \in G$, define f_g by $f_g(x) = f(gx)$. Then a smooth vector field $X \in G$ is *left invariant* if $(Xf)_g = X(f_g)$.

(iii) Let G be a Lie group. If $g \in G$, then $L_g : G \to G$ is defined by $L_g(h) = gh$ is a diffeomorphism and hence induces map $L_{g,\star} : T_h(G) \to T_{gh}(G)$. A vector field X on G is left-invariant if $L_{g,\star}(X_h) = X_{gh}$. The set of all left invariant smooth vector fields forms an algebra $\mathcal{G}$, called the Lie algebra of G.

Remark 179. Vector fields are defined in Chapter 1.

Definition 259 (Lie algebras of Lie groups). [42] Let $T_e(G)$ denote the tangent space to G at the identity. Then for any vector $v \in T_e(G)$, there is a unique left invariant vector field X^v with the property that $X^v_e = v$, defined by

$$X^v_g = (L_{g,\star})(v).$$

The tangent space of G at 'e' is also denoted as G_e. Now for a tangent vector $X \in G_e$, we get one left invariant vector field $\tilde{X}$ on G with $\tilde{X}_e = X$ with $\tilde{X}$ being analytic. Moreover, $\tilde{X}$ can be defined by

Table 6.1 The Lie groups and the corresponding Lie algebras

Lie groups	The corresponding Lie algebras
$GL(n, \mathbb{C})$	$gl(n, \mathbb{C})$
$SL(n, \mathbb{C})$	$sl(n, \mathbb{C})$
$SL(n, \mathbb{R})$	$sl(n, \mathbb{R})$
$U(p, q)$	$u(p, q)$
$U(n)$	$u(n)$
$SU(n)$	$su(n)$
$SO(n, \mathbb{C})$	$so(n, \mathbb{C})$
$SO(p, q)$	$so(p, q)$
$Sp(n, \mathbb{C})$	$sp(n, \mathbb{C})$
$Sp(p, q)$	$sp(p, q)$

$$(\tilde{X}f)(\rho) = \left\{\frac{d}{dt}f(\rho\gamma(t))\right\}_{t=0}$$

for $f \in C^\infty(G)$, $\rho \in G$ and $\gamma(t)$ is any curve in G with tangent vector X at $t = 0$. Now for $X, Y \in G_e$, $[\tilde{X}, \tilde{Y}]$ is left invariant vector field. Denote by $[X, Y]$, the tangent vector $[\tilde{X}, \tilde{Y}]_e$. Now the vector space G_e, the tangent space to G at "e," is denoted by "$\mathcal{G}$" (or $\mathfrak{L}(G)$). This $\mathcal{G}$ becomes a Lie algebra with the rule of composition $(X, Y) \to [X, Y]$. Hence the Lie algebra $\mathcal{G}$ of a Lie group G is the tangent space to G at the identity ($e \in G$) with bracket defined by

$$[v, w] = [X^v, X^w]_e.$$

Remark 180. Let G be a Lie group and $\mathcal{G}$ be a Lie algebra of G. If H is a Lie subgroup of G, then the Lie algebra $\mathfrak{h}$ of H is a subalgebra of $\mathcal{G}$. Each subalgebra of $\mathcal{G}$ is the Lie algebra of exactly one connected Lie subgroup of G.

Example 87. $so(n) = \{X \in sl(n, \mathbb{C}) | X^t + X = 0$ and X has real entries$\}$ is the Lie algebra of SO(n); $u(n) = \{X \in gl(n, \mathbb{C}) \mid \overline{X}^t + X = 0\}$ is the Lie algebra of $U(n)$; $su(n) = \{X \in u(n) \mid TrX = 0\}$ is the Lie algebra of SU(n); $sp(n) = \{X \in u(2n) \mid X^tJ + JX = 0\}$ is the Lie algebra of Sp(n). Here $J = \begin{pmatrix} 0 & I_n \\ -I_n & 0 \end{pmatrix}$.

Definition 260. The Lie algebra $\mathcal{G}$ of any matrix Lie group G is $\{X \mid X$ is a matrix with $e^{tX} \in G$ for all $t \in \mathbb{R}\}$.

Definition 261. A Lie group is called *semisimple,* if its corresponding Lie algebra is semisimple. Similarly, a Lie group is called simple, if its corresponding Lie algebra is simple.

Remark 181. In general, the Lie group $GL(n, \mathbb{C})$ has Lie algebra $gl(n, \mathbb{C})$ and the mapping $\exp : gl(n, \mathbb{C}) \to GL(n, \mathbb{C})$ is surjective (Table 6.1).

Remark 182. The groups $GL(n,\mathbb{C}), SL(n,\mathbb{C}), SL(n,\mathbb{R}), SO(n,\mathbb{C}), SO(n), SU(n), U(n), sp(n,\mathbb{C}), Sp(n)$ are all connected. The group $GL(n,\mathbb{R})$ has two connected components.

Definition 262. Let $\mathcal{G}$ be a Lie algebra over R. Denote by $GL(\mathcal{G})$, the group of all nonsingular linear transformations of $\mathcal{G}$. Now the Lie algebra $gl(\mathcal{G})$ of $GL(\mathcal{G})$, consist of all the endomorphisms of $\mathcal{G}$ with the bracket $[A,B] = AB - BA$. For $X \in \mathcal{G}$, $X \to ad\ X$ is a homomorphism of G onto a subalgebra, $ad(\mathcal{G})$. We denote by $int(\mathcal{G})$, the analytic subgroup of $GL(\mathcal{G})$. Then $int(\mathcal{G})$ is called *adjoint group of* $\mathcal{G}$.

Complex structure on a real vector space and a Lie algebra over R [42]

Definition 263. Let V be a vector space of finite dimension over $\mathbb{R}$. A complex structure on V is a $\mathbb{R}$- linear endomorphism J of V such that $J^2 = -I$, where $I: V \to V$ is the identity mapping. The above vector space V over $\mathbb{R}$ with a complex structure, J will become a vector space $\widetilde{V}$ over $\mathbb{C}$ by taking,

$$(a+ib)x = ax + bJx,$$

for $x \in V; a, b \in \mathbb{R}$. $J^2 = -I$ implies $\alpha(\beta x) = (\alpha\beta)x$ for $\alpha \in \mathbb{C}$ and $x \in V$. Moreover $\dim_c \widetilde{V} = \frac{1}{2}\dim_R V$ and V is even dimension. Here $\tilde{V}$ becomes a complex vector space associated to V. If V is a vector space over $\mathbb{C}$, we can consider V as a vector space $V^{\mathbb{R}}$ over $\mathbb{R}$. The multiplication by i on V gives a complex structure J on $V^{\mathbb{R}}$ and $V = V^{\mathbb{R}}$.

Definition 264. A Lie algebra $\mathcal{G}$ over $\mathbb{R}$ is said to have a *complex structure* J if J is a complex structure on the vector space $\mathcal{G}$ satisfying,

$$[X, JY] = J[X, Y] \text{ for all } X, Y \in \mathcal{G}.$$

Hence $(ad\ X) \circ J = J \circ ad\ X$ for all $X \in V$. So, $ad(JX) = J(ad\ X)$, for all $X \in V$. Also it is clear that

$$[JX, JY] = -[X, Y].$$

Now the corresponding complex vector space $\tilde{\mathcal{G}}$ becomes a Lie algebra over C with the bracket operation is that of $\mathcal{G}$. We have

$$\begin{aligned}[(a+ib)X, (c+id)Y] &= [aX + bJX, cY + dJX] \\ &= ac[X,Y] + bcJ[X,Y] + adJ[X,Y] - bd[X,Y] \\ \Rightarrow [(a+ib)X, (c+id)Y] &= (a+ib)(c+id)[X,Y].\end{aligned}$$

If $\mathcal{G}$ is a Lie algebra over C, then the vector space $\mathcal{G}^R$ has a complex structure J given by the multiplication i on $\mathcal{G}$. Now $\mathcal{G}^R$ becomes a Lie algebra over R with the complex structure J. If $\mathcal{G}_0$ is a Lie algebra over R, then the complex vector space is $\mathcal{G} = (\mathcal{G}_0)^C = \{x + iy | x, y \in \mathcal{G}_0\}$. Now the bracket operation on $\mathcal{G}$ is given by

$$[x_1 + iy_1, x_2 + iy_2] = [x_1, x_2] - [y_1, y_2] + i([y_1, x_2] + [x_1, y_2]).$$

This bracket is a bilinear operation over C. We have $\mathcal{G} = (\mathcal{G}_0)^C$, called the complexification of the Lie algebra $\mathcal{G}_0$. Moreover, $\mathcal{G}^R$ is a Lie algebra over R with complex structure J derived from the multiplication i on $\mathcal{G}$. The following result can be easily established.

Proposition 61. *[42] Let K_0, K, and K^R denote the Killing forms of the Lie algebras $\mathcal{G}_0$, $\mathcal{G}$, and $\mathcal{G}^R$, respectively. Then*

$$K_0(X, Y) = K(X, Y), \text{ for } X, Y \in \mathcal{G}_0 \text{ and}$$
$$K^R(X, Y) = 2 \text{ Real part of } (K(X, Y)), \text{ for } X, Y \in \mathcal{G}^R.$$

Definition 265. Let G be a Lie group of even dimension $2n$ with Lie algebra $\mathcal{G}$. Assume that there exists a real linear map $J : \mathcal{G} \to \mathcal{G}$ such that

(i) $J^2 = -I$,
(ii) $[JX, Y] = J[X, Y]$.

Then J is a *complex structure* on the Lie algebra $\mathcal{G}$. Now we can regard $\mathcal{G}$ as a complex Lie algebra with the multiplication by "i" to be the mapping J. Hence the map J gives $\mathcal{G}$ a complex vector space structure and the bracket is complex bilinear.

Similarly, if G is a complex Lie group, then we can define a map J on its Lie algebra $\mathcal{G}$ so that $\mathcal{G}$ becomes a complex Lie algebra and hence the exponential mapping is holomorphic from $\mathcal{G}$ into G. Conversely, if G is an even dimensional Lie group, we can find a map $J : \mathcal{G} \to \mathcal{G}$ so that G becomes a complex Lie group.

Example 88. Consider $GL(n, \mathbb{C})$. Let G be a subgroup of $GL(n, \mathbb{C})$. The Lie algebra $\mathcal{G}$ of G is a complex subalgebra of $gl(n, \mathbb{C})$ and $\mathcal{G}$ has complex structure.

Definition 266. [42] A *Real form of a Lie algebra $\mathcal{G}$ over* $\mathbb{C}$ is a sub algebra $\mathcal{G}_0$ of the real Lie algebra $\mathcal{G}^R$ such that $\mathcal{G}^R = \mathcal{G}_0 + J\mathcal{G}_0$. Now any element $z \in \mathcal{G}$ can be written uniquely as $z = x + iy$ where $x, y \in \mathcal{G}_0$. So $\mathcal{G}$ is isomorphic to a complexification of $\mathcal{G}_0$. The mapping $\sigma : x + iy \to x - iy$ is called the conjugation of $\mathcal{G}$ with respect to $\mathcal{G}_0$. The mapping σ satisfies the following properties,

$$\sigma(\sigma(X)) = X, \quad \sigma(X + Y) = \sigma(X) + \sigma(Y)$$
$$\sigma(\alpha X) = \overline{\alpha}\, \sigma(X), \quad \sigma[X, Y] = [\sigma X, \sigma Y]$$

for $X, Y \in \mathcal{G}, \alpha \in C$.

We state the following theorem proved in Helgason [42].

Theorem 75. *Every semisimple Lie algebra $\mathcal{G}$ over C has a real form which is compact.*

Let G be a connected Lie group with Lie algebra $\mathcal{G}^R$. Let U be an analytic subgroup of G whose Lie algebra is the compact real form of $\mathcal{G}$. Hence U is a maximal compact subgroup of $\mathcal{G}$. Denote by $\bar{U}$, the universal covering group of U with $Z(\bar{U})$, the center. In Table 6.2, $Z(\bar{U})$ denotes center of $\bar{U}$ with Z_p, a cyclic group of order p.

Table 6.2 **The simple Lie algebras $\mathcal{G}$ over C, their Lie groups G and the corresponding compact real forms U (real forms which are compact) and their dimensions [42]**

$\mathcal{G}$	G	U	dim U	$Z(\bar{U})$
$A_n(n \geq 1)$	$SL(n+1, C)$	$SU(n+1)$	$n(n+2)$	Z_{n+1}
$B_n(n \geq 2)$	$SO(2n+1, C)$	$SO(2n+1)$	$n(2n+1)$	Z_2
$C_n(n \geq 3)$	$Sp(n, C)$	$Sp(n)$	$n(2n+1)$	Z_2
$D_n(n \geq 4)$	$SO(2n, C)$	$SO(2n)$	$n(2n-1)$	Z_4 if $n =$ odd $Z_2 + Z_2$ if $n =$ even
E_6	E_6^C	E_6	78	Z_3
E_7	E_7^C	E_7	133	Z_2
E_8	E_8^C	E_8	248	Z_1
F_4	F_4^C	F_4	52	Z_1
G_2	G_2^C	G_2	14	Z_1

Here A_n, B_n, C_n, D_n are classical structures and E_6, E_7, E_8, F_4, G_2 are exceptional structures.

Definition 267. Let G be a closed linear group. For such a group, we get C^∞ curves $c(t), t \in \mathbb{R}$ with the property that $c(0)$ is the identity 1. Now we define, $\mathcal{G} = \{c'(0) \mid c : \mathbb{R} \to G$ with $c(0) = 1$ and c is smooth function from R to the group of matrices$\}$. This $\mathcal{G}$ defined above is closed under scalar multiplication. Moreover, G is closed under group conjugation $x \mapsto gxg^{-1}$. Again for $x \in \mathcal{G}$ and $c(t)$, a smooth curve in G with $c(0) = 1$, the map $R \to G$ sending $t \mapsto Ad(c(t))x$ is a smooth function in $\mathcal{G}$.

Here "*Ad*" denotes adjoint representation of R in G corresponding to adjoint representation "*ad*" of R in $\mathcal{G}$, the Lie algebra of G.

Definition 268. Let X be an $n \times n$ real or complex matrix. We define the *exponential of X (exp X)* by, $e^X = \sum_{m=0}^{\infty} \frac{X^m}{m!}$. It can be directly proved that, for any real and complex matrix X, the series $\exp X$ is convergent and it is a continuous function of X. Let $\mathcal{G}$ be a Lie algebra. For each $X \in \mathcal{G}$, there exists a unique analytic homomorphism θ of R into G such that $\dot{\theta}(0) = X$. Let $\exp X = \theta(1)$ such that $\dot{\theta}(0) = X$. The mapping $X \to \exp X$ of $\mathcal{G}$ into G ($\exp : gl(n, \mathbb{C}) \to GL(n, \mathbb{C})$), defined by

$$\exp X = e^X = I + X + \frac{1}{2}X^2 + \frac{1}{6}X^3 + \cdots,$$

is called the *exponential mapping*. It can be directly verified that $\exp(t+s)X = \exp tX \cdot \exp sX$ for $s, t \in \mathbb{R}$ and $X \in \mathcal{G}$.

Basic properties of exponential mapping

Let A be an $n \times n$ complex matrix. Then, e^A is a convergent series, which satisfies the following basic properties:

(i) $e^{X+Y} = e^X e^Y$, if X and Y commute.
(ii) e^X is nonsingular.
(iii) $t \mapsto e^{tX}$ is a smooth curve mapping C or R into $GL(n, \mathbb{C})$ which becomes 1 at $t = 0$.
(iv) $\frac{d}{dt}(e^{tX}) = Xe^{tX}$.
(v) $\det e^X = e^{Tr(X)}$.
(vi) $X \mapsto e^X$ is a C^∞ mapping from matrix space $(= (R^{2n})^2)$ into itself.

Example 89.

(1) Consider $G = U(n)$ and the corresponding algebra $u(n)$. For $X \in u(n)$, $X + X^* = 0$. But

$$(e^{tX})(e^{tX})^* = e^{tX} e^{tX^*} = e^{tX} e^{-tX} = 1.$$

Hence, $e^{tX} \in U(n)$, where as $X \in u(n)$. So exponential maps $u(n) \mapsto U(n)$.
(2) Let $G = SL(n, R)$ and $\mathcal{G} = sl(n, R)$. For $X \in sl(n, R)$, Trace $(X) = 0$, which implies that determinant of $e^{tX} = 1$. Hence $e^{tX} \in SL(n, R)$. So here exponential maps $sl(n, R)$ to $SL(n, R)$.

Definition 269. A *one-parameter subgroup* of a Lie group G is an analytic homomorphism of R into G.

Definition 270. Let G be a Lie group and $\mathcal{G}$ be the corresponding algebra. For every $X \in \mathcal{G}$, there exists an one-dimensional subgroup with Lie algebra generated by X. Hence there is a curve $\gamma : R \to G$ such that

(i) $\gamma(s + t) = \gamma(s)\gamma(t)$
(ii) $X(\gamma(s)) = \gamma^*(s)$

Here $\gamma(R)$ is called the *one-parameter subgroup* corresponding to X. Here $\gamma^* : R \to \mathcal{G}$. A function $A : R \to GL(n, C)$ is called one-parameter subgroup of $GL(n, C)$, if

(i) A is continuous
(ii) $A(0) = I$
(iii) $A(t + s) = A(t)A(s)$ for all $t, s \in R$. For any one-parameter subgroup A of $GL(n, C)$, it can be verified that there exists a unique complex $n \times n$ matrix such that $A(t) = e^{tX}$.

We have seen above that the one-parameter subgroups are the mappings $t \to \exp(tX)$ where X is the element of the Lie algebra. We denote this by $t \mapsto gt$. Since tX and uX commute, for $X \in gl(n, C)$, the map $t \to \exp(tX)$ is a one-parameter subgroup.

Example 90.

(i) If $G = U(1)$, then $\gamma(t) = e^{it}$.
(ii) If $G = SU(2)$, then $\gamma(t) = \begin{pmatrix} \cos t & \sin t \\ -\sin t & \cos t \end{pmatrix}$.

Definition 271. A *Cartan subgroup* T of G is a maximal connected abelian subgroup. Such subgroups are closed and are exactly the analytic subgroups of G, corresponding to Cartan subalgebras of $\mathcal{G}$, the Lie algebra of G.

6.2 Kac-Moody groups, supergroups, and some applications

Algebraic group associated to a Kac-Moody algebra

Let $\mathcal{G} = \mathcal{G}(A)$ be the Kac-Moody algebra associated with a symmetrizable generalized Cartan matrix $A = (a_{ij})_{i,j=1}^{n}$. Let $\mathcal{G}' = [\mathcal{G}, \mathcal{G}]$ be the commutator subalgebra of $\mathcal{G}$. Let (V, π) be a $\mathcal{G}'$-module. It is called integrable if $\pi(x)$ is nilpotent for every $x \in \mathcal{G}_\alpha$. Denote by G^*, the free product of the additive groups $\{\mathcal{G}_\alpha\}_{\alpha \in \Delta^{\text{re}}}$. One can take an inclusion map $i_\alpha : \mathcal{G}_\alpha \to G^*$. For any integrable $\mathcal{G}'$-module (V, π), one can define a homomorphism $\pi^* : G^* \to Aut_{\mathbb{C}} V$ by $\pi^*(i_\alpha(x)) = \exp(\pi(x))$, for $x \in \mathcal{G}_\alpha$. Let $N^* =$ Intersection of all $Ker\pi^*$ and $G = G^*/N^*$. Consider the canonical homomorphism, q: $G^* \to G$. Let $\exp(x) = q(i_\alpha x)$, where $\alpha \in \Delta^{\text{re}}$ and $x \in \mathcal{G}_\alpha$. We have $U_\alpha = \exp \mathcal{G}_\alpha$, an additive one-parameter subgroup of G. Denote by U, the subgroup of G generated by the U_α's with $\alpha \in \Delta_+^{\text{re}}$ and U^-, the subgroup of G generated by the U_α's with $\alpha \in \Delta_-^{\text{re}}$. One can define a topology on G in the usual sense. Now G, normally infinite dimensional, is an affine algebraic group in the sense of "Safarevic" with Lie algebra $\mathcal{G}$. Then G is a group (algebraic group) associated to the Kac-Moody algebra $\mathcal{G}$.

Remark 183. For more details on algebraic groups associated to Kac-Moody algebras, one can refer Kostant and Kumar [164].

Lie group module associated to an integrable $\mathcal{G}$-module

Let A be a complex commutative associative algebra with unity and $\mathcal{G}$ be a complex finite dimensional semisimple Lie algebra. Consider the Lie algebra $\mathcal{G}_A = A \otimes_{\mathbb{C}} \mathcal{G}$. This is an integrable algebra over $\mathbb{C}$, because one can choose "*ad*" locally nilpotent operators for $\mathcal{G}$. For any arbitrary Lie algebra $\mathcal{G}$, one can construct the Lie algebra $\mathcal{G}_A$. Here for the Lie algebra $\mathcal{G}_A$ to be integrable, the generators of $\mathcal{G}$ can be chosen to be *ad* locally nilpotent. To each integrable Lie algebra $\mathcal{G}$, one can associate a group G as follows:

Let G^* be the free group on the set $F_{\mathcal{G}} = \mathcal{G} \otimes_{\mathbb{C}} F$. One can define for an integrable $\mathcal{G}$-module $(V, d\pi)$, G^*-module such that

$$\tilde{\pi} = \exp d\pi(x) = \sum_{nn \geq 0} \frac{d\pi(x)^n}{n!}, \ x \in F_{\mathcal{G}}.$$

Let $G = G^*/\cap Ker\tilde{\pi}$, where the intersection taken over all integrable $\mathcal{G}$-modules $d\pi$. So G^*-module with $(V, \tilde{\pi})$ is a G-module (V, π). Here the integrable $\mathcal{G}$-module $(V, d\pi)$ is a differential of (V, π). Here G is called the group associated to the Lie algebra $\mathcal{G}$ and (V, π) being the G-module associated integrable $\mathcal{G}$-module. For any element $x \in F_{\mathcal{G}}$, one can denote $(\exp x)$ its image in G under the canonical homomorphism by $\exp x$. Hence

$$\pi(\exp x) = \exp d\pi(x), \ \ x \in F_{\mathcal{G}}$$

for an integrable $\mathcal{G}$-module $(V, d\pi)$. Here $\{\exp tx \mid t \in \mathbb{C}\}$ is a one-parameter subgroup of G. Similarly, one can define $F_G = \{\exp x \mid x \in F_{\mathcal{G}}\} \subset G$. A G-module (V, π) is called differentiable if all elements of F_G act locally finitely on V and $\exp tx$ restricted to any invariant finite dimensional subspace is analytic in t. Moreover, G-module (V, π) associated to an integrable $\mathcal{G}$-module $(V, d\pi)$ is differentiable.

A homomorphism $d\phi : \mathcal{G}_1 \to \mathcal{G}$ between integrable Lie algebras is called integrable if $d\phi(F_{\mathcal{G}_1}) \subset F_{\mathcal{G}}$. Here $d\phi(F_{\mathcal{G}_1})$ is called integrable subalgebra of $\mathcal{G}$. Moreover, for an integrable homomorphism $d\phi : \mathcal{G}_1 \to \mathcal{G}$ between integrable Lie algebras, one can define a homomorphism of associated groups by $\phi : G_1 \to G$ such that $d(\pi \circ \phi)$ exists and it is equal to $d\pi \circ d\phi$.

Remark 184. For more details regarding Lie group module associated to an integrable $\mathcal{G}$-module, one can refer Kac [8, 165] and Varadarajan [166].

An example to show the structure of the group G(A) associated to the Lie algebra $\mathcal{G}$(A)

This is also an example for connection between the integrable homomorphism of Lie algebras and the corresponding homomorphism between their groups.

Let $\mathcal{G}(A)$ be a Lie algebra with the Cartan matrix A and $G(A)$ be the corresponding group associated to it. We denote by $G_i, H_i, H, U_+, U_-, B_+$, and B_-, the subgroups of $G(A)$ corresponding to the integrable subalgebras $\mathcal{G}_i, \mathbb{C}h_i, \mathfrak{h}, \eta_+, \eta_-, \mathfrak{h} \oplus \eta_+$, and $\mathfrak{h} \oplus \eta_-$, respectively, of $\mathcal{G}(A)$. In the usual notation, define an integrable homomorphism $d\phi_i : sl_2(\mathbb{C}) \to \mathcal{G}(A)$ by

$$d\phi_i \begin{pmatrix} a & b \\ c & -a \end{pmatrix} = ah_i + be_i + cf_i.$$

Let $\phi_i : SL_2(\mathbb{C}) \to G(A)$ be the corresponding homomorphism of groups. Let

$$H_i(t) = \phi_i \begin{pmatrix} t & 0 \\ 0 & t^{-1} \end{pmatrix}.$$

Now the homomorphisms ϕ_i are injective and one has

$$G_i = \phi_i(SL_2(\mathbb{C})); \quad H_i = \{H_i(t) \mid t \in \mathbb{C}^X\} \text{ and}$$

$$\exp te_i = \phi_i \begin{pmatrix} 1 & t \\ 0 & 1 \end{pmatrix}, \quad \exp tf_i = \phi_i \begin{pmatrix} 1 & 0 \\ t & 1 \end{pmatrix}, \quad t \in \mathbb{C}.$$

Moreover, H is an abelian group equal to the (finite) direct product of the subgroups H_i. We have $B_\pm = U_\pm \rtimes H$.

Kac-Moody algebras and corresponding Kac-Moody groups

Let A be the generalized Cartan matrix and $\mathcal{G}(A)$ be the corresponding complex Lie algebra with Harish-Chandra-Serre relations. Let $\eta_{\pm}$ denote the positive and negative nilpotent subalgebras of $\mathcal{G}(A)$ respectively with $\mathfrak{b}_{\pm} = \mathfrak{h} \oplus \eta_{\pm}$ denoting the corresponding Borel subalgebras. Using the structure theory of $\mathcal{G}(A)$, one can construct the topological group $G(A)$ called the (minimal, split) Kac-Moody group over the complex numbers. This group $G(A)$ supports a canonical anti-linear involution ω, and one can define the unitary form $K(A)$ as the fixed group $G(A)^{\omega}$.

Remark 185. For a generalized Cartan matrix of finite type, $K(A)$ is the unitary form of the corresponding complex semisimple Lie group $G(A)$. These groups $K(A)$ are infinite dimensional analogs of compact Lie groups. One can identify the type of Kac-Moody group by that of its generalized Cartan matrix.

(i) If A is a finite type and $\mathcal{G}(A)$ is the corresponding finite dimensional semisimple algebra, then the group $G(A)$ will be the corresponding simply connected semisimple complex Lie group.
(ii) Another subglass of groups of $G(A)$ correspond to Cartan matrices of affine type. These matrices are nonfinite type which are diagonalisable with nonnegative real eigen values. The group of polynomial loops on a complex simply connected semisimple Lie group is a Kac-Moody group of affine type.
(iii) Kac-Moody groups that are not finite or affine types are said to be of indefinite type.

Moreover, one can say a GCM, A is of compact type if every proper subset $J \subset I$, the corresponding sub-matrix $(a_{i,j})_{i,j\in J}$ is of finite type. It can be shown that every indecomposable generalized Cartan matrix of affine type will be of compact type. Similarly, one can introduce an extended compact type as a generalized Cartan matrix A for which there is a unique decomposition $I = I_0 \cup J_0$, with the property that the Cartan matrix $(a_{i,j})_{i,j\in J}$ is of nonfinite type if and only if $I_0 \subseteq J$.

For more details regarding Kac-Moody algebras corresponding Kac-Moody groups, one can refer Kitchloo [167].

Definition 272 (Supergroups). We have already defined Grassman algebra. Consider the complex Grassman algebra $\Lambda(n)$ of order n with n generators 1 (unit element), $\theta_1, \ldots, \theta_n$ (Grassman variables) satisfying $\{\theta_i, \theta_j\} = 0$. The element

$$\eta = \sum_{m\geq 0} \sum_{i_1<\cdots<i_m} \eta_{i_1\ldots i_m}\theta_{i_1}\ldots\theta_{i_m}$$

is called even if each complex coefficient $\eta_{i_1,\ldots,i_m}$ in the above expression corresponds to an even value of m. Similarly, the element

$$\eta = \sum_{m\geq 0} \sum_{i_1<\cdots<i_m} \eta_{i_1\ldots i_m}\theta_{i_1}\ldots\theta_{i_m}$$

is called odd if each complex coefficient $\eta_{i_1,\ldots,i_m}$ in the above expression corresponds to an odd value of m. As a vector space, $\Lambda(n)$ decomposes into $\Lambda(n) = \Lambda_{\bar{0}} \oplus \Lambda_{\bar{1}}$ with $\Lambda_{\bar{0}}$ made of homogeneous even elements and $\Lambda_{\bar{1}}$ made of homogeneous odd elements. Now consider simple superalgebra $A = A_{\bar{0}} \oplus A_{\bar{1}}$. The Grassman envelope $A(\Lambda)$ of A consists of formal linear combinations $\sum_i \eta_i a_i$ where $\{a_i\}$ is a basis of A and $\eta_i \in \Lambda(n)$ such that for a fixed index i, the elements a_i and n_i are both even and odd. For $X = \sum_i \eta_i a_i$ and $Y = \sum_j \eta'_j a_j$, define the bracket by $[X, Y] = \sum_{ij} \eta_i \eta'_j [[a_i, a_j]]$, where $[[a_i, a_j]]$ is, the supercommutator in A. This gives $A(\Lambda)$ a Lie algebra structure. Supergroup A associated to superalgebra $\mathcal{A}$ is the exponential mapping of the Grassman envelope $A(\Lambda)$ of $\mathcal{A}$. The even generators of the superalgebra $\mathcal{A}$ correspond to even parameters, that is even elements of the Grassman algebra and odd generators of $\mathcal{A}$ correspond to odd parameters, that is odd elements of the Grassman algebra.

For more details of supergroups and supergroups of linear transformations, one can refer Frappat et al. [129].

Definition 273 ([129] Supergroups of Linear transformations). Let $\Lambda = \Lambda_{\bar{0}} \oplus \Lambda_{\bar{1}}$ be the Grassman algebra over the field $\mathbb{R}$ or $\mathbb{C}$. Consider the set of $(m+n) \times (m+n)$ supermatrices $M = \begin{pmatrix} A & B \\ C & D \end{pmatrix}$, where A, B, C, and D are $m \times m, m \times n, n \times m$, and $n \times n$ submatrices, respectively, with even entries in $\Lambda_{\bar{0}}$ for A, D and odd entries in $\Lambda_{\bar{1}}$ for B, C. Here M is called complex (resp. real) supermatrix if its entries have values in a complex (resp. real) Grassman algebra $\Lambda = \Lambda_{\bar{0}} \oplus \Lambda_{\bar{1}}$. More precisely, consider the set of $(m+n) \times (p+q)$ supermatrices $M = \begin{pmatrix} A & B \\ C & D \end{pmatrix}$, where A, B, C, and D are $m \times p, n \times p, m \times q$, and $n \times q$ submatrices, respectively. The submatrix M is called even (or degree 0) if $A, D \in \Lambda_{\bar{0}}$ and $B, C \in \Lambda_{\bar{1}}$, while it is called odd (of degree 1) if $A, D \in \Lambda_{\bar{1}}$ and $B, C \in \Lambda_{\bar{0}}$.

The general linear group $GL(m|n; K)$ is the supergroup of even invertible supermatrices M, the product law being product of supermatrices, the usual matrix multiplication.

The transposition and adjoint operations allow us to define the classical subsupergroups of $GL(m|n; K)$ corresponding to the classical superalgebras. The special linear supergroup $SL(m|n; \mathbb{K})$ is the subsupergroup of supermatrices $M \in GL(m|n; \mathbb{K})$ such that $\text{sdet}(M) = 1$. The unitary and superunitary supergroups $U(m|n)$ and $SU(m|n)$ are the subsupergroups of supermatrices $M \in GL(m|n; \mathbb{C})$ such that $M(M^t)* = 1$ and $M(M^{st})^\sharp = 1$, respectively.

Remark 186.

(1) For basic theory of Kac-Moody groups, representation theory and generalized flag varieties etc., one can refer Kumar [168].

(2) There are number of applications for Kac-Moody groups. For some applications, one can refer:

- **(i)** Dominant K-theory and integrable highest weight representations of Kac-Moody groups: Kitchloo [167].
- **(ii)** Constructing groups associated to infinite dimensional Lie algebras: Kac [165].
- **(iii)** Regular functions on certain infinite-dimensional groups: Kac and Peterson [169].
- **(iv)** Generalized invariants of groups generated by reflections: Kac and Peterson [170].
- **(v)** On geometric invariant theory for infinite-dimensional groups: Kac and Peterson [171].
- **(vi)** The nil Hecke ring and cohomology of G/P for a Kac-Moody group G: Kostant and Kumar [164].

(vii) Rings of generalized and stable invariants of pseudo reflections and pseudo reflection groups: Neumann et al. [173].

(viii) Lectures on Chevalley groups: Steinberg [174].

(3) For Monster simple group, one can refer [236].

6.3 Homogeneous spaces, corresponding Lie algebras, and spectra of some differential operators on homogeneous spaces

Definition 274. Let G be a Lie group and H be a closed subgroup. Then there exists an unique analytical structure on H such that H is a topological subgroup of G. Denote by $\mathcal{G}$ and $\mathfrak{h}$, the algebras of G and H, respectively. Also denote by $\mathfrak{p}$, the vector subspace of G such that $\mathcal{G} = \mathfrak{p} + \mathfrak{h}$. Now we can consider G/H the space of all left cosets gH $(g \in G)$ with natural topology. This G/H has a unique analytic structure such that G is a Lie transformation group of G/H. Now G/H is a homogeneous space.

Below we give two examples.

Example 91. **(1)** *Complex quadric hypersurface*, $SO(n+2)/SO(2) \times SO(n)$, $(n \geq 3)$: $SO(n)$ is $\{\sigma \in SL(n, \mathbb{C}) | \sigma = \overline{\sigma} = \sigma^*\}$, where $SL(n, \mathbb{C})$ is the group of all complex $n \times n$ matrices with determinant 1, $\overline{\sigma}$ is the complex conjugate of σ and σ^* is the matrix $(\sigma^t)^{-1}$, the inverse of the transpose of σ. Let $G = SO(n+2)$. The subgroup $H = SO(2) \times SO(n)$ is imbedded in G as

$$(A, B) \to \left(\begin{array}{c|c} A & 0 \\ \hline 0 & B \end{array}\right) \quad \text{for } A \in SO(2),\ B \in SO(n).$$

The space, $P = SO(n+2)/SO(2) \times SO(n)$ is a compact Riemannian symmetric space [175], called *complex quadric hypersurface.* The Riemannian metric on P is the metric induced by the negative of the Killing form in $\mathcal{G}$.

Corresponding Lie algebras of G, H, and P: If $\mathfrak{h}$ is the Lie algebra of H, then

$$\mathfrak{h} = \left\{ \left(\begin{array}{c|c} A & 0 \\ \hline 0 & B \end{array}\right) | A \in so(2),\ B \in so(n) \right\}.$$

We have, $\mathcal{G} = \mathfrak{h} \oplus \mathfrak{p}$ where

$$p = \left\{ \left(\begin{array}{c|c} 0 & -X^t \\ \hline X & 0 \end{array}\right) | X \text{ is a real } n \times\ 2 \text{ matrix} \right\}.$$

By direct computations, we will have $[\mathfrak{h}, \mathfrak{p}] \subseteq \mathfrak{p}$ and $[\mathfrak{p}, \mathfrak{p}] \subseteq \mathfrak{h}$. If $\mathcal{G}_{\mathbb{C}}, \mathfrak{h}_{\mathbb{C}}$ and $\mathfrak{p}_{\mathbb{C}}$ are the complexifications of $\mathcal{G}$,$\mathfrak{h}$, and $\mathfrak{p}$, respectively, then $\mathfrak{h}_{\mathbb{C}} = sp(1, \mathbb{C}) + sp(n, \mathbb{C})$, and

$$\mathcal{G}_{\mathbb{C}} = so(n+2, \mathcal{C}) = \mathfrak{h}_{\mathcal{C}} \oplus \mathfrak{p}_c = so(2, \mathcal{C}) \oplus so(n, \mathcal{C}) \oplus \mathfrak{p}_{\mathbb{C}}.$$

Put $f_{ij} = e_{ij} - e_{ji}$, where e_{ij} is the $(n+2)\times(n+2)$ coordinate matrix. It is clear that the set $\{f_{12}, f_{ij}\, for\ 3 \leq i < j \leq n+2\}$ is a basis for h_C and the set $\{f_{j1}, f_{j2}$ for $3 \leq i < j \leq n+2\}$ forms a basis for $\mathfrak{p}_C$. The element f_{12} is a center of $\mathfrak{h}_C$.

(2) *Quaternionic Projective space*, $Sp(n+1)/Sp(1) \times Sp(n), (n \geq 1)$: The group $Sp(n+1, \mathbb{C})$ is the group of all matrices in $GL(2n+2, \mathbb{C})$, which leave invariant the bilinear form $\sum_{r=1}^{n+1}(X_r Y_{n+r+1} - X_{r+n+1} Y_n)$. If I_n denotes the unit matrix of degree n, then the matrix of the coefficients of this bilinear form is $J = \left(\begin{array}{c|c} 0 & I_n \\ \hline -I_n & 0 \end{array}\right)$. Consequently, as defined earlier $Sp(n+1, \mathbb{C}) = \{A \in GL(2n+2, \mathbb{C}); A^t JA = J\}$. If $U(2n+2)$ is the group of all unitary matrices in $Sp(n+1, \mathbb{C})$, then the symplectic group $G = Sp(n+1) = Sp(n+1, \mathbb{C}) \cap U(2n+2)$. The compact subgroup $H = Sp(1) \times Sp(n)$ is imbedded in $Sp(n+1)$ as

$$\left(\begin{pmatrix} a & b \\ c & d \end{pmatrix}\begin{pmatrix} A & B \\ C & D \end{pmatrix}\right) \to \left(\begin{array}{cc|cc} a & 0 & b & 0 \\ 0 & A & 0 & B \\ \hline c & 0 & d & 0 \\ 0 & C & 0 & D \end{array}\right).$$

Moreover $Sp(1) \times Sp(n)$ is an isotropy subgroup of $Sp(n+1)$ and hence $P = Sp(n+1)/Sp(1) \times Sp(n)$ is a homogeneous space, called *quaternionic projective space.*

Corresponding Lie algebras of G, H, and P: Let $\mathcal{G}$ be the Lie algebra of G and h be the corresponding Lie algebra of H. Using the notations, we have $\mathcal{G} = sp(n+1), \mathfrak{h} = sp(1) + sp(n)$ and $\mathcal{G} = \mathfrak{h} \oplus \mathfrak{p}$ (direct sum). Any element of $\mathcal{G}$ is $\left(\begin{array}{cc|cc} p & y_1 & q & y_2 \\ -y_1^t & x_1 & y_2^t & x_2 \\ \hline -\overline{q} & -\overline{y_2} & \overline{p} & \overline{y_1} \\ -\overline{y_2^t} & -\overline{x_2} & -\overline{y_1^t} & \overline{x_1} \end{array}\right)$. Here $y_i \in M_{1,n}(\mathbb{C}), x_i \in M_{n,n}(\mathbb{C}), p \in i\mathbb{R}$, $q \in i\mathbb{R}$, $\overline{x_1} + x_1^t = 0, x_2 = x_2^t$. Moreover, t is the transpose of the corresponding matrix, $\mathfrak{h}$ is the set of elements of $\mathcal{G}$ such that $y_1 = y_2 = 0$ and p is the set of elements of $\mathcal{G}$ such that $p = q = 0$ and $x_1 = x_2 = 0$. By direct computation, it can be easily established that $[\mathfrak{p}, \mathfrak{p}] \subseteq \mathfrak{h}$ and $[\mathfrak{h}, \mathfrak{p}] \subseteq \mathfrak{p}$. If $\mathcal{G}_\mathbb{C}, \mathfrak{h}_\mathbb{C}$, and $\mathfrak{p}_\mathbb{C}$ are the complexifications of $\mathcal{G}, \mathfrak{h}$, and $\mathfrak{p}$, respectively, then $\mathfrak{h}_\mathbb{C} = sp(1, \mathbb{C}) + sp(n, \mathbb{C})$. Here

$$\mathcal{G}_\mathbb{C} = \left\{ \left(\begin{array}{cc|cc} a & A & b & C^t \\ B & x & C & z \\ \hline c & D^t & -a & -B^t \\ D & y & -A^t & -x^t \end{array}\right), a, b, c \in \mathbb{C}, x, y, z \in M_{n,n}(\mathbb{C}) \right.$$

$$\left. \text{with } y = y^t, z = z^t, A \in M_{1,n}(\mathbb{C}), B, C, D \in M_{n,1}(\mathbb{C}) \right\},$$

$\mathfrak{h}_\mathbb{C}$ is the set of elements of $\mathcal{G}_\mathbb{C}$ with $A = B = C = D = 0$ and $\mathfrak{p}_\mathbb{C}$ is the set of elements of $\mathcal{G}_\mathbb{C}$ with $a = b = c = 0$ and $x = y = z = 0$. We denote by, (A, B, C) the matrix $\left(\begin{array}{c|c} A & B \\ \hline C & -A^t \end{array}\right)$. Herein $(A, B, C) \in M_{n+1,n+1}(\mathbb{C})$. Usually, if (e_{ij}) stands for the $(n+1) \times (n+1)$ unit coordinate matrix, whose ij th element is 1 and zeros elsewhere, we denote by g_{pq}, the matrix $(e_{pq} + e_{qp})$.

(3) *De Rham Hodge operator* $\Box$ [176–178, 182, 183]: Suppose G is a semisimple Lie group and H is a compact subgroup of it with $P = G/H$, the space of left cosets. If $\mathcal{G}$ and $\mathfrak{h}$ are Lie algebras of G and H, respectively, we assume the decomposition of $\mathcal{G} = \mathfrak{h} \oplus \mathfrak{p}$ (direct sum) with $[\mathfrak{h}, \mathfrak{p}] \subseteq \mathfrak{p}$ and $[\mathfrak{p}, \mathfrak{p}] \subseteq \mathfrak{h}$. Here $\mathfrak{p}$ can be identified with the tangent space to P at the identity coset eH of G/H. Denoting by the subscript $\mathbb{C}$, the complexification of the corresponding vector spaces, assume that φ_0^* is the adjoint representation of H in $\mathfrak{p}_{\mathbb{C}}^*$, the dual of the vector space $\mathfrak{p}_{\mathbb{C}}$ and φ_0^{*r} is the corresponding representation of H in $\Lambda^r \mathfrak{p}_{\mathbb{C}}^*$, the exterior rth power of $\mathfrak{p}_{\mathbb{C}}^*$. If φ is an irreducible representation of H in an inner product space F, $\mu = \varphi_0^{*r} \otimes \varphi$ is the representation of H in $\Lambda^r \mathfrak{p}_{\mathbb{C}}^* \otimes F = \Lambda^r T_x(P)_{\mathbb{C}}^* \otimes F = F_1$, for $x \in P$. Suppose F^φ is the associated G homogenous vector bundle of ϕ, then $\Lambda^r T(P)_{\mathbb{C}}^* \otimes F^\varphi$ is the associated G homogeneous vector bundle of μ. If $\mathfrak{C}^\mu(G; F_1) = \{f : f(gh) = \mu(h^{-1})f(g)$, where $g \in G, h \in H$ and f is a C^∞ function on G with values in $F_1\}$, then $\mathfrak{C}^\mu(G; F_1)$ is the C^∞ sections of the vector bundle $\Lambda^r T(P)_{\mathbb{C}}^* \otimes F^\varphi$. But the C^∞ sections of this vector bundle is the set $\Lambda^r(P; F^\varphi)$ of F^φ valued r-forms on P. By this identification, for $\omega \in \Lambda^r(P; F^\varphi)$ and $x \in P$, $\omega_x \in F$ and $f_\omega = \sum_I a_I \zeta^I$, where $I = (i_1, i_2, \ldots, i_r)$, an ordered multi index and $\zeta^I = \zeta^{i_1} \wedge \zeta^{i_2} \wedge \zeta^{i_3} \wedge \cdots \wedge \zeta^{i_\gamma}$ with $f_\omega \in \mathfrak{C}_\mu(G; F_1)$. We define d on ω by $f_{d\omega} = \sum_{I,K} (\partial_K a_1)\zeta^K \wedge \zeta^1$, where ∂_K is the Lie derivative on G. Put $\|\omega_x\|^2 = \mathit{Trace}\ \omega_x^* \omega_x$, where ω_x^* is the adjoint defined relative to the Hermitian structure of $\mathfrak{p}_{\mathbb{C}}$. Putting $\|\omega\|^2 = \int_P \|\omega_x\|^2 dx$, denote the corresponding inner product by $(\omega|\pi)$. For $\omega \in \Lambda^r(P; F^\varphi)$ and $\pi \in \Lambda^{r+1}(P; F^\varphi)$, define d^* on forms with compact support by

$$(d\omega|\pi)_{L^2} = (\omega|d^*\pi)_{L^2}.$$

Put $\Box\omega = (dd^* + d^*d)\omega$, for $\omega \in \Lambda^r(P; F^\varphi)$. Here $\Box$ is called the *De Rham Hodge operator (acting on forms of degree r).* Then the spectrum of $\Box$ = The set of all eigen values of $\Box = \{\lambda \in \mathbb{R}|$ for $\omega(\neq 0)\ \Box\omega = \lambda\omega\}$.

(4) *Operator* Δ_G [176–178, 182, 183]: To compute the spectrum of $\Box$, we introduce another *differential operator* Δ_G, $\Delta_G = \sum_{i=1}^n \mathcal{L}_{x_i}^2$, where $\mathcal{L}_{x_i}$ denotes the Lie derivative with respect to x_i and $\{x_1, x_2, \ldots, x_n\}$ forms an orthonormal basis for $\mathfrak{p}$ with respect to negative Killing form in $\mathcal{G}$. Suppose τ is a representation of $\mathcal{G}$ in $\mathfrak{C}^\mu(G; F_1)$ defined by $(\tau(\gamma)f)(g) = f(\gamma^{-1}g)$, for $\gamma \in G$, then $\tau = \mathit{Ind} \nearrow_H^G (\mu)$ for $\gamma, g \in G$ and $f \in \mathcal{C}^\mu(G; F_1)$. Suppose $\mu_1, \mu_2, \ldots, \mu_q$ are irreducible components of μ such that $\mu = \mu_1 \oplus \mu_2 \oplus \cdots \oplus \mu_q$, then the irreducible components are the representations ρ of G such that $\rho|H$ contains one of μ_i. It was proved in Lévy-Bruhl-Laperrière [176, 177] and Sthanumoorthy [182, 183] that, spectrum of

$$\Delta_G = \cup_i^q \left\{ -\|\Lambda_\rho + \delta_G\|_G^2 + \|\delta_G\|_G^2 - \sum_{j=n+1}^{n+r} \beta_{ij} + \frac{1}{C}(\|\Lambda\mu_i + \delta_{H'}\|_{H'}^2 - \|\delta_{H'}\|_{H'}^2) \right\},$$

where Λ_ρ is the highest weight [172] of ρ, Λ_{μ_i} is the highest weight of μ_i, δ_G is half the sum of the positive roots in $G, \delta_{H'}$ is half the sum of the positive roots in H' and β_{ij}. $Id = \overline{\mu}_i(x_j^2)$ for $n+1 \leq j \leq n+r$ and $1 \leq i \leq q$. Here $\{X_{n+1}, \ldots, X_{n+r}\}$ is an orthonormal basis of the center of $\mathfrak{h}$ in $\mathfrak{h}$ with respect to negative Killing form in $B_\mathcal{G}$ in $\mathcal{G}$, $\{X_{n+r+1}, \ldots, X_{n+t}\}$ is an orthonormal basis for the supplement $\mathfrak{h}'$ of the center of $\mathfrak{h}$ in $\mathfrak{h}$ and $C \in \mathbb{R}$ such that $B_{\mathfrak{h}'} = CB_\mathcal{G}$, $B_{\mathfrak{h}'}$ being the Killing form in $\mathfrak{h}'$. When $J \in End_\mathbb{C}(\mathfrak{p}_\mathbb{C}^*)$ and $A \in \mathfrak{p}_\mathbb{C}^*, B = JA$ [181] will become $B(Z) = -\sum_{i=1}^n A([[Z, x_i], x_i])$ [182]. Herein $\{x_1, x_2, \ldots, x_n\}$ is an orthonormal basis for $\mathfrak{p}_\mathbb{C}$ and $Z \in \mathfrak{p}_\mathbb{C}$. Whenever $J = K \cdot Id (K \in \mathbb{R})$, we use the formula of Weizenbock [181] to conclude that spectrum of $\Box = -$Spectrum of $\Delta_G + K$ [182]. For the complete details regarding the computation of spectrum Δ_G and $\Box$, one can refer Sthanumoorthy [182, 183].

Remark 187. The eigen values of the operator are called spectrum of the corresponding operator. Spectrum of $\Box$ for the case of s forms ($1 \leq s \leq$ n) on the complex projective space $P^n(C)$ and for the case of 1 forms on the sphere S^n was considered in [177, 178, 180].

Example 92. **(1)** *Spectrum of Δ_G for the space $SO(n+2)/SO(2) \times SO(n)$*: For complete details regarding the computation of spectra of Δ_G for the above space, one can refer Sthanumoorthy [182, 183]. For this space, we state the following final results, which were proved in Sthanumoorthy [182, 183].

Theorem. *The spectrum of operator Δ_G acting on forms of degree 1 on the complex quadric hypersurface $SO(n+2)/SO(2) \times SO(n)$(n is even or odd with $n \geq 3$) is the set*

$$\left\{-\frac{1}{2n}(m_1^2 + m_2^2 + m_1 n + m_2 n - 2m_2) + \frac{1}{2};\right.$$
$$\left.(m_1, m_2) \in 2N^* \times \{0\} \ or \ m_1 - m_2 \in 2N, m_1 \in N^*\right\}$$

or

$$\{-(1/2n)(m_1^2 + m_2^2 + m_1 n + m_2 n - 2m_2 + n - 3) + 1/2;$$
$$m_1 - m_2 \in 2N + 1, m_2 \in N^*\}.$$

(2) *Spectrum of $\Box$ for the space $SO(n+2)/SO(2) \times SO(n)$*:

Theorem. *The spectrum of De Rham Hodge operator $\Box$ acting on forms of degree 1 on the complex quadric hypersurface $SO(n+2)/SO(2) \times SO(n)$ (n is even or odd with $n \geq 3$) is the set*

$$\left\{\frac{1}{2n}(m_1^2 + m_2^2 + m_1 n + m_2 n - 2m_2);\right.$$
$$\left.(m_1, m_2) \in 2N^* \times \{0\} \ or \ m_1 - m_2 \in 2N, m_1 \in N^*\right\} \cup$$
$$\{(1/2n)(m_1^2 + m_2^2 + m_1 n + m_2 n - 2m_2 + n - 3); m_1 - m_2 \in 2N + 1, m_2 \in N^*\}.$$

(3) *Spectrum of Δ_G for the space $Sp(n+1)/Sp(1) \times Sp(n)$*: We give below the following results proved for $Sp(n+1)/Sp(1) \times Sp(n)$. For the details of the proof, one can refer Sthanumoorthy [183].

Theorem. *For $G = Sp(n+1)$ and $H = Sp(1) \times Sp(n)$, the spectrum of Δ_G is one of the following sets:*

(i) $\left\{\frac{-1}{2(n+2)}(m_1^2 + 2m_1 n + m_1) + \frac{1}{2}; m_1 \in N^*\right\}$,

(ii) $\left\{\frac{-1}{2(n+2)}(m_1^2 + 2m_1 n + 3m_1 + 2n + 4) + \frac{1}{2}; m_1 \in N\right\}$, *(or)*

(iii) $\left\{\frac{-1}{2(n+2)}(m_1^2 + 2m_1 n + 2m_1 + 2n + 1) + \frac{1}{2}; m_1 \in N^*\right\}$.

(4) *Spectrum of $\square$ for the space $Sp(n+1)/Sp(1) \times Sp(n)$:*

Theorem. *The spectrum of De Rham Hodge operator $\square$ acting on forms of degree 1 on $Sp(n+1)/Sp(1) \times Sp(n)$ $(n \geq 1)$ is one of the following sets:*

(i) $\left\{\frac{1}{2(n+2)}(m_1^2 + 2m_1 n + m_1); m_1 \in N^*\right\}$

(ii) $\left\{\frac{1}{2(n+2)}(m_1^2 + 2m_1 n + 3m_1 + 2n + 4); m_1 \in N\right\}$

(iii) $\left\{\frac{1}{2(n+2)}(m_1^2 + 2m_1 n + 2m_1 + 2n + 1); m_1 \in N^*\right\}$.

Remark 188.

(i) For the above spaces $Sp(n+1)/Sp(1) \times Sp(n)$, $SO(n+2)/SO(2) \times SO(n)$, it was proved in Sthanumoorthy [182, 183], spectrum of $\square = -$spectrumof $\Delta_G + 1/2$.

(ii) For plancherel's measure for the factor space $SL(n, R)/SO(n, R)$, one can refer [1].

(iii) For the asymptotic behavior of zonal spherical functions on the Siegel upper half plane, one can refer [2].

6.4 Spectral invariants of zeta function of the Laplace-Beltrami operator

Definition 275 (Laplace-Beltrami operator Δ). When G is a compact connected Lie group and K is a closed subgroup, we consider the quotient space $M = G/K$. The *Laplace-Beltrami operator* [184] *or Laplacian* is $\Delta = d\delta + \delta d$, where d is exterior differentiation and δ is operator adjoint to d. Δ is a self-adjoint and elliptic differential operator on $C^\infty(\Lambda^p M)$ for each p. Here $C^\infty(\Lambda^p M)$ is the vector space of smooth sections of the vector bundle $\Lambda^p M$, the pth exterior power of the complexified cotangent bundle of M. The set of eigen values of Δ on $C^\infty(\Lambda^p M)$ is a discrete set of real numbers. When G is a compact semisimple Lie group and B is the Killing form of the Lie algebra $\mathcal{G}$ of G, the Casimir element is $C = \sum_{1 \leq i,j \leq N} c^{ij} X_i.X_j$, where $\{X_1, \ldots, X_N\}$ is a basis of $\mathcal{G}$ and $c_{ij} = (B(X_i, X_j))^{-1}$. When (G, K) is a compact symmetric pair with a compact connected semisimple Lie group G, we have the Cartan decomposition

$$\mathcal{G} = \mathcal{K} \oplus \mathcal{M},$$

where $\mathcal{K}$ is the Lie algebra of K and $\mathcal{M}$ is the orthogonal complement to $\mathcal{K}$ in $\mathcal{G}$ with respect to the Killing form. Restricting the Killing form sign changed to $\mathcal{M}$, we get a G-invariant Riemannian metric on $M = G/K$ and $\Delta = -C$ [185].

Definition 276. Let A be a positive elliptic self-adjoint pseudo- differential operator of order $m(> 0)$ acting on a compact n-dimensional manifold X. Its eigenvalues are $\lambda > 0$ and discrete. In analogy with the classical Riemann-zeta function [273, 274], we define, for $Re(s)$ large, zeta function [186] of the differential operator A as

$$\zeta(A, s) = Trace\, A^{-s} = \sum_{\lambda>0} \lambda^{-2}.$$

Here each eigen value λ is repeated as many times as its multiplicity. This series converges for $Re(s) > n/m$ and gives a holomorphic function of the complex variable s in this half plane [187]. Again this series $\zeta(A, s)$ can be analytically continued to the whole of s-plane as a meromorphic function with simple poles. The real valued invariants of the metric satisfying the above equation and satisfying the condition that it is a continuous function of the metric can be obtained by evaluating ζ at some point where it is known to be finite. By the results [188], $\zeta(A, s)$ at $s = 0$ is finite. It may be useful to mention here that for the mean-value and omega-theorems for the Riemann zeta function, one can refer Ramachandra [52]. For more general results on the method of analytic continuation used here, one can refer Mahler [189].

Spectral invariants of the zeta function of the Laplacian for S^{4r-1}

In Sthanumoorthy [190], a general method of analytic continuation was used to compute $\zeta(\Delta, s)$ at $s = 0$ for the Laplacian Δ and computed it for Δ acting on 2-forms on S^{4r-1}. Analytic continuation was used for the function of the form $\sum_{n=1}^{\infty} g(n)\{f(n)\}^{-s}$, where $f(n)$ and $g(n)$ are two polynomials of finite degrees. Using the results in Ikeda and Taniguchi [185], the eigen values for Δ can be written for $S^n = SO(n+1)/SO(n)$. Hence eigen values and their multiplicities can be written as shown below for the Laplacian Δ acting on 2-forms on $S^{4r-1} = SO(4r)/SO(4r-1)$. The following table gives the eigen values and their multiplicities for Laplacian Δ on S^{4r-1}.

Λ_ρ	Eigen value	Multiplicity
$(k+1)\lambda_1 + \lambda_2$	$(k+2)(k+4r-2)$	$\frac{2(k+1)(k+4r-1)(k+2r)}{(4r-3)(k+2)} \binom{k+4r-3}{4r-4}$
$(k+1)\lambda_1 + \lambda_2 + \lambda_3$	$(k+3)(k+4r-3)$	$\frac{(k+1)(k+2r)(k+4r-1)(4r-3)}{(k+3)(k+4r-3)} \binom{k+4r-2}{4r-3}$

Here $\wedge_\rho$ is the highest weight of the irreducible representation ρ intervening in $C^\infty(\wedge^2)S^{4r-1}$. Here K runs over all nonnegative integers. Using the above eigen values and multiplicities, one can write zeta function of Laplacian of Δ as follows [190]: Zeta function of the Laplacian Δ for S^{4r-1} is

$$\zeta(\Delta, s) = \sum_{n=0}^{\infty} \frac{2(n+1)(n+4r-1)(n+2r)}{(4r-3)(n+2)} \binom{n+4r-3}{4r-4} \times \{(n+2)(n+4r-2)\}^{-s} + \sum_{n=0}^{\infty} \frac{(n+1)(n+2r)(n+4r-1)(4r-3)}{(n+3)(n+4r-3)} \binom{n+4r-2}{4r-3} \times \{(n+3)(n+4r-3)\}^{-s}.$$

Following Seeley [188], the value of $\zeta(A, 0)$ can be computed as an explicit integral over the manifold. But in Sthanumoorthy [190] a general method of analytic continuation was used to compute $\zeta(A, 0)$ for the Laplace-Beltrami operator Δ acting on 2-forms on S^{4r-1}. As we consider the positive operator Δ acting on 2-forms on S^{4r-1}, we get $\eta(\Delta, s) = \zeta(\Delta, s)$. Let

$$\frac{2(n+1)(n+4r-1)(n+2r)}{(4r-3)(n+2)} \binom{n+4r-3}{4r-4} = \frac{2}{(4r-3)!} g_1(n).$$

Here $g_1(n)$ can be written as $g_1(n) = \sum_{t=0}^{4r-2} a_t n^t$ with $a_{4r-2} = 1$. Let $P_1(n) = 4rn + 8r - 4$. Then $g_1(n)(P_1(n))^t$ can be written as

$$g_1(n)(P_1(n))^t = \sum_{i=0}^{4r+t-2} a_{(t,i)} n^i \quad \text{for} \quad 1 \le t \le 4r-1.$$

Similarly assuming

$$\frac{(4r-3)(n+1)(n+2r)(n+4r-1)}{(n+3)(n+4r-3)} \binom{n+4r-2}{4r-3} = \frac{1}{(4r-4)!} g_2(n),$$

$g_2(n)$ can be written as $g_2(n) = \sum_{t=0}^{4r-2} b_t n^t$ with $b_{4r-2} = 1$. Let $P_2(n) = 4rn + 12r - 9$. Then $g_2(n)(P_2(n))^t$ can be written as

$$g_2(n)(P_2(n))^t = \sum_{i=0}^{4r+t-2} b_{(t,i)} n^t \quad \text{for} \quad 1 \le t \le 4r-1.$$

Using the above notations, we state the following result which was proved in Sthanumoorthy [190], using the method of analytic continuation.

Theorem 76. *A spectral invariant of the zeta function $\zeta(\Delta, s)$ at $s = 0$ of the Laplace-Beltrami operator Δ acting on 2-forms on $S^{4r-1}(r \ge 2)$ is*

$$\frac{2}{(4r-3)!} \left\{ \sum_{t=1}^{4r-1} \frac{(-1)^t a_{(t,2^t-1)}}{2t} + \sum_{t=1}^{2r-1} \frac{(-1)^t a_{2t-1} B_t}{2t} - \frac{a_0}{2} \right\} + \frac{1}{(4r-4)!}$$

$$\left\{\sum_{t=1}^{4r-1}\frac{(-1)^t b_{(t,2t-1)}}{2t}+\sum_{t=1}^{2r-1}\frac{(-1)^t b_{2t-1}B_t}{2t}-\frac{b_0}{2}\right\}+\frac{2r}{3}(16r^2-1).$$

Here $B_1, B_2, \ldots, B_t$ are Bernoulli's numbers [192] and other coefficients a_t, b_t, $a_{(t,2t-1)}, b_{(t,2t-1)}$ are as explained above.

Spectral invariants of the zeta function of Δ acting on 1-forms on $SO(r+2)/SO(2) \times SO(r)$

The operator Δ is positive and $\eta(\Delta, s) = \zeta(\Delta, s)$. Using the results in Sthanumoorthy [191] on spectra of Δ and multiplicities of the eigen values, one can write the Zeta function of the Laplacian Δ acting on 1-forms on the nonsingular quadric hypersurface $SO(r+2)/SO(2) \times SO(r)$ for $r > 5$ as

$$\zeta(\Delta, s) = S_1 + S_2 + S_3,$$

where

$$S_i = \sum_{n=1}^{\infty} g_i(n)(f_i(n))^{-s} \quad \text{for } i = 1, 2, 3.$$

Here,

$$g_1(n) = \frac{(2n+r-1)(2n+r)}{r(r-1)}\binom{n+r-2}{r-2},$$

$$g_2(n) = \left\{\frac{(2k+1)(2n-2k+r-1)(2n+r)(2n-4k-r-2)}{r(r-1)(r-2)(n+1)}\right.$$

$$\left.\binom{n+r-2}{r-2}\binom{n-2k+r-3}{r-3}\right\}, \text{ and}$$

$$g_3(n) = \frac{(r-2)(2k+2)n(2n+2k+r)(2n+4k+r+2)(2n+r-2)}{r(n+2k+3)(n+1)(n+2k+r-1)(n+r-3)}$$

$$\binom{n+2k+r-1}{r-2}\binom{n+r-3}{r-3}.$$

$$f_1(n) = \left(\frac{2n^2+nr}{r}\right),$$

$$f_2(n) = \left(\frac{n^2 + n(r-2k-1) + 2k^2 + 2k - kr}{r}\right), \text{ and}$$

$$f_3(n) = \left(\frac{1}{r}(n^2 + n(2k+r) + (k+1)(r+2k) - 1)\right).$$

We write the following result, proved in Sthanumoorthy [191]. Let $g_1(n) = \sum_{i=0}^{r} a_i n^i$, $g_2(n) = \sum_{i=0}^{2r-3} b_i^{(k)} n^i$, and $g_3(n) = \sum_{i=0}^{2r-5} c_i^{(k)} n^i$. Also let $P_2(n) = n(r-2k-1)+2k^2+2k-kr$, $P_3(n) = n(2k+r)+(r+2k)(k+1)-1$. Then $g_2(n)(P_2(n))^t$, $g_3(n)(P_3(n))^t$ can be written as $g_2(n)(P_2(n))^t = \sum_{i=0}^{2r+t-3} b_{(t,i)}^{(k)} n^t$ for $1 \le t \le 2r-2$ and $g_3(n)(P_3(n))^t = \sum_{i=0}^{2r-5+i} c_{(t,i)}^{(k)} n^i$ for $1 \le t \le 2r-4$. Using the above notations, we write the following result from Sthanumoorthy [191].

Theorem 77. *Spectral invariant of the Zeta function $\zeta(\Delta, s)$ at $s = 0$ of the Laplace-Beltrami operator Δ acting on 1-forms on the nonsingular quadric hyper-surface $SO(r+2)/SO(2) \times SO(r)(r \ge 5)$ is the set*

$$\left\{\sum_{t=0}^{r} \frac{(-1)^{t+1} r^{r+1}}{2^{t+1}.2(t+1)} a_t - \frac{a_0}{2} + \sum_{t=1}^{p} \frac{(-1)^t a_{2t-1} B_t}{2t} + \sum_{t=1}^{2r-2} \frac{(-1)^t}{2t} b_{(t,2t-1)}^{(k)} - \frac{b_0}{2}\right.$$
$$\left. + \sum_{t=1}^{r-1} \frac{(-1)^t b_{(2t-1)}^{(k)} B_t}{2t} + \sum_{t=1}^{2r-4} \frac{(-1)^t}{2t} c_{(t,2t-1)}^{(k)} - \frac{c}{2} + \sum_{t=1}^{r-2} \frac{(-1)^t c_{(2t-1)}^{(k)} B_t}{2t};\ k \in N\right\}.$$

Here $p = r/2$ or $p = (r+1)/2$ according as r is even or odd and B_m are Bernoulli's numbers [192].

Spectral invariants of the zeta function of the Laplacian on $Sp(r+1)/Sp(1) \times Sp(r)$

We write the following expressions for $\zeta(\Delta, s)$ as proved in Sthanumoorthy [193]. We have, $\eta(\Delta, s) = \zeta(\Delta, s) = S_1 + S_2 + S_3$, where $S_i = \sum_{n=1}^{\infty} g_i(n)\{f_i(n)\}^{-s}$ for $i = 1, 2, 3$. Here,

$$g_1(n) = \frac{2n+2r+1}{(2r+1)(n+1)} \binom{n+2r}{2r} \binom{n+2r-1}{2r-1},$$

$$g_2(n) = \frac{3(2n+2r+3)}{(2r+1)(n+3)} \binom{n+2r-1}{2r-1} \binom{n+2r+2}{2r},$$

$$g_3(n) = \frac{4nr(2r-2)(2n+2r+2)}{(n+1)(n+3)(n+2r-1)(n+2r+1)} \binom{n+2r+2}{2r+1} \binom{n+2r}{2r},$$

$$f_1(n) = \left\{\frac{n^2 + 2nr + n}{2(r+2)}\right\},$$

$$f_2(n) = \left\{ \frac{n^2 + 2nr + 3n + 2r + 4}{2(r+2)} \right\},$$

$$f_3(n) = \left\{ \frac{n^2 + 2nr + 2n + 2r + 1}{2(r+2)} \right\}.$$

Let $g_1(n) = \sum_{i=0}^{4r-1} a_i n^i$, $g_2(n) = \sum_{i=0}^{4r-1} b_i n^i$, and $g_3(n) = \sum_{i=0}^{4r-1} c_i n^i$. Also let $P_2(n) = 2nr + 3n + 2r + 4$ and $P_3(n) = 2nr + 2n + 2r + 1$. Then $g_2(n)(P_2(n))^t$ and $g_3(n)(P_3(n))^t$, can be written as $g_2(n)(P_2(n))^t = \sum_{i=0}^{4r+t-1} b_{(t,i)} n^t$ for $1 \leq t \leq 4r$ and $g_3(n)(P_3(n))^t = \sum_{i=0}^{4r+t-1} c_{(t,i)} n^i$ for $1 \leq t \leq 4r$.

Remark 189. The functions $g_i(n)$, for $i = 1, 2$, or 3 for S^{4r-1} (3.3 and 3.4), $SO(r+2)/SO(2) \times SO(r)$ (3.5 and 3.6), $Sp(r+1)/Sp(1) \times Sp(r)$ (3.7 and 3.8) are different.

Using the above notations, we state below the following theorem proved in Sthanumoorthy [193].

Theorem 78. *A spectral invariant of the Zeta function $\zeta(\Delta, s)$ at $s = 0$ of the Laplace-Beltrami operator Δ acting on forms of degree 1 on $Sp(r+1)/Sp(1) \times Sp(r)$ is*

$$\sum_{t=1}^{4r} \frac{(-1)^t (2r+1)^t}{2t} a_{t-1} - \frac{a_0}{2} + \sum_{t=0}^{2r} \frac{(-1)^t a_{2t-1} B_t}{2t}$$
$$+ \sum_{t=1}^{4r} \frac{(-1)^t}{2t} b_{(t,2t-1)} - \frac{b_0}{2} + \sum_{t=1}^{2r} \frac{(-1)^t b_{2t-1} B_t}{2t}$$
$$+ \sum_{t=1}^{4r} \frac{(-1)^t}{2t} c_{(t,2t-1)} - \frac{c_0}{2} + \sum_{t=1}^{2r} \frac{(-1)^t c_{2t-1} B_t}{2t} + (r+1)(2r+3).$$

Here in B_t are Bernoulli's numbers.

Definition 277 (Eta function and spectral asymmetry). When the operator A is not necessarily positive, the corresponding eta function $\eta(A, s)$ is defined as

$$\eta(A, s) = \sum_{\lambda \neq 0} sign\ \lambda |\lambda|^{-s},$$

where each eigen value λ of A repeats as many times as its multiplicity. When the operator A is positive, this value of ζ at $s = 0$ is the value of η at $s = 0$. The finite value $\zeta(A, s)$ at $s = 0$ is the measure of spectral asymmetry. In Atiyah et al. [186, 187, 194, 195], some information about the spectral asymmetry arising in Riemannian geometry were given. Moreover for the self-adjoint operator $B = \pm(*d - d*)$ acting on even forms on the boundary Y of a $4k$ dimensional compact oriented manifold X, an expression for $\eta(B, 0)$ was given in Atiyah et al. [186]. Here B^2, the square of B, is the usual Laplace-Beltrami operator Δ. For positive self-adjoint elliptic operators, the zeta function defined above have finite values at $s = 0$ by the results in Seeley [188]. Atiyah et al. [195] gave two methods to prove the finiteness of η at $s = 0$ for the special operator $B_{\nu}^{ev} = i'(-1)^{p+1}(*d - d*)$. Using Cobordism theory one can prove the finiteness of $\eta(0)$ and by using invariant theory, $\eta(0)$ can be computed explicitly for B_{ν}^{ev}. If the operator B is self-adjoint and not positive its eigenvalues are real but can be positive or negative. In this case we define, for $Re(s)$ large, its eta-function [186]

$$\eta(B,s) = \sum_{\lambda \neq 0} (sign\ \lambda)|\lambda|^{-s}.$$

Here also λ runs over the eigen values of A and each eigen value repeats as many times as its multiplicity. Again it is true that this $\eta(B,s)$ extends meromorphically to the whole of s-plane and $\eta(B,0)$ is finite [195]. The value of $\eta(B,0)$ is a measure of asymmetry of the spectrum of the operator B. Atiyah et al. [195] had shown that $R(B) = Re_{s=0} s\eta(B,s)$ is a homotopy invariant of the self-adjoint elliptic operator B and that $\eta(B,0)$ is finite for the special operator

$$B_{\nu}^{ev}\phi = i'(-1)^{p+1}(*d - d*)\phi \ [194].$$

($\deg\phi = 2p$, dim of Riemannian manifold$=2\ell - 1$). This finiteness was proved by Atiyah et al. [195] using Cobordism theory and invariant theory. We know that the Laplace-Beltrami operator (or total Laplace operator) Δ is the square of a self-adjoint operator $B = \pm(*d - d*)$. So the eigen values of Δ are of the form λ^2 where λ is an eigen value of B. Thus the operator B is not positive and hence its eigen values can be positive or negative [195]. Without loss of generality, here we assume that the eigen values of B are alternatively positive and negative in each of the two discrete series separately.

Definition 278 (Eta function of the self-adjoint operator B for S^{4r-1}). One can use a simple method of analytic continuation [190] and compute the η-invariant (the spectral asymmetry) of $B = \pm(*d - d*)$ acting on 2-forms on S^{4r-1}. We give the following expressions for S^{4r-1} [113]. Hence as per the assumption that the eigen values of B are alternatively positive and negative in the two series, we write the η function of B as

$$\begin{aligned}\eta(B,s) = &\sum_{n=0}^{\infty} \frac{(-1)^n 2(n+1)(n+4r-1)(n+2r)}{(4r-3)(n+2)} \binom{n+4r-3}{4r-4} \\ &\{(n+2)(n+4r-2)\}^{-s/2} \\ &+ \sum_{n=0}^{\infty} \frac{(-1)^n (n+1)(n+4r-1)(n+2r)(4r-3)}{(n+3)(n+4r-3)} \binom{n+4r-2}{4r-3} \\ &\{(n+3)(n+4r-3)\}^{-s/2}. \qquad \text{(a)}\end{aligned}$$

Each series on the right hand side of (a) can be split into two series, one with all positive terms and the other with all negative terms. So we make the following assumptions:

$$\eta(B,s) = S_1 - S_2 + S_3 - S_4, \text{ where}$$

$$S_i = \sum_{n=1}^{\infty} g_i(n)(f_i(n))^{-s/2}$$

$$g_1(n) = \frac{2(2n-1)(2n+4r-3)(2n+2r-2)}{(4r-3)(2n)} \binom{2n+4r-5}{4r-4}$$

$$= \sum_{i=0}^{4r-2} a_i n^i,$$

$$g_2(n) = \frac{2(2n)(2n+4r-2)(2n+2r-1)}{(4r-3)(2n+1)} \binom{2n+4r-4}{4r-4}$$

$$= \sum_{i=0}^{4r-2} b_i n^i,$$

$$g_3(n) = \frac{(2n-1)(2n+2r-2)(2n+4r-3)(4r-3)}{(2n+1)(2n+4r-5)} \binom{2n+4r-4}{4r-3}$$

$$= \sum_{i=0}^{4r-2} c_i n^i,$$

$$g_4(n) = \frac{(2n)(2n+2r-1)(2n+4r-2)(4r-3)}{(2n+2)(2n+4r-4)} \binom{2n+4r-3}{4r-3}$$

$$= \sum_{i=0}^{4r-2} d_i n^i,$$

$$f_1(n) = 2n(2n+4r-4),$$

$$f_2(n) = (2n+1)(2n+4r-3),$$

$$f_3(n) = (2n+1)(2n+4r-5),$$

$$f_4(n) = (2n+2)(2n+4r-4).$$

Here B_t are Bernoulli's numbers.

Using the above notations, the following theorem was proved in Sthanumoorthy [113].

Theorem 79. *The spectral asymmetry of the operator $B = \pm({}^*d - d^*)$ acting on 2-forms on $S^{4r-1}(r \geq 2)$ (assuming the eigen values with alternate signs) is*

$$\frac{b_0 + d_0 - a_0 - c_0}{2} + \sum_{t=1}^{4r-1} \frac{(-1)^t}{2t} \left\{(2r-2)^t a_{t-1} + c_{(t,2t-1)} - b_{(t,2t-1)} - d_{(t,2t-1)}\right\}$$

$$+ \sum_{t=1}^{2r-1} \frac{(-1)^t B_t}{2t} \{a_{2t-1} + c_{2t-1} - b_{2t-1} - d_{2t-1}\}.$$

Herein, the coefficients $a_t, b_t, c_t, b_{(t,2t-1)}, c_{(t,2t-1)}$ *are given in the following expressions:*

$$g_2(n)(P_2(n))^t = \sum_{i=0}^{4r+t-2} b_{(t,i)} n^i \textit{ for } 1 \leq t \leq 4r-1$$

$$g_3(n)(P_3(n))^t = \sum_{i=0}^{4r+t-2} c_{(t,i)} n^i \textit{ and}$$

$$g_4(n)(P_4(n))^t = \sum_{i=0}^{4r+t-2} d_{(t,i)} n^i,$$

where $P_2(n), P_3(n), P_4(n)$ *are*

$$P_2(n) = n(2r-1) + (r - 3/4),$$
$$P_3(n) = n(2r-2) + r - 5/4, \textit{ and}$$
$$P_4(n) = n(2r-1) + 2r - 2.$$

6.5 Generalization of Macdonald's identities for some Kac-Moody algebras

Let $\eta(q)$ denote Dedekind's η-function. Macdonald [128, 196] obtained a formula for $\eta(q)^{\dim \mathcal{G}}$, for every complex simple Lie algebra $\mathcal{G}$, which gives a generalization of the Jacobi's expansion for $\eta(q)^3$. These formulas are some specializations of the Macdonald's multivariable identities [196]. Many other identities involving Dedekind's η-function were also obtained by Lepowsky [197, 198] and Lepowsky and Primc [199]. For exact formulas for root space multiplicities in Cartan matrix, Lie algebras and their universal enveloping algebras, one can refer Berman and Moody [61]. We explain below some preliminaries needed to find some eta function identities. Let $\mathcal{G}$ denote the simple finite dimensional Lie algebra of the type A_l or B_l and $\mathfrak{h}$ denote its Cartan subalgebra. Let Δ, Δ_+, and $\Delta_+^\vee$ be the root system, the set of positive roots, and the positive dual roots, respectively, of $\mathcal{G}$ and ρ (respectively $\rho^\vee$) be the half sum of the roots in Δ_+ (resp. $\Delta_+^\vee$). W and M are the Weyl group of $\mathcal{G}$ and the lattice spanned over Z by the long roots of $\mathcal{G}$, respectively. Let h and g denote the Coxeter and dual Coxeter numbers of $\mathcal{G}$, respectively. Let $\langle .,. \rangle$ denote the pairing of the elements in $\mathfrak{h}$ and $\mathfrak{h}^*$. We introduce the following notations:

$$\Delta_m^\vee = \{\alpha \in \Delta_+^\vee | \langle \rho, \alpha \rangle \equiv 0 (\mathrm{mod}\ m)\}$$

and

$$d_m(\lambda) = \prod_{\alpha \in \Delta_m^\vee} \frac{\langle \lambda + \rho, \alpha \rangle}{\langle \rho, \alpha \rangle}, (\lambda \in \mathfrak{h}^*).$$

Here we briefly explain some results which were proved in Sthanumoorthy and Tamba [200].

Remark 190. The above formula for $d_m(\lambda)$ is actually the Weyl formula for the dimension of irreducible module with highest weight λ.

Explanation of the techniques involved

We will use the following form of the Macdonald's identity for the affine Lie algebra $\mathcal{G}$ of the type $A_l^{(1)}$ or $B_l^{(1)}$ (the identity is true for any $X_l^{(1)}$) [8]:

$$e\left(-\frac{|\rho|^2}{2g}\delta\right)\Pi_{n\geq 1}((1-e(-n\delta))^l\cdot\Pi_{\alpha\in\Delta}(1-e(-n\delta+\alpha))$$

$$=\sum_{\alpha\in M}\chi(g\alpha)e\left(-\frac{1}{2g}|\rho+g\alpha|^2\delta\right), \tag{6.1}$$

where for $\lambda\in\mathfrak{h}^*$. Here

$$\chi(\lambda)=\frac{\sum_{w\in W}\epsilon(w)e(w(\lambda+\rho)-\rho)}{\Pi_{\alpha\in\Delta_+}(1-e(-\alpha))} \tag{6.2}$$

Let $\alpha_1,\ldots,\alpha_l$ denote the simple roots of $\mathcal{G}$. For any divisor m of g, where g stands for the dual Coxeter number of $\mathcal{G}$, let ϕ_m denote the specialization $\phi_m(e(-\delta))=q$, $\phi_m(e(-\alpha_i))=w$ $(i=1,2,\ldots,l)$. Here w denotes the primitive root of degree m of unity. Using the above notations, we state the following Lemma 16 along with Theorems 80, 81 and Corollaries 7–10 proved in Sthanumoorthy and Tamba [200].

Lemma 16. *For* $\alpha\in M,\phi_m(\chi(g\alpha))=d_m(g\alpha)$.

Let $[x]$ denote the greatest integer contained in x. Let η_p denote the number of roots in Δ_+ with height p [203] and for $0\leq j\leq m$, let $N_j(m)$ denote the number of roots in Δ with height congruent to $j(\text{mod } m)$. One can easily see that

$$N_0(m)=2\sum_{k=1}^{[k/m]}\eta_{km}\quad(\text{for } j=0) \tag{6.3}$$

and

$$N_j(m)=\eta_j+\sum_{k=1}^{[h/m]-1}(\eta_{km-j}+\eta_{km+j})+\eta_{[k/m]m-j'}\quad(\text{for } 1\leq j<\text{m}). \tag{6.4}$$

Now applying ϕ_m to Equation (6.1) and using Equations (6.2)–(6.4) along with the strange formula of Frendental de varies [8],

$$\frac{|\rho|^2}{2g} = \frac{\dim \mathcal{G}}{24}, \tag{6.5}$$

we obtain

$$q^{\dim\ \mathcal{G}/24}\Pi_{n\geq 1}\left((1-q^n)^l \Pi_{j=0}^{m-1}(1-q^n\omega^j)^{N_j(m)}\right) = \sum_{\alpha\in M} d_m(g\alpha)q^{(1/2g)|\rho+g\alpha|^2}. \tag{6.6}$$

One can use the known fact about η_p [203] and can compute η_p: we have

for $\mathcal{G} = A_l$, $\eta_p = l + 1 - p$ and

$$\text{for } \mathcal{G} = B_l,\ \eta_p = \begin{cases} l - p/2 & \text{if } p \text{ is even} \\ l - (p-1)/2 & \text{if } p \text{ is odd;} \end{cases} \quad .$$

We shall discuss three cases separately.

Case (i): $\mathcal{G}$ is of type A_l. By Equations (6.3) and (6.4), we have:

$$N_0(m) = ((l+1)^2/m) - (l+1) \text{ and} \tag{6.7}$$

$$N_j(m) = (l+1)^2/m \text{ for } 1 \leq j \leq m. \tag{6.8}$$

Case (ii): $\mathcal{G}$ is of type B_l and $m = 1$. In this case $[\frac{h}{m}] = 2l$. Hence by Equation (6.3), one can prove that

$$N_0(1) = 2l^2. \tag{6.9}$$

Case (iii): $\mathcal{G}$ is of type B_l and $m > 1$. In this case $[\frac{h}{m}] = \frac{(2l-1)}{m}$. Hence by Equation (6.3), one can directly verify that

$$N_0(m) = (2l^2 + l - 1)/m - (l-1). \tag{6.10}$$

Moreover

$$\eta_{km-j} + \eta_{km+j} = \begin{cases} 2l - km \text{ if } j \text{ is even and } k \text{ is even (or) } j \text{ is odd and } k \text{ is odd} \\ 2l - km + 1 \text{ if } j \text{ is even and } k \text{ is odd (or) } j \text{ is odd and } k \text{ is even.} \end{cases}$$

Hence we have by Equation (6.4), that for $1 \leq j < m$ and j even,

$$N_j(m) = (2l^2 + l - 1)m. \tag{6.11}$$

Similarly, one can prove that for $1 \leq j < m$ and j odd,

$$N_j(m) = (2l^2 + l - 1)/m. \tag{6.12}$$

Now Theorems 80 and 81 given below are the direct consequences of Equations (6.3) and (6.4) along with Equations (6.8), (6.9), (6.11), and (6.12) and also the following identities,

$$\prod_{j=1}^{n-1}(1 - a\omega^j) = (1 - a^n)(1 - a)^{-1}, (a \neq 1).$$

Theorem 80. *For $l \geq 1$, let $m \leq l + 1$ be any divisor of $l + 1$. Then*

$$\eta(q^m)^{(l+1)^2/m}\eta(q)^{-1} = \sum_{\alpha \in M} d_m((l + 1)\alpha)q^{(1/2(l+1))|\rho+(l+1)\alpha|^2},$$

where $d_m(\lambda)$ was defined earlier.

Theorem 81. *For $l \geq 3$, let m be any divisor of $(2l - 1)$. Then*

$$\eta(q)\eta(q^m)^{(2l^2+l-1)/m} = \sum_{\alpha \in M} d_m((2l - 1)\alpha)q^{(1/2(2l-1))|\rho+(2l-1)\alpha|^2}.$$

The above identities are generalizations of Macdonald's formula for $\eta(q)^{(l+1)^2-1}$ and $\eta(q)^{2l^2+l}$. The consequence of the above theorems are the following corollaries. Furthermore, using the following identity due to Euler [204], namely

$$\eta(q) = \sum_{r \in Z}(-1)^r q^{(1/24)(6r+1)^2}$$

and by replacing q by $q^{24/(l+1)^2}$ in Theorem 80, we obtain:

$$\eta(q^{24m/(l+1)^2})^{(l+1)^2/m} = \sum_{\alpha \in M,\ r \in Z} (-1)^r d_m((l + 1)\alpha)q^{(12/(l+1)^3)\{(|\rho+(l+1)\alpha|^2)+1/(l+1)^2(6r+1)^2\}}. \quad (6.13)$$

Now Corollary 7 follows by taking $n = 12$ and $m = 6$, Corollary 8 follows by taking $n = 4$ and $m = 2$ and Corollary 9 follows by taking $n = 6$ and $m = 3$, respectively. Corollary 10 follows by taking $m = 2\ell - 1$, by Theorem 80.

Corollary 7.

$$\eta(q)^{24} = \sum_{\substack{r_1,\ldots,r_{12}; \\ r \in \mathbb{Z}_{r_1+\cdots+r_{12}=0}}} \left\{(-1)^r.D_1(r_1, \ldots, r_{12}).q^{(1/44)(6r+1)^2} + \sum_{i=1}^{12}(13 - 2i + 24r_i)^2/4\right\},$$

where

$$D_1(r_1,\dots,r_{12}) = \prod_{i=1}^{6}(2(r_i - r_{i+6}) + 1).$$

Corollary 1 gives new formula for the generating function $\eta(q)^{24}$ of the Ramanujan's τ-function [237]. This is different from that of Dyson [205] and Lepowsky [198, 203].

Corollary 8.

$$q\{(1-q^3)(1-q^6)\dots\}^8 = \sum_{\substack{r_1,\dots,r_4;\\ r\in\mathbb{Z}_{r_1+\dots+r_4=0}}} \left\{(-1)^r.D_2(r_1,\dots,r_4).q^{(1/16)(6r+1)^2+(3/4)} \sum_{i=1}^{4}(5-2i+8r_i)^2\right\},$$

where

$$D_2(r_1,\dots,r_4) = (2(r_1 - r_3) + 1)(2(r_2 - r_4) + 1).$$

Corollary 9.

$$q\{(1-q)^2(1-q^4)\dots\}^{12} = \sum_{\substack{r_1,\dots,r_6;\\ r\in\mathbb{Z}_{r_1+\dots+r_6=0}}} \left\{(-1)^r.D_3(r_1,\dots,r_6).q^{(1/36)(6r+1)^2+(1/2)} \sum_{i=1}^{6}(7-2i+12r_i)^2\right\},$$

where

$$D_3(r_1,\dots,r_6) = (2(r_1 - r_4) + 1)(2(r_2 - r_5) + 1)(2(r_3 - r_6) + 1).$$

Corollaries 8 and 9 give formulas for the generating function G_k of the Ramanujan's ψ_α-functions [77, 206].

Corollary 10. *For $l \geq 3$, we have*

$$\eta(q)\eta(q^{2l-1})^{l+1} = \sum_{\alpha\in M} d_{2l-1}((2l-1)\alpha)q^{(1/2(2l-1))|\rho+(2l-1)\alpha|^2}.$$

6.6 Some special infinite dimensional Lie algebras

Definition 279. The *oscillator (Heisenberg) algebra $\mathcal{A}$* is a complex Lie algebra with a basis $\{a_n, n \in \mathbb{Z}; \mathfrak{h}\}$, satisfying the commutation relations,

$$[\mathfrak{h}, a_n] = 0 \ (n \in \mathbb{Z}) \text{ and}$$
$$[a_m, a_n] = \delta_{m,-n}\mathfrak{h} \ (m, n \in \mathbb{Z}).$$

In particular, $[a_0, a_n] = 0$ for $n \in \mathbb{Z}$. Alternatively, a *Heisenberg algebra* $\mathcal{A}$ can also be defined as a Lie algebra with basis p_i, q_i ($i = 1, 2, \ldots$), and c, with the following commutation relations:

$$[p_i, q_i] = c \ (i = 1, 2, \ldots),$$

all the other brackets are zero.

In general, one can define a *finite dimensional Heisenberg Lie algebra* as follows:

Definition 280. A *finite dimensional Heisenberg Lie algebra* $\mathcal{A}$ of order n has a basis $\{e_i, f_i, z, i = 1, 2, \ldots, n\}$ such that $[e_i, f_i] = \delta_{ij}z$ for $i = 1, 2, \ldots, n$ and all other brackets are zero. Introduce the following commutative subalgebras $\mathcal{A}_-, \mathcal{A}_+$ of $\mathcal{A}$ ($\mathcal{A} = \mathcal{A}_- \oplus \mathcal{A}_+$) with

$$\mathcal{A}_+ = \sum_{j \geq 1} \mathbb{C}p_j; \quad \mathcal{A}_- = \sum_{j<1} \mathbb{C}q_j.$$

A vector v of an $\mathcal{A}$ module is called a *vacuum vector* with eigenvalue $\lambda \in \mathbb{C}$, if $\mathcal{A}_+(v) = 0$ and $c(v) = \lambda v$.

Example 93. Lie algebra of all matrices $\begin{pmatrix} 0 & a & c \\ 0 & 0 & b \\ 0 & 0 & 0 \end{pmatrix}$ where $a, b, c \in \mathbb{R}$, is a nilpotent Lie algebra. This Lie algebra is actually a Heisenberg Lie algebra.

Definition 281 (Heisenberg group). H is the set of all 3×3 matrices of the form $A = \begin{pmatrix} 1 & a & b \\ 0 & 1 & c \\ 0 & 0 & 1 \end{pmatrix}$ where $a, b, c \in \mathbb{R}$. It forms a group under matrix multiplication such that *inverse* of $\begin{pmatrix} 1 & a & b \\ 0 & 1 & c \\ 0 & 0 & 1 \end{pmatrix} = \begin{pmatrix} 1 & -a & ac-b \\ 0 & 1 & -c \\ 0 & 0 & 1 \end{pmatrix}$. This Heisenberg group H is a subgroup of $Gl(3, \mathbb{R})$. Hence it is a matrix Lie group.

Definition 282. Let $\mathcal{A}$ be a Heisenberg algebra generated by p_i, q_i, and c. For every complex number a, the Lie algebra $\mathcal{A}$ has an irreducible representation σ_a, called the *canonical commutation relations representation*, on the space $R = \mathbb{C}[x_1, x_2, \ldots]$ of polynomials in infinitely many indeterminates x_i defined by

$$\sigma_a(p_i) = \frac{\partial}{\partial x_i}; \quad \sigma_a(q_i) = x_i; \quad \sigma_a(c) = aI_R.$$

We denote this $\mathcal{A}$-module by R_a.

Definition 283 (Witt algebra). Let M be a differentiable manifold. If X is a vector field on M, then $X(p)$ is a tangent vector at p of M. For example if M is a circle S^1 then we can define vector field on S^1 by $X = f(\phi)\frac{d}{d\phi}$. Here f is a periodic function such that $f(\phi) = f(\phi + 2\pi)$. If $X = f(\phi)\frac{d}{d\phi}$, $Y = g(\phi)\frac{d}{d\phi}$, then we can define a Lie bracket $[.,.]$ by

$$[X, Y] = \left(f(\phi)\frac{dg}{d\phi} - g(\phi)\frac{df}{d\phi}\right)\frac{d}{d\phi}.$$

If we define f by Fourier decomposition,

$$f(\phi) = a + \sum_{k=1}^{\infty} b_k \cos k\phi + \sum_{k=1}^{\infty} c_k \sin k\phi,$$

then one can write a basis $\left\{\frac{d}{d\phi}, \cos k\phi \frac{d}{d\phi}, \sin k\phi \frac{d}{d\phi} \mid k = 1, 2, 3, \ldots\right\}$ for the real Lie algebra on vector field S^1 denoted by Vect S^1. Now we can define a complex Lie algebra called *Witt algebra* denoted by W as

$$W = Vect\, S^1 + \mathrm{i}\; Vect\, S^1.$$

This Witt algebra W has a basis

$$d_n = \mathrm{i}\exp(\mathrm{i}\, n\phi)\frac{d}{d\phi} = -z^{n+1}\frac{d}{dz}(n \in Z),$$

where $z = e^{\mathrm{i}\phi}$. One can verify that

$$[d_l, d_m] = d_l d_m - d_m d_l = (l - m)d_{l+m} = \sum_n C^n_{lm} d_n, (l, m \in Z)$$

with $C^n_{lm} = (l - m)\delta^n_{l+m}$. Define a two cocycle ϕ on W as $\phi : W \times W \to C$. Now define $W_{\text{ext}} := W \oplus Cc$ with

$$[d_l + \mu c, d_m + \nu c]_{W_{\text{ext}}} := [d_l, d_m]W + \phi(d_l, d_m)c.$$

W_{ext} is the central extension of W with one-dimensional Lie algebra $K = \mathbb{C}c$.

Definition 284 (Virasoro algebra).

(1) *Construction of Virasoro algebra by one-dimensional central extension of Witt algebra*: Let W be the Witt algebra with basis d_n. Then every Lie algebra which is a nontrivial one-dimensional central extension of the Witt algebra is called the Virasoro algebra generated by $\{d_n, n \in \mathbb{Z}, c\}$ with commutation relations: $[d_m, d_n] = (m - n)d_{m+n} + a(m, n)c$, and $[d_m, c] = 0$. Using certain conditions, one can establish that $a(m, n) = \delta_{m,-n}\frac{(m^3-m)}{12}c$. Now we get a Virasoro algebra '*Vir*' with a basis $\{d_m, m \in Z; c\}$ satisfying the following commutation relations:

$$[d_m, c] = 0$$

and

$$[d_m, d_n] = (m-n)d_{m+n} + \delta_{m,-n}\frac{(m^3-m)}{12}c.$$

(2) *Virasoro algebra can be constructed directly as follows*: Let $\mathcal{A}$ be a complex Lie algebra with a basis,

$$\left\{d_n = \mathrm{i}\exp(\mathrm{i}\,n\theta)\frac{d}{d\theta} = -z^{n+1}\frac{d}{dz}(n \in \mathbb{Z}),\ z = e(\mathrm{i}\theta)\right\},$$

where $z = \exp(\mathrm{i}\,\theta)$. These elements satisfy the commutation relations:

$$[d_m, d_n] = (m-n)d_{m+n}(m, n \in \mathbb{Z}). \tag{6.14}$$

Let $\widehat{\mathcal{A}}$ denote the extensions of $\mathcal{A}$ by a one-dimensional center $\mathbb{C}c$. This means that $\widehat{\mathcal{A}} = \mathcal{A} \oplus \mathbb{C}c$.

This $\widehat{\mathcal{A}}$ will become an *infinite dimensional Lie algebra* with the commutation relations as in Equation (6.14).

Taking $a(m,n) = \frac{(m^3-m)}{12}c$, we arrive at a Lie algebra called the Virasoro algebra denoted by Vir with a basis, $\{d_m, m \in \mathbb{Z}; c\}$.

Remark 191. For Virasoro algebras in basic representations, one can refer Wakimoto [4].

Definition 285 (Triangular decomposition of Vir and a Vir module). We have the following triangular decomposition of Vir:

$$Vir = Vir_- \oplus Vir_0 \oplus Vir_+.$$

Here $Vir_\pm = \oplus_{j>0}\mathbb{C}d_{\pm j}$; $Vir_0 = \mathbb{C}c \oplus \mathbb{C}d_0$.

Given $c, h \in \mathbb{C}$, define a Vir module V with highest weight (c, h) as follows:

There exists a nonzero vector $v = v_{c,h}$ such that

$$Vir_+(v) = 0,\ \mathcal{U}(Vir_-)v = V,\ d_0(v) = hv,\ c(v) = cv.$$

Here $\mathcal{U}$ denotes the corresponding universal enveloping algebra.

Definition 286. Let $V = \bigoplus_{j\in\mathbb{Z}} \mathbb{C}v_j$ with a basis $\{v_j | j \in \mathbb{Z}\}$, where v_j is a column vector with 1 in the jth place, 0 elsewhere. Any vector in V has only a finite, but arbitrary number of nonzero coordinates $(a_i)_{i\in Z}$. Hence $V = C^\infty$, the space of column vectors. Now we define the Lie algebra gl_∞ by

$$gl_\infty = \{(a_{ij})_{i,j\in Z}|\ \text{all but a finite number of the } a_{ij}'s \text{ are } 0\}$$

with the Lie bracket being the ordinary commutator.

The elements $E_{ij}(i, j \in \mathbb{Z})$ form a basis for gl_∞ with

$$E_{ij}v_k = \delta_{jk}v_i,$$
$$E_{ij}E_{mn} = \delta_{jm}E_{in}.$$

We have the following commutation relations:

$$[E_{ij}, E_{mn}] = \delta_{jm}E_{in} - \delta_{ni}E_{mj.}$$

The Lie algebra gl_∞ is the Lie algebra of the group defined by,

$$GL_\infty = \{A = (a_{ij})_{i,j \in Z} |\ A \text{ is invertible and all but a finite number of } a_{ij} - \delta_{ij} \text{ are } 0\}$$

with group operation being the matrix multiplication.

Definition 287 (An alternative approach to $gl(\infty)$ using fermions). Let A be an associative algebra generated by the elements $\psi_i,\ \psi_n{}^*$ $(n \in \mathbb{Z})$, satisfying the canonical anti-commutation relations,

$$[\psi_m, \psi_n]_+ = [\psi_m^*, \psi_n^*]_+ = 0 \ \text{ and } [\psi_m^*, \psi_n]_+ = \delta_{m,n}.$$

The elements ψ_n and ψ_n^* are called fermions. Let

$$\mathcal{G} = \{\sum c_m \psi_m \psi_m^* |\ \psi_m, \psi_m^* \in A\}.$$

Then $\mathcal{G}$ is a Lie algebra with the bracket,

$$[\psi_m\psi_n^*, \psi_p\psi_q^*] = \delta_{np}\psi_m\psi_q^* - \delta_{mq}\psi_p\psi_n^*.$$

Here $\psi_m\psi_n^*$ can be identified with the matrix E_{mn} of infinite rank given by

$$(E_{mn})_{pq} = \delta_{mp}\delta_{nq}.$$

Consider an expression of the form,

$$X = \sum a_{mn} : \psi_m\psi_n^* : +ac,$$

where c is the central element and $a_{mn},\ a \in C$ are such that

$$a_{mn} = 0 \ if \ |m - n| \gg 0.$$

Define a Lie bracket

$$[: \psi_m\psi_n^* :, : \psi_p\psi_q^* :] = \delta_{mp} : \psi_m\psi_q^* : -\delta_{mq} : \psi_p\psi_n^* : +\delta_{np}\delta_{mq}(Y_-(q) - Y_-(m))c,$$

where

$$Y_-(m) = \begin{cases} 0 \text{ if } m \geq 0 \\ 1 \text{ if } m < 0 \end{cases}.$$

Then, $gl(\infty)$ is the Lie algebra generated by the elements of the form X satisfying the above relations.

6.7 Hirota bilinear differential operators and soliton solutions for KdV equation

A connection between infinite dimensional Lie algebras and integrable systems first appeared in the two dimensional Ising model solved by Onsagar [207]. Hirota [208] introduced a new calculus to construct n-soliton solutions.

Definition 288 (Hirota derivative or Hirota differential operator). One can define a new operator D_x on an ordered pairs of functions $\sigma(x)$ and $\tau(x)$ as

$$D_x(\sigma \cdot \tau) = \lim_{\epsilon \to 0} \frac{\partial}{\partial \epsilon} \sigma(x+\epsilon)\tau(x-\epsilon).$$

Here D_x is called the *Hirota derivative.* Alternatively, if $f(x)$ and $g(x)$ are functions in one variable $x \in \mathbb{R}$, define D_x on f and g by

$$\begin{aligned} D_x(f \cdot g) &= \frac{\partial}{\partial y}(f(x+y)g(x-y))|_{y=0} \\ &= f_x g - f g_x. \end{aligned}$$

Here D_x is called the *First order Hirota bilinear differential operator.*

"Higher order" Hirota bilinear differential operators $D_x^k (k \in \mathbb{Z}_{\geq 0})$ are defined on two functions f and g as follows:

$$\begin{aligned} D_x^k(f \cdot g) &:= \left(\frac{\partial}{\partial y}\right)^k f(x+y)g(x-y)|_{y=0}. \\ &= \sum_{j=0}^{k} (-1)^j \binom{k}{j} \partial_x^{k-j} f \cdot \partial_x^j g. \end{aligned}$$

In particular, $D_x^2 f \cdot g = \dfrac{\partial^2 f}{\partial x^2} g - 2\dfrac{\partial f}{\partial x}\dfrac{\partial g}{\partial x} + f\dfrac{\partial^2 g}{\partial x^2}.$

Moreover, $D_t D_x f \cdot f = 2\left(\dfrac{\partial^2 f}{\partial t \partial x} - \dfrac{\partial f}{\partial t}\dfrac{\partial f}{\partial x}\right).$

In general, given two functions $f(x)$ and $g(x)$ of a single variable x, we can write the Taylor expansion of $f(x+y)g(x-y)$ around $y=0$ in the form,

$$f(x+y)g(x-y) = \sum_{j=0}^{\infty} \frac{1}{j!} D_x^j(f \cdot g) y^j.$$

The operator $(f,g) \to D_x^j(f \cdot g)$ is the jth *Hirota derivative* with respect to the variable x on the pair (f,g). Hirota derivatives in many variables can be defined similarly.

This definition can be extended to the functions $\sigma(x_1, x_2, \ldots)$ and $\tau(x_1, x_2, \ldots)$ of infinitely many variables and higher power of the operator as follows:

$$D_{x_1}^{\alpha_1} D_{x_2}^{\alpha_2}, \ldots, D_{x_n}^{\alpha_n} \sigma \cdot \tau = \prod_{r=1}^{n} \lim_{\epsilon_r \to 0} \frac{\partial^{\alpha_r}}{\partial \epsilon_r^{\alpha_r}} \sigma(x_r + \epsilon_r)\tau(x_r - \epsilon_r).$$

This is same as

$$D_x^i D_y^j, \ldots, D_z^k \sigma \cdot \tau = \left(\frac{\partial}{\partial x} - \frac{\partial}{\partial x'}\right)^i \left(\frac{\partial}{\partial y} - \frac{\partial}{\partial y'}\right)^j, \ldots, \left(\frac{\partial}{\partial z} - \frac{\partial}{\partial z'}\right)^k$$
$$\sigma(x, y, \ldots, z)\tau(x', y', \ldots, z')|x = x', y = y', \ldots, z = z'.$$

Following can be directly established

$$\begin{aligned}
\partial_x D_x^k &= D_x^k \partial_x, \\
D_x^k(f \cdot g) &= (-1)^k D_x^k(g \cdot f), \\
D_x^k(f \cdot 1) &= \partial_x^k f, \\
D_x^k(1 \cdot g) &= (-1)^k \partial_x^k g, \text{ here 1 is the Identity function}, \\
D_x^k(\mathrm{e}^{ax} \cdot \mathrm{e}^{bx}) &= (a-b)^k \mathrm{e}^{ax}\mathrm{e}^{bx}, \\
D_x^k(f \cdot f) &= 0 \text{ if } k \text{ is odd.}
\end{aligned}$$

In fact, we will have

$$D_x^i D_y^j \ldots D_z^k \sigma \cdot \tau = (-1)^{i+j+\cdots+k} D_x^i D_y^j \ldots D_z^k \tau \cdot \sigma.$$

Here $D_x, D_y, \ldots$ are Hirota derivatives.

Hirota differential operator on n-variables

In general, Hirota bilinear differential operators acting on functions with n-variables $(x_1, \ldots, x_n)$, where n is any positive integer and fixed, can be explained as follows: Let

$$P \in \mathbb{C}[x_1, \ldots, x_n].$$

Hirota bilinear differential operator $P(D_1,\ldots,D_n)(f\cdot g) := P(\frac{\partial}{\partial y_1},\ldots,\frac{\partial}{\partial y_n})(f(x_1+y_1,\ldots,x_n+y_n)\times g(x_1-y_1,\ldots,x_n-y_n))|_{y_1=\cdots=y_n=0}$.

Remark 192. As in the one-dimensional case,

$$P(D_1,\ldots,D_n)f\cdot f = 0 \text{ if } \deg(P) \text{ is odd}$$

(Degree of polynomial functions $P(x_1\cdots x_n)$ is the degree with respect to usual gradation $\deg(x_1)=\cdots=\deg(x_n)=1$.)

Hirota derivatives using Taylor expansion

Let

$$f=f(x_1,x_2,\ldots) \quad \text{and} \quad g=g(x_1,x_2,\ldots.).$$

Now, we have,

$$(e^{y_1D_1+y_2D_2+\cdots})f\cdot g = f(x_1+y_1,x_2+y_2,\ldots)g(x_1-y_1,x_2-y_2,\ldots). \tag{6.15}$$

In particular,

$$(e^{a_1D_1+a_2D_2+\cdots})f\cdot g = f(x_1+a_1,x_2+a_2,\ldots)g(x_1-a_1,x_2-a_2,\ldots)$$

from which, we can get all Hirota derivatives of all orders by expanding L.H.S. as a series in $(a_1,a_2,\ldots)$ and comparing with R.H.S. Expanding L.H.S in Equation (6.15), Taylor series, we get

$$f\cdot g + y_1(D_1f\cdot g) + y_2(D_2f\cdot g) + \cdots + \frac{1}{2}y_1^2(D_1^2f\cdot g) + \cdots.$$

Hence comparing with R.H.S., we can get all Hirota derivatives.

Transformation of KdV equation in Hirota form and its solution

Consider the KdV equation,

$$u_t - \frac{3}{2}uu_x - \frac{1}{4}u_{xxx} = 0.$$

After making the transformation

$$u(x,t) = 2\frac{\partial^2}{\partial x^2}\log f(x,t),$$

one can transform the KdV equation as,

$$2(ff_{xxxx} - 4f_xf_{xxx} + 3f_{xx}^2) - 4 \times 2(f_{tx}f - f_tf_x) = 0.$$

This can be transformed into Hirota form as

$$(D_x^4 - 4D_xD_t)(f \circ f) = 0.$$

Naturally $u = 2\frac{\partial^2}{\partial x^2} \log f(x; t)$ is a solution of KdV equation.

A solution $f(x,t)$ of the above Hirota bilinear differential equation is called a τ-function. Alternatively, the KdV equation can be written in the form,

$$q_t + 6qq_x + q_{xxx} = 0.$$

After making the transformation,

$$q(x,t) = 2\frac{\partial^2}{\partial x^2} \log \tau,$$

one can transform the above KdV equation to

$$\tau\tau_{xt} - \tau_x\tau_t + \tau\tau_{xxx} - 4\tau_x\tau_{xxx} + 3\tau_{xx}^3 = 0.$$

This can be transformed to $(D_xD_t - D_x^4)\tau \cdot \tau = 0$ in terms of Hirota derivatives.

Remark 193.

(1) For the above KdV equation, $q_t + q_{xxx} + 6qq_x = 0$, an ordinary solution is $q(x,t) = (1/2)sech^2(\sqrt{c/2})(x - ct - a)$.

(2) For Solitons, Vertex operators, etc., one can also refer M. Jimbo and T. Miwa [222].

Soliton solutions for KdV equation can also be constructed as follows:

Remark 194 (n-soliton solution for KdV equation). For the above KdV equation,

$$\tau\tau_{xt} - \tau_x\tau_t + \tau\tau_{xxx} - 4\tau_x\tau_{xxx} + 3\tau_{xx}^3 = 0,$$

consider the corresponding polynomial,

$$P(k_1, k_2) = 4k_1k_2 - k_1^4.$$

Clearly $(2k, 2k^3)$ is a solution set for $P = 0$ and hence $\tau = 1 + ce^{2kx+2k^3t}$ is a 1-*soliton solution.* Assuming two solution sets $(2k_1, 2k_1^3)$ and $(2k_2, 2k_2^3)$, we get

$$\tau = 1 + \epsilon(c_1e^{2k_1x+2k_1^3t} + c_2e^{2k_2x+2k_2^3t})\epsilon^2\left(\frac{k_1 - k_2}{k_1 + k_2}\right)^2 c_1c_2e^{2(k_1+k_2)x+2(k_1^3+k_2^3)t}$$

is a 2-*soliton solution*. In general, the n-soliton solution of the KdV equation can be constructed as follows: Consider the infinite number of variables $x_1, x_3, x_5, x_7, \ldots$ with the notation

$$\xi(x,k) = \sum_{i=1}^{\infty} x_i k^i.$$

Introduce the parameters $c_1, \ldots, c_n$ and $k_1, \ldots, k_n$. Define the exponents ξ_i by

$$\xi_i = \xi(x,k_i) - \xi(x,-k_i) = 2\sum_{j=0}^{\infty} k_i^{2j+1} x_{2j+1}.$$

Let $a_{ii'} = \left(\frac{k_i - k_i'}{k_i + k_i'}\right)^2$. If $I = \{1, \ldots, n\}$, let

$$\tau(x_1, x_3, \ldots) = \sum_{J \subset I} \left(\prod_{i \in J} c_i\right) \left(\prod_{\substack{i,i' \in J \\ i < i'}} a_{ii'}\right) \exp\left(\sum_{i \in J} \xi_i\right),$$

where the sum is taken over all subsets J of I. Then τ is the *n-soliton solution* for the KdV equation.

6.8 Principal vertex operator construction of basic representation and homogeneous vertex operator construction of the basic representation

(i) *Preliminaries*:

Definition 289 (Algebra of formal power series and Vertex operators). Let $R = \mathbb{C}[x_1, x_2, \ldots]$ be the algebra of polynomials in infinitely many indeterminates x_i and let $\widehat{R} = \mathbb{C}[[x_1, x_2, \ldots]]$ be the algebra of formal power series, that is the algebra of all linear combinations of finite monomial in x_i. So $\hat{R}$ is the formal completion of R. In two variables x_1 and x_2,

$$\mathbb{C}[x_1, x_2] = \left\{ \sum_{m,n \in \mathbb{N}} v_{mn} x_1^m x_2^n |\ v_{mn} \in \mathbb{C};\ \text{all but finitely many } v_{mn} = 0 \right\}.$$

The corresponding $\hat{R}$ is:

$$\mathbb{C}[[x_1, x_2]] = \left\{ \sum_{m,n \in \mathbb{N}} v_{mn} x_1^m x_2^n |\ v_{mn} \in \mathbb{C} \right\}.$$

This above space is the *algebra of formal power series* in x_1, x_2. A differential on R is, in general, an infinite sum of the form

$$\sum_{r\geq 0}\sum_{1\leq i_1\leq\cdots\leq i_r} P_{i_1\ldots i_r}\frac{\partial}{\partial x_{i_1}},\ldots,\frac{\partial}{\partial x_{i_r}},$$

where $P_{i_1\ldots i_r}\in\hat{R}$. This operator defines a linear map $D: R\to\hat{R}$. Now one can define an operator T_λ by

$$(T_\lambda f)(x_1,x_2,\ldots)=f(x_1+\lambda_1,x_2+\lambda_2,\ldots)$$

By Taylor's formula, it is clear that,

$$T_\lambda=\underset{i\geq 1}{\exp}\lambda_i\frac{\partial}{\partial x_i}=e^{\lambda_i\frac{\partial}{\partial x_i}}.$$

For the proof of the following lemma, one can refer Kac [8].

Lemma 17.

(1) *If D satisfies $[x_i,D]=\lambda_i D, i=1,2,\ldots$, then $D=D(1)\exp\left(-\sum_i\lambda_i\frac{\partial}{\partial x_i}\right)$.*

(2) *If D satisfies $[\frac{\partial}{\partial x_i},D]=\mu_i D, i=1,2,\ldots$, then $D(1)=c\exp(\sum_i\mu_i x_i), c\in\mathbb{C}$.*

(3) *If D satisfies $[x_i,D]=\lambda_i D, i=1,2,\ldots$ and $[\frac{\partial}{\partial x_i},D]=\mu_i D, i=1,2,\ldots$ then $D=c\exp(\sum_i\mu_i x_i)\exp(-\sum_i\lambda_i\frac{\partial}{\partial x_i})$ for some $c\in\mathbb{C}$.*

Definition 290. Let $f(x)$ be real or complex valued function and consider $e^{\lambda\frac{d}{dx}}$, a operator acting on $f(x)$ for $\lambda\in\mathbb{R}$ or $\mathbb{C}$.

$$\begin{aligned}e^{\lambda\frac{d}{dx}}f(x)&:=\sum_{j=0}^{\infty}\frac{\lambda^j}{j!}\left(\frac{d}{dx}\right)^j f(x)\\&=f(x+\lambda).\end{aligned}$$

Again

$$\begin{aligned}e^{\lambda\frac{d}{dx}}(g(x)f(x))&=g(x+\lambda)f(x+\lambda),\\&=g(x+\lambda)e^{\lambda\frac{d}{dx}}f(x).\end{aligned}$$

In particular, $e^{\lambda\frac{d}{dx}}(e^{\mu x}f(x))=e^{\lambda\mu}e^{\mu x}e^{\lambda\frac{d}{dx}}f(x)$.

Consider the space $\mathbb{C}[x_1,x_2,\ldots]=\mathbb{C}(x_j,\ (j\in\mathbb{N}))$.

Let $V_\lambda^+(Z):=\exp\left(-\sum_{j=1}^{\infty}\frac{\lambda}{j}\frac{\partial}{\partial x_j}Z^{-j}\right)$ and

$$V_\lambda^-(Z):=\exp\left(\sum_{j=1}^{\infty}\lambda x_j Z^j\right).$$

Here z is a formal variable and $\lambda \in \mathbb{C}$. Operator of the form,

$$V_\lambda(Z) := V_\lambda^-(Z)V_\lambda^+(z)$$
$$= \exp\left(\sum_{j=1}^{\infty} \lambda x_j Z^j\right) \exp\left(-\sum_{j=1}^{\infty} \frac{\lambda}{j} \frac{\partial}{\partial x_j} Z^{-j}\right)$$

is called a vertex operator.

In general, operators of the form $\exp(\sum_j \lambda_j x_j)\exp(-\sum_j \mu_j \frac{\partial}{\partial x_j})$ are called vertex operators.

Example 94 (KP equation, vertex operators and n-soliton solution). In the case of KP equation,

$$\frac{3}{4}\frac{\partial^2 u}{\partial x_2^2} = \frac{\partial}{\partial x}\left(\frac{\partial u}{\partial x_3} - \frac{3}{2}u\frac{\partial u}{\partial x} - \frac{1}{4}\frac{\partial^3 x}{\partial x^3}\right),$$

the transformed equation in Hirota's form [209] is

$$(D_1^4 + 3D_2^2 - 4D_1 D_3)\tau \cdot \tau = 0.$$

The vertex operator here is

$$X(p,q) = \exp\left(\sum_{j=1}^{\infty} \lim(p^j - q^j)x_j\right) \exp\left(-\sum_{j=1}^{\infty} \frac{1}{j}(p^{-j} - q^{-j})\frac{\partial}{\partial x_j}\right),$$

where $\xi_i = \xi(X,p_i) - \xi(X,q_i)$. $\tau = e^{c_1 X(p_1,q_1)}, \ldots, e^{c_n X(p_n q_n)}$ is an n-soliton solution. If we write

$$a_{ii'} = \frac{(p_i - p_{i'})(q_i - q_{i'})}{(p_i - q_{i'})(q_i - p_{i'})}$$

then

$$\tau(x_1, x_2, \ldots) = \tau(x_1, x_3, \ldots) = \sum_{J \subset I}\left(\prod_{i \in j} c_i\right)\left(\prod_{i,i' \in J, i<i'} a_{ii'}\right)\exp\left(\sum_{i \in j} \xi_i\right),$$

where $I = \{1, \ldots, n\}$ and J is a subset of I. In particular, taking $p_i = -q_i$ this τ becomes the τ-function for KdV equation. These vertex operators will be viewed as elements of the corresponding algebras $gl(\infty)$ or $A_1^{(1)}$ as the case may be. (For vertex operators, τ functions of transformation groups for soliton equations, one can refer Date et al. [210].)

(ii) *Principal realization of the basic representation*:

The highest weight module $L(\wedge_0)$ over an affine Lie algebra $\mathcal{G}(A)$ is called the basic representation. Let $\mathcal{G}(A)$ be the affine Lie algebra of type $X_N^{(r)}$ of rank $(l+1)$.

$\mathcal{G}(A) = \mathcal{G}'(A) \oplus \mathbb{C}d$, where $\mathcal{G}'(A)$ is the derived subalgebra, generated by the Chevalley generators $e_i, f_i (i = 0, \ldots, l)$ and d is the scaling element.

Center of $\mathcal{G}'(A)$ coincides with the central element c. Let $\overline{\mathcal{G}}(A) = \mathcal{G}'(A) \oplus \mathbb{C}c$, so that we have an exact sequence,

$$0 \to \mathbb{C}c \to \mathcal{G}'(A) \xrightarrow{\Pi} \overline{\mathcal{G}}(A) \to 0,$$

where $\deg e_i = -\deg f_i = 1$ $(i = 0, \ldots, l)$.

Define the principal gradation,

$$\mathcal{G}'(A) = \oplus_{j \in \mathbb{Z}} \mathcal{G}_j(A).$$

It induces the principal gradation,

$$\overline{\mathcal{G}}(A) = \oplus_{j \in \mathbb{Z}} \overline{\mathcal{G}}_j(A) \text{ for } j \neq 0$$

so that $\dim \bar{\mathcal{G}}_j(1) = \dim \mathcal{G}_j(A)$ for $j \neq 0$. The element $\bar{e} = \sum_{i=0}^{l} \Pi(e_i) \in \bar{\mathcal{G}}_1(1)$ is called the cyclic element of $\overline{\mathcal{G}}(A)$. Let $\bar{s} = \{x \in \bar{\mathcal{G}} | [x, \bar{e}] = 0\}$ be the centralizer of $\bar{e}$ in $\bar{\mathcal{G}}(A)$. It is clear that $\bar{s}$ is graded with respect to principal gradation of $\bar{\mathcal{G}}(A)$,

$$\bar{s} = \bigoplus_{j \in \mathbb{Z}} \bar{s}_j.$$

The subalgebra $s = \Pi^{-1}(\bar{s})$ is called the principal subalgebra of $\mathcal{G}'(A)$ (or $\mathcal{G}(A)$). It is graded with respect to the principal gradation of $\mathcal{G}'(A)$, $s = \oplus_{j \in \mathbb{Z}} s_j$. Note that $\dim s_j = \dim \bar{s}_j$ for $j \neq 0$, $s_0 = \mathbb{C}K$.

The nonzero integers of the set which contains j with multiplicity $\dim s_j$ are called the exponents of the affine algebra $\mathcal{G}(A)$.

One can construct the basic representation $L(\Lambda_0)$ of the affine Lie algebra $\mathcal{G}'(A)$ with the Cartan matrix A of type $X_N^{(r)}$, where either $r = 1$ and A is symmetric or $r > 1$. The basic representation $L(\Lambda_0)$ of an affine Lie algebra $\mathcal{G}'(A)$ with the Cartan matrix A of type $X_N^{(r)}$ (where either $r = 1$ and $X_N = A, D, E$ or $r > 1$ and rank $l + 1$) can be constructed in terms of differential operators. This is called the principal realization of the basic representation.

(iii) *Principal vertex operator construction of the basic representation*:

Let $\mathcal{G}$ be a simple finite dimensional Lie algebra of type X_N and μ be a diagram automorphism of $\mathcal{G}$ of order $r(= 1,\ 2 \text{ or } 3)$. Let $\mathcal{G}(A)$ be an affine algebra of type $X_N^{(r)}$ and rank $l + 1$, such that $\mathcal{G}(A) = \mathcal{G}'(A) + \mathbb{C}d$

Let E_i, F_i and H_i $(0 \leq i \leq l)$ be the generators of $\mathcal{G}$. Center of $\mathcal{G}(A)$ is spanned by $\sum_{i=0}^{l} a_i^{\vee} H_i$ and here $\sum_{i=0}^{l} a_i^{\vee} H_i = 0$.

Let $a_i^{\vee}$ be the labels of the diagram of the transpose of the affine matrix of type $X_N^{(r)}$. $E_i's (0 \leq i \leq l)$ generate Lie algebra $\mathcal{G}$.

Let a_i be the labels of the diagram of the affine matrix X_N^r. The integer $h^{(r)} = r\sum_{i=0}^{l} a_i$ is called the rth Coxeter number of $\mathcal{G}$. We know that $\deg E_i = -\deg F_i = 1, \deg H_i = 0 (i = 0, \ldots, l)$. Define a $\mathbb{Z}/h^{(r)}\mathbb{Z}$ gradation $\mathcal{G} = \oplus_j \mathcal{G}_j(1, r)$ called the r-principal gradation of $\mathcal{G}$. The element $E = \sum_{i=0}^{l} E_i$ is called the cyclic element of $\mathcal{G}$. Denote by $S^{(r)}$, the centralizer of E in $\mathcal{G}$. It is graded with respect to r-principal gradation.

$$S^{(r)} = \bigoplus_{j \in \mathbb{Z}/h^{(r)}Z} S_j^{(r)}.$$

For the proof of the following proposition, one can refer Kac [8].

Proposition 62.

(a) $\dim \mathcal{G}_j(1, r) = l + \dim S_j^{(r)} \quad (j \in \mathbb{Z}/h^{(r)}(\mathbb{Z}))$

(b) *$S^{(r)}$ is a Cartan subalgebra of $\mathcal{G}$.*

(c) *The subspaces $S_i^{(r)}$ and $S_j^{(r)}$ are orthogonal (respectively nondegenerately paired) with respect to a nondegenerate invariant bilinear form on $\mathcal{G}$ if $i+j \neq 0 \bmod h^{(r)}$ (respectively $i + j \equiv 0 \bmod h^{(r)}$).*

Definition 291. The nondecreasing sequences of integers $m_1^{(r)} \leq m_2^{(r)} \leq \cdots$ from the interval $[1, h^{(r)-1}]$, in which j appears with multiplicity $\dim S_j^{(r)}$, is called the set of *r-exponents* of $\mathcal{G}$.

We state the following proposition, the proof of which is in Kac et al. [211]. The space $\hat{L}(\mathcal{G}, \sigma)$ in the following proposition was already explained in Section 2.7 in Chapter 2.

Proposition 63. *Let $\mathcal{G}$ be a simple finite dimensional Lie algebra and let X_n denote its Dynkin diagram. Let $k = 1$, 2 or 3 and let $\mathcal{G}(A)$ denote the Euclidean Lie algebra corresponding to the Cartan matrix $A = X_n^{(k)}$, with canonical generators $e_i, f_i, h_i (i = 0, \ldots, n)$. Then the mapping*

$$\Pi : e_i \mapsto \widetilde{e}_i; \;\; f_i \mapsto \widetilde{f}_i; \;\; h_i \mapsto \widetilde{h}_i, \quad i = 0, \ldots, n$$

defines an epimorphism $\Pi : \mathcal{G}(A) \rightarrow \widehat{L}(\mathcal{G}, \sigma)$ such that $\mathrm{Ker}\Pi$ is one-dimensional and is the center of $\mathcal{G}(A)$ spanned by, $c = \sum_{i=0}^{n} a_i^{\vee} \alpha_i^{\vee}$.

Let $\Delta \subset S^{(r)^*}$ be the set of roots of $\mathcal{G}$ with respect to $S^{(r)}$. For $\beta \in R$, let $A_\beta \in \mathcal{G}$ be a root vector and A_β can also be decomposed with respect to the principal gradation as

$$A_\beta = \sum_{j \in \mathbb{Z}/h^{(r)}\mathbb{Z}} A_{\beta, j}.$$

Therefore the root vectors $A_{\beta_1}, \ldots, A_{\beta_l}$ corresponding to the l roots $\beta_1, \ldots, \beta_l \in S^{(r)^*}$ can be chosen such that their projections on $\mathcal{G}_0(1; r)$ form a basis of this space. For all β and j, we have

$$[E, A_{\beta,j}] = \langle \beta, E \rangle A_{\beta,j+1} \text{ with } \langle \beta, E \rangle \neq 0.$$

Let $T_j (j = 1, \ldots, N)$ be a basis of $S^{(r)}$ such that $T_j \in S^{(r)}_{m_j^{(r)}}$ and $(T_i | T_{N+1-j}) = h\delta_{ij}$ for all $i, j = 1, \ldots, N$. Let $\lambda_{\beta_j} = \langle \beta, T_j \rangle$. Hence $[T_k, A_{\beta,j}] = \lambda_{\beta_k} A_{\beta,j+m_k}(r)$. The set of elements $\{A_{\beta_{i,j}}, T_k\}$ form a basis of $\mathcal{G}$. Note also that $(A_{\beta,j} | T_k) = 0$.

Now the affine algebra $\mathcal{G}'(A)$ can be defined by

$$\mathcal{G}'(A) = \bigoplus_{j \in \mathbb{Z}} (t^j \otimes \mathcal{G}_{j \bmod h^{(r)}}(1; r) \oplus \mathbb{C}c).$$

The Lie bracket in $\hat{\underline{\mathfrak{g}}}'(A)$ is given by

$$[p_1(t) \otimes g_1 \oplus \lambda_1 c, p_2(t) \otimes g_2 \oplus \lambda_2 c] = p_1(t) p_2(t) \otimes [g_1, g_2] \oplus \frac{1}{3} Res \left(\frac{dp_1(t)}{dt} p_2(t) \right) (g_1 | g_2) c,$$

where $g_1, g_2 \in \underline{\mathfrak{g}}$, $p_1(t), p_2(t) \in \mathbb{C}[t, t^{-1}]$, $\lambda_1, \lambda_2 \in \mathbb{C}$. Let $E_+ = \{b_1, b_2, \ldots\}$ be the sequence of positive exponents of $\mathcal{G}(A)$ arranged in nondecreasing order. Now let

$$S_j = t^j \otimes S^{(r)}_{j \bmod h^{(r)}} \oplus \mathbb{C}c \quad \text{and} \quad S = \bigoplus_{j \in \mathbb{Z}} S_j.$$

Then S is called the principal subalgebra. The basis elements of $\mathfrak{s}$ is the set $\{c, p_i = t^{b_i} \otimes T_{i'}, q_i = b_i^{-1} t^{-b_i} \otimes T_{N+1-i'}$ where $i' \in \{1, \ldots, N\}$ such that $i' \equiv i \bmod N$ and $i = 1, 2, \ldots\}$. We have,

$$[p_i, q_j] = \delta_{ij} c \text{ for all } i, j = 1, 2, \ldots$$

Now one can consider the basic module $L(\Lambda_0)$. Since $L(\Lambda_0)$ has level 1, the element c can be represented by the identity operator. But as $L(\Lambda_0)$ is irreducible as an $\mathfrak{s}$ module, and we can identify $L(\Lambda_0)$ with the space $R = \mathbb{C}[x_1, x_2, \ldots]$. Hence c operates as an identity operator, p_i acts as $\frac{\partial}{\partial x_i}$ and q_i acts as multiplication of $x_i (i = 1, 2, \ldots)$. In order to extend the realization of the basic representation from S to $\mathcal{G}'(A)$, we extend the identification of $L(\Lambda_0)$ with R to that of $\widehat{L}(\Lambda_0)$ with $\widehat{R}$, where $\widehat{L}(\Lambda_0)$ denotes the formal completion of $L(\Lambda_0)$. Let $X^\beta(z)$ be an element of this completion, depending on a parameter $z \in \mathbb{C}^\times$:

$$X^\beta(z) = \sum_{j \in \mathbb{Z}} z^{-j} (t^j \otimes A_{\beta,j}), \quad \beta \in \Delta.$$

Here

$$[p_s, X^\beta(z)] = [t^{b_s} \otimes T_{s'}, \sum_{j \in \mathbb{Z}} z^{-j} (t^j \otimes A_{\beta,j})]$$

$$= \sum z^{-j} t^{j+b_s} \otimes [T_{s'}, A_{\beta,j}]$$
$$= \lambda_{\beta s'} z^{b_s} X^{\beta}(z).$$

Similarly $\left[q_s, X^{\beta}(z)\right] = \lambda_{\beta,N+1-s'} b_s^{-1} z^{-b_s} X^{\beta}(z).$

Therefore using Lemma 17 and the above relations, it is clear that the operator $X^{\beta}(z) : R \to \widehat{R}$ acts as the following vertex operator:

$$\Gamma^{\beta}(z) = \langle \Lambda_0, A_{\beta,0} \rangle \exp\left(\sum_{j=1}^{\infty} \lambda_{\beta j'} z^{b_j} x_j\right) \exp\left(-\sum_{j=1}^{\infty} \lambda_{\beta,N+1-j'} b_j^{-1} z^{-b_j} \frac{\partial}{\partial x_j}\right).$$

We expand the vertex operator $\Gamma^{\beta}(z) = \sum_{j \in \mathbb{Z}} \Gamma_{\beta,j} z^j$.

Then $\Gamma_{\beta,j}$ are infinite-order differential operators which map $L(\Lambda_0)$ into itself. Hence the following theorem is proved.

Theorem 82 ([8]). *Let $\mathcal{G}'(A)$ be an affine algebra such that either A is symmetric or A is from tables Aff 2 or 3. Let $R = \mathbb{C}[x_1, x_2, \ldots]$. Then the identity operator, the operators x_j, $\frac{\partial}{\partial x_j}(j = 1, 2, \ldots)$ and $\Gamma_{\beta_i,j}(i = 1, \ldots, l, j \in \mathbb{Z})$ form a basis of a Lie subalgebra of differential operators preserving R. This subalgebra is isomorphic to $\mathcal{G}'(A)$, and the representation of it on R is equivalent to the basic representation of $\mathcal{G}'(A)$.*

(iv) *The homogeneous vertex operator construction of the basic representation*:

This construction is based on considering the homogeneous Heisenberg subalgebra. Let $\underline{\mathfrak{g}} = \underline{\mathfrak{h}} \oplus (\bigoplus_{\alpha \in \Delta} \mathbb{C}E_{\alpha})$ be a simple finite dimensional Lie algebra of type A_l, D_l, or E_l. Let $(.|.)$ be the normalized invariant form on $\underline{\mathfrak{g}}$. Let

$$\hat{\mathfrak{L}}(\underline{\mathfrak{g}}) = \mathbb{C}[t, t^{-1}] \otimes_{\mathbb{C}} \underline{\mathfrak{g}} \oplus \mathbb{C}c \oplus \mathbb{C}d$$

be the associated affine algebra of type $A_l^{(1)}, D_l^{(1)}$, and $E_l^{(1)}$, respectively. Let us consider the complex commutative associative algebra

$$V = \mathfrak{s}(\bigoplus_{j<0} (t^j \otimes \underline{\mathfrak{h}})) \otimes_{\mathbb{C}} \mathbb{C}[Q],$$

where $\mathfrak{s}$ stands for the symmetric algebra and $\mathbb{C}[Q]$ stands for the group algebra of the root lattice $Q \subset \underline{\mathfrak{h}}$ of $\mathfrak{g}(X_l)$. Let $\alpha \mapsto e^{\alpha}$ denote the inclusion $Q \subset \mathbb{C}[Q]$. Let $u^{(n)} = t^n \otimes u \quad (n \in \mathbb{Z},\ u \in \underline{\mathfrak{g}})$. For $n > 0,\ u \in \underline{\mathfrak{h}}$, denote by $u(-n)$ the operator on V of multiplication by $u^{(-n)}$. For $n \geq 0,\ u \in \underline{\mathfrak{h}}$ denote by $u(n)$ the derivation of the algebra V defined by

$$u(n)(v^{(-m)} \otimes e^{\alpha}) = n\delta_{n,-m}(u|v) \otimes e^{\alpha} + \delta_{n,0}(\alpha|u) v^{(-m)} \otimes e^{\alpha}.$$

Choosing dual bases u_i and u^i of $\underline{\mathfrak{h}}$, define the operator D_0 on V by the formula

$$D_0 = \sum_{i=1}^{l} \left(\frac{1}{2} u_i(0) u^i(0) + \sum_{n \geq 1} u_i(-n) u^i(n) \right).$$

Let $\alpha \in Q$ and define c_α as $c_\alpha(f \otimes e^\beta) = \epsilon(\alpha, \beta)(f \otimes e^\beta)$, where $\epsilon : Q \times Q \to \{\pm 1\}$ is a function satisfying the following properties:

$$\epsilon(\alpha + \alpha', \beta) = \epsilon(\alpha, \beta)\epsilon(\alpha', \beta)$$
$$\epsilon(\alpha, \beta + \beta') = \epsilon(\alpha, \beta)\epsilon(\alpha, \beta') \text{ for } \alpha, \alpha', \beta, \beta' \in Q$$

Here we note that, $\epsilon(\alpha, \alpha) = (-1)^{\frac{1}{2}(\alpha|\alpha)}$ for $\alpha \in Q$. Also for $\alpha \in \Delta \subset Q$, the vertex operator is given as

$$\Gamma_\alpha(z) = \exp\left(\sum_{j \geq 1} \frac{z^j}{j} \alpha(-j) \right) \exp\left(-\sum_{j \geq 1} \frac{z^j}{j} \alpha(-j) \right) e^\alpha x^{\alpha(0)} c_\alpha$$
$$\Gamma_\alpha(z) = \sum_{j \in \mathbb{Z}} \Gamma_\alpha^{(j)} z^{-j-1},$$

which is a sequence of operators $\Gamma_\alpha^{(j)}$ on V.

For the proof of the following theorem, one can refer Kac [8].

Theorem 83. *The map* $\sigma : \hat{\mathfrak{L}}(\underline{g}) \to End\ V$ *given by,*

$$K \mapsto 1; \quad u^{(n)} \mapsto u(n) \text{ for } u \in \underline{\mathfrak{h}}, n \in \mathbb{Z}$$
$$E_\alpha^{(n)} \mapsto \Gamma_\alpha^{(n)} \text{ for } \alpha \in \Delta, n \in \mathbb{Z};$$
$$\text{and } d \mapsto -D_0$$

defines the basic representation of the affine algebra $\hat{\mathfrak{L}}(\underline{\mathfrak{g}})$ *on* V.

Moreover, the following theorem given in Kac and Peterson [26] will be used for the construction of Hirota bilinear equation.

Theorem 84. *Let* $\mathfrak{g}$ *be a Kac-Moody algebra with a symmetrizable Cartan matrix and let* G *be the associated group. Let* $\{u_j\}$ *and* $\{u^j\}$ *be bases of* $\mathfrak{g}$ *dual with respect to a nondegenerate invariant bilinear form* $(\cdot|\cdot)$ *on* $\mathfrak{g}$ *and consistent with a triangular decomposition of* $\mathfrak{g}$. *Let* $L(\Lambda)$ *be an integrable representation of* $\mathfrak{g}$ *with highest weight* Λ *and let* v_Λ *be its highest weight vector. Then*

(a) *A nonzero vector* v *of* $L(\Lambda)$ *lies in the orbit* $G \cdot v_\Lambda$ *if and only if* $\sum u_j(v) \otimes u^j(v) = (\Lambda|\Lambda)(v \otimes v)$ *in* $L(\Lambda) \otimes L(\Lambda)$.

(b) *A vector* v *of* $L(\Lambda)$ *satisfies the equation* $(*)$ *if and only if* $v \otimes v$ *lies in the highest component of* $L(\Lambda) \otimes L(\Lambda)$.

Below we discuss briefly the vertex operator construction needed for the homogeneous hierarchy:

Let $\underline{\mathfrak{g}}$ be a finite dimensional Lie algebra of rank l with a symmetric Cartan matrix. Let Q be its root lattice. $\underline{\mathfrak{g}} = \underline{\mathfrak{h}} \oplus \sum_{\alpha\in\Delta} \mathbb{C}E_\alpha$ where $\Delta = \{\alpha \in Q \mid (\alpha|\alpha) = 2\}$. Let $R : Q \times Q \to \mathbb{Z}$ be a bilinear form such that $R(\alpha, \beta) + R(\beta, \alpha) = (\alpha|\beta)$. Let $\epsilon(\alpha, \beta) = (-1)^{R(\alpha,\beta)}$. Then one can have the usual description on $\underline{\mathfrak{g}}$.

Let $\hat{\underline{\mathfrak{g}}}$ be the associated affine Kac-Moody algebra. That is,

$$\hat{\underline{\mathfrak{g}}} = \mathbb{C}[t, t^{-1}] \otimes \underline{\mathfrak{g}} \oplus \mathbb{C}c \oplus \mathbb{C}d$$

with communication relations

$$[u(j), v(k)] = [u, v](j + k) + j(u|v)\delta_{j,-k}c, \quad [d, u(j)] = ju(j), \quad [c, \hat{\underline{\mathfrak{g}}}] = 0.$$

Here $u(j) = t^j \otimes u$. The homogeneous realization of the basic representation $L(\Lambda_0)$ of $\hat{\underline{\mathfrak{g}}}$ is constructed in the space

$$\mathbb{C}[x] \otimes \mathbb{C}[Q] = \mathbb{C}[x_k^{(j)}; i \le j \le l,\ k \in \mathbb{N}] \otimes \sum_{\alpha\in Q} \mathbb{C}e^\alpha.$$

Here $\hat{\underline{\mathfrak{g}}}$ acts on the above space as follows:

$$u_j(-k) = kx_k^{(j)},\ u^j(k) = \frac{\partial}{\partial x_k^{(j)}} \quad (1 \le j \le l, k \in \mathbb{N})$$

$$h(0)(f \otimes e^\beta) = \beta(h)f \otimes e^\beta \quad \text{for } h \in \mathfrak{h}$$

$$d(f \otimes e^\beta) = -\left(\sum_{k\ge1}\sum_{i=1}^{l} kx_k^{(j)} \frac{\partial f}{\partial x_k^{(j)}} + \frac{1}{2}|\beta|^2 f\right) \otimes e^\beta$$

$$E_\gamma(-k)(f \otimes e^\beta) = \epsilon(\gamma, \beta)X_k(\gamma)(f \otimes e^\beta) \quad \text{for } \gamma \in \Delta, \text{ where}$$

$$X(\gamma, z) = \sum_{k\in\mathbb{Z}} X_k(\gamma)z^k$$

$$= z^{|\gamma|^2/2}\left(\exp\sum_{j\ge1}\frac{\gamma(-j)z^j}{j}\right)\left(\exp-\sum_{j\ge1}\frac{\gamma(j)z^{-j}}{j}\right) \otimes e^\gamma z^{\partial_\gamma}$$

is the vertex operator defined for $\gamma \in Q$. Here, $z^{\partial_\gamma}(f \otimes e^\beta) = z^{(\beta|\gamma)}(f \otimes e^\beta)$.

The operator $S = \sum(u_j \otimes u^j)$ is defined on the space

$$L(\Lambda_0) \otimes L(\Lambda_0) = (\mathbb{C}[x'] \otimes \mathbb{C}[Q']) \otimes (\mathbb{C}[x''] \otimes \mathbb{C}[Q'']).$$

The following will give a dual basis for $\hat{\underline{\mathfrak{g}}}$: For $1 \le j \le l$,

$$u_k : \{u^j(k)\ (k > 0),\ u_j(-k)\ (k \ge 0), E_\gamma(k)(\gamma \in \Delta, k \in \mathbb{Z}), c, d\}$$

$$u^k : \{u^j(-k)\ (k > 0),\ u^j(k)\ (k \ge 0), -E_{-\gamma}(-k)(\gamma \in \Delta, k \in \mathbb{Z}), d, c\}.$$

Then one can compute $S(f \otimes g)$ as in the principal vertex operator construction of basic modules and prove that $S(f \otimes g)$ becomes a hierarchy of Hirota bilinear equations. Thus we can get the following theorem due to Kac and Wakimoto [229].

Theorem 85. *An element $\tau = \sum_{\beta \in Q} \tau_\beta \otimes e^\beta$ of $L(\Lambda_0) = \mathbb{C}[x] \otimes \mathbb{C}[Q]$ is contained in the G-orbit of the vacuum vector $1 \otimes e^0$ if and only if it satisfies the following hierarchy of Hirota bilinear differential equations:*

$$-\left\{2\sum_{k\geq 1}\sum_{j=1}^{l} k y_k^{(j)} D_k^j + \frac{1}{2}|\alpha-\beta|^2\right\} e^{\sum y_k^{(j)} D_k^j} \tau_\alpha \cdot \tau_\beta$$
$$+\sum_{\gamma\in\Delta}\epsilon(\gamma,\alpha-\beta)\sum_{n\geq 0} Q_n^\gamma(2y) P^\gamma_{n-2+(\gamma|\alpha-\beta)}(-\tilde{D}) e^{\sum y_k^{(j)} D_k^{(j)}} \tau_{\alpha-\gamma}\cdot\tau_{\beta+\gamma} = 0.$$

Here the hierarchy is called the homogeneous hierarchy of type $X_l^{(1)}$. Here

$$\sum_{n\geq 0} P_n^\gamma(x) z^n = \exp\sum_{k\geq 1}\sum_{j=1}^{l}\langle\gamma, u_j\rangle x_k^{(j)} z^k \text{ and}$$

$$\sum_{n\geq 0} Q_n^\gamma(x) z^n = \exp\sum_{k\geq 1}\sum_{j=1}^{l}\langle\gamma, u^j\rangle x_k^{(j)} z^k.$$

(v) *Construction of homogeneous hierarchy and the corresponding nonlinear differential equations for $A_1^{(1)}$.*

First we describe the structure of $\mathfrak{g} = A_1^{(1)}$.

In the usual notation,

$$A_1^{(1)} = \widehat{sl}(2,\mathbb{C}) = \bigoplus_{j\in\mathbb{Z}} sl(2,\mathbb{C}) \bigotimes t^j + \mathbb{C}k + \mathbb{C}d,$$

$$\begin{aligned} \text{with } \left[u_{(j)}, v_{(k)}\right] &:= [u,v]_{(j+k)} + j\cdot tr(uv)\delta_{j+k,0}\cdot k,\\ \left[d, u_{(j)}\right] &:= j u_{(j)},\\ \left[k, \tilde{g}\right] &:= \{0\} \end{aligned}$$

for $u, v \in \overline{\mathcal{G}}, j, k \in \mathbb{Z}, u_{(j)} = u \bigotimes t^j$.

In $sl(2,\mathbb{C})$, one can define an involution σ of $sl(2,\mathbb{C})$, $\sigma x = -x^t$, for all $x \in sl(2,\mathbb{C})$.

Consider eigen spaces

$$\overline{\mathcal{G}}_{\bar{0}} := \{x \in sl(2,\mathbb{C}) | \sigma(x) = x\} = \{x \in sl(2,\mathbb{C}) | x^t = -x\},$$
$$\mathcal{G}_{\bar{1}} := \{x \in sl(2,\mathbb{C}) | \sigma(x) = -x\} = \{x \in sl(2,\mathbb{C}) | x^t = x\}.$$

Then $\overline{\mathcal{G}}_{\bar{0}} = \mathbb{C}X_{\bar{0}}$ and $\overline{\mathcal{G}}_{\bar{1}} = \mathbb{C}H \bigoplus \mathbb{C}X_1$, with $\overline{\mathcal{G}} = \overline{\mathcal{G}}_{\bar{0}} \oplus \overline{\mathcal{G}}_{\bar{1}}$.

We have $X_{\bar{0}} := E - F = \begin{pmatrix} 0 & 1 \\ -1 & 0 \end{pmatrix}, X_{\bar{1}} := E + F = \begin{pmatrix} 0 & 1 \\ 1 & 0 \end{pmatrix}$.
Here $E = \begin{pmatrix} 0 & 1 \\ 0 & 0 \end{pmatrix}$, $F = \begin{pmatrix} 0 & 0 \\ 1 & 0 \end{pmatrix}, H = \begin{pmatrix} 1 & 0 \\ 0 & -1 \end{pmatrix}$.

$$\tilde{\mathcal{G}} := \left(\bigoplus_{j\in\mathbb{Z}_{\text{even}}} \tilde{\mathcal{G}}_{\bar{0}} \bigotimes t^j\right) \bigoplus \left(\bigoplus_{j\in\mathbb{Z}_{\text{odd}}} \tilde{\mathcal{G}}_{\bar{1}} \bigotimes t^j\right) + Ck + \mathbb{C}d_0.$$

With the same above relations for $j, k \in \mathbb{Z}$, $u \in \overline{g}_{\bar{j} \bmod 2}$ and $v \in \overline{\mathcal{G}}_{\bar{k} \bmod 2}$.

Here $Q = \mathbb{Z}\alpha_1$ and $\epsilon(m\alpha_1, n\alpha_1) = (-1)^{mn}$. Let $u_1 = \alpha_1, u^1 = \frac{1}{2}\alpha_1$ and $\tau_n(x) = \tau_{n\alpha_1}(x)$. We also have $P_n^{\pm\alpha_1}(x) = P_n(\pm 2x)$, $Q_n^{\pm\alpha_1}(x) = P_n(\pm x)$. Then the above homogeneous hierarchy as given in the above theorem looks as follows:

$$\begin{aligned}
&\left(2\sum_{k\in\mathbb{N}} ky_k D_k + (m-n)^2\right) e^{\sum y_k D_k}\tau_n \cdot \tau_m \\
&\quad + (-1)^{m-n} \sum_{k\in\mathbb{Z}_+} P_k(2y)P_{k-2(m-n+1)}(2\tilde{D})e^{\sum y_k D_k}\tau_{n-1} \cdot \tau_{m+1} \\
&\quad + (-1)^{m-n} \sum_{k\in\mathbb{Z}_+} P_k(-2y)P_{k+2(m-n+1)}(2\tilde{D})e^{\sum y_k D_k}\tau_{n+1} \cdot \tau_{m-1} = 0.
\end{aligned}$$

Taking $m - n = 0, -1$ or 1 and considering the coefficients of y_1^2, we get the following equations on $\tau_n(x), \tau_{n-1}(x)$, and $\tau_{n+1}(x)$:

$$\begin{aligned}
D_1^2\tau_n \cdot \tau_n + 2\tau_{n-1} \cdot \tau_{n+1} &= 0 \\
(D_1^2 + D_2)\tau_n \cdot \tau_{n+1} &= 0 \\
(D_1^2 + D_2)\tau_{n-1} \cdot \tau_n &= 0
\end{aligned}$$

Let $x = x_1, t = x_2, q(x,t) = \tau_1/\tau_0, q^*(x,t) = \tau_{-1}/\tau_0$ and $u(x,t) = \log \tau_0$. Let us consider the first equation, $D_1^2\tau_n \cdot \tau_n + 2\tau_{n-1} \cdot \tau_{n+1} = 0$. If $n = 0$, then this equation becomes

$$D_1^2\tau_0 \cdot \tau_0 + 2\tau_{-1} \cdot \tau_1 = 0 \tag{i}$$

Using $D_1^n f \cdot g = \sum_{k=0}^{n}(-1)^k \begin{pmatrix} n \\ k \end{pmatrix} \frac{\partial^k f}{\partial x_1^k}\frac{\partial^{n-k} g}{\partial x_1^{n-k}}$, we get from (i), $D_1^2\tau_0 \cdot \tau_0 = 2\tau_0\frac{\partial^2\tau_0}{\partial x^2} - 2\left(\frac{\partial\tau_0}{\partial x}\right)^2$. Here equation (i) becomes

$$-\tau_{-1} \cdot \tau_1 = \tau_0\frac{\partial^2\tau_0}{\partial x^2} - \left(\frac{\partial\tau_0}{\partial x}\right)^2 \tag{ii}$$

Taking $u(x,t) = \log\ \tau_0$, we get

$$u_{xx} = -\frac{1}{\tau_0^2}\left(\frac{\partial \tau_0}{\partial x}\right)^2 + \frac{1}{\tau_0}\frac{\partial^2 \tau_0}{\partial x^2} \tag{iii}$$

Substituting the value of $\left(\frac{\partial \tau_0}{\partial x}\right)^2$ from (ii) in (iii), we get

$$u_{xx} = -qq^*.$$

Similarly, from $(D_1^2 + D_2)\tau_n \cdot \tau_{n+1} = 0$, $(D_1^2 + D_2)\tau_{n-1} \cdot \tau_n = 0$, one can arrive at the following two equations $-q_t + q_{xx} + 2qu_{xx} = 0$ and $q_t^* = -q_{xx}^* + 2qq^{*2}$. Substituting the value of $u_{xx} = -qq^*$ in the above equations, we get the following equations

$$q_t = q_{xx} - 2q^2q^*, \quad q_t^* = -q_{xx}^* + 2qq^{*2}.$$

Let $q^*(x,it) = \pm\overline{q(x,it)} = g(x,t)$, the complex conjugate of $q(x,it)$. Then $ig_t = -g_{xx} \pm 2|g|^2 g$ which is the classical nonlinear Schrodinger equation. Hence the above hierarchy is called NLS hierarchy.

Remark 195. For the homogeneous realization of the basic representation of $A_1^{(1)}$ and the Toda lattice, one can refer Tenkroode and Berg Velt [212]. Homogeneous vertex operator construction can be used for other Kac-Moody algebras and many other bilinear identities can be deduced. Homogeneous vertex operators and corresponding Hirota bilinear equations for $B_2^{(1)}$ were constructed in Sthanumoorthy and Kiruba Bagirathi [214], using the structure of standard modules in the homogeneous picture for $A_1^{(1)}$ [215].

6.9 Principal vertex operators and the corresponding Hirota bilinear equations for $A_2^{(1)}$, $A_2^{(2)}$, $C_2^{(1)}$, and $B_3^{(1)}$

6.9.1 Construction of principal vertex operators and Hirota bilinear equations for $A_2^{(1)}$ [213]

Consider the finite dimensional Lie algebra $\underline{g} = sl(3,\mathbb{C})$. Let $(x|y) = \text{trace}(xy) = \text{tr}(xy)$ be a nondegenerate symmetric bilinear form defined on $\underline{g}$, for $x, y \in \mathcal{G}$.

In the usual notation, let $\{E_0 = e_{31}, E_1 = e_{12}, E_2 = e_{23}, F_0 = e_{13}, F_1 = e_{21}, F_2 = e_{32}, H_0 = e_{33} - e_{11}, H_1 = e_{11} - e_{22}$, and $H_2 = e_{22} - e_{33}\}$ be a set of canonical generators for $\underline{g}$. $\underline{\mathfrak{h}} = \text{span}\ \{h_i;\ 1 \le i \le 2\}$ is a Cartan subalgebra of $\underline{g}$. Let Δ be a simple root system. The bilinear form $(.|.)$ is normalized so that $(\alpha|\alpha) = 2$ for all $\alpha \in \Delta$. Let $\Pi = \{\alpha_1, \alpha_2\}$ be a basis of Δ. We can now consider the corresponding affine algebra $A_2^{(1)}$ with labels $(1,1,1)$. Then, $h^{(1)} = \sum_{i=0}^{2} a_i = 3$ is the Coxeter number. We have, $\deg E_i = -\deg F_i = 1$, $\deg H_i = 0$, $i = 0,1,2$. This defines a $\mathbb{Z}/3\mathbb{Z}$ gradation of $\underline{g} = sl(3,\mathbb{C})$ which is the principal gradation of $\underline{g}$. We have, the following decomposition of $\underline{g}$ due to the above principal gradation:

$$\underline{g} = \bigoplus_{j\in\mathbb{Z}/3\mathbb{Z}} \underline{g}_{[j]}, \text{ where } \underline{g}_{[j]} = \{x \in \underline{g} \mid \deg x = j\}.$$

We have

$$\underline{g}_{[0]} = \mathbb{C}H_1 \oplus \mathbb{C}H_2,\ \underline{g}_{[1]} = \mathbb{C}E_0 \oplus \mathbb{C}E_1 \oplus \mathbb{C}E_2,\ \underline{g}_{[2]} = \mathbb{C}F_0 \oplus \mathbb{C}F_1 \oplus \mathbb{C}F_2.$$

Let $E = E_0 + \sum_{i=1}^{2} E_i$. Here E_0 is the lowest root vector of A_2 with respect to $\underline{\mathfrak{h}}$. $E = E_0 + E_1 + E_2 = e_{31} + e_{12} + e_{23}$ is the cyclic element of A_2. Let $S = \{x \in \underline{g} \mid [x, E] = 0\}$. Then S is the principal Cartan subalgebra of $\underline{g}$. The basis elements of S are $E = e_{31} + e_{12} + e_{23}$ and $F = e_{13} + e_{21} + e_{32}$. Hence $S = \mathbb{C}E \oplus \mathbb{C}F$. The principal Cartan subalgebra S of $\underline{g}$ is also graded with respect to the principal gradation as,

$$S = \bigoplus_{j\in\mathbb{Z}/3\mathbb{Z}} S_j; \quad \text{with } S_j = S \cap \underline{g}_j \text{ where}$$

$S_0 = \phi,\ S_1 = \mathbb{C}E$ and $S_2 = \mathbb{C}F$.

Hence, the exponents of $sl(3,\mathbb{C})$ are $\{1, 2\}$. Let us denote the basis elements E, F of S by T_1 and T_2 respectively so that $(T_i|T_{3-j}) = 3\delta_{ij},\ i,j = 1,2$. Let $\beta_1 = (\epsilon, 1)$ and $\beta_2 = (\epsilon^2, 1)$ be the simple roots of A_2 with respect to the Cartan subalgebra S and X_{β_1} and X_{β_2} be the corresponding root vectors. Here $\langle \beta_j, T_i\rangle = \beta_j(T_i)$ for $i,j = 1,2$. Let $\epsilon = \frac{-1+I\sqrt{3}}{2}$, $\epsilon^2 = \frac{-1-I\sqrt{3}}{2}$ where $1, \epsilon, \epsilon^2$ are the primitive cubic roots of unity, with $I = \sqrt{-1}$. Now,

$$\begin{aligned}
X_{\beta_1} &= \text{matrix}(\epsilon^r 1^{-s})_{r,s=1}^{3} \\
&= \epsilon H_1 - H_2 + E_0 + F_2 + \epsilon(F_0 + E_1) + \epsilon^2(F_1 + E_2) \\
X_{\beta_2} &= \text{matrix}(\epsilon^{2r} 1^{-s})_{r,s=1}^{3} \\
&= \epsilon^2 H_1 - H_2 + \epsilon^2(E_1 + F_0) + E_0 + F_2 + \epsilon(E_2 + F_1)
\end{aligned}$$

are root vectors with respect to S. Moreover,

$$\begin{aligned}
[T_1, X_{\beta_1}] &= [E, X_{\beta_1}] = \langle \beta_1, T_1\rangle X_{\beta_1} = (\epsilon - 1)X_{\beta_1}, \\
[T_2, X_{\beta_1}] &= [F, X_{\beta_1}] = \langle \beta_1, T_2\rangle X_{\beta_2} = (\epsilon^2 - 1)X_{\beta_1}, \\
[T_1, X_{\beta_2}] &= [E, X_{\beta_2}] = \langle \beta_2, T_1\rangle X_{\beta_2} = (\epsilon^2 - 1)X_{\beta_2}, \\
[T_2, X_{\beta_2}] &= [F, X_{\beta_2}] = \langle \beta_2, T_2\rangle X_{\beta_2} = (\epsilon^4 - 1)X_{\beta_2}.
\end{aligned}$$

One can decompose $X_{\beta_1}, X_{\beta_2} \in \underline{g}$ with respect to the principal gradation and we get $X_{\beta_1} = \sum_{j=0}^{2} X_{\beta_{1,j}}$, with $X_{\beta_{1,0}} = \epsilon H_1 - H_2$; $X_{\beta_{1,1}} = E_0 + \epsilon E_1 + \epsilon^2 E_2$; and $X_{\beta_{1,2}} = \epsilon F_0 + \epsilon^2 F_1 + F_2$. Similarly, $X_{\beta_2} = \sum_{j=0}^{2} X_{\beta_{2,j}}$, where $X_{\beta_{2,0}} = \epsilon^2 H_1 - H_2$; $X_{\beta_{2,1}} = E_0 + \epsilon^2 E_1 + \epsilon E_2$; $X_{\beta_{2,2}} = \epsilon^2 F_0 + \epsilon F_1 + F_2$. Clearly, the set $\{X_{\beta_{1,j}}, X_{\beta_{2,j}}, T_1, T_2;\ j = 0, 1, 2\}$ forms a basis of $\underline{g} = sl(3,\mathbb{C})$.

Let $\hat{\underline{g}}(A)$ be the corresponding affine algebra, $\hat{\underline{g}}(A) = \hat{\underline{g}}'(A) \oplus \mathbb{C}d$, where $\hat{\underline{g}}'(A)$ is the derived subalgebra defined by,

$$\hat{\underline{g}}'(A) = \bigoplus_{j\in\mathbb{Z}} t^j \otimes \underline{g}_{j \bmod 3} \oplus \mathbb{C}c$$

and the scaling element, d. The canonical generators for $A_2^{(1)}$ are given by

$$\hat{E}_j = 1 \otimes E_j, \quad \hat{E}_0 = t \otimes E_0,$$
$$\hat{F}_j = 1 \otimes F_j, \quad \hat{F}_0 = t^{-1} \otimes F_0,$$
$$\hat{H}_j = \alpha_j \text{ and } \hat{H}_0 = \frac{2}{(\theta|\theta)}c - \theta^v.$$

For $j \in \{1,2\}$, θ is the highest root of $sl(3,\mathbb{C})$. The Lie bracket in $\hat{\underline{g}}'(A)$ is given by

$$[p_1(t) \otimes g_1 \oplus \lambda_1 c, p_2(t) \otimes g_2 \oplus \lambda_2 c] = p_1(t)p_2(t) \otimes [g_1, g_2] \oplus \frac{1}{3}Res\left(\frac{dp_1(t)}{dt}p_2(t)\right)(g_1|g_2)c,$$

where $g_1, g_2 \in \underline{g}$, $p_1(t), p_2(t) \in \mathbb{C}[t, t^{-1}], \lambda_1, \lambda_2 \in \mathbb{C}$.

For $\underline{g} = sl(3,\mathbb{C})$, we have

$$\begin{aligned}\hat{\underline{g}}'(A) &= \bigoplus_{j\in\mathbb{Z}} t^j \otimes \underline{g}_{j \bmod 3} \oplus \mathbb{C}c \\ &= t^0\, \mathbb{C}[t^3, t^{-3}] \otimes \underline{g}_0 \oplus t^1\, \mathbb{C}[t^3, t^{-3}] \otimes \underline{g}_1 \oplus t^2\, \mathbb{C}[t^3, t^{-3}] \otimes \underline{g}_2 \oplus \mathbb{C}c.\end{aligned}$$

The principal Cartan subalgebra S of $sl(3,\mathbb{C})$ has $\{E, F\}$ as basis elements.

Let

$$\begin{aligned}S &= \bigoplus_{j\in\mathbb{Z}}(t^j \otimes S_{j \bmod 3}) \oplus \mathbb{C}c \\ &= t^1\, \mathbb{C}[t^3, t^{-3}] \otimes S_1 \oplus t^2\, \mathbb{C}[t^3, t^{-3}] \otimes S_2 \oplus \mathbb{C}c.\end{aligned}$$

This is the principal subalgebra of $\hat{\underline{g}}'(A) = A_2^{(1)}$. The set of its positive exponents is $E_+ = \{1, 2, 4, 5, 7, 8, 10, 11, \ldots\}$. Let us denote these elements of E_+ by $\{b_1, b_2, \ldots\}$. The generators of S are

$$c,\ p_i = t^i \otimes T_{i'}, \quad \text{and} \quad q_i = i^{-1}t^{-i} \otimes T_{3-i'},$$

where $i \in E_+$ such that $i' \equiv i \bmod 3$ with $i' \in \{1,2\}$. They satisfy the relation $[p_i, q_j] = c\delta_{ij}$. Hence S is isomorphic to the infinite Heisenberg algebra. Let $L(\Lambda_0)$ be the basic module of $A_2^{(1)}$. It is irreducible as a S module [8]. Hence, $L(\Lambda_0)$ is identified with $\mathbb{C}[x_i,\ i \in E_+]$. Here the generators of S acts on $\mathbb{C}[x_i, i \in E_+]$ as $c \longmapsto Id.\ p_i \longmapsto \frac{\partial}{\partial x_i}$ $q_i \longmapsto x_i,\ i \in E_+$. The representation can be extended to the whole algebra $\underline{\hat{\mathfrak{g}}}(A)$, using the vertex operators as follows:

Let us consider the simple roots $\beta_i \in \Delta_S,\ i \in \{1,2\}$. Let

$$X_{\beta_i}(z) = \sum_{j \in \mathbb{Z}} z^{-j}\ (t^j \otimes X_{\beta_{i,j \bmod 3}}).$$

Then $X_{\beta_i}(z)$ maps $L(\Lambda_0)$ into its formal completion. In particular, for the simple root $\beta_1 \in \Delta_S$, we consider the following commutation relations which can be directly verified that:

$$[p_i, X_{\beta_1}(z)] = z^i(\epsilon^{i'} - 1)\ X_{\beta_1}(z),$$

where $i' \in \{1,2\}$ such that $i' \equiv i \bmod 3$ with $i \in E_+$.

Similarly,

$$[q_i, X_{\beta_1}(z)] = z^{-i}(\epsilon^{3-i'} - 1)X_{\beta_1}(z).$$

The principal vertex operator corresponding to the root β_1 is

$$\Gamma^{\beta_1}(z) = \frac{\epsilon}{(\epsilon - 1)} \exp\left(\sum_{j \in E_+} z^j(\epsilon^{j'} - 1)x_j\right) \exp\left(\sum_{j \in E_+} z^{-j}\left(\frac{1 - \epsilon^{-j'}}{j}\right)\frac{\partial}{\partial x_j}\right).$$

Here $\langle \Lambda_0, X_{\beta_1,0}\rangle = \frac{\epsilon}{(\epsilon-1)}$. Moreover, $X_{\beta_1}(z) = \sum_{j\in\mathbb{Z}} z^{-j}(t^j \otimes X_{\beta_1,j \bmod 3})$ acts as the vertex operator $\Gamma^{\beta_1}(z)$.

Similarly, the principal vertex operator corresponding to the root β_2 is

$$\Gamma^{\beta_2}(z) = \frac{\epsilon^2}{\epsilon^2 - 1} \exp\left(\sum_{j \in E_+} z^j(\epsilon^{2j'} - 1)x_j\right) \exp\left(\sum_{j \in E_+} z^{-j}\left(\frac{1 - \epsilon^{-2j'}}{j}\right)\frac{\partial}{\partial x_j}\right),$$

where $\langle \Lambda_0, X_{\beta_2,0}\rangle = \frac{\epsilon^2}{(\epsilon^2-1)}$.

Herein, $X_{\beta_2}(z) = \sum_{j\in\mathbb{Z}} z^{-j}(t^j \otimes X_{\beta_{2,j \bmod 3}})$ acts as the vertex operator $\Gamma^{\beta_2}(z)$.

Therefore the representation of $A_2^{(1)}$ on

$$\mathbb{C}[x_i;\ i \in E_+] = \mathbb{C}[x_i; i \in \mathbb{Z}_+,\ i \not\equiv 0 \bmod 3]$$

is given by

$$\begin{aligned} p_j &\mapsto \frac{\partial}{\partial x_j};\ q_j \mapsto jx_j \\ c &\mapsto Id;\ d \mapsto \frac{1}{3}\sum_{x_j\in E_+} jx_j\frac{\partial}{\partial x_j} \\ X_{\beta_i}(z) &\mapsto \Gamma^{\beta_i}(z),\ i\in\{1,2\}. \end{aligned}$$

We choose the following dual bases of $A_2^{(1)}$.

$$v_j:\left\{\frac{1}{\sqrt{3}}p_j,\ \frac{1}{\sqrt{3}}q_j,\ \frac{1}{\sqrt{3}}\,t^j\otimes X_{\beta_1,j \bmod 3},\ \frac{1}{\sqrt{3}}\,t^j\otimes X_{\beta_2,j \bmod 3},\ c,\ d\right\}$$
$$v^j:\left\{\frac{1}{\sqrt{3}}q_j,\ \frac{1}{\sqrt{3}}p_j,\ \frac{1}{\sqrt{3}}t^{-j}\otimes X_{-\beta_1,-j \bmod 3},\ \frac{1}{\sqrt{3}}t^{-j}\otimes X_{-\beta_2,-j \bmod 3},\ d,\ c\right\}.$$

Using these bases, the operator $S=\sum(v_j\otimes v^j)$ on $L(\Lambda_0)\otimes L(\Lambda_0)=\mathbb{C}[x']\otimes\mathbb{C}[x'']$ (Theorem 84(b)) is computed as follows:

Let $'$ and $''$ denote the operators acting on the first and second factors of the tensor product respectively. So

$$S=\frac{1}{3}\sum_{j\in E_+}\left(jx_j'\otimes\frac{\partial}{\partial x_j''}+\frac{\partial}{\partial x_j'}\otimes jx_j''\right)+\text{Coeff. of } z^0 \text{ in } \frac{1}{3}\left(\Gamma^{\beta_1}(z)\otimes\Gamma^{-\beta_1}(z)\right)$$
$$+\text{Coeff. of } z^0 \text{ in } \frac{1}{3}\left(\Gamma^{\beta_2}(z)\otimes\Gamma^{-\beta_2}(z)\right)-1\otimes\frac{1}{3}\sum_{j\in E_+} jx_j''\frac{\partial}{\partial x_j''}-\frac{1}{3}\sum_{j\in E_+} jx_j'\frac{\partial}{\partial x_j'}\otimes 1.$$

Hence,

$$S=-\frac{1}{3}\sum_{j\in E_+}\left(j(x_j'-x_j'')\right)\left(\frac{\partial}{\partial x_j'}-\frac{\partial}{\partial x_j''}\right)+\text{Coeff. of } z^0 \text{ in}$$
$$\left\{\frac{\epsilon^2}{3(\epsilon-1)^2}\exp\left(\sum_{j\in E_+}(\epsilon^{j'}-1)(x_j'-x_j'')z^j\right)\exp\left(\sum_{j\in E_+}\frac{(1-\epsilon^{-j'})}{j}\left(\frac{\partial}{\partial x_j'}-\frac{\partial}{\partial x_j''}\right)z^{-j}\right)\right\}$$
$$+\text{Coeff. of } z^0 \text{ in}$$
$$\left\{\frac{\epsilon^4}{3(\epsilon^2-1)^2}\exp\left(\sum_{j\in E_+}(\epsilon^{2j'}-1)(x_j'-x_j'')z^j\right)\exp\left(\sum_{j\in E_+}\frac{(1-\epsilon^{-2j'})}{j}\left(\frac{\partial}{\partial x_j'}-\frac{\partial}{\partial x_j''}\right)z^{-j}\right)\right\}.$$

Assume

$$x_j = \frac{1}{2}(x'_j + x''_j); \qquad y_j = \frac{1}{2}(x'_j - x''_j),$$
$$\frac{\partial}{\partial x_j} = \frac{\partial}{\partial x'_j} + \frac{\partial}{\partial x''_j} \quad \text{and} \quad \frac{\partial}{\partial y_j} = \frac{\partial}{\partial x'_j} - \frac{\partial}{\partial x''_j}.$$

Hence,

$$S = -\frac{2}{3}\sum_{j\in E_+} jy_j\frac{\partial}{\partial y_j} + \text{Coeff. of } z^0 \text{ in}$$
$$\left\{\frac{\epsilon^2}{3(\epsilon-1)^2}\exp\left(\sum_{j\in E_+}(2(\epsilon^{j'}-1)y_j)z^j\right)\exp\left(\sum_{j\in E_+}\left(\frac{1-\epsilon^{-j'}}{j}\right)\frac{\partial}{\partial y_j}z^{-j}\right)\right\}$$
$$+ \text{Coeff. of } z^0 \text{ in}$$
$$\left\{\frac{\epsilon^4}{3(\epsilon^2-1)^2}\exp\left(\sum_{j\in E_+}(2(\epsilon^{2j'}-1)y_j)z^j\right)\exp\left(\sum_{j\in E_+}\left(\frac{1-\epsilon^{-2j'}}{j}\right)\frac{\partial}{\partial y_j}z^{-j}\right)\right\}.$$

The usual Schur polynomials, $P_k(x)$, $k \in \mathbb{Z}_+$ with $x = (x-0, x_1, \ldots)$ are defined by

$$\sum_{k\geq 0} P_k(x)z^k = \exp\left(\sum_{k\geq 0} x_k z^k\right).$$

Similarly define the Schur polynomial is $P_k^{E_+}(x)$ by

$$\sum_{k\geq 0} P_k^{E_+}(x)z^k = \exp\left(\sum_{k\in E_+} x_k z^k\right).$$

So S can be rewritten in terms of Schur polynomials as,

$$S(\tau\otimes\tau) = \left\{-\frac{2}{3}\sum_{j\in E_+} jy_j\frac{\partial}{\partial y_j} + \frac{\epsilon^2}{3(\epsilon-1)^2}\sum_{n\geq 0} P_n^{E_+}(2(\epsilon^{j'}-1)y_j)P_n^{E_+}\left(\frac{1-\epsilon^{-j'}}{j}\frac{\partial}{\partial y_j}\right)\right.$$
$$\left. + \frac{\epsilon^4}{3(\epsilon^2-1)^2}\sum_{n\geq 0} P_n^{E_+}(2(\epsilon^{2j'}-1)y_j)P_n^{E_+}\left(\frac{1-\epsilon^{-2j'}}{j}\frac{\partial}{\partial y_j}\right)\right\}\tau(x+y)\tau(x-y).$$

Using Taylor's expansion, we have $f(x+y)g(x-y) = \left(\exp \sum_{j\in E_+} y_j \frac{\partial}{\partial u_j}\right) f(x+u)g(x-u)|_{u=0}$. Hence,

$$S(\tau \otimes \tau) = \left\{ -\frac{2}{3} \sum_{j\in E_+} j y_j \frac{\partial}{\partial u_j} + \frac{\epsilon^2}{3(\epsilon-1)^2} \sum_{n\geq 0} P_n^{E_+}(2(\epsilon^{j'}-1)y_j) P_n^{E_+}\left(\frac{1-\epsilon^{-j'}}{j}\frac{\partial}{\partial u_j}\right) \right.$$
$$\left. + \frac{\epsilon^4}{3(\epsilon^2-1)^2} \sum_{n\geq 0} P_n^{E_+}(2(\epsilon^{2j'}-1)y_j) P_n^{E_+}\left(\frac{1-\epsilon^{-2j'}}{j}\frac{\partial}{\partial u_j}\right) \right\}.$$
$$\exp\left(\sum_{j\in E_+} y_j \frac{\partial}{\partial u_j}\right) \tau(x+u)\tau(x-u)\,|_{u=0}\,.$$

We conclude, by using Theorem 84(b), that the element τ of $L(\Lambda_0)$ lies in the orbit $G.1$, where G is the group associated to $A_2^{(1)}$, if and only if $S(\tau\otimes\tau) = 0$. Hence τ satisfies the following hierarchy of Hirota's bilinear equations:

$$\left\{ -\frac{2}{3} \sum_{j\in E_+} j y_j D_j + \frac{\epsilon^2}{3(\epsilon-1)^2} \sum_{n\geq 0} P_n^{E_+}(2(\epsilon^{j'}-1)y_j) P_n^{E_+}\left(\frac{1-\epsilon^{-j'}}{j} D_j\right) \right.$$
$$\left. + \frac{\epsilon^4}{3(\epsilon^2-1)^2} \sum_{n\geq 0} P_n^{E_+}(2(\epsilon^{2j'}-1)y_j) P_n^{E_+}\left(\frac{1-\epsilon^{-2j'}}{j} D_j\right) \right\}.$$
$$\exp\left(\sum_{j\in E_+} y_j D_j\right) \tau(x+u)\tau(x-u) = 0$$

In particular, taking the coefficient of y_4, we get the equation,

$$(D_1^4 + D_2^2)\,\tau\cdot\tau = 0.$$

This is a soliton equation called Boussinesq equation in Hirota's form. The above hierarchy of equations is called Boussinesq hierarchy [209, 210, 216–223]. Similarly, one can get the other equations in the above hierarchy, by taking the coefficients of other variables.

6.9.2 Construction of principal vertex operators and Hirota bilinear equations for $A_2^{(2)}$ [213]

(1) *Construction of principal vertex operators* $A_2^{(2)}$:

Consider the finite dimensional Lie algebra $\underline{g} = sl(3,\mathbb{C})$. Let $(x|y) = \text{trace}\,(xy) = \text{tr}(xy)$ be a nondegenerate symmetric bilinear form defined on $\underline{g}$.

Let

$$\begin{array}{lll} E_0 = e_{31}, & E_1 = e_{12}, & E_2 = e_{23}, \\ F_0 = e_{13}, & F_1 = e_{21}, & F_2 = e_{32}, \\ H_0 = e_{33} - e_{11}, & H_1 = e_{11} - e_{22} \text{ and } & H_2 = e_{22} - e_{33} \end{array}$$

be a set of canonical generators for $\underline{g}$. Let $\underline{\mathfrak{h}} = \text{span}\ \{h_i;\ 1 \leq i \leq 2\}$ be a Cartan subalgebra of $\underline{g}$ and Δ be a simple root system. The bilinear form $(\cdot|\cdot)$ is normalized so that $(\alpha|\alpha) = 2$ for all $\alpha \in \Delta$. Let $\Pi = \{\alpha_1, \alpha_2\}$ be a basis of Δ. Let θ be the Dynkin diagram automorphism of order 2 given by

$$\theta(\alpha_1) = \alpha_2 \text{ and } \theta(\alpha_2) = \alpha_1.$$

The automorphism θ is extended to $\underline{g}$ by defining $\theta(x_\alpha) = x_{\theta(\alpha)}$ for all $\alpha \in \Delta$. The decomposition of $\underline{g}$ due to θ is as follows:

$$\underline{g} = \bigoplus_{j \in \mathbb{Z}/2\mathbb{Z}} \underline{g}_{[j]}, \text{ where } \underline{g}_{[j]} = \{x \in \underline{g} \mid \theta(x) = \epsilon^j x\} \text{ with } \epsilon = e^{\pi I} \text{ where } I = \sqrt{-1}.$$

We have

$$\underline{g}_{[0]} = \mathbb{C}(e_{12} + e_{23}) \oplus \mathbb{C}(e_{21} + e_{32}) \oplus \mathbb{C}(h_1 + h_2),$$
$$\underline{g}_{[1]} = \mathbb{C}(e_{31}) \oplus \mathbb{C}(e_{13}) \oplus \mathbb{C}(e_{12} - e_{23}) \oplus \mathbb{C}(e_{21} - e_{32}) \oplus \mathbb{C}(h_1 - h_2).$$

Let

$$\begin{array}{lll} \bar{H}_0 = 2(h_1 + h_2); & \bar{E}_0 = 2^{1/2}(e_{12} + e_{23}); & \bar{F}_0 = 2^{1/2}(e_{21} + e_{32}) \\ \bar{H}_1 = -(h_1 + h_2)\ ; & \bar{E}_1 = e_{31}; & \bar{F}_1 = -e_{13}. \end{array}$$

The 2-Coxeter number of $\underline{g}$ is $h^{(2)} = 2\sum_{i=0}^{1} a_i = 6$ with $(a_0, a_1) = (2, 1)$. Let $\nu : A_2 \to A_2$ be the sixth order automorphism defined by

$$\nu(\bar{E}_j) = \omega \bar{E}_j,$$
$$\nu(\bar{F}_j) = \omega^{-1} \bar{F}_j \text{ and}$$
$$\nu(\bar{H}_j) = \bar{H}_j \text{ (for } j = 0, 1), \text{ where } \omega = e^{\frac{2\pi I}{6}}.$$

Then $\underline{g}$ is decomposed with respect to this automorphism ν as follows:

$$\underline{g} = \bigoplus_{j \in \mathbb{Z}/6\mathbb{Z}} \underline{g}_j,$$

where $\underline{g}_j = \{x \in \underline{g} \mid \nu x = \omega^j x, \omega = e^{2\pi I/6}$ and $j = 0, \ldots, 5.\}$ This is called the 2-principal gradation of $\underline{g}$. Now one can have the following explicit descriptions of each graded subspace $\underline{g}_j$ of $\underline{g}$ with respect to the above 2-principal gradation: $\underline{g}_0 = \mathbb{C}\bar{H}_0$, $\underline{g}_1 = \mathbb{C}\bar{E}_0 \oplus \mathbb{C}\bar{E}_1$, $\underline{g}_2 = \mathbb{C}(F_1 - F_2)$, $\underline{g}_3 = \mathbb{C}(H_1 - H_2)$, $\underline{g}_4 = \mathbb{C}(E_1 - E_2)$, and $\underline{g}_5 = \mathbb{C}\bar{F}_0 \oplus \mathbb{C}\bar{F}_1$.

Now $E = \sum_{i=0}^{1} \bar{E}_i = 2^{1/2}(E_1 + E_2) + E_0$ is the 2-cyclic element of $\underline{g} = sl(3, \mathbb{C})$. Let

$$S = \{x \in \underline{g} \mid [x, E] = 0\}.$$

Then S is the 2-principal Cartan subalgebra of A_2. The basis elements of S are,

$$E = e_{31} + 2^{1/2}(e_{12} + e_{23}) \text{ and}$$
$$F = 2e_{13} + 2^{1/2}(e_{21} + e_{32}).$$

Hence $S = \mathbb{C}E \oplus \mathbb{C}F$. The principal Cartan subalgebra S of A_2 is also graded with respect to the principal gradation as

$$S = \bigoplus_{j \in \mathbb{Z}/6\mathbb{Z}} S_j; \quad S_j = S \cap \underline{g}_j \quad \text{with}$$

$S_0 = \phi$, $S_1 = \mathbb{C}E$, $S_2 = \phi$, $S_3 = \phi$, $S_4 = \phi$, and $S_5 = \mathbb{C}F$.

Therefore, the 2-exponents of $sl(3, \mathbb{C})$ are $\{1, 5\}$. For the sake of convenience, let us denote the basis elements $E = \bar{e}_0 + \bar{e}_1$ and $F = \bar{f}_0 - 2\bar{f}_1$ of S by T_1 and T_2, respectively, so that $(T_i|T_{3-j}) = 6\,\delta_{ij}$, $i, j = 1, 2$. The root vectors A_{β_i} corresponding to the simple roots are chosen such that their projections on the space $\underline{g}_0$ form a basis of $\underline{g}_0$. Here, $\underline{g}_0$ is an one-dimensional space. Therefore, let us take A_β to be the root vector corresponding to the simple root β with respect to the Cartan subalgebra S.

Let $\epsilon = \frac{1+I\sqrt{3}}{2}$, $\epsilon^2 = \frac{-1+I\sqrt{3}}{2}$, where $1, \epsilon, \epsilon^2$ are sixth roots of unity. We find that,

$$X_\beta = 2^{1/3}e_{12} + \frac{(-I\sqrt{3}2^{1/3} - 2^{1/3})}{2}e_{23} + \frac{(I\sqrt{6} - \sqrt{2})}{2}e_{13} + \frac{(-2^{1/6} - I\sqrt{3}2^{1/6})}{2}h_1$$
$$- 2^{1/6}h_2 + e_{21} + \frac{(-1 - I\sqrt{3})}{2}e_{32} + \frac{(I2^{5/6}\sqrt{3} - 2^{5/6})}{4}e_{31}$$

is a root vector with respect to S. We have,

$$\left[T_1, X_\beta\right] = [E, X_\beta] = 2^{1/3}(\epsilon + \epsilon^2)X_\beta = 2^{1/3}I\sqrt{3}X_\beta \text{ and}$$
$$\left[T_2, X_\beta\right] = [F, X_\beta] = -2^{2/3}(\epsilon + \epsilon^2)X_\beta = -2^{2/3}I\sqrt{3}X_\beta$$

Decomposing $X_\beta \in \underline{g}$ with respect to the principal gradation, we get

$$X_\beta = \sum_{j=0}^{6} X_{\beta,j}$$
$$= X_{\beta,0} + X_{\beta,1} + X_{\beta,2} + X_{\beta,3} + X_{\beta,4} + X_{\beta,5}.$$

Here

$$X_{\beta,0} = \left(\frac{-32^{1/6} - I\sqrt{3}2^{1/6}}{4}\right)(H_1 + H_2),\ X_{\beta,1} = \left(\frac{2^{5/6}I\sqrt{3} - 2^{5/6}}{4}\right)E_0 + \left(\frac{2^{1/3} - I\sqrt{3}2^{1/3}}{4}\right)(E_1 + E_2)$$
$$X_{\beta,2} = \left(\frac{3 + I\sqrt{3}}{4}\right)(F_1 - F_2), \qquad X_{\beta,3} = \left(\frac{2^{1/6} - I\sqrt{3}2^{1/6}}{4}\right)(H_1 - H_2)$$
$$X_{\beta,4} = \left(\frac{32^{1/3} + I\sqrt{3}2^{1/3}}{4}\right)(E_1 - E_2), \quad X_{\beta,5} = \left(\frac{1 - I\sqrt{3}}{4}\right)(F_1 + F_2) + \left(\frac{I\sqrt{6} - 2}{2}\right)(F_0).$$

Now the set $\left\{ X_{\beta,j}\ (j = 0 \text{ to } 5),\ \ T_1,\ T_2 \right\}$ forms a basis of $\underline{\mathfrak{g}} = sl(3, \mathbb{C})$.

Similarly, we find the element $X_{-\beta} \in \underline{\mathfrak{g}}$ such that it satisfies the following conditions:

$$\left[T_1, X_{-\beta}\right] = [E, X_{-\beta}] = -2^{1/3}(\epsilon + \epsilon^2)X_{-\beta} = (-2^{1/3}I\sqrt{3})X_{-\beta},$$
$$\left[T_2, X_{-\beta}\right] = [F, X_{-\beta}] = 2^{2/3}(\epsilon + \epsilon^2)X_{-\beta} = (2^{2/3}I\sqrt{3})X_{-\beta}.$$

Let $\hat{\underline{\mathfrak{g}}}(A)$ be the corresponding affine algebra. That is, $\hat{\underline{\mathfrak{g}}}(A) = \hat{\underline{\mathfrak{g}}}'(A) \oplus \mathbb{C}d$ where

$$\hat{\underline{\mathfrak{g}}}'(A) = \bigoplus_{j \in \mathbb{Z}} t^j \otimes \underline{\mathfrak{g}}_{j \bmod 6} \oplus \mathbb{C}c$$

is the derived subalgebra and $d = t\frac{d}{dt}$ is the scaling element. The canonical generators for $A_2^{(2)}$ are given by

$$\begin{aligned} \hat{E}_0 &= 1 \otimes \bar{E}_0, & \hat{E}_1 &= t \otimes \bar{E}_1, \\ \hat{F}_0 &= 1 \otimes \bar{F}_0, & \hat{F}_1 &= t^{-1} \otimes \bar{F}_1, \\ \hat{H}_0 &= 1 \otimes \bar{H}_0 \text{ and } & \hat{H}_1 &= c + 1 \otimes \bar{H}_1. \end{aligned}$$

The Lie bracket in $\hat{\underline{\mathfrak{g}}}'$ is given by

$$[p_1(t) \otimes g_1 \oplus \lambda_1 c, p_2(t) \otimes g_2 \oplus \lambda_2 c] = p_1(t)p_2(t) \otimes [g_1, g_2]$$
$$\oplus \frac{1}{6} Res\left(\frac{dp_1(t)}{dt} p_2(t)\right)(g_1|g_2)c,$$

where $g_1, g_2 \in \underline{\mathfrak{g}}$, $p_1(t), p_2(t) \in \mathbb{C}[t, t^{-1}]$, $\lambda_1, \lambda_2 \in \mathbb{C}$. For $\hat{\underline{\mathfrak{g}}}' = sl(3, \mathbb{C})$, we have

$$\begin{aligned} \hat{\underline{\mathfrak{g}}}'(A) &= \bigoplus_{j \in \mathbb{Z}} t^j \otimes \underline{\mathfrak{g}}_{j \bmod 6} \oplus \mathbb{C}c \\ &= t^0\mathbb{C}[t^6, t^{-6}] \otimes \underline{\mathfrak{g}}_0 \oplus t^1\mathbb{C}[t^6, t^{-6}] \otimes \underline{\mathfrak{g}}_1 \oplus t^2\mathbb{C}[t^6, t^{-6}] \otimes \underline{\mathfrak{g}}_2 \oplus t^3\mathbb{C}[t^6, t^{-6}] \otimes \underline{\mathfrak{g}}_3 \\ &\quad \oplus t^4\mathbb{C}[t^6, t^{-6}] \otimes \underline{\mathfrak{g}}_4 \oplus t^5\mathbb{C}[t^6, t^{-6}] \otimes \underline{\mathfrak{g}}_5 \oplus \mathbb{C}c. \end{aligned}$$

The principal Cartan subalgebra S of $sl(3,\mathbb{C})$ has $\{E,F\}$ as basis elements. Let

$$\begin{aligned}\mathfrak{s} &= \bigoplus_{j\in\mathbb{Z}}(t^j \otimes S_{j \bmod 6}) \oplus \mathbb{C}c \\ &= t^1\mathbb{C}[t^6,t^{-6}]\otimes\ S_1 \oplus t^5\mathbb{C}[t^6,t^{-6}]\otimes S_5 \oplus \mathbb{C}c.\end{aligned}$$

This is the principal Heisenberg subalgebra of $\hat{\underline{\mathfrak{g}}}'(A) = A_2^{(2)}$ and the set of its positive exponents is $E_+ = \{1,5,7,11,13,17,19\cdots\}$. Let us denote these elements of E_+ by $\{b_1,b_2,\ldots\}$. Now the following are the generators of S:

$$c,\ p_i = t^i \otimes T_{i'},\ q_i = i^{-1}t^{-i} \otimes T_{3-i'},$$

where $i \in E_+$ such that $i' = i \bmod 3$ with $i' \in \{1,2\}$. They satisfy the relation $[p_i,q_j] = c\delta_{ij}$. Hence $\mathfrak{s}$ is isomorphic to the infinite Heisenberg algebra.

Let $L(\Lambda_0)$ be the basic module of $A_2^{(2)}$. It is irreducible as a $\mathfrak{s}$ module. Hence $L(\Lambda_0)$ is identified with $\mathbb{C}[x_i,\ i \in E_+]$. Here the generators of $\mathfrak{s}$ act on $\mathbb{C}[x_i,\ i \in E_+]$ as follows:

$$c \longmapsto Id,\ \ p_i \longmapsto \frac{\partial}{\partial x_i}\ \ q_i \longmapsto x_i,\ i \in E_+.$$

The representation is extended to the whole algebra $\hat{\underline{\mathfrak{g}}}(A)$ using the vertex operators constructed below:

Let us consider the simple root $\beta \in \Delta_S$. Let

$$X_\beta(z) = \sum_{j\in\mathbb{Z}} z^{-j}\ (t^j \otimes X_{\beta,j \bmod 6}).$$

Then $X_\beta(z)$ maps $L(\Lambda_0)$ into the space of formal series $L(\Lambda_0)[[z,z^{-1}]]$ with coefficients in $L(\Lambda_0)$. In particular, for the simple root $\beta \in \Delta_S$, we consider the following commutation relations:

$$\begin{aligned}[p_i, X_\beta(z)] &= [t^i \otimes T_{i'}, \sum_{j\in\mathbb{Z}} z^{-j}(t^j \otimes X_{\beta,j \bmod 6})] \\ &= \sum_{j\in\mathbb{Z}} z^{-j}\ t^{i+j} \otimes [T_{i'}, X_{\beta,j \bmod 6}] \\ &= z^i\ (\ (-1)^{(i'-1)}\ 2^{\frac{i'}{3}}\ I\sqrt{3}\ ) \sum_{j\in\mathbb{Z}} z^{-(i+j)}t^{i+j} \otimes X_{\beta,(j+i') \bmod 6} \\ &= z^i\ ((-1)^{(i'-1)}\ 2^{\frac{i'}{3}}\ I\sqrt{3}\ )X_\beta(z),\end{aligned}$$

where $i' \in \{1,2\}$ such that $i' \equiv i \bmod 3$ with $i \in E_+$.

Also,

$$\begin{aligned}
[q_i, X_\beta(z)] &= i^{-1}[t^{-i} \otimes T_{3-i'}, \sum_{j\in\mathbb{Z}} z^{-j}(t^j \otimes X_{\beta,j \bmod 6})] \\
&= i^{-1} \sum_{j\in\mathbb{Z}} z^{-j} t^{-i+j} \otimes [T_{3-i'}, X_{\beta,j \bmod 6}] \\
&= z^{-i}\,(\,(-1)^{(3-i'-1)}\, 2^{\frac{(3-i')}{3}}\, I\sqrt{3}) X_\beta(z).
\end{aligned}$$

Therefore by Kac [8] and the above relations, the principal vertex operator corresponding to the root β is

$$\begin{aligned}
\Gamma^\beta(z) = \frac{(3+I\sqrt{3})2^{1/6}}{12} \exp\left(\sum_{j\in E_+} z^j((-1)^{(j'-1)} 2^{\frac{j'}{3}} I\sqrt{3}) x_j\right) \cdot \\
\exp\left(-\sum_{j\in E_+} \frac{z^{-j}}{j}\left(((-1)^{(3-j'-1)} 2^{\frac{(3-j')}{3}} I\sqrt{3})\right) \frac{\partial}{\partial x_j}\right),
\end{aligned}$$

where $\langle \Lambda_0, X_{\beta,0}\rangle = \frac{(3+I\sqrt{3})\, 2^{1/6}}{12}$ and j' is such that $j' \in \{1,2\}$ and $j' \equiv j \bmod 3$ with $j \in E_+$. Moreover, $X_\beta(z) = \sum_{j\in\mathbb{Z}} z^{-j}(t^j \otimes X_{\beta,j \bmod 6})$ acts as the vertex operator $\Gamma^\beta(z)$.

Therefore the representation of $A_2^{(2)}$ on

$$\mathbb{C}[x_i;\ i \in E_+] = \mathbb{C}[x_i; i \in \mathbb{Z}_+,\ i \not\equiv 0 \bmod 3,\ i \equiv 1 \bmod 2]$$

is given by

$$\begin{aligned}
&p_j \mapsto \frac{\partial}{\partial x_j}; \quad q_j \mapsto x_j; \quad c \mapsto Id.; \\
&d \mapsto -\frac{1}{3}\sum_{j\in E_+} j x_j \frac{\partial}{\partial x_j}; \quad X_\beta(z) \mapsto \frac{1}{\langle \Lambda_0, X_{\beta,0}\rangle}\Gamma^\beta(z).
\end{aligned}$$

(2) *Hirota bilinear equations for* $A_2^{(2)}$:

We choose the following dual bases of $A_2^{(2)}$.

$$\begin{aligned}
v_j &: \left\{\frac{1}{\sqrt{3}} p_j,\ \frac{1}{\sqrt{3}} q_j,\ \frac{1}{3} t^j \otimes A_{\beta,j \bmod 6},\ c,\ d\right\} \\
v^j &: \left\{\frac{1}{\sqrt{3}} q_j,\ \frac{1}{\sqrt{3}} p_j,\ \frac{1}{3} t^{-j} \otimes A_{-\beta,-j \bmod 6},\ d,\ c\right\}.
\end{aligned}$$

Using these bases, the operator $S = \sum(v_j \otimes v^j)$ on $L(\Lambda_0) \otimes L(\Lambda_0) = \mathbb{C}[x'] \otimes \mathbb{C}[x'']$ (Theorem 84(b)) is computed as follows:

Therefore,

$$S = \frac{1}{3}\sum_{j\in E_+}\left(jx_j' \otimes \frac{\partial}{\partial x_j''} + \frac{\partial}{\partial x_j'} \otimes jx_j''\right) + \text{Coeff. of } z^0 \text{ in } \frac{1}{9}\left(\Gamma^{\beta}(z) \otimes \Gamma^{-\beta}(z)\right)$$
$$- 1 \otimes \frac{1}{3}\sum_{j\in E_+} jx_j'' \frac{\partial}{\partial x_j''} - \frac{1}{3}\sum_{j\in E_+} jx_j' \frac{\partial}{\partial x_j'} \otimes 1.$$

Hence,

$$S = -\frac{1}{3}\sum_{j\in E_+}\left(j(x_j' - x_j'')\right)\left(\frac{\partial}{\partial x_j'} - \frac{\partial}{\partial x_j''}\right) + \text{Coeff. of } z^0 \text{ in}$$
$$\frac{1}{9}\exp\left(\sum_{j\in E_+}\left(\left(-1\right)^{(j'-1)} 2^{\frac{j'}{3}} I\sqrt{3}\,\right)(x_j' - x_j'')z^j\right) \times$$
$$\exp\left(-\sum_{j\in E_+}\frac{1}{j}\left(\left(-1\right)^{(3-j'-1)} 2^{\frac{(3-j')}{3}} I\sqrt{3}\right)\left(\frac{\partial}{\partial x_j'} - \frac{\partial}{\partial x_j''}\right)z^{-j}\right)$$

Now assume

$$x_j = \frac{1}{2}(x_j' + x_j''); \quad y_j = \frac{1}{2}(x_j' - x_j'')$$
$$\frac{\partial}{\partial x_j} = \frac{\partial}{\partial x_j'} + \frac{\partial}{\partial x_j''} \quad \text{and} \quad \frac{\partial}{\partial y_j} = \frac{\partial}{\partial x_j'} - \frac{\partial}{\partial x_j''}.$$

So

$$S = -\frac{2}{3}\sum_{j\in E_+} jy_j\frac{\partial}{\partial y_j} + \text{Coeff. of } z^0 \text{ in}$$
$$\frac{1}{9}\exp\left(\sum_{j\in E_+} z^j((-1)^{(j'-1)}2^{1+\frac{j'}{3}}I\sqrt{3})y_j\right)$$
$$\times \exp\left(-\sum_{j\in E_+}\frac{z^{-j}}{j}\left((-1)^{(3-j'-1)}2^{\frac{(3-j')}{3}}I\sqrt{3}\right)\frac{\partial}{\partial y_j}\right).$$

The usual Schur polynomials, $P_k(x),\ k \in \mathbb{Z}_+$ with $x = (x_0, x_1, \ldots)$ are defined by

$$\sum_{k\geq 0} P_k(x) z^k = \exp\left(\sum_{k\geq 0} x_k z^k\right).$$

So similarly define the Schur polynomial $P_k^{E_+}(x)$ by

$$\sum_{k\geq 0} P_k^{E_+}(x) z^k = \exp\left(\sum_{k\in E_+} x_k z^k\right).$$

Hence for $\tau \in L(\Lambda_0)$, the operator S can be rewritten in terms of Schur polynomials as,

$$S(\tau\otimes\tau) = \left\{-\frac{2}{3}\sum_{j\in E_+} j y_j \frac{\partial}{\partial y_j} + \frac{1}{9}\sum_{j\geq 0} P_j^{E_+}\left(((-1)^{(j'-1)}\, 2^{1+\frac{j'}{3}} I\sqrt{3}\,) y_j\right)\right.$$
$$\left. P_j^{E_+}\left(-\frac{1}{j}((-1)^{(3-j'-1)}\, 2^{\frac{(3-j')}{3}} I\sqrt{3}\,\frac{\partial}{\partial y_j}\right)\right\} \tau(x+y)\tau(x-y).$$

By Taylor's expansion, we have,

$$f(x+y)g(x-y) = \left(\exp \sum_{j\in E^+} y_j \frac{\partial}{\partial u_j}\right)(f(x+u)g(x-u))|_{u=0}.$$

So

$$S(\tau\otimes\tau) = \left\{-\frac{2}{3}\sum_{j\in E_+} j y_j \frac{\partial}{\partial u_j} + \frac{1}{9}\sum_{j\geq 0} P_j^{E_+}\left(((-1)^{(j'-1)}\, 2^{1+\frac{j'}{3}}\, I\sqrt{3}\,) y_j\right)\right.$$
$$\left. \cdot\, P_j^{E_+}\left(-\frac{1}{j}((-1)^{(2-j')}\, 2^{\frac{(3-j')}{3}}\, I\sqrt{3})\frac{\partial}{\partial u_j}\right)\right\} e^{(\sum_{j\in E_+} y_j \frac{\partial}{\partial u_j})}\, \tau(x+u)\tau(x-u)\,|_{u=0}.$$

Hence one can conclude, using Theorem 84(b), that the element τ of $L(\Lambda_0)$ lies in the orbit $G.1$, where G is the group associated to $A_2^{(2)}$, if and only if, $S(\tau\otimes\tau) = 0$. Hence τ satisfies the following hierarchy of Hirota's bilinear equations.

$$\left\{-\frac{2}{3}\sum_{j\in E_+} jy_jD_j + \frac{1}{9}\sum_{j\geq 0} P_j^{E_+}\left(((-1)^{(j'-1)}\, 2^{1+\frac{j'}{3}}\, I\sqrt{3})y_j\right)\right.$$

$$\left. P_j^{E_+}\left(-\frac{1}{j}((-1)^{(2-j')}\, 2^{\frac{(3-j')}{3}}\, I\sqrt{3})D_j\right)\right\} e^{(\sum_{s\in E_+} y_sD_s)}\tau(x+u)\tau(x-u)|_{u=0} = 0.$$

In particular, taking the coefficient of y_1y_5 and making suitable transformations to the Hirota derivatives, we get the equation,

$$(D_1^6 + 9D_1D_5)\ \tau\cdot\tau = 0.$$

This is the Sawada-Kotera equation in Hirota's form corresponding to the 3-reduction type of BKP [209, 210, 216–222, 224].

Remark 196. The above equation was obtained in Goddard and Olive [227, 228], as the lowest degree even equation for $A_2^{(2)}$ by obtaining the principal hierarchy of $A_2^{(2)}$ from that of $D_4^{(1)}$. Here it is to be mentioned that for Affine algebra $D_4^{(1)}$ having diagram automorphism of order 4, the fixed point set will be the affine algebra $A_2^{(2)}$. Similarly, we can get other equations in the above hierarchy, by taking the coefficients of other variables.

6.9.3 Principal vertex operators and super Hirota bilinear equations for $C_2^{(1)}$

Now we use the method followed in Kac and Wakimoto [229], to write down the Hirota bilinear equations for $C_2^{(1)}$. For detailed explanation, one can refer Sthanumoorthy and Kiruba Bagirathi [225].

(1) Let $\underline{g}$ be a finite dimensional Lie algebra of type A_3. Let $(x|y) = \frac{1}{2}tr(xy)$ be a nondegenerate symmetric bilinear form on $\underline{g}$.

Let $e_0 = e_{41}, e_1 = e_{12}, e_2 = e_{23}, e_3 = e_{34}, f_0 = e_{14}, f_1 = e_{21}, f_2 = e_{32}, f_3 = e_{43}$, $h_0 = e_{44} - e_{11}, h_1 = e_{11} - e_{22}, h_2 = e_{22} - e_{33}$, and $h_3 = e_{33} - e_{44}$ be the set of all canonical generators for $\underline{g}$. Here (e_{ij}) is the 4×4 matrix with 1 in the ijth place and 0 elsewhere.

Let $\underline{\mathfrak{h}} = span\{h_i,\ \ 1\leq i\leq 3\}$ be a Cartan subalgebra of $\underline{g}$, Δ be a simple root system. The bilinear form $(.|.)$ is normalized so that $(\alpha|\alpha) = 2$ for all $\alpha\in\Delta$. Let $\Pi = \{\alpha_1, \alpha_2, \alpha_3\}$ be a basis of Δ.

Let θ be the Dynkin diagram automorphism of order 2 given by $\theta(\alpha_1) = \alpha_3, \theta(\alpha_2) = \alpha_2$ and $\theta(\alpha_3) = \alpha_1$. The automorphism θ is extended to $\underline{g}$ by defining $\theta(x_\alpha) = x_{\theta(\alpha)}$ for all $\alpha\in\Delta$. Using this automorphism of Dynkin diagram of A_3, we can construct C_2 as a subalgebra of fixed points of this automorphism.

We have the following decomposition of $\underline{g}$ with respect to θ :

$$\underline{g} = \bigoplus_{j \in Z/2Z} \underline{g}_{[j]},$$

where $\underline{g}_{[j]} = \{x \in \underline{g} \mid \theta(x) = \epsilon^j x\}, \epsilon = e^{\pi I}$ and $I = \sqrt{-1}$. We have

$$\underline{\mathfrak{g}}_{[0]} = \mathbb{C}(e_{12} + e_{34}) \oplus \mathbb{C}e_{23} \oplus \mathbb{C}e_{14} \oplus \mathbb{C}e_{41} \oplus \mathbb{C}e_{32} \oplus \mathbb{C}(e_{21} + e_{43}) \oplus$$
$$\mathbb{C}(h_1 + h_3) \oplus \mathbb{C}h_2 \oplus \mathbb{C}(e_{13} - e_{24}) \oplus \mathbb{C}(e_{31} - e_{42}) \text{ and}$$
$$\underline{\mathfrak{g}}_{[1]} = \mathbb{C}(e_{12} - e_{34}) \oplus \mathbb{C}(e_{21} - e_{43}) \oplus \mathbb{C}(h_1 - h_3) \oplus \mathbb{C}(e_{13} + e_{24}) \oplus \mathbb{C}(e_{31} + e_{42}).$$

We omit the proof of following proposition which is similar to proposition in Lepowsky and Primc [199].

Proposition 64.

(i) $\underline{\mathfrak{g}}_{[0]}$ *is of the type* C_2 *and* $\underline{\mathfrak{h}}_{[0]}$ *is a Cartan subalgebra of* $\underline{\mathfrak{g}}_{[0]}$.

(ii) *The set* $\left\{\frac{\alpha + \theta\alpha}{2};\ \alpha \in \Delta\right\}$ *is a root system with respect to* $\underline{\mathfrak{h}}_{[0]}$ *with basis* $\left\{\frac{\alpha_i + \theta\alpha_i}{2};\ 1 \le i \le 2\right\}$ *and the systems of long and short roots are respectively*

$$(\Delta_{[0]})_l^+ = \{\alpha_2, \alpha_1 + \alpha_2 + \alpha_3\} \text{ and}$$
$$(\Delta_{[0]})_l^+ = \left\{\frac{\alpha_1 + \alpha_3}{2}, \frac{\alpha_1 + 2\alpha_2 + \alpha_3}{2}\right\}.$$

The corresponding generating elements of $\underline{\mathfrak{g}}_{[0]} = C_2$ *are* $H_2 = h_2, E_1 = e_{12} + e_{34}$, $F_1 = e_{21} + e_{43}, E_2 = e_{23}, F_2 = e_{32}$, *and* $H_1 = h_1 + h_3$. *To construct principal vertex operators for* C_2, *we consider* C_2 *as the subalgebra of fixed points of the automorphism* θ *of* A_3. *The basis elements of* $\mathfrak{a}$ *are* $E = e_{41} + e_{12} + e_{23} + e_{34}$, $F = e_{14} + e_{21} + e_{32} + e_{43}$, *and* $G = e_{13} + e_{24} + e_{31} + e_{42}$, *with* $\mathfrak{a} = \mathbb{C}E \oplus \mathbb{C}F \oplus \mathbb{C}G$.

Let $E = e_0 + \sum_{i=1}^{3} e_i$, *where* e_0 *is the lowest root vector of* A_3 *with respect to* $\mathfrak{h}$. *Then* $E = e_{41} + e_{12} + e_{23} + e_{34}$ *is the cyclic element of* A_3. *Let* $\mathfrak{a} = \{x \in \underline{\mathfrak{g}} \mid [x, E] = 0\}$. *Then* $\underline{\mathfrak{a}}$ *is the principal Cartan subalgebra of* A_3.

Let $\overline{\eta} = \{x \in \underline{\mathfrak{g}}_{[0]} \mid [x, E] = 0\}$. *Then* $\underline{\eta}$ *is the principal Cartan subalgebra of* C_2. *A basis of* $\underline{\eta}$ *is* $\{E, F\}$.

The following proposition can be directly proved.

Proposition 65. $\underline{\eta}$ *is the subalgebra of* $\underline{\mathfrak{a}}$ *of* θ *fixed points.*

We have decomposition of $C_2 = \underline{\mathfrak{g}}_{[0]}$ *as* $\underline{\mathfrak{g}}_{[0]} = \oplus_{j \in \mathbb{Z}/4\mathbb{Z}} (\underline{\mathfrak{g}}_{[0]})_j$, *where*

$$(\underline{\mathfrak{g}}_{[0]})_0 = \mathbb{C}(h_1 + h_3) \oplus \mathbb{C}h_2; (\underline{\mathfrak{g}}_{[0]})_1 = \mathbb{C}(e_{12} + e_{34}) \oplus \mathbb{C}e_{23} \oplus \mathbb{C}e_{41},$$
$$(\underline{\mathfrak{g}}_{[0]})_2 = \mathbb{C}(e_{13} - e_{24}) \oplus \mathbb{C}(e_{31} - e_{42}) \text{ and } (\underline{\mathfrak{g}}_{[0]})_3 = \mathbb{C}(e_{21} + e_{43}) \oplus \mathbb{C}e_{32} \oplus \mathbb{C}e_{14}.$$

The principal Cartan subalgebra $\underline{\eta}$ of C_2 is also graded with respect to the principal gradation:

$$\underline{\eta} = \oplus_{j\in\mathbb{Z}/4\mathbb{Z}}\underline{\eta}_j; \underline{\eta}_j = \eta \cap (\underline{\mathfrak{g}}_{[0]})_j \text{ where } \underline{\eta}_0 = \{0\}, \underline{\eta}_1 = \mathbb{C}E, \underline{\eta}_2 = \{0\} \text{ and } \underline{\eta}_3 = \mathbb{C}F.$$

Let $\Delta_{\underline{\mathfrak{a}}}$ be the set of roots for A_3 with respect to $\underline{\mathfrak{a}}$.

We omit the proofs of the following theorem and propositions [225].

Theorem 86. *The set $\Delta_{\underline{\eta}}$ of roots of C_2 with respect to $\underline{\eta}$ is*

$$\Delta_{\underline{\eta}} = \left\{\underline{\beta} = \frac{\beta + \theta\beta}{2}; \beta \in \Delta_{\underline{\mathfrak{a}}}\right\}.$$

Proposition 66. *Let $\{\gamma_1, \gamma_2, \gamma_3\}$ be the set of simple roots of A_3 with respect to $\Delta_{\underline{\mathfrak{a}}}$ such that*

$$\langle\gamma_i, \gamma_i\rangle = 2, \langle\gamma_1, \gamma_2\rangle = \langle\gamma_2, \gamma_3\rangle = -1 \quad \text{and} \quad \langle\gamma_1, \gamma_3\rangle = 0.$$

Then it can be directly verified that $X_{\pm\gamma_i}$, $(1 \le i \le 3)$ are the root vectors for the simple roots $\pm\gamma_i$, $(1 \le i \le 3)$ satisfying the relations:

$$\begin{aligned}
&[E, X_{\pm\gamma_1}] = \pm(1 + I)X_{\pm\gamma_1}, [F, X_{\pm\gamma_1}] = \pm(1 - I)X_{\pm\gamma_1}, [G, X_{\pm\gamma_1}] = \mp 2X_{\pm\gamma_1},\\
&[E, X_{\pm\gamma_2}] = \pm(-I + 1)X_{\pm\gamma_2}, [F, X_{\pm\gamma_2}] = \pm(1 + I)X_{\pm\gamma_2}, [G, X_{\pm\gamma_2}] = \mp 2X_{\pm\gamma_2},\\
&[E, X_{\pm\gamma_3}] = \mp(1 + I)X_{\pm\gamma_3}, [F, X_{\pm\gamma_3}] = \pm(I - 1)X_{\pm\gamma_3}, [G, X_{\pm\gamma_3}] = \mp 2X_{\pm\gamma_3}.
\end{aligned}$$

Decomposing $X_{\gamma_i} \in \underline{\mathfrak{g}}, (1 \le i \le 3)$ with respect to the principal gradation, we get, $X_{\gamma_i} = \sum_{j=0}^{3} X_{\gamma_{i,j}}, X_{i,j} \in \underline{\mathfrak{g}}_j$, we observe that the elements $\{E_i, X_{\gamma_{i,j}}, 1 \le i \le 3;\ 0 \le j \le 3\}$ form a basis of A_3. Now the corresponding affine algebra is defined by $\hat{\underline{\mathfrak{g}}}' = \oplus_{j\in\mathbb{Z}} t^j \otimes \underline{\mathfrak{g}}_{j \bmod 4} \oplus \mathbb{C}c$, where c is the canonical central element and the generators for $A_3^{(1)}$ are given by

$$\begin{aligned}
&\hat{e}_j = 1 \otimes e_j, \quad \hat{e}_0 = t \otimes e_0,\\
&\hat{f}_j = 1 \otimes f_j, \quad \hat{f}_0 = t^{-1} \otimes f_0,\\
&\hat{h}_j = \alpha_j, \quad \hat{h}_0 = \frac{2}{(\theta|\theta)}c - \theta^\vee,
\end{aligned}$$

where $j \in \{1, 2, 3\}$ and $\theta^\vee$ is the coroot corresponding to the highest root θ of A_3. The Lie bracket in $\hat{\underline{\mathfrak{g}}}'(A)$ is given by,

$$\begin{aligned}
[p_1(t) \otimes g_1 \oplus \lambda_1 c, p_2(t) \otimes g_2 \oplus \lambda_2 c] &= p_1(t)p_2(t) \otimes [g_1, g_2]\\
&\oplus \frac{1}{4} Res\left(\frac{dp_1(t)}{dt} p_2(t)\right)(g_1|g_2)c,
\end{aligned}$$

for all $p_1(t), p_2(t) \in \mathbb{C}[t, t^{-1}]$ and $g_1, g_2 \in \underline{\mathfrak{g}}$.
Explicitly,

$$\begin{aligned}\hat{\underline{\mathfrak{g}}}' &= \bigoplus_{j\in\mathbb{Z}} t^j \otimes \underline{\mathfrak{g}}_{j \bmod 4} \oplus \mathbb{C}c \\ &= t^0\, \mathbb{C}[t^4, t^{-4}] \otimes \underline{\mathfrak{g}}_0 \oplus t\, \mathbb{C}[t^4, t^{-4}] \otimes \underline{\mathfrak{g}}_1 \oplus t^2\, \mathbb{C}[t^4, t^{-4}] \otimes \underline{\mathfrak{g}}_2 \oplus t^3\, \mathbb{C}[t^4, t^{-4}] \otimes \underline{\mathfrak{g}}_3 \oplus \mathbb{C}c.\end{aligned}$$

The principal Cartan subalgebra $\underline{\mathfrak{a}}$ of $\underline{\mathfrak{g}}$ has $\{E, F, G\}$ as basis elements. Let us rename these basis elements as E_1, E_2, E_3, where $E_1 = E, E_2 = G, E_3 = F$, so that $E_i \in \underline{\mathfrak{a}}_i$ and $(E_i|E_{4-j}) = 4\delta_{ij}$ for $j = 1, 3$. Now $\mathfrak{a} = \bigoplus_{j\in\mathbb{Z}}(t^j \otimes \underline{a}_{j \bmod 4}) \oplus \mathbb{C}c = t\mathbb{C}[t^4, t^{-4}] \otimes \underline{\mathfrak{a}}_1 \oplus t^2\mathbb{C}[t^4, t^{-4}] \otimes \underline{\mathfrak{a}}_2 \oplus t^3\mathbb{C}[t^4, t^{-4}] \otimes \underline{\mathfrak{a}}_3 \oplus \mathbb{C}c$.

This is the principal Heisenberg-Cartan subalgebra of $\hat{\underline{\mathfrak{g}}}'$ and we observe that $E_+ = \{n \in \mathbb{Z}_+ \mid n \neq 0 \bmod 4\} = \{1, 2, 3, 5, 6, 7, 9, 10, \ldots\}$ is the set of exponents of $A_3^{(1)}$. Its principal subalgebra $\mathfrak{a}$ has basis elements $c, p_i = t^i \otimes E_{i'}, q_i = \frac{1}{i} t^{-i} \otimes E_{4-i'}$ where $i \in E_+$ and $i' \in \{1, 2, 3\}$ such that $i' = i \bmod 4$. These generators satisfy the relation $[p_i, q_j] = c\delta_{ij}$.

The principal Cartan subalgebra of C_2 is η with $\{E, F\}$ as a basis. Let us rename the basis elements of $\underline{\eta}$ as E_1, E_3, where $E_1 = \bar{E}, E_3 = F$ so that $E_i \in \underline{\eta}_i$ and $(E_i|E_{4-j}) = 4\delta_{ij}$ for $i, j = 1, 3$. Let γ_i $(1 \leq i \leq 3)$ be the simple roots of A_3 with respect to $\underline{\mathfrak{a}}$. The root system of C_2 with respect to the Cartan subalgebra $\underline{\eta}$ of C_2 is

$$\Delta_{\underline{\eta}} = \pm\{\bar{\gamma}_1, \bar{\gamma}_2, \bar{\gamma}_1 + \bar{\gamma}_2, 2\bar{\gamma}_1 + \bar{\gamma}_2\},$$

with $(\Delta_{\underline{\eta}})_l^+ = \{2\bar{\gamma}_1 + \bar{\gamma}_2, \bar{\gamma}_2\}$ as the simple root of the system of long roots and $(\Delta_{\underline{\eta}})_s^+ = \{\bar{\gamma}_1, \bar{\gamma}_1 + \bar{\gamma}_2\}$ as the simple root of the system of short roots. Here $\overline{\gamma}_1 = \frac{1}{2}(\gamma_1 - \gamma_3), \overline{\gamma}_{-1} = (\gamma_2 + \gamma_3)$ are the simple roots and $X_{\bar{\gamma}_i}$ are the corresponding root vectors of C_2 which are computed using the relations $\bar{\gamma}_i = \frac{1}{2}(\gamma_i + \theta\gamma_i)$; $X_{\bar{\gamma}_i} = X_{\gamma_i} + X_{\theta\gamma_i}$ $(i = 1, 2)$. The corresponding root vectors satisfy the following relations:

$$\begin{aligned}&[E_i, X_{\bar{\gamma}_1}] = (1 + I^i)X_{\bar{\gamma}_1}; \quad [E_i, X_{-\bar{\gamma}_1}] = -(1 + I^i)X_{\bar{\gamma}_1};\\ &[E_i, X_{\bar{\gamma}_2}] = (-2)I^i X_{\bar{\gamma}_2}; \quad \text{and} \quad [E_i, X_{-\bar{\gamma}_2}] = 2I^i X_{-\bar{\gamma}_2},\end{aligned}$$

where $i = 1, 2$. Decomposing $X_{\overline{\gamma}_i} \in \underline{\mathfrak{g}}$, $(1 \leq i \leq 2)$ with respect to the principal gradation of C_2, we get, $X_{\overline{\gamma}_i} = \sum_{j=0}^{3} X_{\overline{\gamma}_i, j}$, $X_{\overline{\gamma}_i, j} \in (\underline{\mathfrak{g}}_{[0]})_j$. We observe that the elements $\{E_i, X_{\overline{\gamma}_i, j},\ 1 \leq i \leq 2, 0 \leq j \leq 3\}$ form a basis of C_2. The corresponding affine algebra is given by $\hat{\underline{\mathfrak{g}}}_{[0]} = \oplus_{j\in\mathbb{Z}} t^j \otimes (g_{[0]})_{j \bmod 4} \oplus \mathbb{C}c$, with generators $\hat{E}_0 = \hat{e}_0, \hat{E}_1 = \hat{e}_1 + \hat{e}_3, \hat{E}_2 = \hat{2}_2, \hat{F}_0 = \hat{f}_0, \hat{F}_1 = \hat{f}_1 + \hat{f}_3, \hat{F}_2 = \hat{f}_2, \hat{H}_0 = \hat{h}_0, \hat{H}_1 = \hat{h}_1 + \hat{h}_3$, and $\hat{H}_2 = \hat{h}_2$.

Let $\hat{\underline{\mathfrak{g}}} = \hat{\underline{\mathfrak{g}}}' \oplus \sum_{j\in\mathbb{Z}} \mathbb{C}d_j$ and $u(j) = u \otimes t^j$ be with brackets defined by

$$[d_n, u(j)] = -ju(j + n),\ [d_n, c] = 0,\ \text{ and } [d_m, d_n] = (m - n)d_{m+n} + \delta_{m,-n}\frac{m^3 - m}{12}c.$$

Similarly we can define, $\hat{\underline{\mathfrak{g}}}_{[0]} = \hat{\underline{\mathfrak{g}}}'_{[0]} \bigoplus \sum_{j\in\mathbb{Z}} \mathbb{C}d_j$ with $[d_n, u(j)] = -ju(j+n)$, $[d_n, c] = 0$ and $[d_m, d_n] = (m-n)d_{m+n} + \delta_{m,-n}\frac{m^3-m}{12}c$. Let $X_{\bar{\gamma}}(z) = \sum_{j\in\mathbb{Z}} z^{-j}(t^j \otimes X_{\bar{\gamma},j \bmod 4})$. Then $[d_n, X_{\bar{\gamma}}(z)] = z^n\left(n + z\frac{d}{dz}\right)X_{\bar{\gamma}}(z)$. Let us consider the principal Cartan subalgebra $\underline{\eta}$ of C_2 and its basis $\{E_1, E_3\}$. Let $\eta = \oplus_{j\in\mathbb{Z}}(t^j \otimes \underline{\eta}_{j \bmod 4}) \oplus \mathbb{C}c$. This is the principal subalgebra of $\hat{\underline{\mathfrak{g}}}'_{[0]}$. Here we observe that $\mathbb{Z}_+^{\text{odd}}$ is the set of exponents for $C_2^{(1)}$, which we denote by $\underline{E}_+$. We also note that $\underline{E}_+ \subset E_+$. It can also be proved that $\eta = \{x \in \mathfrak{a} \mid \theta x = x\}$. Here c, $p_i = t^i \otimes E_{i'}$, $q_i = i^{-1}t^{-i} \otimes E_{4-i'}$ where $i \in \underline{E}_+$ such that $i' = i \bmod 4$ with $i' \in \{1,3\}$ are the generators of η. They satisfy the relation $[p_i, q_j] = c\delta_{ij}$. Hence η is isomorphic to the infinite Heisenberg algebra.

Let us consider the space $\mathbb{C}[x_i;\ i \in \mathbb{Z}_{+\text{odd}}]$ and consider it as an η module on which p_i acts as $\frac{\partial}{\partial x_i}$, q_i acts as multiplication by ix_i and c acts as Id.

Let us take $\bar{\gamma}_1 \in \Delta_{\underline{\eta}}$. Then $[p_i, X_{\bar{\gamma}_1}(z)] = [t^i \otimes E_{i'}, \sum_{j\in\mathbb{Z}_{+\text{odd}}} z^{-j}(t^j \otimes X_{\bar{\gamma}_1,j \bmod 4})] = z^i(1 + I^{i'})X_{\bar{\gamma}_1}(z)$, where $i' \in \{1,3\}$ such that $i' \equiv i \bmod 4$ with $i \in \underline{E}_+$.

Similarly, $[q_i, X_{\bar{\gamma}_1}(z)] = z^{-i}(1 + I^{4-i'})X_{\bar{\gamma}_1}(z)$. Using Kac et al. [211] and Kac [8], the principal vertex operator corresponding to the root $\bar{\gamma}_1$ is

$$\Gamma^{\bar{\gamma}_1}(z) = \exp\left(\sum_{j\in\mathbb{Z}_{+\text{odd}}} (1 + I^{j'})z^j x_j\right) \exp\left(-\sum_{j\in\mathbb{Z}_{+\text{odd}}} (1 + I^{4-j'})z^{-j}\frac{\partial}{\partial x_j}\right)$$

(where $j' \in \{1,3\}$ with $j' \equiv \bmod 4$). Similarly, we can have the vertex operator expressions for all the roots $\bar{\gamma} \in (\Delta_{\underline{\eta}})$.

Consider the fermionic oscillators ψ_n satisfying the following anticommutation relation:

$$[\psi_m, \psi_n]_+ = \psi_m\psi_n + \psi_n\psi_m = \delta_{m,-n} \quad (m, n \in \delta + \mathbb{Z}),$$

where either $\delta = 0$ (or) $\delta = \frac{1}{2}$ with $\psi_0^2 = \frac{1}{2}$. We denote by A_δ, the algebra generated by $\{\psi_n | n \in \delta + \mathbb{Z}\}$. Consider the space

$$V_\delta = \wedge[\xi_i \mid i \geq 0,\ i \in \delta + \mathbb{Z}],$$

where $\wedge$ means the exterior algebra generated by ξ_j.

The algebra A_δ can be represented in V_δ by $\psi_n \mapsto \frac{\partial}{\partial \xi_n}$; $\psi_{-n} \mapsto \xi_n$ for $n > 0$. $\psi_0 \mapsto \frac{1}{\sqrt{2}}(\xi_0 + \frac{\partial}{\partial \xi_0})$, for $n = 0$.

The proof of the following proposition is in Kac and Raina [226].

Proposition 67. *Let l_k ($k \in \mathbb{Z}$) be the operators in V_δ defined by*

$$l_k = \delta_{k,0}\frac{1-2\delta}{16} + \frac{1}{2}\sum_{j\in\delta+\mathbb{Z}} j : \psi_{-j}\psi_{j+k} :,$$

$$where: \psi_m\psi_n := \begin{cases} \psi_m\psi_n, if\ n \geq m \\ -\psi_n\psi_m,\ if\ n < m\ \ (m,n \in \delta + \mathbb{Z}). \end{cases}$$

Then

$$[l_m, l_n] = (m-n)l_{m+n} + \delta_{m,-n}\left(\frac{m^3 - m}{24}\right).$$

Here

$$l_0 = \frac{1-2\delta}{16} + \sum_{j>0} j\psi_{-j}\psi_j\ for\ j\ \in \delta + \mathbb{Z}\ and\ [\psi_m, l_k] = \left(m + \frac{k}{2}\right)\psi_{m+k}.$$

The operators l_n $(n \in \mathbb{Z})$ along with identity operator, form a Virasoro algebra *Vir* with central charge $\frac{1}{2}$ [4, 211, 227, 228]. The representation of *Vir* in V_δ has central charge $c = \frac{1}{2}$. The subspaces V_δ^+ and V_δ^- of even and odd elements respectively are sub representations of V_δ. If c is the central charge and k is the eigen value of the energy operator [8, 226] then the corresponding Virasoro module can be denoted by $V(c,k)$.

Using this notation we can write:

$$V_{\frac{1}{2}}^+ = V\left(\frac{1}{2}, 0\right) \quad \text{and} \quad V_{\frac{1}{2}}^- = V\left(\frac{1}{2}, \frac{1}{2}\right), \text{ with}$$

$$V_{\frac{1}{2}} = V_{\frac{1}{2}}^+ \oplus V_{\frac{1}{2}}^- \quad \text{and} \quad V_0^+ = V_0^- = V\left(\frac{1}{2}, \frac{1}{16}\right).$$

Let $a_j = \frac{\partial}{\partial x_j}, a_{-j} = jx_j$ for $j \in \mathbb{N}_{\text{odd}}$ be the operators defined on $\mathbb{C}[x]$ and $\dot{L}_n^{(h)} = \frac{1}{2h}\sum_{j\in\mathbb{N}_{\text{odd}}} : a_{nh-j}a_j :$, for $n \in \mathbb{Z}$. Here h is the coxeter number of $C_2^{(1)}$.

Using the construction given in Goddard and Olive [228] and also using Kac [8] and Kac and Raina [226], one can directly prove the following theorem (we omit the proof) which gives the coset Virasoro algebra representation of $C_2^{(1)}$.

Theorem 87. *The representation of* $\hat{\underline{\mathfrak{g}}}'_{[0]} + Vir = C_2^{(1)} \oplus Vir$ *on the space* $\mathbb{C}[x_i,\ i \in \mathbb{N}_{\text{odd}}] \otimes \wedge[\xi_i \mid i \in \frac{1}{2} + \mathbb{Z}_+]$ *can be given by*

$$p_j \mapsto \frac{\partial}{\partial x_j};\quad q_j \mapsto jx_j;\quad K \mapsto Id;\quad d_j \mapsto \dot{L}_j^{(\mathrm{h})} + l_j;$$

$$X_{\bar{\gamma}_i}(z) \mapsto \Gamma^{\bar{\gamma}_i}(z)\psi(z),\quad if\ (\bar{\gamma}_i|\bar{\gamma}_i) = 1\ for\ i = 1,2\quad and$$

$$X_{\bar{\gamma}_i}(z) \mapsto \Gamma^{\bar{\gamma}_i}(z),\quad if\ (\bar{\gamma}_i|\bar{\gamma}_i) = 2\ for\ \ i = 1,2.$$

Super Hirota bilinear equations for $C_2^{(1)}$ from those of $A_3^{(1)}$

We use Theorem 84 (b) (proved in Peterson and Kac [26]) and Taylor expansion for the G-module $L(\wedge)$ and conclude that the element τ of $L(\wedge)$ lies in the orbit $G.1$ if and only if, $S(\tau \otimes \tau) = 0$, where

$$S = \sum u_j \otimes u^j.$$

Let us choose dual bases for $C_2^{(1)}$ as follows:

$$u_j : \{\frac{1}{\sqrt{8}}p_j, \quad \frac{1}{\sqrt{8}}q_j, \quad t^j \otimes X_{\bar{\gamma}_1, j \bmod 4}, \quad t^j \otimes X_{\bar{\gamma}_2, j \bmod 4}, \ c, \ d\}$$
$$u^j : \{\frac{1}{\sqrt{8}}q_j, \quad \frac{1}{\sqrt{8}}p_j, \quad t^{-j} \otimes X_{-\bar{\gamma}_1, -j \bmod 4}, \quad t^{-j} \otimes X_{-\bar{\gamma}_2, -j \bmod 4}, \ d, \ c\}$$

Let $S = \sum (u_j \otimes u^j)$ be an operator acting on the module

$$\mathbb{C}[x'_i,\ i \in \mathbb{Z}_{+\mathrm{odd}}] \otimes \wedge[\xi'_i \mid i \in \frac{1}{2} + \mathbb{Z}_+] \bigotimes \mathbb{C}[x''_i,\ i \in \mathbb{Z}_{+\mathrm{odd}}] \otimes \wedge[\xi''_i \mid i \in \frac{1}{2} + \mathbb{Z}_+]$$
$$= \mathbb{C}[x'_i,\ x''_i,\ i \in \mathbb{Z}_{+\mathrm{odd}}] \bigotimes \wedge[\xi'_i,\ \xi''_i \ \mid \ i \in \frac{1}{2} + \mathbb{Z}_+].$$

Here $'$ and $''$ denote the operators acting on the first and second factors of the tensor product respectively. Hence,

$$S = \frac{1}{8} \sum_{j \in \underline{E}_+} \left(\frac{\partial}{\partial x'_j} \otimes j x''_j + j x'_j \frac{\partial}{\partial x''_j} \right) + \text{Coeff. of } z^0 \text{ in}$$
$$\left(\sum_{i=1}^{2} X_{\bar{\gamma}_i}(z) \otimes X_{-\bar{\gamma}_i}(z) \right) - 1 \otimes (\dot{L}_0^{(h)''} + l''_0) - (\dot{L}_0^{(h)'} + l'_0) \otimes 1.$$
$$= \sum_{j \in \underline{E}_+} \frac{j}{8} \left(\frac{\partial}{\partial x'_j} \otimes x''_j + x'_j \otimes \frac{\partial}{\partial x''_j} \right) + \text{Coeff. of } z^0 \text{ in} \sum_{i=1}^{2} X_{\bar{\gamma}_i}(z) \otimes X_{-\bar{\gamma}_i}(z)$$
$$- \frac{1}{8} \sum_{j \in \underline{E}_+} \frac{\partial}{\partial x'_j} j x'_j \otimes 1 - 1 \otimes \frac{1}{8} \sum_{j \in \underline{E}_+} \frac{\partial}{\partial x''_j} j x''_j - \sum_{j \in \frac{1}{2} + \mathbb{Z}_+} j \psi''_{-j} \psi''_j - \sum_{j \in \frac{1}{2} + \mathbb{Z}_+} j \psi'_{-j} \psi'_j$$
$$S = \frac{-1}{8} \sum_{j \in \underline{E}_+} j (x'_j - x''_j) \left(\frac{\partial}{\partial x'_j} - \frac{\partial}{\partial x''_j} \right) - \sum_{j \in \frac{1}{2} + \mathbb{Z}_+} j (\psi'_{-j} \psi'_j + \psi''_{-j} \psi''_j)$$
$$+ \text{Coeff. of } z^0 \text{ in} \sum_{i=1}^{2} (X_{\bar{\gamma}_i}(z) \otimes X_{-\bar{\gamma}_i}(z)).$$

Here $\bar{\gamma}_1$ is a short root and $\bar{\gamma}_2$ is a long root. Therefore,

$$\text{Coeff. of } z^0 \text{ in } X_{\bar{\gamma}_1}(z) \otimes X_{-\bar{\gamma}_1}(z) = \text{Coeff. of } z^0 \text{ in } \Gamma^{\bar{\gamma}_1}(z) \psi'_{1/2}(z) \otimes \Gamma^{-\bar{\gamma}_1}(z) \psi''_{1/2}(-z)$$

It can be proved that, Coeff. of z^0 in $\Gamma^{\bar{\gamma}_1}(z)\psi'_{1/2}(z) \otimes \Gamma^{-\bar{\gamma}_1}(z)\psi''_{1/2}(-z)$

$$= \text{Coeff. of } \mathrm{z}^0 \text{ in}$$

$$\exp\left(\sum_{j\in \underline{E}_+}(1+I^{j'})(x'_j - x''_j)z^j\right)\exp\left(-\sum_{j\in \underline{E}_+}(1+I^{4-j'})\left(\frac{\partial}{\partial x'_j} - \frac{\partial}{\partial x''_j}\right)z^{-j}\right)$$

$$\otimes\, \psi'_{1/2}(z)\psi''_{1/2}(-z).$$

Similarly Coeff. of z^0 in $X_{\bar{\gamma}_2}(z) \otimes X_{-\bar{\gamma}_2}(z) =$ Coeff. of z^0 in $\Gamma^{\bar{\gamma}_2}(z) \otimes \Gamma^{-\bar{\gamma}_2}(z)$.
Hence, coeff. of z^0 in $\Gamma^{\bar{\gamma}_2}(z) \otimes \Gamma^{-\bar{\gamma}_2}(z)$

$$= \text{Coeff. of } \mathrm{z}^0 \text{ in}$$

$$\left\{\exp\left(\sum_{j\in \underline{E}_+}(-2)I^{j'}(x'_j - x''_j)z^j\right)\exp\left(\sum_{j\in \underline{E}_+}2I^{4-j'}\left(\frac{\partial}{\partial x'_j} - \frac{\partial}{\partial x''_j}\right)z^{-j}\right)\right\}.$$

Assume

$$x_j = \frac{1}{2}(x'_j + x''_j); \quad y_j = \frac{1}{2}(x'_j - x''_j)$$

$$\frac{\partial}{\partial x_j} = \frac{\partial}{\partial x'_j} + \frac{\partial}{\partial x''_j} \quad \text{and} \quad \frac{\partial}{\partial y_j} = \frac{\partial}{\partial x'_j} - \frac{\partial}{\partial x''_j}.$$

Therefore Coeff. of z^0 in $X_{\bar{\gamma}_1}(z) \otimes X_{-\bar{\gamma}_1}(z) =$ Coeff. of z^0 in

$$\exp\left(\sum_{j\in \underline{E}_+}2(1+I^{j'})y_j z^j\right)\exp\left(-\sum_{j\in \underline{E}_+}(1+I^{4-j'})\frac{\partial}{\partial y_j}z^{-j}\right) \otimes \psi'_{1/2}(z)\psi''_{1/2}(-z).$$

Coeff. of z^0 in $X_{\bar{\gamma}_2}(z) \otimes X_{-\bar{\gamma}_2}(z)$

$$= \text{Coeff. of } z^0 \text{ in } \exp\left(\sum_{j\in \underline{E}_+}(-2)I^{j'}2y_j z^j\right)\exp\left(\sum_{j\in \underline{E}_+}2I^{4-j'}\frac{\partial}{\partial y_j}z^{-j}\right).$$

We note that $\psi_{1/2}(-z) = -\,\psi_{1/2}(z)$. Therefore, we have

$$S = -\sum_{j\in \underline{E}_+}\frac{j}{4}y_j\frac{\partial}{\partial y_j} - \sum_{j\in \frac{1}{2}+\mathbb{Z}_+} j(\psi'_{-j}\psi'_j + \psi''_{-j}\psi''_j) - \text{ Coeff. of } \mathrm{z}^0 \text{ in}$$

$$\left\{\exp\left(\sum_{j\in\underline{E}_+} 2(1+I^{j'})y_j z^j\right)\exp\left(-\sum_{j\in\underline{E}_+}(1+I^{4-j'})\frac{\partial}{\partial y_j}z^{-j}\right)\right.$$
$$\left.\otimes\ \psi'_{1/2}(z)\psi''_{1/2}(z)\right\}$$
$$+\ \text{Coeff. of } z^0 \text{ in}\left\{\exp\left(\sum_{j\in\underline{E}_+} -4I^{j'}y_j z^j\right)\exp\left(\sum_{j\in\underline{E}_+} 2I^{4-j'}\frac{\partial}{\partial y_j}z^{-j}\right)\right\}.$$

Let $P_k^{E_+}(x)$, $k \in E_+$ be the Schur polynomials defined as follows:

$$\sum_{k\geq 0} P_k^{E_+}(x)z^k = \exp\left(\sum_{k\in\underline{E}_+} x_k z^k\right),\quad \text{where } P_k^{E_+}(x) = P_k(x_k),\quad k\in\underline{E}_+.$$

Finally, it can be proved that,

$$S(\tau\otimes\tau) = \left\{-\sum_{j\in\underline{E}_+}\frac{j}{4}y_j\frac{\partial}{\partial y_j} - \sum_{j\in\frac{1}{2}+\mathbb{Z}_+} j(\psi'_{-j}\psi'_j + \psi''_{-j}\psi''_j)\right.$$
$$-\sum_{n\in\mathbb{Z}_+,\ r\in\mathbb{Z}} P_n^{E_+}(2(1+I^{n'})y_n)P_{n-r}^{E_+}\left(-(1+I^{4-n'})\frac{\partial}{\partial y_n}\right)\sum_{\substack{j+k=r\\ j,k\in\frac{1}{2}+\mathbb{Z}}}\psi'_j\psi''_k$$
$$\left.+\sum_{n\in\mathbb{Z}_+} P_n^{E_+}(-4I^{n'}y_n)P_n^{E_+}\left(2I^{4-n'}\frac{\partial}{\partial y_n}\right)\right\}\tau(x+y,\xi')\cdot\tau(x-y,\xi'').$$

Now the module is $\mathbb{C}[x',x'']\otimes V_{\frac{1}{2}}(\xi',\xi'')$.

As in Kac [8], let us change the variables ξ',ξ'' into ξ,η by using

$$\xi_j = \frac{1}{2}(\xi'_j + \xi''_j) \text{ and } \eta_j = \frac{1}{2}(\xi' - \xi'').$$

Therefore,

$$f(x+y,\xi')\cdot g(x-y,\xi'') = f(x+y,\xi+\eta)\cdot g(x-y,\xi-\eta).$$

Moreover the operators ψ'_j, ψ''_j act on $V_{\frac{1}{2}}(\xi,\eta)$. Let $\psi_j = \frac{1}{2}(\psi'_j + \psi''_j)$ and $\theta_j = \frac{1}{2}(\psi'_j - \psi''_j)$, for $j \in \frac{1}{2}+\mathbb{Z}_+$.

The action of the above elements is given by

$$\psi_j \mapsto \frac{1}{2}\frac{\partial}{\partial \xi_j} \quad \text{and} \quad \theta_j \mapsto \frac{1}{2}\frac{\partial}{\partial \eta_j}$$

$\psi_{-j} \mapsto \xi_j$ and $\theta_{-j} \mapsto \eta_j$ for $j \in \frac{1}{2} + \mathbb{Z}_+$. They satisfy the following relations:

$$[\psi_j, \psi_k]_+ = \frac{1}{2}\delta_{j,-k}; \quad [\theta_j, \theta_k]_+ = \frac{1}{2}\delta_{j,-k} \quad \text{and} \quad [\psi_j, \theta_k]_+ = 0.$$

Therefore

$$\sum_{j\in\frac{1}{2}+\mathbb{Z}_+} j(\psi'_{-j}\psi'_j + \psi''_{-j}\psi''_j) = 2\sum_{j\in\frac{1}{2}+\mathbb{Z}_+} j(\psi_{-j}\psi_j + \theta_{-j}\theta_j) \text{ and}$$

$$\sum_{\substack{i,j\in\frac{1}{2}+\mathbb{Z}\\ i+j=s}} \psi'_i\psi''_j = 2\sum_{\substack{j,k\in\frac{1}{2}+\mathbb{Z}_+\\ j+k=s}} \psi_j\theta_k + 2\sum_{j\in\frac{1}{2}+\mathbb{Z}_+} (\psi_{-j}\psi_{j+s} - \theta_{-j}\theta_{j+s}).$$

Let us use here the Taylor's expansion:

$$e^{\sum y_j \frac{\partial}{\partial u_j}} f(x+u,\xi)\cdot g(x-u,\eta)|_{u=0} = f(x+y,\xi)\cdot g(x-y,\eta)$$
$$e^{\sum \xi_j \frac{\partial}{\partial \alpha_j}} f(\alpha)|_{\alpha=0} = f(\xi) \quad \text{and} \quad e^{\sum \eta_j \frac{\partial}{\partial \beta_j}} f(\beta)|_{\beta=0} = f(\eta).$$

We use Theorem 84(b). Hence τ satisfies the following hierarchy of Hirota's bilinear equations corresponding to $C_2^{(1)}$:

$$S(\tau\otimes\tau) = \Bigg\{\frac{1}{2} - \sum_{j\geq 0}\frac{j}{4}y_j\frac{\partial}{\partial u_j} - 2K$$
$$- \sum_{n\geq 0,\ r\in\mathbb{Z}} P_n^{E_+}(2(1+I^{n'})y_n)P_{n-r}^{E_+}(-(1+I^{4-n'})\frac{\partial}{\partial u_n})K_r$$
$$+ \sum_{n\geq 0} P_n^{E_+}(-4I^{n'}y_n)P_n^{E_+}(2I^{4-n'}\frac{\partial}{\partial u_n})\Bigg\} e^{\sum_{j\geq 0} y_j\frac{\partial}{\partial u_j}} \cdot e^{\sum_{j\in\frac{1}{2}+\mathbb{Z}_+} \xi_j\frac{\partial}{\partial \alpha_j}}.$$
$$e^{\sum_{j\in\frac{1}{2}+\mathbb{Z}_+} \eta_j\frac{\partial}{\partial \beta_j}} \tau(x+u,\alpha+\beta)\cdot\tau(x-u,\alpha-\beta)\big|_{u=\alpha=\beta=0} = 0,$$

where

$$K = \sum_{j\in\frac{1}{2}+\mathbb{Z}_+} j\left(\xi_j\frac{\partial}{\partial\alpha_j} + \eta_j\frac{\partial}{\partial\beta_j}\right)$$

$$K_r = \frac{1}{2}\sum_{\substack{j,k\in\frac{1}{2}+\mathbb{Z}_+\\ j+k=r}} \left(\frac{\partial}{\partial\alpha_j}\frac{\partial}{\partial\beta_k}\right) + \sum_{j\in\frac{1}{2}+\mathbb{Z}_+}\left(\xi_j\frac{\partial}{\partial\alpha_{j+r}} - \eta_j\frac{\partial}{\partial\beta_{j+r}}\right) \text{ and}$$

$$K_{-r} = 2\sum_{\substack{j,k\in\frac{1}{2}+\mathbb{Z}_+\\ j+k=r}} \eta_j\xi_k + \sum_{j\in\frac{1}{2}+\mathbb{Z}_+}\left(\xi_{j+r}\frac{\partial}{\partial\alpha_j} - \eta_{j+r}\frac{\partial}{\partial\beta_j}\right).$$

Let us consider some particular coefficients from the above hierarchy.

Coeff. of $\xi_j\eta_k$:

$$\left\{\frac{1}{2} - 2(j+k)\frac{\partial}{\partial\alpha_j}\frac{\partial}{\partial\beta_k}\right.$$
$$- \left(2P^{E_+}_{j+k}\left(-(1+I^{4-n'})\frac{\partial}{\partial u_n}\right) + \sum_{0<s\leq j} P^{E_+}_{j-s}\left(-(1+I^{4-n'})\frac{\partial}{\partial u_n}\right)\frac{\partial}{\partial\alpha_s}\frac{\partial}{\partial\beta_k}\right.$$
$$\left.\left. - \sum_{0<s\leq k} P^{E_+}_{k-s}\left(-(1+I^{4-n'})\frac{\partial}{\partial u_n}\right)\frac{\partial}{\partial\beta_s}\frac{\partial}{\partial\alpha_j}\right)\right\}\tau\cdot\tau\bigg|_{u=\alpha=\beta=0} = 0.$$

After making suitable proportionate transformation to the Hirota operators in the above hierarchy, we get the following:

Coeff. of $\xi_{\frac{3}{2}}\eta_{\frac{3}{2}}$:

$$\left\{\frac{\partial}{\partial\alpha_{\frac{3}{2}}}\cdot\frac{\partial}{\partial\beta_{\frac{3}{2}}}\right\}\tau\cdot\tau = 0. \tag{6.16}$$

Coeff. of $\xi_{\frac{3}{2}}\eta_{\frac{5}{2}}$:

$$\left\{D_1^4 - 12D_1D_3 + \frac{3}{2}D_1^2\left(\frac{\partial}{\partial\beta_{\frac{1}{2}}}\frac{\partial}{\partial\alpha_{\frac{3}{2}}}\right) - \frac{39}{2}\frac{\partial}{\partial\alpha_{\frac{3}{2}}}\frac{\partial}{\partial\beta_{\frac{5}{2}}}\right\}\tau\cdot\tau = 0. \tag{6.17}$$

Coeff. of $\xi_{\frac{7}{2}}\eta_{\frac{1}{2}}$:

$$\left\{D_1^4 - 4D_1D_3 - D_1^2\left(\frac{\partial}{\partial\alpha_{\frac{3}{2}}}\cdot\frac{\partial}{\partial\beta_{\frac{1}{2}}}\right) - \frac{3}{2}\frac{\partial}{\partial\alpha_{\frac{7}{2}}}\cdot\frac{\partial}{\partial\beta_{\frac{1}{2}}}\right\}\tau\cdot\tau = 0. \tag{6.18}$$

Coeff. of $\xi_{\frac{7}{2}}\eta_{\frac{5}{2}}$:

$$\left\{D_1^6 - 20D_1^3D_3 - 8D_3^2 + 18D_1D_5 - \frac{D_1^2}{2}\left(\frac{\partial}{\partial\alpha_{\frac{3}{2}}}\cdot\frac{\partial}{\partial\beta_{\frac{5}{2}}} - \frac{\partial}{\partial\beta_{\frac{1}{2}}}\cdot\frac{\partial}{\partial\alpha_{\frac{7}{2}}}\right)\right.$$
$$\left. -\frac{9}{2}\frac{\partial}{\partial\alpha_{\frac{7}{2}}}\cdot\frac{\partial}{\partial\beta_{\frac{5}{2}}}\right\}\tau\cdot\tau = 0. \tag{6.19}$$

Here we recall the definition of a super bilinear equation from Kac and Wakimoto [229] which is defined as follows:

$$P(D_x, D^{\alpha}_{\frac{1}{2}}, D^{\beta}_{\frac{1}{2}})\tau\cdot\tau = 0.$$

That is,

$$P\left(\frac{\partial}{\partial u}, \frac{\partial}{\partial\alpha_{\frac{1}{2}}}, \frac{\partial}{\partial\beta_{\frac{1}{2}}}\right)\tau(x+u,\alpha+\beta)\cdot\tau(x-u,\alpha-\beta)|_{u=\alpha=\beta=0} = 0.$$

Hence Equations (6.16)–(6.19) can be rewritten as follows:

Coefficient of $\xi_{\frac{3}{2}}\eta_{\frac{3}{2}}$:

$$\left\{D^{\alpha}_{\frac{3}{2}}D^{\beta}_{\frac{3}{2}}\right\}\tau\cdot\tau = 0.$$

Coefficient of $\xi_{\frac{3}{2}}\eta_{\frac{5}{2}}$:

$$\left\{D_1^4 - 12D_1D_3 + \frac{3}{2}D_1^2\left(D^{\alpha}_{\frac{3}{2}}D^{\beta}_{\frac{1}{2}}\right) - \frac{39}{2}D^{\alpha}_{\frac{3}{2}}D^{\beta}_{\frac{5}{2}}\right\}\tau\cdot\tau = 0.$$

Coefficient of $\xi_{\frac{7}{2}}\eta_{\frac{1}{2}}$:

$$\left\{D_1^4 - 4D_1D_3 - D_1^2\left(D^{\alpha}_{\frac{3}{2}}D^{\beta}_{\frac{1}{2}}\right) - \frac{3}{2}D^{\alpha}_{\frac{7}{2}}D^{\beta}_{\frac{1}{2}}\right\}\tau\cdot\tau = 0.$$

Coefficient of $\xi_{\frac{7}{2}}\eta_{\frac{5}{2}}$:

$$\left\{D_1^6 - 20D_1^3D_3 - 8D_3^2 + 18D_1D_5 - \frac{D_1^2}{2}\left(D^{\alpha}_{\frac{3}{2}}\cdot D^{\beta}_{\frac{5}{2}} - D^{\beta}_{\frac{1}{2}}\cdot D^{\alpha}_{\frac{7}{2}}\right)\right.$$
$$\left. -\frac{9}{2}D^{\alpha}_{\frac{7}{2}}\cdot D^{\beta}_{\frac{5}{2}}\right\}\tau\cdot\tau = 0.$$

6.9.4 Principal vertex operators and super Hirota bilinear equations for $B_3^{(1)}$ [232]

In this section, we consider $B_3^{(1)}$ as a subalgebra of $D_4^{(1)}$ and construct the principal vertex operators and the corresponding hierarchy of super Hirota bilinear equations for $B_3^{(1)}$. One can use Mandia [230] to construct principal Cartan subalgebras of D_4 and B_3 and also can use Kac et al. [211] to construct principal vertex operators for $B_3^{(1)}$ corresponding to the simple roots of B_3 which are written in terms of simple roots of D_4.

(1) *Construction of B_3 as a subalgebra of D_4 using a fixed point automorphism θ of the Dynkin diagram Γ of D_4*

Let $\underline{g}$ be a finite dimensional Lie algebra of type D_4. Let $(x|y) = \frac{1}{2}tr(xy)$ be a nondegenerate symmetric bilinear form defined on $\underline{g}$. One can take

$$\mu_{pq} = e_{pq} - e_{q+4,p+4} \ \ (1 \le p, q \le 4)$$
$$\nu_{pq} = e_{p,q+4} - e_{q,p+4} \ \ (1 \le p < q \le 4)$$

and

$$\rho_{pq} = e_{p+4,q} - e_{q+4,p} \ \ (1 \le p < q \le 4)$$

as a set of basis elements for this Lie algebra. Let

$$\begin{array}{llll} e_1 = \mu_{12} & e_2 = \mu_{23} & e_3 = \mu_{34} & e_4 = \nu_{34}, \\ f_1 = \mu_{12}, & f_2 = \mu_{32}, & f_3 = \mu_{43}, & f_4 = -\rho_{34}, \\ h_1 = \mu_{11} - \mu_{22}, & h_2 = \mu_{22} - \mu_{33}, & h_3 = \mu_{33} - \mu_{44}, & \text{and } h_4 = \mu_{33} + \mu_{44} \end{array}$$

be a set of canonical generators of $\underline{g}$. Let $\underline{\mathfrak{h}} = span\{h_i, 1 \le i \le 4\}$ be a Cartan subalgebra of $\underline{g}$, Δ a simple root system. The bilinear form $(.|.)$ is normalized so that $(\alpha|\alpha) = 2$ for all $\alpha \in \Delta$. Let $\Pi = \{\alpha_1, \alpha_2, \alpha_3, \alpha_4\}$ be a base of Δ and let Γ be the Dynkin diagram of Δ. One can identify $\underline{\mathfrak{h}}$ with $\underline{\mathfrak{h}}^*$ via $(.|.)$. Let θ be the Dynkin diagram automorphism of order 2 given by

$$\theta\alpha_1 = \alpha_1, \quad \theta\alpha_2 = \alpha_2, \quad \theta\alpha_3 = \alpha_4 \quad \text{and} \quad \theta\alpha_4 = \alpha_3.$$

The automorphism θ of Γ is extended to D_4 so that $\theta(x_\alpha) = x_{\theta(\alpha)}$ for all $\alpha \in \Delta$. Hence we have the following decomposition of $\underline{g}$ with respect to θ :

$$\underline{g} = \bigoplus_{j \in Z/2Z} \underline{g}_{[j]},$$

where $\underline{g}_{[j]} = \{x \in \underline{g} | \theta x = \epsilon^j x\}, \epsilon = e^{2\pi I/2}$ and $I = \sqrt{-1}$.

We have

$$\underline{g}_{[0]} = \mathbb{C}h_1 \oplus \mathbb{C}h_2 \oplus \mathbb{C}(h_3 + h_4) \oplus \mathbb{C}\mu_{12} \oplus \mathbb{C}\mu_{23} \oplus \mathbb{C}(\mu_{34} + \nu_{34}) \oplus$$
$$\mathbb{C}\mu_{21} \oplus \mathbb{C}\mu_{32} \oplus \mathbb{C}(\mu_{34} - \rho_{34}) \oplus \mathbb{C}\mu_{13} \oplus \mathbb{C}(\mu_{24} + \nu_{24}) \oplus$$
$$\mathbb{C}(\mu_{14} + \nu_{14}) \oplus \mathbb{C}\nu_{23} \oplus \mathbb{C}\nu_{13} \oplus \mathbb{C}\gamma_{12} \oplus \mathbb{C}\mu_{31} \oplus$$
$$\mathbb{C}(\mu_{42} - \rho_{24}) \oplus \mathbb{C}(\mu_{41} - \rho_{14}) \oplus \mathbb{C}\rho_{23} \oplus \mathbb{C}\rho_{13} \oplus \mathbb{C}\rho_{12}$$

and

$$\underline{g}_{[1]} = \mathbb{C}(h_3 - h_4) \oplus \mathbb{C}(\mu_{34} - \nu_{34}) \oplus \mathbb{C}(\mu_{43} + \rho_{34}) \oplus \mathbb{C}(\mu_{24} - \nu_{24}) \oplus$$
$$\mathbb{C}(\mu_{14} - \nu_{14}) \oplus \mathbb{C}(\mu_{42} + \rho_{24}) \oplus \mathbb{C}(\mu_{41} + \rho_{14}).$$

Moreover

$$\underline{g}_{[0]} = \{x \in \underline{g} \mid \theta x = x\}.$$

For the proof of the following two propositions, one can refer Mandia [230].
Proposition 68.

(i) *$\underline{g}_{[0]}$ is of type B_3, $\underline{\mathfrak{h}}_{[0]}$ is its Cartan subalgebra.*
(ii) *The set $\{\frac{\alpha+\theta\alpha}{2};\ \alpha \in \Delta\}$ is a root system with respect to $\underline{\mathfrak{h}}_{[0]}$ with base $\{\frac{\alpha_i+\theta\alpha_i}{2};\ 1 \le i \le 4\}$.*

Corresponding generators of $\underline{\mathfrak{g}}_0 = B_3$ are

$$\begin{array}{lll} E_1 = \mu_{12} & E_2 - \mu_{23} & E_3 = \mu_{34} + \nu_{34} \\ F_1 = \mu_{21} & F_2 = \mu_{32} & F_3 = \mu_{43} - \rho_{34} \\ H_1 = \mu_{11} - \mu_{22} = h_1 & H_2 = \mu_{22} - \mu_{33} = h_2 \ \textit{and} & H_3 = 2\mu_{23} = h_3 + h_4. \end{array}$$

Principal Cartan subalgebras of D_4 and B_3

Let $E = e_0 + \sum_{i=1}^{4} e_i$, where e_0 is the lowest root vector of D_4 with respect to $\mathfrak{h}$. Then $E = \mu_{12}+\mu_{23}+\mu_{34}+\nu_{34}-\rho_{12}$ is the cyclic element of D_4. Let $\underline{\mathfrak{a}} = \{x \in \underline{\mathfrak{g}} | [x, \overline{E}] = 0\}$. Then the $\underline{\mathfrak{a}}$ is the principal Cartan sub algebra of D_4. The basis elements of $\underline{\mathfrak{a}}$ are

$$E = \mu_{12} + \mu_{23} + \mu_{34} + \nu_{34} - \rho_{12},$$
$$F = \mu_{21} + 2\mu_{32} + \mu_{43} - \rho_{34} + \nu_{12},$$
$$G = -\mu_{41} - \rho_{14} + \nu_{14} - \mu_{14}, \ \textit{and}$$
$$H = \rho_{14} - \rho_{23} + \nu_{23} - \nu_{14}.$$

The lowest root of B_3 with respect to $\underline{\mathfrak{h}}$ is $-(\alpha_1 + 2\alpha_2 + \alpha_3 + \alpha_4)$. The corresponding root vector $E_0 = -\rho_{12}$.

Then $E_0 + E_1 + E_2 + E_3 = \mu_{12} + \mu_{23} + \mu_{34} + \nu_{34} - \rho_{12}$ is the cyclic element of $\underline{\mathfrak{g}}_{[0]}$. Let $\underline{\eta} = \{x \in \underline{\mathfrak{g}}_{[0]} | [x, E] = 0\}$. Then $\underline{\eta}$ is the principal Cartan subalgebra of B_3. A basis of $\underline{\eta}$ is $\{E, F, G + 2H\}$.

Proposition 69. *$\underline{\eta}$ is the subalgebra of $\underline{\mathfrak{a}}$ of θ fixed points.*

The Coxeter number for D_4 and B_3 is 6. Let $\nu : D_4 \to D_4$ be the principal automorphism defined by

$$\nu(h_j) = h_j, \quad \nu(e_j) = we_j, \quad \nu(f_j) = w^{-1} f_j, j = 1, 2, 3, 4 \text{ and } w = e^{\frac{2\Pi I}{6}}.$$

Then $\underline{\mathfrak{g}}$ can be decomposed with respect to this automorphism. This is called the principal gradation of $\underline{\mathfrak{g}}$.

We have the following decomposition of $\underline{\mathfrak{g}}$, with respect to the principal gradation $\underline{\mathfrak{g}} = \bigoplus_{j \in \mathbb{Z}/6\mathbb{Z}} \underline{\mathfrak{g}}_j$ where $\underline{\mathfrak{g}}_j = \{x \in \underline{\mathfrak{g}} \mid \nu(x) = \epsilon^j x\}, \epsilon = e^{\frac{2\pi I}{6}}, 0 \leq j \leq 5$, where

$$\underline{\mathfrak{g}}_0 = \mathbb{C}h_1 \oplus \mathbb{C}h_2 \oplus \mathbb{C}h_3 \oplus \mathbb{C}h_4$$

$$\underline{\mathfrak{g}}_1 = \mathbb{C}\mu_{12} \oplus \mathbb{C}\mu_{23} \oplus \mathbb{C}\mu_{34} \oplus \mathbb{C}\nu_{34} \oplus \mathbb{C}\rho_{12}$$

$$\underline{\mathfrak{g}}_2 = \mathbb{C}\mu_{13} \oplus \mathbb{C}\mu_{24} \oplus \mathbb{C}\nu_{24} \oplus \mathbb{C}\rho_{13}$$

$$\underline{\mathfrak{g}}_3 = \mathbb{C}\mu_{14} \oplus \mathbb{C}\mu_{41} \oplus \mathbb{C}\nu_{14} \oplus \mathbb{C}\nu_{23} \oplus \mathbb{C}\rho_{24} \oplus \mathbb{C}\rho_{23}$$

$$\underline{\mathfrak{g}}_4 = \mathbb{C}\mu_{31} \oplus \mathbb{C}\mu_{42} \oplus \mathbb{C}\rho_{24} \oplus \mathbb{C}\nu_{13} \textit{ and}$$

$$\underline{\mathfrak{g}}_5 = \mathbb{C}\mu_{21} \oplus \mathbb{C}\mu_{32} \oplus \mathbb{C}\mu_{43} \oplus \mathbb{C}\rho_{34} \oplus \mathbb{C}\nu_{12}.$$

We have the decomposition of $\underline{\mathfrak{a}}$ as follows:

$$\underline{\mathfrak{a}} = \bigoplus_{j \in \mathbb{Z}/6\mathbb{Z}} \underline{\mathfrak{a}}_j, \text{ where } \underline{\mathfrak{a}}_j = \underline{\mathfrak{a}} \cap \underline{\mathfrak{g}}_j,$$

with $\underline{\mathfrak{a}}_0 = 0, \ \underline{\mathfrak{a}}_1 = \mathbb{C}E, \ \underline{\mathfrak{a}}_2 = 0, \underline{\mathfrak{a}}_3 = \mathbb{C}G \oplus \mathbb{C}H, \ \underline{\mathfrak{a}}_4 = 0,$ *and* $\underline{\mathfrak{a}}_5 = \mathbb{C}F.$

Finally, we have the decomposition of $B_3 = \underline{\mathfrak{g}}_{[0]}$ as follows:

$$\underline{\mathfrak{g}}_{[0]} = \bigoplus_{j \in \mathbb{Z}/6\mathbb{Z}} (\underline{\mathfrak{g}}_{[0]})_j, \textit{ where}$$

$$(\underline{\mathfrak{g}}_{[0]})_0 = \mathbb{C}h_1 \oplus \mathbb{C}h_2 \oplus \mathbb{C}(h_3 + h_4)$$

$$(\underline{\mathfrak{g}}_{[0]})_1 = \mathbb{C}\mu_{12} \oplus \mathbb{C}\mu_{23} \oplus \mathbb{C}(\mu_{34} + \nu_{34}) \oplus \mathbb{C}\rho_{12}$$

$$(\underline{\mathfrak{g}}_{[0]})_2 = \mathbb{C}\mu_{13} \oplus \mathbb{C}(\mu_{24} + \nu_{24}) \oplus \mathbb{C}\rho_{13}$$

$$(\underline{\mathfrak{g}}_{[0]})_3 = \mathbb{C}(\mu_{14} + \nu_{14}) \oplus \mathbb{C}\nu_{23} \oplus \mathbb{C}(\mu_{41} - \rho_{14}) \oplus \mathbb{C}\rho_{23}$$

$$(\underline{\mathfrak{g}}_{[0]})_4 = \mathbb{C}\nu_{13} \oplus \mathbb{C}\mu_{31} \oplus \mathbb{C}(\mu_{42} - \rho_{24}) \textit{ and}$$

$$(\underline{\mathfrak{g}}_{[0]})_5 = \mathbb{C}\mu_{21} \oplus \mathbb{C}\mu_{32} \oplus \mathbb{C}(\mu_{43} - \rho_{34}) \oplus \mathbb{C}\nu_{12}.$$

The principal Cartan subalgebra $\underline{\eta}$ of B_3 is also graded with respect to the principal gradation:

$$\underline{\eta} = \bigoplus_{j\in\mathbb{Z}/6\mathbb{Z}} \underline{\eta}_j, \quad \underline{\eta}_j = \underline{\eta} \cap (\underline{\mathfrak{g}_{[0]}})_j$$

$$\text{with } \underline{\eta}_0 = \phi, \quad \underline{\eta}_1 = \mathbb{C}E, \quad \underline{\eta}_2 = 0, \quad \underline{\eta}_3 = \mathbb{C}(G+2H), \quad \underline{\eta}_4 = 0$$
$$\text{and } \quad \underline{\eta}_5 = \mathbb{C}F.$$

Definition 292. Let h be the Coxeter number of $\underline{\mathfrak{g}}$ and $\underline{\mathfrak{g}} = \bigoplus_{j\in\mathbb{Z}/h\mathbb{Z}} \underline{\mathfrak{g}}_j$ be the principal gradation of $\underline{\mathfrak{g}}$. Let E be the cyclic element and $S = \oplus_{j\in\mathbb{Z}/h\mathbb{Z}} S_j$ be the principal Cartan subalgebra. The nondecreasing sequence of integers $m_1 \leq m_2 \leq m_3 \leq \ldots$ from the interval $[1, h-1]$ in which j appears with multiplicity $\dim S_j$ is called the exponents of $\underline{\mathfrak{g}}$.

For the proof of the following theorem, one can refer Mandia [230].

Theorem 88. *The set $\Delta_{\underline{\eta}}$ of roots of B_3 with respect to $\underline{\eta}$ is $\Delta_{\underline{\eta}} = \{\frac{\beta+\theta\beta}{2}, \ \beta \in \Delta_{\underline{\mathfrak{a}}}\}$.*

For the proof of the following proposition, one can refer Sthanumoorthy and Kiruba Bagirathi [232].

Proposition 70. *Let γ_i $(1 \leq i \leq 4)$ be the simple roots of D_4 with respect to $\underline{\mathfrak{a}}$. Then $A_{\pm\gamma_i}$, $(1 \leq i \leq 4)$ are the root vectors for the roots $\pm\gamma_i, (1 \leq i \leq 4)$ satisfying the following relations:*

$$[E, A_{\pm\gamma_1}] = \pm 2^{1/3} A_{\pm\gamma_1}, [F, A_{\pm\gamma_1}] = \pm 2^{2/3} A_{\pm\gamma_1}, [G, A_{\pm\gamma_1}] = 0, [H, A_{\pm\gamma_1}] = \pm 2A_{\pm\gamma_1},$$
$$[E, A_{\pm\gamma_2}] = \pm k_3 A_{\pm\gamma_2}, [F, A_{\pm\gamma_2}] = \mp k_3^2 A_{\pm\gamma_2}, [G, A_{\pm\gamma_2}] = 0, [H, A_{\pm\gamma_2}] = \mp 2A_{\pm\gamma_2},$$
$$[E, A_{\pm\gamma_3}] = \pm k_1 A_{\pm\gamma_3}, [F, A_{\pm\gamma_3}] = \mp k_1^2 A_{\pm\gamma_2}, [G, A_{\pm\gamma_3}] = \pm 2A_{\pm\gamma_3}, [H, A_{\pm\gamma_3}] = 0,$$
$$[E, A_{\pm\gamma_4}] = \pm k_1 A_{\pm\gamma_4}, [F, A_{\pm\gamma_4}] = \mp k_1^2 A_{\pm\gamma_4}, [G, A_{\pm\gamma_4}] = \mp 2A_{\pm\gamma_4}, [H, A_{\pm\gamma_4}] = \pm 2A_{\pm\gamma_4}.$$

Here, $k_1 = \frac{2^{1/3} + I\sqrt{32}^{1/3}}{2}, k_2 = -k_1, k_3 = \frac{2^{1/3} - I\sqrt{32}^{1/3}}{2}$, *and* $k_4 = -k_3$.

Remark 197. For Coxeter-Killing transformation, one can refer Kostant [231].

Construction of principal vertex operators for $B_3^{(1)}$

We observe that the set of elements $\{E, F, H, G, A_{\gamma_{i,j}}, 1 \leq i \leq 4, 1 \leq j \leq 6\}$ is a basis of D_4. Now the corresponding affine Lie algebra is defined by

$$\underline{\hat{\mathfrak{g}}}' = \bigoplus_{j\in\mathbb{Z}} t^j \otimes \underline{\mathfrak{g}}_{j \bmod 6} \oplus \mathbb{C}K,$$

where K is the canonical central element and the generators of $D_4^{(1)}$ are given by

$$\hat{e}_j = 1 \otimes e_j, \ \ \hat{e}_0 = t \otimes e_0,$$
$$\hat{f}_j = 1 \otimes f_j, \ \ \hat{f}_0 = t^{-1} \otimes f_0,$$
$$\hat{h}_j = \alpha_j, \ \ \hat{h}_0 = \frac{2}{(\theta|\theta)} c - \theta^\vee,$$

where $\theta^\vee$ is the coroot corresponding to the highest root θ. The Lie bracket in $\hat{\underline{\mathfrak{g}}}'(A)$ is given by,

$$[p_1(t)\otimes g_1 \oplus \lambda_1 K, p_2(t)\otimes g_2 \oplus \lambda_2 K] = p_1(t)p_2(t)\otimes [g_1,g_2] \oplus \frac{1}{6}\left(Res\frac{dp_1(t)}{dt}p_2(t)\right)(g_1|g_2)K,$$

for all $p_1(t), p_2(t) \in \mathbb{C}[t,t^{-1}]$ and $g_1, g_2 \in \underline{\mathfrak{g}}$. For the proof of the following proposition, one can refer Sthanumoorthy and Kiruba Bagirathi [232].

Proposition 71. *The root vectors of B_3 corresponding to the simple roots $\overline{\gamma}_i$, $(1 \leq i \leq 3)$ with respect to the Cartan subalgebra $\underline{\eta}$ are $A_{\overline{\gamma}_i}$. Here γ_i $(1 \leq i \leq 4)$ are the simple roots of D_4 with respect to $\underline{\mathfrak{a}}$. The root system of B_3 with respect to the Cartan subalgebra of B_3 is*

$$\begin{aligned}\Delta_{\underline{\eta}} = \pm\{&\overline{\gamma}_1,\overline{\gamma}_2,\overline{\gamma}_3,\overline{\gamma}_1+\overline{\gamma}_2,\overline{\gamma}_2+\overline{\gamma}_3,\overline{\gamma}_2+2\overline{\gamma}_3,\overline{\gamma}_1+\overline{\gamma}_2+\overline{\gamma}_3,\overline{\gamma}_1+\overline{\gamma}_2\\ &+2\overline{\gamma}_3,\overline{\gamma}_1+2\overline{\gamma}_2+2\overline{\gamma}_3\},\end{aligned}$$

where $\overline{\gamma}_1 = \gamma_1, \overline{\gamma}_2 = \gamma_2$ and $\overline{\gamma}_3 = \frac{\gamma_3+\gamma_4}{2}$ with

$$\begin{aligned}(\Delta_{\underline{\eta}})_l &= \pm\{\overline{\gamma}_1,\overline{\gamma}_2,\overline{\gamma}_1+\overline{\gamma}_2,\overline{\gamma}_2+2\overline{\gamma}_3,\overline{\gamma}_1+\overline{\gamma}_2+2\overline{\gamma}_3,\overline{\gamma}_1+2\overline{\gamma}_2+2\overline{\gamma}_3\},\\ (\Delta_{\underline{\eta}})_s &= \pm\{\overline{\gamma}_3,\overline{\gamma}_2+\overline{\gamma}_3,\overline{\gamma}_1+\overline{\gamma}_2+2\overline{\gamma}_3\}.\end{aligned}$$

Here $(\Delta_{\underline{\eta}})_l$ and $(\Delta_{\underline{\eta}})_s$ are the sets of long and short roots, respectively.

The principal Cartan subalgebra $\underline{\mathfrak{a}}$ of $\underline{\mathfrak{g}} = D_4$ has $\{E,F,G,H\}$ as basis elements. Now $\mathfrak{a} = \underset{j\in\mathbb{Z}\oplus}{(}\, t^j \otimes \underline{\mathfrak{a}}_{j \bmod 4}) \oplus \mathbb{C}c$. This is the principal Heisenberg-Cartan subalgebra of $\hat{\underline{\mathfrak{g}}}'$ and we observe that $E_+ = \{n \in \mathbb{Z}_+ \mid n \neq 0 \bmod 4\} = \{1,3,3,5,\ldots\}$ is the set of exponents of D_4. The principal Cartan subalgebra of B_3 is $\underline{\eta}$ with $\{E_1 = E,\ E_3 = \frac{1}{\sqrt{2}}(G+2H),\ E_5 = F\}$ as a basis.

Decomposing $A_{\overline{\gamma}_i} \in \underline{\mathfrak{g}}$, $(1 \leq i \leq 2)$ with respect to the principal gradation of B_3, we get, $A_{\overline{\gamma}_i} = \sum_{j=0}^{3} X_{\overline{\gamma}_i j}$, $A_{\overline{\gamma}_i j} \in (\underline{\mathfrak{g}}_{[0]})_j$. We observe that the elements $\{E_1, E_3, E_5, A_{\overline{\gamma}_i j}, 1 \leq i \leq 2, 0 \leq j \leq 3\}$ form a basis of B_3. The corresponding affine algebra is given by $\hat{\underline{\mathfrak{g}}}_{[0]} = \oplus_{j\in\mathbb{Z}} t^j \otimes (g_{[0]})_{j \bmod 6} \oplus \mathbb{C}c$, with generators $\hat{E}_0 = \hat{e}_0, \hat{E}_1 = \hat{e}_1, \hat{E}_2 = \hat{e}_2, \hat{E}_3 = \hat{e}_3 + \hat{e}_4, \hat{F}_0 = \hat{f}_0, \hat{F}_1 = \hat{f}_1, \hat{F}_2 = \hat{f}_2, \hat{F}_3 = \hat{f}_3 + \hat{f}_4$, $\hat{H}_0 = \hat{h}_0, \hat{H}_1 = \hat{h}_1, \hat{H}_2 = \hat{h}_2, \hat{H}_3 = \hat{h}_3 + \hat{h}_4$.

Let $\hat{\underline{\mathfrak{g}}} = \hat{\underline{\mathfrak{g}}}' \oplus \sum_{j\in\mathbb{Z}} \mathbb{C}d_j$ be defined by $[d_n, u(j)] = -ju(j+n)$, $[d_n, c] = 0$ and $[d_m, d_n] = (m-n)d_{m+n} + \delta_{m,-n}\frac{m^3-m}{12}c$. Similarly, we can define, $\hat{\underline{\mathfrak{g}}}_{[0]} = \hat{\underline{\mathfrak{g}}}'_{[0]} \bigoplus \sum_{j\in\mathbb{Z}} \mathbb{C}d_j$ with $[d_n, u(j)] = -ju(j+n)$, $[d_n, c] = 0$ and $[d_m, d_n] = (m-n)d_{m+n} + \delta_{m,-n}\frac{m^3-m}{12}c$. Let $X_{\bar{\gamma}}(z) = \sum_{j\in\mathbb{Z}} z^{-j}(t^j \otimes A_{\bar{\gamma},j \bmod 6})$. Then it is clear that,

$[d_n, A_{\bar\gamma}(z)] = z^n\left(n + z\frac{d}{dz}\right)X_{\bar\gamma}(z)$. Let us consider the principal Cartan subalgebra $\underline{\eta}$ of B_3 and its basis $\{E_1 = E,\ E_3 = \frac{1}{\sqrt{2}}(G+2H), E_5 = F\}$. Let $\eta = \oplus_{j\in\mathbb{Z}}(t^j \otimes \underline{\eta}_{j \bmod 6}) \oplus \mathbb{C}c$. This is the principal subalgebra of $\hat{\underline{\mathfrak{g}}}'_{[0]} = B_3^{(1)}$. Here we observe that $\mathbb{Z}_+^{\text{odd}}$ is the set of exponents for $B_3^{(1)}$, which we denote by $\underline{E}_+$. We also note that $\underline{E}_+ \subset E_+$. It can also be proved that $\eta = \{x \in \mathfrak{a} \mid \theta x = x\}$. Here c, $p_i = t^i \otimes E_{i'}$, $q_i = i^{-1}t^{-i} \otimes E_{4-i'}$ where $i \in \underline{E}_+$ such that $i' = i \bmod 4$ with $i' \in \{1, 3\}$ are the generators of η. They satisfy the relation $[p_i, q_j] = c\delta_{ij}$. Hence η is isomorphic to the infinite Heisenberg algebra.

Let us consider the space

$$\mathbb{C}[x_i;\ i \in \mathbb{Z}_{+\text{odd}}]$$

and consider it as an η module on which p_i acts as $\frac{\partial}{\partial x_i}$, q_i acts as multiplication by ix_i and c acts as Id.

Let us take $\bar\gamma_1 \in \Delta_{\underline{\eta}}$. Then

$$\begin{aligned}[p_s,\ X_{\bar\gamma_1}(z)] &= [t^s \otimes E_{s'},\ \sum_{j\in\mathbb{Z}} z^{-j}(t^j \otimes A_{\bar\gamma_{1,j}})] \\ &= \sum_{j\in\mathbb{Z}} z^{-j}t^{s+j} \otimes [E_{s'},\ A_{\bar\gamma_{1,j}}]\end{aligned}$$

where $s \in \mathbb{Z}_{+odd} = (6\mathbb{N}+1) \cup (6\mathbb{N}+3) \cup (6\mathbb{N}+5)$.

We use Proposition 71 in all the following commutation relations:

If $s \in 6\mathbb{N}+1$, then $s' = 1$, $[p_s, X_{\bar\gamma_1}(z)] = 2^{1/3}z^s X_{\bar\gamma_1}(z)$ and $[q_s, X_{\bar\gamma_1}(z)] = 2^{2/3}s^{-1}z^{-s}X_{\bar\gamma_1}(z)$.

If $s \in 6\mathbb{N}+3$, then $s' = 3$, $[p_s, X_{\bar\gamma_1}(z)] = 2.2^{1/3}z^s X_{\bar\gamma_1}(z)$ and $[q_s, X_{\bar\gamma_1}(z)] = 2.2^{1/3}s^{-1}z^{-s}X_{\bar\gamma_1}(z)$.

If $s \in 6\mathbb{N}+5$, then $s' = 5$, $[p_s, X_{\bar\gamma_1}(z)] = 2^{2/3}z^s X_{\bar\gamma_1}(z)$ and $[q_s, X_{\bar\gamma_1}(z)] = 2^{1/3}s^{-1}z^{-s}X_{\bar\gamma_1}(z)$.

Using Kac et al. [211] and Kac [8] it can be verified that the principal vertex operator corresponding to the root $\bar\gamma_1$ is

$$\begin{aligned}\Gamma^{\bar\gamma_1}(z) = \exp&\left(\sum_{j\in 6\mathbb{N}+1} 2^{1/3}z^j x_j + \sum_{j\in 6\mathbb{N}+3} 2\sqrt{2}z^j x_j + \sum_{j\in 6\mathbb{N}+5} 2^{2/3}z^j x_j\right) \\ \exp&\left(-\left(\sum_{j\in\mathbb{N}+1} 2^{2/3}z^{-j}\frac{\partial}{\partial x_j} + \sum_{j\in\mathbb{N}+3} 2\sqrt{2}z^{-j}\frac{\partial}{\partial x_j} + \sum_{j\in\mathbb{N}+5} 2^{1/3}z^{-j}\frac{\partial}{\partial x_j}\right)\right).\end{aligned}$$

Similarly, we can have the vertex operator expressions for all the roots $\bar\gamma \in (\Delta_{\underline{\eta}})$.

Remark 198. Oscillator representation for Virasoro algebra with central charge 1/2 was discussed in the previous section.

We state as a theorem the following results proved in Sthanumoorthy and Kiruba Bagirathi [232]. Using the construction given in Goddard and Olive [228] and also using Kac [8] and Kac and Raina [226], one can directly prove the following theorem (we omit the proof) which gives the coset Virasoro algebra representation of $B_3^{(1)}$.

Theorem 89. *The representation of* $\hat{\mathfrak{g}}'_{\underline{[0]}} + Vir = B_3^{(1)} \oplus Vir$ *on the space* $\mathbb{C}[x_i, i \in \mathbb{N}_{\text{odd}}] \otimes \wedge[\xi_i \mid i \geq 0]$ *is given by*

$$p_j \mapsto \frac{\partial}{\partial x_j};\ \ q_j \mapsto jx_j;\ \ c \mapsto Id;\ \ d_j \mapsto \dot{L}_j^{(h)} + l_j;$$

$$X_{\bar{\gamma}_i}(z) \mapsto \Gamma^{\bar{\gamma}_i}(z)\psi(z), \quad \textit{if } (\bar{\gamma}_i|\bar{\gamma}_i) = 1 \textit{ for } i = 1, 2, 3 \textit{ and}$$

$$X_{\bar{\gamma}_i}(z) \mapsto \Gamma^{\bar{\gamma}_i}(z), \ \textit{if } (\bar{\gamma}_i|\bar{\gamma}_i) = 2 \textit{ for } i = 1,\ 2,\ 3.$$

Hirota bilinear equations for $B_3^{(1)}$

For complete construction of Hirota bilinear equations and super bilinear equations for $B_3^{(1)}$, one can refer Sthanumoorthy and Kiruba Bagirathi [232]. We give below the final results.

We get the following Hirota differential equations:

(1) $\{D_1^6 - 5D_1^3 D_3 - 5D_3^2 + 9D_1D_5\}\tau_0 \cdot \tau_1 = 0.$

(2) $\left\{D_1^2 + \frac{\partial}{\partial\alpha_1} \cdot \frac{\partial}{\partial\beta_1}\right\} \tau_0 \cdot \tau_1|_{\alpha=\beta=0} = 0.$

(3) $\left\{D_1^4 - 4D_1D_3 + 2D_1(\frac{\partial}{\partial\alpha_1} \cdot \frac{\partial}{\partial\beta_2} - \frac{\partial}{\partial\beta_1} \cdot \frac{\partial}{\partial\alpha_2}) - \frac{\partial}{\partial\alpha_2} \cdot \frac{\partial}{\partial\beta_2}\right\} \tau_0 \cdot \tau_1|_{\alpha=\beta=0} = 0.$

(4) $\left\{D_1^6 + 9D_1D_5 - 5D_1^3D_3 - 5D_3^2 + D_1(\frac{\partial}{\partial\alpha_2} \cdot \frac{\partial}{\partial\beta_3} - \frac{\partial}{\partial\beta_2} \cdot \frac{\partial}{\partial\alpha_3}) - 5\frac{\partial}{\partial\alpha_3} \cdot \frac{\partial}{\partial\beta_3}\right\}$
$\tau_0 \cdot \tau_1|_{\alpha=\beta=0} = 0.$

Defining a super bilinear equation

$$P(D_x, D^\alpha, D^\beta)\tau \cdot \tau = 0 \text{ by}$$

$$P\left(\frac{\partial}{\partial u}, \frac{\partial}{\partial\alpha}, \frac{\partial}{\partial\beta}\right) \tau_0(x+u, \alpha+\beta) \cdot \tau_1(x-u, \alpha-\beta)|_{u=\alpha=\beta=0},$$

we can write the above equations (2), (3), (4) as follows:

$$(2')\ \{D_1^2 + D_1^\alpha D_1^\beta\}\tau_0 \cdot \tau_1 = 0.$$

$$(3')\ \{D_1^4 - 4D_1D_3 + 2D_1(D_1^\alpha D_2^\beta - D_2^\alpha D_1^\beta) - D_2^\alpha D_2^\beta\}\tau_0 \cdot \tau_1 = 0.$$

$$(4')\ \{D_1^6 + 9D_1D_5 - 5D_1^3D_3 - 5D_3^2 + D_1(D_2^\alpha D_3^\beta - D_2^\beta D_3^\alpha) - 5D_3^\alpha D_3^\beta\}\tau_0 \cdot \tau_1 = 0.$$

Remark 199. For vertex operator algebras, vertex algebras, vertex operator superalgebras, one can refer Adamovic [233].

6.10 Fermionic Fock space, Clifford algebra, Bosonic Fock space, and Boson-Fermion correspondence

Definition 293 (Fermionic Fock space). Let F be complex vector space with a basis consisting of all semi-infinite monomials of the form $i_1 \wedge i_2 \wedge i_3 \wedge \dots$, where $i_1, i_2, \dots$ are integers such that $i_1 > i_2 > i_3, \dots$, and $i_n = i_{n-1} - 1$ for $n \gg 0$.

Let $|m> = m \wedge m-1 \wedge m-2 \wedge \dots$, denote the vacuum vector of charge m and $F^{(m)}$ denote the linear span of all semi-infinite monomials of charge m, that is those differ from $|m>$ only at finite number of places. We have the charge decomposition as $F = \bigoplus_{m\in\mathbb{Z}} F^{(m)}$. Here F is called Fermionic Fock space.

Definition 294 (Clifford algebra). For $j \in \mathbb{Z}$, we define ψ_j, the wedging operators on F and $\psi_j^\star$, the contracting operators on F as follows:

$$\psi_j(i_1 \wedge i_2 \wedge \cdots) = \begin{Bmatrix} 0 & \text{if} \quad j = i_s \quad \text{for some } s \\ (-1)^s & (i_1 \wedge i_2 \wedge \cdots \wedge i_{s-1} \wedge j \wedge i_{s+1} \wedge \cdots \text{if } i_s > j > i_{s+1}) \end{Bmatrix}.$$

$$\psi_j^*(i_1 \wedge i_2 \wedge \cdots) = \begin{Bmatrix} 0 & \text{if} \quad j \neq i_s \quad \text{for all } s \\ (-1)^{s+1} & (i_1 \wedge i_2 \wedge \cdots \wedge i_{s-1} \wedge i_{s+1} \wedge \cdots \text{if } j = i_s) \end{Bmatrix}.$$

Actually we can define a Hermitian form $H(,.,)$ on F (for which the orthonormal basis consist of all semi-infinite monomials). Then ψ_j and $\psi_j^\star$ are adjoint with respect to Hermitian form H.

We have, $\psi_i\psi_j^\star + \psi_j\psi_i^\star = [\psi_i, \psi_j^\star]_+ = \delta_{ij}$, $\psi_i\psi_j + \psi_j\psi_i = [\psi_i, \psi_j]_+ = 0$, and $\psi_i^\star\psi_j^\star + \psi_j^\star\psi_i^\star = [\psi_i^\star, \psi_j^\star]_+ = 0$.

Moreover, $\psi_j, \psi_j^\star$ generate a Clifford algebra, denoted as Cl. In other words, the Clifford algebra, Cl, can also be defined as follows. The Clifford algebra is the C-algebra with basis $\{\psi(i), \psi^*(i)\}_{i\in\mathbb{Z}}$ satisfying the relations,

$$\{\psi(i), \psi^*(j)\} = \delta_{ij} \text{ and } \{\psi(i), \psi(j)\} = \{\psi^*(i), \psi^*(j)\} = 0,$$

for all $i, j \in \mathbb{Z}$, where $\{x, y\} = xy + yx$.

Define the normal ordering prescription

$$: \psi(i)\psi^*(j) := \begin{cases} \psi(i)\psi^*(j), & \text{if } j > 0, \\ -\psi^*(j)\psi(i), & \text{if } j \leq 0. \end{cases}$$

Here the operators ψ_j and $\psi_j^\star$ are called free fermions. We know that $GL_\infty = \{a \in (a_{ij})_{i,j\in\mathbb{Z}} | a$ is invertible and all but a finite number of $a_{ij} - \delta_{ij}$ are $0\}$ and gl_∞ is the Lie algebra of the group GL_∞.

We have an embedding, $r : gl_\infty \to \text{Cl}$ defined by $r(E_{ij}) = \psi_i\psi_j^\star$ defines a *representation* r *of* gl_∞ *on* F, producing thereby a representation r_m of gl_∞ on $F^{(m)}$ for each $m \in \mathbb{Z}$.

The group Gl_∞ of automorphisms of the space $\Psi = \sum_{j\in\mathbb{Z}} \mathbb{C}\psi_j$, that leave all but a finite number of ψ_j, fixed acts also on $\Psi_j^\star = \sum_{j\in\mathbb{Z}} \mathbb{C}\psi_j^\star$, which is identified with a subspace in the dual of Ψ via $\langle \psi_i, \psi_j^\star \rangle = \delta_{ij}$.

For $n \in \mathbb{Z}\backslash\{0\}$, let

$$\alpha_n = \sum_{j\in\mathbb{Z}} \psi_j \psi_{j+n}^\star.$$

Then one can prove that $[\alpha_m, \alpha_n] = m\delta_{m,-n}$.

Remark 200. In general, for a given vector space V, with a symmetric bilinear form $(. | .)$, one can define the Clifford algebra $Cl(V)$ as follows:

$$Cl(V) = T(V)/J,$$

where $T(V)$ is the tensor algebra over V and J is its 2-sided ideal generated by elements of the form $x \otimes y - (x|y)$, $(x, y \in V)$.

Definition 295 (Bosonic Fock space). The Bosonic Fock space B is $B := \mathbb{C}[x_1, x_2, \ldots; q, q^{-1}]$, in the usual notation, with $B = \oplus_{m\in Z} B^m$, where the charge m part of B is $B^{(m)} := q^m \mathbb{C}[x_1, x_2, \ldots]$.

Definition 296 (Boson-Fermion correspondence). We get a so called Boson-Fermion correspondence as follows: $\sigma_m : F^{(m)} \overset{\sim}{\rightarrow} B^{(m)} = \mathbb{C}[x_1, x_2 \cdots]$, polynomial algebra on infinitely many indeterminates $x = (x_1, x_2, \ldots)$, is defined by $\sigma_m(|m>) = 1$, $\sigma_m \alpha_n \sigma_m^{-1} = \frac{\partial}{\partial x_n}$, $\sigma_m \alpha_{-n} \sigma_m^{-1} = n x_n$ for $n = 1, 2, \ldots$.

The essential part of Boson-Fermion correspondence is the calculation of operators $\sigma_{m+1}\ \psi_j\ \sigma_m^{-1} : B^{(m)} \rightarrow B^{(m+1)}$.

Remark 201. Fermionic Fock space F can also be defined, using a diagram named as Maya diagram.

6.11 Remark on Quantum groups, String theory and Mathematical Physics with some references

Quantum groups

The term "quantum groups" first appeared in the theory of quantum integrable systems, which was then formalized by Vladimir Drinfeld and Michio Jimbo as a particular class of Hopf algebras which are in fact, universal enveloping algebras of some semisimple or affine Lie algebras. In mathematics and theoretical physics, quantum group represents various kinds of noncommutative algebra with some additional structures. For Quantum groups and related topics, one can refer:

1) Hopf algebras and quantum Yang-Baxter equation: V.G. Drinfeld [242].
2) Quantum groups: V.G. Drinfeld [243].
3) From quantum affine Kac-Moody algebras to Drinfeldians and Yangians: V. Tolstoy [53].
4) A q-difference analogue of U(g) and the Yang-Baxter equation: M. Jimbo [245].

5) A q-analogue of $U_q(gl(N+1))$, Hecke algebra and the Yang-Baxter equation: M. Jimbo [246].
6) Quantum R-matrix for the generalized Toda system: M. Jimbo [247].
7) Lectures on Quantum groups: J.C. Jantzen [248].
8) Quantum physics: a functional integral point of view (2nd ed.): Glimm, James, Jaffe Arthur [249].
9) Affine Lie algebras and Quantum groups: Fuchs [258].
10) A guide to Quantum Groups: V. Chari and A. Pressley [259].
11) Quantum Groups: Kassel [260].
12) Introduction to quantum groups: P. Podles; E. Muller [250].
13) Some introductory notes on quantum groups, quantum algebras, and their applications: R. Jagannathan [244].

String theory

String theory is a theoretical framework in which the point-like particles of particle physics are replaced by one-dimensional objects called strings. The original version of string theory was bosonic string theory, which describes only bosons, a class of particles which transmit forces between the matter particles or fermions. Bosonic string theory was superseded by theories called superstring theories, which describe both bosons and fermions, and also a theoretical idea called supersymmetry. This is a mathematical relation that exists in certain physical theories between the bosons and fermions.

For string theory and related topics, one can refer:

1) Monstrous moonshine and Lie superalgebras: Borcherds, Richard [150].
2) Monstrous moonshine, Conway, John; Norton, Simon [28].
3) String theory and M-theory: A modern introduction: Becker, Katrin; Becker, Melanie; Schwarz, John [251].
4) Supermembranes and eleven-dimensional supergravity: Bergshoeff, Eric; Sezgin, Ergin; Townsend, Paul [252].
5) Noncommutative Geometry: Connes, Alain [253].

Mathematical Physics

Mathematical Physics is the development of mathematical methods for application to problems in physics such as ordinary differential equations, symplectic geometry (purely mathematical disciplines), dynamical systems and Hamiltonian mechanics (mathematical physics).

For topics related to Mathematical Physics, one can refer:

1) Methods of mathematical physics: Courant, Richard; Hilbert, David [254].
2) Methods of modern mathematical physics 4: Reed, Michael; Simon, Barry [255].
3) Mathematical Physics: Menzel, Donald Howard [256].

4) Methods of Mathematical Physics: M. Reed; B. Simon [257].
5) The definition of conformal field theory, in: Differential Geometrical Methods in Theoretical Physics: G. Segal [179].

Exercises

6.1 Using the theorems (results) on the spectra of the operator Δ_G, De Rham Hodge operator $\Box$, compute the same for the following spaces:
- **(i)** $SO(5)/SO(2) \times SO(3)$,
- **(ii)** $SO(6)/SO(2) \times SO(4)$,
- **(iii)** $Sp(3)/Sp(1) \times Sp(2)$,
- **(iv)** $Sp(4)/Sp(2) \times Sp(2)$.

6.2 Using Theorem 76 on a spectral invariant of the zeta function $\zeta(\Delta, s)$ at $s = 0$ of the Laplace-Beltrami operator Δ acting on 2-forms, compute the same for the following spaces:
(a) S^7; (b) S^{11}.

6.3 Using Theorem 77 on a spectral invariant of the zeta function $\zeta(\Delta, s)$ at $s = 0$ of the Laplace-Beltrami operator Δ acting on 1-form, compute the same for the following space: $SO(7)/SO(2) \times SO(5)$.

6.4 Using Theorem 78 on a spectral invariant of the zeta function $\zeta(\Delta, s)$ at $s = 0$ of the Laplace-Beltrami operator Δ acting on 1-form, compute the same for the following space: $Sp(5)/Sp(1) \times Sp(4)$.

6.5 Using Theorem 79 on the spectral asymmetry of the operator $B = \pm(\star d - d\star)$, compute the same for the following space: S^{15}.

6.6 Using the theorem on the generalization of Macdonald formula, obtain the same for $\eta(q)^{(4+1)^2-1}(=\eta(q)^{24})$, namely the value of $\eta(q^m)^{5^2/m}\eta(q)^{-1}$ in terms of a series in q.

6.7 Using the theorem on the generalization of Macdonald formula, obtain the same for $\eta(q)^{33}$, namely, the value of $\eta(q)\eta(q^m)^{(2\times 4^2+4-1)}(= \eta(q)\eta(q^m)^{35})$ in terms of a series in q.

6.8 Find the Hirota derivatives D_x^k on $e^{ax}e^{bx}e^{cx}e^{dx}$ namely the value of $D_x^k(e^{ax}e^{bx}e^{cx}e^{dx})$.

6.9 Find the value of $D_x^2(e^x \sin x)$ in terms of ordinary partial derivatives.

6.10 Find the value $D_t D_x(e^t \cos x)$.

6.11 For the KP equation, find the transformed equation in Hirota's form.

6.12 Prove that in the usual notation for any $k \in R$, $e^{y\frac{d}{dx}}f(x) = f(x+y)$, $e^{5\frac{d}{dx}}f(x) = f(x+5)$.

6.13 Show that the Lie algebra of left invariant vector fields on a Lie group is isomorphic to the Lie algebra of right invariant vector fields.

Appendix

Abstract

In this appendix, definitions of differentiable manifold, product manifold, C^∞ -functions, pseudo Riemannian manifold, Riemann surface, Lie algebra realized as operators, Maya diagram, Fermionic Fock space, Young tableau, presheaf, sheaf, symmetric group, generalized Cartan matrices of quasi finite type, quasi affine type and quasi hyperbolic type with the corresponding Dynkin diagrams are given with examples wherever necessary.

Definition 297 (Differentiable manifold). Let M be a Hausdorff space. An *open chart* is a pair (U, ϕ), where U is an open subset of M and ϕ is a homeomorphism of U onto an open subset of $\mathbb{R}^m$. A differentiable structure on M of dimension m is a collection of open charts $(U_\alpha, \varphi_\alpha)_{\alpha\in A}$, where A is the index set of all open charts on M and $\varphi_\alpha(U_\alpha)$ is an open subset of $\mathbb{R}^m$ such that the following conditions are satisfied:

(i) $M = \bigcup_{\alpha\in A} U_\alpha$.

(ii) For each pair $\alpha, \beta \in A$, the mapping $\varphi_\beta \circ \varphi_\alpha^{-1}$ is a differentiable mapping of $\varphi_\alpha(U_\alpha \cap U_\beta)$ onto $\varphi_\beta(U_\beta \cap U_\alpha)$.

(iii) The collection $(U_\alpha, \varphi_\alpha)_{\alpha\in A}$ is a maximal family of open charts for which (i) and (ii) hold.

A differentiable manifold (C^∞ manifold or simply manifold) of dimension m is a Hausdorff space with a differentiable structure of dimension m.

If M is a manifold, a local chart on M (or a local coordinate system on M) is by definition, a pair $(U_\alpha, \varphi_\alpha)$, where $\alpha \in A$. If $p \in U_\alpha$ and $\varphi_\alpha(p) = (x_1(p), \ldots, x_m(p))$, the set U_α is called a coordinate neighborhood of p and the numbers $x_i(p)$ are called local coordinates of p. The mapping φ_α: $q \to (x_1(q), \ldots, x_m(q)), q \in U_\alpha$ is often denoted by $\{x_1, \ldots, x_m\}$.

Remark 202. A manifold is connected if and only if it is pathwise connected.

Remark 203. A analytical structure of dimension m is defined in a similar fashion. In (ii) (the above definition of differentiable manifold), if we replace "differentiable" by "analytic," then M is called an analytic manifold.

Example 95.

(1) The n-dimensional sphere $S^n = \left\{(x_1, x_2, \ldots, x_{n+1}) \in \mathbb{R}^{n+1} | \sum_{n+1}^{i=1} (x_i)^2 = 1\right\} = U_+ \cup U_-$, where $U_+ = S^n - (0, 0, \ldots, -1)$ (northern hemisphere) and $U_- = S^n - (0, 0, \ldots, 1)$ (southern hemisphere). In particular, $S^2 = \{(x, y, z) \in R^3 | x^2 + y^2 + z^2 = 1\} = U_+ \cup U_-$, where $U_+ = \{(x, y, z) \in S^2; z \neq -1\}, \phi_+: (x, y, z) \to \left(\frac{x}{1+z}, \frac{y}{1+z}\right)$ $U_- = \{(x, y, z) \in S^2; z \neq 1\}, \phi_-: (x, y, z) \to \left(\frac{x}{1-z}, \frac{y}{1-z}\right)$.

(2) Real projective space $RP(n)$: The real n-dimensional projective space $RP(n)$ is the set of straight lines through the origin in R^{n+1} space. $RP(n)$ is equivalent to manifold S^n/Z_2, in which the antipodal points on sphere S^n are taken to be one point. For example,

$$RP(3) \simeq \frac{S^3}{Z_2} \simeq \frac{SU(2)}{Z_2} \simeq SO(3),$$

where $Z_2 \simeq S_0$ is a group consisting of two elements.

Remark 204. One-dimensional connected manifolds are R^1 and S^1.

A two-dimensional manifold is called a surface.

Definition 298 (Product manifold). If M and N are two C^∞ manifolds then one can define a C^∞ manifold, called the *product manifold*. The examples are

(i) Cylinder = $R \times S^1$.
(ii) Two-dimensional (ordinary) torus = $S_1 \times S_1 = T^2$.
(iii) d-dimensional torus = $S_1 \times \cdots \times S_1$ (d factors) = T^d.

Definition 299 ($C^\infty(M)$ functions on a differentiable manifold). Let f be a real-valued function on a C^∞ manifold M. The function f is called *differentiable at a point* $p \in M$, if there exists a *local chart* $(U_\alpha, \varphi_\alpha)$ with $p \in U_\alpha$ such that the composite function $f \circ \varphi_\alpha^{-1}$ is a differentiable function on $\varphi_\alpha(U_\alpha)$. The function is called differentiable if it is differentiable at each point $p \in M$. The function f is said to be analytic at $p \in M$ if there exists a local chart $(U_\alpha, \varphi_\alpha)$ with $p \in U_\alpha$ such that $f \circ \varphi_\alpha^{-1}$ is an analytic function on the set $\varphi_\alpha(U_\alpha)$. We shall often write $C^\infty(M)$ for the set of all functions on M, differentiable at p for all $p \in M$. The set $C^\infty(M)$ is an algebra over R with the following operations:

(i) $(\lambda f)(p) = \lambda f(p)$,
(ii) $(f + g)(p) = f(p) + g(p)$, and
(iii) $(fg)(p) = f(p)g(p)$,

for $\lambda \in R; p \in M; f, g \in C^\infty(M)$.

Definition 300 (Lie group SO(3) and the corresponding Lie algebra so(3)). Orthogonal matrices SO(3) corresponding to counter clockwise rotations of $\mathbb{R}^3$ with determinant 1 are

$$R_1(\alpha) = \begin{pmatrix} 1 & 0 & 0 \\ 0 & \cos\alpha & -\sin\alpha \\ 0 & \sin\alpha & \cos\alpha \end{pmatrix}, R_2(\beta) = \begin{pmatrix} \cos\beta & 0 & \sin\beta \\ 0 & 1 & 0 \\ -\sin\beta & 0 & \cos\beta \end{pmatrix},$$
$$\text{and } R_3(\gamma) = \begin{pmatrix} \cos\gamma & -\sin\gamma & 0 \\ \sin\gamma & \cos\gamma & 0 \\ 0 & 0 & 1 \end{pmatrix}.$$

Corresponding Lie algebra is generated by $L_1 = \frac{d}{d\alpha}R_1(\alpha), L_2 = \frac{d}{d\beta}R_2(\beta)$, $L_3 = \frac{d}{d\gamma}R_3(\gamma)$ at $\alpha, \beta, \gamma = 0$. So,

$$L_1 = \begin{pmatrix} 0 & 0 & 0 \\ 0 & 0 & -1 \\ 0 & 1 & 0 \end{pmatrix}, L_2 = \begin{pmatrix} 0 & 0 & 1 \\ 0 & 0 & 0 \\ -1 & 0 & 0 \end{pmatrix}, \text{ and } L_3 = \begin{pmatrix} 0 & -1 & 0 \\ 1 & 0 & 0 \\ 0 & 0 & 0 \end{pmatrix}.$$

This set $\{L_1, L_2, L_3\}$ forms a basis of $so(3) = \{3 \times 3$ matrices A over R trace zero satisfying $A + A^t = 0\}$. This is a Lie algebra.

Remark 205. The above example also gives an example for the construction of a Lie algebra from a Lie group.

Definition 301 (Pseudo-Riemannian manifold). A pseudo-Riemannian structure on a C^∞ manifold M is a tensor field G of type $(0, 2)$ such that

(1) $g(X, Y) = g(Y, X)$ for all $X, Y \in D^1(M)$ and
(2) for each $p \in M$, g_p is a nondegenerate bilinear form on $M_p \times M_p$.

A pseudo-Riemannian manifold is a connected C^∞ manifold with the pseudo-Riemannian structure. If in addition, g_p is positive definite for each $p \in M$, then we get a Riemannian structure instead of pseudo-Riemannian structure and the corresponding manifold is a *Riemannian manifold*. Let M be a Riemannian manifold with Riemannian structure Q. This Riemannian manifold is called analytic Riemannian manifold if both M and Q are analytic.

Example 96 (Some examples of Riemannian symmetric space). Riemannian symmetric spaces are an important class of homogeneous spaces. We give below some examples:

1. Positive curvature:
  (i) Sphere (orthogonal group): $S^{n-1} \cong O(n)/O(n-1)$.
  (ii) Oriented sphere (special orthogonal group): $S^{n-1} \cong SO(n)/SO(n-1)$.
  (iii) Projective space (projective orthogonal group): $P^{n-1} \cong PO(n)/PO(n-1)$.
2. Flat (zero curvature):
  (i) Euclidean space (Euclidean group, point stabilizer is orthogonal group): $A^n \cong E(n)/O(n)$.
3. Negative curvature:
  (i) Anti-de Sitter space: $AdS_{n+1} = O(2, n)/O(1, n)$.
4. Others:
  (i) Grassmannian: $Gr(r, n) = O(n)/(O(r) \times O(n-r))$.

Remark 206. For details regarding $SL(n, R)/SO(n)$ and zonal spherical functions on the Siegel upper half plane, one can refer Bhanumurti [1, 2].

Definition 302 (Riemann surface). Riemann surface is an one-dimensional complex manifold. Riemann surfaces can also be thought of *as deformed versions of the complex plane*. Hence, neighborhood of every point of a Riemann surface looks like patches of the complex plane; for example, they look like a sphere or a torus or several sheets glued together. Main point of Riemann surfaces is that the holomorphic functions may be defined between them.

Definition 303. Symmetries of the heat equation $u_t = u_{xx}$ are generated by the following vector fields:

$$
\begin{aligned}
X_1 &= \frac{\partial}{\partial x}, X_2 = \frac{\partial}{\partial t}, X_3 = u\frac{\partial}{\partial u}, \\
X_4 &= \frac{\partial}{\partial x} + 2t\frac{\partial}{\partial t}, X_5 = 2t\frac{\partial}{\partial x} - xu\frac{\partial}{\partial u}, \quad \text{and} \\
X_6 &= 4tx\frac{\partial}{\partial x} + 4t^2\frac{\partial}{\partial t} - (x^2 + 2t)u\frac{\partial}{\partial u}
\end{aligned}
$$

They form a six-dimensional Lie algebra.

Definition 304 (Lie algebra realized as operators). The following gives an example of a Lie algebra, realized as operators on a three-dimensional vector space. That is, they are

$$R_k = x^j \frac{\partial}{\partial x^i} - x^i \frac{\partial}{\partial x^j}, \quad i, j, k = 1, 2, 3.$$

In particular,

$$R_1 = x^3 \frac{\partial}{\partial x^2} - x^2 \frac{\partial}{\partial x^3}$$
$$R_2 = x^3 \frac{\partial}{\partial x^1} - x^1 \frac{\partial}{\partial x^3}$$
$$R_3 = x^2 \frac{\partial}{\partial x^1} - x^1 \frac{\partial}{\partial x^2}$$

with $[R_j, R_k] = a^l_k R_l$.

These operators act on the infinite-dimensional vectorspace $C^\infty(R^3)$, so this gives an infinite-dimensional representation of so(3).

Definition 305 (Maya diagram and Fermionic Fock space [234]). A Maya diagram is a placement of a white or black bead at each position in $\mathbb{Z} + 1/2$, subject to the condition that all but finitely many positions $m < 0$ are filled with a black bead and all but finitely many positions $m > 0$ are filled with a white bead. For instance, we get the following diagram:

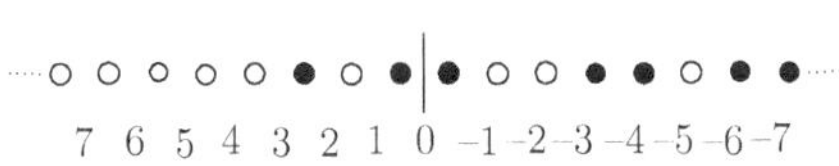

Note that we label the real line from right to left.

We can define Fermionic Fock space F to be the free span over $\mathbb{C}$ of all Maya diagrams. The basis of F consisting of Maya diagrams is called the standard basis. The standard basis of F is also indexed by charged partitions or regular normally ordered semi-infinite wedges. The charge m part of F is $F^{(m)} :=$ span {charged partitions with charge m}. In particular, $F^{(0)}$ is just spanned by ordinary positions. Clearly $F = \oplus_{m \in \mathbb{Z}} F^{(m)}$.

Definition 306 (Young tableau). A Young tableau is constituted by a certain number of "boxes" set on one or more rows such that, if the rows are numbered from top to the bottom, the number of boxes in the ith row is bigger or equal to the number of boxes in the $(i+1)$-th row. In general, let $[\alpha]$ denote set of numbers $[\alpha_1, \alpha_2, \ldots, \alpha_n]$ such that

$$\alpha_1 \geq \alpha_2 \geq \cdots \geq \alpha_n \geq 0$$

with

$$\alpha_1 + \alpha_2 + \cdots + \alpha_n = n.$$

This partition of $[\alpha]$ can be denoted by Young shape consisting of n boxes, α_r-th will occur in the rth row. If $r\,\alpha$'s are equal, we denote it as α^r. Thus a Young tableau is a Young shaped in which boxes will be numbered in some order. A standard Young tableau is a tableau with the property that if all but the boxes labeled $1, \ldots, h$ with $h < n$ are erased, then these remaining h boxes by themselves form a Young tableau. An example of a standard Young tableau $[43^2 2]$ is

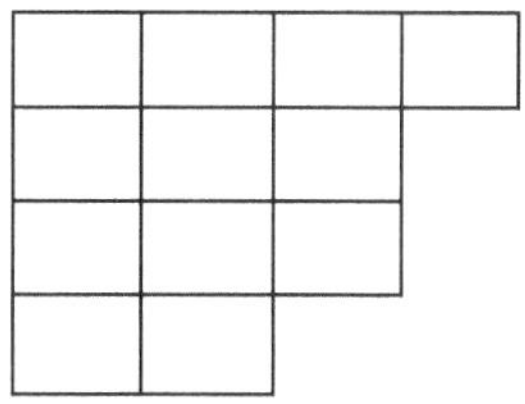

The standard Young tableau corresponding to a given shape $[\alpha]$ can be labeled as $[\alpha]_r$, where $r = 1, \ldots, f_\alpha$. The particular method of numbering the standard tableau is irrelevant and just a matter of convention. Here the number of standard tableaux for a given shape $[\alpha]$ is f_α.

Remark 207. For basic properties of Young tableau, one can refer Belinfante et al. [154, 156] and for further details, one can refer Frappat et al. [129] and Fulton [276].

Definition 307. Let X be a topological space. An assignment to every open subset U of X, of a set $\mathcal{F}(U)$ and to every pair of open sets U, V with $V \subset U$, of a map (to be called restriction map), $res_{UV}: \mathcal{F}(U) \to \mathcal{F}(V)$ satisfying

$$res_{VW} \circ res_{UV} = res_{UW}$$

for every triple $W \subset V \subset U$ of open sets, is called a *presheaf* of sets.

Remark 208. If $\mathcal{F}(U)$ are all abelian groups, rings, vector spaces, etc., and the restriction maps are homomorphism of the respective structures, then we say that $\mathcal{F}$ is a presheaf of abelian groups, rings, vector spaces, etc.

Definition 308. A presheaf will become a sheaf if the following additional conditions are satisfied.

Let $U = \cup_{i \in I} U_i$ be any open covering of open sets U. Then

(S_1) Two elements $s, t \in \mathcal{F}(U)$ are equal if $res_{UU_i} s = res_{UU_i} t$ for all $i \in I$.

(S_2) If $s_i \in \mathcal{F}(U_i)$ satisfy $res_{U_iU_i} \cap U_j s_i = res_{U_jU_i} \cap U_j s_j$ for all $i, j \in I$, then there exists an element $s \in \mathcal{F}(U)$ with $res_{UU_i} s = s_i$ for all i. We will assume that $\mathcal{F}(\phi)$ consists of single point.

Remark 209. For the above definition of presheaf and sheaf, one can refer Ramanan [235].

Definition 309. Symmetric groups are some of the most essential types of finite groups. The symmetric group S_n of degree n and of order $n!$ is the group of all permutations on n symbols. Cayley's theorem proves that any finite group G is isomorphic to a subgroup of a symmetric group. For groups and symmetry, one can refer [271] and [272].

Example 97 (An example of symmetric group). Consider $\mathbb{R}^4$. Let L be the group of all linear transformations of $\mathbb{R}^4$, which leave the symmetric bilinear form

$$((x_1, x_2, x_3, x_4), (y_1, y_2, y_3, y_4)) \mapsto -x_1y_1 + x_2y_2 + x_3y_3 + x_4y_4,$$

invariant. Then L is called homogeneous Lorentz group. Let L' be the group generated by this L and translations. Then L' is called as a homogenous Lorentz group. This is an example of symmetric group.

Definition 310 (Quasi-finite type [264]). An indecomposable generalized Cartan matrix (GCM) A of order n (or the associated Dynkin diagram $S(A)$) is said to be of quasi-finite type if $S(A)$ has a proper connected subdiagram of finite type of order $n-1$.

 is an example of quasi-finite Dynkin diagram.

Like extended hyperbolic type of Kac Moody algebras (Chapter 2), quasi-hyperbolic type is a more generalized extension of the hyperbolic type.

Definition 311 (Quasi-hyperbolic type [265]). A Dynkin diagram $S(A)$, associated with the indefinite, indecomposable GCM A of order n is said to be of quasi-hyperbolic type if $S(A)$ contains at least one proper, connected subdiagram of hyperbolic type of order $n-1$. In this case, the GCM, A is also said to be of quasi-hyperbolic type. The Kac Moody algebra associated with a quasi-hyperbolic GCM is called quasi-hyperbolic Kac Moody algebra.

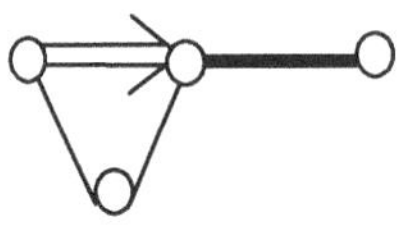 is an example of quasi-hyperbolic Dynkin diagram.

Definition 312 (Quasi-affine type [266]). A Dynkin diagram $S(A)$, associated with the indefinite, indecomposable GCM A of order n is said to be of quasi-affine type if $S(A)$ contains at least one proper, connected subdiagram of affine type of order $n-1$. In this case, the GCM, A is also said to be of quasi-affine type. The Kac Moody algebra associated with a quasi-affine GCM is called quasi-affine Kac Moody algebra.

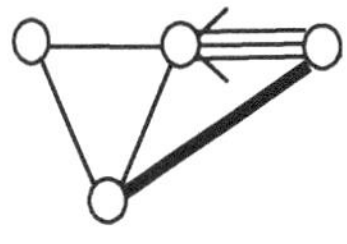 is an example of quasi-affine Dynkin diagram.

For the results on the root systems, Dynkin diagrams and the structure of some specific classes of quasi-affine and quasi-hyperbolic type of Kac Moody algebras, one can refer [265–268].

Remark 210.

(1) It is to be noted that every extended hyperbolic diagram is quasi-hyperbolic but every quasi-hyperbolic type need not be extended hyperbolic. Also, there are indefinite Kac Moody algebras which are neither extended hyperbolic nor quasi-hyperbolic.

(2) Quasi-affine family is a broader class in the indefinite type, which contains the hyperbolic class. Every hyperbolic diagram is of quasi-affine type, but the converse is not true in general.

Remark 211. For many mathematical definitions, formulas, figures, tabulations and references, one can refer Weisstein [275].

Bibliography

[1] T.S. Bhanu Murti, Plancherel's measure for the factor space $SL(n, R)/SO(n, R)$, Soviet Math. Doklady 1 (1961a) 860–862.
[2] T.S. Bhanu Murti, The asymptotic behavior of zonal spherical functions on the Siegel upper half-plane, Soviet Math. Doklady 1 (1961b) 1325–1329.
[3] Z. Wan, An Introduction to Kac-Moody Algebras, World Scientific Publ. Co. Ltd, Singapore, 1990.
[4] M. Wakimoto, Affine Lie algebras and Virosoro algebras I, Jpn J. Math. 12 (2) (1986) 379–400.
[5] S. Lie, Gruppenregister, in: Gesammelte Abhandlungen, vol. 5, B.G. Teubner, Leipzig, 1924, pp. 767–773.
[6] S. Lie, Theorie der Transformationsgruppen, Math. Ann. 6 (1880) 441–528, 16, Gesammelte Abhandlungen, B.G. Teubner, Leipzig, pp. 1–94.
[7] S. Lie, Theorie der Transformationsgruppen, vol. 3, B.G. Teubner, Leipzig, 1893.
[8] V.G. Kac, Infinite Dimensional Lie Algebras, III ed., Cambridge University Press, Cambridge, 1990.
[9] E. Cartan, Premiere These: Sur la Structure des Groupes de Transformations Finis et Continues, Nony, Paris, 1894.
[10] E.B. Dynkin, The classification of simple Lie groups, Rec. Math. [Mat. Sbornik] XVIII (60) (1946) 347–352.
[11] E.B. Dynkin, The structure of semisimple Lie algebras, Uspehi Mat. Nank 2 (1947) 59–127.
[12] C. Chevalley, Sur la classification des algebras de Lie simples et de leur representations, C. R. Acad. Sci., Paris 227 (1948a) 1136–1138.
[13] Harish-Chandra, On some applications of the universal enveloping algebra of a semisimple Lie algebra, Trans. Amer. Math. Soc. 70 (1951) 28–96.
[14] J.P. Serre, Divisibilite de certaines fonctions arithmetiques, Enseignement Math. 2 (22) (1976) 227–260.
[15] J.P. Serre, Algebres de Lie Semi-Simple Complexes, Mathematics Lecture Note Series, Benjamin, New York and Amsterdam, 1966.
[16] J.P. Serre, Lie Algebras and Lie Groups: 1964 Lectures Given at Harvard University, vol. 2: Mathematics Lecture Note Series, Benjamin, New York and Amsterdam, 1965.
[17] R.V. Moody, A new class of Lie algebras, J. Algebra 10 (1968) 211–230.
[18] R.V. Moody, Euclidean Lie algebras, Canad. J. Math. 21 (1969) 1432–1454.
[19] V.G. Kac, Simple graded Lie algebras of finite growth, Funct. Anal. Appl. 1 (1967) 328–329.
[20] V.G. Kac, Highest weight representations of infinite dimensional Lie algebras, in: Proceedings of ICM, Helsinki, 1978, 1980, pp. 299–304.
[21] V.G. Kac, Simple irreducible graded Lie algebras of finite growth, Izv. Akad. Nauk USSR 32 (2) (1968). Eng. Trans. Math of USSR-Izvestija 2 (1968) 1271–1311.
[22] N. Jacobson, Lie Algebras, Interscience, New York, 1962.

[23] H. Garland, J. Lepowsky, Lie algebra homology and the Macdonald-Kac formulas, Invent. Math. 34 (1976) 37–76.
[24] N. Sthanumoorthy, A. Uma Maheswari, Root multiplicities of extended-hyperbolic Kac-Moody algebras, Comm. Algebra (USA) 24 (14) (1996) 4495–4512.
[25] S. Berman, K.H. Parshall, Victor Kac and Robert Moody: their paths to Kac-Moody Lie algebras, Math. Intell. 24 (1) (2002) 50–60.
[26] V.G. Kac, D.H. Peterson, Infinite flag varieties and conjugacy theorems, Proc. Nat Acad. Sci. USA 80 (1983) 1778–1782.
[27] R.E. Borcherds, Generalized Kac-Moody algebras, J. Algebra. 115 (III) (1988) 502–512.
[28] J.H. Conway, S. Norton, Monstrous moonshine, Bull. Lond. Math. Soc. 11 (1979) 308–339.
[29] I.B. Frenkel, V.G. Kac, Basic representations of affine Lie algebras and dual resonance models, Invent. Math. 62 (1980) 23–66.
[30] R.E. Borcherds, Vertex algebras, Kac-Moody algebras and the Monster, Proc. Natl Acad. Sci. USA 83 (1986) 3068–3071.
[31] I.B. Frenkel, J. Lepowsky, J. Meurman, Vertex Operator Algebras and the Monster, Pure and Appl. Math, vol. 134, Academic Press, London, 1988.
[32] I.B. Frenkel, Representations of Kac-Moody algebras and dual resonance models, Lect. Appl. Math. 21 (1985) 325–353.
[33] I.B. Frenkel, Two constructions of affine Lie algebra representations and Boson-Fermion correspondence in quantum field theory, J. Funct. Anal. 44 (3) (1981) 259–327.
[34] V.G. Kac, Lie superalgebras, Adv. Math. 26 (1977) 8–96.
[35] M. Scheunert, The Theory of Lie Superalgebras, vol. 716, Lecture Notes in Mathematics, Springer-Verlag, 1979.
[36] V.G. Kac, M. Wakimoto, Integrable highest weight modules over affine superalgebras and number theory in "Lie Theory and Geometry", in: Progress in Mathematics, vol. 123, Birkhauser, Boston, 1994, pp. 415–456.
[37] S.J. Kang, M.H. Kim, Dimension formula for graded Lie algebras and its applications, Trans. Amer. Math. Soc. 351 (1999) 4281–4336.
[38] Yu.A. Bahturin, A.A. Mikhalev, V.M. Petrogradsky, M.V. Zaicev, Infinite dimensional Lie superalgebras [J], in: Expositions in Mathematics, vol. 7, Walter de Gruyter, New York, 1992.
[39] V.G. Kac, On the classification of simple Lie algebras over a field of non-zero characteristic, Math. USSR Izvestija 4 (1970) 391–413.
[40] J.E. Humphreys, Introduction to Lie Algebras and Representations Theory, Springer-Verlag, Berlin, Heidelberg, New York, 1972.
[41] K. Erdmann, M.J. Wildon, Introduction to Lie Algebras, Springer, New York, 2009.
[42] S. Helgason, Differential Geometry, Lie Groups and Symmetric Spaces, Academic Press, New York, 1978.
[43] H. Samelson, Notes on Lie algebras, Springer-Verlag, 1969.
[44] N. Shapovalov, On a bilinear form on the Universal enveloping algebra of a complex semisimple Lie algebra, Funct. Anal. Appl. 6 (4) (1972) 307–312.
[45] L. Bianchi, Lezioni Sulla Teoria Dei Gruppi Continui Finiti di Trasformazioni, Enrico Spoerri, Editore, Pisa, 1918.
[46] J.E. Cambpell, Introductory Treatise on Lie's Theory of Finite Continuous Transformation Groups, The Clarendon Press, Oxford, 1903.
[47] R. Hermann, M. Ackerman, Sophus Lie's 1880 Transformation Group Paper, Math. Sci. Press, Brookline, MA, 1975.

[48] A. Gonzalez-Lopez, N. Kamran, P.J. Olver, Lie algebras of vector fields in real plane, Proc. Lond. Math. Soc. 3 (64) (1992) 339–368.
[49] A. Gonzalez-Lopez, N. Kamran, P.J. Olver, Lie algebras of differential operators in two complex variables, Amer. J. Math. 114 (1992) 1163–1185.
[50] M.S. Narasiman, C.S. Seshadri, Stable and unitary vector bundle on a compact Riemann surface, Anal. Math. Sec. Ser. 82 (3) (1965) 540–567.
[51] N. Kamran, P.J. Olver, Lie algebras of differential operators and Lie algebraic potentials, J. Math. Anal. Appl. 145 (1990) 342–356.
[52] K. Ramachandra, On the Mean-Value and Omega-Theorems for the Riemann Zeta-Function, Lectures on Mathematics and Physics 85, Published for the Tata Institute of Fundamental Research, Bombay, 1995, ISBN 978-3-540-58437-7, MR 1332493.
[53] V.N. Tolstoy, From quantum affine Kac-Moody algebras to Drinfeldians and Yangians, Contemp. Math. 343 (2004) 349–370.
[54] C. Bennett, Imaginary roots of a Kac-Moody algebra whose reflections preserve root multiplicities, J. Algebra 158 (1993) 244–267.
[55] R.V. Moody, Root systems of hyperbolic types, Adv. Math. 33 (1979) 144–160.
[56] V.G. Kac, Infinite dimensional algebras, Dedekind's eta function, classical Mobius function and the very strange formula, Adv. Math. 30 (1978) 85–136.
[57] D. Casperson, Strictly imaginary roots of Kac-Moody algebras, J. Algebra 168 (1994) 90–122.
[58] M. Jimbo, K.C. Misra, T. Miwa, M. Okado, Combinatorics of representations of $U_q(sl(n))$ at $q = 0$, Commn. Math. Phys. 136 (1991) 543–566.
[59] M. Primc, Standard representations of $A_n^{(1)}$, in: Infinite-Dimensional Lie Algebras and Groups, vol. 7, World Scientific, Singapore, 1989, pp. 273–282, Adv. Ser. Math. Phys.
[60] N. Sthanumoorthy, A. Uma Maheswari, Structure and root multiplicities for two classes of extended hyperbolic Kac-Moody algebras $EHA_{(1)}^1$ and $EHA_{(2)}^2$ for all cases, Commun. Algebra 40 (2) (2012) 632–665.
[61] S. Berman, R.V. Moody, Multiplicities in Lie algebras, Proc. Amer. Math. Soc. 76 (1979) 223–228.
[62] S.J. Kang, Root multiplicities of the hyperbolic Kac-Moody Lie algebra $HA_1^{(1)}$, J. Algebra 160 (1993) 492–523.
[63] S.J. Kang, Root multiplicities of Kac-Moody algebras, Duke Math. J. 74 (1994) 635–666.
[64] A.J. Feingold, I.B. Frenkal, A hyperbolic Kac-Moody algebra and the theory of Siegel modular forms of genus 2, Math. Ann. 263 (1983) 87–144.
[65] S.J. Kang, Kac-Moody Lie algebras, spectral sequences, and the Witt formula, Trans. Amer. Math. Soc. 339 (1993) 463–495.
[66] S.J. Kang, Root multiplicities of hyperbolic Kac-Moody Lie algebra $HA_1^{(1)}$, J. Algebra 160 (1993) 492–523.
[67] V.G. Kac, S.J. Kang, Trace formula for graded Lie algebras and Monstrous Moonshine, in: B. Allison, G. Cliff (Eds.), Representations of Groups, Canad. Math. Soc. Conf. Proc., vol. 16, 1995, pp. 141–154, MR 96k:17044.
[68] S.J. Kang, D.J. Melville, Root multiplicities of the Kac-Moody algebras $HA_n^{(1)}$, J. Algebra 170 (1994) 277–299.
[69] J. Hontz, K.C. Misra, On root multiplicities of $HA_n^{(1)}$, Int. J. Algebra Comp. 12 (2002) 477–508.
[70] V.G. Kac, R.V. Moody, M. Wakimoto, On E_{10}, Differential Geometrical Methods in Theoretical Physics, Kluwer, Dordrecht, Boston, 1988, pp. 109–128.
[71] S. Kass, R.V. Moody, J. Patera, R. Slansky, Affine Lie Algebras, Weight Multiplicities and Branching Rules, vol. 1, University of California Press, Berkeley, 1990.

[72] S.J. Kang, On the hyperbolic Kac-Moody Lie algebra $HA_1^{(1)}$, Trans. Amer. Math. Soc. 341 (1994) 623–638.

[73] S.J. Kang, Root multiplicities of graded Lie algebras, in: S.J. Kang, M.H. Kim, I.S. Lee (Eds.), Lie Algebras and Their Representations, Contemp. Math., vol. 194, 1996, pp. 161–176.

[74] A.J. Feingold, J. Lepowsky, The Weyl-Kac character formula and power series identities, Adv. in. Math. 29 (1978) 271–309.

[75] A.J. Feingold, Tensor product of certain module for the generalized Cartan matrix Lie algebra $A_1^{(1)}$, Comm. Algebra 12 (9) (1981) 1323–1341.

[76] J. Hontz, K.C. Misra, Root multiplicities of the indefinite Kac-Moody algebras $HD_4^{(3)}$ and $HG_2^{(1)}$, Comm. Algebra 30 (2002) 2941–2959.

[77] K.C. Misra, Structure of certain modules for $A_n^{(1)}$ and the Rogers-Ramanujan identities, J. Algebra 88 (1984) 196–227.

[78] K.C. Misra, Structure of certain modules for $C_n^{(1)}$, J. Algebra 90 (1984) 385–409.

[79] N. Sthanumoorthy, A. Uma Maheswari, Purely imaginary roots of Kac-Moody algebras, Comm. Algebra 24 (2) (1996) 677–693.

[80] N. Sthanumoorthy, P.L. Lilly, A. Uma Maheswari, Root multiplicities of some classes of extended-hyperbolic Kac-Moody and extended hyperbolic generalized Kac-Moody algebras, Contemp. Math. 343 (2004) 315–347.

[81] N. Sthanumoorthy, A. Uma Maheswari, P.L. Lilly, Extended Hyperbolic Kac-Moody Algebras $EHA_2^{(2)}$: structure and root multiplicities, Comm. Algebra (USA) 32 (6) (2004) 2457–2476.

[82] N. Sthanumoorthy, P.L. Lilly, Complete classifications of generalized Kac-Moody algebras possessing special imaginary roots and strictly imaginary property, Comm. Algebra 35 (8) (2007) 2450–2471.

[83] O. Gabber, V.G. Kac, On defining relations of certain infinite-dimensional Lie algebras, Bull. Amer. Math. Soc (NS) 5 (2) (1981) 185–189.

[84] J. Lepowsky, R.L. Wilson, Construction of the affine Lie algebra $A_1^{(1)}$, Comm. Math. Phys. 62 (1978) 43–53.

[85] S. Eswara Rao, K. Zhao, On integrable representations of toroidal Lie superalgebras, Contemp. Math. 243 (2004) 243–261.

[86] S. Eswara Rao, Classification of irreducible integrable modules for toroidal Lie algebras with finite dimensional weight spaces, J. Algebra 277 (2004) 318–348.

[87] S. Eswara Rao, On representations of Toroidal Lie algebras, 2005, arXiv. Math/0503629v1, MathRT.

[88] G.M. Benkart, S.J. Kang, K.C. Misra, Graded Lie algebras of Kac-Moody type, Adv. Math. 97 (1993) 154–190.

[89] G.M. Benkart, S.J. Kang, K.C. Misra, Indefinite Kac-Moody algebras of classical type, Adv. Math. 105 (1994) 76–110.

[90] G.M. Benkart, S.J. Kang, K.C. Misra, Indefinite Kac-Moody algebras of classical linear type, Pacific J. Math. 170 (1995) 379–404.

[91] L.S. Liu, Kostant's formula for Kac-Moody Lie algebras, J. Algebra 149 (1992) 155–178.

[92] S.J. Kang, M. Kashiwara, K.C. Misra, Crystal bases of Verma modules for quantum affine Lie algebras, Comp. Math. 92 (1994) 299–325.

[93] S.J. Kang, M. Kashiwara, K.C. Misra, T. Miwa, T. Nakashima, A. Nakayashiki, Affine crystals and vertex models, Int. J. Modern Phys. A7 (1992) 449–484.

[94] R. Carter, Lie Algebras of Finite and Affine Type, Cambridge University Press, Cambridge, 2005.

[95] V. Marotta, A. Sciarrino, Realization of Borcherds algebras, Int. J. Mod. Phys. A 27 (10) (1995) 3921–3936.
[96] S. Naito, The strong Bernstein-Gelfand-Gelfand resolution for generalized Kac-Moody algebras I: the existence of the resolution, Publ. RIMS, Kyoto Univ. 29 (2000) 81–103.
[97] J. Fuchs, U. Ray, C. Schweigert, Some automorphisms of generalized Kac-Moody algebras, J. Algebra 191 (1997) 518–540.
[98] Z.H. Li, Y.D. Jin, R.P. Bai, Relationship between reflections determined by imaginary roots and the Weyl group for a special GKM algebra, J. Math. Res. Exp. 3 (19) (1999) 501–507.
[99] O. Barwald, R. Gebert, H. Nicolai, On the imaginary simple roots of the Borcherds algebras $gII_{9,1}$, Nucl. Phy. B 510 (1998) 721–738.
[100] N. Sthanumoorthy, P.L. Lilly, A note on purely imaginary roots of generalized Kac-Moody algebras, Comm. Algebra (USA) 31 (11) (2003) 5467–5480.
[101] N. Sthanumoorthy, P.L. Lilly, On the root systems of generalized Kac-Moody algebras, J. Madras Univ. (WMY-2000, Special Issue) Sect. B Sci. 52 (2000) 81–103.
[102] N. Sthanumoorthy, P.L. Lilly, Special imaginary roots of generalized Kac-Moody algebras, Comm. Algebra (USA) 10 (2002) 4771–4787.
[103] N. Sthanumoorthy, P.L. Lilly, On some classes of root systems of generalized Kac-Moody algebras, Contemp. Math., AMS, 343 (2004) 289–313.
[104] R.E. Borcherds, A characterization of generalized Kac-Moody algebras, J. Algebra 174 (1995) 1073–1079.
[105] R.E. Borcherds, The Monster Lie algebra, Adv. Math. 83 (III) (1990) 30–47.
[106] S.J. Kang, Generalized Kac-Moody algebras and the modular function, j Mat. Ann. 298 (1994) 373–384.
[107] S. Naito, Kazhdan-Luszting conjecture for generalized Kac-Moody algebras II. Proof of the conjecture, Trans. Amer. Math. Soc. 347 (10) (1995) 3891–3919.
[108] E. Jurisich, Generalized Kac-Moody Lie algebras, free Lie algebras and the structure of the Monster Lie algebra, J. Pure Appl. Algebra 126 (1998) 233–266.
[109] E. Jurisich, An exposition of generalized Kac-Moody algebras, Lie algebras and their representations, Contemp. Math. 194 (1996) 121–159.
[110] E. Jurisich, J. Lepowsky, R.L. Wilson, Realizations of the Monster Lie algebra, Sel. Math. New Ser. 1 (III) (1995) 121–161.
[111] M. Miyamoto, A generalization of Borcherds algebra and denominator formula, the recursive dimension formula, J. Algebra 180 (1996) 631–651.
[112] N. Sthanumoorthy, P.L. Lilly, Root multiplicities of some generalized Kac-Moody algebras, Indian J. Pure Appl. Math. 38 (2) (2007) 45–78.
[113] N. Sthanumoorthy, Eta function and spectral asymmetry for $S^{(4r-1)}$, Indian J. Pure Appl. Math. 21 (11) (1990) 971–976.
[114] R.W. Gebert, J. Teschner, On the fundamental representation of Borcherds algebras with one simple imaginary root, 1993, arXiv:hep-th/9310139v1.
[115] S.J. Kang, M.H. Kim, Borcherds superalgebras and monstrous Lie superalgebras, Math. Ann. 307 (1997) 677–694.
[116] V.G. Kac, D.H. Peterson, Affine Lie algebras and Hecke modular forms, Bull. Amer. Math. Soc. 3 (1980) 1057–1061.
[117] K. Kim, D.U. Shin, The recursive dimension formula for graded Lie algebras and its applications, Comm. Algebra 27 (1999) 2627–2652.
[118] U. Ray, Automorphic Forms and Lie Superalgebras, Springer, Berlin, 2007.
[119] S. Govindarajan, Brewing Moonshine for Mathieu, (2010) arxiv.1012, 5732v1[math.NT].
[120] Z. Nijenhuis, Proc. K. Ned. Adad. Wet. A58,390, (1955).

[121] A. Frolicher, Z. Nijenhuis, A theorem on stability of complex structures, Proc. Natl Acad. Sci. USA 43 (2) (1957) 239–241.

[122] F.A. Berezin, G.I. Kats, Lie groups with commuting and anticommuting parameters, Math. USSR. Sb. 11 (3) (1970) 311–325.

[123] F.A. Berezin, D.A. Leites, Supermanifolds, Soviet Math. Dokl. 16 (1975) 1218–1222.

[124] D.A. Llites, Cohomology of Lie superalgebras, Funct. Anal. Appl. 9 (4) (1975) 75–76.

[125] M. Gerstenhaber, The cohomology structure of an associative ring, Ann. Math. 78 (2) (1963) 267–288.

[126] S.J. Kang, Graded Lie superalgebras and the superdimensional formula, J. Algebra 204 (2) (1998) 597–655.

[127] V.G. Kac, classification of simple Lie superalgebras, Fund. Anal. Appl. 9 (1975) 263–265.

[128] I.G. Macdonald, Kac-Moody algebras, in: CMS Conference Proceedings. 1984 Conference on Lie Algebras and Related Topics, vol. 5 June 26–July 6, 1984, pp. 69–109.

[129] L. Frappat, A. Sciarrino, P. Sorba, Dictionary on Lie Algebras and Superalgebras, Academic Press, Elsevier, London, 1999.

[130] I.M. Musson, Lie Superalgebras and Enveloping Algebras, in: Graduate Text in Mathematics, vol. 131, AMS, 2012.

[131] A. Paris, V. Rittenberg, Semisimple graded Lie algebras, J. Math. Phys. 16 (1975) 2062–2073.

[132] D.Z. Djokovic, Classification of some 2-graded Lie algebras, J. Pure Appl. Algebra 7 (1976) 217–230.

[133] P.G.O. Freund, I. Kaplansky, Simple Supersymmetries, J. Math. Phys. 17 (1976) 228–231.

[134] W. Nahm, V. Rittenberg, M. Scheunert, The classification of graded Lie algebras, Phys. Lett. B 61 (1976) 383–385, (later all classical ones in J. Math. Phys. 17 (1976) 1626–1640).

[135] M. Gorelik, V. Serganova, On representations of the affine superalgebra $q(n)^{(2)}$, Mosc. Math. J. 8 (1) (2008) 91–109.

[136] C. Hoyt, V. Serganova, Classification of finite-growth general Kac-Moody superalgebras, Comm. Algebra 35 (3) (2007) 851–874.

[137] J. Brunden, Representations of the general linear Lie superalgebra in the BGG category O, in: Retrospective Volumes of the West Coast Lie Theory Seminar, 22 pages, (expository article).

[138] J. Brunden, Tilting modules for Lie superalgebras, Commun. Algebra 32 (2004) 2251–2268.

[139] J. Brunden, Kazhdan-Lusztig polynomials and character formulae for the Lie superalgebra $gl(m|n)$, 505–508 (in Russian); English translation: Soviet Math. Doklady 16 (5) (1975), 1218–1222. J. Am. Math. Soc. 16 (2003) 185–231.

[140] F.A. Berezin, D.A. Leites, Supermanifolds, Dokl. Akad. Nauk SSSR. 224 (3) (1975) 505–508 (Russian).

[141] S. Cheng, V.G. Kac, Structure of some $\mathbb{Z}$-graded Lie superalgebras of vector fields, Transformation Groups 4 (2–3) (1999) 219–272.

[142] M. Wakimoto, Infinite Dimensional Lie Algebras, Translation of Mathematical Monographs, American Mathematical Society, vol. 195, 2001.

[143] V.K. Dobrev, Canonical q-deformation of noncompact Lie (super)algebras, J. Phys. A Math. Gen. 26 (6) (1993) 1317.

[144] N. Sthanumoorthy, P. Lilly, A.N. Basha, Special imaginary roots of BKM Lie superalgebras, Int. J. Pure Appl. Math. 38 (4) (2007) 513–542.

[145] N. Sthanumoorthy, P. Lilly, A.N. Basha, Strictly imaginary roots and purely imaginary roots of BKM Lie superalgebras, Comm. Algebra 37 (7) (2009) 2357–2390.

[146] S.J. Kang, J.H. Kwon, Graded Lie superalgebras, supertrace formula and orbit Lie superalgebras, Proc. Lond. Math. Soc. 81 (3) (2000) 675–724.
[147] N. Sthanumoorthy, K. Priyadharsini, Complete classification of BKM Lie superalgebras possessing strictly imaginary roots, Appl. Math. 2 (4) (2012) 100–115.
[148] N. Sthanumoorthy, K. Priyadharsini, Root supermultiplicities and corresponding combinatorial identities for some Borcherds superalgebras, Glasnik Math. 49 (69) (2014) 53–81.
[149] N. Sthanumoorthy, P. Lilly, A.N. Basha, Root multiplicities of some Borcherds superalgebras, Commun. Algebra 37 (5) (2009) 1353–1388.
[150] R.E. Borcherds, Monstrous moonshine and monstrous Lie superalgebras, Invent. Math. 109 (III) (1992) 405–444.
[151] U. Ray, Generalized Kac-Moody algebras and some related topics, Bull. Am. Math. Soc. 38 (2001) 1–42.
[152] N. Sthanumoorthy, K. Priyadharsini, Complete classification of BKM Lie superalgebras possessing special imaginary roots, Commun. Algebra 41 (1) (2013) 367–400.
[153] N. Sthanumoorthy, K. Priyadharsini, Combinatorial identities related to root supermultiplicities in some Borcherds superalgebras, Int. J. Eng. Sci. (IJES) 4 (6) (2015) 63–77.
[154] J.G.F. Belinfante, B. Kolmen, H.A. Smith, An introduction to Lie groups and Lie algebras with applications, II: the basic methods and results of representation theory, SIAM Rev. 10 (1968) 160–195.
[155] J.G.F. Belinfante, B. Kolman, An introduction to Lie groups and Lie algebras with applications, III: computational methods and applications of representation theory, SIAM Rev. 4 (11) (1969) 510–543.
[156] J.G.F. Belinfante, B. Kolmen, H.A. Smith, An introduction to Lie groups and Lie algebras with applications, SIAM Rev. 8 (1966) 11–46.
[157] V.S. Varadarajan, Lie Groups, Lie Algebras and Their Representations, Prentice-Hall, Englewood Cliffs, 1974, pp. 41–148.
[158] A. Knapp, Lie groups: beyond an introduction, in: Progress in Mathematics, vol. 140, second ed. 2002, ISBN 0-8176-4259-5 (Originally publ. 1996).
[159] G.G.A Baurle, E.A. De Kerf, Lie Algebras Part 1 Finite and Infinite Dimensional Lie Algebras and Applications in Physics, vol. II, North Holland Publ., Amsterdam, 1999.
[160] N. Bourbaki, Groupes et Algebres de Lie. Chapitres IV, V et VT, Hermann, Paris, 1968.
[161] H. Boerner, Representations of Groups with special consideration for the needs of modern physics, North-Holland Publishing Company, Amsterdam, 1962, Series in Physics.
[162] J. Lepowsky, Multiplicity formula for certain semisimple Lie groups, Bull. Am. Math. Soc. 77 (II) (1971) 601–605.
[163] M.S. Raghunathan, Discrete Subgroups of Lie Groups, Springer Verlag, Germany, 1972.
[164] B. Kostant, S. Kumar, The Nil Hecke ring and cohomology of G/P for a Kac-Moody group G^*, Adv. Math. 62 (1986) 187–237.
[165] V.G. Kac, Constructing Groups Associated to Infinite Dimensional Lie Algebras, vol. 4, MSRI Publications, 1985.
[166] V.S. Varadarajan, Lie Groups, Lie Algebras and Their Representations, Springer-Verlag, Dordrecht, 1984.
[167] N. Kitchloo, Dominant K-theory and integrable highest weight representations of Kac-Moody groups, Adv. Math. 221 (2009) 1191–1226.
[168] S. Kumar, Kac-Moody groups, their flag varieties and representation theory, Progress in Mathematics, vol. 204, Birkhauser, 2002.
[169] V.G. Kac, D.H. Peterson, Regular functions on certain infinite-dimensional groups, in: Arithmetic and Geometry, Progress in Mathematics, vol. 36, Birkhauser, Boston, 1983, pp. 141–166.

[170] V.G. Kac, D.H. Peterson, Generalized invariants of groups generated by reflections, in: Progress in Math., vol. 60, Birkhauser, Boston, 1985, pp. 276–298.
[171] V.G. Kac, D.H. Peterson, On geometric invariant theory for infinite-dimensional groups, in: Algebraic Groups, Lecture Notes in Math., Utrecht, vol. 1271, Springer-Verlag, 1987, 1986, pp. 102–142.
[172] B. Kostant, Lie algebra cohomology and the generalized Borel-Weil theorem, Ann. Math. II Ser. 74 (1961) 329–387.
[173] F. Neumann, M.D. Neusel, L. Smith, Rings of generalized and stable invariants of pseudo reflections and pseudo reflection groups, J. Algebra 182 (1996) 85–122.
[174] R. Steinberg, Lectures on Chevalley Groups, 1967.
[175] J.A. Wolf, Spaces of Constant Curvature, third ed., McGraw-Hill Book Company, Boston, Publish or Perish, 1974.
[176] A. Levy-Bruhl-Lapperiere, Spectre du De Rham-Hodge sur les formes de degre 1 des spheres de $R^n (n \geq 6)$, Bull. Sci. Math. 99 (1975) 213–240.
[177] A. Levy-Bruhl-Lapperiere, Spectre du De Rham-Hodge sur l'espace projectif complexe, Bull. Sci. Math. 104 (1980) 135–143.
[178] A. Levy-Bruhl-Lapperiere, Spectre du De Rham-Hodge sur l'espace projectif quaternionique, in: Seminaire d'Algebre Paul Dubreil et Marie-Paule Malliavin, 33e annee, Paris, Springer-Verlag, Berlin, Heidelberg, New York, 1980, pp. 98–129, Lecture Notes in Mathematics, No. 867.
[179] G. Segal, The definition of conformal field theory, in: Differential Geometrical Methods in Theoretical Physics, Kluwer Acad. Publ., New York, 1988, pp. 165–172.
[180] A. Lévy-Bruhl-Laperrière, Spectre du De Rham-Hodge sur l'espace projectif complexe Seminaire d'Algebre Paul Dubreil, in: Proceedings 30e Annee, Paris, Springer-Verlag, 1978, pp. 163–188 (Lecture Notes in Mathematics, 641).
[181] M.P. Malliavin, P. Malliavin, Diagonalisation du Systeme de Rham-Hodge au Dessus D'un Espace Riemannien Homogene Non-Commutative Harmonic Analysis, Springer-Verlag, Marseille-Luminy, Berlin, Heidelberg, New York, 1974, pp. 135–146, 1975 (Lecture Notes in Math, No. 466).
[182] N. Sthanumoorthy, Spectra of de Rham Hodge operator on $SO(n+2)/SO(2) \times SO(n)$, Bull. Sci. Math. (France) Sec. Ser. 108 (1984) 297–320.
[183] N. Sthanumoorthy, Spectra of de Rham Hodge operator on $Sp(n+1)/Sp(1) \times Sp(n)$, Bull. Sci. Math. (France) Sec. Ser. 111 (1987) 201–227.
[184] C. Tsukamoto, Spectra of Laplace-Beltrami operators on $SO(n+2)/SO(2) \times SO(n)$ and $Sp(n+1)/Sp(1) \times Sp(n)$, Osaka Math. J. 18 (1981) 407–426.
[185] A. Ikeda, Y. Taniguchi, Spectra of eigen forms of the Laplacian on S^n and $P^n(C)$, Osaka J. Math. 15 (1978) 515–546.
[186] M.F. Atiyah, V.K. Patodi, I.M. Singer, Spectral asymmetry and Riemannian geometry, Bull. Lond. Math. Soc. 5 (1973) 229–234.
[187] M.F. Atiyah, V.K. Patodi, I.M. Singer, Spectral asymmetry and Riemannian geometry: III, Math. Proc. Cambridge Philos. Soc. 79 (1976) 71–99.
[188] R.T. Seeley, Complex powers of an elliptic operator, Proc. Symp. Pure Math. Am. Math. Soc. 10 (1967) 288–307.
[189] K. Mahler, Uber einen Satz von Mellin, Math. Ann. 100 (1928) 384–398.
[190] N. Sthanumoorthy, Spectral invariants of the zeta function of the Laplacian on $S^{(4r-1)}$, Indian J. Pure Appl. Math. 19 (5) (1988) 407–414.
[191] N. Sthanumoorthy, Spectral invariants for the non singular quadric hypersurface, Bull. Sci. Math. (France) Sec. Ser. 114 (1990) 361–371.
[192] E.C. Titchmarsh, The Theory of the Riemann Zeta Function, Clarendon Press, Oxford, 1967.

[193] N. Sthanumoorthy, Spectral invariant of the Zeta function of the Laplacian on $Sp(r+1)/Sp(1) \times Sp(r)$, Arch. Math. (BRNO) (Czech) 27a (1991) 95–104, tomus.
[194] M.F. Atiyah, V.K. Patodi, I.M. Singer, Spectral asymmetry and Riemannian geometry: I, Math. Proc. Cambridge Philos. Soc. 77 (1975) 43–69.
[195] M.F. Atiyah, V.K. Patodi, I.M. Singer, Spectral asymmetry and Riemannian geometry: II, Math. Proc. Cambridge Philos. Soc. 78 (1975) 405–432.
[196] I.G. Macdonald, Affine root systems and Dedekind's η-function, Invent. Math. 15 (1972) 91–143.
[197] J. Lepowsky, Lectures on Kac-Moody Lie Algebras, Universite Paris VI, Springer, 1978.
[198] J. Lepowsky, Mcdonald type identities, Adv. Math. 27 (1978) 230–234.
[199] J. Lepowsky, M. Primc, Standard modules for type one affine Lie algebras, Number theory, in: Lecture Notes in Mathematics, vol. 1052, Springer-Verlag, 1984, pp. 194–251.
[200] N. Sthanumoorthy, M. Tamba, A note on generalization of Macdonald's identities for A_l and B_l, Proc. Indian Acad. Sci. (Math. Sci.) 104 (1994) 377–383.
[201] V.G. Kac, An algebraic definition of compact Lie groups, Trudy MIEM 5 (1969) 36–47.
[202] V.G. Kac, Root systems, representations of quivers and invariant theory, Lecture Notes in Mathematics, vol. 996, 2006, pp. 74–108.
[203] J. Lepowsky, Generalized Verma modules, loop space cohomology and MacDonold type identities, Ann. Sci. Ecol. Norm. Sup. 12 (1979) 169–234.
[204] J. Lepowsky, Affine Lie algebras and combinatorial identities, Lie algebras and related topics, in: Lecture Notes in Mathematics, vol. 933, Springer-Verlag, 1982, pp. 130–156.
[205] F.J. Dyson, Missed opportunities, Bull. Math. Soc. 78 (1972) 635–652.
[206] S. Ramanujan, On certain arithmetical functions, Trans. Cambridge Philos. Soc. 22 (1916) 159–184.
[207] L. Onsagar, Crystal statistics. I. A two dimensional model with an order-disorder transition, Phys. Rev. 65 (1944) 117–149.
[208] R. Hirota, Direct method of finding exact solutions of non linear evolution equations, in: Lecture Notes in Mathematics, vol. 515, Springer-Verlag, Berlin and New York, 1976, pp. 40–68.
[209] E. Date, M. Jimbo, T. Miwa, Solitons Differential Equations, Symmetries and Infinite Dimensional Algebras, vol. 135, Cambridge University Press, Cambridge, 2000.
[210] E. Date, M. Kashiwara, T. Miwa, Vertex operators and τ functions transformation groups for Soliton equations II, Proc. Jpn Acad. Ser. A Math. Sci. 57 (8) (1981) 387–392.
[211] V.G. Kac, D. Kazhdan, J. Lepowsky, R. Wilson, Realization of the basic representations of the Euclidean Lie algebras, Adv. Math. 42 (1981) 83–112.
[212] A.P.E. Tenkroode, M.J.B. Velt, The homogeneous realization of the basic representation of $A_1^{(1)}$ and the Toda lattice, Lett. Math. Phys. 12 (2) (1986) 139–147.
[213] N. Sthanumoorthy, Hirota derivatives, vertex operators and Plucker coordinates, in: Proceedings of VIII Ramanujan Symposium on "Nonlinear Systems", Narosa Publishing House, New Delhi, 2002, pp. 71–84.
[214] N. Sthanumoorthy, C.K. Bagirathi, Homogeneous vertex operators and Hirota bilinear equations for $B_2^{(1)}$, Indian J. Pure Appl. Math. 33 (11) (2002) 1649–1664.
[215] J. Lepowsky, M. Primc, Structure of the standard modules for the affine Lie algebras $A_1^{(1)}$ in the homogeneous picture, in: Vertex Operators in Mathematics and Physics, Publication of the Mathematical Sciences Research Institute #3, Springer-Verlag, Berlin, 1983, pp. 143–162.
[216] E. Date, M. Jimbo, M. Kashiwara, T. Miwa, Operator approach to the KP-equation. Transformation groups for Soliton equations. III, J. Phys. Soc. Jpn 50 (1981) 3806–3812.

[217] E. Date, M. Jimbo, M. Kashiwara, T. Miwa, A new hierarchy of soliton equations of KP-type, Transformation groups for soliton equations. IV. Phys. D 4 (1982) 343–365.
[218] E. Date, M. Jimbo, M. Kashiwara, T. Miwa, Euclidean Lie algebras and reduction of the KP hierarchy. Transformation groups for Soliton equations, Publ. Res. Inst. Math. Sci. 18 (1982) 1077–1110.
[219] R. Hirota, Reduction of soliton equations in bilinear form, Physica D: Nonlinear Phenomena 18 (1–3) (1986) 161–170.
[220] E. Date, M. Jimbo, M. Kashiwara, T. Miwa, Nonlinear Integrable Systems—Classical Theory and Quantum Theory, World Scientific, Singapore, 1983.
[221] E. Date, M. Jimbo, M. Kashiwara, T. Miwa, Transformation groups for Soliton equations, in: Proc. Res. Inst. Math. Sci. Symp., World Scientific, Singapore, 1983.
[222] M. Jimbo, T. Miwa, Solitons and infinite dimensional Lie algebras, Publ. RIMS Kyoto Univ. 19 (1983) 943–1001.
[223] T. Miwa, Soliton Theory: A Survey of Results, Manchester University Press, Manchester, 1990.
[224] M. Jimbo, T. Miwa, Monodromy, Solitons and Infinite Dimensional Lie Algebra, Proceedings Conference: Vertex operators in mathematics and physics, Publication Math. Sci. Inst. 3, Springer-Verlag, 1983, pp. 275–290.
[225] N. Sthanumoorthy, C.K. Bagirathi, Principal vertex operators and super Hirota bilinear equations for a non-simply laced affine Kac-Moody Lie algebra, Int. Electron. J. Pure Appl. Math. 1 (4) (2010) 429–444.
[226] V.G. Kac, A. Raina, Bombay lectures on highest weight representations of infinite dimensional Lie algebras, Advanced Series in Mathematical Physics, vol. 2, World Scientific, Singapore, 1987.
[227] P. Goddard, D. Olive, Kac Moody and Virosoro algebras in relation to quantum physics, Int. J. Mod. Phys. A 1 (1980) 23–66.
[228] P. Goddard, D. Olive, Kac Moody and Virosoro algebras in relation to quantum physics, Int. J. Mod. Phys. A 1 (1986) 303–414.
[229] V.G. Kac, M. Wakimoto, Exceptional hierarchies of Soliton equations, in: Proceedings of Symposia in Pure Mathematics, Part I, vol. 49, American Mathematical Society, Providence, RI, 1989, pp. 191–237.
[230] M. Mandia, Structure of the level one standard modules for the affine Lie algebras $B_l^{(1)}, F_4^{(1)}$ and $G_2^{(1)}$, Amer. Math. Soc. Memoirs, 362 (1987).
[231] B. Kostant, The three dimensional subgroup and the Betti numbers of a complex simple Lie group, Amer. Jour. of Math. 81 (1959) 973–1032.
[232] N. Sthanumoorthy, C.K. Bagirathi, Principal vertex operators and super Hirota bilinear equations for $B_3^{(1)}$, Contemp. Math. 343 (2004) 263–288.
[233] D. Adamovic, Regularity of certain vertex operators superalgebras, Contemp. Math. 343 (2004) 1–16.
[234] P. Tingley, Notes on Focks Space, Cambridge University Press, Cambridge, 2011.
[235] S. Ramanan, Global Calculus, vol. 65, Graduate Studies in Mathematics, AMS, 2005.
[236] B. Fischer, D. Livingstone, M.P. Thorne, The Characters of the "Monster" Simple Group, Birmingham, 1978.
[237] G.H. Hardy, Ramanujan, twelve lectures on subjects suggested by his life and work, Chelsea Publishing Company, New York, 1940.
[238] V.G. Kac, A description of filtered Lie algebras whose associated graded Lie algebras are of Cartan types, Math. of USSR-Izvestijia 8 (1974) 801–835.
[239] A.L. Onishchik, Lie superalgebras of vector fields, Amer. Math. Soc. Transl. 148 (2) (1991) 87–97.

[240] N. Sthanumoorthy, K.C. Misra (Eds.), Kac-Moody Lie algebras and related topics, in: Proceedings of the Ramanujan International symposium on Kac-Moody Lie algebras and applications, Contemporary Mathematics, vol. 343, American Mathematical Society, 2004.
[241] I.M. Singer, J.A. Thorpe, Lecture Notes on Elementary Topology and Geometry, Springer-Verlag, New York, 1967.
[242] V.G. Drinfeld, Hopf algebras and quantum Yang-Baxter equation, Sov. Math. Dokl. 283 (1985) 1060–1064.
[243] V.G. Drinfeld, Quantum groups, in: Proc. ICM-1986 (Berkeley, USA), vol. 1, American Mathematical Society, Providence, RI, 1987, pp. 798–820.
[244] R. Jagannathan, Some introductory notes on quantum groups, quantum algebras, and their applications, 2001, arXiv: math-ph/0105002v1.
[245] M. Jimbo, A q-difference analogue of $U(g)$ and the Yang-Baxter equation, Lett. Math. Phys. 10 (1985) 63–69.
[246] M. Jimbo, A q-analogue of $U_q(gl(N+1))$, Hecke algebra and the Yang-Baxter equation, Lett. Math. Phys. 11 (1986) 247–252.
[247] M. Jimbo, Quantum R-matrix for the generalized Toda system, Commun. Math. Phys. 102 (4) (1986) 537–547.
[248] J.C. Jantzen, Lectures on quantum groups, in: Graduate Studies in Mathematics, vol. 6, American Mathematical Society, Providence, RI, 1996.
[249] J. Glimm, A. Jaffer, Quantum Physics: A Functional Integral Point of View, second ed., Springer-Verlag, New York, 1987.
[250] P. Podles, E. Muller, Introduction to Quantum Groups, 1997, p. 4002.
[251] K. Becker, M. Becker, J. Schwarz, String Theory and M-Theory: A Modern Introduction, Cambridge University Press, Cambridge, 2007.
[252] E. Bergshoeff, E. Sezgin, P. Townsend, Supermembranes and eleven-dimensional supergravity, Phys. Lett. B 189 (1) (1987) 75–78.
[253] A. Connes, Noncommutative Geometry, Academic Press, Boston, MA, 1994.
[254] R. Courant, D. Hilbert, Methods of Mathematical Physics, Interscience Publishers, New York, 1989.
[255] M. Reed, B. Simon, Methods of modern mathematical physics, in: Functional Analysis, vol. 1, Academic Press, New York, 1980.
[256] D.H. Menzel, Mathematical Physics, Dover Publications, New York, 1961.
[257] M. Reed, B. Simon, Methods of Mathematical Physics, vols. 1–3, Academic Press, New York, 1972–1977.
[258] J. Fuchs, Affine Lie Algebras and Quantum Groups: An Introduction, With Applications in Conformal Field Theory, Cambridge University Press, Cambridge, 1992.
[259] V. Chari, A. Pressley, A Guide to Quantum Groups, Cambridge University Press, Cambridge, 1995.
[260] C. Kassel, Quantum groups, in: Graduate Text in Mathematics, Springer-Verlag, New York, 1995.
[261] M. Hausner, J.T. Schwartz, Lie Groups; Lie Algebras, Gordon and Breach, New York, 1968.
[262] V.G. Kac, Vertex algebras for beginners, in: University Lecture Series 10, second ed., American Mathematical Society, Providence, RI, 1998.
[263] S. Eswara Rao, Complete reducibility of integrable modules for the affine Lie (super)algebras, J. Algebra 264 (2003) 269–278.
[264] H. Zhang, K. Zhao, Root systems of a class of Kac-Moody algebras, Algebra Colloq. 1 (4) (1994) 359–368.

[265] A. Uma Maheswari, Imaginary roots and Dynkin diagrams of quasi hyperbolic Kac-Moody algebras of rank 3, Int. J. Math. Comput. Appl. Res. 4 (2) (2014) 19–28, ISSN(P): 2249-6955.
[266] A. Uma Maheswari, Quasi-affine Kac-Moody algebras: classification of Dynkin diagrams & realization for a class of QAG2(1), Int. J. Math. Sci. 34 (2) (2014) 1590–1609, ISSN: 2051-5995.
[267] A. Uma Maheswari, A study on the structure of indefinite quasi-affine Kac-Moody algebras QAC2(1), Int. Math. Forum 9 (32) (2014) 1595–1609.
[268] A. Uma Maheswari, Structure of indefinite quasi-affine Kac-Moody algebras QAD3(2), Int. J. Pure Appl. Math. 102 (1) (2015) 57–72, ISSN: 1311-8080 (printed version); ISSN: 1314-3395 (on-line version).
[269] G. Arfken, Infinite Groups, Lie Groups, Mathematical Methods for Physicists, third ed., Academic Press, Orlando, FL, 1985, pp. 251–252.
[270] C. Chevalley, Theory of Lie Groups, Princeton University Press, Princeton, NJ, 1946.
[271] J.S. Lomont, Symmetric Groups, in: Applications of Finite Groups, Dover, New York, 1987, pp. 258–273 (Chapter 7).
[272] D. Earner, Groups and Symmetry, American Mathematical Society, Providence, RI, 1995.
[273] H.M. Edwards, Riemann's Zeta Function, Academic Press, New York, 1974.
[274] A. Ivic, The Riemann Zeta Function, Wiley, New York, 1985.
[275] E.W. Weisstein, CRC Concise Encyclopedia of Mathematics, Chapman and Hall/CRC Press, New York/Washington, D.C., 1999.
[276] W. Fulton, Young Tableaux With Applications to Representation Theory and Geometry, Cambridge University Press, New York, 1996.

Index

Note: Page numbers followed by *t* indicate tables.

A

Abelian Lie algebra, 3
Abelian Lie superalgebra, 209–210, 215
Affine KM matrix extension, 274, 278–279
Affine Lie algebras
  GCM of, 70
  nontwisted, 86–90
  properties of matrices of, 69–70
  realization of, 86–94
  root systems of, 83–84
  twisted, 91–94
  Weyl groups of, 85–86
Algebraic group, to Kac-Moody algebra, 386
Alien type imaginary roots, in BKM Lie superalgebras, 272–281
Associative superalgebra, 207

B

Basic Lie superalgebra, 233, 233*t*, 235
BCM. *See* Borcherds-Cartan matrix (BCM)
Bilinear forms, 216–217
  extension of, 90
  invariant, 75–78
BKM algebras, 136–139, 150–163, 168, 169, 258–260, 268, 297, 309, 314
Borcherds algebra, 144, 183–186
Borcherds-Cartan matrix (BCM), 136
  indecomposable, 139
Borcherds Kac-Moody (BKM) Lie superalgebras, 258–264
  classification of, 285–308, 315–325
  domestic type and alien type imaginary roots in, 272–281
  Dynkin diagrams of, 264–272
  even and odd roots, 260–264
  as extensions of KM algebras, 273–274
  with finite number of real roots, 323–325
  fundamental results on real roots, 269–270
  imaginary roots in, strictly and purely, 298–308
  with infinite number of imaginary roots, 323–325
  possessing purely imaginary property, 308–314
  possessing purely imaginary roots, 298–308
  possessing special imaginary roots, 285–298
  possessing strictly imaginary property, 315–325
  root space decomposition in, 268–269
  special imaginary roots in, 281–298
Borcherds Kac-Moody superalgebras
  classes of, 315
  extended-hyperbolic, 305–308
  strictly imaginary roots and purely imaginary roots of, 308
Borcherds Kac-Moody supermatrices, 258–264
  Dynkin diagrams of, 264–266
  even and odd roots, 260–264
Borcherds superalgebra
  extended-hyperbolic, 351–368
  of root supermultiplicities, 325–368, 336*t*, 339*t*
Boson-Fermion correspondence, 460
Bosonic Fock space, 460

C

Cartan group, 385
Cartan matrices
  of hyperbolic types, 71–72

Cartan matrices *(Continued)*
Kac-Moody algebras associated with, 105–107
of simple Lie algebras, 36–38
Cartan subalgebras, 230–237
of Lie algebras, 230, 277*t*
of Lie superalgebras, 230–237
Cartan-Weyl basis, 51–52
Category $\mathcal{O}$-modules, 100
Chevalley generators, 62
Chevalley involution, 91
Chevalley's normalization, 34
Classical Lie superalgebras, 244*t*
classes of, 220–224
classification of, 241–242, 243*t*
properties of, 224
rank and root systems of, 232–237
with zero Killing form, 243–244
Z-gradation of, 247*t*
Z_2-gradation of, 245*t*
Classical simple Lie algebras
Cartan matrices of, 37–38, 38*t*
long roots and short roots of, 44*t*
root systems of, 40–44
Classical simple Lie superalgebras, finite dimensional, 244–245
Clifford algebra, 459
θ-colored Lie superalgebras, 212–215, 224–230
Coloring matrices, 212–215
Combinatorial identity, for extended-hyperbolic Borcherds superalgebra, 351–368
Commutative Lie superalgebra, 209–210
Commutative superalgebra, 206
Compact groups, finite dimensional representations of, 379–382
Complex general linear group GL(n, C), 378–379
Complex quadric hypersurface, 390
Consistently graded Lie superalgebra, 210
Contragradient Lie superalgebras construction, 224
Coxeter graphs, 34–36
Coxeter groups, 78–81
Coxeter system, 145
Crystal base theory, quantum affine Lie algebras, 109

D

De Rham Hodge operator, 392
Diagram automorphism, 64
Distinguished Cartan matrix, 232
Distinguished simple root system, 230, 232
Distinguished Z-gradation, 231
Domestic imaginary roots, in BKM Lie superalgebras, 272–281
Dynkin diagrams, 34–36, 66–75, 139–144
for affine types, 68–69
of BKM Lie superalgebras, 264–272
of Borcherds Kac-Moody supermatrices, 264–266
of extended-hyperbolic types, 73
of GCM of extended-hyperbolic types, 72–73
of hyperbolic types, 70–71
of Lie superalgebras, 239–241
properties, 67, 70–71
of A and A^t, 70

E

Eta function, 399–400, 402–403
Euclidean spaces/root diagram, root system in, 30–34
Exponential mapping property, 384–385
Extended-hyperbolic algebra, 74–75
completely connected, 73
Dynkin diagrams of, 73
Dynkin diagrams of GCM of, 72–73
Extended-hyperbolic BKM superalgebras, 305–308
Extended-hyperbolic Borcherds superalgebra, combinatorial identity for, 351–368
Extended-hyperbolic KM algebra, structure and root multiplicities for, 110–134

F

Fermionic Fock space, 459
Filtration, 102, 211
Finite dimensional classical simple Lie superalgebras, 244–245

Finite dimensional contragradient Lie superalgebras, 223
Finite dimensional Heisenberg Lie algebra, 407–411
Finite-dimensional Lie algebras, 368–373
  construction of nontwisted affine Lie algebras from, 87–90, 93–94
  construction of twisted affine Lie algebras from, 91–94
  Weyl groups of, 85–86
Finite-dimensional Lie superalgebras, 368–373
Finite dimensional representation, of compact groups/Lie groups, 379–382
Finite and infinite dimensional Lie algebras and Lie superalgebras
  description, 368–373
  different classes of root systems, 368–373
Finite KM matrix extension, 274–278
Finite Lie algebras, properties of matrices of, 69–70
Finite number of real roots, BKM Lie superalgebras with, 323–325
Formal character, 102–103

G

Generalized Cartan matrix (GCM), 63, 285
  of affine type, 70
  construction of, 86–87
  Dynkin diagrams of, 72–73
  expression of symmetrizable, 76–78
  of invariant bilinear form, 76–78
Generalized generalized Cartan matrix (GGCM), 136
Generalized Generalized Cartan supermatrix, 285
Generalized Kac-Moody (GKM) algebras
  affine GKM, 174–175, 369–370
  Dynkin diagrams, 139–144
  extension of finite type, 146
  finite GKM, 369–370
  homology modules, 187–202
  indefinite GKM, 174–175
  nonsymmetric affine type, 147
  nonsymmetric hyperbolic type, 148
  properties of real roots, 149
  purely imaginary roots, 169–179
  real and imaginary roots, 146–148
  representation, 180–186
  root multiplicities, simple imaginary root, 190–202
  root systems and Weyl groups, 144–149
  strictly imaginary roots, 164–168
  weight $\mathcal{G}$-module, 181–182
General linear algebra, 3
General linear Lie superalgebras, 216
GGCM. *See* Generalized generalized Cartan matrix (GGCM)
GKM algebras. *See* Generalized Kac-Moody (GKM) algebras
Gradation
  bitransitive, 211
  transitive, 211
Graded Jacobi identity, 207, 218
Graded Lie algebras
  construction of, 105–107
  of Kac-Moody algebras, 100–101
Grassmann algebra, 208–209
Grassmann superalgebra, 208–209

H

$\mathfrak{h}$-diagonalizable $\mathcal{G}$-module, 180–181, 327
Heisenberg algebra, 406
Heisenberg superalgebra, 210
Highest long and short roots, 40–44
Hirota bilinear differential equations, 424
Hirota bilinear differential operators, super, 411–415
Hirota bilinear equation, 426–453, 458
Hirota derivatives, using Taylor expansion, 413
Hirota differential equation, 458
Hirota differential operator, on n-variables, 412–413
Hirota form and solution, KdV equation in, 413–415
Homogeneous hierarchy, 422, 424–425

Homogeneous spaces, differential operators on, 390–394
Homogeneous vertex operator construction, 415–426
Homology
of Lie algebra, 107–108
modules and root multiplicities, 187–202
Homomorphism
Lie algebra, 3
Lie superalgebras., 216
Horchschild-Serre five-term exact sequence, 107–108
Hyperbolic algebra
Cartan matrices of, 71–72
Dynkin diagrams of, 70–71
extended (*see* Extended-hyperbolic algebra)
Hyperbolic KM matrix extension, 275, 280–281

I

Ideals in Lie algebras, 10–12
Imaginary roots, in Kac-Moody algebras, 81–84, 94–98
Induced module, 187, 219, 328
Infinite dimensional Lie algebras, 368–373
Infinite dimensional Lie superalgebras, 368–373
Infinite number of imaginary root, BKM Lie superalgebras with, 323–325
Integrable $\mathcal{G}$-module, Lie group module associated to, 386–387
Integrable highest-weight module, 103–105, 183–184
Invariant bilinear form, 75–78
Invariant nondegenerate symmetric bilinear form property, 75–76
Ising model, 411–412
Isomorphism
of Lie algebra, 3
of Lie superalgebras, 216
theorem, 14–19

K

Kac-Moody algebras, 388–390
algebraic group associated to, 386
associated with Cartan matrix, 105–107
basic concepts in, 60–66
classification, 99–100
definition for, 81
extended-hyperbolic, 110–134
extensions of, 351–368
graded Lie algebras of, construction of, 100–101
Kostant's formula for symmetrizable, 108–109
Macdonald's identities for, 402–406
possessing purely imaginary roots, 98–100
properties of, 98–99
real roots of, 81–84
representation of, 100–101
root systems of, 83
and some combinatorial identities, 351–368
special imaginary roots of, 81–84
strictly imaginary roots in, 94–98
Kac-Moody groups, 388–390
applications, 386–390
Kac Moody Lie superalgebras, 264–272
types, 270–272
KdV equation
in Hirota form and solution, 413–415
soliton solutions for, 411–415
transformation of, 413–415
Killing forms, 14–19
on Lie superalgebras, 237–239
root space decomposition and properties of, 30
Kostant's formula, for symmetrizable Kac-Moody algebras, 108–109

L

Laplace-Beltrami operator, 394–402
Lie algebras, 381*t*
character of finite dimensional representation, 52–53
definition, 2–8
derivation of, 19–21
derived sub Lie algebra, 10–12
direct sum, 10–12
G and real forms *U*, 384*t*
group *G(A)* associated to, 387
homology of, 107–108
of Lie groups, 376–385

over C, 384*t*
over R, 382–384
representations of, 21–28
special infinite dimensional, 406–411
subalgebras of, 8–10
of vector fields, 53–57
Lie groups, 376–385, 381*t*
finite dimensional representations of, 379–382
module associated to integrable $\mathcal{G}$-module, 386–387
Lie superalgebras
Abelian, 215
basic, 231
basic concepts in, 204–212
consistently graded, 210
definition, 204
Dynkin diagrams of, 239–241
general linear, 216
ideals of, 215
killing forms on, 237–239
nilpotent, 215
representations of, 218–220
root systems of, 230–237
solvable, 215
strange, 231, 244*t*
supercommutative, 207
universal enveloping algebras of, 224–230
of vector fields, 248–255

M

Macdonald's identities
generalization of, 402–406
for Kac-Moody algebras, 402–406
Mathematical Physics, 460–462
Monster Lie algebra, 163, 168
Monster Lie superalgebras, 323

N

Nilpotent Lie algebras, 12–14
Nilpotent Lie superalgebras, 215
Non-classical Lie superalgebras classification, 245–248
Non compact group, 379
Non-degenerate killing form, Lie superalgebras with, 243
Nontwisted affine Lie algebras, 86–87
construction of, 87–90

O

One-dimensional Lie algebra, 7
One parameter subgroup, 385
Oscillator algebra, 406

P

Poincare-Birkhoff-Witt (PBW) theorem, 46–47, 229
Principal Cartan subalgebras, 453–455
Principal vertex operator, 440–445, 452–453
construction, 415–432, 455–458
Purely imaginary roots
BKM Lie superalgebras possessing, 298–308
Dynkin diagram, 174
GGCM, 173
GKM algebras, 174–175
and strictly, GKM algebra, 175–179

Q

Quantum affine Lie algebras, crystal base theory of, 109
Quantum groups, 460–461
Quasi affine type, 468
Quasi-Dynkin diagram, 264
Quasi finite type, 468
Quasi hyperbolic type, 468
Quaternionic projective space, 391
Quotient Lie algebras, 10–12

R

Real roots of Kac-Moody algebras, 81–84
Real vector space, complex structure on, 382–384
Reflection in Euclidean space, 30
Riemannian symmetric space, 390

Root multiplicities
GKM algebras, 190–197
homology modules, 187–202
Rootspace decomposition
Killing form, 30
of semisimple Lie algebras, 28–30
Root supermultiplicities, of Borcherds superalgebras, 325–368, 336*t*, 339*t*
Root system
of affine Lie algebras, 83–84
of classical simple Lie algebras, 40–44
in Euclidean spaces and root diagrams, 30–34
of Kac-Moody algebras, 83
properties of, 29–30

S

Semisimple Lie algebras, 12–14
construction of, 50–51
representation theory of, 49–50
rootspace decomposition of, 28–30
Semisimple Lie superalgebras, 216–217
Serre's theorem, 64
Simple Lie algebras, 12–14
Cartan matrices, ranks, and dimensions of, 36–38, 38*t*
classical (*see* Classical simple Lie algebras)
properties of, 66
Weyl groups of, 39, 40*t*
Simple Lie superalgebras, 216–217
classification of, 243, 248
Soliton solutions, for KdV equation, 411–415
Solvable Lie algebras, 12–14
Solvable Lie superalgebras, 215
Special imaginary roots
BKM, infinite number, 163
BKM Lie superalgebras possessing, 285–298
extension, 282–283
GGCMs, 150–151
GKM, finite index set, 150–163
non-existence, 283–284
Special infinite dimensional Lie algebras, 406–411
Spectral invariants, of zeta function, 394–402
asymmetry, 399, 401
Strictly imaginary roots
BKM algebras, infinite number, 168
GKM algebras, finite number, 164–168
String theory, 460–461
Structure constants, 2–8, 47
Subalgebras
Cartan, 230–237
of gl(n,$\mathcal{C}$), 8–10
of Lie algebras, 8–10
toral, 28–29
Sub Lie algebras, 10–12
Subsuperalgebras, 215
Superalgebras
associative, 207
θ-colored Lie, 212–215
commutative, 206
on differential forms, 209–212
extended-hyperbolic BKM, 305–308
Grassmann, 208–209
Heisenberg, 210
Supercharacter, 219
Supercommutative Lie superalgebra, 207
Superdimension formula, 351–368
Supergroups, applications, 386–390
Super Hirota bilinear equation, 440–453
Symmetric bilinear form, invariant nondegenerate, 75–76
Symmetrizable Kac-Moody algebras, Kostant's formula for, 108–109

T

Tangent bundle, 53
Tangent vector, 53, 380
Taylor expansion, Hirota derivatives using, 413
Three-dimensional Lie algebra, 8
Tits cone, 81
Toral subalgebra, 28–29
Triangular decomposition, 90–91
Two-dimensional Lie algebra, 7

U

Universal enveloping algebras
of Lie algebras, 44–49

of Lie superalgebras, 224–230
θ-colored Lie superalgebras, 212–215, 224–230
Universal mapping property, 45–49

V

Vector bundle, 56, 392
Vector fields
Lie algebras of, 53–57
Lie superalgebras of, 248–255
Verma module, 102
character formula, 100–105
highest weight $\mathcal{G}$-module, 187
properties, 182
Vinberg's classification theorem, 66
Virasoro algebra, 408–409

W

Weight module, 100–105, 417
W-equivariant bijection, 82
Weyl chamber, 81
fundamental, 40, 80–81
Weyl dimension formula, 52–53
Weyl groups, 78–81, 79*t*, 144–149
of affine Lie algebras, 85–86
elements of, 79*t*
elements of $SBGB_3$, 276*t*, 277*t*
of finite-dimensional Lie algebras, 85–86
properties of, 79–81
of simple Lie algebras, 39, 40*t*
Weyl-Kac-Borcherds character formula
GKM algebra, 182–183
imaginary simple root, 183–186
Weyl reflection, 39
Witt algebra, 407

Z

Zero Killing form, classical Lie superalgebras with, 243–244
Zeta function, spectral invariants of, 394–402
Z-gradation, of classical Lie superalgebras, 247*t*
Z_2-graded vector space
differential manifold, 463
heat equation, 465
Maya diagram, 466
presheaf, 467
product manifold, 464
pseudo Riemannian manifold, 465
Riemannian manifold, 465
Riemannian symmetric space, 465
Riemann surface, 465
sheaf, 467
symmetric group, 467
symmetries, 465
Young tableau, 466